D1709780

Exact Methods in Low-Dimensional Statistical Physics and Quantum Computing

Lecturers who contributed to this volume

Ian Affleck
Fabien Alet
John Cardy
Nigel R. Cooper
Bertrand Duplantier
Mikhail V. Feigelman
Thierry Jolicoeur
Richard Kenyon
Alexei Kitaev
Ivan Kostov
Werner Krauth
Satya N. Majumdar
Grégoire Misguich
Bernard Nienhuis
Nicolai Reshetikhin
Hubert Saleur
David Thouless
Paul Zinn-Justin

École d'été de Physique des Houches

Session LXXXIX, 30 June–1 August 2008

École thématique du CNRS

Exact Methods in Low-Dimensional Statistical Physics and Quantum Computing

Edited by

Jesper Jacobsen, Stéphane Ouvry,
Vincent Pasquier, Didina Serban
and Leticia F. Cugliandolo

OXFORD
UNIVERSITY PRESS

OXFORD
UNIVERSITY PRESS

Great Clarendon Street, Oxford OX2 6DP

Oxford University Press is a department of the University of Oxford.
It furthers the University's objective of excellence in research, scholarship,
and education by publishing worldwide in

Oxford New York

Auckland Cape Town Dar es Salaam Hong Kong Karachi
Kuala Lumpur Madrid Melbourne Mexico City Nairobi
New Delhi Shanghai Taipei Toronto

With offices in

Argentina Austria Brazil Chile Czech Republic France Greece
Guatemala Hungary Italy Japan Poland Portugal Singapore
South Korea Switzerland Thailand Turkey Ukraine Vietnam

Oxford is a registered trade mark of Oxford University Press
in the UK and in certain other countries

Published in the United States
by Oxford University Press Inc., New York

© Oxford University Press 2010

The moral rights of the authors have been asserted
Database right Oxford University Press (maker)

First published 2010

All rights reserved. No part of this publication may be reproduced,
stored in a retrieval system, or transmitted, in any form or by any means,
without the prior permission in writing of Oxford University Press,
or as expressly permitted by law, or under terms agreed with the appropriate
reprographics rights organization. Enquiries concerning reproduction
outside the scope of the above should be sent to the Rights Department,
Oxford University Press, at the address above

You must not circulate this book in any other binding or cover
and you must impose the same condition on any acquirer

British Library Cataloguing in Publication Data
Data available

Library of Congress Cataloging in Publication Data
Data available

Printed in Great Britain
on acid-free paper by
CPI Antony Rowe, Chippenham, Wiltshire

ISBN 978–0–19–957461–2

1 3 5 7 9 10 8 6 4 2

École de Physique des Houches
Service inter-universitaire commun à l'Université Joseph Fourier de Grenoble et à l'Institut National Polytechnique de Grenoble

Subventionné par le Ministère de l'Éducation Nationale, de l'Enseignement Supérieur et de la Recherche, le Centre National de la Recherche Scientifique, le Commissariat à l'Énergie Atomique

Membres du conseil d'administration:
Farid Ouabdesselam (président), Paul Jacquet (vice-président), Jacques Deportes, Cécile DeWitt, Thérèse Encrenaz, Bertrand Fourcade, Luc Frappat, Jean-François Joanny, Michèle Leduc, Jean-Yves Marzin, Giorgio Parisi, Eva Pebay-Peyroula, Michel Peyrard, Luc Poggioli, Jean-Paul Poirier, François Weiss, Jean Zinn-Justin

Directeur:
Leticia F. Cugliandolo, Laboratoire de Physique Théorique et Hautes Énergies, Université Pierre et Marie Curie – Paris VI, France

Directeurs scientifiques de la session LXXXIX:
Jesper Jacobsen, Laboratoire de Physique Théorique, École Normale Supérieure, Paris
Stéphane Ouvry, Laboratoire de Physique Théorique et Modèles Statistiques, Centre National de la Récherche Scientifique, Université Paris-Sud – Centre Scientifique d'Orsay
Vincent Pasquier, Institut de Physique Théorique, Commissariat à l'Énergie Atomique, Saclay
Didina Serban, Institut de Physique Théorique, Commissariat à l'Énergie Atomique, Saclay

Previous sessions

I	1951	Quantum mechanics. Quantum field theory
II	1952	Quantum mechanics. Statistical mechanics. Nuclear physics
III	1953	Quantum mechanics. Solid state physics. Statistical mechanics. Elementary particle physics
IV	1954	Quantum mechanics. Collision theory. Nucleon-nucleon interaction. Quantum electrodynamics
V	1955	Quantum mechanics. Non equilibrium phenomena. Nuclear reactions. Interaction of a nucleus with atomic and molecular fields
VI	1956	Quantum perturbation theory. Low temperature physics. Quantum theory of solids. Ferromagnetism
VII	1957	Scattering theory. Recent developments in field theory. Nuclear and strong interactions. Experiments in high energy physics
VIII	1958	The many body problem
IX	1959	The theory of neutral and ionized gases
X	1960	Elementary particles and dispersion relations
XI	1961	Low temperature physics
XII	1962	Geophysics; the earth's environment
XIII	1963	Relativity groups and topology
XIV	1964	Quantum optics and electronics
XV	1965	High energy physics
XVI	1966	High energy astrophysics
XVII	1967	Many body physics
XVIII	1968	Nuclear physics
XIX	1969	Physical problems in biological systems
XX	1970	Statistical mechanics and quantum field theory
XXI	1971	Particle physics
XXII	1972	Plasma physics
XXIII	1972	Black holes
XXIV	1973	Fluids dynamics
XXV	1973	Molecular fluids
XXVI	1974	Atomic and molecular physics and the interstellar matter
XXVII	1975	Frontiers in laser spectroscopy
XXVIII	1975	Methods in field theory
XXIX	1976	Weak and electromagnetic interactions at high energy
XXX	1977	Nuclear physics with heavy ions and mesons
XXXI	1978	Ill condensed matter
XXXII	1979	Membranes and intercellular communication
XXXIII	1979	Physical cosmology

XXXIV	1980	Laser plasma interaction
XXXV	1980	Physics of defects
XXXVI	1981	Chaotic behaviour of deterministic systems
XXXVII	1981	Gauge theories in high energy physics
XXXVIII	1982	New trends in atomic physics
XXXIX	1982	Recent advances in field theory and statistical mechanics
XL	1983	Relativity, groups and topology
XLI	1983	Birth and infancy of stars
XLII	1984	Cellular and molecular aspects of developmental biology
XLIII	1984	Critical phenomena, random systems, gauge theories
XLIV	1985	Architecture of fundamental interactions at short distances
XLV	1985	Signal processing
XLVI	1986	Chance and matter
XLVII	1986	Astrophysical fluid dynamics
XLVIII	1988	Liquids at interfaces
XLIX	1988	Fields, strings and critical phenomena
L	1988	Oceanographic and geophysical tomography
LI	1989	Liquids, freezing and glass transition
LII	1989	Chaos and quantum physics
LIII	1990	Fundamental systems in quantum optics
LIV	1990	Supernovae
LV	1991	Particles in the nineties
LVI	1991	Strongly interacting fermions and high T_c superconductivity
LVII	1992	Gravitation and quantizations
LVIII	1992	Progress in picture processing
LIX	1993	Computational fluid dynamics
LX	1993	Cosmology and large scale structure
LXI	1994	Mesoscopic quantum physics
LXII	1994	Fluctuating geometries in statistical mechanics and quantum field theory
LXIII	1995	Quantum fluctuations
LXIV	1995	Quantum symmetries
LXV	1996	From cell to brain
LXVI	1996	Trends in nuclear physics, 100 years later
LXVII	1997	Modeling the earth's climate and its variability
LXVIII	1997	Probing the Standard Model of particle interactions
LXIX	1998	Topological aspects of low dimensional systems
LXX	1998	Infrared space astronomy, today and tomorrow
LXXI	1999	The primordial universe
LXXII	1999	Coherent atomic matter waves
LXXIII	2000	Atomic clusters and nanoparticles
LXXIV	2000	New trends in turbulence
LXXV	2001	Physics of bio-molecules and cells
LXXVI	2001	Unity from duality: Gravity, gauge theory and strings

LXXVII	2002	Slow relaxations and nonequilibrium dynamics in condensed matter
LXXVIII	2002	Accretion discs, jets and high energy phenomena in astrophysics
LXXIX	2003	Quantum entanglement and information processing
LXXX	2003	Methods and models in neurophysics
LXXXI	2004	Nanophysics: Coherence and transport
LXXXII	2004	Multiple aspects of DNA and RNA
LXXXIII	2005	Mathematical statistical physics
LXXXIV	2005	Particle physics beyond the Standard Model
LXXXV	2006	Complex systems
LXXXVI	2006	Particle physics and cosmology: the fabric of spacetime
LXXXVII	2007	String theory and the real world: from particle physics to astrophysics
LXXXVIII	2007	Dynamos
LXXXIX	2008	Exact methods in low-dimensional statistical physics and quantum computing
XC	2008	Long-range interacting systems

Publishers

- Session VIII: Dunod, Wiley, Methuen
- Sessions IX and X: Herman, Wiley
- Session XI: Gordon and Breach, Presses Universitaires
- Sessions XII–XXV: Gordon and Breach
- Sessions XXVI–LXVIII: North Holland
- Session LXIX–LXXVIII: EDP Sciences, Springer
- Session LXXIX–LXXXVIII: Elsevier
- Session LXXXIX– : Oxford University Press

Organizers

JACOBSEN Jesper, Laboratoire de Physique Théorique, École Normale Supérieure, Paris, France
OUVRY Stéphane, Laboratoire de Physique Théorique et Modèles Statistiques, Centre Nationale de la Récherche Scientifique, Université Paris-Sud – Centre Scientifique d'Orsay, France
PASQUIER Vincent, Institut de Physique Théorique, Commissariat à l'Énergie Atomique, Saclay, France
SERBAN Didina, Institut de Physique Théorique, Commissariat à l'Énergie Atomique, Saclay, France
CUGLIANDOLO Leticia, Laboratoire de Physique Théorique et Hautes Énergies, Université Pierre et Marie Curie, Paris VI, France

Preface

The interest in exact solutions in low-dimensional theoretical physics comes from several directions.

First, low-dimensional (and, more specifically, two-dimensional) models provide reliable descriptions of a wide array of physical phenomena, ranging from condensed matter (disordered electron gases, quantum liquids) to string theory (D-brane physics, AdS/CFT duality). Many interdisciplinary applications also find a representation within this class of models, such as problems in quantum information theory, to give but one important example.

Second, in recent years an extraordinary convergence of the methods used to study such models has been achieved. The fruitful interplay between conformal field theory and quantum inverse scattering has been supplemented by mathematical methods coming from combinatorics, representation theory, and probability theory. One striking mathematical development, initiated by the late O. Schramm, is stochastic Loewner evolution, which has provided a fundamentally new way of thinking about conformal field theory.

Third, the very existence of exact solutions of nontrivial statistical models has profound implications. A few decades ago, such solutions were often, if not frowned upon, at least considered with some suspicion: if a model turned out to be exactly solvable, it must certainly be somehow pathological and not representative of the physical phenomenon it purported to describe. The advent of ideas of universality and the explosion in the number and diversity of exact solutions have taught us otherwise. If a physical phenomenon is considered interesting, the tendency nowadays is to look for an exactly solvable model describing it. Needless to say, an exact solution is a measuring stick by which the validity of an approximation or of a numerical method can be judged.

The purpose of this School was to treat this plethora of models and techniques in a transverse way, emphasizing the common underlying structures. To this end, we have brought together a group of carefully selected young scientists with the leading experts in a variety of fields gravitating around exactly solvable models in low-dimensional condensed matter and statistical physics.

The interplay between conformal field theory (CFT) and stochastic Loewner evolution was emphasized at the beginning of the School, through the lectures by J. Cardy and W. Werner. It was made clear that the description of conformally invariant two-dimensional curves can be developed in parallel within the two frameworks. The treatment of such models by Coulomb gas techniques was covered by B. Nienhuis, who also presented tiling models, which provide a different class of random geometries. The special case of dimers was discussed by R. Kenyon. The definition of models of random curves on a background of random triangulations is also possible, and was presented by I. Kostov, with a special emphasis on boundary conditions.

Another field that links up closely with CFT is that of quantum inverse scattering (QIS), treated by N. Reshetikhin. The prototype model here is that of an integrable quantum spin chain. The interplay between such integrable models and combinatorics was discussed by

P. Zinn-Justin. H. Saleur presented the case where the spin chains possess target space supersymmetry. In a rather different vein, mass transport models of condensation in real space were covered by S. Majumdar.

A certain class of numerical methods belongs to the realm of exact solutions, in the sense that they provide completely unbiased sampling and/or are made possible by an underlying exact solution. These aspects were discussed by W. Krauth and F. Alet, who also made the connection with topics in condensed matter physics.

The techniques of CFT and QIS can be combined to provide exact solutions of various quantum impurity problems. This subject was treated by I. Affleck. Still within the domain of condensed matter, G. Misguich discussed the physics of spin liquids, and the subject of the quantum Hall effect was covered jointly by F. D. M. Haldane and J. Fröhlich. Finally, N. Cooper lectured on models of rotating Bose–Einstein condensates, and A. Kitaev presented models of topological order with applications to quantum computing.

Seminars on special topics were given by B. Duplantier, M. Feigel'man, T. Jolicoeur, and D. Thouless.

This Les Houches session was made possible by generous financial support from NATO, the ESF network INSTANS, Triangle de la Physique, CNRS, and CEA.

The organizers wish to express their gratitude to C. LeVaou for her invaluable assistance throughout the organization of this session, and to the Les Houches administrative staff M. Gardette, I. Lelievre, and B. Rousset for their organizational skills during the session itself. We are indebted to V. Degat and J. Dubail for their help in typesetting this volume. And, finally, our thanks go to the staff members at Les Houches for solving all kinds of practical problems and providing the participants with tasty memories of French cuisine.

A few of the lecturers (F. D. M. Haldane, W. Werner, and J. Fröhlich) were unfortunately not able to provide a written contribution to this volume. We have nevertheless included a page containing their titles and affiliations at the place where the text of their lectures would have appeared.

Paris, Orsay, and Saclay, June 2009

J. L. Jacobsen
S. Ouvry
V. Pasquier
D. Serban
L. F. Cugliandolo

Contents

List of participants

ORGANIZERS

MICHAEL FEIGEL'MAN
Landau Institute for Theoretical Physics, Kosygina Street 2, 119334 Moscow, Russia

JESPER JACOBSEN
LPTMS, CNRS, Université Paris Sud, Orsay, France

STÉPHANE OUVRY
LPTMS, CNRS, Université Paris Sud, Orsay, France

VINCENT PASQUIER
IPhT, CEA, Saclay, France

DIDINA SERBAN
IPhT, CEA, Saclay, France

LECTURERS

IAN AFFLECK
Department of Physics and Astronomy, University of British Columbia, Vancouver, B.C. V6T 1Z1, Canada

FABIEN ALET
Laboratoire de Physique Théorique, Université Paul Sabatier, 118 route de Narbonne, F-31062 Toulouse Cedex 4, France

JOHN CARDY, FRS
Rudolf Peierls Centre for Theoretical Physics, University of Oxford, 1 Keble Road, Oxford OX1 3NP, United Kingdom

NIGEL COOPER
Theory of Condensed Matter Group, Cavendish Laboratory, JJ Thomson Avenue, Cambridge CB3 0HE, United Kingdom

BERTRAND DUPLANTIER
DSM/IphT, Orme des Merisiers, CEA Saclay, F-91191 Gif-sur-Yvette Cedex, France

JÜRG FRÖHLICH
Department of Physics HPF G9.5, ETH Zurich, CH-8093 Zurich, Switerland

F. DUNCAN M. HALDANE
Department of Physics, Princeton University, 327 Jadwin Hall, Princeton, NJ 08544-0708, USA

THIERRY JOLICOEUR
LPTMS, CNRS UMR8626, Bat. 100, Université Paris Sud, F-91405 Orsay Cedex, France

RICHARD KENYON
Mathematics Department, Brown University, Providence, RI, USA

ALEXEI KITAEV
Department of Physics and Computer Science, California Institute of Technology, Pasadena, CA 91125, USA

IVAN KOSTOV
Institute for Nuclear Research and Nuclear Energy, 72 Tsarigradsko Chaussee, BG-1784 Sofia, Bulgaria

WERNER KRAUTH
Laboratoire de Physique Statistique, École Normale Supérieure, 24 rue Lhomond, F-75231 Paris Cedex 05, France

SATYA MAJUMDAR
LPTMS, CNRS UMR8626, Bat. 100, Université Paris Sud, F-91405 Orsay Cedex, France

GRÉGOIRE MISGUICH
DSM/IphT, Orme des Merisiers, CEA Saclay, F-91191 Gif-sur-Yvette Cedex, France

BERNARD NIENHUIS
Institute for Theoretical Physics, University of Amsterdam, Valckenierstraat 65, 1018XE Amsterdam, The Netherlands

NICOLAI RESHETIKHIN
Department of Mathematics, University of California, Berkeley, CA 94720-3840, USA

HUBERT SALEUR
Institute for Nuclear Research and Nuclear Energy, 72 Tsarigradsko Chaussee, BG-1784 Sofia, Bulgaria

DAVID THOULESS
Department of Physics, Box 351560, University of Washington, Seattle, WA 98195, USA

WENDELIN WERNER
Laboratoire de Mathématiques, Bat. 425, Université Paris Sud, F-91405 Orsay Cedex, France

PAUL ZINN-JUSTIN
LPTMS, CNRS UMR8626, Bat. 100, Université Paris Sud, F-91405 Orsay Cedex, France

STUDENTS AND AUDITORS

VINCENZO ALBA
Scuola Normale Superiore, Piazza del Cavalieri 7, 56126 Pisa, Italy

S. ASHOK PARAMESWARAN
Department of Physics, Princeton University, Jadwin Hall, Washington Road, Princeton, NJ 08540, USA

ANDREA BEDINI
DSM/IphT, CEA Saclay, Orme des Merisiers, F-91191 Gif-sur-Yvette Cedex, France

Adel BENLAGRA
DSM/IphT, CEA Saclay, Orme des Merisiers, F-91191 Gif-sur-Yvette Cedex, France

Jean-Émile BOURGINE
DSM/IphT, CEA Saclay, Orme des Merisiers, F-91191 Gif-sur-Yvette Cedex, France

Michele BURELLO
International School for Advanced Studies (SISSA), Via Beirut 2-4, 34014 Trieste, Italy

Luigi CANTINI
LPTHE, Tour 24-25, 5e étage, Boite 126, 4 place Jussieu, F-75252 Paris Cedex 05, France

Daniel CHARRIER
LPT, CNRS UMR5152, Université Paul Sabatier, 118 Route de Narbonne, F-31062 Toulouse Cedex 04, France

Olga DIMITROVA
Landau Institute for Theoretical Physics, Kosygina Street 2, 119334 Moscow, Russia

Jérôme DUBAIL
DSM/IphT, CEA Saclay, Orme des Merisiers, F-91191 Gif-sur-Yvette Cedex, France

Vitalie EREMEEV
Institute of Applied Physics of Academy of Sciences of Moldova, str. Academiei 5, Chisinau MD-2028, Moldova

Benoit ESTIENNE
LPTHE, Tour 24-25, 5e étage, Boite 126, 4 place Jussieu, F-75252 Paris Cedex 05, France

Davide FIORETTO
International School for Advanced Studies (SISSA), Via Beirut 2-4, 34014 Trieste, Italy

Tiago FONCESA
LPTHE, Tour 24-25, 5e étage, Boite 126, 4 place Jussieu, F-75252 Paris Cedex 05, France

Azat GAINUTDINOV
P. N. Lebedev Physical Institute (I. E. Tamm Theory Department), 53 Leninsky prospect, Moscow 119991, Russia

Oleksandr GAMAYUN
Institute of Physics and Technology of the National Technical University, Kyiv Polytechnic Institute, Peremogy AV, Ukraine

Nikolay GROMOV
St. Petersburg State University, Faculty of Physics, Peterhof, 198 904 St. Petersburg, Russia

Jutho HAEGEMAN
Ghent University, Department of Mathematical Physics and Astronomy, Krijgslaan 281, S9, 9000 Gent, Belgium

Maria HERMANNS
Stockholm University, Roslagstullsbacken 21, 11421 Stockholm, Sweden

Chang-Yu HOU
Department of Physics, Boston University, 590 Commonwealth Ave., Room 255, Boston, MA 02215, USA

BENJAMIN HSU
University of Illinois, Urbana-Champaign, Department of Physics, 1110 W. Green Street, Urbana, IL 61801, USA
LIZA HUIJSE
Institute for Theoretical Physics of the University of Amsterdam, Valckenierstraat 65, 1018 XE Amsterdam, The Netherlands
YACINE IKHLEF
Rudolf Peierls Centre for Theoretical Physics, University of Oxford, 1 Keble Road, Oxford OX1 3NP, United Kingdom
THOMAS JACKSON
Department of Physics, Yale University, P. O. Box 208120, New Haven, CT 06520-8120, USA
LUDOVIC JAUBERT
Laboratoire de Physique, École Normale Supérieure, 46 allée d'Italie, F-69364 Lyon Cedex 07, France
ANTOINE KLAUSER
Institut Lorentz for Theoretical Physics, Niels Bohrweg 2, Leiden, NL-233 CA, The Netherlands
CHRISTOPHER LAUMANN
Princeton University, Department of Physics, Washington Road, Princeton, NJ 08544, USA
IVAN LEVKIVSKYI
Bogolyubov Institute for Theoretical Physics, 14-b Metrologichna Str, Kyiv 03143, Ukraine
DMYTRO MAKOGON
Faculty of Radiophysics of Taras Shevchenko Kyiv University, Vladimirskaya, 64, Kyiv 252601, Ukraine
CHIHIRO MATSUI
University of Tokyo, 7-3-1, Hongo, Bunkyo-ku, Tokyo 113-0033, Japan
LAURA MESSIO
LPTMC, Tour 24, 4 place Jussieu, F-75252 Paris Cedex 05, France
VICTOR MIKHAYLOV
Moscow Institute of Physics and Technology, Institutsky per. 9, Dolgopruchy, Moscow Region, 141700 Russia
NIALL MORAN
Department of Mathematical Physics, NUI Maynooth, Co. Kildare, Ireland
MICHAEL MULLIGAN
Department of Physics, Stanford University, 382 Via Pueblo Mall, Stanford, CA 94305, USA
ZLATKO PAPIC
Institut of Physics, Pregrevica 118, 11000 Belgrade, Serbia
OVIDIU IONEL PATU
C. N. Yang Institute for Theoretical Physics, State University at Stony Brook, Stony Brook, NY 11790-3840, USA
BALAZS POZSGAI
Institute for Theoretical Physics, Eotvos Lorand University, Pazmani Peter Setany 1/a 1117, Budapest, Hungary

Sylvain PROLHAC
DSM/IphT, CEA Saclay, Orme des Merisiers, F-91191 Gif-sur-Yvette Cedex, France

Armin RAHMANI-SISAN
Boston University, Physics Department, 590 Commonwealth Ave., Boston, MA 02215, USA

Enrique RICO ORTEGA
Innsbruck University, Institut für Theoretische Physik, Room 2/24, Technikstrasse 25, Innsbruck, A-6020 Austria

Maksym SERBYN
Landau Institute for Theoretical Physics, Kosygina Street 2, 119334 Moscow, Russia

Igor SHENDEROVICH
St. Petersburg State University, Physics Department, Ulianovskaya Street 3, Petrodvorets, St. Petersburg 198504, Russia

Keiichi SHIGECHI
LMPT, UFR Sciences et Technologie, Université de Tours, Parc de Grandmont, F-37200 Tours, France

Jacob SIMMONS
Rudolf Peierls Centre for Theoretical Physics, University of Oxford, 1 Keble Road, Oxford OX1 3NP, United Kingdom

Damien SIMON
Laboratoire de Physique Statistique, École Normale Supérieure, 24 rue Lhomond, F-75231 Paris Cedex 05, France

Andrea SPORTIELLO
Dipartimento di Fisica, Universita degli Studi di Milano, Via Celoria 16, 20133 Milano, Italy

Brigitte SURER
Institut for Theoretical Physics, ETH Zurich, Schafmattstr. 32, CH-8093 Zurich, Switzerland

Ronny THOMALE
Institut für Theorie der Kondensierten Materie, Universität Karlsruhe, Physikhochhaus 76128 Karlsruhe, Germany

Creighton THOMAS
Department of Physics, Syracuse University, 201 Physics Building, Syracuse, NY 13244, USA

Konstantin TIKHONOV
Landau Institute for Theoretical Physics, Kosygina Street 2, 119334 Moscow, Russia

Fabien TROUSSELET
LPT, CNRS UMR5152, Université Paul Sabatier, 118 Route de Narbonne, F-31062 Toulouse Cedex 04, France

Jacopo VITI
University of Florence, Department of Physics, Via Sansone 1, Sesto Fiorentino, Italy

Dmytro VOLIN
Bogolyubov Institute for Theoretical Physics, 14b Metrolohichna Str., Kyiv 03143, Ukraine

Mikhail ZVONAREV
Petersburg Department of Steklov Institute of Mathematics, 27 Fontanka, St. Petersburg 191023, Russia

All participants.

The organizers of the session, here seen at the Les Houches workshop entitled 'Physics in the plane: From condensed matter to string theory', in March 2010. From left to right: Jesper L. Jacobsen, Stephane Ouvry, Vincent Pasquier, and Didina Serban.

On the summit of Mont Blanc.

Part I

Long lectures

1
Quantum impurity problems in condensed matter physics

I. AFFLECK

Ian Affleck,
Department of Physics and Astronomy,
University of British Columbia,
Vancouver, B.C., Canada, V6T 1Z1.

1.1 Quantum impurity problems and the renormalization group

A remarkable property of nature that has intrigued physicists for many years is universality at critical points (see, for example, Ma 1976). An impressive example is the critical point of water. By adjusting the temperature and pressure, one can reach a critical point where the correlation length diverges and the long-distance physics becomes the same as that of the Ising model. A microscopic description of water is very complicated and has very little connection with the Ising model; in particular, there is no lattice, no spin operators, and not even any Z_2 symmetry. Nonetheless, various experimentally measured critical exponents appear to be exactly the same as those of the Ising model. Furthermore, the best description of this universal long-distance behavior is probably provided by the φ^4 field theory at its critical point. Our understanding of universality is based upon the renormalization group (RG). For a system at or near a critical point with a diverging correlation length, it is convenient to consider an effective free energy (or Hamiltonian), used only to describe long-distance properties, which is obtained by integrating out short-distance degrees of freedom. It is found that the same long-distance Hamiltonian, characterizing an RG fixed point, is obtained from many different microscopic models. These fixed-point Hamiltonians are universal attractors for all microscopic models (Ma 1976).

The same features hold for quantum models of many-body systems at low temperature. In many cases such models can exhibit infinite correlation lengths and vanishing excitation energy gaps. In this situation one again expects universality. Important examples are the Fermi liquid fixed point for interacting electrons in three dimensions ($D = 3$) (see, for example, the RG treatment of Landau Fermi liquid theory in Shankar 1994), its cousin, the Luttinger liquid fixed point in $D = 1$, and various models of interacting quantum spins (Affleck 1990, Giamarchi 2004).

In these lectures, I will be concerned with a single quantum impurity embedded in such a critical system. The quantum impurity can be of quite a general form, possibly comprising several nearby impurities. If we study its behavior at long length scales (compared with all microscopic lengths, including the spatial extent of the impurity and the range of its interactions with the host) and at low energies compared with all microscopic energy scales, then universality again emerges. These single-impurity models, while simplified, have the attractive feature that such powerful methods as boundary conformal field theory can be used to tackle them. They provide quite nontrivial examples of quantum critical phenomena and, in some cases, appear to be good descriptions of experimental reality.

An important example is provided by the simplest version of the Kondo model (Kondo 1964, Hewson 1997, Affleck 1995). This is a model invented to describe a single magnetic impurity (such as an iron atom) in a nonmagnetic metal (such as copper). Traditional experiments in this field always involve a finite density of impurities, but if this density is low enough, we may consider only one of them; technically, this gives the first term in a virial expansion in the impurity density. Furthermore, these models can be applied to situations where the single impurity is a nanostructure device such as a quantum dot. A simple Hamiltonian to describe this system can be written

$$H = \int d^3k\, \psi^{\dagger\alpha}_{\vec{k}} \psi_{\vec{k}\alpha} \epsilon(k) + J \int \frac{d^3k\, d^3k'}{(2\pi)^3} \psi^{\dagger\alpha}_{\vec{k}} \frac{\vec{\sigma}^{\beta}_{\alpha}}{2} \psi_{\vec{k}'\beta} \cdot \vec{S}\,. \tag{1.1}$$

Here $\psi_{\vec{k}\alpha}$ annihilates an electron of wave vector $\vec{k}$ and spin α and is normalized so that

$$\{\psi_{\vec{k}}^{\dagger\alpha}, \psi_{\vec{k}'\beta}\} = \delta^{\alpha}_{\beta}\delta^3(\vec{k} - \vec{k}') \,. \tag{1.2}$$

Repeated spin indices are summed over. $\epsilon(\vec{k})$ is the dispersion relation for the electrons, which we will usually approximate by the free-electron form

$$\epsilon(\vec{k}) = \frac{k^2}{2m} - \epsilon_F \,, \tag{1.3}$$

where ϵ_F is the Fermi energy. (So this is actually $H - \mu\hat{N}$.) $\vec{S}$ is an impurity spin operator, of magnitude S. J measures the strength of a Heisenberg-type exchange interaction between the electron spin density and the impurity spin. Usually, $J > 0$. Note that this form of interaction is a δ-function in position space:

$$H = \int d^3r \left[\psi^\dagger(\vec{r})\left(-\frac{\nabla^2}{2m} - \epsilon_F\right)\psi(\vec{r}) + J\delta^3(r)\psi^\dagger\frac{\vec{\sigma}}{2}\psi\cdot\vec{S}\right] . \tag{1.4}$$

Here we have suppressed the spin indices completely. Actually, this model is ultraviolet divergent unless we truncate the integral over $\vec{k}$, $\vec{k}'$ in the interaction term. Such a truncation is assumed here but its details will not be important in what follows. The dimensionless measure of the strength of the Kondo interaction is

$$\lambda \equiv J\nu \,, \tag{1.5}$$

where ν is the density of states, per unit energy per unit volume per spin. For free electrons this is

$$\nu = \frac{mk_F}{\pi^2} \,, \tag{1.6}$$

where k_F is the Fermi wave vector. Typically, $\lambda \ll 1$.

This model is a considerable simplification of reality. In particular, electrons in metals interact with each other via the Coulomb interaction and this is neglected. This can be justified using Fermi liquid ideas. Since $\lambda \ll 1$, the Kondo interaction affects only electrons close to the Fermi surface. The Coulomb interactions become increasingly ineffective for these electrons, as can be seen from phase space arguments, after taking into account screening of the Coulomb interactions. The free-electron Hamiltonian (with an appropriate effective mass) represents the fixed-point Hamiltonian, valid at low energies. Treating the Kondo interaction as a δ-function is another approximation; a more realistic model would give it a finite range. Again, if we are concerned with the long-distance, low-energy physics, we might expect this distinction to be unimportant. The spherical symmetry of the dispersion relation and Kondo interaction will considerably simplify our analysis, but again can be seen to be inessential.

Owing to the absence of bulk interactions, the δ-function form of the Kondo interaction, and the spherical symmetry of $\epsilon(k)$, we may usefully expand the electron operators in spherical harmonics, finding that only the s-wave harmonic interacts with the impurity. (See, for example, Appendix A of Affleck and Ludwig (1991).) This gives us an effectively one-dimensional problem, defined on the half-line $r > 0$, with the impurity sitting at the beginning of the line, $r = 0$. Thus we write

$$\psi_{\vec{k}} = \frac{1}{\sqrt{4\pi}k}\psi_0(k) + \text{ higher harmonics} \,. \tag{1.7}$$

Next we restrict the integral over k in the Hamiltonian to a narrow band around the Fermi wave vector:

$$-\Lambda < k - k_F < \Lambda\,. \tag{1.8}$$

This is justified by the fact that $\lambda \ll 1$. To be more accurate, we should integrate out the Fourier modes further from the Fermi surface, renormalizing the Hamiltonian in the process. However, for small λ this only generates small corrections to H, which we simply ignore. Actually, this statement is true only if Λ is chosen to be small but not *too* small. We want it to be $\ll k_F$. However, if it becomes arbitrarily small, eventually the renormalized λ starts to blow up, as we discuss below. Thus, we assume that Λ is chosen judiciously to have an intermediate value. We can then approximate the dispersion relation by

$$\epsilon(k) \approx v_F(k - k_F)\,. \tag{1.9}$$

We now define the following position space fields:

$$\psi_{L/R}(r) \equiv \int_{-\Lambda}^{\Lambda} dk\, e^{\pm ikr}\psi_0(k_F + k)\,. \tag{1.10}$$

Note that these obey the boundary condition

$$\psi_L(t, r = 0) = \psi_R(t, r = 0)\,. \tag{1.11}$$

Furthermore, they obey approximately the anticommutation relations

$$\{\psi_{L/R}(x), \psi^\dagger_{L/R}(x')\} = 2\pi\delta(x - x')\,. \tag{1.12}$$

(This is only approximately true at long distances. The Dirac δ-function is actually smeared over a distance of order $1/\Lambda$. Note also the unconventional normalization in eqn (1.12).) The Hamiltonian can then be written

$$H = \frac{v_F}{2\pi} i \int_0^\infty dr \left(\psi^\dagger_L \frac{d}{dr}\psi_L - \psi^\dagger_R \frac{d}{dr}\psi_R\right) + v_F\lambda\psi^\dagger_L(0)\frac{\vec{\sigma}}{2}\psi_L(0)\cdot\vec{S} + \text{higher harmonics}\,. \tag{1.13}$$

The "higher harmonics" terms in H are noninteracting and we will generally ignore them. This is a (1+1)-dimensional massless Dirac fermion (with 2 "flavors" or spin components) defined on a half-line, interacting with the impurity spin. The velocity of light is replaced by the Fermi velocity, v_F. We shall generally set $v_F = 1$. Note that in a space–imaginary-time representation, the model is defined on the half-plane and there is an interaction with the impurity spin along the edge, $r = 0$. Since the massless Dirac fermion model is a conformal field theory, this is a type of conformal field theory (CFT) with a boundary. However, it is a much more complicated boundary than that discussed in John Cardy's lectures at this summer school. There, he considered CFTs with conformally invariant boundary conditions. If we set $\lambda = 0$ then we have such a model since the boundary condition (BC) of eqn (1.11) is conformally invariant. More precisely, we have a boundary conformal field theory (BCFT) and a decoupled spin, sitting at the boundary. However, for $\lambda \neq 0$, we do not have merely a BC but a boundary interaction with an impurity degree of freedom. Nonetheless, as we shall see, the low-energy fixed-point Hamiltonian is just a standard BCFT.

Although the form of the Hamiltonian in eqn (1.13) makes the connection with BCFT most explicit, it is often convenient to make an "unfolding" transformation. Since $\psi_L(t,x)$ is a function of $(t+x)$ only and $\psi_R(t,x)$ is a function of $(t-x)$ only, the boundary condition of eqn (1.11) implies that we may think of ψ_R as the analytic continuation of ψ_L to the negative r-axis:

$$\psi_R(r) = \psi_L(-r) \quad (r > 0)\,. \tag{1.14}$$

The Hamiltonian can be rewritten

$$H = \frac{v_F}{2\pi} i \int_{-\infty}^{\infty} dr\, \psi_L^\dagger \frac{d}{dr}\psi_L + v_F \lambda \psi_L^\dagger(0)\frac{\vec{\sigma}}{2}\psi_L(0)\cdot\vec{S}\,. \tag{1.15}$$

We have reflected the outgoing wave to the negative r-axis. In this representation, the electrons move to the left only, interacting with the impurity spin as they pass the origin.

The phrase "Kondo problem", as far as I know, refers to the infrared-divergent property of perturbation theory, in λ, discovered by Kondo in the mid 1960s. In the more modern language of the RG, this simply means that the renormalized coupling constant, $\lambda(E)$, increases as the characteristic energy scale, E, is lowered. The "problem" is how to understand the low-energy behavior given this failure of perturbation theory, a failure which occurs despite the fact that the original coupling constant $\lambda \ll 1$. The β-function may be calculated using Feynman diagram methods; the first few diagrams are shown in Fig. 1.1. The dotted line represents the impurity spin. The simplest way to deal with it is to use time-ordered real-time perturbation theory and to explicitly evaluate the quantities

$$\mathcal{T}\left\langle 0|S^a(t_1)S^b(t_2)S^c(t_3)\ldots|0\right\rangle. \tag{1.16}$$

(For a detailed discussion of this approach and some third-order calculations, see, for example, Barzykin and Affleck (1998).) Since the noninteracting part of the Hamiltonian is independent of $\vec{S}$, these products are actually time-independent, up to some minus signs arising from the time-ordering. For instance, for the $S = 1/2$ case,

$$\mathcal{T}\langle 0|S^x(t_1)S^y(t_2)|0\rangle = \theta(t_1-t_2)S^xS^y + \theta(t_2-t_1)S^yS^x = \mathrm{sgn}(t_1-t_2)iS^z\,. \tag{1.17}$$

Here $\mathrm{sgn}(t)$ is the sign function, ± 1 for t positive or negative, respectively. Using the spin commutation relations and $\vec{S}^2 = S(S+1)$, it is possible to explicitly evaluate the expectation

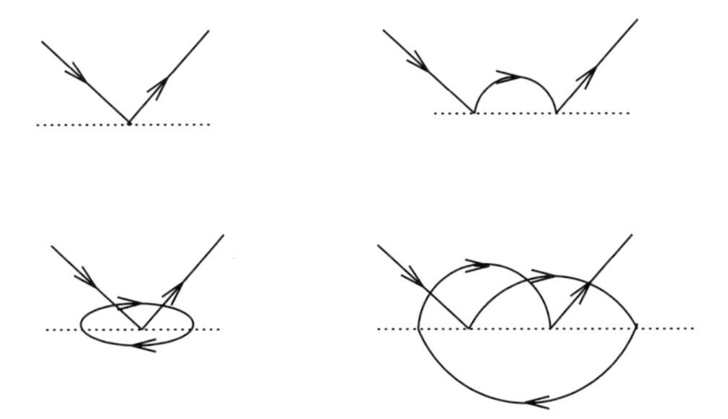

Fig. 1.1 Feynman diagrams contributing to renormalization of the Kondo coupling constant to third order.

values of any of the spin products occurring in perturbation theory. We are then left with standard fermion propagators. These are simplified by the fact that all fermion fields occurring in the Kondo interaction are at $r = 0$; we suppress the spatial labels in what follows. For instance, the second-order diagram is

$$-\frac{\lambda^2}{2}\int dt\, dt'\, T(S^a(t)S^b(t'))\cdot T\left[\psi^\dagger(t)\frac{\sigma^a}{2}\psi(t)\psi^\dagger(t')\frac{\sigma^b}{2}\psi(t')\right], \tag{1.18}$$

which can be reduced, using Wick's theorem, to

$$\frac{\lambda^2}{2}\int dt\, dt'\, \psi^\dagger\frac{\vec{\sigma}}{2}\psi\cdot\vec{S}\,\mathrm{sgn}(t-t')\,\langle 0|\psi(t)\psi^\dagger(t')|0\rangle\,. \tag{1.19}$$

The free-fermion propagator is simply

$$G(t) = \frac{-i}{t}\,. \tag{1.20}$$

This gives a correction to the effective coupling constant

$$\delta\lambda = \frac{\lambda^2}{2}\int dt\frac{\mathrm{sgn}(t)}{t}\,. \tag{1.21}$$

Integrating symmetrically, eqn (1.19) would give zero if the factor sgn (t), coming from the impurity spin Green's function, were absent. This corresponds to a cancelation between particle and hole contributions. This is as it should be. If we have a simple nonmagnetic scatterer, with no dynamical degrees of freedom, there is no renormalization. The Kondo problem arises entirely from the essentially quantum mechanical nature of the impurity spin. Including the $\mathrm{sgn}(t)$ factor, the integral in eqn (1.19) is infrared and ultraviolet log-divergent. In an RG transformation, we integrate over only a restricted range of wave vectors, integrating out modes with $D' < |k| < D$. We then obtain

$$\delta\lambda = \lambda^2\ln(D/D')\,. \tag{1.22}$$

The corresponding β-function, to this order, is then

$$\frac{d\lambda}{d\ln D} = -\lambda^2 + \ldots\,. \tag{1.23}$$

Solving for the effective coupling at scale D, in terms of its bare value λ_0 at scale D_0, we obtain

$$\lambda(D) \approx \frac{\lambda_0}{1-\lambda_0\ln(D_0/D)}\,. \tag{1.24}$$

If the bare coupling is *ferromagnetic*, $\lambda_0 < 0$, then $\lambda(D)$ is well behaved, getting smaller in magnitude at lower energy scales. However, if it is antiferromagnetic, $\lambda(D)$ continues to increase as we reduce the energy scale until it gets so large that lowest-order perturbation theory for the β-function breaks down. We may estimate the energy scale where this happens as

$$T_K \approx D_0\exp(-1/\lambda_0)\,. \tag{1.25}$$

The scale D_0, which plays the role of the ultraviolet cutoff, is of order the band width or Fermi energy, and $D_0 = v_F/\Lambda$, where Λ is the cutoff in momentum units.

After many years of research by many theorists, a very simple picture emerged for the low-energy behavior of the Kondo model, due in large part to the contributions of Anderson (1970), Wilson (1975), Nozières (1975), and their collaborators. We may think of λ as renormalizing to ∞. What is perhaps surprising is that the infinite-coupling limit is actually very simple. To see this, it is very convenient to consider a lattice model,

$$H = -t \sum_{i=0}^{\infty} (\psi_i^\dagger \psi_{i+1} + \psi_{i+1}^\dagger \psi_i) + J \psi_0^\dagger \frac{\vec{\sigma}}{2} \psi_0 \cdot \vec{S}. \tag{1.26}$$

The strong-coupling limit corresponds to $J \gg t$. It is quite easy to solve this limit exactly. One electron sits at site 0 and forms a spin singlet with the impurity, which I assume to have $S = 1/2$ for now. $| \Uparrow\downarrow\rangle - | \Downarrow\uparrow\rangle$. (Here the double arrow refers to the impurity spin and the single arrow to the spin of the electron at site zero.) The other electrons can do anything they like, as long as they don't go to site 0. Thus, we say the impurity spin is "screened," or, more accurately, has formed a spin singlet. To understand the low-energy effective Hamiltonian, we are more interested in what the other electrons are doing, on the other sites. If we now consider a small but finite t/J, the other electrons will form the usual free-fermion ground state, filling a Fermi sea, but with a modified boundary condition that they cannot enter site 0. It is as if there were an infinite repulsion at site 0. The single-particle wave functions are changed from $\sin k(j+1)$ to $\sin kj$. In the particle–hole (PH) symmetric case of a half-filled band, $k_F = \pi/2$, the phase shift at the Fermi surface is $\pi/2$. In this one-dimensional case, we take the continuum limit by writing

$$\psi_j \approx e^{ik_F j} \psi_R(j) + e^{-ik_F j} \psi_L(j) . \tag{1.27}$$

For $\lambda = 0$, in the PH symmetric case, the open boundary condition for the lattice model corresponds to

$$\psi_L(0) = \psi_R(0) \tag{1.28}$$

in the continuum model, just as in $D = 3$. On the other hand, the strong-coupling BC is

$$\psi_L(0) = -\psi_R(0) . \tag{1.29}$$

The strong-coupling fixed point is the same as the weak-coupling fixed point except for a change in boundary conditions (and the removal of the impurity). We describe the strong-coupling fixed point by the conformally invariant BC of eqn (1.29).

This simple example illustrates the main ideas of the BCFT approach to quantum impurity problems. In general, we consider systems whose long-distance, low-energy behavior, in the absence of any impurities, is described by a CFT. Examples include noninteracting fermions in any dimension, and interacting fermions (Luttinger liquids) in $D = 1$. We then add some interactions, involving impurity degrees of freedom, localized near $r = 0$. Despite the complicated, interacting nature of the boundary in the microscopic model, the low-energy, long-distance physics is always described by a conformally invariant BC. The impurity degrees of freedom always either get screened or decouple, or some combination of both. Why should this be true in general? Some insight can be gained by considering the behavior of arbitrary Green's functions at space–(imaginary-)time points $z_1 = \tau_1 + ir_1$, $z_2 = \tau_2 + ir_2$, ... Very close to the boundary, we expect nonuniversal behavior. If all points r_i are very far from the boundary, $r_i \gg |z_j - z_k|$, then we expect to recover the bulk behavior, unaffected by the boundary interactions. This behavior is conformally invariant. However, if the time separations of some of the points are larger than or of the order of the distances from the boundary, which are themselves large compared with microscopic scales, then the boundary still affects the Green's functions. We expect it to do so in a conformally invariant way. We have a sort of conformally invariant termination of the bulk conformal behavior, which is influenced by the universality class of the boundary, encoded in a conformally invariant boundary condition. Note that the RG flow being discussed here is entirely restricted to boundary interactions. The bulk terms in the effective Hamiltonian do not renormalize, in our description; they sit at a bulk critical point. We do not expect finite-range interactions, localized near $r = 0$, to produce any renormalization of the bulk behavior. All of the RG flows, which play such an important role in these lectures, are boundary RG flows.

It actually turns out to be extremely important to go slightly beyond merely identifying the low-energy fixed point, and to consider in more detail how it is approached, as the energy is lowered. As is usual in RG analyses, this is controlled by the leading irrelevant operator (LIO) at this fixed point. This is a boundary operator, defined in the theory with the conformally invariant boundary condition (CIBC) characteristic of the fixed point. It is important to realize that, in general, the set of boundary operators which exist depends on the CIBC.

In the case at hand, the simplest version of the Kondo model, the boundary operators are simply constructed out of the fermion fields, which now obey the BC of eqn (1.29). It is crucial to realize that the impurity spin operator cannot appear in the low-energy effective Hamiltonian, because it is screened. Thus the LIO is constructed from fermion fields only. It must be SU(2) invariant. In general, the operator $\psi^{\dagger\alpha}(0)\psi_\alpha(0)$ might appear. This has dimension 1 and is thus marginal. Note that 1 is the marginal dimension for boundary operators in a (1+1)-dimensional CFT since these terms in the action are integrated over time only, not space. If we restrict ourselves to the PH symmetric case, then this operator cannot appear, since it is odd under the PH transformation, $\psi \rightarrow \psi^\dagger$. Thus we must turn to four-fermion operators, of dimension 2, which are irrelevant. There are two operators allowed by SU(2) symmetry, $J(0)^2$ and $\vec{J}(0)^2$, where the charge and spin currents are defined as

$$J \equiv \psi^{\dagger\alpha}\psi_\alpha\,, \quad \vec{J} \equiv \psi^{\dagger\alpha}\frac{\vec{\sigma}_\alpha^\beta}{2}\psi_\beta\,. \tag{1.30}$$

Here I am suppressing L, R indices. I work in the purely left-moving formalism, so all operators are left-movers. I will argue below that, since the Kondo interaction involves the spin degrees of freedom, the $\vec{J}^2$ term in the effective Hamiltonian has a much larger coefficient that does the J^2 term. More precisely, we expect that coefficient of the $\vec{J}^2$ term to be of order $1/T_K$. By power counting, it must have a coefficient with dimensions of inverse energy. $1/T_K$ is the largest possible coefficient (corresponding to the lowest characteristic energy scale) that could occur, and there is no reason why it should not occur in general. Another way of looking at this is that the low-energy effective Hamiltonian has a reduced cutoff of order T_K (or T_K/v_F in wave vector units). The coefficient of $\vec{J}^2$ is of the order of the inverse cutoff. By contrast, I shall argue below that the coefficient of J^2 is of order $1/D_0 \ll 1/T_K$. Thus, this term can be ignored. The precise value of the coefficient of $\vec{J}^2$ is not known; but neither have I yet given a precise definition of T_K. Unfortunately, there are a large number of definitions of characteristic energy scales in use, referred to as T_K among other things. One possibility is to fix a definition of T_K from the coupling constant of the LIO:

$$H = \frac{v_F}{2\pi} i \int_{-\infty}^{\infty} dr\, \psi_L^\dagger \frac{d}{dr} \psi_L - \frac{1}{6T_K} \vec{J}_L(0)^2 . \tag{1.31}$$

The factor of $1/6$ is inserted for convenience. The fact that the sign is negative has physical significance and, in principle, can only be deduced from comparison with other calculations (or experiments). Note that this Hamiltonian is defined with the low-energy effective BC of eqn (1.29). This means that the unfolding transformation used to write eqn (1.31) is changed by a minus sign, i.e. eqn (1.14) is modified to

$$\psi_R(r) = -\psi_L(-r) \quad (r > 0) . \tag{1.32}$$

With this effective Hamiltonian in hand, we may calculate various physical quantities in perturbation theory in the LIO, i.e. perturbation theory in $1/T_K$. This will be discussed in detail later, but basic features follow from power counting. An important result is for the impurity magnetic susceptibility. The susceptibility is

$$\chi(T) \equiv \frac{1}{T} \left\langle (S_T^z)^2 \right\rangle , \tag{1.33}$$

where $\vec{S}_T$ is the *total* spin operator including both impurity and electron spin operators. The impurity susceptibility is defined, motivated by experiments, as the difference in susceptibilities of samples with and without the impurity. In practice, for a finite density of impurities, it is the term in the virial expansion of the susceptibility of first order in the impurity density n_i:

$$\chi = \chi_0 + n_i \chi_{\rm imp} + \dots . \tag{1.34}$$

If we ignore the LIO, $\chi_{\rm imp}$ vanishes at low T. This follows because the Hamiltonian of eqn (1.31) is translationally invariant. A simple calculation, reviewed later, shows that, to first order in the LIO,

$$\chi_{\rm imp} \to \frac{1}{4T_K} . \tag{1.35}$$

This is the leading low-T result, corrected by a power series in T/T_K. On the other hand, the high-T result at $T \gg T_K$, in the scaling limit of small λ_0, is

$$\chi \to \frac{1}{4T} , \tag{1.36}$$

the result for a decoupled impurity spin. Our RG, BCFT methods give the susceptibility only in the low-T and high-T limits. More powerful machinery is needed to also calculate it throughout the crossover, when T is of order T_K. Such a calculation has been done accurately using the Bethe ansatz solution (Andrei 1980, Weigmann 1980), giving

$$\chi(T) = \frac{1}{4T_K} f\left(\frac{T}{T_K}\right), \tag{1.37}$$

where $f(x)$ is a universal scaling function. The asymptotic results of eqns (1.35) and (1.36) are obtained. While the RG, BCFT methods are generally restricted to low-energy and high-energy limits (near the RG fixed points), they have the advantages of relative simplicity and generality (i.e. they are not restricted to integrable models). The impurity entropy (or, equivalently, the impurity specific heat) has a similar behavior, with a contribution in first order in the LIO:

$$S_{\text{imp}} \rightarrow \frac{\pi^2 T}{6T_K} . \tag{1.38}$$

Again, S_{imp} is a universal scaling function of T/T_K, approaching $\ln 2$, the result for a decoupled impurity, at high T. It is also possible to calculate the impurity contribution to the electrical resistivity due to scattering off a dilute random array of impurities, using the Kubo formula. In this case, there is a contribution from the fixed-point Hamiltonian itself (i.e. from the modified BC of eqn (1.29)), even without including the LIO. This modified BC is equivalent to a $\pi/2$ phase shift in the s-wave channel (in the PH symmetric case). We may simply take over standard formulas for scattering from nonmagnetic impurities, which make a contribution to the resistivity expressed entirely in terms of the phase shift at the Fermi energy. A phase shift of $\pi/2$ gives the maximum possible resistivity, the so-called unitary limit:

$$\rho_u = \frac{3n_i}{(ev_F \nu)^2} . \tag{1.39}$$

The correction to the unitary limit can again be calculated in perturbation theory in the LIO. In this case the leading correction is second order:

$$\rho(T) \approx \rho_u \left[1 - \frac{\pi^4 T^2}{16 T_K^2}\right] \quad (T \ll T_K) . \tag{1.40}$$

Again, at $T \gg T_K$ we can calculate $\rho(T)$ in perturbation theory in the Kondo interaction:

$$\rho(T) \approx \rho_u \frac{3\pi^2}{16} \frac{1}{\ln^2(T/T_K)} \quad (T \gg T_K) . \tag{1.41}$$

In between, a scaling function of T/T_K occurs. It has not so far been possible to calculate this from the Bethe ansatz, but fairly accurate results have been obtained using numerical renormalization group methods. For a review, see Bulla *et al.* (2008).

This perturbation theory in the LIO is referred to as "Nozières local Fermi liquid theory (FLT)." This name is highly appropriate owing to the close parallels with Fermi liquid theory for bulk (screened) Coulomb interactions.

1.2 Multichannel Kondo model

In this lecture, I will generalize the BCFT analysis of the simplest Kondo model to the multichannel case (Nozières and Blandin 1980). I will give a fairly sketchy overview of this subject here; more details are given in my previous summer school lecture notes (Affleck 1995). One now imagines k identical "channels" all interacting with the same impurity spin, preserving an $\mathrm{SU}(k)$ symmetry. In fact, it is notoriously difficult to find physical systems with such $\mathrm{SU}(k)$ symmetry, so the model is an idealization. Much effort has gone into finding (or creating) systems realizing the $k = 2$ case, with recent success. Jumping immediately to the continuum limit, the analogue of eqn (1.13) is

$$H = \frac{v_F}{2\pi} i \int_0^\infty dr \left(\psi_L^{\dagger j} \frac{d}{dr} \psi_{Lj} - \psi_R^{\dagger j} \frac{d}{dr} \psi_{Rj} \right) + v_F \lambda \psi_L^{\dagger j}(0) \frac{\vec{\sigma}}{2} \psi_{Lj}(0) \cdot \vec{S}\,. \tag{1.42}$$

The repeated index j is summed from 1 to k; the spin indices are not written explicitly. The same "bare" BC, eqn (1.11) is used. The RG equations are only trivially modified:

$$\frac{d\lambda}{d \ln D} = -\lambda^2 + \frac{k}{2}\lambda^3 + \ldots\,. \tag{1.43}$$

The factor of k in the $O(\lambda^3)$ term follows from the closed loop in the third diagram in Fig. 1.1. We again conclude that a small bare coupling gets larger under the RG. However, for general choices of k and the impurity spin magnitude S, it can be readily seen that the simple strong-coupling BC of eqn (1.29) does not occur at the infrared fixed point. This follows from Fig. 1.2. If the coupling flowed to infinity then we would expect k electrons, one from each channel, to go into a symmetric state near the origin (at the first site in the limit of a strong bare coupling). They would form a total spin $k/2$. The antiferromagnetic coupling to the impurity of spin S would lead to a ground state of size $|S - k/2|$. It is important to distinguish three cases: $S < k/2$, overscreened; $S > k/2$, underscreened; and $S = k/2$, exactly screened. It turns out that this strong-coupling fixed point is stable in the underscreened and exactly screened cases only. In the exactly screened case, this follows from the same considerations as for $S = 1/2$, $k = 1$, discussed in the previous lecture, which is the simplest exactly screened case. Otherwise, further considerations of this strong-coupling fixed point are necessary. This effective spin will itself have a Kondo coupling to the conduction electrons, obeying the strong coupling BC of eqn (1.29). This is clearest in the lattice model discussed in the previous lecture. The effective spin is formed between the impurity spin and the electrons on site 0. If $J \gg t$, electrons from site 1 can make virtual transitions onto site 0, producing a high-energy state. Treating these in second order gives an effective Kondo coupling to site 1 with

$$J_{\text{eff}} \propto \frac{t^2}{J}\,. \tag{1.44}$$

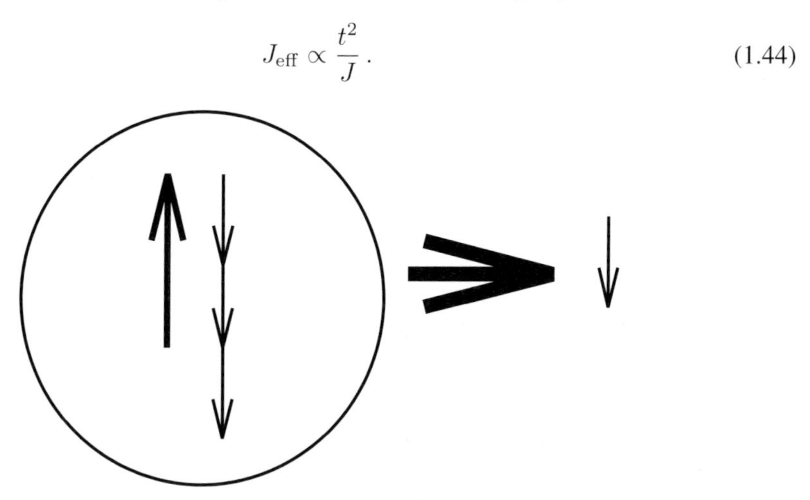

Fig. 1.2 Formation of an effective spin at strong Kondo coupling. $k = 3$, $s = 1$, and $s_{\text{eff}} = 1/2$.

The sign of J_{eff} is clearly crucial; if it is ferromagnetic, then it would renormalize to zero. In this case, the strong-coupling fixed point is stable. (Note that $J_{\text{eff}} \to 0$ corresponds to the original $J \to \infty$.) However, if it is antiferromagnetic then J_{eff} gets larger as we lower the energy scale. This invalidates the assumption that the strong-coupling fixed point is stable. (J_{eff} getting larger corresponds to J getting smaller.) It is not hard to see that $J_{\text{eff}} > 0$ for the overscreened case and $J_{\text{eff}} < 0$ for the underscreened case. (This calculation can be done in two steps. First, consider the sign of the exchange interaction between the electron spins on site 0 and on site 1. This is always antiferromagnetic, as in the Hubbard model. Then consider the relative orientation of the electron spin on site 0 and $\vec{S}_{\text{eff}}$. These are parallel in the overscreened case but antiparallel in the underscreened case.) So, in the underscreened case, we may apply the results of the previous lecture with minor modifications. There is an impurity spin of size S_{eff} in the low-energy Hamiltonian but it completely decouples at the infrared fixed point. The overscreened case is much more interesting. It is a "non-Fermi liquid (NFL)."

To solve this case, I introduce the idea of a conformal embedding (CE). This is actually useful for various other BCFT problems. It is a generalization of the idea of bosonization, a powerful technique in (1+1) dimensions.

We start by considering a left-moving single component (no channels, no spin) fermion field with Hamiltonian density

$$\mathcal{H} = \frac{1}{2\pi}\psi_L^\dagger i\frac{d}{dx}\psi_L\,. \tag{1.45}$$

We define the current (= density) operator,

$$J_L(t+x) =: \psi_L^\dagger\psi_L : (x,t) = \lim_{\epsilon\to 0}[\psi_L^\dagger(x)\psi_L(x+\epsilon) - \langle 0|\psi_L^\dagger(x)\psi_L(x+\epsilon)|0\rangle]\,. \tag{1.46}$$

(Henceforth we shall generally drop the subscripts L and the time argument. The double dots denote normal ordering: creation operators on the right.) We will reformulate the theory in terms of currents (the key to bosonization). Consider

$$\begin{aligned} J(x)J(x+\epsilon) &= :\psi^\dagger(x)\psi(x)\psi^\dagger(x+\epsilon)\psi(x+\epsilon): \\ &\quad + [:\psi^\dagger(x)\psi(x+\epsilon): + :\psi(x)\psi^\dagger(x+\epsilon):]G(\epsilon) + G(\epsilon)^2\,, \\ G(\epsilon) &= \langle 0|\psi(x)\psi^\dagger(x+\epsilon)|0\rangle = \frac{1}{-i\epsilon}\,. \end{aligned} \tag{1.47}$$

By Fermi statistics, the four-fermion term vanishes as $\epsilon \to 0$:

$$:\psi^\dagger(x)\psi(x)\psi^\dagger(x)\psi(x): = -:\psi^\dagger(x)\psi^\dagger(x)\psi(x)\psi(x): = 0\,. \tag{1.48}$$

The second term becomes a derivative:

$$\begin{aligned} \lim_{\epsilon\to 0}\left[J(x)J(x+\epsilon) + \frac{1}{\epsilon^2}\right] &= \lim_{\epsilon\to 0}\frac{1}{-i\epsilon}[:\psi^\dagger(x)\psi(x+\epsilon): - :\psi^\dagger(x+\epsilon)\ \psi(x):] \\ &= 2i:\psi^\dagger\frac{d}{dx}\psi:\,, \\ \mathcal{H} &= \frac{1}{4\pi}J(x)^2 + \text{constant}\,. \end{aligned} \tag{1.49}$$

Now consider the commutator $[J(x), J(y)]$. The quartic and quadratic terms cancel. We must be careful about the divergent c-number part,

$$
\begin{aligned}
[J(x), J(y)] &= -\frac{1}{(x-y-i\delta)^2} + \frac{1}{(x-y+i\delta)^2} \qquad (\delta \to 0^+) \\
&= \frac{d}{dx}\left[\frac{1}{x-y-i\delta} - \frac{1}{x-y+i\delta}\right] \\
&= 2\pi i \frac{d}{dx}\delta(x-y)\,.
\end{aligned} \tag{1.50}
$$

Here δ is an ultraviolet cutoff.

Now we consider the free-massless-boson theory with Hamiltonian density (setting $v_F = 1$)

$$
\mathcal{H} = \frac{1}{2}\left(\frac{\partial \varphi}{\partial t}\right)^2 + \frac{1}{2}\left(\frac{\partial \varphi}{\partial x}\right)^2, \quad [\varphi(x), \frac{\partial}{\partial t}\varphi(y)] = i\delta(x-y)\,. \tag{1.51}
$$

We can again decompose this into left- and right-moving parts,

$$
\begin{aligned}
(\partial_t{}^2 - \partial_x{}^2)\varphi &= (\partial_t + \partial_x)(\partial_t - \partial_x)\varphi\,, \\
\varphi(x,t) &= \varphi_L(x+t) + \varphi_R(x-t)\,, \\
(\partial_t - \partial_x)\varphi_L \equiv \partial_-\varphi_L &= 0,\ \ \partial_+\varphi_R = 0\,, \\
H = \frac{1}{4}(\partial_-\varphi)^2 + \frac{1}{4}(\partial_+\varphi)^2 &= \frac{1}{4}(\partial_-\varphi_R)^2 + \frac{1}{4}(\partial_+\varphi_L)^2\,,
\end{aligned} \tag{1.52}
$$

where

$$
\partial_\pm \equiv \partial_t \pm \partial_x\,. \tag{1.53}
$$

Consider the Hamiltonian density for a left-moving boson field,

$$
\begin{aligned}
\mathcal{H} &= \frac{1}{4}(\partial_+\varphi_L)^2\,, \\
[\partial_+\varphi_L(x), \partial_+\varphi_L(y)] &= [\dot\varphi + \varphi', \dot\varphi + \varphi'] = 2i\frac{d}{dx}\delta(x-y)\,.
\end{aligned} \tag{1.54}
$$

Comparing this with the fermionic case, we see that

$$
J_L = \sqrt{\pi}\partial_+\varphi_L = \sqrt{\pi}\partial_+\varphi\,, \tag{1.55}
$$

since the commutation relations and the Hamiltonian are the same. That means the operators are the same, with appropriate boundary conditions.

This equivalence becomes especially powerful when the fermions have several components. Consider the case at hand with two spin components and k channels. Clearly, we can write the free-fermion Hamiltonian in this case as

$$
\mathcal{H}(x) = \frac{1}{4\pi}\lim_{\epsilon\to 0}\sum_{\alpha j} :\psi^{\dagger\alpha j}\psi_{\alpha j}:(x)\ :\psi^{\dagger\alpha j}\psi_{\alpha j}:(x+\epsilon)\ + \text{constant}\,. \tag{1.56}
$$

It turns out to be very useful to use simple algebraic identities to rewrite this in terms of charge, spin, and channel current operators:

$$J(x) \equiv :\psi^{\dagger\alpha i}\psi_{\alpha i}:,$$
$$\vec{J} \equiv \psi^{\dagger\alpha i}\frac{\vec{\sigma}_\alpha^\beta}{2}\psi_{\beta i},$$
$$J^A \equiv \psi^{\dagger\alpha i}\left(T^A\right)_i^j\psi_{\alpha i}. \tag{1.57}$$

Here the T^A's are generators of SU(k), i.e. a set of traceless Hermitian matrices obeying the orthonormality condition

$$\text{Tr } T^A T^B = \frac{1}{2}\delta^{AB} \tag{1.58}$$

and hence the completeness relation

$$\sum_A \left(T^A\right)_a^b \left(T^A\right)_c^d = \frac{1}{2}\left[\delta_c^b\delta_a^d - \frac{1}{k}\delta_a^b\delta_c^d\right]. \tag{1.59}$$

They obey the commutation relations

$$[T^A, T^B] = i\sum_C f^{ABC}T^C, \tag{1.60}$$

where the numbers f^{ABC} are the structure constants of SU(k). In the $k = 2$ case, we may choose

$$T^a \to \frac{\sigma^a}{2}. \tag{1.61}$$

It is now straightforward to prove the following identity:

$$\mathcal{H} = \frac{1}{8\pi k}J^2 + \frac{1}{2\pi(k+2)}\vec{J}^2 + \frac{1}{2\pi(k+2)}J^A J^A. \tag{1.62}$$

The coefficients of each term are chosen so that the normal-ordered products $:\psi^{\dagger\alpha a}\psi_{\alpha a}\psi^{\dagger\beta b}\psi_{\beta b}:$ and $:\psi^{\dagger\alpha a}\psi_{\alpha b}\psi^{\dagger\beta b}\psi_{\beta a}:$ have zero coefficients. The current operators now obey the current algebras

$$[J(x), J(y)] = 4\pi i k\delta'(x-y),$$
$$[J^a(x), J^b(y)] = 2\pi i\delta(x-y)\epsilon^{abc}J^c + \pi i k\delta^{ab}\delta'(x-y),$$
$$[J^A(x), J^B(y)] = 2\pi i\delta(x-y)f^{ABC}J^C + 2i\pi\delta^{AB}\delta'(x-y). \tag{1.63}$$

The currents of different types (charge, spin, and flavor) commute with each other. Thus the Hamiltonian is a sum of three commuting terms, for charge, spin, and flavor, each of which is quadratic in currents and each of which is fully characterized by the current commutation relations. Upon including the right-moving degrees of freedom, we may define three independent field theories, for spin, charge, and flavor with the corresponding Hamiltonians. The charge Hamiltonian is simply a free boson, with

$$J = \sqrt{2\pi k}\partial_+\varphi. \tag{1.64}$$

The spin and channel Hamiltonians are Wess–Zumion–Witten (WZW) nonlinear σ-models (NLσM) (Witten 1984, Knizhnik and Zamolodchikov 1984), labeled SU$(2)_k$ and SU$(k)_2$.

These can be written in terms of SU(2) and SU(k) bosonic matrix fields $g^{\alpha}_{\beta}(t,x)$ and $h^{i}_{j}(t,x)$, respectively. (These fields are Lorentz scalars, i.e. they have zero conformal spin.) The corresponding current operators can be written in a form quadratic in these matrix fields. In the particular case $k=1$, the corresponding $\mathrm{SU}(2)_1$ WZW model is simply equivalent to a free boson. The connection between the multicomponent free-fermion model and the sum of spin, charge, and channel bosonic models is an example of a conformal embedding. All conformal towers in the free-fermion finite size spectrum (FSS), with various BCs, can be written as sums of products of conformal towers from the three constituent models. Likewise, each local operator in the free-fermion model is equivalent to a product of charge, spin, and flavor local operators in the bosonic models. (Actually, this statement needs a little qualification. It is literally true for free-fermion operators which contain even numbers of fermion fields and have zero conformal spin. It is only true for the fermion fields themselves if we are allowed to define chiral components of the WZW matrix fields.)

We now adopt the purely left-moving representation of the Kondo model, in eqn (1.15). This has the advantage that the Kondo interaction can be written in terms of the spin current operators at the origin only. Thus, remarkably, the Kondo interaction is entirely in the spin sector:

$$H=\frac{1}{2\pi(k+2)}\int_{-\infty}^{\infty}dr\vec{J}(r)^2+\lambda\vec{J}(0)\cdot\vec{S}+\dots . \tag{1.65}$$

Here the ... represents the charge and channel parts of the Hamiltonian which are noninteracting, decoupled from the impurity. An immediate consequence of this spin–charge–channel-separated form of the Kondo Hamiltonian is that the Kondo interaction appears only in the spin sector. It is then reasonable to expect that the LIO at the Kondo fixed point involves the spin operators only, corresponding to eqn (1.31). If we took eqn (1.65) at face value this would appear to be exactly true. In fact, since eqn (1.65) is only a low-energy effective Hamiltonian, we can generate other operators, in the charge (and channel) sectors during intermediate stages of the RG. However, we expect all such operators to have much smaller coefficients, with the scale set by D_0 rather than T_K. This is the reason we ignored the irrelevant operator J^2 in the previous lecture. The marginal operator which could be added to the effective Hamiltonian when particle–hole symmetry is broken is now seen to be purely a charge operator: $J_L=\sqrt{2\pi k}\partial_+\phi$. Because it is linear in the charge boson, its effects are easy to include and it is strictly marginal. Assuming it is small, we can simply ignore it. It leads to a line of fixed points.

It is interesting to observe that this Hamiltonian can be formally diagonalized, for a special value of $\lambda_c=2/(k+2)$, by redefining the spin currents:

$$\tilde{J}^a(r)\equiv J^a(r)+2\pi\delta(r)S^a\,. \tag{1.66}$$

It can readily be checked that the $\tilde{J}^a(r)$ obey the same commutation relations as in eqn (1.63). Furthermore, for $\lambda=\lambda_c$, the interacting Hamiltonian reduces to

$$H=\frac{1}{2\pi(k+2)}\int_{-\infty}^{\infty}dr\,[\tilde{J}^a(r)]^2+\dots . \tag{1.67}$$

This suggests that there might be an infrared stable fixed point of the RG at an intermediate value of λ corresponding to a BCFT. To make this idea more quantitative, we must use the full apparatus of BCFT. See Cardy's lecture notes from this Summer School for a review of this subject.

A central idea of Cardy's BCFT is that one should represent CIBCs by *boundary states* (Cardy 1989). These states contain all low-energy information about a BCFT. From them one can construct the finite-size spectrum (with any pair of CIBCs at the two ends of a finite system), operator product expansion (OPE) coefficients, boundary operator content, and all other universal properties. For each CFT there is a set of possible boundary states (i.e. a set of possible CIBCs). In general, a complete classification of *all* (conformally invariant) boundary states is not available. However, in the case of rational CFTs, with a finite number of conformal towers, a complete classification is available. The boundary states are in one-to-one correspondence with the conformal towers, i.e. with the primary operators. One can obtain the complete set of boundary states from a reference state by a process of fusion with primary operators. Our strategy for finding the low-energy fixed point of the general Kondo models (and various other quantum impurity problems) is to first identify the boundary state corresponding to the trivial boundary conditions of eqn (1.11). We then obtain the CIBC corresponding to the low-energy fixed point by fusion with an appropriate primary operator. The choice of primary operator is inspired by the mapping in eqn (1.66).

The conformal towers of the $\mathrm{SU}(2)_k$ WZW model are labeled by the spin of the "highest-weight state" (i.e. the lowest-energy state) (Knizhnik and Zamolodchikov 1984, Gepner and Witten 1986, Di Francesco *et al.* 1997). There is one conformal tower for each spin j with

$$j = 0, 1/2, 1, \ldots, k/2\,. \tag{1.68}$$

These primary fields have zero conformal spin and left and right scaling dimensions:

$$\Delta = \frac{j(j+1)}{k+2}\,. \tag{1.69}$$

We may associate them with the spin part of the fermion operators $(\psi_L)^n$. The largest possible spin we can get this way, from $n = k$, antisymmetrized with respect to flavor, is $j = k/2$. The fusion rules are

$$j \otimes j' = |j - j'|, |j - j'| + 1, |j - j'| + 2, \ldots, \min\{j + j', k - j - j'\}\,. \tag{1.70}$$

Note that this generalizes the ordinary angular-momentum addition rules in a way which is consistent with the conformal tower structure of the theories (i.e. the fact that primaries only exist with $j \leq k/2$).

Based on eqn (1.66), we expect that the infrared stable fixed point of the k-channel Kondo model with a spin-S impurity corresponds to fusion with the spin $j = S$ primary operator, whenever $S \leq k/2$. Note that the spin quantum numbers of the conformal towers match nicely with the over/underscreening paradigm. For $S > k/2$, we obtain the infrared stable fixed point by fusion with the maximal-spin primary operator of spin $k/2$. Thus the boundary state at the infrared ("Kondo") fixed point is related to the free-fermion boundary state by (see Cardy's chapter in this book and Cardy (1989))

$$\langle j0|\mathrm{Kondo}\rangle = \langle j0|\mathrm{free}\rangle\, \frac{S^j_S}{S^j_0}\,. \tag{1.71}$$

Here the boundary states are expanded in the Ishibashi states corresponding to the Kac–Moody conformal towers of spin j, and $|j0\rangle$ labels the ground state of the spin-j conformal tower. $S^j_{j'}$ is the modular S-matrix (Kac and Peterson 1984) for $\mathrm{SU}(2)_k$:

$$S^j_{j'}(k) = \sqrt{\frac{2}{2+k}} \sin\left[\frac{\pi(2j+1)(2j'+1)}{2+k}\right]. \tag{1.72}$$

This "fusion rule hypothesis" leads immediately to various predictions about the low-energy behavior which can be tested against numerical simulations, Bethe ansatz calculations, and experiments. One important comparison involves the finite-size spectrum. With the free BCs of eqn (1.11), the FSS can be written as a sum of direct products of conformal towers from spin, charge, and channel sectors, (Q, j, j_c). Here Q is the charge of the highest-weight state (measured from the charge of the ground state), and j_c is a shorthand notation for the $\mathrm{SU}(k)$ quantum numbers of the highest-weight state of the $\mathrm{SU}(k)_2$ WZW model for the channel degrees of freedom. In the important example $k = 2$, it corresponds literally to a second set of $\mathrm{SU}(2)$ "pseudo-spin" quantum numbers. To obtain the spectrum at the infrared fixed point, one replaces the spin-j conformal tower by a set of spin conformal towers using the $\mathrm{SU}(2)_k$ fusion rules of eqn (1.70) with j' replaced by S, the impurity spin magnitude, in the overscreened and exactly screened cases. (In the underscreened case, S should be replaced by $k/2$.) Since the full spectrum of each conformal tower is easily constructed, the "fusion rule hypothesis" predicts an infinite number of finite-size energy levels. These can be compared with the results of numerical renormalization group (NRG) calculations. These calculations give the spectrum of a finite chain of length l, with the impurity spin at one end, like the tight-binding model of eqn (1.26). These spectra reveal an interesting crossover behavior. For a weak Kondo coupling and a relatively short chain length, the spectrum is essentially that of the zero-Kondo-coupling model, i.e. the conformal spectrum with the BC of eqn (1.11) factored with the decoupled impurity spin. However, as the chain length increases, this spectrum shifts. The characteristic crossover length is

$$\xi_K \equiv \frac{v_F}{T_K} \propto \exp\left[\frac{1}{\lambda_0}\right]. \tag{1.73}$$

For longer chain lengths, the FSS predicted by our BCFT methods is observed, for the low-energy part of the spectrum. (In principle, the smaller the bare Kondo coupling and the longer the chain length, the more states in this conformal BCFT spectrum are observed.) In Affleck *et al.* (1992), for example, the first six energy levels (most of which are multiply degenerate) were compared, for the $k = 2$ case, obtaining excellent agreement.

1.2.1 Impurity entropy

We define the impurity entropy as

$$S_{\mathrm{imp}}(T) \equiv \lim_{l\to\infty} \left[S(l,T) - S_0(l,T)\right], \tag{1.74}$$

where $S_0(l, T)$ is the free-fermion entropy, proportional to l, in the absence of the impurity. Note that, for zero Kondo coupling, $S_{\mathrm{imp}} = \ln[2s + 1]$, simply reflecting the ground state degeneracy of the free spin. In the case of exact screening ($k = 2s$), $S_{\rm imp}(0) = 0$. For underscreening,

$$S_{\rm imp}(0) = \ln[2s' + 1] \,, \tag{1.75}$$

where $s' \equiv s - k/2$. What happens for overscreening? Surprisingly, we will obtain, in general, the log of a noninteger, implying a sort of "noninteger ground state degeneracy."

To proceed, we show how to calculate $S_{\rm imp}(0)$ from the boundary state. All calculations are done in the scaling limit, ignoring irrelevant operators, so that $S_{\rm imp}(T)$ is a constant, independent of T, and characterizes the particular boundary condition. It is important, however, that we take the limit $l \to \infty$ first, as specified in eqn (1.74), at fixed, nonzero T, i.e. we are interested in the limit $l/\beta \to \infty$. Thus it is convenient to use the following expression for the partition function (see Cardy's chapter in this book and Cardy (1989)) Z_{AB}:

$$Z_{AB} = \sum_a \langle A|a0\rangle\langle a0|B\rangle \chi_a(e^{-4\pi l/\beta}) \to e^{\pi lc/6\beta}\langle A|00\rangle\langle 00|B\rangle \,. \tag{1.76}$$

Here $|a0\rangle$ labels the ground state in the conformal tower of the operator O_a, and χ_a is the corresponding character. c is the conformal anomaly. Thus the free energy is

$$F_{AB} = -\frac{\pi c T^2 l}{6} - T\ln\langle A|00\rangle\langle 00|B\rangle \,. \tag{1.77}$$

The first term gives the specific heat,

$$C = \frac{\pi c T l}{3} \,, \tag{1.78}$$

and the second gives the impurity entropy,

$$S_{\rm imp} = \ln\langle A|00\rangle\langle 00|B\rangle \,. \tag{1.79}$$

This is a sum of contributions from the two boundaries,

$$S_{\rm imp} = S_A + S_B \,. \tag{1.80}$$

Thus we see that the "ground state degeneracy" g_A associated with boundary condition A is

$$\exp[S_{{\rm imp}A}] = \langle A|00\rangle \equiv g_A \,. \tag{1.81}$$

Here we have used our freedom to choose the phase of the boundary state so that $g_A > 0$. For our original, antiperiodic boundary condition, $g = 1$. For the Kondo problem, we expect the low-T impurity entropy to be given by the value at the infrared fixed point. Since this is obtained by fusion with the spin-s (or $k/2$) operator, we obtain from eqn (1.72)

$$g = \frac{S_s^0}{S_0^0} = \frac{\sin[\pi(2s+1)/(2+k)]}{\sin[\pi/(2+k)]} \,. \tag{1.82}$$

This formula agrees exactly with the Bethe ansatz result (Tsvelik 1985) and has various interesting properties. Recall that in the case of exact screening or underscreening ($s \geq k/2$) we must replace s by $k/2$ in this formula, in which case it reduces to 1. Thus the ground state degeneracy is 1 for exact screening. For underscreening, we must multiply g by $(2s' + 1)$ to account for the decoupled, partially screened impurity. Note that in the overscreened case, where $s < k/2$, we have

$$\frac{1}{2+k} < \frac{2s+1}{2+k} < 1 - \frac{1}{2+k}, \tag{1.83}$$

so $g > 1$. In the case $k \to \infty$ with s held fixed, $g \to 2s + 1$, i.e. the entropy of the impurity spin is hardly reduced at all by the Kondo interaction, corresponding to the fact that the critical point occurs at weak coupling. In general, for underscreening,

$$1 < g < 2s + 1, \tag{1.84}$$

i.e. the free-spin entropy is somewhat reduced, but not completely eliminated. Furthermore, g is not, in general, an integer. For instance, for $k = 2$ and $s = 1/2$, $g = \sqrt{2}$. Thus we may say that there is a noninteger "ground state degeneracy". Note that in all cases the ground state degeneracy is reduced under renormalization from the zero-Kondo-coupling fixed point to the infrared stable fixed point. This is a special case of a general result: *the ground state degeneracy always decreases under renormalization.* This is related to Zamolodchikov's c-theorem (Zamolodchikov 1986), which states that the conformal anomaly parameter, c, always decreases under renormalization. The intuitive explanation of the c-theorem is that as we probe lower energy scales, degrees of freedom which appeared approximately massless start to exhibit a mass. This freezes out their contribution to the specific heat, the slope of which can be taken as the definition of c. In the case of the "g-theorem," the intuitive explanation is that as we probe lower energy scales, approximately degenerate levels of impurities exhibit small splittings, reducing the degeneracy.

A "perturbative" proof of the g-theorem was given in Affleck and Ludwig (1993), where RG flow between two "nearby" boundary RG fixed points with almost the same values of g was considered. A general proof was given in Friedan and Konechny (2004).

1.2.2 Resistivity/conductance

In this subsection, I consider the resistivity due to scattering from a dilute array of k-channel Kondo impurities (Affleck and Ludwig 1993), and the closely related conductance through a single k-channel impurity. This latter quantity, in the $k = 2$ case, was recently measured in quantum dot experiments, as I discuss in the next lecture. Using the Kubo formula, these quantities can be expressed in terms of the single-electron Green's function. Owing to the δ-function nature of the Kondo interaction, the exact retarded Green's function (in one, two, or three dimensions) with a single impurity at $r = 0$ can be written as

$$G(\vec{r}, \vec{r}'; \omega) = G_0(|\vec{r} - \vec{r}'|, \omega) + G_0(r, \omega)\mathcal{T}(\omega)G_0(r', \omega). \tag{1.85}$$

Here G_0 is the noninteracting Green's function. The function $\mathcal{T}$, which depends on the frequency only, not the spatial coordinates, is known as the $\mathcal{T}$-matrix. Note that I am using a mixed space–frequency representation of the Green's function, which is invariant under time translations, but not space translations. The only thing which distinguishes the dimensionality of space is G_0.

In the case of a dilute random array of impurities in $D = 3$, the Green's function, to first order in the impurity concentration n_i, can be written exactly as

$$G(|\vec{r} - \vec{r}'|, \omega) = \frac{1}{G_0^{-1}(\vec{r} - \vec{r}'|, \omega) - \Sigma(\omega)}, \tag{1.86}$$

where the self-energy is given by

$$\Sigma(\omega) = n_i \mathcal{T}(\omega). \tag{1.87}$$

(Translational invariance is restored after averaging over impurity positions.) The single-electron lifetime is given by

$$\tau^{-1}(\omega) = \text{Im}\, \Sigma(\omega) \tag{1.88}$$

and the finite-temperature resistivity, $\rho(T)$, can be expressed in terms of this lifetime by the standard formula

$$\frac{1}{\rho(T)} = \frac{2e^2 k}{3m^2} \int \frac{d^3 p}{(2\pi)^3} \left[\frac{-dn_F}{d\epsilon_p}\right] \vec{p}^2 \tau(\epsilon_p), \tag{1.89}$$

where n_F is the Fermi distribution function,

$$n_F \equiv \frac{1}{\exp[\epsilon_p/T] + 1}. \tag{1.90}$$

At low temperatures, this integral is dominated by low energies so our field theory results can be used. A similar calculation, reviewed in the next lecture, expresses also the conductance through a single impurity in terms of Im $\mathcal{T}$.

Thus, our task is to calculate the electron Green's function in the low-energy one-dimensional effective field theory. In the zero-temperature limit, we may simply evaluate it at the Kondo fixed point. At low finite temperatures, we consider the correction from the LIO. At the fixed point, the chiral Green's functions $\langle \psi_L^\dagger(r+i\tau)\psi_L(r'+i\tau')\rangle$ and $\langle \psi_R^\dagger(r-i\tau)\psi_R(r'-i\tau')\rangle$ are unaffected by the Kondo interaction. We only need to consider $\langle \psi_L^\dagger(r,\tau)\psi_R(r',\tau')\rangle$. By general methods of BCFT, this behaves as a two-point function of left-movers with the right-mover reflected to the negative axis, $(-r', \tau')$:

$$\langle 0|\psi_L^{\dagger i\alpha}(\tau, r)\psi_{Rj\beta}(r', \tau')|0\rangle = \frac{S_{(1)}\delta^\alpha_\beta \delta^i_j}{(\tau - \tau') + i(r + r')}. \tag{1.91}$$

Only the constant, $S_{(1)}$, depends on the particular CIBC. For instance, if the BC is of free-fermion type, $\psi_R(0) = e^{i\delta}\psi_L(0)$, then $S_{(1)} = e^{i\delta}$. In general, $S_{(1)}$ can be expressed in terms of the boundary state (Cardy and Lewellen 1991). Since the fermion field has spin $j = 1/2$, the general expression is

$$S_{(1)} = \frac{\langle 1/2, 0|A\rangle}{\langle 00|A\rangle}. \tag{1.92}$$

By comparing this with the free-fermion BC where $S_{(1)} = 1$ and obtaining the Kondo BC by fusion, using eqn (1.71), it follows that $S_{(1)}$ is given in terms of the modular S-matrix:

$$S_{(1)} = \frac{S_S^{1/2} S_S^0}{S_0^{1/2} S_S^0} = \frac{\cos[\pi(2S+1)/(2+k)]}{\cos[\pi/(2+k)]}. \tag{1.93}$$

The zero-temperature $\mathcal{T}$-matrix can be expressed directly in terms of $S_{(1)}$:

$$\mathcal{T}(\omega) = \frac{-i}{2\pi\nu}[1 - S_{(1)}]\,. \tag{1.94}$$

In this limit, $\mathcal{T}$ is independent of ω and purely imaginary. This result follows from the definition, (1.85), of the $\mathcal{T}$-matrix upon using the free Green's function:

$$G_0(r, \omega_n) = 2\pi i e^{\omega_n r}\left[\theta(-\omega_n)\theta(r) - \theta(\omega_n)\theta(-r)\right], \tag{1.95}$$

where $\theta(x)$ is the step function, together with the analytic continuation to real frequency,

$$\theta(\omega_n) \to \theta(\delta - i\omega) = 1\,. \tag{1.96}$$

In the exactly screened or underscreened case, where we set $S = k/2$, eqn (1.93) gives $S_{(1)} = -1$, the free-fermion result with a $\pi/2$ phase shift. This is the unitary limit resistivity of eqn (1.39), which I now define divided by a factor of k since we have k parallel channels. In general, at the non-Fermi-liquid fixed points,

$$\rho(0) = \rho_U\left[\frac{1 - S_{(1)}}{2}\right] \leq \rho_U\,. \tag{1.97}$$

To calculate the leading corrections at low temperature (or frequency), we must do perturbation theory in the LIO. The LIO must be a boundary operator which exists under the CIBCs characterizing the fixed point and which, furthermore, respects all symmetries of the Hamiltonian. The set of boundary operators (for *any* CIBC) is a subset of the set of chiral operators in the bulk theory. This follows from the "method of images" approach to BCFT, which expresses any local operator with left- and right-moving factors as a bilocal product of left-movers. In the limit where the operator is taken to the boundary, we may use the OPE to express it in terms of local left-moving operators. The set of boundary operators which actually exist for a given CIBC is in one-to-one correspondence with the set of conformal towers in the finite-size spectrum where the corresponding boundary condition is imposed *at both ends* of a finite system. This can be obtained by "double fusion" from the operator content with free-fermion BCs. The boundary operators with free BCs all have integer dimensions and include Kac–Moody descendants of the identity operator, such as the current operators. Double fusion, starting with the identity operators, corresponds to applying eqn (1.70) *twice*, starting with $j = 0$ and $j' = S$. This gives operators of spin $j = 0, 1, \ldots, \min\{2S, k - 2S\}$. While this only gives back the identity operator $j = 0$ for the exactly screened and underscreened cases, where $S = k/2$ it always gives $j = 1$ (and generally higher integer spins) for the overscreened case. We see from the dimensions, eqn (1.69), that the spin-1 primary is the lowest-dimension one that occurs, with dimension

$$\Delta = \frac{2}{2 + k}\,. \tag{1.98}$$

None of these nontrivial primary operators can appear directly in the effective Hamiltonian, since they are not rotationally invariant, having nonzero spin. However, we may construct descendant operators of spin 0. The lowest-dimension spin-zero boundary operator for all overscreened cases is $\vec{J}_{-1}\cdot\vec{\varphi}$, where $\vec{\varphi}$ is the spin-1 primary operator. This is a first descendant, with scaling dimension $1+\Delta$. This is less than 2, the dimension of the Fermi liquid operator, $\vec{J}^2$, which can also occur. This is the LIO in the exactly screened and underscreened cases, eqn (1.31). Thus the effective Hamiltonian, in the overscreened case, may be written

$$H = H_0 - \frac{1}{T_K^{\Delta}}\vec{J}_{-1}\cdot\vec{\varphi}. \tag{1.99}$$

Here H_0 is the WZW Hamiltonian with the appropriate BC. As usual, we assume that the dimensional coupling constant multiplying the LIO has its scale set by T_K, the crossover scale determined by the weak-coupling RG. We may take eqn (1.99) as our precise definition of T_K (with the operator normalized conventionally). As in the Fermi liquid case, many different physical quantities can be calculated in lowest-order perturbation theory in the LIO, giving various generalized "Wilson ratios" in which T_K cancels. One of the most interesting of these perturbative calculations is for the single-fermion Green's function, giving the $\mathcal{T}$-matrix. In the Fermi liquid case, the first-order perturbation theory in the LIO gives a correction to the $\mathcal{T}$-matrix which is purely real. Only in second order do we get a correction to Im $\mathcal{T}$, leading to the correction to the resistivity of $O(1/T_K^2)$ in eqn (1.40). On the other hand, a detailed calculation shows that first-order perturbation theory in the non-Fermi liquid LIO of eqn (1.99) gives a correction to the $\mathcal{T}$-matrix with both real *and* imaginary parts, and hence a correction to the resistivity of the form

$$\rho(T) = \rho_U\left[\frac{1-S_{(1)}}{2}\right]\left[1-\alpha\left(\frac{T}{T_K}\right)^{\Delta}\right]. \tag{1.100}$$

Here α is a constant which was obtained explicitly from the detailed perturbative calculation, having the value $\alpha = 4\sqrt{\pi}$ for the two-channel $S=1/2$ case (for which $S_{(1)}=0$). Also note that the *sign* of the coupling constant in eqn (1.99) is *not* determined a priori. If we assumed the opposite sign, the T-dependent term in the resistivity in eqn (1.100) would switch. It is reasonable to expect this negative sign, for a weak bare coupling, since the resistivity is also a decreasing function of T at $T \gg T_K$, where it can be calculated perturbatively in the Kondo coupling. An assumption of monotonicity of $\rho(T)$ leads to the negative sign in eqn (1.99). In fact, this negative sign has recently been confirmed by experiments, as I will discuss in the next lecture.

A number of other low-energy properties of the non-Fermi-liquid Kondo fixed points have been calculated by these methods, including the T dependence of the entropy and susceptibility and space- and time-dependent Green's function of the spin density, but I will not take the time to review them here.

1.3 Quantum dots: Experimental realizations of one- and two-channel Kondo models

In this lecture, I will discuss theory and experiments on quantum dots, as experimental realizations of both single- and two-channel Kondo models.

1.3.1 Introduction to quantum dots

Experiments on gated semiconductor quantum dots begin with two-dimensional electron gases (2DEGs) in semiconductor heterostructures, usually GaAs–AlGaAs. (These are the same types of semiconductor wafers as are used for quantum Hall effect experiments.) A low areal density of electrons is trapped in an inversion layer between the two different bulk semiconductors. Great effort goes into making these 2DEGs very clean, with long scattering lengths. Because the electron density is so low compared with the interatomic distance, the dispersion relation near the Fermi energy is almost perfectly quadratic, with an effective mass much lower than that of free electrons. The inversion layer is located quite close to the upper surface of the wafer (typically around 100 nm below it). Leads are attached to the edges of the 2DEG to allow conductance measurements. In addition, several leads are attached to the upper surface of the wafer, to apply gate voltages to the 2DEG, which can vary over distances of order 0.1 microns. Various types of quantum dot structures can be built on the 2DEG using gates. A simple example is a single quantum dot, a roughly circular puddle of electrons, with a diameter of around 0.1 μm. The quantum dot is separated from the left and right regions of the 2DEG by large electrostatic barriers so that there is a relatively small rate at which electrons tunnel from the dot to the left and right regions of the 2DEG. In simple devices, the only appreciable tunneling path for electrons from the left to right 2DEG regions is through the quantum dot. Because the electron transport in the 2DEG is essentially ballistic, the current is proportional to the voltage difference V_{sd} (source–drain voltage) between the leads for small V_{sd}, rather than to the electric field. The linear conductance

$$I = GV_{sd} \tag{1.101}$$

(and also the nonlinear conductance) is measured versus T and the various gate voltages.

An even simpler device of this type does not have a quantum dot, but just a single point contact between the two leads. (The quantum dot devices have essentially two point contacts, from the left side to the dot and from the dot to the right side.) As the barrier height of the point contact is raised, so that it is nearly pinched off, it is found that the conductance at sufficiently low T has sharp plateaus and steps, with the conductance on the plateaus being $2ne^2/h$, for integer n; $2e^2/h$ is the conductance of an ideal noninteracting one-dimensional wire, with the factor of 2 arising from the electron spin. This can be seen from a Landauer approach. Imagine left and right reservoirs at different chemical potentials, μ and $\mu - eV_{sd}$, with each reservoir emitting electrons to the left and right into wires with equilibrium distributions characterized by different chemical potentials:

$$\begin{aligned} I &= -2e \int_0^\infty \frac{dk}{2\pi} v(k)[n_F(\epsilon_k - \mu + eV_{sd}) - n_F(\epsilon_k - \mu)], \\ G &= 2e^2 \int_0^\infty \frac{d\epsilon}{2\pi} \frac{dn_F}{d\mu}(\epsilon_k - \mu) = \frac{2e^2}{h} n_F(-\mu), \end{aligned} \tag{1.102}$$

where I have inserted a factor of $\hbar$, previously set equal to one, in the last step. Thus, provided that $\mu \gg k_B T$, $G = 2e^2/h$ for an ideal one-dimensional conductor. A wider noninteracting wire would have n partially occupied bands and a conductance of $2ne^2/h$. As the point contact is progressively pinched off, it is modeled as a progressively narrower quantum wire with fewer channels, thus explaining the plateaus. Because the gate voltage varies gradually in the 2DEG, backscattering at the constriction is ignored; otherwise the conductance of a single channel would be $2e^2 T_r/h$, where T_r is the transmission probability.

The tunnel barriers separating the quantum dot from the left and right 2DEG regions are modeled as single-channel point contacts. Nonetheless, as the temperature is lowered, the conductance through a quantum dot often tends towards zero. This is associated with the Coulomb interactions between the electrons in the quantum dot. Although it may be permissable to ignore Coulomb interactions in the leads, this is not permissable in the quantum dot itself. A simple and standard approach is to add a term to the Hamiltonian of the form $Q^2/(2C)$, where Q is the charge on the quantum dot and C is its capacitance. In addition, a gate voltage V is applied to the dot, so that the total dot Hamiltonian may be written

$$H_d = \frac{U}{2}(\hat{n} - n_0)^2 \,, \tag{1.103}$$

where $\hat{n}$ is the number operator for electrons on the dot and $n_0 \propto V$. An important dimensionless parameter is $t^2\nu/U$, where t is the tunneling amplitude between leads and dot. If $t^2\nu/U \ll 1$ then the charge on the dot is quite well defined and will generally stay close to n_0, with virtual fluctuations into higher-energy states with $n = n_0 \pm 1$. (An important exception to this is when n_0 is a half-integer.) It is possible to actually observe changes in the behavior of the conductance as n_0 is varied by a single step. For n_0 close to an integer value, the conductance tends to become small at low T. This is a consequence of the fact that for an electron to pass through the dot it must go temporarily into a high-energy state with $n = n_0 \pm 1$, an effect known as the Coulomb blockade. At the special values of the gate voltage where n_0 is a half-integer, the Coulomb blockade is lifted and the conductance is larger.

1.3.2 Single-channel Kondo effect

At still lower temperatures, a difference emerges between the cases where n_0 is close to an even or an odd integer; this is due to the Kondo effect. When n_0 is near an odd integer, the dot must have a nonzero (half-integer) spin, generally $1/2$. At energy scales small compared with U, we may disregard charge fluctuations on the dot and consider only its spin degrees of freedom. Virtual processes, of second order in t, lead to a Kondo exchange interaction between the spin on the quantum dot and the spin of the mobile electrons on the left and right sides of the 2DEG. A simplified and well-studied model is obtained by considering only a single energy level on the quantum dot, the one nearest the Fermi energy. Then there are only four states available to the quantum dot: zero or two electrons, or one electron with spin up or down. The corresponding model is known as the Anderson model (AM):

$$H = \int dk\, \psi_k^{\dagger\alpha}\psi_{k\alpha}\epsilon(k) + \Gamma \int dk\, [\psi_k^{\dagger\alpha} d_\alpha + h.c.] + \frac{U}{2}(\hat{n}_d - n_0)^2 \,, \tag{1.104}$$

where

$$\hat{n}_d \equiv d^{\dagger\alpha} d_\alpha \,. \tag{1.105}$$

$d^{\dagger\alpha}$ creates an electron in the single energy level under consideration on the dot. Note that I am now treating the conduction electrons as one-dimensional. This is motivated by the fact that the point contacts between the 2DEG regions and the dot are assumed to be single-channel. However, the actual wave functions of the electrons created by $\psi_k^{\dagger\alpha}$ are extended in two dimensions on the left and right sides of the dot. The conventional label k doesn't really label a wave vector anymore. An important assumption is being made here that there is a set of energy levels near the Fermi energy which are equally spaced and have equal hybridization amplitudes t with the level d on the quantum dot. This is expected to be reasonable for small quantum dots with weak tunneling amplitudes t and smooth point contacts. I assume, for convenience, that these wave functions are parity symmetric between the left and right 2DEGs. As the gate voltage is varied, n_0 passes through 1. Provided that $\Gamma^2\nu \ll U$, where ν is the (one-dimensional) density of states, we can obtain the Kondo model as the low-energy effective theory at scales small compared with U, with an effective Kondo coupling

$$J = \frac{2\Gamma^2}{U(2n_0 - 1)(3 - 2n_0)}. \tag{1.106}$$

Thus we again expect the spin of the quantum dot to be screened at $T \ll T_K$ by the conduction electrons in the leads.

A crucial and perhaps surprising point is how the Kondo physics affects the conductance through the dot. This is perhaps best appreciated by considering a tight-binding version of the model, where we replace the left and right leads by 1D tight-binding chains:

$$H = -t\sum_{j=-\infty}^{-2}(c_j^\dagger c_{j+1} + h.c.) - t\sum_1^\infty (c_j^\dagger c_{j+1} + h.c.) - t'[(c_{-1}^\dagger + c_1^\dagger)d + h.c.)] + \frac{U}{2}(\hat{n}_d - n_0)^2. \tag{1.107}$$

The Kondo limit gives

$$H = -t\sum_{j=-\infty}^{-2}(c_j^\dagger c_{j+1} + h.c.) - t\sum_1^\infty (c_j^\dagger c_{j+1} + h.c.) + J(c_{-1}^\dagger + c_1^\dagger)\frac{\vec{\sigma}}{2}(c_{-1} + c_1)\cdot\vec{S}, \tag{1.108}$$

with

$$J = \frac{2t'^2}{U(2n_0 - 1)(3 - 2n_0)}. \tag{1.109}$$

Note that the only way electrons can pass from the right to the left lead is via the Kondo interaction. Thus if the Kondo interaction is weak, the conductance should be small. The renormalization of the Kondo coupling to large values at low energy scales implies a dramatic characteristic *increase* of the conductance upon lowering the temperature. In the particle–hole-symmetric case of half-filling, it is easy to understand the low-temperature limit by simply taking the bare Kondo coupling to infinity, $J \gg t$. Now the spin of the quantum dot (at site 0) forms a singlet with an electron in the parity-symmetric state on sites 1 and -1:

$$\left[d^{\dagger\uparrow}\frac{(c_{-1}^{\dagger\downarrow} + c_1^\downarrow)}{\sqrt{2}} - d^{\dagger\downarrow}\frac{(c_{-1}^{\dagger\uparrow} + c_1^\uparrow)}{\sqrt{2}}\right]|0\rangle. \tag{1.110}$$

The parity-antisymmetric orbital

$$c_a \equiv \frac{c_{-1} - c_1}{\sqrt{2}} \tag{1.111}$$

is available to conduct current past the screened dot. The resulting low-energy effective Hamiltonian,

$$H=-t\sum_{j=-\infty}^{-3}(c_j^\dagger c_{j+1}+h.c.)-t\sum_{2}^{\infty}(c_j^\dagger c_{j+1}+h.c.)-\frac{t}{\sqrt{2}}[(-c_{-2}^\dagger+c_2)c_a+h.c.]\,, \quad (1.112)$$

has resonant transmission, $T_r = 1$, at the Fermi energy in the particle–hole-symmetric case of half-filling. This leads to ideal $2e^2/h$ conductance from the Laudauer formula. Thus we expect the conductance to increase from a small value of order J^2 at $T \gg T_K$ to the ideal value at $T \ll T_K$. Thus, the situation is rather inverse to the case of the resistivity due to a dilute random array of Kondo scatterers in three (or lower) dimensions. For the quantum dot geometry discussed here, lowering T leads to an increase in conductance rather than an increase in resistivity. Actually, it is easy to find another quantum dot model, referred to as "side-coupled," where the behavior is like the random-array case. In the side-coupled geometry, the tight-binding Hamiltonian is

$$H=-t\sum_{j=-\infty}^{\infty}(c_j^\dagger c_{j+1}+h.c.)+Jc_0^\dagger\frac{\vec{\sigma}}{2}c_0\cdot\vec{S}\,. \quad (1.113)$$

Now there is perfect conductance at $J = 0$ due to the direct hopping from sites -1 to 0 to 1. On the other hand, in the strong-Kondo-coupling limit, an electron sits at site 0 to form a singlet with the impurity. This completely blocks transmission, since an electron cannot pass through without destroying the Kondo singlet.

At arbitrary temperatures, the conductance through the quantum dot may be expressed exactly in terms of the $\mathcal{T}$-matrix, $\mathcal{T}(\omega, T)$. This is precisely the same function which determines the resistivity for a dilute random array of Kondo scatterers. To apply the Kubo formula, it is important to carefully distinguish conduction electron states in the left and right leads. Thus we write the Kondo Hamiltonian in the form

$$H=\sum_{L/R}\int dk\,\psi_{L/R,k}^\dagger\psi_{L/R,k}\epsilon(k)+\frac{J}{2}\int\frac{dk\,dk'}{2\pi}(\psi_{L,k}^\dagger+\psi_{R,k}^\dagger)\frac{\vec{\sigma}}{2}(\psi_{L,k'}+\psi_{R,k'})\cdot\vec{S}\,. \quad (1.114)$$

The Kondo interaction involves only the symmetric combination of the left and right leads. On the other hand, the current operator which appears in the Kubo formula for the conductance is

$$j=-e\frac{d}{dt}[N_L-N_R]=-ie[H,N_L-N_r]\,, \quad (1.115)$$

where $N_{L/R}$ are the number operators for electrons in the left and right leads:

$$N_{L/R}\equiv\int dk\,\psi_{L/R,k}^\dagger\psi_{L/R,k}\,. \quad (1.116)$$

If we introduce symmetric and antisymmetric combinations

$$\psi_{s/a}\equiv\frac{\psi_L\pm\psi_R}{\sqrt{2}}\,, \quad (1.117)$$

the Kondo interaction only involves ψ_s but the current operator is

$$j = \frac{d}{dt} \int dk \left[\psi_s^\dagger \psi_a + h.c.\right], \tag{1.118}$$

which contains a product of symmetric and antisymmetric operators. The Kubo formula

$$G = \lim_{\omega \to 0} \frac{1}{\omega} \int_0^\infty e^{i\omega t} \langle [j(t), j(0)] \rangle \tag{1.119}$$

then may be expressed as a product of the free Green's function for ψ_a and the interacting Green's function for ψ_s. Expressing the ψ_s Green's function in terms of the $\mathcal{T}$-matrix by eqn (1.85), it is not hard to show that the conductance is given by

$$G(T) = \frac{2e^2}{h} \int d\epsilon \left[-\frac{dn_F}{d\epsilon}(T) \right] \left[-\pi\nu \, \mathrm{Im} \, \mathcal{T}(\epsilon, T) \right]. \tag{1.120}$$

For $T \gg T_K$, a perturbative calculation of the $\mathcal{T}$-matrix gives

$$-2\pi\nu\mathcal{T} \to -\frac{3\pi^2}{8}\lambda^2 + \dots . \tag{1.121}$$

It can be checked that the higher-order terms replace the Kondo coupling λ by its renormalized value at a scale ω or T (whichever is higher), leading to the conductance

$$G \to \frac{2e^2}{h} \frac{3\pi^2}{16 \ln^2(T/T_K)} \qquad (T \gg T_K). \tag{1.122}$$

On the other hand, at $T, \omega \to 0$, we find that $-2\pi\nu\mathcal{T} \to -2i$, corresponding to a $\pi/2$ phase shift, leading to ideal conductance. By doing second-order perturbation theory in the LIO, Nozières Fermi liquid theory gives

$$G \to \frac{2e^2}{h} \left[1 - \left(\frac{\pi^2 T}{4T_K} \right)^2 \right]. \tag{1.123}$$

The calculation of $\mathcal{T}$ at intermediate temperatures and frequencies of order T_K is a difficult problem. It goes beyond the scope of our RG methods, which apply only near the high- and low-energy fixed points. It is also not feasible using the Bethe ansatz solution of the Kondo model. The most accurate results at present come from the numerical renormalization group method.

1.3.3 Two-channel Kondo effect

Quite recently, the first generally accepted experimental realization of an overscreened Kondo effect, in the two-channel, $S = 1/2$ case, was obtained in a quantum dot device. To understand the difficulty in obtaining a two-channel situation, consider the case discussed above. In a sense, there are two channels in play, corresponding to the left and right leads. However, the problem is that only the even channel actually couples to the spin of the quantum dot. In eqn (1.114), it is the left–right cross terms in the Kondo interaction that destroy the two-channel behavior. If such terms could somehow be eliminated, we would obtain a two-channel model.

On the other hand, the only thing which is readily measured in a quantum dot experiment is the conductance, and this is trivially zero if the left–right Kondo couplings vanish.

A solution to this problem, proposed by Oreg and Goldhaber-Gordon (2003), involves a combination of a small dot, in the Kondo regime, and a "large dot," i.e. another, larger puddle of conduction electrons with only weak tunneling from it to the rest of the system, as shown in Fig. 1.3. This was then realized experimentally by Goldhaber-Gordon's group (Potok *et al.* 2007). The key feature is to adjust the size of this large dot to be not too large and not too small. It should be chosen to be large enough that the finite-size level spacing is negligibly small compared with the other relevant energy scales T_K and T, which may be in the millikelvin to kelvin range. On the other hand, the charging energy of the dot, effectively the U parameter discussed above, must be relatively large compared with these other scales. In this case, the charge degrees of freedom of the large dot are frozen out at low energy scales. An appropriate Kondo-type model could be written as

$$H = \sum_{j=1}^{3} \int dk\, \psi_{j,k}^{\dagger} \psi_{j,k} \epsilon(k) + \frac{2}{U_s} \sum_{i,j=1}^{3} \Gamma_i \Gamma_j \int \frac{dk\, dk'}{2\pi} \psi_{i,k}^{\dagger} \frac{\vec{\sigma}}{2} \psi_{j,k'} \cdot \vec{S} + \frac{U_l}{2} (\hat{n}_3 - n_0)^2 . \tag{1.124}$$

Here $j = 1$ corresponds to the left lead, $j = 2$ corresponds to the right lead, and $j = 3$ corresponds to the large dot. $U_{s/l}$ is the charging energy for the small/large dot, respectively, the Γ_i are the corresponding tunneling amplitudes onto the small dot, and $\hat{n}_3$ is the total number of electrons on the large dot. n_0 is the lowest-energy electron number for the large dot, which is now a rather large number. (It is actually unimportant here whether n_0 is an integer or half-integer, because the Kondo temperature for the large dot is assumed to be negligibly small.) If U_l is sufficiently large, the 1–3 and 2–3 cross terms in the Kondo interaction can be ignored, since they take the large dot from a low-energy state with $n_3 = n_0$ to a high-energy state with $n_3 = n_0 \pm 1$. Dropping these cross terms, assuming $\Gamma_1 = \Gamma_2$ as before, and replacing $(\psi_1 + \psi_2)/\sqrt{2}$ by ψ_s as before, we obtain a two-channel Kondo model, but with different Kondo couplings for the two channels:

$$\begin{aligned} J_1 &\equiv \frac{4\Gamma_1^2}{U_s} , \\ J_2 &\equiv \frac{2\Gamma_3^2}{U_s} . \end{aligned} \tag{1.125}$$

Finally, by fine-tuning Γ_3 it is possible to make the two Kondo couplings equal, obtaining precisely the standard two-channel Kondo Hamiltonian.

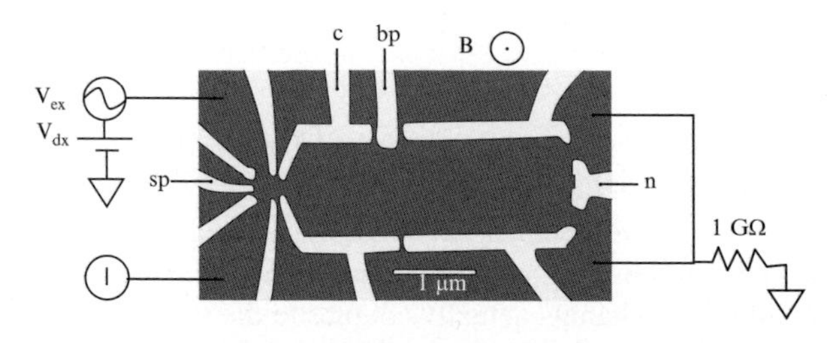

Fig. 1.3 Device for realizing the two-channel Kondo effect.

Using our BCFT methods, it is easily seen that this type of channel anisotropy, with $J_1 \neq J_2$, is a relevant perturbation (Affleck *et al.* 1992). The relevant operator which now appears at the low-energy fixed point is

$$\delta H \propto (J_1 - J_2)\varphi_c^3 \,. \tag{1.126}$$

Here $\vec{\varphi}_c$ is a primary field in the channel sector. Since the associated channel WZW model is also $\mathrm{SU}(2)_2$ for $k = 2$, we may label channel fields by their corresponding pseudo-spin. This primary field has pseudo-spin 1, and scaling dimension $1/2$. It must be checked that it occurs in the boundary operator spectrum at the Kondo fixed point. This follows from the fact that $\varphi_c^a \varphi_s^b$, the product of the channel and spin $j = 1$ primaries, occurs for free-fermion BCs. This dimension-1 operator occurs in the non-Abelian bosonization formula

$$\psi_L^{\dagger j\alpha}(\sigma^a)_\alpha^\beta(\sigma^b)_j^i \psi_{Li\beta} \propto \varphi_s^a \varphi_c^b \,. \tag{1.127}$$

(Note that both sides of this equation have the same scaling dimensions and the same symmetries.) We get the boundary operator spectrum at the Kondo fixed point by double fusion with the $j = 1/2$ primary in the spin sector. The first fusion operation maps the $j_s = 1$ spin primary into $j_s = 1/2$ and the second one maps $j_s = 1/2$ into $j_s = 0$ (and $j_s = 1$). Therefore, the $j_c = 1$, $j_s = 0$ primary is in the boundary operator spectrum at the (overscreened) Kondo fixed point. Since this operator exists and is allowed by all symmetries once the channel $\mathrm{SU}(2)$ symmetry is broken, we expect it to be generated in the low-energy effective Hamiltonian. It thus destabilizes the fixed point since it is relevant. It is not hard to guess what stable fixed point the system flows to. Suppose $J_1 > J_2$. The stable fixed point corresponds to J_2 flowing to zero and J_1 flowing to large values. The more strongly coupled channel 1 screens the $S = 1/2$ impurity, while the more weakly coupled channel 2 decouples. This fixed point is characterized by simple phase shifts of $\pi/2$ for channel 1 and 0 for channel 2. Such behavior is consistent with the weak-coupling RG equations

$$\begin{aligned}
\frac{d\lambda_1}{d\ln D} &= -\lambda_1^2 + \frac{1}{2}\lambda_1(\lambda_1^2 + \lambda_2^2) + \dots \,, \\
\frac{d\lambda_2}{d\ln D} &= -\lambda_2^2 + \frac{1}{2}\lambda_2(\lambda_1^2 + \lambda_2^2) + \dots \,.
\end{aligned} \tag{1.128}$$

Once λ_1^2 gets larger than $2\lambda_2$, these equations predict that the growth of λ_2 is arrested and starts to decrease. This RG flow is also consistent with the g-theorem: $g = (1/2)\ln 2$ at the symmetric fixed point but $g = 0$ at the stable fixed point. The implications of this RG flow for the conductance in the quantum dot system is also readily deduced (Pustilnik *et al.* 2004). From eqn (1.94), using $S_{(1)} = 0$ for the $k = 2$, $S = 1/2$ Kondo fixed point, we see that $\mathcal{T}(\omega = T = 0)$ has half the value it has in the Fermi liquid case, and therefore the conductance through the quantum dot has half the value for the single-channel fixed point, $G(0) = e^2/h$. Let us call the Kondo coupling to the large dot J_l and the coupling to the symmetric combination of left and right leads J_s. The $T = 0$ conductance when $J_s > J_l$ is $2e^2/h$, owing to the $\pi/2$ phase shift in the $\mathcal{T}$-matrix for the s channel. On the other hand, if $J_l > J_s$, the $T = 0$ conductance is zero since the s-channel phase shift is zero. For bare couplings that are close to each other, $\lambda_1 \approx \lambda_2$, the system will flow towards the NFL critical point, before diverging from it at low T, i.e. there is a "quantum critical region" at finite T for λ_1 near λ_2. The basic scaling properties follow from the fact the relevant operators destabilizing the NFL critical point have dimension 1/2, together with the fact that the LIO at the NFL critical point has dimension 3/2. Right at the critical point, the finite-T correction to the conductance, of first order in the LIO, is

$$G(T) \to \frac{e^2}{h}\left[1 - \left(\frac{\pi T}{T_K}\right)^{1/2}\right]. \tag{1.129}$$

(The prefactor was also determined exactly here, but this is useful only if some independent measurement of T_K can be made experimentally.) We define T_c as the crossover scale at which the RG flow starts to deviate from the NFL critical point. This defines the energy scale occurring in the relevant perturbation in the effective Hamiltonian

$$H = H_{NFL} \pm T_c^{1/2}\varphi_c^3(0) - \frac{1}{T_K^{1/2}}\vec{J}_{-1}\cdot\vec{\varphi}_s\,, \tag{1.130}$$

where I have included both the LIO (last term) and the relevant operator, which is present when $\lambda_1 \neq \lambda_2$ with a sign for the coupling constant $\propto \lambda_1 - \lambda_2$. For almost equal bare couplings, this will be much less than T_K, the scale at which the renormalized couplings become large. At $T_c \ll T \ll T_K$, we can calculate the correction to the NFL conductance to first order in φ_c^c, giving

$$G(T) \approx \frac{e^2}{h}\left[1 + \text{constant}\cdot\text{sgn}(\Delta)\left(\frac{T_c}{T}\right)^{1/2}\right]. \tag{1.131}$$

Here,

$$\Delta \equiv \lambda_1 - \lambda_2\,. \tag{1.132}$$

We may also estimate (Pustilnik *et al.* 2004) T_c in terms of Δ and the average bare Kondo coupling,

$$\bar{\lambda} \equiv \frac{\lambda_1 + \lambda_2}{2}\,. \tag{1.133}$$

The weak-coupling RG equations (to second order only) are

$$\begin{aligned} \frac{d\bar{\lambda}}{d\ln D} &= -\bar{\lambda}^2\,, \\ \frac{d\Delta}{d\ln D} &= -2\,\Delta\bar{\lambda}\,. \end{aligned} \tag{1.134}$$

The solution to the first of these can be written

$$\bar{\lambda}(D) = \frac{1}{\ln(D/T_K)}\,, \tag{1.135}$$

for $D \gg T_K$. The second of these RG equations can then be written

$$\frac{d\Delta}{d\ln D} = -\frac{2}{\ln(D/T_K)}\Delta\,. \tag{1.136}$$

Integrating this equation gives

$$\Delta(T_K) \propto \frac{\Delta_0}{\bar{\lambda}_0^2}, \tag{1.137}$$

where $\bar{\lambda}_0$ and Δ_0 are the bare couplings. At energy scales below T_K, we may write the RG equation for $\Delta(D)$,

$$\frac{d\Delta}{d\ \ln D} = \frac{1}{2}\Delta, \tag{1.138}$$

reflecting the fact that Δ has scaling dimension $1/2$ at the NFL fixed point. Thus,

$$\Delta(D) = \left(\frac{T_K}{D}\right)^{1/2} \Delta(T_K) \tag{1.139}$$

for $D < T_K$. By definition, the crossover scale T_c is the energy scale where $\Delta(D)$ becomes of order 1. Thus

$$1 \propto \Delta(T_K)\left(\frac{T_K}{T_C}\right)^{1/2}. \tag{1.140}$$

Using eqn (1.137) for $\Delta(T_K)$, we finally determine the crossover scale in terms of T_K and the bare parameters:

$$T_c \propto T_K \frac{\Delta_0^2}{\bar{\lambda}_0^4}. \tag{1.141}$$

Another interesting quantity is the dependence of the conductance at $\Delta_0 = 0$ on the temperature and the source–drain voltage. V_{sd} defines another energy scale, in addition to T, so we expect

$$G \equiv \frac{dI}{dV_{sd}} = \frac{e^2}{h}\left[1 - \left(\frac{\pi T}{T_K}\right)^{1/2} F\left(\frac{eV_{sd}}{T}\right)\right], \tag{1.142}$$

where $F(x)$ is some universal scaling function. A theoretical calculation of F remains an open problem. These theoretical predictions, in particular the occurrence of the critical exponent 1/2, are in good agreement with the experiments of the Goldhaber-Gordon group (Potok *et al.* 2007).

1.4 Quantum impurity problems in Luttinger liquids

The Kondo models considered so far in these lectures all have the property that the electrons are assumed to be noninteracting, except with the impurity. The validity of this approximation is based on Fermi liquid theory ideas, as mentioned in the first lecture. Although our model becomes one-dimensional after s-wave projection, it is probably important that it was originally two- or three-dimensional, to justify ignoring these interactions, since in the one-dimensional case, Fermi liquid theory definitely fails. Now interactions are important, leading, at low energies, to "Luttinger liquid" (LL) behavior. We will now find interesting boundary RG phenomena for a potential scatterer, even without any dynamical degrees of freedom at the impurity (Kane and Fisher 1992, Eggert and Affleck 1992). The physical applications of this theory include point contacts in quantum wires, carbon nanotubes, constrictions in quantum Hall bars, and impurities in spin chains.

I will just give a lightning review of LL theory here, since it has been reviewed in many other places (for example Affleck (1990) and Giamarchi (2004)) and is not the main subject of these lectures. We are generally interested in the case of fermions with spin, but no additional "channel" quantum numbers ($k = 1$). A typical microscopic model is the Hubbard model,

$$H = -t \sum_j [(\psi_j^\dagger \psi_{j+1} + h.c.) + U \hat{n}_j^2], \tag{1.143}$$

where $\hat{n}_j$ is the total number operator (summed over spin directions) on site j. (More generally, we might consider "ladder" models, in which case we would also get several "channels" and a plethora of complicated interactions.) Non-Abelian bosonization is again useful, leading to a separation of spin and charge degrees of freedom. But now we must consider the various bulk interactions. These fall into several classes:

- $g_c J_L J_R$. An interaction term of this form, which is proportional to $(\partial_\mu \varphi)^2$, in the Lagrangian density can be eliminated by rescaling the charge boson field: $\varphi \rightarrow \sqrt{g}\varphi$. Here the Luttinger parameter g has the value 1 in the noninteracting case. (Unfortunately, there are numerous different conventions for the Luttinger parameter. I follow here the notation of Oshikawa *et al.* (2006).) This leaves the Hamiltonian in noninteracting form, but the rescaling changes the scaling dimensions of various operators.
- $\psi_R^{\dagger\uparrow} \psi_R^{\dagger\downarrow} \psi_{L\uparrow} \psi_{L\downarrow} + h.c.$ This can be bosonized as a pure charge operator. Depending on the value of the Luttinger parameter, it can be relevant, in which case it produces a gap for charge excitations. However, this "Umklapp" term is accompanied by oscillating factors $e^{\pm 2ik_F x}$, so it can usually be ignored unless $k_F = \pi/2$, corresponding to half-filling. It produces a charge gap in the repulsive Hubbard model at half-filling. The low-energy Hamiltonian then involves the spin degrees of freedom only. In particular, it may correspond to the $\mathrm{SU}(2)_1$ WZW model. In the large-U limit of the Hubbard model, we obtain the $S = 1/2$ Heisenberg model, with antiferromagnetic coupling $J \propto t^2/U$, as a low-energy ($E \ll U$) lattice model. The low-energy Hamiltonian for the Heisenberg model is again the $\mathrm{SU}(2)_1$ WZW model.
- *Marginal terms of nonzero conformal spin.* The only important effect of these is assumed to be to change the velocities of the spin and charge degrees of freedom, making them, in general, different.
- $-(g_s/2\pi)\vec{J}_L \cdot \vec{J}_R$. g_s has a quadratic β-function at weak coupling; it flows to zero logarithmically if it is initially positive, as occurs for the repulsive ($U > 0$) Hubbard model. It is often simply ignored but, in fact, it leads to important logarithmic corrections to all quantities.
- *Spin-anisotropic interactions of zero conformal spin.* Often $\mathrm{SU}(2)$ spin symmetry is a good approximation in materials, but it is generally broken to some extent, owing to spin–orbit couplings. If a $\mathrm{U}(1)$ spin symmetry is preserved then, depending on the parameters, the spin degrees of freedom can remain gapless. It is then usually convenient to use ordinary Abelian bosonization. The spin boson then also gets rescaled so that $\varphi_s \rightarrow g_s \varphi_s$, where $g_s = 1$ in the isotropic case. This leads to further changes in the scaling dimensions of various operators.
- *Various higher-dimensional operators of nonzero conformal spin.* Some of these have very interesting and nontrivial effects and are the subject of current research. However, these effects generally go away at low energies.

Let us begin with an interacting spinless fermion model with impurity scattering at the origin only, corresponding to a point contact in a quantum wire. A corresponding lattice model could be, for example,

$$H = \left[-t \sum_{j=-\infty}^{-1} \psi_j^\dagger \psi_{j+1} - t' \psi_0^\dagger \psi_1 - t \sum_{j=1}^{\infty} \psi_j^\dagger \psi_{j+1} + h.c. \right] + U \sum_{j=-\infty}^{\infty} \hat{n}_j \hat{n}_{j+1} . \quad (1.144)$$

The hopping term between sites 0 and 1 has been modified from t to t'; we might expect $t' \ll t$ for a point contact or constriction. Upon bosonizing and rescaling the boson, the bulk terms in the action just give

$$S_0 = \frac{g}{4\pi} \int_{-\infty}^{\infty} dx\, d\tau\, (\partial_\mu \varphi)^2 . \quad (1.145)$$

The impurity term, in terms of continuum limit fermions,

$$\sqrt{2\pi}\psi_j \approx e^{ik_F j} \psi_R(j) + e^{-ik_F j} \psi_R(j) , \quad (1.146)$$

is

$$H_{\text{int}} \approx \frac{t - t'}{2\pi} [J_L(0) + J_R(0) + (\psi_L^\dagger(0)\psi_R(0)e^{ik_F} + h.c.)] . \quad (1.147)$$

(Note that we ignore the small variation of the continuum limit fields over one lattice spacing here. Including this effect leads only to irrelevant operators. This continuum limit Hamiltonian is appropriate for small $|t' - t|$, since we have taken the continuum limit assuming $t' = t$.) Using the bosonization formulas

$$\psi_{L/R} \propto e^{i(\varphi \pm \theta)/\sqrt{2}} , \quad (1.148)$$

this becomes

$$H_{\text{int}} = -(t' - t)\sqrt{2}\partial_x \theta(0) - \text{constant} \cdot (t' - t) \cos[\sqrt{2}(\theta(0) - \alpha)] , \quad (1.149)$$

for a constant α depending on k_F. While the first term is always exactly marginal, the second term, which arises from "backscattering" ($L \leftrightarrow R$), has dimension

$$x = g . \quad (1.150)$$

It is marginal for free fermions, where $g = 1$, but is relevant for $g < 1$, corresponding to repulsive interactions, $U > 0$. It is convenient to go to a basis of even and odd channels,

$$\theta_{e/o}(x) \equiv \frac{\theta(x) \pm \theta(-x)}{\sqrt{2}} . \quad (1.151)$$

The $\theta_{e/o}$ fields obey Neumann (N) and Dirichlet (D) BCs, respectively:

$$\begin{aligned} \partial_x \theta_e(0) &= 0 , \\ \theta_o(0) &= 0 . \end{aligned} \quad (1.152)$$

Then the action separates into even and odd parts, $S = S_e + S_o$, with

$$S_e = \frac{1}{4\pi g}\int_{-\infty}^{\infty} d\tau \int_0^{\infty} dx\,(\partial_\mu \theta_e)^2 - V_b \cos(\theta_e(0) - \alpha)\,,$$
$$S_o = \frac{1}{4\pi g}\int_{-\infty}^{\infty} d\tau \int_0^{\infty} dx\,(\partial_\mu \theta_o)^2 - V_f \partial_x \theta_o(0)\,, \tag{1.153}$$

where $V_{f/b}$, the forward and backward scattering amplitudes, are both $\propto t' - t$. The interaction term can be eliminated from S_o by the transformation

$$\theta_o(x) \to \theta_o(x) - 2\pi V_f g\,\mathrm{sgn}(x)\,. \tag{1.154}$$

On the other hand, S_e is the well-known boundary sine–Gordon model, which is not so easily solved. It is actually integrable (Zamolodchikov and Goshal 1994), and a great deal is known about it, but here I will just discuss simple RG results. For $g < 1$, when backscattering is relevant, it is natural to assume that V_b renormalizes to infinity, thus changing the N boundary condition on θ_e to D, $\theta_e(0) = \alpha$. This has the effect of severing all communication between the left and right sides of the system, corresponding to a cut chain. If the forward scattering, V_f, is equal to 0 then we have independent D boundary conditions on the left and right sides:

$$\theta(0^{\pm}) = \frac{\alpha}{\sqrt{2}}\,. \tag{1.155}$$

For nonzero V_f, the left and right side are still severed but the D BCs are modified to

$$\theta(0^{\pm}) = \frac{\alpha}{\sqrt{2}} \mp \sqrt{2}\pi V_f g\,. \tag{1.156}$$

The simple D BCs of eqns (1.155) and (1.156) correspond, in the original fermion language, to

$$\psi_L(0^{\pm}) \propto \psi_R(0^{\pm})\,. \tag{1.157}$$

The right-moving excitations on the $x < 0$ axis are reflected at the origin, picking up a phase shift which depends on V_f, and likewise for the left-moving excitations on $x > 0$.

Of course, we have made a big assumption here, that V_b renormalizes to infinity, giving us this simple D BC. It is important to at least check the self-consistency of the assumption. This can be done by checking the stability of the D fixed point. Thus we consider the Hamiltonian of eqn (1.144) with $t' \ll t$. To take the continuum limit, we must carefully take into account the boundary conditions when $t' = 0$. Consider the chain from $j = 1$ to ∞ with open boundary conditions (OBCs). This model is equivalent to one where a hopping term, of strength t, to site 0 is included but then a BC $\psi_0 = 0$ is imposed. From eqn (1.146) we see that this corresponds to $\psi_L(0) = -\psi_R(0)$. Using the bosonization formulas of eqn (1.148), we see that this corresponds to a D BC, $\theta(0) =$ constant, as we would expect from the previous discussion. A crucial point is that imposing a D BC changes the scaling dimension of the fermion fields at the origin. Setting $\theta(0) =$ constant, (1.148) reduces to

$$\psi_{L/R}(0) \propto e^{i\varphi(0)/\sqrt{2}}\,. \tag{1.158}$$

The dimension of this operator is itself affected by the D BC. Decomposing $\varphi(t, x)$ and θ into left- and right-moving parts,

$$\begin{aligned} \varphi(t,x) &= \frac{1}{\sqrt{g}}[\varphi_L(t+x) + \varphi_R(t-x)]\,, \\ \theta &= \sqrt{g}[\varphi_L - \varphi_R]\,, \end{aligned} \tag{1.159}$$

we see that the D BC implies

$$\varphi_R(0) = \varphi_L(0) + \text{constant}\,. \tag{1.160}$$

Thus, to evaluate correlation functions of $\varphi(0)$ with a D BC, we can use

$$\varphi(0) \to \frac{2}{\sqrt{g}}\varphi_L(0) + \text{constant}\,. \tag{1.161}$$

The bulk correlation functions of exponentials of φ decay as

$$\begin{aligned} &\langle e^{ia\varphi(t,x)}e^{-ia\varphi(0,0)}\rangle \\ &= \langle e^{ia\varphi_L(t+x)/\sqrt{g}}e^{-ia\varphi_L(0,0)/\sqrt{g}}\rangle\langle e^{ia\varphi_R(t-x)/\sqrt{g}}e^{-ia\varphi_R(0,0)/\sqrt{g}}\rangle \\ &= \frac{1}{(x+t)^{a^2/2g}(x-t)^{a^2/2g}}\,. \end{aligned} \tag{1.162}$$

On the other hand, at a boundary with a D BC,

$$\langle e^{ia\varphi(t,0)}e^{-ia\varphi(0,0)}\rangle = \langle e^{2ia\varphi_L(t)/\sqrt{g}}e^{-2ia\varphi_L(0,0)/\sqrt{g}}\rangle = \frac{1}{t^{(2a)^2/2g}}\,. \tag{1.163}$$

The RG scaling dimension of the operator $e^{ia\varphi}$ doubles at a boundary with a D BC to $\Delta = a^2/g$. Thus the fermion field at a boundary with a D BC, eqn (1.158), has a scaling dimension $1/2g$. The weak tunneling amplitude t' in eqn (1.144) couples together two independent fermion fields from the left and right sides, both obeying D BCs. Therefore the scaling dimension of this operator is obtained by adding the dimensions of each independent fermion field, and has the value $1/g$. This is relevant when $g > 1$, the case where the weak backscattering is irrelevant, and is irrelevant for $g < 1$, the case where the weak backscattering is relevant. Thus our bold conjecture that the backscattering V_b renormalizes to ∞ for $g < 1$ has passed an important consistency test. The infinite-backscattering, cut-chain D BC fixed point is indeed stable for $g < 1$. On the other hand, and perhaps even more remarkably, it seems reasonable to hypothesize that even a weak tunneling t' between two semi-infinite chains flows to the N BC at low energies. This is a type of "healing" phenomenon: translational invariance is restored in the low-energy, long-distance limit.

The conductance is clearly zero at the D fixed point. At the N fixed point, we may calculate it using a Kubo formula. One approach is to apply an AC electric field to a finite region, L, in the vicinity of the point contact:

$$G = \lim_{\omega\to 0}\frac{-e^2}{h}\frac{1}{\pi\omega L}\int_{-\infty}^{\infty}d\tau\, e^{i\omega\tau}\int_0^L dx\, T\langle J(y,\tau)J(x,0)\rangle \tag{1.164}$$

(independent of x). Here the current operator is $J = -i\partial_\tau \theta$. Using

$$\langle \theta(x,\tau)\theta(0,0)\rangle = -\frac{g}{2}\ln(\tau^2 + x^2)\,, \tag{1.165}$$

for the infinite-length system, it is straightforward to obtain

$$G = g\frac{e^2}{h}\,. \tag{1.166}$$

(While this is the conductance predicted by the Kubo formula, it is apparently not necessarily what is measured experimentally (Tarucha *et al.* 1995). For various theoretical discussions of this point, see Maslov and Stone (1995), Ponomarenko (1995), Safi and Shulz (1995), Chamon and Fradkin (1997), and Imura *et al.* (2002).) Low-temperature corrections to this conductance can be obtained by doing perturbation theory in the LIO, as usual. At the cut-chain, D fixed point, for $g < 1$, the LIO is the tunneling term, $\propto t'$ in eqn (1.144). This renormalizes as

$$t'(T) \approx t'_0 \left(\frac{T}{T_0}\right)^{1/g-1}\,. \tag{1.167}$$

Since the conductance is second order in t', we predict

$$G(T) \propto t'^2_0 T^{2(1/g-1)} \quad (g < 1, T \ll T_0)\,. \tag{1.168}$$

Here T_0 is the lowest characteristic energy scale in the problem. If the bare t'_0 is small then T_0 will be of the order of the band width, t. However, if the bare model has only a weak backscattering V_b then T_0 can be much smaller, corresponding to the energy scale where the system crosses over between the N and D fixed points, analogous to the Kondo temperature. (Note that there is no Kondo impurity spin in this model, however.) On the other hand, near the N fixed point, where the backscattering is weak, we may do perturbation theory in the renormalized backscattering; again the contribution to G is second order. Now, for $g > 1$,

$$V_b(T) = V_{b0}\left(\frac{T}{T_0}\right)^{g-1} \tag{1.169}$$

and hence

$$G - \frac{ge^2}{h} \propto V_b^2 T^{2(g-1)} \quad (g > 1, T \ll T_0)\,. \tag{1.170}$$

Again T_0 is the lowest characteristic energy scale; it is a small crossover scale if the microscopic model has only a small tunneling t'.

A beautiful application (Moon *et al.* 1993, Fendley *et al.* 1995) of this quantum impurity model is to tunneling through a constriction in a quantum Hall bar, as illustrated in Fig. 1.4. Consider a 2DEG in a strong magnetic field at the fractional-quantum-Hall-effect plateau of filling factor $\nu = 1/3$. Owing to the bulk excitation gap in the Laughlin ground state, there is no current flowing in the bulk of the sample. However, there are gapless edge states which behave as a chiral Luttinger liquid. Now the currents are chiral, with right-movers restricted to the lower edge and left-movers to the upper edge in the figure. Nonetheless, we may apply our field theory to this system, and the Luttinger parameter turns out to have the

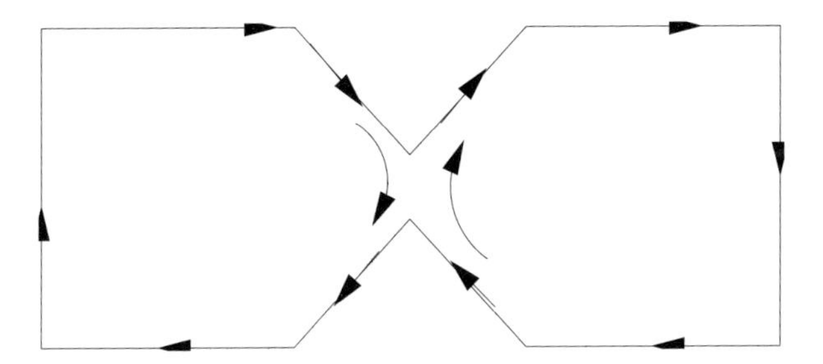

Fig. 1.4 Quantum Hall bar with a constriction. Edge currents circulate clockwise and can tunnel from the upper to the lower edge at the constriction.

value $g = \nu < 1$. (Furthermore, the edge states are believed to be spin-polarized, making the spinless model discussed here appropriate.) Right- and left-movers interact at the constriction, with a finite probability of backscattering, which in this case takes quasiparticles between the upper and lower edges. The entire crossover function for the conductance can be calculated, either by quantum Monte Carlo methods (together with a delicate analytic continuation to zero frequency) or using the integrability of the model. The results agree fairly well with experiments.

A number of other interesting quantum impurity problems have been studied in Luttinger liquids. These include the generalization of the model discussed above to include electron spin (Kane and Fisher 1992). Four simple fixed points are found, which are obvious generalizations of the two discussed in the spinless case. We may now have D or N BCs for both charge and spin bosons, corresponding to perfect reflection/transmission for charge/spin. Interestingly, additional fixed points occur for certain ranges of the charge and spin Luttinger parameters, which have charge and spin conductances which are universal, nontrivial numbers. A general solution for these fixed points remains an open problem. A simpler situation occurs in spin chains (Eggert and Affleck 1992). Again, I give only a telegraphic reminder of the field theory approach to the $S = 1/2$ Heisenberg antiferromagnetic chain with Hamiltonian

$$H = J \sum_j \vec{S}_j \cdot \vec{S}_{j+1} \, . \tag{1.171}$$

One method is to start with the Hubbard model at half-filling, where the charge excitations are gapped owing to the Umklapp interaction. We may simply drop the charge boson from the low-energy effective Hamiltonian, which then contains only the spin boson or, equivalently, an $\mathrm{SU}(2)_1$ WZW model. The low-energy degrees of freedom of the spin operators occur at wave vectors 0 and π:

$$\vec{S}_j \approx \frac{1}{2\pi}[\vec{J}_L(j) + \vec{J}_R(j)] + (-1)^j \vec{n}(j) \, , \tag{1.172}$$

where the staggered component, of scaling dimension 1/2, can be written either in terms of a free boson φ and its dual θ, with $g = 1/2$, or else in terms of the primary field g^α_β of the WZW model:

$$\vec{n} \propto \mathrm{tr}\, g\vec{\sigma} \propto \begin{pmatrix} \cos(\varphi/\sqrt{2}) \\ \sin(\varphi/\sqrt{2}) \\ \cos(\sqrt{2}\theta) \end{pmatrix} . \tag{1.173}$$

The spin boson Hamiltonian contains the marginally irrelevant interaction $-(g_s/2\pi)\vec{J}_L \cdot \vec{J}_R$, with a bare coupling constant g_s, of O(1). By including a second-neighbor coupling, J_2, in the microscopic Hamiltonian, the bare value of g_s can be varied. At $J_2 = J_{2c} \approx 0.2411J$, a phase transition occurs, with the system going into a gapped, spontaneously dimerized phase. In the low-energy effective Hamiltonian, the phase transition corresponds to the bare g_s passing through zero and becoming marginally relevant rather than marginally irrelevant.

A semi-infinite spin chain, $j \geq 0$, with a free BC corresponds to a D BC on θ, just as for the fermionic model discussed above (Eggert and Affleck 1992). Then the staggered spin operator at zero becomes

$$\vec{n}(0) \propto \begin{pmatrix} \cos[\sqrt{2}\varphi_L(0)] \\ \sin[\sqrt{2}\varphi_L(0)] \\ \partial_x \varphi_L(0) \end{pmatrix} . \tag{1.174}$$

All three components now have scaling dimension 1, and, it is easily seen, taking into account $\mathrm{SU}(2)$ symmetry, that

$$\vec{n}(0) \propto \vec{J}_L(0) . \tag{1.175}$$

The D BC also implies $\vec{J}_L(0) = \vec{J}_R(0)$, so that both the uniform and the staggered spin components at $x = 0$ reduce to $\vec{J}_L(0)$. Now consider the effect of a Kondo-type coupling between a spin chain and one additional "impurity spin." In the simplest case, where the impurity spin is also of size $S = 1/2$, it makes an enormous difference exactly how it is coupled to the other spins. The simplest case is where it is coupled at the end of a semi-infinite chain:

$$H = J'\vec{S}_1 \cdot \vec{S}_2 + J\sum_{i=2}^{\infty} \vec{S}_i \cdot \vec{S}_{i+1} , \tag{1.176}$$

with the impurity coupling $J' \ll J$. For small J', a low-energy Hamiltonian description is appropriate, and since $\vec{S}_2 \propto \vec{J}_L(0)$, we obtain the continuum limit of the Kondo model with a bare Kondo coupling $\lambda \propto J'$ (Eggert and Affleck 1992, Laflorencie *et al.* 2008). Thus we can take over immediately all the RG results for the Kondo effect, except that we must beware of logarithmic corrections arising from the bulk marginal coupling constant g_s, which are absent for the free-fermion Kondo model. The correspondence with the free-fermion Kondo model becomes nearly perfect when a bulk second-neighbor interaction J_2 is added to the Hamiltonian and fine-tuned to the critical point where this bulk marginal interaction vanishes. Then, only truly irrelevant bulk interactions (of dimension 4 or greater) distinguish the two models. The strong-coupling fixed point of this Kondo model corresponds simply to the impurity spin being absorbed into the chain, and corresponds to a renormalized $J' \to J$ at low energies. A more interesting model involves an impurity spin coupled to two semi-infinite chains:

$$H = J\sum_{-\infty}^{-2} \vec{S}_j \cdot \vec{S}_{j+1} + J\sum_{j=1}^{\infty} \vec{S}_j \cdot \vec{S}_{j+1} + J'\vec{S}_0 \cdot (\vec{S}_{-1} + \vec{S}_1) . \tag{1.177}$$

The continuum limit is now the two-channel Kondo model (Eggert and Affleck 1992), with the left and right sides of the impurity corresponding to the two channels. Again the Kondo fixed point corresponds simply to a "healed chain," with J' renormalizing to J and a restoration of translational invariance at low energies. Other possibilities involve a "side-coupled" impurity spin. For example, we may couple the impurity spin $\vec{S}'$ to one site on a uniform chain:

$$H = J \sum_{j=-\infty}^{\infty} \vec{S}_j \cdot \vec{S}_{j+1} + J'\vec{S}' \cdot \vec{S}_0 . \tag{1.178}$$

Now the correspondence to the ordinary free-fermion Kondo model fails dramatically because the boundary interaction $\propto J'\vec{S}' \cdot \vec{n}(0)$ appears in the effective Hamiltonian, where $\vec{n}$ is the staggered spin operator introduced in eqn (1.173). This is a strongly relevant dimension-1/2 boundary interaction. It renormalizes to infinity. It is easy to understand the low-energy fixed point in this case by imagining an infinite bare J'. The impurity spin forms a singlet with $\vec{S}_0$, and the left and right sides of the chain are decoupled. The stability of such a fixed point is verified by the fact that the spins at the ends of the open chains, $\vec{S}_{\pm 1}$, have dimension 1 so that an induced weak "bridging" coupling $J_{\text{eff}}\vec{S}_{-1} \cdot \vec{S}_1$ has dimension 2 and is thus an irrelevant boundary interaction. We expect even a small J' to renormalize to such a strong-coupling fixed point but, in general, the screening cloud will be spread over longer distances. Nonetheless, the left and right sides decouple at low energies and long distances. Other examples, including larger-spin impurities, were discussed in Eggert and Affleck (1992).

We may also couple an impurity spin to a Hubbard type model with gapless spin *and* charge degrees of freedom. The various situations closely parallel the spin chain case. In particular, the cases of the impurity spin at the end of the chain and embedded in the middle still correspond essentially to the simple Kondo model. This follows because the D BC on both the spin and the charge bosons has the effect of reducing both the uniform *and the staggered* spin density operators at the boundary to $\vec{J}_L(0)$. This model, which can be applied to a quantum dot coupled to a quantum wire in a semiconductor heterostructure, was analyzed in detail in Pereira *et al.* (2008).

1.5 Quantum impurity entanglement entropy

Quantum entanglement entropy has become a popular subject in recent years because of its connection with black holes, quantum computing, and the efficiency of the density matrix renormalization group method and its generalizations for calculating many-body ground states (on a classical computer). In this lecture I will discuss the intersection of this subject with quantum impurity physics (Sorensen *et al.* 2007). After some generalities, I will focus on the simple example of the single-channel Kondo model, obtaining a novel perspective on the nature of the Kondo ground state and the meaning of the characteristic length scale ξ_K. In the second lecture, I discussed and defined the zero-temperature impurity entropy, showing that it is a universal quantity, characterizing the BCFT fixed point, and always decreases under boundary RG flow. Quantum entanglement entropy is, in general, quite distinct from thermodynamic entropy, being a property of a quantum ground state and depending on an arbitrary division of a system into two different spatial regions. Nonetheless, as we shall see, the thermodynamic impurity entropy, in the $T = 0$ limit, also appears as a term in the entanglement entropy, in a certain limit.

Consider first a CFT with central charge c on a semi-infinite interval, $x > 0$, with a CIBC, labeled A, at $x = 0$. We trace out the region $x' \geq x$ to define the density matrix, and hence the entanglement entropy $S_A(x) = -\text{tr}\rho \ln \rho$ for the region $0 \leq x' \leq x$. (Note that I am using the natural logarithm in my definition of entanglement entropy. Some authors define S using the logarithm to base 2, which simply divides S by $\ln 2$.) Calabrese and Cardy (2004) (C&C) showed, generalizing earlier results of Holzhey, Larsen, and Wilczek (1994), that this entanglement entropy is given by

$$S_A(x) = \frac{c}{6} \ln\left(\frac{x}{a}\right) + c_A . \tag{1.179}$$

Here a is a nonuniversal constant. c_A is another constant, which could have been absorbed into a redefinition of a. However, $S_A(x)$ is written this way because, by construction, the constant a is independent of the choice of the CIBC A, while the constant c_A depends on it. C&C showed that the generalization of $S_A(x)$ to a finite inverse temperature β is given by a standard conformal transformation:

$$S_A(x,\beta) = \frac{c}{6} \ln\left[\left(\frac{\beta}{\pi a}\right) \sinh\left(\frac{2\pi x}{\beta}\right)\right] + c_A . \tag{1.180}$$

$S_A(x,\beta)$ is defined by beginning with the Gibbs density matrix for the entire system, $e^{-\beta H}$, and then again tracing out the region $x' > x$. Now consider the high-temperatures, long-length limit, $\beta \ll x$:

$$S_A \to \frac{2\pi c x}{\beta} + \frac{c}{6} \ln\left(\frac{\beta}{2\pi a}\right) + c_A + O(e^{-4\pi x/\beta}) . \tag{1.181}$$

The first term is the extensive term (proportional to x) in the thermodynamic entropy for the region $0 < x' < x$. The reason that we recover the thermodynamic entropy when $x \gg \beta$ is because, in this limit, we may regard the region $x' > x$ as an "additional reservoir" for the region $0 \leq x' \leq x$. That is, the thermal density matrix can be defined by integrating out degrees of freedom in a thermal reservoir, which is weakly coupled to the entire system. On the other hand, the region $x' > x$ is quite strongly coupled to the region $x' < x$. Although this coupling is quite strong, it occurs only at one point, x. When $x \gg \beta$, this coupling only weakly perturbs the density matrix for the region $x' < x$. Only low-energy states, with energies of order $1/x$, and a negligible fraction of the higher-energy states (those localized near $x' = x$) are affected by the coupling to the region $x' > x$. The thermal entropy for the system, with the boundary at $x = 0$ in the limit $x \gg \beta$, is

$$S_{A,th} \to \frac{2\pi c x}{\beta} + \ln g_A + \text{constant} , \tag{1.182}$$

with corrections that are exponentially small in x/β. The only dependence on the CIBC, in this limit, is through the constant term, $\ln g_A$, the impurity entropy. Thus it is natural to identify the BC-dependent term in the entanglement entropy with the BC-dependent term in the thermodynamic entropy:

$$c_A = \ln g_A . \tag{1.183}$$

This follows since, in the limit $x \gg \beta$, we don't expect the coupling to the region $x' > x$ to affect the thermodynamic entropy associated with the boundary $x' = 0$, c_A. Note that the entanglement entropy, eqn (1.181), contains an additional large term not present in the thermal entropy. We may ascribe this term to a residual effect of the strong coupling to the region $x' > x$ on the reduced density matrix. However, this extra term does not depend on the CIBC, as we would expect in the limit $x \gg \beta$, in which the "additional reservoir" is far from the boundary. Now, passing to the opposite limit $\beta \to \infty$, we obtain the remarkable result that the only term in the (zero-temperature) entanglement entropy depending on the BC is precisely the impurity entropy, $\ln g_A$.

Since the impurity entropy $\ln g_A$ is believed to be a universal quantity characterizing boundary RG fixed points, it follows that the boundary-dependent part of the ($T = 0$) entanglement entropy also possesses this property. In particular, we might then expect this quantity to exhibit an RG crossover as we increase x. That is, consider the entanglement entropy $S(x)$ for the type of quantum impurity model discussed in these lectures, that is, a model which is described by a conformal field theory in the bulk (at low energies) and has more or less arbitrary boundary interactions. As we increase x, we might expect $S(x)$ to approach the CFT value, eqn (1.179), with the value of c_A corresponding to the corresponding CIBC. More interestingly, consider such a system which is flowing between an unstable and a stable CIBC, A and B respectively, such as a general Kondo model with weak bare couplings. Then, as discussed in lectures 1 and 2, the impurity part of the thermodynamic entropy crosses over between two values, $\ln g_A$ and $\ln g_B$. In the general k-channel Kondo case, $g_A = (2S + 1)$, the degeneracy of the decoupled impurity spin. g_B is determined by the Kondo fixed point, having the values

$$\begin{aligned} g_B &= 2S' + 1 \quad (S' \equiv S - 2k, k \leq 2S) \\ &= \frac{\sin[\pi(2s+1)/(2+k)]}{\sin[\pi/(2+k)]} \quad (k > 2S)\,. \end{aligned} \tag{1.184}$$

This *change* in $\ln g_A$ ought to be measurable in numerical simulations or possibly even experiments.

To make this discussion more precise, it is convenient to define a "quantum impurity entanglement entropy" as the difference between the entanglement entropies with and without the impurity. Such a definition parallels the definition of impurity thermodynamic entropy (and impurity susceptibility, impurity resistivity, etc.) used in lectures 1 and 2. To keep things simple, consider the Kondo model in three dimensions and consider the entanglement of a spherical region of radius r containing the impurity at its center, with the rest of the system (which we take to be of infinite size, for now). This entanglement entropy could be measured both before and after adding the impurity, the difference giving the impurity part. Because of the spherical symmetry, it is not hard to show that the entanglement entropy reduces to a sum of contributions from each angular-momentum channel (l, m). For a δ-function Kondo interaction, only the s-wave harmonic is affected by the interaction. Therefore, the impurity entanglement entropy is determined entirely by the s-wave harmonic and can thus be calculated in the usual 1D model. Thus we may equivalently consider the entanglement between a section of the chain $0 < r' < r$, including the impurity, and the rest of the chain. It is even more convenient, especially for numerical simulations, to use the equivalent model, discussed in the previous lecture, of an impurity spin weakly coupled at the end of a Heisenberg $S = 1/2$ spin chain, eqn (1.176). Region A is the first r sites of the chain, which in general has a finite total length R. The corresponding entanglement entropy is written as $S(J'_K, r, R)$, where we set $J = 1$ and now refer to the impurity (Kondo) coupling as J'_K. The long-distance behavior of $S(r)$ is found to have both uniform and alternating parts:

$$S(r) = S_U(r) + (-1)^r S_A(r)\,, \tag{1.185}$$

where both S_U and S_A are slowly varying. I will just focus here on S_U, which we expect to have the same universal behavior as in other realizations of the Kondo model, including the 3D free-fermion one. We define the impurity part of S precisely as

$$S_{\text{imp}}(J'_K, r, R) \equiv S_U(J'_K, r, R) - S_U(1, r-1, R-1) \quad (r > 1)\,. \tag{1.186}$$

Note that we are subtracting the entanglement entropy when the first spin, at site $j = 1$, is removed. This removal leaves a spin chain of length $R - 1$, with all couplings equal to 1. After the removal, region A contains only $r - 1$ sites. If we start with a weak Kondo coupling $J' \ll J = 1$, we might expect to see crossover between weak- and strong-coupling fixed points as r is increased past the size ξ_K of the Kondo screening cloud. Ultimately, this behavior was found, but with a surprising dependence on whether R is even or odd. (Note that this is a separate effect from the dependence on whether r is even or odd, which I have already removed by focusing on the uniform part, defined in eqn (1.186).)

As usual, I focus first on the behavior near the fixed points, where perturbative RG methods can be used. Strong-coupling perturbation theory in the LIO at the Kondo fixed point turns out to be very simple. The key simplifying feature is that the Fermi liquid LIO is proportional to the energy density itself, at $r = 0$. It follows from eqns (1.31) and (1.62) (in the single-channel $k = 1$ case) that the low-energy effective Hamiltonian including the LIO at the Kondo fixed point can be written, in the purely left-moving formalism, as

$$H = \frac{1}{6\pi}\int_{-R}^{R} d\vec{J}_L(x)^2 - \frac{\xi_K}{6}\vec{J}_L^2(0) = \int_{-R}^{R} d\mathcal{H}(x) - \pi\xi_K\mathcal{H}(0)\,, \tag{1.187}$$

where $\mathcal{H}(x) \equiv (1/6\pi)\vec{J}_L^2(x)$ is the energy density. (I have set $v = 1$ and used $\xi_K = 1/T_K$.) The method of Holzhey and Wilczek and of Calabrese and Cardy for calculating the entanglement entropy is based on the "replica trick." That is to say, the partition function Z_n is calculated on an n-sheeted Riemann surface $\mathcal{R}_n$, with the sheets joined along region A, from $r' = 0$ to $r' = r$. The trace of the n-th power of the reduced density matrix can be expressed as

$$\text{Tr}\,\rho(r)^n = \frac{Z_n(r)}{Z^n}\,, \tag{1.188}$$

where Z is the partition function on the normal complex plane, $\mathcal{C}$. The entanglement entropy is obtained from the formal analytic continuation in n,

$$S = -\lim_{n\to 1}\frac{d}{dn}[\text{Tr}\,\rho(r)^n]\,. \tag{1.189}$$

The correction to $Z_n(r)$ of first order in ξ_K is

$$\delta Z_n = (\xi_K\pi)n\int_{-\infty}^{\infty} d\tau\langle\mathcal{H}(\tau,0)\rangle_{\mathcal{R}_n}\,. \tag{1.190}$$

C&C related this expectation value of the energy density on $\mathcal{R}_n$ to a three-point correlation function on the ordinary complex plane:

$$\langle \mathcal{H}(\tau,0)\rangle_{\mathcal{R}_n} = \frac{\langle \mathcal{H}(\tau,0)\varphi_n(r)\varphi_{-n}(-r)\rangle_{\mathcal{C}}}{\langle \varphi_n(r)\varphi_{-n}(-r)\rangle_{\mathcal{C}}}\,. \tag{1.191}$$

Here the primary operators $\varphi_{\pm n}$ sit at the branch points $\pm r$ (in the purely left-moving formulation) and have scaling dimension

$$\Delta_n = \frac{c}{24}\left[1-\left(\frac{1}{n}\right)^2\right], \tag{1.192}$$

where, in this case, the central charge is $c = 1$. Thus eqn (1.191) gives

$$\langle \mathcal{H}(\tau,0)\rangle_{\mathcal{R}_n} = \frac{[1-(1/n)^2]}{48\pi}\frac{(2ir)^2}{(\tau-ir)^2(\tau+ir)^2}\,. \tag{1.193}$$

Doing the τ-integral in eqn (1.190) gives

$$\delta Z_n = -\frac{\xi_K \pi}{24r} n[1-(1/n)^2]\,. \tag{1.194}$$

Since there is no correction to Z, to first order in ξ_K, inserting this result in eqns (1.188) and (1.189) gives

$$S_{\rm imp} = \frac{\pi \xi_K}{12r}\,. \tag{1.195}$$

Here we have used the fact that $g_A = 1$, $c_A = 0$ at the Kondo fixed point. C&C also observed that the entanglement entropy for a finite total system size R can be obtained by a conformal transformation. For a conformally invariant system, this generalizes eqn (1.179) to

$$S(r,R) = \frac{c}{6}\ln\left[\left(\frac{R}{\pi a}\right)\sin\left(\frac{\pi r}{R}\right)\right] + c_A\,. \tag{1.196}$$

Our result for $S_{\rm imp}$ can also be extended to finite R by the same conformal transformation of the two-point and three-point functions in eqn (1.191). The three-point function now has a disconnected part, but this is canceled by the correction to Z^n of first order in ξ_K, leaving

$$\frac{\delta Z^n}{Z^n} = \frac{\xi_K \pi n[1-(1/n)^2]}{48\pi}\int_{-\infty}^{\infty} d\tau \left[\frac{(\pi/R)\sinh 2i\pi r/R}{\sinh[\pi(\tau+ir)/R]\sinh[\pi(\tau-ir)/R]}\right]^2\,. \tag{1.197}$$

Doing the integral and taking the replica limit now gives

$$S_{\rm imp}(r,R) = \frac{\pi\xi_K}{12R}\left[1+\pi\left(1-\frac{r}{R}\right)\cot\left(\frac{\pi r}{R}\right)\right]. \tag{1.198}$$

We emphasize that these results can only be valid at long distances where we can use FLT, i.e. $r \gg \xi_K$. They represent the first term in an expansion in ξ_K/r.

The thermodynamic impurity entropy is known to be a universal scaling function of T/T_K. It seems reasonable to hypothesize that the impurity entanglement entropy (for infinite system size) is a universal scaling function of r/ξ_K. At finite R, we might then expect it to be a universal scaling function of the two variables r/ξ_K and r/R. Our numerical results bear out this expectation, with one perhaps surprising feature: while the scaling function is independent of the total size of the system at $R \to \infty$, there is a large difference between integer and half-integer total spin (i.e. even and odd R in the spin chain version of the Kondo model) for finite R, i.e. we must define two universal scaling functions $S_{\text{imp},e}(r/\xi_K, r/R)$ and $S_{\text{imp},o}(r/\xi_K, r/R)$ for even and odd R. These become the same at $r/R \to \infty$.

We have calculated the impurity entanglement entropy numerically using the density matrix renormalization group (DMRG) method. In this approach, a chain system is built up by adding pairs of additional sites near the center of the chain and systematically truncating the Hilbert space at a manageable size (typically around 1000) at each step. The key feature of the method is the choice of which states to keep during the truncation. It has been proven that the optimum choice is the eigenstates of the reduced density matrix with the largest eigenvalues. Since the method, by construction, calculates eigenvalues of ρ_A, it is straightforward to calculate the corresponding entanglement entropy. Some of our results for $S_{\text{imp}}(r, R, J_K')$ are shown in Figs. 1.5 and 1.6. The second of these figures shows that the large-r result of eqn (1.198) works very well. This is rather remarkable confirmation of the universality of the quantum impurity entanglement entropy, since we have obtained results for the microscopic model using a continuum field theory. It supports the idea that the impurity entanglement entropy would be given by the same universal functions for other realizations of the Kondo model, including the standard 3D free-fermion one. Note that the even and odd scaling functions have very different behavior when $R < \xi_K$. For the integer spin case, S_{imp} increases monotonically with decreasing r and appears to be approaching $\ln 2$ in the limit $r \ll \xi_K, R$. This is what is expected from the general C&C result, since $g_A = \ln 2$ at the weak-coupling

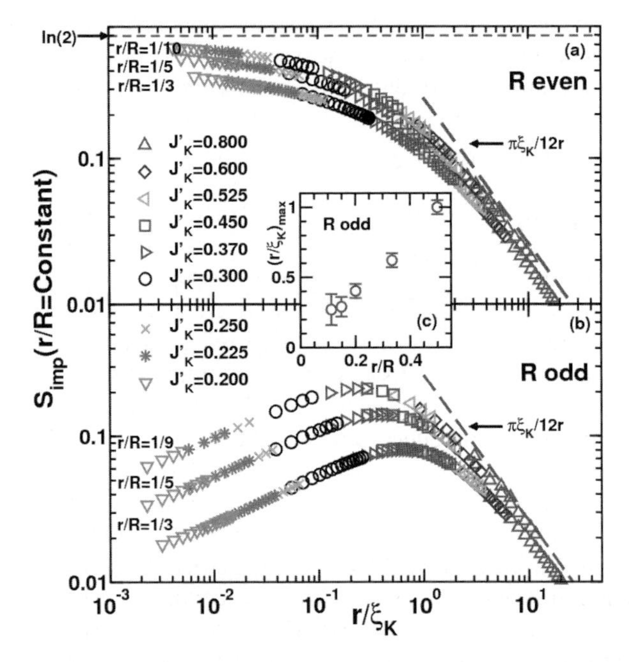

Fig. 1.5 Impurity entanglement entropy for fixed r/R plotted versus r/ξ_K for both of the cases R even and R odd. FLT predictions for large r/ξ_K are shown. Inset: location of the maximum, $(r/\xi_K)_{\text{max}}$ for odd R plotted versus r/R.

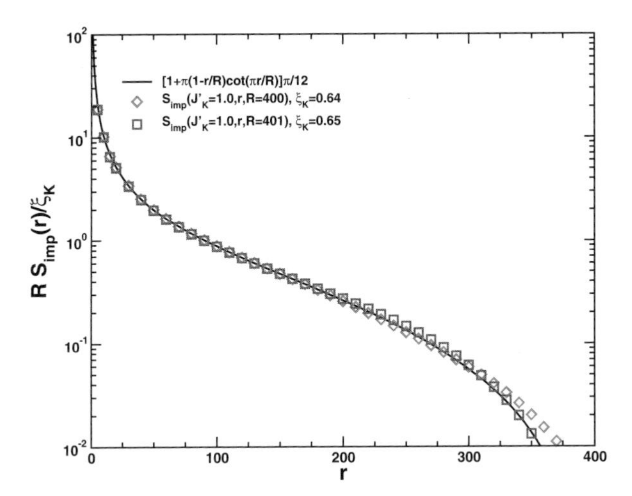

Fig. 1.6 Impurity entanglement entropy for the Kondo spin chain model calculated by DRMG compared with the Fermi liquid calculation of eqn (1.198).

fixed point with a decoupled impurity spin. On the other hand, for the half-integer spin case S_{imp} initially increases with decreasing r/ξ_K but eventually goes through a maximum and starts to decrease. The maximum occurs when $r/\xi_K \approx r/R$, i.e. when $\xi_K \approx R$.

These results can be understood heuristically from a resonating-valence-bond (RVB) picture of the ground spin in the Kondo spin chain model. Let us first consider the case of integer total spin, R even. Then the ground state is a spin singlet. It is important to realize that when $J'_K \to 0$, the singlet ground state becomes degenerate with a triplet state. This occurs since the ground state of the other $R-1$ sites, $j = 2, 3, \ldots, R$, has spin 1/2, as does the impurity spin and the two are coupled. However, by continuity in J'_K, it is the singlet state which is considered. This singlet state has strong entanglement of the impurity spin with the rest of the system, despite the fact that there is no term in the Hamiltonian coupling them together when $J'_K = 0$. Any singlet state, and hence the ground state for any value of J'_K, can be written as some linear combination of "valence bond states," i.e. states in which pairs of spins form a singlet. (We can always restrict ourselves to "noncrossing" states such that if we draw lines connecting every pair of contracted spins, none of these lines cross each other.) If we consider a "frozen" valence bond state, i.e. any particular basis state, then the entanglement entropy is simply $\ln 2$ times the number of valence bonds from region A to B. There will always be a valence bond connecting the impurity spin to some other spin in the system; we refer to this as the impurity valence bond (IVB). Intuitively, if the IVB connects the impurity to a spin outside of region A then we think of this as resulting in an impurity entanglement of $\ln 2$; however, this picture is certainly naive because the valence bond basis, while complete, is not orthogonal. We may think of the typical length of the IVB as being ξ_K, since the spin screening the impurity is precisely the one forming the IVB. This picture makes it quite clear why S_{imp} is a decreasing function of r and why ξ_K is the characteristic scale for its variation.

Now consider the case of half-integer spin, R odd. In this case, when $J'_K = 0$ there is zero entanglement between the impurity spin and the rest of the system. In this case the other $R-1$ sites have a spin-zero ground state, decoupled from the impurity, which is unpaired.

For R odd and any J'_K, the ground state always has spin 1/2. This can again be represented as an RVB state but each basis valence bond state has precisely one unpaired spin, which may or may not be the impurity. At $J'_K = 0$ it *is* the impurity with probability 1, but consider what happens as we increase J'_K from zero, corresponding to decreasing $\xi_K \propto \exp[\text{constant}/J'_K]$ from infinity. The probability of having an IVB increases. On the other hand, the typical length of the IVB when it is present is ξ_K, which decreases. These two effects trade off to give a peak in S_{imp} when ξ_K is approximately R.

Our most important conclusion is probably that the quantum impurity entanglement entropy appears to exhibit a universal crossover between boundary RG fixed points, with the size r of region A acting as an infrared cutoff.

1.6 Y-junctions of quantum wires

Now I consider three semi-infinite spinless Luttinger liquid quantum wires, all with the same Luttinger parameter g, meeting at a Y-junction (Oshikawa *et al.* 2006), as shown in Fig. 1.7. By imposing a magnetic field near the junction, we can introduce a nontrivial phase ϕ into the tunneling terms between the three wires. A corresponding lattice model is

$$H = \sum_{n=0}^{\infty}\sum_{j=1}^{3}[-t(\psi^\dagger_{n,j}\psi_{n+1,j} + h.c.) + \tilde{V}\hat{n}_{n,j}\hat{n}_{n+1,j}] - \frac{\tilde{\Gamma}}{2}\sum_{j=1}^{3}[e^{i\phi/3}\psi^\dagger_{0,j}\psi_{0,j-1} + h.c.]\,. \tag{1.199}$$

(In general, we can also introduce potential scattering terms at the end of each wire.) Here $j = 1, 2, 3$ labels the three chains cyclically so that we identify $j = 3$ with $j = 0$. We bosonize, initially introducing a boson $\varphi_j(x)$ for each wire:

$$\psi_{j,L/R} \propto \exp[i(\varphi_j \pm \theta_j)/\sqrt{2}]\,. \tag{1.200}$$

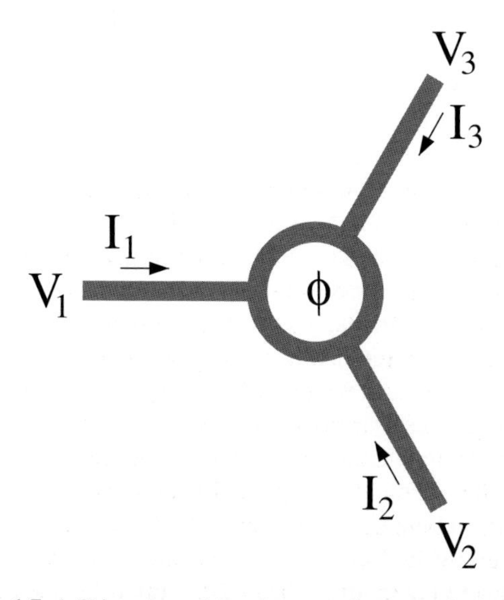

Fig. 1.7 A Y-junction with voltages and currents indicated.

It follows immediately from the discussion of the two-wire case in the fourth lecture that $\tilde{\Gamma}$ is irrelevant for $g < 1$, the case of repulsive interactions. Thus we expect the decoupled-wire, zero-conductance fixed point to be the stable one in that case. On the other hand, if $g > 1$ the behavior is considerably more complex and interesting. This case corresponds effectively to an attractive interaction between electrons, as can arise from phonon exchange.

It is convenient to make a basis change, analogous to the even and odd channels introduced in Lecture 4 for the two-wire junction:

$$\begin{aligned}
\Phi_0 &= \frac{1}{\sqrt{3}}(\varphi_1 + \varphi_2 + \varphi_3)\,, \\
\Phi_1 &= \frac{1}{\sqrt{2}}(\varphi_1 - \varphi_2)\,, \\
\Phi_2 &= \frac{1}{\sqrt{6}}(\varphi_1 + \varphi_2 - 2\varphi_3)\,,
\end{aligned} \tag{1.201}$$

and similarly for the θ_i's. It is also convenient to introduce three unit vectors, at angles $2\pi/3$ with respect to each other, acting on the (Φ_1, Φ_2) space:

$$\begin{aligned}
\vec{K}_1 &= (-1/2, \sqrt{3}/2)\,, \\
\vec{K}_2 &= (-1/2, -\sqrt{3}/2)\,, \\
\vec{K}_3 &= (1, 0)\,.
\end{aligned} \tag{1.202}$$

We will also use three other unit vectors, rotated by $\pi/2$ relative to these ones, which we write as $\hat{z} \times \vec{K}_i = (-K_{iy}, K_{ix})$. The various boundary interactions are now written in this basis:

$$\begin{aligned}
T^{RL}_{21} &= \psi^\dagger_{2R}\psi_{1L} \propto e^{i\vec{K}_3\cdot\vec{\Phi}} e^{i(1/\sqrt{3})\hat{z}\times\vec{K}_3\cdot\vec{\Theta}} e^{i\sqrt{2/3}\Theta_0}\,, \\
T^{RL}_{12} &= \psi^\dagger_{1R}\psi_{2L} \propto e^{-i\vec{K}_3\cdot\vec{\Phi}} e^{i(1/\sqrt{3})\hat{z}\times\vec{K}_3\cdot\vec{\Theta}} e^{i\sqrt{2/3}\Theta_0}\,, \\
T^{LL}_{21} &= \psi^\dagger_{2L}\psi_{1L} \propto e^{i\vec{K}_3\cdot\vec{\Phi}} e^{i\vec{K}_3\cdot\vec{\Theta}}\,, \\
T^{RL}_{11} &= \psi^\dagger_{R1}\psi_{L1} \propto e^{-i(2/\sqrt{3})\hat{z}\times\vec{K}_1\cdot\Theta} e^{i\sqrt{2/3}\Theta_0}\,,
\end{aligned} \tag{1.203}$$

etc. Note that we have not yet imposed any particular BCs on the fields at the origin. A more standard approach would probably be to start with the $\tilde{\Gamma} = 0$ BC, $\Theta_i(0) =$ constant, and then allow for the possibility that these BCs renormalize owing to the tunneling. We call the current approach the method of "delayed evaluation of boundary conditions." However, we expect that the "center of mass" field, θ_0, is always pinned, $\Theta_0(0) =$ constant. Since Φ_0 carries a nonzero total charge, and hence doesn't appear in any of the boundary interactions, it would not make sense for any other type of boundary condition to occur in the "0" sector. We may thus simply drop the factor involving Θ_0 from all of the boundary interactions; it makes no contribution to scaling dimensions. However, we must consider other possible BCs in the 1, 2 sector, since, for $g > 1$, the simple D BC on $\vec{\Theta}$ is not stable, as mentioned above. Another simple possibility would be a D BC on $\vec{\Phi}$: $\Phi_i(0) = c_i$ for two constants c_i. To check the stability of this BC under the RG, we must consider the LIO. The various candidates are the tunneling and backscattering terms in eqn (1.203). Imposing the BC $\vec{\Phi} =$ constant, we can evaluate the scaling dimension of the exponential factors involving $\vec{\Theta}(0)$ by the same method as that used in Lecture 4. The BC implies that we should replace $\vec{\Theta}(0)$ by $2\sqrt{g}\vec{\Phi}_L(0)$. The dimensions of these operators can then be read off:

$$\begin{aligned}\Delta_{21}^{RL} &= g/3\,,\\ \Delta_{21}^{LL} &= g\,,\\ \Delta_{11}^{RL} &= 4g/3\,,\end{aligned} \tag{1.204}$$

etc. The T_{ij}^{RL}, for $i \neq j$, are the LIOs. They are relevant for $g < 3$. Thus we see that this cannot be the stable fixed point for $1 < g < 3$.

Stable fixed points for $1 < g < 3$ can be identified from the form of the $T_{j,j\pm1}^{RL}$. If $\tilde{\Gamma}$ grows large under renormalization, it is plausible that one or the other of this set of operators could develop an expectation value. Note that if

$$\langle T_{j,j+1}\rangle \neq 0\,, \tag{1.205}$$

this would correspond to strong tunneling from j to $j+1$. On the other hand, if

$$\langle T_{j,j-1}\rangle \neq 0\,, \tag{1.206}$$

this would correspond to strong tunneling from j to $j-1$. Breaking time reversal by adding a nonzero magnetic flux ϕ favors one or the other of these tunneling paths, depending on the sign of ϕ. On the other hand, in the time-reversal-invariant case $\phi = 0$ (or π), we do not expect such an expectation value to develop. These fixed points obey the BCs

$$\pm\vec{K}_i\cdot\vec{\Phi}(0) + \sqrt{\frac{1}{3}}(\hat{z}\times\vec{K}_i)\cdot\vec{\Theta}(0) = \vec{C} \quad (i = 1,2,3)\,, \tag{1.207}$$

for some constants C_i. Note that since $\sum_{i=1}^{3}\vec{K}_i = 0$, these are only two independent BCs. Introducing left- and right-moving fields

$$\begin{aligned}\vec{\Phi} &= \frac{1}{\sqrt{g}}(\vec{\Phi}_L + \vec{\Phi}_R)\,,\\ \vec{\Theta} &= \sqrt{g}(\vec{\Phi}_L - \vec{\Phi}_R)\,,\end{aligned} \tag{1.208}$$

we may write these BCs as

$$\vec{\Phi}_R(0) = \mathcal{R}\vec{\Phi}_L(0) + \vec{C}'\,, \tag{1.209}$$

where $\vec{C}'$ is another constant vector and $\mathcal{R}$ is an orthogonal matrix, which we parameterize as

$$\mathcal{R} = \begin{pmatrix}\cos\xi & -\sin\xi\\ \sin\xi & \cos\xi\end{pmatrix}, \tag{1.210}$$

with

$$\xi = \pm 2\arctan\left(\frac{\sqrt{3}}{g}\right). \tag{1.211}$$

We refer to these as the "chiral" fixed points, $\chi_\pm$, with the $+$ or $-$ corresponding to the sign in eqns (1.207) and (1.211). Note that if $\xi = 0$ we obtain the usual D BC on $\vec{\Theta}$, and if $\xi = \pi$

we obtain a D BC on $\vec{\Phi}$. The chiral BCs are, in a sense, intermediate between these other two BCs.

As usual, we check their stability by calculating the scaling dimension of the LIO. Once we have the BCs in the form of eqn (1.209), it is straightforward to calculate the scaling dimension of any vertex operators of the general form

$$\begin{aligned}\mathcal{O} = \exp\left(i\sqrt{g}\vec{a}\cdot\vec{\Phi} + i\frac{1}{\sqrt{g}}\vec{b}\cdot\vec{\Theta}\right) &= \exp[i(\vec{a}-\vec{b})\cdot\vec{\Phi}_R + i(\vec{a}+\vec{b})\cdot\vec{\Phi}_L] \\ &\propto \exp\{i[\mathcal{R}^{-1}(\vec{a}-\vec{b}) + (\vec{a}+\vec{b})]\cdot\vec{\Phi}_L\}\,,\end{aligned} \tag{1.212}$$

$$\Delta_{\mathcal{O}} = \frac{1}{4}|\mathcal{R}(\vec{a}+\vec{b}) + (\vec{a}-\vec{b})|^2\,. \tag{1.213}$$

Applying this formula to the $\chi_\pm$ fixed points, we find that all of the operators listed in eqn (1.203) have the same dimension,

$$\Delta = \frac{4g}{3+g^2}\,. \tag{1.214}$$

Since $\Delta > 1$ for $1 < g < 3$, we conclude that the chiral fixed points are stable for this intermediate range of g only. Thus we hypothesize that the system renormalizes to these chiral fixed points for this range of g whenever there is a nonzero flux, $\phi \neq 0$. However, these fixed points are presumably not allowed owing to time reversal symmetry when $\phi = 0$, and there must be some other stable fixed point to which the system renormalizes. This fixed point, which we have referred to as "M," appears to be of a less trivial type than the "rotated D" states. An attractive possibility is that the M fixed point is destabilized by an infinitesimal flux, leading to a flow to one of the chiral fixed points. Alternatively, it is possible that there is a critical value of the flux $|\phi_c| \neq 0$ necessary to destabilize the M fixed point. In that case there would presumably be two more, as yet undetermined, CIBCs corresponding to these critical points.

1.6.1 Conductance

Once we have identified the CIBCs, it is fairly straightforward to calculate the conductance using the Kubo formula. For a Y-junction, the conductance is a tensor. If we apply voltages V_i far from the junction on leads i and let I_i be the current flowing towards the junction in lead i, then, for small V_i,

$$I_i = \sum_{j=1}^{3} G_{ij}V_j\,. \tag{1.215}$$

Since there must be no current when all the V_i are equal to each other and since the total current flowing into the junction is always zero, it follows that

$$\sum_i G_{ij} = \sum_j G_{ij} = 0\,. \tag{1.216}$$

Taking into account also the Z_3 symmetry of the model, we see that the most general possible form of G_{ij} is

$$G_{ij} = \frac{G_S}{2}(3\delta_{ij} - 1) + \frac{G_A}{2}\epsilon_{ij}\,. \tag{1.217}$$

Here ϵ_{ij} is the 3×3 antisymmetric, Z_3-symmetric tensor with $\epsilon_{12} = 1$. G_A, which is odd under time reversal, can only be nonzero when there is a non-zero flux.

In the noninteracting case, $\tilde{V} = 0$ in eqn (1.199), $g = 1$, we may calculate the conductance by a simple generalization of the Landauer formalism. We imagine that the three leads are connected to distant reservoirs at chemical potentials $\mu - eV_j$. Each reservoir is assumed to emit a thermal distribution of electrons down the lead and to perfectly absorb electrons heading towards it. The conductance can then be expressed in terms of the S-matrix for the Y-junction. This is defined by solutions of the lattice Schrödinger equation

$$\begin{aligned} - t(\Phi_{n+1,j} + \Phi_{n-1,j}) &= E\Phi_{n,j} \quad (n \geq 1)\,, \\ -t\Phi_{1,j} - \frac{\tilde{\Gamma}}{2}(e^{i\phi/3}\Phi_{0,j-1} + e^{-i\phi/3}\Phi_{0,j+1}) &= E\phi_{o,j}\,. \end{aligned} \tag{1.218}$$

The wave functions are of the form

$$\Phi_{n,j} A_{\text{in},j} e^{-ikn} + A_{\text{out},j} e^{ikn}\,, \tag{1.219}$$

for all $n \geq 0$ and $j = 1, 2, 3$, with energy eigenvalues $E = -2t\cos k$. The 3×3 S-matrix is defined by

$$\vec{A}_{\text{out}} = S\vec{A}_{\text{in}}\,. \tag{1.220}$$

The most general Z_3-symmetric form is

$$\begin{aligned} S_{ij} &= S_0 \ \ (i = j) \\ &= S_- \ \ (i = j - 1) \\ &= S_+ \ \ (i = j + 1)\,. \end{aligned} \tag{1.221}$$

In the time-reversal-invariant case, $S_+ = S_-$. It is then easy to see that unitarity of S implies $|S_\pm| \leq 2/3$. In this case, for any wave vector k, $|S_\pm|$ reaches $2/3$ for some value of $\tilde{\Gamma}$ of O(1). It can also be checked that when $k - \phi = \pi$ and $\tilde{\Gamma} = 2$, $S_+ = 0 = S_0 = 0$ and $|S_-| = 1$. In this case an electron incident on lead j is transmitted with unit probability to lead $j - 1$. To calculate the Landauer conductance, we observe that the total current in lead j is the current emitted by reservoir j, minus the reflected current plus the current transmitted from leads $j \pm 1$:

$$I_j = e\int \frac{dk}{2\pi} v(k) \left[(1 - |S_{jj}|^2) n_F(\epsilon_k - eV_j) - \sum_{\pm} |S_{j,j\pm1}|^2 n_F(\epsilon_k - eV_{j\pm1})\right]. \tag{1.222}$$

This gives a conductance tensor of the form of eqn (1.217), with

$$G_{S/A} = \frac{e^2}{h}(|S_+|^2 \pm |S_-|^2)\,, \tag{1.223}$$

where $S_\pm$ are now evaluated at the Fermi energy. Thus the maximum possible value of G_S, in the zero-flux case, is $(8/9)e^2/h$, when $|S_\pm| = 2/3$. On the other hand, for nonzero flux, when $S_+ = S_0 = 0$, $G_S = -G_A = e^2/h$, i.e. $G_{ii} = -G_{i,i+1} = e^2/h$ but $G_{i,i-1} = 0$. This implies that if a voltage is imposed on lead 1 only, a current $I = (e^2/h)V_1$ flows from lead 1 to lead 2 but zero current flows into lead 3. We refer to this as a perfectly chiral conductance tensor. Of course, if we reverse the sign of the flux then the chirality reverses, with V_1 now inducing a current $(e^2/h)V_1$ from lead 1 to lead 3.

Now consider the conductance in the interacting case, for the three fixed points that we have identified. From the Kubo formula, we may write the DC linear conductance tensor as

$$G_{jk} = \lim_{\omega \to 0} \frac{-e^2}{h} \frac{1}{\pi \omega L} \int_{-\infty}^{\infty} d\tau \, e^{i\omega\tau} \int_0^L dx \, T\langle J_j(y,\tau) J_k(x,0) \rangle \,, \tag{1.224}$$

where $J_j = -i\partial_\tau \theta_j$. (At zero temperature, which I consider here, it is straightforward to take the zero-frequency limit, in the imaginary (Matsubara) formulation. At finite T it is necessary to do an analytic continuation to real frequency first.) We first transform from the ϕ_j basis to Φ_μ, inverting the transformation in eqn (1.201):

$$\phi_j = \sum_\mu v_{j\mu} \Phi_\mu \,. \tag{1.225}$$

Φ_0 makes no contribution to the conductance, since Θ_0 always obeys a D BC as discussed above. Therefore

$$G_{jk} = \lim_{\omega \to 0} \frac{-e^2}{h} \frac{1}{\pi \omega L} \sum_{\mu,\nu=1,2} v_{j\mu} v_{k\nu} \int_{-\infty}^{\infty} d\tau \, e^{i\omega\tau} \int_0^L dx \, T\langle J_\mu(y,\tau) J_\nu(x,0) \rangle \,, \tag{1.226}$$

where $J_\mu = -i\partial_\tau \Theta_\mu$, and the result is independent of $y > 0$. Here

$$v_{j\mu} = \sqrt{\frac{2}{3}} (\hat{z} \times \vec{K}_j)_\mu = -\sqrt{\frac{2}{3}} \sum_\nu \epsilon_{\mu\nu} K_j^\nu \,. \tag{1.227}$$

To proceed, we use the following decomposition:

$$\vec{J} = -i\partial_\tau \vec{\Phi}_L + i\partial_\tau \vec{\Phi}_R \equiv -\vec{J}_L + \vec{J}_R \,. \tag{1.228}$$

The general type of BC of eqn (1.209) allows us to regard the $\Phi_{R\mu}(x)$'s as the analytic continuation of the $\Phi_{L\nu}(x)$'s to the negative x-axis:

$$\vec{\Phi}_R(x) = \mathcal{R}\vec{\Phi}_L(-x) + \vec{C}' \,, \tag{1.229}$$

and thus

$$\vec{J}_R(x) = \mathcal{R}\vec{J}_L(-x) \,. \tag{1.230}$$

The Green's function for $\vec{J}_L$ is unaffected by the BC:

$$\langle J_{L\mu}(\tau + iy) J_{L\nu}(\tau' + ix) \rangle = \frac{g\delta_{\mu\nu}}{2[(\tau - \tau') + i(y - x)]^2} \,. \tag{1.231}$$

The τ integral in eqn (1.224) gives

$$\int_{-\infty}^{\infty} d\tau \, e^{i\omega\tau} T\langle J_{L\mu}(y,\tau) J_{L\nu}(x,0) \rangle = -2\pi\omega H(x - y) e^{\omega(y - x)} \,. \tag{1.232}$$

Here $H(x)$ is the Heaviside step function, often written $\theta(x)$, but I avoid that notation here since $\theta(x)$ has another meaning. Thus we obtain

$$\int_{-\infty}^{\infty} d\tau\, e^{i\omega\tau} \langle J_\mu(\tau, y) J_\nu(0, x)\rangle = -2\pi\omega[\delta_{\mu\nu} H(x-y) e^{\omega(y-x)} + \delta_{\mu\nu} H(y-x) e^{\omega(x-y)} - \mathcal{R}_{\mu\nu} H(y+x) e^{-\omega(y+x)} - \mathcal{R}_{\nu\mu} H(-y-x) e^{\omega(y+x)}]. \tag{1.233}$$

Observing that $H(x-y) + H(y-x) = 1$ and that $H(x+y) = 1$, $H(-x-y) = 0$ since x and y are always positive in eqn (1.224), we obtain

$$G_{ij} = g\frac{e^2}{h} \sum_{\mu,\nu=1,2} v_{j\mu} v_{k\nu} [\delta_{\mu\nu} - R_{\nu\mu}]\,. \tag{1.234}$$

For the D BC on $\vec{\Phi}$, $\mathcal{R} = -I$, so

$$G_{ij} = 2g\frac{e^2}{h} \vec{v}_j \cdot \vec{v}_k = 2g\frac{e^2}{h}\left(\delta_{jk} - \frac{1}{3}\right), \tag{1.235}$$

corresponding to $G_S = (4/3)g(e^2/h)$, $G_A = 0$. We observed above that this is a stable fixed point for $g > 3$, with $G_S > 4e^2/h$. This exceeds the unitary bound on the conductance in the noninteracting case. An intuitive way of understanding why increasing g leads to enhanced transmission is that attractive interactions can lead to pairing, and either coherent pair tunneling or Andreev-type processes (where an incident electron in one lead reflects as a hole while a pair is transmitted to a different lead) could lead to enhanced conductance.

We can now also obtain the conductance tensor for the chiral fixed points, which are stable for $1 < g < 3$. Inserting eqns (1.210 and (1.211) in (1.234), we obtain

$$G_S = \frac{e^2}{h}\frac{4g}{g^2+3}\,, \qquad G_A = \pm\frac{e^2}{h}\frac{4g^2}{g^2+3}\,. \tag{1.236}$$

For $g = 1$, this reduces to the chiral conductance tensor discussed above in the noninteracting case with $G_S = \pm G_A = e^2/h$. However, for $g > 1$, $G_A > G_S$, implying that a voltage on lead 1 not only leads to all current from 1 flowing to 2 but some additional current also flows from 3 to 2. Intuitively, we might think that as the electrons pass from lead 1 to 2 they attract some electrons from lead 3. If our hypothesis discussed above that the zero-flux "M" fixed point is unstable is correct, then presumably an infinitesimal flux could lead to an RG flow to these stable chiral fixed points. Such a device would have an interesting switching property. Even a small magnetic field could switch the current completely from lead 2 to lead 3, at low enough temperatures and currents.

1.7 Boundary-condition-changing operators and the X-ray edge singularity

There are some situations in condensed matter physics where we are interested in the response of a system to a sudden change in the Hamiltonian. A well-known example is the "X-ray edge singularity" in the absorbtion intensity for X-rays in a metal, plotted versus X-ray energy. The X-ray dislodges an electron from a core level. This is assumed to suddenly switch on a localized impurity potential which acts on the conduction electrons. Since I have argued that, quite generally, the low-energy properties of quantum impurity problems are described by CIBCs, we might expect that the low-energy response to a sudden change in impurity interactions might be equivalent to a response to a sudden change in CIBCs. Very fortunately, Cardy has also developed a theory of BC-changing operators which can be applied to this situation. In this lecture, I will show how this theory can be applied to the usual X-ray edge problem and to a multichannel Kondo version (Affleck and Ludwig 1994).

1.7.1 The X-ray edge singularity

When an X-ray is absorbed by a metal it can raise an electron from a deep core level, several keV below the Fermi surface, up to the conduction band. Let E_0 be this large energy difference between the Fermi energy and the core level and let ω be the energy of the X-ray. At $T = 0$, ignoring electron–electron interactions, this transition is possible only for $\omega \geq \tilde{E}_0$. Here $\tilde{E}_0$ is a "renormalized" value of E_0. I am assuming that the core level has a distinct energy, rather than itself being part of an energy band. This may be a reasonable approximation, since core levels are assumed to be tightly bound to nuclei and to have very small tunneling matrix elements to neighboring nuclei. Presumably the excited electron will eventually relax back to the core level, possibly emitting phonons or electron–hole pairs. This is ignored in the usual treatment of X-ray edge singularities. Thus we are effectively ignoring the finite width of the excited electron states. Thus the X-ray absorption intensity, $I(\omega)$, will be strictly zero for $\omega \leq \tilde{E}_0$ in this approximation. When the core electron is excited into the conduction band, it leaves behind a core hole, which interacts with all the electrons in the conduction band. Note that the only interaction being considered here is the one between the core hole and the conduction electrons. The X-ray edge singularity at $\omega = \tilde{E}_0$, in this approximation, is determined by the response of the conduction electrons to the sudden appearance of the core hole potential, at the instant that the X-ray is absorbed. It turns out that, for ω only slightly larger then $\tilde{E}_0$, very close to the threshold, $I(\omega)$ is determined only by the conduction electron states very close to ϵ_F; this fact allows us to apply low-energy effective-Hamiltonian methods. The difference between $\tilde{E}_0$ and E_0 arises from the energy shift of the filled Fermi sea due to the core hole potential. Not including the interaction with the external electromagnetic field, which I turn to momentarily, the Hamiltonian is simply

$$H = \int d^3r \left[\psi^\dagger(\vec{r})\left(-\frac{\nabla^2}{2m} - \epsilon_F\right)\psi(\vec{r}) + \tilde{V}bb^\dagger\delta^3(r)\psi^\dagger\psi\right] + E_0 b^\dagger b\,, \tag{1.237}$$

where b annihilates an electron in the core level at $\vec{r} = 0$. I have assumed, for simplicity, that the core hole potential $\tilde{V}\delta^3(r)$ is a spherically symmetric δ-function. These assumptions can be easily relaxed. Following the same steps as in Section 1.1, we can reduce the problem to a one-dimensional one, with left-movers only:

$$H = \frac{1}{2\pi}i\int_{-\infty}^{\infty} dr\,\psi_L^\dagger\frac{d}{dr}\psi_L + \frac{V}{2\pi}bb^\dagger\psi_L^\dagger(0)\psi_L(0) + E_0 b^\dagger b\,. \tag{1.238}$$

(I have set $v_F = 1$ and $V \propto \tilde{V}$.) It is convenient to bosonize. We may introduce a left-moving boson only:

$$\psi_L \propto e^{i\sqrt{4\pi}\phi_L}\,, \tag{1.239}$$

$$H = \int_{-\infty}^{\infty} (\partial_x \phi_L)^2 - \frac{V}{\sqrt{\pi}} bb^\dagger \partial_x \phi_L + E_0 bb^\dagger\,. \tag{1.240}$$

The solubility of this model hinges on the fact that $b^\dagger b$ commutes with H. Thus the Hilbert space breaks up into two parts, in which the core level is either empty or occupied. The spectrum of the Hamiltonian, in each sector of the Hilbert space, is basically trivial. In the sector where the core level is occupied, $b^\dagger b = 1$, we get the spectrum of free electrons with no impurity:

$$H_0 = \int_{-\infty}^{\infty} (\partial_x \phi_L)^2\,. \tag{1.241}$$

In the sector with $b^\dagger b = 0$, we get the spectrum with a potential scatterer present:

$$H_1 = \int_{-\infty}^{\infty} (\partial_x \phi_L)^2 - \frac{V}{\sqrt{\pi}} \partial_x \phi_L + E_0\,. \tag{1.242}$$

What makes this problem somewhat nontrivial is that, to obtain the edge singularity, we must calculate the Green's function of the operator which couples to the electromagnetic field associated with the X-rays, $\psi^\dagger(t, r = 0)b(t)$. This operator, which excites an electron from the core level into the conduction band, mixes the two sectors of the Hilbert space. There is also some interest in calculating the Green's function of the operator $b(t)$ itself; this is associated with photoemission processes in which the core electron is ejected from the metal by the X-ray. Again this operator mixes the two sectors of the Hilbert space.

In what may have been the first paper on bosonization, in 1969, Schotte and Schotte (1969) observed that these Green's functions can be calculated by taking advantage of the fact that the two Hamiltonians H_0 and H_1 are related by a canonical transformation (and a shift of the ground state energy). To see this, note that we may write H_1 in the form

$$H_1 = \int_{-\infty}^{\infty} (\partial_x \tilde{\phi}_L)^2 + \text{constant}\,, \tag{1.243}$$

where

$$\tilde{\phi}_L(x) \equiv \phi_L(x) - \frac{V}{4\sqrt{\pi}} \text{sgn}(x)\,. \tag{1.244}$$

Using the commutator

$$[\partial_y \phi_L(y), \phi_L(y)] = \frac{-i}{2} \delta(x - y)\,, \tag{1.245}$$

we see that

$$H_1 = U^\dagger H_0 U + \text{constant}\,, \tag{1.246}$$

with the unitary transformation

$$U = \exp\left[\frac{-iV\phi_L(0)}{\sqrt{\pi}}\right]. \tag{1.247}$$

(U can only be considered a unitary operator if we work in the extended Hilbert space which includes states with *all* possible BCs. U maps whole sectors on the Hilbert space, with particular BCs, into each other.) Consider the core electron Green's function, $\langle b(t)^\dagger b(0)\rangle$. We may write

$$b^\dagger(t) = e^{iHt}b^\dagger e^{-iHt} = e^{iH_0 t}b^\dagger e^{-iH_1 t}\,. \tag{1.248}$$

This is valid because, owing to Fermi statistics, the core level must be vacant before $b^\dagger$ acts, and occupied after it acts, i.e. we can replace the factor $bb^\dagger$ in H by 1 on the right-hand side and by 0 on the left. But H_0 and H_1 commute with b, so we have

$$\langle 0|b^\dagger(t)b(0)|0\rangle = \langle 1|b^\dagger b|1\rangle\langle\tilde{0}|e^{iH_0 t}e^{-iH_1 t}|\tilde{0}\rangle\,. \tag{1.249}$$

Here $|0\rangle$ is the ground state of the system, including the core level and the conduction electrons. This state can be written $|0\rangle = |1\rangle|\tilde{0}\rangle$, where $|1\rangle$ is the state with the core level occupied and $|\tilde{0}\rangle$ is the filled Fermi sea ground state of the conduction electrons, with no impurity potential. Using eqn (1.246), we see that

$$\langle\tilde{0}||e^{iH_0 t}e^{-iH_1 t}|\tilde{0}\rangle = e^{-i\tilde{E}_0 t}\langle\tilde{0}|e^{iH_0 t}U^\dagger e^{-iH_0 t}U|\tilde{0}\rangle = e^{-i\tilde{E}_0 t}\langle\tilde{0}|U(t)^\dagger U(0)|\tilde{0}\rangle\,. \tag{1.250}$$

Using our explicit expression, eqn (1.247), for U, we have reduced the calculation to one involving only a free-boson Green's function:

$$\langle 0|b^\dagger(t)b(0)|0\rangle = e^{-i\tilde{E}_0 t}\langle e^{iV\phi_L(t,0)/\sqrt{\pi}}e^{-iV\phi_L(0,0)/\sqrt{\pi}}\rangle \propto \frac{e^{-i\tilde{E}_0 t}}{t^{V^2/4\pi^2}}\,. \tag{1.251}$$

Similarly, to get the Green's function of $b^\dagger(t)\psi_L(t,0)$, I use

$$b^\dagger(t)\psi_L(t,0) = e^{iH_0 t}b^\dagger e^{-iH_1 t}\psi_L(t,0) \tag{1.252}$$

and thus

$$\begin{aligned}
&\langle 0|b^\dagger(t)\psi_L(t,0)b(0)\psi_L(0,0)|0\rangle \\
&= e^{-i\tilde{E}_0 t}\langle\tilde{0}|U^\dagger(t)U(0)\psi_L(t,0)\psi_L^\dagger(0,0)|\tilde{0}\rangle \\
&\propto e^{-i\tilde{E}_0 t}\langle e^{i\sqrt{4\pi}(1+V/2\pi)\phi_L(t,0)}e^{-i\sqrt{4\pi}(1+V/2\pi)\phi_L(0,0)}\rangle \\
&\propto \frac{e^{-i\tilde{E}_0 t}}{t^{(1+V/2\pi)^2}}\,.
\end{aligned} \tag{1.253}$$

Finally, we Fourier transform to get the X-ray edge singularity:

$$\begin{aligned}
&\int_{-\infty}^{\infty} dt\, e^{i\omega t}\langle 0|b^\dagger(t)\psi_L(t,0)b(0)\psi_L^\dagger(0,0)|0\rangle \\
&\propto \int_{-\infty}^{\infty} dt\frac{e^{i(\omega-\tilde{E}_0)t}}{(t-i\delta)^{(1+V/2\pi)^2}} \\
&\propto \frac{\theta(\omega-\tilde{E}_0)}{(\omega-\tilde{E}_0)^{1-(1+V/2\pi)^2}}\,.
\end{aligned} \tag{1.254}$$

This result is conventionally written in terms of a phase shift at the Fermi surface, rather than the potential strength V. The connection can be readily seen from eqn (1.244) and the bosonization formula:

$$\psi_L(x) \propto e^{i\sqrt{4\pi}\phi_L(x)} e^{iV\cdot \mathrm{sgn}(x)/2} . \tag{1.255}$$

Since $\psi_L(x)$ for $x < 0$ represents the outgoing field, we see that

$$\psi_{\mathrm{out}} = e^{2i\delta}\psi_{\mathrm{in}} , \tag{1.256}$$

where the phase shift is

$$\delta = -\frac{V}{2} . \tag{1.257}$$

In fact, the parameter V appearing in the bosonized Hamiltonian should be regarded as a renormalized one. Its physical meaning is the phase shift at the Fermi surface induced by the core hole. Only for small $\tilde{V}$ is it linearly related to the bare potential. Even if the core hole potential has a finite range, we expect the formulas for the X-ray edge singularity to still be correct, when expressed in terms of the phase shift at the Fermi surface, δ. More generally, if the core hole potential is not a δ-function, but is still spherically symmetric, a similar expression arises for the X-ray edge singularity, with the exponent involving a sum over phase shifts at the Fermi surface δ_l in all angular-momentum channels l.

The connection with a boundary-condition-changing operator (BCCO) (see Cardy's chapter in this book and Cardy (1989)) is now fairly evident. If we revert to the formulation of the model on the semi-infinite line $r > 0$, then the boundary condition is

$$\psi_R(0) = e^{2i\delta}\psi_L(0) . \tag{1.258}$$

The operator, U or b, which creates the core hole potential in the Hamiltonian can be viewed as changing the boundary condition, by changing the phase shift δ. It is interesting to consider the relationship between the finite-size spectrum with various BCs and the scaling dimensions of b and $\psi_L(0)^\dagger b$. We consider the system on a line of length l with a fixed BC $\psi_R(l) = -\psi_L(l)$ at the far end. Equivalently, in the purely left-moving formulation, for $\delta = 0$ we have antiperiodic BCs on a circle of circumference $2l$, $\psi_L(x+2l) = -\psi_L(x)$. This corresponds to periodic BCs on the left-moving boson field,

$$\phi_L(x+2l) = \phi_L(x) + \sqrt{\pi}Q, \quad (Q = 0, \pm 1, \pm 2, \ldots) . \tag{1.259}$$

The mode expansion is

$$\phi_L(t,x) = \sqrt{\pi}\frac{(t+x)}{2l}Q + \sum_{m=1}^{\infty} \frac{1}{\sqrt{2\pi m}} \left[\exp\left(\frac{-i\pi m(t+x)}{l} \right) a_m + h.c. \right] . \tag{1.260}$$

The finite-size spectrum is

$$E = \int_{-l}^{l} (\partial_x \phi_L)^2 = \frac{\pi}{l} \left[-\frac{1}{24} + \frac{1}{2}Q^2 + \sum_{m=1}^{\infty} m n_m \right] . \tag{1.261}$$

The universal ground state energy term, $-\pi/(24l)$, has been included. We see that Q can be identified with the charge of the state (measured relative to the filled Fermi sea). There is a one-to-one correspondence between the states in the FSS with excitation energy $\pi x/l$ and operators with dimension x in the free-fermion theory. For example, $Q = \pm 1$ corresponds to $\psi_L^\dagger$ and ψ_L, respectively, of dimension $x = 1/2$. If we impose the "same" BC at both ends,

$$\begin{aligned} \psi_R(0) &= e^{2i\delta}\psi_L(0)\,, \\ \psi_R(l) &= -e^{2i\delta}\psi_L(l)\,, \end{aligned} \tag{1.262}$$

then this corresponds to the same antiperiodic BC on $\psi_L(x)$ in the purely left-moving formulation, so the spectrum is unchanged. On the other hand, if the phase shift is inserted at $x = 0$ only, then $\psi_L(x + 2l) = -e^{-2i\delta}\psi_L(x)$, corresponding to

$$\phi_L(x + 2l) - \phi_L(0) = \sqrt{\pi}\left(n - \frac{\delta}{\pi}\right), \tag{1.263}$$

corresponding to the replacement $Q \to Q - \delta/\pi$ in eqn (1.260). The FSS is now modified to

$$E = \int_{-l}^{l} (\partial_x\phi_L)^2 = \frac{\pi}{l}\left[-\frac{1}{24} + \frac{1}{2}(Q - \delta/\pi)^2 + \sum_{m=1}^{\infty} m n_m\right]. \tag{1.264}$$

In particular, the change in ground state energy due to the phase shift is

$$E_0(\delta) - E_0(0) = \frac{\pi}{l}\frac{1}{2}\left(\frac{\delta}{\pi}\right)^2. \tag{1.265}$$

Actually, adding the potential scattering also changes the ground state energy by a nonuniversal term of O(1), which was absorbed into $\tilde{E}_0$ in the above discussion. It is the term of $O(1/l)$ which is universal and determines scaling exponents. The corresponding scaling dimension, $x = (\delta/\pi)^2/2$, is precisely the scaling dimension of the BCCO b (or U). The energy of the excited state with $Q = 1$ obeys

$$E(Q, \delta) - E_0(0) = \frac{\pi}{l}\frac{1}{2}\left(1 - \frac{\delta}{\pi}\right)^2. \tag{1.266}$$

The corresponding scaling dimension, $x = (1-\delta/\pi)^2/2$, is the scaling dimension of $\psi_L(0)^\dagger b$.

This is all in accord with Cardy's general theory of BCCOs. An operator which changes the BCs from A to B generally has a scaling dimension x, which gives the ground state energy on a finite strip of length l with BCs A at one end of the strip and B at the other end, measured relative to the absolute ground state energy with the same BCs A at both ends. This follows by making a conformal transformation from the semi-infinite plane to the finite strip. Explicitly, consider acting with a primary BCCO $\mathcal{O}$ at time τ_1 at the edge $x = 0$ of a semi-infinite plane. Assume $\mathcal{O}$ changes the BCs from A to B. Then, at time τ_2, we change the BCs back to A with the Hermitian conjugate operator $\mathcal{O}^\dagger$, also acting at $x = 0$, as shown in Fig. 1.8. Then the Green's function on the semi-infinite plane is

$$\langle A|\mathcal{O}(\tau_1)\mathcal{O}^\dagger(\tau_2)|A\rangle = \frac{1}{(\tau_2 - \tau_1)^{2x}}, \tag{1.267}$$

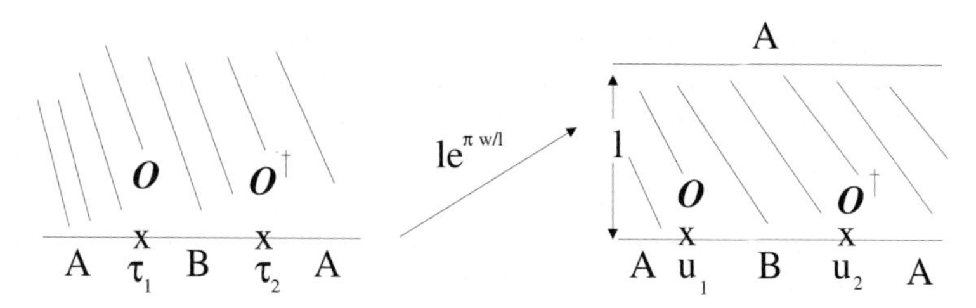

Fig. 1.8 BCCOs act at times τ_1 and τ_2 on a semi-infinite plane. This is conformally mapped to an infinite strip, with the BCCOs acting on the lower boundary.

where x is the scaling dimension of $\mathcal{O}$. Now we make a conformal transformation from the semi-infinite plane $z = \tau + ix$ ($x \geq 0$) to the finite-width strip $w = u + iv$, $0 < v < l$:

$$z = le^{\pi w/l}\,. \tag{1.268}$$

Note that the positive real axis, $x = 0$, $\tau > 0$, maps onto the bottom of the strip, $v = 0$. If we choose τ_1, $\tau_2 > 0$, both points map onto the bottom of the strip. Note that, on the strip, the BCs are A at all times at the upper boundary $v = l$ but change from A to B and then back to A on the lower strip, at times u_1, u_2. The Green's function on the strip is given by

$$\langle AA|\mathcal{O}(u_1)\mathcal{O}^\dagger(u_2)|AA\rangle = \left(\frac{\pi}{2l\sinh[(\pi(u_1-u_2)/(2l)]}\right)^{2x}. \tag{1.269}$$

Here $|AA\rangle$ denotes the ground state of the system on a strip of length l with the same BC, A, at both ends of the strip. We may insert a complete set of states:

$$\langle AA|\mathcal{O}(u_1)\mathcal{O}^\dagger(u_2)|AA\rangle = \sum_n |\langle AA|\mathcal{O}|n\rangle|^2 e^{-E_n(u_2-u_1)}\,. \tag{1.270}$$

However, the states $|n\rangle$ must all be states in the Hilbert space with *different* BCs A at $v = 0$ and B at $v = l$. The corresponding energies E_n are the energies of the states with these different BCs measured relative to the absolute ground state energy with the same BCs A at both ends. Taking the limit of large $u_2 - u_1$ in eqns (1.269) and (1.270), we see that

$$\frac{\pi x}{l} = (E^{\mathcal{O}}_{AB} - E_0)\,, \tag{1.271}$$

where $E^{\mathcal{O}}_{AB}$ is the lowest-energy state with BCs A and B produced by the primary operator $\mathcal{O}$. The lowest-dimension BCCO will simply produce the ground state with BCs A and B. This corresponds to the operator b in the X-ray edge model. On the other hand, the lowest-energy state produced could be an excited state with BCs A and B as in the example of $b\psi^\dagger_L(0)$, which produces the primary excited state with $Q = 1$ and BCs twisted by the phase δ.

1.7.2 X-ray edge singularities and the Kondo model

This BCCO approach has various other applications that go beyond the Schotte and Schotte results. One of them is to the Kondo model. So far, we have been ignoring electron spin.

If the core level is doubly occupied in the ground state, then it would have spin 1/2 after one electron is ejected from it by the X-ray. An initially spin-0 ion would then acquire a net spin 1/2. In addition to the potential-scattering interaction with the localized core hole, there would also be a Kondo interaction. Thus, it is interesting to consider the effect of suddenly turning on a Kondo interaction. We might expect that the Kondo effect could dominate the X-ray edge exponent, at least at low enough temperatures $T \ll T_K$ and frequencies such that $\omega - \tilde{E}_0 \ll T_K$. Thus, we consider the Hamiltonian

$$H = \frac{i}{2\pi}\int_{-\infty}^{\infty} dr\, \psi_L^{i\alpha\dagger}\frac{d}{dr}\psi_{Li\alpha} + \lambda\psi_L^{i\gamma\dagger}(0)\frac{\vec{\sigma}_\gamma^\delta}{2}\psi_{Li\delta}(0)\cdot b^{\alpha\dagger}\frac{\vec{\sigma}_\alpha^\beta}{2}b_\beta + E_0 b^{\alpha\dagger}b_\alpha\,. \tag{1.272}$$

Now $b^{\alpha\dagger}$ creates a core electron with spin α, and we have considered the general case of k channels of conduction electrons, $i = 1, 2, 3, \ldots$. Again we are interested in Green's functions for the core electron operator $b_\alpha(t)$ and also the operators $b^{\alpha\dagger}(t)\psi_{Li\beta}(t, 0)$. These Green's functions should exhibit a nontrivial crossover with frequency or time, but at long times (frequencies very close to the threshold) we expect to be able to calculate them using properties of the Kondo fixed point. Since the operator b_α creates an impurity spin, thus turning on the Kondo effect, it is again a BCCO. In this case, it should switch the BC from free to Kondo. Thus we expect its infrared scaling dimension to be given by the energy of the ground state with a free BC at one end of the finite system and a Kondo BC at the other. This spectrum is given by fusion with the $j = 1/2$ primary in the spin sector. The ground state with these BCs is always the $j = 1/2$ primary itself, of dimension

$$x = \frac{(3/4)}{2+k}\,. \tag{1.273}$$

This follows because the free spectrum includes the charge-zero, spin $j = 0$ flavor singlet, $(0, 0, I)$. Fusion with the spin-j primary always gives $(0, j, I)$, among other operators. This appears to have the lowest dimension of all fusion products. Note that this operator *does not* occur in the operator spectrum considered earlier at the Kondo fixed point. There, we only considered operators produced by double fusion, corresponding to the FSS with Kondo BCs at both ends of the finite system. This gives the operator spectrum with a fixed Kondo BC. But for the Kondo X-ray problem, we must consider the corresponding BCCO. In general, for a CIBC obtained by fusion with some operator $\mathcal{O}$ from a noninteracting BC, we may expect that the BCCO will be $\mathcal{O}$ itself. We may check this result by a more elementary method in the single-channel case. There, the Kondo fixed point is equivalent to a phase shift of magnitude $\pi/2$ for both spin up and spin down, $\delta_{\uparrow,\downarrow}$. The energy of this state, from eqn (1.265), is simply $(1/2\pi l)[(\delta_\uparrow/\pi)^2 + (\delta_\downarrow/\pi)^2]$, implying a dimension $x = 1/4$. This agrees with eqn (1.273) in the special case $k = 1$. We may also consider the dimensions of the operators $\psi_L^{i\alpha\dagger}(0)b_\beta$. This operator has $Q = 1$ (one extra electron added to the conduction band), transforms under the fundamental representation of flavor, of dimension k, and has spin either $j = 0$ or $j = 1$ depending on how we sum over the spin indices α and β. The free spectrum always contains the operator corresponding to the fermion field itself, $(Q = 1, j = 1/2, k)$ (where k now denotes the k-dimensional fundamental representation of SU(k)), and fusion with $j = 1/2$ gives $(Q = 1, j = 0, k)$ for all k and $(Q = 1, j = 1, k)$ for $k \geq 2$. These operators have dimension

$$x_j = \frac{1}{4k} + \frac{k^2-1}{2k(2+k)} + \frac{j(j+1)}{2+k} \quad (j = 0, 1)\,. \tag{1.274}$$

(It can be seen than $x_{1/2} = 1/2$, corresponding to the free-fermion operator.) Again, we may check the case $k = 1$ by more elementary arguments. We may find the unitary operators corresponding to b_α as

$$b_\alpha \propto \exp[2i(\delta_\uparrow \phi_{\uparrow L} + \delta_\downarrow \phi_{\downarrow L}]\,, \tag{1.275}$$

where we have introduced separate bosons for spin-up and spin-down electrons. $\delta_\uparrow$, $\delta_\downarrow$ can depend on α. It is convenient to switch to charge and spin bosons,

$$\phi_{c/s} \equiv \frac{\phi_\uparrow \pm \phi_\downarrow}{\sqrt{2}}\,. \tag{1.276}$$

By choosing $(\delta_\uparrow, \delta_\downarrow) = (\pi/2, -\pi/2)$ for $b_\uparrow$ and $(\delta_\uparrow, \delta_\downarrow) = (-\pi/2, \pi/2)$ for $b_\downarrow$, we obtain

$$b_{\uparrow/\downarrow} \propto \exp(\pm i\sqrt{2\pi}\phi_{Ls})\,. \tag{1.277}$$

These have the correct S^z quantum numbers, as can be seen by comparing with the standard bosonization formula for $\psi_{L\alpha}$:

$$\psi_{\uparrow/\downarrow L} \propto \exp(i\sqrt{2\pi}\phi_{Lc}) \exp(\pm i\sqrt{2\pi}\phi_{Ls})\,. \tag{1.278}$$

The factor $\exp(\pm i\sqrt{2\pi}\phi_{Ls})$ can be identified with $g_{L\uparrow/\downarrow}$, the chiral component of the WZW model fundamental field. We then see that the spin singlet operator, $\exp(i\sqrt{2\pi}\phi_{Lc})$, has $x = 1/4$, in agreement with eqn (1.274). On the other hand, the triplet operators have dimension $x = 5/4$. Since there is no $j = 1$ primary for $k = 1$, they contain Kac–Moody descendants, i.e. the spin current operator.

1.8 Conclusions

Apart from the examples discussed in these lectures, BCFT techniques have been applied to a number of other quantum impurity problems, including the following. We can consider a local cluster of impurities. At distances large compared with the separation between the impurities, the same methods can be applied. The two-impurity Kondo model exhibits an NFL fixed point, which can be obtained (Affleck *et al.* 1995) by a conformal embedding, which includes an Ising sector in which the fusion is performed. The three-impurity Kondo model also exhibits a novel NFL fixed point. This was obtained (Ingersent *et al.* 2005) by a different conformal embedding with fusion in a Z_8 parafermion CFT sector. Impurities in $\mathrm{SU}(3)$ spin chains (Affleck *et al.* 2001*a*) and quantum Brownian motion (Affleck *et al.* 2001*b*) were also solved by these techniques. These techniques were even applied (Affleck and Sagi 1994) to a high-energy physics model associated with Callan and Rubukaov. This describes a superheavy magnetic monopole interacting with the k-flavors of effectively massless fermions (quarks and leptons). The monopole is actually a dyon, having a set of electric charge states as well as a magnetic charge. When the fermions scatter off the dyon, they can exchange electric charge. In this case fusion takes place in the charge sector.

The assumption that essentially arbitrary impurity interactions, possibly involving localized impurity degrees of freedom, interacting with a gapless continuum, renormalize at low energies to a CIBC has worked in numerous examples. It appears to be generally valid and will likely find many other applications in the future.

Acknowledgments

I would like to thank my collaborators in the work discussed here, including Andreas Ludwig, Sebastian Eggert, Masaki Oshikawa, Claudio Chamon, Ming-Shyang Chang, Erik Sorensen, and Nicolas Laflorencie.

References

I. Affleck (1990). In *Fields, Strings and Critical Phenomena*, eds. E. Brézin and J. Zinn-Justin, North-Holland, Amsterdam, pp. 563–640.
I. Affleck (1995). *Acta Phys. Polon. B* **26**, 1869, cond-mat/9512099.
I. Affleck and A. W. W. Ludwig (1991). *Nucl. Phys. B* **360**, 641.
I. Affleck and A. W. W. Ludwig (1993). *Phys. Rev. B* **48**, 7297.
I. Affleck and A. W. W. Ludwig (1994). *J. Phys. A* **27**, 5375.
I. Affleck and J. Sagi (1994). *Nucl. Phys. B* **417**, 374.
I. Affleck, A. W. W. Ludwig, H.-B. Pang, and D. L. Cox (1992). *Phys. Rev. B* **45**, 7918.
I. Affleck, A. W. W. Ludwig, and B. A. Jones (1995). *Phys. Rev. B* **52**, 9528.
I. Affleck, M. Oshikawa, and H. Saleur (2001*a*). *J. Phys. A* **34**, 1073.
I. Affleck, M. Oshikawa, and H. Saleur (2001*b*). *Nucl. Phys. B* **594**, 535.
P. W. Anderson (1970). *J. Phys. C* **3**, 2346.
N. Andrei (1980). *Phys. Rev. Lett.* **45**, 379.
V. Barzykin and I. Affleck (1998). *Phys. Rev. B* **57**, 432.
R. Bulla, T. A. Costi, and T. Pruschke (2008). *Rev. Mod. Phys.* **80**, 395.
P. Calabrese and J. L. Cardy (2004). *J. Stat. Mech.* 04010.
J. L. Cardy (1989). *Nucl. Phys. B* **324**, 581.
J. L. Cardy and D. Lewellen (1991). *Phys. Lett. B* **259**, 274.
C. Chamon and E. Fradkin (1997). *Phys. Rev. B* **56**, 2012.
P. Di Francesco, P. Mathieu, and D. Senechal (1997). *Conformal Field Theory*, Springer, New York.
S. Eggert and I. Affleck (1992). *Phys. Rev. B* **46**, 10866.
P. Fendley, A. W. Ludwig, and H. Saleur (1995). *Phys. Rev. Lett.* **74**, 3005.
D. Friedan and A. Konechny (2004). *Phys. Rev. Lett.* **93**, 030402.
D. Gepner and E. Witten (1986). *Nucl. Phys. B* **278**, 493.
T. Giamarchi (2004). *Quantum Physics in One Dimension*, Oxford University Press, Oxford.
A. Hewson (1997). *The Kondo Model to Heavy Fermions*, Cambridge University Press, Cambridge.
K.-I. Imura, K.-V. Pham, P. Lederer, and F. Piechon (2002). *Phys. Rev. B* **66**, 035313.
K. Ingersent, A. W. W. Ludwig, and I. Affleck (2005). *Phys. Rev. Lett.* **95**, 257204.
V. G. Kac and K. Peterson (1984). *Adv. Math.* **53**, 125.
C. L. Kane and M. P. A. Fisher (1992). *Phys. Rev. B* **46**, 15233.
V. G. Knizhnik and A. B. Zamolodchikov (1984). *Nucl. Phys. B* **247**, 83.
J. Kondo (1964). *Prog. Theor. Phys.* **32**, 37.
N. Laflorencie, E. S. Sorensen, and I. Affleck (2008). *J. Stat. Mech.* P02007.
S.-K. Ma (1976). *Modern Theory of Critical Phenomena*, Benjamin-Cummings, Reading, MA.
D. L. Maslov and M. Stone (1995). *Phys. Rev. B* **52**, R5539.

K. Moon, H. Yi, C. L. Kane, S. M. Girvin, and M. P. A. Fisher (1993). *Phys. Rev. Lett.* **71**, 4381.
P. Nozières (1975). In *Proc. of 14th Int. Conf. on Low Temp. Phys.*, eds. M. Krusius and M. Vuorio, Vol. 5, p. 339.
P. Nozières and A. Blandin (1980). *J. de Physique* **41**, 193.
Y. Oreg and D. Goldhaber-Gordon (2003). *Phys. Rev. Lett.* **90**, 136602.
M. Oshikawa, C. Chamon, and I. Affleck (2006). *J. Stat. Mech.* P02008, 102.
R. G. Pereira, N. Laflorencie, I. Affleck, and B. I. Halperin (2008). *Phys. Rev. B* **77**, 125327.
V. V. Ponomarenko (1995). *Phys. Rev. B* **52**, R8666.
R. M. Potok, I. G. Rau, H. Shtrikman, Y. Oreg, and D. Goldhaber-Gordon (2007). *Nature* **446**, 167.
M. Pustilnik, L. Borda, L. Glazman, and J. von Delft (2004). *Phys. Rev. B* **69**, 115316.
I. Safi and H. J. Schulz (1995). *Phys. Rev. B* **52**, R17040.
K. D. Schotte and U. Schotte (1969). *Phys. Rev.* **182**, 479.
R. Shankar (1994). *Rev. Mod. Phys.* **66**, 129.
E. S. Sorensen, M.-C. Chang, N. Laflorencie, and I. Affleck (2007). *J. Stat. Mech.* P08003.
S. Tarucha, T. Honda, and T. Saku (1995). *Solid State Commun.* **94**, 413.
A. M. Tsvelick (1985). *J. Phys. C* **18**, 159.
P. B. Weigmann (1980). *Sov. Phys. JETP Lett.* **31**, 392.
K. G. Wilson (1975). *Rev. Mod. Phys.* **47**, 773.
E. Witten (1984). *Commun. Math. Phys.* **92**, 455.
A. B. Zamolodchikov (1986). *Pis'ma Zh. Eksp. Teor. Fiz.* **43**, 565 [*JETP Lett.* **43**, 730].
A. B. Zamalodchikov and S. Ghoshal (1994). *Int. J. Mod. Phys. A* **9**, 3841.

2
Conformal field theory and statistical mechanics

J. CARDY

John Cardy,
Rudolf Peierls Centre for Theoretical Physics,
1 Keble Road, Oxford OX1 3NP, UK
and All Souls College, Oxford.

2.1 Introduction

It is twenty years ago almost to the day that the Les Houches school on *Fields, Strings and Critical Phenomena* took place. It came on the heels of a frenzied period of five years or so following the seminal paper of Belavin, Polyakov, and Zamolodchikov (BPZ) in which the foundations of conformal field theory (CFT) and related topics had been laid down, and featured lectures on CFT by Paul Ginsparg, Ian Affleck, Jean-Bernard Zuber, and myself, related courses and talks by Hubert Saleur, Robert Dijkgraaf, Bertrand Duplantier, Sasha Polyakov, and Daniel Friedan, and lectures on several other topics. The list of young participants, many of whom have since gone on to make their own important contributions, is equally impressive.

Twenty years later, CFT is an essential item in the toolbox of many theoretical condensed matter physicists and string theorists. It has also had a marked impact in mathematics, in algebra, geometry, and, more recently, probability theory. It is the purpose of these lectures to introduce some of the tools of CFT. In some ways they are an updated version of those I gave in 1988. However, there are some important topics there which, in order to include new material, I will omit here, and I would encourage the diligent student to read both versions. I should stress that the lectures will largely be about conformal field theory, rather than conformal field theor<u>ies</u>, in the sense that I'll be describing rather generic properties of CFT rather than discussing particular examples. That said, I don't want the discussion to be too abstract, and in fact I will have a very specific idea about what are the kind of CFTs I will be discussing: the scaling limit (in a sense to be described) of critical lattice models (either classical, in two dimensions, or quantum, in 1+1 dimensions.) This will allow us to have, I hope, more concrete notions of the mathematical objects which are being discussed in CFT. However, there will be no attempt at mathematical rigor. Despite the fact that CFT can be developed axiomatically, I think that for this audience it is more important to understand the physical origin of the basic ideas.

2.2 Scale invariance and conformal invariance in critical behavior

2.2.1 Scale invariance

The prototype lattice model we have at the back of our minds is the ferromagnetic Ising model. We take some finite domain $\mathcal{D}$ of d-dimensional Euclidean space and impose a regular lattice, say hypercubic, with lattice constant a. With each node r of the lattice we associate a binary-valued spin $s(r) = \pm 1$. Each configuration $\{s\}$ carries a relative weight $W(\{s\}) \propto \exp(\sum_{rr' \in \mathcal{D}} J(r-r')s(r)s(r))$, where $J(r-r') > 0$ is some short-ranged interaction (i.e. it vanishes if $|r-r'|$ is larger than some fixed multiple of a.)

The usual kinds of local observable $\phi_j^{\rm lat}(r)$ considered in this lattice model are sums of products of nearby spins over some region of size $O(a)$, for example the local magnetization $s(r)$ itself, or the energy density $\sum_{r'} J(r-r')s(r)s(r')$. However, it will become clear later on that there are other observables, also labeled by a single point r, which are functions of the whole configuration $\{s\}$. For example, we shall show that this Ising model can be mapped onto a gas of nonintersecting loops. An observable then might depend on whether a particular loop passes through the given point r. From the point of view of CFT, these observables are equally valid objects.

Correlation functions, that is, expectation values of products of local lattice observables, are then given by

$$\langle\phi_1^{\rm lat}(r_1)\phi_2^{\rm lat}(r_2)\ldots\phi_n^{\rm lat}(r_n)\rangle = Z^{-1}\sum_{\{s\}}\phi_1^{\rm lat}(r_1)\ldots\phi_n^{\rm lat}(r_n)W(\{s\})\,,$$

where $Z = \sum_{\{s\}} W(\{s\})$ is the partition function. In general, the connected pieces of these correlation functions fall off over the same distance scale as the interaction $J(r - r')$, but, close to a critical point, this correlation length ξ can become large, $\gg a$.

The *scaling limit* is obtained by taking $a \to 0$ while keeping ξ and the domain $\mathcal{D}$ fixed. In general, the correlation functions as defined above do not possess a finite scaling limit. However, the theory of renormalization (based on studies in exactly solved models, as well as perturbative analysis of cutoff quantum field theory) suggests that in general there are particular linear combinations of local lattice observables which are *multiplicatively renormalizable*. That is, the limit

$$\lim_{a\to 0} a^{-\sum_{j=1}^n x_j}\langle\phi_1^{\rm lat}(r_1)\phi_2^{\rm lat}(r_2)\ldots\phi_n^{\rm lat}(r_n)\rangle \tag{2.1}$$

exists for certain values of the $\{x_j\}$. We usually denote this by

$$\langle\phi_1(r_1)\phi_2(r_2)\ldots\phi_n(r_n)\rangle\,, \tag{2.2}$$

and we often think of it as the expectation value of the product of the random variables $\phi_j(r_j)$, known as *scaling fields* (sometimes scaling operators, to be even more confusing), with respect to some "path integral" measure. However, it should be stressed that this is only an occasionally useful fiction, which ignores all the wonderful subtleties of renormalized field theory. The basic objects of QFT are the correlation functions. The numbers $\{x_j\}$ in eqn (2.1) are called the *scaling dimensions*.

One important reason why this is not true in general is that the limit in eqn (2.1) in fact only exists if the points $\{r_j\}$ are noncoincident. The correlation functions in eqn (2.2) are singular in the limits when $r_i \to r_j$. However, the nature of these singularities is prescribed by the *operator product expansion* (OPE)

$$\langle\phi_i(r_i)\phi_j(r_j)\ldots\rangle = \sum_k C_{ijk}(r_i - r_j)\langle\phi_k((r_i + r_j)/2)\ldots\rangle\,. \tag{2.3}$$

The main point is that, in the limit when $|r_i - r_j|$ is much less than the separation between r_i and all the other arguments in $\ldots$, the coefficients C_{ijk} are independent of what is in the dots. For this reason, eqn (2.3) is often written as

$$\phi_i(r_i)\cdot\phi_j(r_j) = \sum_k C_{ijk}(r_i - r_j)\phi_k((r_i + r_j)/2)\,, \tag{2.4}$$

although it should be stressed that this is merely a shorthand for eqn (2.3).

So far, we have been talking about how to get a continuum (Euclidean) field theory as the scaling limit of a lattice model. In general, this will be a massive QFT, with a mass scale given by the inverse correlation length ξ^{-1}. In general, the correlation functions will depend on this scale. However, at a (second-order) critical point the correlation length ξ diverges, that is, the mass vanishes, and there is no length scale in the problem besides the overall size L of the domain $\mathcal{D}$.

The fact that the scaling limit of eqn (2.1) exists then implies that, instead of starting with a lattice model with lattice constant a, we could equally well have started with one with some fraction a/b. This would, however, be identical with a lattice model with the original spacing a, in which all lengths (including the size of the domain $\mathcal{D}$) are multiplied by b. This implies that the correlation functions in eqn (2.2) are *scale covariant*:

$$\langle\phi_1(br_1)\phi_2(br_2)\ldots\phi_n(br_n)\rangle_{b\mathcal{D}} = b^{-\sum_j x_j}\langle\phi_1(r_1)\phi_2(r_2)\ldots\phi_n(r_n)\rangle_{\mathcal{D}}\,. \tag{2.5}$$

Once again, we can write this in the suggestive form

$$\phi_j(br) = b^{-x_j}\phi_j(r)\,, \tag{2.6}$$

as long as what we really mean is eqn (2.5).

In a massless QFT, the form of the OPE coefficients in eqn (2.4) simplifies: by scale covariance,

$$C_{ijk}(r_j - r_k) = \frac{c_{ijk}}{|r_i - r_j|^{x_i+x_j-x_k}}\,, \tag{2.7}$$

where the c_{ijk} are pure numbers, and are *universal* if the two-point functions are normalized so that $\langle\phi_j(r_1)\phi_j(r_2)\rangle = |r_1 - r_2|^{-2x_j}$. (This assumes that the scaling fields are all rotational scalars—otherwise it is somewhat more complicated, at least for general dimension.)

From scale covariance, it is a simple but powerful leap to *conformal covariance*: suppose that the scaling factor b in eqn (2.5) is a slowly varying function of the position r. Then we can try to write a generalization of eqn (2.5) as

$$\langle\phi_1(r_1')\phi_2(r_2')\ldots\phi_n(r_n')\rangle_{\mathcal{D}'} = \prod_{j=1}^{n} b(r_j)^{-x_j}\langle\phi_1(r_1)\phi_2(r_2)\ldots\phi_n(r_n)\rangle_{\mathcal{D}}\,, \tag{2.8}$$

where $b(r) = |\partial r'/\partial r|$ is the local Jacobian of the transformation $r \to r'$.

For what transformations $r \to r'$ do we expect eqn (2.8) to hold? The heuristic argument runs as follows: if the theory is local (that is, the interactions in the lattice model are short-ranged), then as long as the transformation looks *locally* like a scale transformation (plus a possible rotation), eqn (2.8) may be expected to hold. (In Section 2.3 we will make this more precise, based on the assumed properties of the stress tensor, and argue that in fact it holds only for a special class of scaling fields $\{\phi_j\}$ called primary scaling fields.)

It is most important that the underlying lattice does *not* transform (otherwise the statement is a tautology): eqn (2.8) relates correlation functions in $\mathcal{D}$, defined in terms of the limit $a \to 0$ of a model on a regular lattice superimposed on $\mathcal{D}$, to correlation functions defined by a regular lattice superimposed on $\mathcal{D}'$.

Transformations which are locally equivalent to a scale transformation and rotation, that is, have no local components of shear, also locally preserve angles and are called *conformal*.

2.2.2 Conformal mappings in general

Consider a general infinitesimal transformation (in flat space) $r^\mu \to r'^\mu = r^\mu + \alpha^\mu(r)$ (we distinguish between upper and lower indices in anticipation of using coordinates in which the metric is not diagonal.) The shear component is the traceless symmetric part

$$\alpha^{\mu,\nu} + \alpha^{\nu,\mu} - (2/d)\alpha^\lambda{}_{,\lambda}g^{\mu\nu}\,,$$

all $\frac{1}{2}d(d+1)-1$ components of which must vanish for the mapping to be conformal. For general d this is very restrictive, and, in fact, apart from uniform translations, rotations and scale transformations, there is only one other type of solution,

$$\alpha^{\mu}(r)=b^{\mu}r^{2}-2(b\cdot r)r^{\mu}\,,$$

where b^{μ} is a constant vector. These are in fact the composition of the finite conformal mapping of inversion $r^{\mu}\to r^{\mu}/|r|^{2}$, followed by an infinitesimal translation b^{μ}, followed by a further inversion. They are called the special conformal transformations and, together with the others, they generate a group isomorphic to $\mathrm{SO}(d+1,1)$.

These special conformal transformations have enough freedom to fix the form of the three-point functions in $\mathbf{R}^{d}$ (just as scale invariance and rotational invariance fix the two-point functions): for scalar operators,[1]

$$\langle\phi_1(r_1)\phi_2(r_2)\phi_3(r_3)\rangle=\frac{c_{123}}{|r_1-r_2|^{x_1+x_2-x_3}|r_2-r_3|^{x_2+x_3-x_1}|r_3-r_1|^{x_3+x_1-x_2}}\,. \tag{2.9}$$

Comparing this with the OPE in eqns (2.4) and (2.7), and assuming nondegeneracy of the scaling dimensions,[2] we see that c_{123} is the same as the OPE coefficient defined earlier. This shows that the OPE coefficients c_{ijk} are symmetric in their indices.

In two dimensions, the condition that $\alpha^{\mu}(r)$ be conformal imposes only two differential conditions on two functions, and there is a much wider class of solutions. These are more easily seen using *complex coordinates*[3] $z\equiv r^{1}+ir^{2}$, $\bar{z}\equiv r^{1}-ir^{2}$, so that the line element is $ds^{2}=dz\,d\bar{z}$, and the metric is

$$g_{\mu\nu}=\begin{pmatrix}0&\frac{1}{2}\\ \frac{1}{2}&0\end{pmatrix},\qquad g^{\mu\nu}=\begin{pmatrix}0&2\\ 2&0\end{pmatrix}.$$

In this basis, two of the conditions are satisfied identically and the others become

$$\alpha^{z,z}=\alpha^{\bar{z},\bar{z}}=0\,,$$

which means that $\partial\alpha^{z}/\partial\bar{z}=\partial\alpha^{\bar{z}}/\partial z=0$, that is, α^{z} is a holomorphic function $\alpha(z)$ of z, and $\alpha^{\bar{z}}$ is an antiholomorphic function.

Generalizing this to a finite transformation, this means that conformal mappings $r\to r'$ correspond to functions $z\to z'=f(z)$ which are analytic in $\mathcal{D}$. (Note that the only such functions on the whole Riemann sphere are the Möbius transformations $f(z)=(az+b)/(cz+d)$, which are the finite special conformal mappings.)

In passing, let us note that complex coordinates give us a nice way of discussing non-scalar fields: if, for example, under a rotation $z\to ze^{i\theta}$, $\phi_j(z,\bar{z})\to e^{is_j\theta}\phi_j$, we say that

[1] The easiest way to show this it to make an inversion with an origin very close to one of the points, say r_1, and then use the OPE, since its image is then very far from those of the other two points.

[2] This and other properties fail in so-called logarithmic CFTs.

[3] For many CFT computations we may treat z and $\bar{z}$ as independent, imposing only at the end the condition that they should be complex conjugates.

ϕ_j has conformal spin s_j (not related to quantum mechanical spin), and, under a combined transformation $z \to \lambda z$, where $\lambda = be^{i\theta}$, we can write (in the same spirit as in eqn (2.6))

$$\phi_j(\lambda z, \bar{\lambda}\bar{z}) = \lambda^{-\Delta_j}\bar{\lambda}^{-\overline{\Delta}_j}\phi_j(z,\bar{z}),$$

where $x_j = \Delta_j + \overline{\Delta}_j$, $s_j = \Delta_j - \overline{\Delta}_j$. $(\Delta_j, \overline{\Delta}_j)$ are called the complex scaling dimensions of ϕ_j (although they are usually both real, and not necessarily complex conjugates of each other.)

2.3 The role of the stress tensor

Since we wish to explore the consequences of conformal invariance for correlation functions in a *fixed* domain $\mathcal{D}$ (usually the entire complex plane), it is necessary to consider transformations which are *not* conformal everywhere. This brings in the stress tensor $T_{\mu\nu}$ (also known as the stress–energy tensor or the (improved) energy–momentum tensor). This is the object appearing on the right-hand side of Einstein's equations in curved space. In a classical field theory, it is defined in terms of the response of the action S to a general infinitesimal transformation $\alpha^\mu(r)$:

$$\delta S = -\frac{1}{2\pi}\int T_{\mu\nu}\alpha^{\mu,\nu}d^2r \tag{2.10}$$

(the $(1/2\pi)$ avoids awkward such factors later on.) Invariance of the action under translations and rotations implies that $T_{\mu\nu}$ is conserved and symmetric. Moreover, if S is scale invariant, $T_{\mu\nu}$ is also traceless. In complex coordinates, the first two conditions imply that $T_{z\bar{z}}+T_{\bar{z}z} = 0$ and $T_{z\bar{z}} = T_{\bar{z}z}$, so they both vanish, and the conservation equations then read $\partial^z T_{zz} = 2\partial T_{zz}/\partial\bar{z} = 0$ and $\partial T_{\bar{z}\bar{z}}/\partial z = 0$. Thus the nonzero components $T \equiv T_{zz}$ and $\overline{T} \equiv T_{\bar{z}\bar{z}}$ are, respectively, holomorphic and antiholomorphic fields. Now, if we consider a more general transformation for which $\alpha^{\mu,\nu}$ is symmetric and traceless, that is, a conformal transformation, we see that $\delta S = 0$ in this case also. Thus, at least classically, we see that scale invariance and rotational invariance imply conformal invariance of the action, at least if eqn (2.10) holds. However, if the theory contains long-range interactions, for example, this is no longer the case.

In a quantum 2D CFT, it is assumed that the above analyticity properties continue to hold at the level of correlation functions: those of $T(z)$ and $\overline{T}(\bar{z})$ are holomorphic and antiholomorphic functions of z, respectively (except at coincident points.)

Example: Free (Gaussian) scalar field. The prototype CFT is the free, or Gaussian, massless scalar field $h(r)$ (we use this notation for reasons that will emerge later). It will turn out that many other CFTs are basically variants of this. The classical action is

$$S[h] = \frac{g}{4\pi}\int(\partial_\mu h)(\partial^\mu h)\,d^2r\,.$$

Since $h(r)$ can take any real value, we could rescale it to eliminate the coefficient in front, but in later extensions this will have a meaning, so we keep it. In complex coordinates, $S \propto \int(\partial_z h)(\partial_{\bar{z}} h)\,d^2z$, and it is easy to see that this is conformally invariant under $z \to z' =$

$f(z)$, since $\partial_z = f'(z)\partial_{z'}$, $\partial_{\bar{z}} = \overline{f'(z)}\partial_{\bar{z}'}$, and $d^2z = |f'(z)|^{-2}\, d^2z'$. This is confirmed by calculating $T_{\mu\nu}$ explicitly: we find $T_{z\bar{z}} = T_{\bar{z}z} = 0$, and

$$T = T_{zz} = -g(\partial_z h)^2, \qquad \overline{T} = T_{\bar{z}\bar{z}} = -g(\partial_{\bar{z}} h)^2 .$$

These are holomorphic and antiholomorphic, respectively, by virtue of the classical equation of motion $\partial_z \partial_{\bar{z}} h = 0$.

In the quantum field theory, a given configuration $\{h\}$ is weighted by $\exp(-S[h])$. The two-point function is[4]

$$\langle h(z,\bar{z})h(0,0)\rangle = \frac{2\pi}{g}\int \frac{e^{ik\cdot r}}{k^2}\frac{d^2k}{(2\pi)^2} \sim -\frac{1}{2g}\log\left(\frac{z\bar{z}}{L^2}\right).$$

This means that $\langle T\rangle$ is formally divergent. It can be made finite by, for example, point-splitting and subtracting off the divergent piece:

$$T(z) = -g\lim_{\delta\to 0}\left(\partial_z h\left(z + \frac{1}{2}\delta\right)\partial_z h\left(z - \frac{1}{2}\delta\right) - \frac{1}{2g\delta^2}\right). \tag{2.11}$$

This doesn't affect the essential properties of T.

2.3.1 Conformal Ward identity

Consider a general correlation function of scaling fields $\langle\phi_1(z_1,\bar{z}_1)\phi_2(z_2,\bar{z}_2)\ldots\rangle_{\mathcal{D}}$ in some domain $\mathcal{D}$. We want to make an infinitesimal conformal transformation $z \to z' = z + \alpha(z)$ on the points $\{z_j\}$, without modifying $\mathcal{D}$. This can be done by considering a contour C which encloses all the points $\{z_j\}$ but which lies wholly within $\mathcal{D}$, such that the transformation is conformal within C, and the identity $z' = z$ applies outside C (Fig. 2.1). This gives rise to an (infinitesimal) discontinuity on C and, at least classically, to a modification of the action S according to eqn (2.10). Integrating by parts, we find $\delta S = (1/2\pi)\int_C T_{\mu\nu}\alpha^\mu n^\nu\, d\ell$, where n^ν

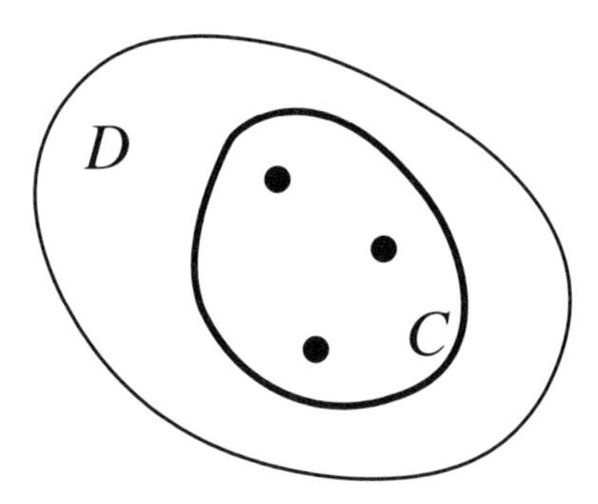

Fig. 2.1 We consider an infinitesimal transformation which is conformal within C, and the identity in the complement in $\mathcal{D}$.

[4]This is cut off at small k by the assumed finite size L, but we are also assuming here that the points are far from the boundary.

is the outward-pointing normal and $d\ell$ is a line element of C. This is more easily expressed in complex coordinates, after some algebra, as

$$\delta S = \frac{1}{2\pi i}\int_C \alpha(z)T(z)\,dz + \text{complex conjugate}\,.$$

This extra factor can then be expanded, to first order in α, out of the weight $\exp(-S[h] - \delta S) \sim (1 - \delta S)\exp(-S[h])$, and the extra piece δS considered as an insertion into the correlation function. This is balanced by the explicit change in the correlation function under the conformal transformation:

$$\delta\langle\phi_1(z_1,\bar{z}_1)\phi_2(z_2,\bar{z}_2)\ldots\rangle = \frac{1}{2\pi i}\int_C \alpha(z)\langle T(z)\phi_1(z_1,\bar{z}_1)\phi_2(z_2,\bar{z}_2)\ldots\rangle\,dz + \text{c.c.} \quad (2.12)$$

Let us first consider the case when $\alpha(z) = \lambda(z - z_1)$, that is, it corresponds to a combined rotation and scale transformation. In that case $\delta\phi_1 = (\Delta_1\lambda + \overline{\Delta}_1\bar{\lambda})\phi_1$ and, therefore, equating coefficients of λ and $\bar{\lambda}$,

$$\int_C (z - z_1)\langle T(z)\phi_1(z_1,\bar{z}_1)\phi_2(z_2,\bar{z}_2)\ldots\rangle\frac{dz}{2\pi i} = \Delta_1\langle\phi_1(z_1,\bar{z}_1)\phi_2(z_2,\bar{z}_2)\ldots\rangle + \cdots\,.$$

Similarly, if we take $\alpha =$ constant, corresponding to a translation, we have $\delta\phi_j \propto \partial_{z_j}\phi_j$, so

$$\int_C \langle T(z)\phi_1(z_1,\bar{z}_1)\phi_2(z_2,\bar{z}_2)\ldots\rangle\frac{dz}{2\pi i} = \sum_j \Delta_j\partial_{z_j}\langle\phi_1(z_1,\bar{z}_1)\phi_2(z_2,\bar{z}_2)\ldots\rangle\,.$$

Using Cauchy's theorem, these two equations tell us about two of the singular terms in the OPE of $T(z)$ with a general scaling field $\phi_j(z_j,\bar{z}_j)$:

$$T(z)\cdot\phi_j(z_j,\bar{z}_j) = \cdots + \frac{\Delta_j}{(z - z_j)^2}\phi_j(z_j,\bar{z}_j) + \frac{1}{z - z_j}\partial_{z_j}\phi_j(z_j,\bar{z}_j) + \cdots\,. \quad (2.13)$$

Note that only integer powers can occur, because the correlation function is a meromorphic function of z.

Example of the Gaussian free field. If we take $\phi_q^{\text{lat}}(r) = e^{iqh(r)}$ then

$$\langle\phi_q^{\text{lat}}(r_1)\phi_{-q}^{\text{lat}}(r_2)\rangle = \exp\Big(-\frac{1}{2}q^2\langle(h(r_1) - h(r_2))^2\rangle\Big) \sim \Big(\frac{a}{|r_1 - r_2|}\Big)^{q^2/g}\,,$$

which means that the renormalized field $\phi_q \sim a^{-q^2/2g}\phi_q^{\text{lat}}$ has scaling dimension $x_q = q^2/2g$. It is then a nice exercise in Wick's theorem to check that the OPE with the stress tensor (2.13) holds with $\Delta_q = x_q/2$. (Note that in this case the multiplicative renormalization of ϕ_q is equivalent to ignoring all Wick contractions between fields $h(r)$ at the same point.)

Now suppose each ϕ_j is such that the terms $O\big((z - z_j)^{-2-n}\big)$ with $n \geq 1$ in eqn (2.13) are absent. Since a meromorphic function is determined entirely by its singularities, we then know the correlation function $\langle T(z)\ldots\rangle$ exactly:

$$\langle T(z)\phi_1(z_1,\bar{z}_1)\phi_2(z_2,\bar{z}_2)\ldots\rangle = \sum_j\left(\frac{\Delta_j}{z - z_j)^2} + \frac{1}{z - z_j}\partial_{z_j}\right)\langle\phi_1(z_1,\bar{z}_1)\phi_2(z_2,\bar{z}_2)\ldots\rangle\,. \quad (2.14)$$

This (as well as a similar equation for an insertion of $\overline{T}$) is the *conformal Ward identity*. We have derived it assuming that the quantum theory could be defined by a path integral and the change in the action δS follows the classical pattern. For a more general CFT, not necessarily "defined" (however loosely) by a path integral, eqn (2.14) is usually assumed as a property of T. In fact, many basic introductions to CFT use this as a starting point.

Scaling fields $\phi_j(z_j, \bar{z}_j)$ such that the most singular term in their OPE with $T(z)$ is $O((z-z_j)^{-2})$ are called *primary*.[5] All other fields like those appearing in the less singular terms in eqn (2.13) are called *descendants*. Once one knows the correlation functions of all the primaries, those of the rest follow from eqn (2.13).[6]

For correlations of such primary fields, we can now reverse the arguments leading to eqn (2.13) for the case of a general infinitesimal conformal transformation $\alpha(z)$ and conclude that

$$\delta\langle\phi_1(z_1,\bar{z}_1)\phi_2(z_2,\bar{z}_2)\ldots\rangle = \sum_j(\Delta_j\alpha'(z_j) + \alpha(z_j)\partial_{z_j})\langle\phi_1(z_1,\bar{z}_1)\phi_2(z_2,\bar{z}_2)\ldots\rangle\,,$$

which may be integrated up to get the result for a *finite* conformal mapping $z \to z' = f(z)$:

$$\langle\phi_1(z_1,\bar{z}_1)\phi_2(z_2,\bar{z}_2)\ldots\rangle_{\mathcal{D}} = \prod_j f'(z_j)^{\Delta_j}\overline{f'(z_j)}^{\bar{\Delta}_j}\langle\phi_1(z_1',\bar{z}_1')\phi_2(z_2',\bar{z}_2')\ldots\rangle_{\mathcal{D}'}\,.$$

This is just the result we wanted to postulate in eqn (2.8), but now we see that it can hold only for correlation functions of *primary* fields.

It is important to realize that T itself is not in general primary. Indeed, its OPE with itself must take the form[7]

$$T(z)\cdot T(z_1) = \frac{c/2}{(z-z_1)^4} + \frac{2}{(z-z_1)^2}T(z_1) + \frac{1}{z-z_1}\partial_{z_1}T(z_1)\cdots\,. \tag{2.15}$$

This is because (taking the expectation value of both sides) the two-point function $\langle T(z)T(z_1)\rangle$ is generally nonzero. Its form is fixed by the fact that $\Delta_T = 2$, $\overline{\Delta}_T = 0$, but, since the normalization of T is fixed by its definition (2.10), its coefficient $c/2$ is fixed. This introduces the *conformal anomaly* number c, which is part of the basic data of the CFT, along with the scaling dimensions $(\Delta_j, \overline{\Delta}_j)$ and the OPE coefficients c_{ijk}.[8]

This means that, under an infinitesimal transformation $\alpha(z)$, there is an additional term in the transformation law of T:

$$\delta T(z) = 2\alpha'(z)T(z) + \alpha(z)\partial_z T(z) + \frac{c}{12}\alpha'''(z)\,.$$

For a finite conformal transformation $z \to z' = f(z)$, this integrates up to

$$T(z) = f'(z)^2 T(z') + \frac{c}{12}\{z', z\}\,, \tag{2.16}$$

where the last term is the Schwarzian derivative

[5] If we assume that there is a lower bound to the scaling dimensions, such fields must exist.

[6] Since the scaling dimensions of the descendants differ from those of the corresponding primaries by positive integers, they are increasingly irrelevant in the sense of the renormalization group.

[7] The $O((z-z_1)^{-3})$ term is absent by symmetry under exchange of z and z_1.

[8] When the theory is placed in a curved background, the trace $\langle T^\mu_\mu\rangle$ is proportional to cR, where R is the local scalar curvature.

$$\{w(z), z\} = \frac{w'''(z)w'(z) - (3/2)w''(z)^2}{w'(z)^2}.$$

The form of the Schwarzian can be seen most easily in the example of a Gaussian free field. In this case, the point-split terms in eqn (2.11) transform properly and give rise to the first term in eqn (2.16), but the fact that the subtraction has to be made separately in the origin frame and the transformed one leads to an anomalous term

$$\lim_{\delta\to 0} g \left(\frac{f'(z+\frac{1}{2}\delta)f'(z-\frac{1}{2}\delta)}{2g\big(f(z+\frac{1}{2}\delta)-f(z-\frac{1}{2}\delta)\big)^2} - \frac{1}{2g\delta^2} \right),$$

which, after some algebra, gives the second term in eqn (2.16) with $c = 1$.[9]

2.3.2 An application—entanglement entropy

Let's take a break from the development of the general theory and discuss how the conformal anomaly number c arises in an interesting (and topical) physical context. Suppose that we have a massless relativistic quantum field theory (QFT) in 1+1 dimensions, whose imaginary-time behavior therefore corresponds to a Euclidean CFT. (There are many condensed matter systems whose large-distance behavior is described by such a theory.) The system is at zero temperature and therefore in its ground state $|0\rangle$, corresponding to a density matrix $\rho = |0\rangle\langle 0|$. Suppose an observer A has access to just part of the system, for example a segment of length ℓ inside the rest of the system, of total length $L \gg \ell$, observed by B. The measurements of A and B are entangled. It can be argued that a useful measure of the degree of entanglement is the *entropy* $S_A = -\mathrm{Tr}_A\, \rho_A \log \rho_A$, where $\rho_A = \mathrm{Tr}_B\, \rho$ is the reduced density matrix corresponding to A.

How can we calculate S_A using CFT methods? The first step is to realize that if we can compute $\mathrm{Tr}\, \rho_A^n$ for positive integer n, and then analytically continue in n, the derivative $\partial/\partial n|_{n=1}$ will give the required quantity. For any QFT, the density matrix at finite inverse temperature β is given by the path integral over some fundamental set of fields $h(x,t)$ (t is imaginary time)

$$\rho\big(\{h(x,0)\},\{h(x,\beta)\}\big) = Z^{-1} \int' [dh(x,t)] e^{-S[h]},$$

where the rows and columns of ρ are labeled by the values of the fields at times 0 and β, respectively, and the path integral is over all histories $h(x,t)$ consistent with these initial and final values (Fig. 2.2). Z is the partition function, obtained by sewing together the edges at these two times, that is, setting $h(x,\beta) = h(x,0)$ and integrating $\int [dh(x,0)]$.

We are interested in the partial density matrix ρ_A, which is similarly found by sewing together the top and bottom edges for $x \in B$, that is, leaving open a slit along the interval A (Fig. 2.3). The rows and columns of ρ_A are labeled by the values of the fields on the edges of the slit. $\mathrm{Tr}\, \rho_A^n$ is then obtained by sewing together the edges of n copies of this slit cylinder

[9]This is a classic example of an anomaly in QFT, which comes about because the regularization procedure does not respect a symmetry.

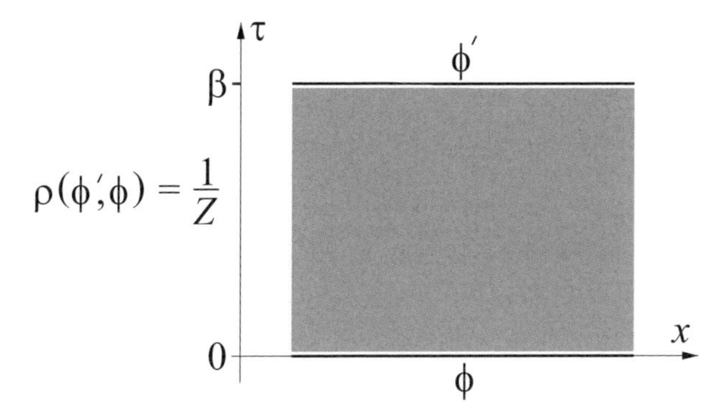

Fig. 2.2 The density matrix is given by the path integral over a space–time region in which the rows and columns are labeled by the initial and final values of the fields.

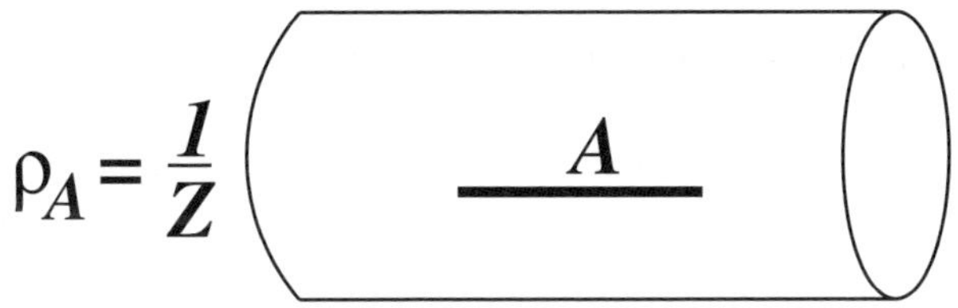

Fig. 2.3 The reduced density matrix ρ_A is given by the path integral over a cylinder with a slit along the interval A.

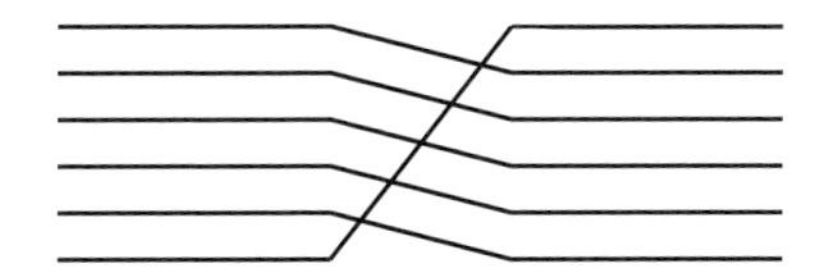

Fig. 2.4 $\mathrm{Tr}\,\rho_A^n$ corresponds to sewing together n copies so that the edges are connected cyclically.

in a cyclic fashion (Fig. 2.4). This gives an n-sheeted surface with branch points, or conical singularities, at the ends of the interval A. If Z_n is the partition function on this surface, then

$$\mathrm{Tr}\,\rho_A^n = \frac{Z_n}{Z_1^n}\,.$$

Let us consider the case of zero temperature, $\beta \to \infty$, when the whole system is in the ground state $|0\rangle$. If the ends of the interval are at (x_1, x_2), the conformal mapping

$$z = \left(\frac{w - x_1}{w - x_2}\right)^{1/n}$$

maps the n-sheeted w-surface to the single-sheeted complex z-plane. We can use the transformation law (2.16) to compute $\langle T(w)\rangle$, given that $\langle T(z)\rangle = 0$ by translational and rotational invariance. This gives, after a little algebra,

$$\langle T(w)\rangle = \frac{(c/12)(1 - 1/n^2)(x_2 - x_1)^2}{(w - x_1)^2(w - x_2)^2}\,.$$

Now suppose we change the length $\ell = |x_2 - x_1|$ slightly, by making an infinitesimal non-conformal transformation $x \to x + \delta\ell\,\delta(x - x_0)$, where $x_1 < x_0 < x_2$. The response of the log of the partition function, by the definition of the stress tensor, is

$$\delta \log Z_n = -\frac{n\,\delta\ell}{2\pi} \int_{-\infty}^{\infty} \langle T_{xx}(x_0, t)\rangle\, dt$$

(the factor n occurs because it has to be inserted on each of the n sheets.) If we write $T_{xx} = T + \overline{T}$, the integration in each term can be carried out by wrapping the contour around either of the points x_1 or x_2. The result is

$$\frac{\partial \log Z_n}{\partial \ell} = -\frac{(c/6)(n - 1/n)}{\ell},$$

so that $Z_n/Z_1^n \sim \ell^{-(c/6)(n-1/n)}$. Taking the derivative with respect to n at $n = 1$, we get the final result,

$$S_A \sim \frac{c}{3} \log \ell\,.$$

2.4 Radial quantization and the Virasoro algebra

2.4.1 Radial quantization

Like any quantum theory, CFT can be formulated in terms of local operators acting on a Hilbert space of states. However, as it is massless, the usual quantization of the theory on the infinite line is not so useful, since it is hard to disentangle the continuum of eigenstates of the Hamiltonian, and we cannot define asymptotic states in the usual way. Instead, it is useful to exploit the scale invariance rather than the time-translation invariance, and quantize on a circle of fixed radius r_0. In the path integral formulation, heuristically the Hilbert space is the space of field configurations $|\{h(r_0, \theta)\}\rangle$ on this circle. The analogue of the Hamiltonian is then the generator $\hat{D}$ of scale transformations. It will turn out that the spectrum of $\hat{D}$ is discrete. In the vacuum state, each configuration is weighted by the path integral over the disk $|z| < r_0$, conditioned on taking the assigned values on r_0 (see Fig. 2.5):

$$|0\rangle = \int [dh(r \le r_0)] e^{-S[h]} |\{h(r_0, \theta)\}\rangle\,.$$

The scale invariance of the action means that this state is independent of r_0, up to a constant. If instead we insert a scaling field $\phi_j(0)$ into the above path integral, we get a different state

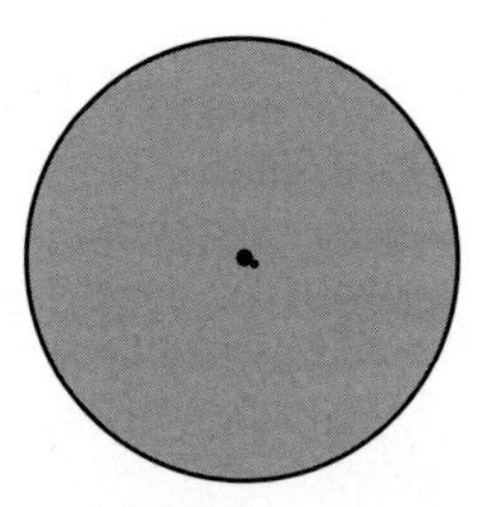

Fig. 2.5 The state $|\phi_j\rangle$ is defined by weighting field configurations on the circle with the path integral inside, with an insertion of the field $\phi_j(0)$ at the origin.

$|\phi_j\rangle$. On the other hand, more general correlation functions of scaling fields are given in this operator formalism by[10]

$$\langle\phi_j(z_1,\bar{z}_1)\phi_j(z_2,\bar{z}_2)\rangle = \langle 0|\mathbf{R}\,\hat{\phi}_j(z_1,\bar{z}_1)\hat{\phi}_j(z_2,\bar{z}_2)|0\rangle\,,$$

where $\mathbf{R}$ means the r-ordered product (like the time-ordered product in the usual case), and $\langle 0|$ is defined similarly in terms of the path integral over $r > r_0$. Thus we can identify

$$|\phi_j\rangle = \lim_{z\to 0}\hat{\phi}_j(z,\bar{z})|0\rangle\,.$$

This is an example of the *operator–state correspondence* in CFT.

Just as the Hamiltonian in ordinary QFT can be written as an integral over the appropriate component of the stress tensor, we can write

$$\hat{D} = \hat{L}_0 + \hat{\bar{L}}_0 \equiv \frac{1}{2\pi i}\int_C z\hat{T}(z)dz + \text{c.c.}\,,$$

where C is any closed contour surrounding the origin. This suggests that we define more generally

$$\hat{L}_n \equiv \frac{1}{2\pi i}\int_C z^{n+1}\hat{T}(z)\,dz\,,$$

and similarly $\hat{\bar{L}}_n$.

If there are no operator insertions inside C it can be shrunk to zero for $n \geq -1$, and thus

$$\hat{L}_n|0\rangle = 0 \qquad \text{for } n \geq -1\,.$$

On the other hand, if there is an operator ϕ_j inserted at the origin, we see from the OPE (2.13) that

$$\hat{L}_0|\phi_j\rangle = \Delta_j|\phi_j\rangle\,.$$

If ϕ_j is *primary*, we see further from the OPE that

$$\hat{L}_n|\phi_j\rangle = 0 \qquad \text{for } n \geq 1\,,$$

while for $n \leq -1$ we get states corresponding to descendants of ϕ_j.

2.4.2 The Virasoro algebra

Now consider $\hat{L}_m\hat{L}_n$ acting on some arbitrary state. In terms of correlation functions, this involves the contour integrals

$$\int_{C_2}\frac{dz_2}{2\pi i}z_2^{m+1}\int_{C_1}\frac{dz_1}{2\pi i}z_1^{n+1}T(z_2)T(z_2)\,,$$

where C_2 lies *outside* C_1, because of the $\mathbf{R}$-ordering. If instead we consider the operators in the reverse order, the contours will be reversed. However, we can then always distort them to

[10] We shall make an effort consistently to denote actual operators (as opposed to fields) with a hat.

restore them to their original positions, as long as we leave a piece of the z_2 contour wrapped around z_1. This can be evaluated using the OPE (2.15) of T with itself:

$$\int_C \frac{dz_1}{2\pi i} z_1^{n+1} \oint \frac{dz_2}{2\pi i} z_2^{m+1} \left(\frac{2}{(z_2 - z_1)^2} T(z_1) + \frac{1}{z_2 - z_1} \partial_{z_1} T(z_1) + \frac{c/2}{(z_2 - z_1)^4} \right)$$
$$= \int_C \frac{dz_1}{2\pi i} z_1^{n+1} \left(2(m+1) z_1^m T(z_1) + z_1^{m+1} \partial_{z_1} T(z_1) + \frac{c}{12} m(m^2-1) z_1^{m-2} \right).$$

The integrals can then be re-expressed in terms of the $\hat{L}_n$. This gives the *Virasoro algebra*

$$[\hat{L}_m, \hat{L}_n] = (m-n)\hat{L}_{m+n} + \frac{c}{12} m(m^2-1)\delta_{m+n,0}, \tag{2.17}$$

with an identical algebra satisfied by the $\hat{\bar{L}}_n$. It should be stressed that eqn (2.17) is completely equivalent to the OPE (2.15), but of course algebraic methods are often more efficient in understanding the structure of QFT. The first term on the right-hand side could have been foreseen if we think of $\hat{L}_n$ as being the generator of infinitesimal conformal transformations with $\alpha(z) \propto z^{n+1}$. Acting on functions of z, this can therefore be represented by $\hat{\ell}_n = z^{n+1}\partial_z$, and it is easy to check that these satisfy eqn (2.17) without the central term (called the Witt algebra.) However, the $\hat{L}_n$ act on states of the CFT rather than functions, which allows for the existence of the second term, the central term. The form of this, apart from the undetermined constant c, is in fact dictated by consistency with the Jacobi identity. Note that there is a closed subalgebra generated by $(\hat{L}_1, \hat{L}_0, \hat{L}_{-1})$, which corresponds to special conformal transformations.

One consequence of eqn (2.17) is

$$[\hat{L}_0, \hat{L}_{-n}] = n\hat{L}_{-n},$$

so that $\hat{L}_0(\hat{L}_{-n}|\phi_j\rangle) = (\Delta_j + n)\hat{L}_{-n}|\phi_j\rangle$, which means that the $\hat{L}_n$ with $n < 0$ act as raising operators for the weight, or scaling dimension, Δ, and those with $n > 0$ act as lowering operators. The state $|\phi_j\rangle$ corresponding to a primary operator is annihilated by all the lowering operators. It is therefore a *lowest-weight state*.[11] By acting with all possible raising operators, we build up a *lowest-weight representation* (called a Verma module) of the Virasoro algebra:

$$\begin{array}{l} \vdots \\ \hat{L}_{-3}|\phi_j\rangle, \hat{L}_{-2}\hat{L}_{-1}|\phi_j\rangle, \hat{L}_{-1}^3|\phi_j\rangle; \\ \hat{L}_{-2}|\phi_j\rangle, \hat{L}_{-1}^2|\phi_j\rangle; \\ \hat{L}_{-1}|\phi_j\rangle; \\ |\phi_j\rangle. \end{array}$$

2.4.3 Null states and the Kac formula

One of the most important issues in CFT is whether, for a given c and Δ_j, this representation is unitary, and whether it is reducible (more generally, decomposable). It turns out that these

[11] In the literature this is often called, confusingly, a highest-weight state.

two are linked, as we shall see later. Decomposability implies the existence of null states in the Verma module, that is, some linear combination of states at a given level is itself a lowest state. The simplest example occurs at level 2, if

$$\hat{L}_n \left(\hat{L}_{-2}|\phi_j\rangle - (1/g)\hat{L}_{-1}^2|\phi_j\rangle\right) = 0\,,$$

for $n > 0$ (a notation using g is chosen to correspond to the Coulomb gas later on.) By taking $n = 1$ and $n = 2$ and using the Virasoro algebra and the fact that $|\phi_j\rangle$ is a lowest-weight state, we get

$$\Delta_j = \frac{3g-2}{4}\,, \qquad c = \frac{(3g-2)(3-2g)}{g}\,.$$

This is the special case $(r, s) = (2, 1)$ of the *Kac formula*: with c parameterized as above, if[12]

$$\Delta_j = \Delta_{r,s}(g) \equiv \frac{(rg-s)^2 - (g-1)^2}{4g}\,, \tag{2.18}$$

then $|\phi_j\rangle$ has a null state at level $r \cdot s$. We will not prove this here, but later will indicate how it is derived from Coulomb gas methods.

Removing all the null states from a Verma module gives an irreducible representation of the Virasoro algebra. Null states (and all their descendants) can consistently be set to zero in a given CFT. (This is no guarantee that they are in fact absent, however.)

One important consequence of the null state is that the correlation functions of $\phi_j(z, \bar{z})$ satisfy linear differential equations in z (or $\bar{z}$) of order rs. The case $rs = 2$ will be discussed as an example in the last lecture. This allows us in principle to calculate all of the four-point functions and hence the OPE coefficients.

2.4.4 Fusion rules

Let us consider the three-point function

$$\langle \phi_{2,1}(z_1)\phi_{r,s}(z_2)\phi_\Delta(z_3)\rangle\,,$$

where the first two fields sit at the indicated places in the Kac table, but the third is a general primary scaling field of dimension Δ. The form of this is given by eqn (2.9). If we insert $\oint_C (z - z_1)^{-1} T(z)\, dz$ into this correlation function, where C surrounds z_1 but not the other two points, this projects out $L_{-2}\phi_{2,1} \propto \partial_{z_1}^2 \phi_{2,1}$. On the other hand, the full expression is given by the Ward identity (2.14). After some algebra, we find that this is consistent only if

$$\Delta = \Delta_{r\pm1,s}\,,$$

otherwise the three-point function, and hence the OPE coefficient of ϕ_Δ in $\phi_{2,1} \cdot \phi_{r,s}$, vanishes.

This is an example of the *fusion rules* in action. It shows that Kac operators compose very much like irreducible representations of SU(2), with the r label playing the role of a spin $\frac{1}{2}(r-1)$. The s labels compose in the same way. More generally, the *fusion rule coefficients*

[12] Note that g and $1/g$ give the same value of c, and that $\Delta_{r,s}(g) = \Delta_{s,r}(1/g)$. This has led to endless confusion in the literature.

N_{ij}^k tell us not only which OPEs can vanish, but which ones actually do appear in a particular CFT.[13] In this simplest case we have (suppressing the s indices for clarity)

$$N_{rr'}^{r''} = \delta_{r'',r+r'-1} + \delta_{r'',r+r'-3} + \cdots + \delta_{r'',|r-r'|+1} \, .$$

A very important thing happens if g is rational $= p/p'$. Then we can write the Kac formula as

$$\Delta_{r,s} = \frac{(rp - sp')^2 - (p-p')^2}{4pp'} \, ,$$

and we see that $\Delta_{r,s} = \Delta_{p'-r,p-s}$, that is, the same primary field sits at two different places in the rectangle $1 \le r \le p'-1, 1 \le s \le p-1$, called the Kac table. If we now apply the fusion rules to these fields we see that we get consistency between the different constraints only if the fusion algebra is *truncated*, and that fields within the rectangle do not couple to those outside.

This shows that, at least at the level of fusion, we can have CFTs with a *finite number* of primary fields. These are called the *minimal models*.[14] However, it can be shown that, among these, the only ones admitting *unitary* representations of the Virasoro algebra, that is, for which $\langle\psi|\psi\rangle \ge 0$ for all states $|\psi\rangle$ in the representation, are those with $|p-p'| = 1$ and $p, p' \ge 3$. Moreover, these are the *only* unitary CFTs with $c < 1$. The physical significance of unitarity will be mentioned shortly.

2.5 CFT on the cylinder and torus

2.5.1 CFT on the cylinder

One of the most important conformal mappings is the logarithmic transformation $w = (L/2\pi)\log z$, which maps the z-plane (punctured at the origin) to an infinitely long cylinder of circumference L (or, equivalently, a strip with periodic boundary conditions.) It is useful to write $w = t + iu$, and to think of the coordinate t running along the cylinder as imaginary time, and u as space. CFT on the cylinder then corresponds to Euclidean QFT on a circle.

The relation between the stress tensors on the cylinder and in the plane is given by eqn (2.16):

$$T(w)_{\rm cyl} = \left(\frac{dz}{dw}\right)^2 T(z) + \frac{c}{12}\{z, w\} = \left(\frac{2\pi}{L}\right)^2 \left(z^2 T(z)_{\rm plane} - \frac{c}{24}\right) ,$$

where the last term comes from the Schwarzian derivative.

The Hamiltonian $\hat{H}$ on the cylinder, which generates infinitesimal translations in t, can be written in the usual way as an integral over the time–time component of the stress tensor

$$\hat{H} = \frac{1}{2\pi}\int_0^L \hat{T}_{tt}(u)\, du = \frac{1}{2\pi}\int_0^L \left(\hat{T}(u) + \hat{\overline{T}}(u)\right) du \, ,$$

which corresponds in the plane to

[13] They can actually take values ≥ 2 if there are distinct primary fields with the same dimension.

[14] The minimal models are examples of *rational* CFTs: those which have only a finite number of fields which are primary with respect to some algebra, more generally an extended one containing Virasoro algebra as a subalgebra.

$$\hat{H} = \frac{2\pi}{L}\left(\hat{L}_0 + \hat{\overline{L}}_0\right) - \frac{\pi c}{6L}\,. \tag{2.19}$$

Similarly, the total momentum $\hat{P}$, which generates infinitesimal translations in u, is the integral of the T_{tu} component of the stress tensor, which can be written as $(2\pi/L)(\hat{L}_0 - \hat{\overline{L}}_0)$.

Equation (2.19), although elementary, is one of the most important results of CFT in two dimensions. It relates the dimensions of all the scaling fields in the theory (which, recall, are the eigenvalues of $\hat{L}_0$ and $\hat{\overline{L}}_0$) to the spectra of $\hat{H}$ and $\hat{P}$ on the cylinder. If we have a lattice model on the cylinder whose scaling limit is described by a given CFT, we can therefore read off the scaling dimensions, up to finite-size corrections in (a/L), by diagonalizing the transfer matrix $\hat{t} \approx 1 - a\hat{H}$. This can be done either numerically for small values of L or, for integrable models, by solving the Bethe ansatz equations.

In particular, we see that the lowest eigenvalue of $\hat{H}$ (corresponding to the largest eigenvalue of the transfer matrix) is

$$E_0 = -\frac{\pi c}{6L} + \frac{2\pi}{L}(\Delta_0 + \overline{\Delta}_0)\,,$$

where $(\Delta_0, \overline{\Delta}_0)$ are the lowest possible scaling dimensions. In many CFTs, and all unitary ones, this corresponds to the identity field, so that $\Delta_0 = \overline{\Delta}_0 = 0$. This shows that c can be measured from the finite-size behavior of the ground state energy.

E_0 also gives the leading term in the partition function $Z = \mathrm{Tr}\, e^{-\ell\hat{H}}$ on a finite cylinder (a torus) of length $\ell \gg L$. Equivalently, the free energy (in units of $k_B T$) is

$$F = -\log Z \sim -\frac{\pi c \ell}{6L}\,.$$

In this equation, F represents the scaling limit of the free energy of a 2D *classical* lattice model.[15] However, we can equally well think of t as being space and u imaginary time, in which case periodic boundary conditions imply a finite inverse *temperature* $\beta = 1/k_B T = L$ in a 1D *quantum* field theory. For such a theory we then predict that

$$F \sim -\frac{\pi c \ell k_B T}{6}\,,$$

or, equivalently, that the low-temperature specific heat, at a quantum critical point described by a CFT (generally, with a linear dispersion relation $\omega \sim |q|$), has the form

$$C_v \sim \frac{\pi c k_B^2 T}{3}\,.$$

Note that the Virasoro generators can be written in terms of the stress tensor on the cylinder as

$$\hat{L}_n = \frac{L}{2\pi}\int_0^L e^{inu}\hat{T}(u,0)\,du\,.$$

[15]This treatment overlooks UV-divergent terms in F of order $(\ell L/a^2)$, which are implicitly set to zero by the regularization of the stress tensor.

In a unitary theory, $\hat{T}$ is self-adjoint, and hence $\hat{L}_n^\dagger = \hat{L}_{-n}$. Unitarity of the QFT corresponds to *reflection positivity* of correlation functions: in general,

$$\langle \phi_1(u_1, t_1)\phi_2(u_2, t_2) \ldots \phi_1(u_1, -t_1)\phi_2(u_2, -t_2)\rangle$$

is positive if the transfer matrix $\hat{T}$ can be made self-adjoint, which is generally true if the Boltzmann weights are positive. Note, however, that any given lattice model (e.g. the Ising model) contains fields which are the scaling limit of lattice quantities in which the transfer matrix can be locally expressed (e.g. the local magnetization and energy density) and for which one would expect reflection positivity to hold, and other scaling fields (e.g. the probability that a given edge lies on a cluster boundary) which are not locally expressible. Within such a CFT, then, one would expect to find a unitary *sector*—in fact, in the Ising model, this corresponds to the $p = 3, p' = 4$ minimal model—but also possible nonunitary sectors in addition.

2.5.2 Modular invariance on the torus

We have seen that unitarity (for $c < 1$) and, more generally, rationality, fix which scaling fields may appear in a given CFT, but they don't fix which ones actually appear. This is answered by considering another physical requirement: that of modular invariance on the torus.

We can make a general torus by imposing periodic boundary conditions on a parallelogram whose vertices lie in the complex plane. Scale invariance allows us to fix the length of one of the sides to be unity: thus we can choose the vertices to be at $(0, 1, 1 + \tau, \tau)$, where τ is a complex number with positive imaginary part. In terms of the conventions of the previous section, we start with a finite cylinder of circumference $L = 1$ and length $\mathrm{Im}\,\tau$, twist one end by an amount $\mathrm{Re}\,\tau$, and sew the ends together (see Fig. 2.6). An important feature of this parameterization of the torus is that it is not unique: the transformations $T : \tau \to \tau + 1$ and $S : \tau \to -1/\tau$ give the same torus (Fig. 2.7). Note that S interchanges space u and imaginary time t in the QFT. S and T generate an infinite discrete group of transformations

$$\tau \to \frac{a\tau + b}{c\tau + d}$$

with (a, b, c, d) all integers and $ad - bc = 1$. This is called $\mathrm{SL}(2, \mathbf{Z})$, or the *modular group*. Note that $S^2 = 1$ and $(ST)^3 = 1$.

Consider the scaling limit of the partition function Z of a lattice model on this torus. Apart from the divergent term in $\log Z$, proportional to the area divided by a^2, which in CFT is set to

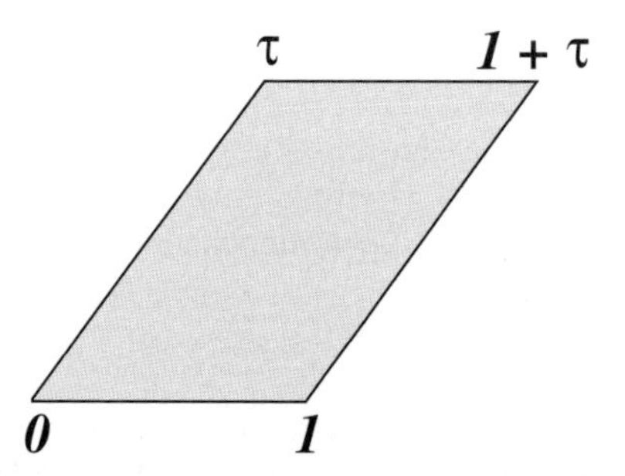

Fig. 2.6 A general torus is obtained by identifying opposite sides of a parallelogram.

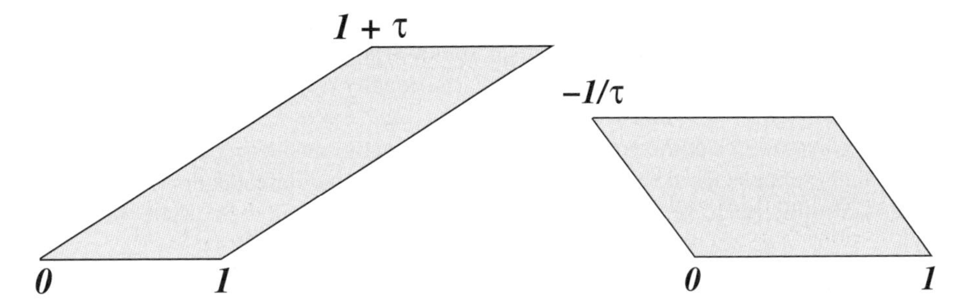

Fig. 2.7 Two ways of viewing the same torus, corresponding to the modular transformations T and S.

zero by regularization, the rest should depend only on the aspect ratio of the torus and thus be *modular invariant*. This would be an empty statement were it not that Z can be expressed in terms of the spectrum of scaling dimensions of the CFT in a manner which is not, manifestly, modular invariant.

Recall that the generators of infinitesimal translations along and around the cylinder can be written as

$$\hat{H} = 2\pi(\hat{L}_0 + \hat{\overline{L}}_0) - \frac{\pi c}{6}, \qquad \hat{P} = 2\pi(\hat{L}_0 - \hat{\overline{L}}_0).$$

The action of twisting the cylinder corresponds to a finite translation around its circumference, and sewing the ends together corresponds to taking the trace. Thus

$$\begin{aligned} Z &= \mathrm{Tr}\, e^{-(\mathrm{Im}\,\tau)\hat{H} + i(\mathrm{Re}\tau)\hat{P}} \\ &= e^{\pi c \mathrm{Im}\,\tau/6}\, \mathrm{Tr}\, e^{2\pi i \tau \hat{L}_0}\, e^{-2\pi i \hat{\overline{L}}_0} \\ &= (q\bar{q})^{-c/24}\, \mathrm{Tr}\, q^{\hat{L}_0}\, \bar{q}^{\hat{\overline{L}}_0}, \end{aligned}$$

where, in the last line, we have defined $q \equiv e^{2\pi i \tau}$.

The trace means that we sum over all eigenvalues of $\hat{L}_0$ and $\hat{\overline{L}}_0$, that is, all scaling fields of the CFT. We know that these can be organized into irreducible representations of the Virasoro algebra, and therefore have the form $(\Delta + N, \overline{\Delta} + \bar{N})$, where Δ and $\overline{\Delta}$ correspond to primary fields and $(N, \bar{N})$ are non-negative integers labeling the levels of the descendants. Thus we can write

$$Z = \sum_{\Delta, \overline{\Delta}} n_{\Delta, \overline{\Delta}} \chi_{\Delta}(q) \chi_{\overline{\Delta}}(\bar{q}),$$

where $n_{\Delta,\overline{\Delta}}$ is the number of primary fields with lowest weights $(\Delta, \overline{\Delta})$, and

$$\chi_{\Delta}(q) = q^{-c/24+\Delta} \sum_{N=0}^{\infty} d_{\Delta}(N) q^N,$$

where $d_{\Delta}(N)$ is the degeneracy of the representation at level N. This is purely a property of the representation, not the particular CFT, and therefore so is $\chi_{\Delta}(q)$. This is called the *character* of the representation.

The requirement of modular invariance of Z under T is rather trivial: it says that all fields must have integer conformal spin.[16] However, the invariance under S is highly nontrivial: it states that Z, which is a power series in q and $\bar{q}$, can equally well be expressed as an identical power series in $\tilde{q} \equiv e^{-2\pi i/\tau}$ and $\bar{\tilde{q}}$.

We can get some idea of the power of this requirement by considering the limit $q \to 1, \tilde{q} \to 0$, with q real. Suppose the density of scaling fields (including descendants) with dimension $x = \Delta + \overline{\Delta}$ in the range $(x, x + \delta x)$ (where $1 \gg \delta x \gg x$) is $\rho(x)\,\delta x$. Then, in this limit, when $q = 1 - \epsilon$, $\epsilon \ll 1$,

$$Z \sim \int^{\infty} \rho(x) e^{x \log q}\, dx \sim \int^{\infty} \rho(x) e^{-\epsilon x}\, dx\,.$$

On the other hand, we know that as $\tilde{q} \to 0$, $Z \sim \tilde{q}^{-c/12+x_0} \sim e^{(2\pi)^2(c/12-x_0)/\epsilon}$, where $x_0 \leq 0$ is the lowest scaling dimension (usually 0). Taking the inverse Laplace transform,

$$\rho(x) \sim \int e^{\epsilon x + (2\pi)^2(c/12-x_0)/\epsilon} \frac{d\epsilon}{2\pi i}\,.$$

Using the method of steepest descents, we then see that, as $x \to \infty$,

$$\rho(x) \sim \exp\left(4\pi \left(\frac{c}{12} - x_0\right)^{1/2} x^{1/2}\right),$$

times a (calculable) prefactor. This relation is of importance in understanding black hole entropy in string theory.

Modular invariance for the minimal models. Let us apply this to the minimal models, where there is a finite number of primary fields, labeled by (r, s). We need the characters $\chi_{r,s}(q)$. If there were no null states, the degeneracy at level N would be the number of states of the form $\ldots \hat{L}_{-3}^{n_3} \hat{L}_{-2}^{n_2} \hat{L}_{-1}^{n_1} |\phi\rangle$ with $\sum_j j n_j = N$. This is just the number of distinct partitions of N into positive integers, and the generating function is $\prod_{k=1}^{\infty}(1 - q^k)^{-1}$.

However, we know that the representation has a null state at level rs, and this, and all its descendants, should be subtracted off. Thus

$$\chi_{rs}(q) = q^{-c/24} \prod_{k=1}^{\infty} (1 - q^k)^{-1} (1 - q^{rs} + \cdots)\,.$$

But, as can be seen from the Kac formula (2.18), $\Delta_{r,s} + rs = \Delta_{p'+r,p-s}$, and therefore the null state at level rs has itself null states in its Verma module, which should not have been subtracted off. Thus we must add these back in. However, it is slightly more complicated than this, because for a minimal model each primary field sits in two places in the Kac rectangle, $\Delta_{r,s} = \Delta_{p'-r,p-s}$. Therefore this primary field also has a null state at level $(p' - r)(p - s)$, and this has the dimension $\Delta_{2p'-r,s} = \Delta_{r,2p-s}$ and should therefore also be added back in

[16] It is interesting to impose other kinds of boundary conditions, e.g. antiperiodic, on the torus, when other values of the spin can occur.

if it has not been included already. A full analysis requires understanding how the various submodules sit inside each other, but fortunately the final result has a nice form

$$\chi_{rs}(q) = q^{-c/24} \prod_{k=1}^{\infty} (1 - q^k)^{-1} \left(K_{r,s}(q) - K_{r,-s}(q) \right) , \tag{2.20}$$

where

$$K_{r,s}(q) = \sum_{n=-\infty}^{\infty} q^{(2npp' + rp - sp')^2/4pp'} . \tag{2.21}$$

The partition function can then be written as the finite sum

$$Z = \sum_{r,s;\bar{r},\bar{s}} n_{r,s;\bar{r},\bar{s}} \chi_{rs}(q) \chi_{\bar{r}\bar{s}}(\bar{q}) = \sum_{r,s;\bar{r},\bar{s}} n_{r,s;\bar{r},\bar{s}} \chi_{rs}(\tilde{q}) \chi_{\bar{r}\bar{s}}(\bar{\tilde{q}}) .$$

The reason this can happen is that the characters themselves transform linearly under $S : q \to \tilde{q}$ (as can be seen after quite a bit of algebra, by applying the Poisson sum formula to eqn (2.21) and Euler's identities to the infinite product):

$$\chi_{rs}(\tilde{q}) = \sum_{r',s'} S_{rs}^{r's'} \chi_{r's'}(q) ,$$

where $\mathbf{S}$ is a matrix whose rows and columns are labeled by (rs) and $(r's')$. Another way to state this is that the characters form a representation of the modular group. The form of $\mathbf{S}$ is not that important, but we give it anyway:

$$S_{rs}^{r's'} = \left(\frac{8}{pp'} \right)^{1/2} (-1)^{1+r\bar{s}+s\bar{r}} \sin \frac{\pi p' r \bar{r}}{p'} \sin \frac{\pi p' s \bar{s}}{p} .$$

The important properties of $\mathbf{S}$ are that it is real and symmetric and $\mathbf{S}^2 = 1$.

This immediately implies that the *diagonal* combination, with $n_{r,s;\bar{r},\bar{s}} = \delta_{r\bar{r}}\delta_{s\bar{s}}$, is modular invariant:

$$\sum_{r,s} \chi_{rs}(\tilde{q}) \chi_{rs}(\bar{\tilde{q}}) = \sum_{r,s} \sum_{r',s'} \sum_{r'',s''} S_{rs}^{r's'} S_{rs}^{r''s''} \chi_{r's'}(q) \chi_{r''s''}(\bar{q}) = \sum_{r,s} \chi_{rs}(q) \chi_{rs}(\bar{q}) ,$$

where we have used $\mathbf{S}\mathbf{S}^T = 1$. This gives the diagonal series of CFTs, in which all possible scalar primary fields in the Kac rectangle occur just once. These are known as the A_n series.

It is possible to find other modular invariants by exploiting symmetries of $\mathbf{S}$. For example, if $p'/2$ is odd, the space spanned by $\chi_{rs} + \chi_{p'-r,s}$, with r odd, is an invariant subspace of $\mathbf{S}$, and is multiplied only by a pure phase under T. Hence the diagonal combination within this subspace

$$Z = \sum_{r \text{ odd},s} |\chi_{r,s} + \chi_{p'-r,s}|^2$$

is modular invariant. Similar invariants can be constructed if $p/2$ is odd. This gives the D_n series. Note that in this case some fields appear with degeneracy 2. Apart from these two infinite series, there are three special values of p and p' (12, 18, and 30), denoted by $E_{6,7,8}$.

The reason for this "ADE" classification will become more obvious in the next section when we construct explicit lattice models which have these CFTs as their scaling limits.[17]

2.6 Height models, loop models, and Coulomb gas methods

2.6.1 Height models and loop models

Although the ADE classification of minimal CFTs with $c < 1$ through modular invariance was a great step forwards, one can ask whether there are in fact lattice models which have these CFTs as their scaling limit. The answer is yes—in the form of the ADE lattice models. These can be analyzed nonrigorously by so-called Coulomb gas methods.

For simplicity, we shall describe only the so-called dilute models, defined on a triangular lattice.[18] At each site r of the lattice is defined a "height" $h(r)$, which takes values on the nodes of some connected graph $\mathcal{G}$. An example is the linear graph called A_m, shown in Fig. 2.8, in which $h(r)$ can be thought of as an integer between 1 and m. There is a restriction in these models that the heights on neighboring sites of the triangular lattice must either be the same or be adjacent on $\mathcal{G}$. It is then easy to see that around a given triangle either all three heights are the same (which carries relative weight 1), or two of them are the same and the other is adjacent on $\mathcal{G}$.[19] In this case, if the heights are (h, h', h'), the weight is $x(S_h/S_{h'})^{1/6}$, where S_h is a function of the height h, to be made explicit later, and x is a positive temperature-like parameter.[20] (A simple example is A_2, corresponding to the Ising model on the triangular lattice.)

The weight for a given configuration of the whole lattice is the product of the weights for each elementary triangle. Note that this model is local and has positive weights if S_h is a positive function of h. Its scaling limit at the critical point should correspond to a unitary CFT.

The height model can be mapped to a loop model as follows: every time the heights in a given triangle are not all equal, we draw a segment of a curve through it, as shown in Fig. 2.9. These segments all link up, and if we demand that all the heights on the boundary are the same, they form a set of nested, nonintersecting closed loops on the dual honeycomb lattice, separating regions of constant height on the original lattice. Consider a loop for which the heights just inside and outside are h and h', respectively. This loop has convex (outward-pointing) and concave (inward-pointing) corners. Each convex corner carries a factor $(S_h/S_{h'})^{1/6}$, and each concave corner the inverse factor. But each loop has exactly six more outward-pointing corners than inward-pointing ones, so it always carries an overall weight S_h/S'_h, times a factor x raised to the power of the length of the loop. Let us now sum over the heights consistent with a fixed loop configuration, starting with the innermost regions. Each sum has the form

Fig. 2.8 The graph A_m, with m nodes.

[17] This classification also arises in the finite subgroups of SU(2) and of simply laced Lie algebras, and in catastrophe theory.

[18] Similar models can be defined on a square lattice. They give rise to critical loop models in the dense phase.

[19] Apart from the pathological case when $\mathcal{G}$ itself has a 3-cycle, in which case we can enforce the restriction by hand.

[20] It is sometimes useful, e.g. for implementing a transfer matrix, to redistribute these weights around the loops.

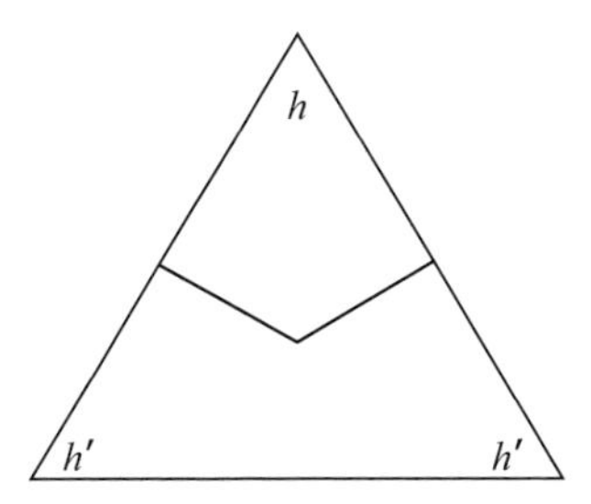

Fig. 2.9 If the heights on the vertices are not all equal, we denote this by a segment of a curve on the dual lattice, as shown.

$$\sum_{h:|h-h'|=1} (S_h/S_{h'}) ,$$

where $|h - h'| = 1$ means that h and h' are adjacent on $\mathcal{G}$. The next stage in the summation will be simple only if this is a constant independent of h'. Thus we assume that the S_h satisfy

$$\sum_{h:|h-h'|=1} S_h = \Lambda S_{h'} ,$$

that is, S_h is an eigenvector of the *adjacency matrix* of $\mathcal{G}$, with eigenvalue Λ. For A_m, for example, these have the form

$$S_h \propto \sin \frac{\pi k h}{m+1} , \tag{2.22}$$

where $1 \leq k \leq m$, corresponding to $\Lambda = 2\cos\big(\pi k/(m+1)\big)$. Note that only the case $k = 1$ gives weights that are all real and positive.

Having chosen the S_h in this way, we can sum out all the heights consistent with a given loop configuration (Fig. 2.10), starting with the innermost and moving outwards, and thereby express the partition function as a sum over loop configurations:

$$Z = \sum_{\text{loop configs}} \Lambda^{\text{number of loops}}\, x^{\text{total length}} . \tag{2.23}$$

When x is small, the heights are nearly all equal (depending on the boundary condition), and the typical loop length and number are small. At a critical point $x = x_c$ we expect these to diverge. Beyond this, we enter the *dense phase*, which is still critical in the loop sense, even though observables which are local in the original height variables may have a finite correlation length. For example, for $x > x_c$ in the Ising model, the Ising spins are disordered but the cluster boundaries are the same, in the scaling limit, as those of critical percolation for site percolation on the triangular lattice.

However, we could have obtained the same expression for Z in several different ways. One is by introducing n-component spins $s_a(R)$ with $a = 1, \dots, n$ on the sites of the dual lattice, and the partition function

$$Z_{\mathrm{O}(n)} = \mathrm{Tr} \prod_{RR'} \left(1 + x \sum_{a=1}^{n} s_a(R) s_a(R') \right) ,$$

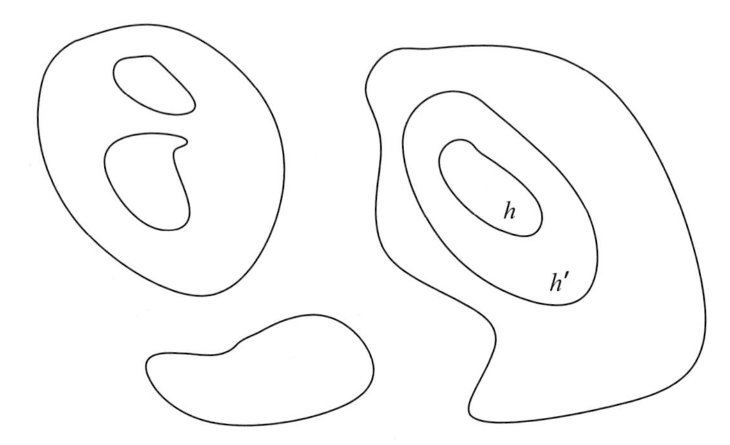

Fig. 2.10 Nested set of loops, separating regions of constant height. We iteratively sum over the heights h, starting with the innermost.

where the product is over edges of the honeycomb lattice, the trace of an odd power of $s_a(R)$ is zero, and $\text{Tr}s_a(R)s_b(R) = \delta_{ab}$. Expanding in powers of x, and drawing in a curve segment each time the term proportional to x is chosen on a given edge, we get the same set of nested loop configurations, weighted as above with $\Lambda = n$. This is the $\mathrm{O}(n)$ model. Note that the final expression makes sense for all real, positive values of n, but it can be expressed in terms of weights local in the original spins only for positive integer n. Only in the latter case do we therefore expect that correlations of the $\mathrm{O}(n)$ spins will satisfy reflection positivity and therefore correspond to a unitary CFT, even though the description in terms of heights is unitary. This shows how different sectors of the "same" CFT can describe rather different physics.

2.6.2 Coulomb gas methods

These loop models can be solved in various ways, for example by realizing that their transfer matrix gives a representation of the Temperley–Lieb algebra, but a more powerful if less rigorous method is to use the so-called Coulomb gas approach. Recalling the arguments of the previous section, we see that yet another way of getting to eqn (2.23) is by starting from a different height model, where now the heights $h(r)$ are defined on the integers (times π, for historical reasons) $\pi\mathbf{Z}$. As long as we choose the correct eigenvalue of the adjacency matrix, this will give the same loop gas weights. That is, we take $\mathcal{G}$ to be A_∞. In this case the eigenvectors correspond to plane wave modes propagating along the graph, labeled by a quasi-momentum χ with $|\chi| < 1$: $S_h \propto e^{i\chi h}$, and the eigenvalue is $\Lambda = 2\cos(\pi\chi)$. Because these modes are chiral, we have to orient the loops to distinguish between χ and $-\chi$. Each oriented loop then gets weighted with a factor $e^{\pm i\pi\chi/6}$ at each vertex of the honeycomb lattice it goes through, depending on whether it turns to the left or the right.

This version of the model, where the heights are unbounded, is much easier to analyze, at least nonrigorously. In particular, we might expect that in the scaling limit, after coarse-graining, we can treat $h(r)$ as taking all real values, and write down an effective field theory. This should have the property that it is local, invariant under $h(r) \to h(r)+$ constant, and with no terms irrelevant under the renormalization group (RG) (that is, entering the effective action with positive powers of a.) The only possibility is a free Gaussian field theory, with action

$$S = \frac{g_0}{4\pi} \int (\nabla h)^2 \, d^2r \, .$$

However, this cannot be the full answer, because we know this corresponds to a CFT with $c = 1$. The resolution of this is most easily understood by considering the theory on a long cylinder of length ℓ and circumference $L \ll \ell$. Noncontractible loops which go around the cylinder have the same number of inside and outside corners, so they are incorrectly counted. This can be corrected by inserting a factor $\prod_t e^{i\chi h(t,0)} e^{-i\chi h(t+1,0)}$, which counts each loop passing between $(t, 0)$ and $(t + 1, 0)$ with just the right factors $e^{\pm i\pi\chi}$ (Fig. 2.11). These factors accumulate to $e^{i\chi h(-\ell/2,0)} e^{-i\chi h(\ell/2,0)}$, corresponding to charges $\pm\chi$ at the ends of the cylinder. This means that the partition function is

$$Z \sim Z_{c=1} \langle e^{i\chi h(\ell/2)} e^{-i\chi h(-\ell/2)} \rangle \, .$$

But we know (Section 2.3.1) that this correlation function decays like r^{-2x_q} in the plane, where $x_q = q^2/2g_0$, and therefore on the cylinder

$$Z \sim e^{\pi\ell/6L} \, e^{-2\pi(\chi^2/2g_0)\ell/L} \, ,$$

from which we see that the central charge is actually

$$c = 1 - \frac{6\chi^2}{g_0} \, .$$

However, we haven't yet determined g_0. This is fixed by the requirement that the *screening fields* $e^{\pm i2h(r)}$, which come from the fact that originally $h(r) \in \pi\mathbf{Z}$, should be marginal, that is, they do not affect the scaling behavior, so that we can add them to the action with impunity. This requires that they have scaling dimension $x_2 = 2$. However, now x_q should be calculated from the cylinder with the charges $e^{\pm i\chi h(\pm\ell/2)}$ at the ends:

$$x_q = \frac{(q \pm \chi)^2}{g_0} - \frac{\chi^2}{g_0} = \frac{q^2 \pm 2\chi q}{g_0} \, . \tag{2.24}$$

Setting $x_2 = 2$, we then find $g_0 = 1 \pm \chi$ and therefore

$$c = 1 - \frac{6(g_0 - 1)^2}{g_0} \, .$$

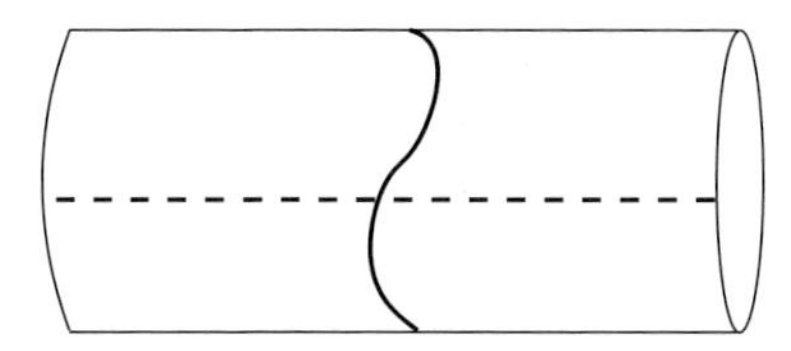

Fig. 2.11 Noncontractible loops on the cylinder can be taken into account by the insertion of a suitable factor along a seam.

2.6.3 Identification with minimal models

The partition function for the height models (at least on a cylinder) depends only on the eigenvalue Λ of the adjacency matrix, and hence the Coulomb gas should work equally well for the models on A_m if we set $\chi = k/(m+1)$. The corresponding central charge is then

$$c = 1 - \frac{6k^2}{(m+1)(m+1\pm k)}\,.$$

If we compare this with the formula for the minimal models

$$c = 1 - \frac{6(p-p')^2}{pp'}\,,$$

we are tempted to identify $k = p - p'$ and $m+1 = p'$. This implies $g_0 = p/p'$, which can therefore be identified with the parameter g introduced in the Kac formula.[21] Moreover, if we compute the scaling dimensions of local fields $\phi_r(R) = \cos\big((r-1)kh(R)/(m+1)\big)$ using eqn (2.24), we find perfect agreement with the leading diagonal $\Delta_{r,r}$ of the Kac table.[22] We therefore have strong circumstantial evidence that the scaling limit of the dilute $A_{p'-1}$ models (choosing the eigenvalue $\Lambda = 2\cos(\pi(p-p')/p'))$ is the (p,p') minimal model with $p > p'$. Note that only if $k = 1$, that is, $p = p'+1$, are these CFTs unitary, and this is precisely the case where the weights of the lattice model are real and positive.

For other graphs $\mathcal{G}$, we can try to make a similar identification. However, this is going to work only if the maximal eigenvalue of the adjacency matrix of $\mathcal{G}$ is strictly less than 2. A famous classification then shows that this restricts $\mathcal{G}$ to be either of the form A_m or D_m, or one of three exceptional cases $E_{6,7,8}$.[23] These other graphs also have eigenvalues of the same form (2.22) as A_m, but with $m+1$ now being the Coxeter number, and the allowed integers k being only a subset of those appearing in the A-series. These correspond to the Kac labels of the allowed scalar operators which appear in the appropriate modular-invariant partition function.

2.7 Boundary conformal field theory

2.7.1 Conformal boundary conditions and Ward identities

So far, we haven't considered what happens at the boundary of the domain $\mathcal{D}$. This is a subject with several important applications, for example to quantum impurity problems (see the lectures by Affleck) and to D-branes in string theory.

In any field theory in a domain with a boundary, one needs to consider how to impose a set of consistent boundary conditions. Since CFT is formulated independently of any particular set of fundamental fields and Lagrangian, this must be done in a more general manner. A natural requirement is that the off-diagonal component $T_{\parallel\perp}$ of the stress tensor parallel/perpendicular to the boundary should vanish. This is called the conformal boundary condition. If

[21] The other solution corresponds to interchanging p' and p.

[22] These are the relevant fields in the RG sense.

[23] $\Lambda = 2$ corresponds to the extended diagrams $\hat{A}_m$, etc., which give interesting rational CFTs with $c = 1$. However, models based on graphs with $\Lambda > 2$ probably have a different kind of transition, at which the mean loop length remains finite.

the boundary is parallel to the time axis, this implies that there is no momentum flow across the boundary. Moreover, it can be argued that, under the RG, any uniform boundary condition will flow into a conformally invariant one. For a given bulk CFT, however, there may be many possible distinct such boundary conditions, and it is one task of boundary CFT (BCFT) to classify these.

To begin with, we take the domain to be the upper half-plane, so that the boundary is the real axis. The conformal boundary condition then implies that $T(z) = \overline{T}(\bar{z})$ when z is on the real axis. This has the immediate consequence that the correlators of $\overline{T}$ are those of T analytically continued into the lower half-plane. The conformal Ward identity now reads

$$\left\langle T(z) \prod_j \phi_j(z_j, \bar{z}_j) \right\rangle = \sum_j \left(\frac{\Delta_j}{(z - z_j)^2} + \frac{1}{z - z_j} \partial_{z_j} + \frac{\overline{\Delta}_j}{(z - \bar{z}_j)^2} + \frac{1}{z - \bar{z}_j} \partial_{\bar{z}_j} \right) \left\langle \prod_j \phi_j(z_j, \bar{z}_j) \right\rangle . \quad (2.25)$$

In radial quantization, in order that the Hilbert spaces defined on different hypersurfaces should be equivalent, one must now choose semicircles centered on some point on the boundary, conventionally the origin. The dilatation operator is now

$$\hat{D} = \frac{1}{2\pi i} \int_S z \hat{T}(z) \, dz - \frac{1}{2\pi i} \int_S \bar{z} \hat{\overline{T}}(\bar{z}) \, d\bar{z} \,, \quad (2.26)$$

where S is a semicircle. Using the conformal boundary condition, this can also be written as

$$\hat{D} = \hat{L}_0 = \frac{1}{2\pi i} \int_C z \hat{T}(z) \, dz \,, \quad (2.27)$$

where C is a complete circle around the origin.

Note that there is now only one Virasoro algebra. This is related to the fact that conformal mappings which preserve the real axis correspond to real analytic functions. The eigenstates of $\hat{L}_0$ correspond to *boundary operators* $\hat{\phi}_j(0)$ acting on the vacuum state $|0\rangle$. It is well known that in a renormalizable QFT, fields at the boundary require a different renormalization from those in the bulk, and this will in general lead to a different set of conformal weights. It is one of the tasks of BCFT to determine these, for a given allowed boundary condition.

However, there is one feature unique to boundary CFT in two dimensions. Radial quantization also makes sense, leading to the same form (2.27) for the dilation operator, if the boundary conditions on the negative and positive real axes are *different*. As far as the structure of BCFT goes, correlation functions with this mixed boundary condition behave as though a local scaling field were inserted at the origin. This has led to the term "boundary-condition-changing" (bcc) operator.

2.7.2 CFT on an annulus and classification of boundary states

Just as consideration of the partition function on a torus illuminates the bulk operator content $n_{\Delta,\overline{\Delta}}$, it turns out that consistency on an annulus helps classify both the allowed boundary conditions and the boundary operator content. To this end, consider a CFT in an annulus formed from a rectangle of unit width and height δ, with the top and bottom edges identified

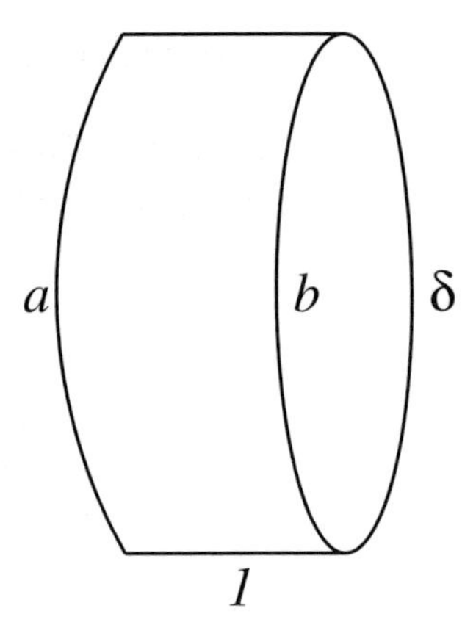

Fig. 2.12 The annulus, with boundary conditions a and b on each boundary.

(see Fig. 2.12). The boundary conditions on the left and right edges, labeled by $a, b, \ldots$, may be different. The partition function with boundary conditions a and b on each edge is denoted by $Z_{ab}(\delta)$.

One way to compute this is by first considering the CFT on an infinitely long strip of unit width. This is conformally related to the upper half-plane (with an insertion of boundary-condition-changing operators at 0 and ∞ if $a \neq b$) by the mapping $z \to (1/\pi)\ln z$. The generator of infinitesimal translations along the strip is

$$\hat{H}_{ab} = \pi\hat{D} - \frac{\pi c}{24} = \pi\hat{L}_0 - \frac{\pi c}{24} \,. \tag{2.28}$$

Thus, for the annulus,

$$Z_{ab}(\delta) = \mathrm{Tr}\, e^{-\delta\, \hat{H}_{ab}} = \mathrm{Tr}\, q^{\hat{L}_0 - \pi c/24} \,, \tag{2.29}$$

with $q \equiv e^{-\pi\delta}$. As before, this can be decomposed into characters

$$Z_{ab}(\delta) = \sum_{\Delta} n_{ab}^{\Delta}\, \chi_{\Delta}(q) \,, \tag{2.30}$$

but note that now the expression is linear. The non-negative integers n_{ab}^{Δ} give the operator content with the boundary conditions (ab): the lowest value of Δ with $n_{ab}^{\Delta} > 0$ gives the conformal weight of the bcc operator, and the others give the conformal weights of the other allowed primary fields which may also sit at this point.

On the other hand, the annulus partition function may be viewed, up to an overall rescaling, as the path integral for a CFT on a circle of unit circumference, propagated for (imaginary) time δ^{-1}. From this point of view, the partition function is no longer a trace, but rather the matrix element of $e^{-\hat{H}/\delta}$ between *boundary states*:

$$Z_{ab}(\delta) = \langle a|e^{-\hat{H}/\delta}|b\rangle \,. \tag{2.31}$$

Note that $\hat{H}$ is the same Hamiltonian that appears on the cylinder, and the boundary states lie in the Hilbert space of states on the circle. They can be decomposed into linear combinations of states in the representation spaces of the two Virasoro algebras, labeled by their lowest weights $(\Delta, \overline{\Delta})$.

How are these boundary states to be characterized? Recalling that on the cylinder $\hat{L}_n \propto \int e^{inu}\hat{T}(u)\,du$ and $\hat{\overline{L}}_n \propto \int e^{-inu}\hat{\overline{T}}(u)\,du$, the conformal boundary condition implies that any boundary state $|B\rangle$ lies in the subspace satisfying

$$\hat{L}_n|B\rangle = \hat{\overline{L}}_{-n}|B\rangle\,. \tag{2.32}$$

This condition can be applied in each subspace. Taking $n = 0$ in eqn (2.32) constrains $\overline{\Delta} = \Delta$. It can then be shown that the solution of eqn (2.32) is unique within each subspace and has the following form. The subspace at level N has dimension $d_\Delta(N)$. Denote an orthonormal basis by $|\Delta, N; j\rangle$, with $1 \leq j \leq d_\Delta(N)$, and the same basis for the representation space of $\overline{\text{Vir}}$ by $\overline{|\Delta, N; j\rangle}$. The solution to eqn (2.32) in this subspace is then

$$|\Delta\rangle\rangle \equiv \sum_{N=0}^{\infty}\sum_{j=1}^{d_\Delta(N)} |\Delta, N; j\rangle \otimes \overline{|\Delta, N; j\rangle}\,. \tag{2.33}$$

These are called Ishibashi states. One way to understand this is to note that eqn (2.32) implies that

$$\langle B|\hat{L}_n|B\rangle = \langle B|\hat{\overline{L}}_{-n}|B\rangle = \langle B|\hat{\overline{L}}_n|B\rangle\,,$$

where we have used $\hat{\overline{L}}^{\dagger}_{-n} = \hat{\overline{L}}_n$ and assumed that the matrix elements are all real. This means that acting with the raising operators $\hat{\overline{L}}_n$ on $|B\rangle$ has exactly the same effect as the $\hat{L}_n$, so, starting with $N = 0$, we must build up exactly the same state in the two spaces.

The matrix elements of the translation operator along the cylinder between Ishibashi states are simple:

$$\langle\langle\Delta'|e^{-\hat{H}/\delta}|\Delta\rangle\rangle \tag{2.34}$$

$$= \sum_{N'=0}^{\infty}\sum_{j'=1}^{d_{\Delta'}(N')}\sum_{N=0}^{\infty}\sum_{j=1}^{d_\Delta(N)} \langle\Delta', N'; j'| \otimes \overline{\langle\Delta', N'; j'|} e^{-(2\pi/\delta)(\hat{L}_0+\hat{\overline{L}}_0-c/12)}$$

$$\times\, |\Delta, N; j\rangle \otimes \overline{|\Delta, N; j\rangle} \tag{2.35}$$

$$= \delta_{\Delta'\Delta}\sum_{N=0}^{\infty}\sum_{j=1}^{d_\Delta(N)} e^{-(4\pi/\delta)\left(\Delta+N-(c/24)\right)} = \delta_{\Delta'\Delta}\,\chi_\Delta(e^{-4\pi/\delta})\,. \tag{2.36}$$

Note that the characters which appear are related to those in eqn (2.30) by the modular transformation S.

The *physical* boundary states satisfying eqn (2.30) are linear combinations of these Ishibashi states:

$$|a\rangle = \sum_{\Delta}\langle\langle\Delta|a\rangle\,|\Delta\rangle\rangle\,. \tag{2.37}$$

Equating the two different expressions in eqns (2.30) and (2.31) for Z_{ab}, and using the modular transformation law for the characters and their linear independence, gives the (equivalent) conditions

$$n_{ab}^{\Delta} = \sum_{\Delta'} S_{\Delta'}^{\Delta} \langle a|\Delta'\rangle\rangle\langle\langle\Delta'|b\rangle\,, \tag{2.38}$$

$$\langle a|\Delta'\rangle\rangle\langle\langle\Delta'|b\rangle = \sum_{\Delta} S_{\Delta}^{\Delta'} n_{ab}^{\Delta}\,. \tag{2.39}$$

The requirements that the right-hand side of eqn (2.38) should give a non-negative integer and that the right-hand side of eqn (2.39) should factorize in a and b give highly nontrivial constraints on the allowed boundary states and their operator content.

For the diagonal CFTs considered here (and for the nondiagonal minimal models), a complete solution is possible. Since the elements S_0^{Δ} of $\mathbf{S}$ are all non-negative, one may choose $\langle\langle\Delta|\tilde{0}\rangle = \left(S_0^{\Delta}\right)^{1/2}$. This defines a boundary state

$$|\tilde{0}\rangle \equiv \sum_{\Delta} \left(S_0^{\Delta}\right)^{1/2} |\Delta\rangle\rangle\,, \tag{2.40}$$

and a corresponding boundary condition such that $n_{\tilde{0}\tilde{0}}^{\Delta} = \delta_{\Delta 0}$. Then, for each $\Delta' \neq 0$, one may define a boundary state

$$\langle\langle\Delta|\tilde{\Delta}'\rangle \equiv \frac{S_{\Delta'}^{\Delta}}{\left(S_0^{\Delta}\right)^{1/2}}\,. \tag{2.41}$$

From eqn (2.38), this gives $n_{\tilde{\Delta}'\tilde{0}}^{\Delta} = \delta_{\Delta'\Delta}$. For each allowed Δ' in the torus partition function, there is therefore a boundary state $|\tilde{\Delta}'\rangle$ satisfying eqns (2.38) and (2.39). However, there is a further requirement, that

$$n_{\tilde{\Delta}'\tilde{\Delta}''}^{\Delta} = \sum_{\ell} \frac{S_{\ell}^{\Delta} S_{\Delta'}^{\ell} S_{\Delta''}^{\ell}}{S_0^{\ell}} \tag{2.42}$$

should be a non-negative integer. Remarkably, this combination of elements of $\mathbf{S}$ occurs in the *Verlinde formula*, which follows from considering consistency of the CFT on the torus. This states that the right-hand side of eqn (2.42) is equal to the fusion rule coefficient $N_{\Delta'\Delta''}^{\Delta}$. Since these coefficients are non-negative integers, the above ansatz for the boundary states is consistent. The appearance of the fusion rules in this context can be understood by the following argument, illustrated in Fig. 2.13. Consider a very long strip. At "time" $t \to -\infty$, the boundary conditions on both sides are those corresponding to $\tilde{0}$, so that only states in the representation 0 propagate. At time t_1, we insert the bcc operator $(0|\Delta')$ on one edge: the states Δ' then propagate. This can be thought of as the fusion of 0 with Δ'. At some much later time, we insert the bcc operator $(0|\Delta'')$ on the other edge: by the same argument, this

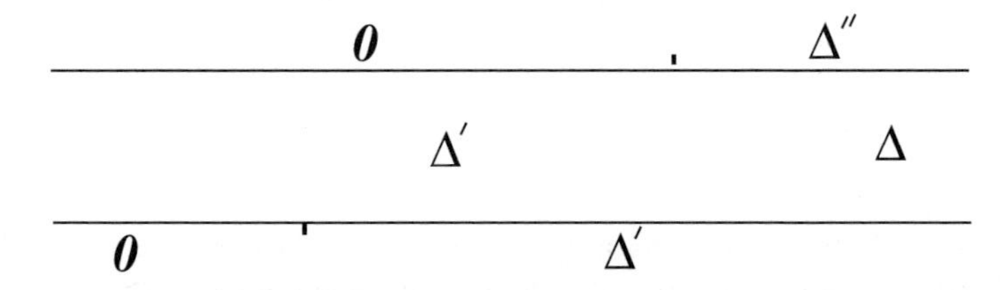

Fig. 2.13 Argument illustrating the fusion rules.

should correspond to the fusion of Δ' and Δ'', which gives all states Δ with $N^{\Delta}_{\Delta',\Delta''} = 1$. But, by definition, these are those with $n^{\Delta}_{\Delta',\Delta''} = 1$.

We conclude that, at least for the diagonal models, there is a bijection between the allowed primary fields in the bulk CFT and the allowed conformally invariant boundary conditions. For the minimal models, with a finite number of such primary fields, this correspondence has been followed through explicitly.

Example. The simplest example is the diagonal $c = \frac{1}{2}$ unitary CFT corresponding to $p = 4, p' = 3$. The allowed values of the conformal weights are $h = 0$, 1/2, and 1/16, and

$$\mathbf{S} = \begin{pmatrix} \frac{1}{2} & \frac{1}{2} & \frac{1}{\sqrt{2}} \\ \frac{1}{2} & \frac{1}{2} & -\frac{1}{\sqrt{2}} \\ \frac{1}{\sqrt{2}} & -\frac{1}{\sqrt{2}} & 0 \end{pmatrix}, \tag{2.43}$$

from which one finds the allowed boundary states

$$|\tilde{0}\rangle = \frac{1}{\sqrt{2}}|0\rangle\rangle + \frac{1}{\sqrt{2}}\left|\frac{1}{2}\right\rangle\rangle + \frac{1}{2^{1/4}}\left|\frac{1}{16}\right\rangle\rangle, \tag{2.44}$$

$$\left|\frac{\tilde{1}}{2}\right\rangle = \frac{1}{\sqrt{2}}|0\rangle\rangle + \frac{1}{\sqrt{2}}\left|\frac{1}{2}\right\rangle\rangle - \frac{1}{2^{1/4}}\left|\frac{1}{16}\right\rangle\rangle, \tag{2.45}$$

$$\left|\frac{\tilde{1}}{16}\right\rangle = |0\rangle\rangle - \left|\frac{1}{2}\right\rangle\rangle. \tag{2.46}$$

The nonzero fusion rule coefficients of this CFT are

$$N^0_{0,0} = N^{1/16}_{0,1/16} = N^{1/2}_{0,1/2} = N^0_{1/16,1/16} = N^{1/2}_{1/16,1/16} = N^0_{1/2,1/2} = N^{1/16}_{1/16,1/2} = 1\,.$$

The $c = \frac{1}{2}$ CFT is known to describe the continuum limit of the critical Ising model, in which spins $s = \pm 1$ are localized on the sites of a regular lattice. The above boundary conditions may be interpreted as the continuum limit of the lattice boundary conditions $s = 1$, $s = -1$, and free (f), respectively. Note that there is a symmetry of the fusion rules which means that one could equally well have reversed the first two. This shows, for example, that for the (ff) boundary conditions the states with lowest weights 0 (corresponding to the identity operator) and $\frac{1}{2}$ (corresponding to the magnetization operator at the boundary) can propagate. Similarly, the scaling dimension of the $(f|\pm 1)$ bcc operator is $1/16$.

2.7.3 Boundary operators and stochastic Loewner evolution

Let us now apply the above ideas to the A_m models. There should be a set of conformal boundary states corresponding to the entries of first row $(r, 1)$ of the Kac table, with $1 \le r \le m$. It is an educated guess (confirmed by exact calculations) that these in fact correspond to lattice boundary conditions where the heights on the boundary are fixed to be at a particular node r of the graph A_m. What about the boundary-condition-changing operators? These are given by the fusion rules. In particular, since (suppressing the index $s = 1$)

$$N_{r,2}^{r'} = \delta_{|r-r'|,1}\,,$$

we see that the bcc operator between r and $r \pm 1$, corresponding to a single cluster boundary intersecting the boundary of the domain, must be a $(2, 1)$ operator in the Kac table.[24] This makes complete sense: if we want to go from r_1 to r_2 we must bring together at least $|r_1 - r_2|$ cluster boundaries, showing that the leading bcc operator in this case is at $(|r_1 - r_2|, 1)$, which is consistent once again with the fusion rules. If the bcc operators corresponding to a single curve are $(2, 1)$, this means that the corresponding states satisfy

$$\left(2\hat{L}_{-2} - \frac{2}{g}\hat{L}_{-1}^2\right)|\phi_{2,1}\rangle = 0\,. \tag{2.47}$$

We are now going to argue that eqn (2.47) is equivalent to the statement that the cluster boundary starting at this boundary point is described by stochastic Loewner evolution (SLE). In order to avoid being too abstract initially, we'll first show how the calculations of a particular observable agree in the two different formalisms.

Let ζ be a point in the upper half-plane and let $P(\zeta)$ be the probability that the curve, starting at the origin, passes to the left of this point (of course, it is not holomorphic). First we'll give the physicist's version of the SLE argument (assuming familiarity with Werner's lectures). We imagine performing the exploration process for a small Loewner time δt, and then continuing the process to infinity. Under the conformal mapping $f_{\delta t}(z)$ which removes the first part of the curve, we get a new curve with the same measure as the original one, but the point ζ is mapped to $f_{\delta t}(\zeta)$. But this will lie to the right of the new curve if and only if the original point lies to the right of the original curve. Also, by integrating the Loewner equation starting from $f_0(z) = z$, we have approximately

$$f_{\delta t}(z) \approx z + \frac{2\delta t}{z} + \sqrt{\kappa}\,\delta B_t\,,$$

at least for $z \gg \delta t$. Thus we can write down an equation[25]

$$P(\zeta) = \mathbf{E}\left[P\left(\zeta + \frac{2\delta t}{\zeta} + \sqrt{\kappa}\,\delta B_t\right)\right]_{\delta B_t}\,,$$

where $\mathbf{E}[\ldots]_{\delta B_t}$ means an average over all realizations of the Brownian motion up to time δt. Expanding the right-hand side to $O(\delta t)$, and remembering that $\mathbf{E}[\delta B_t] = 0$ and $\mathbf{E}[(\delta B_t)^2] = \delta t$, we find (with $\zeta = x + iy$), the linear partial differential equation (PDE)

$$\left(\frac{2x}{x^2+y^2}\frac{\partial}{\partial x} - \frac{2y}{x^2+y^2}\frac{\partial}{\partial y} + \frac{\kappa}{2}\frac{\partial^2}{\partial x^2}\right)P(x,y) = 0\,. \tag{2.48}$$

By scale invariance, $P(x, y)$ depends in fact only on the ratio y/x, and therefore this can be reduced to a second-order ordinary differential equation, whose solution, with appropriate

[24] If, instead of the dilute lattice model, we consider the dense phase, which corresponds, for example, to the boundaries of the FK clusters in the Potts model, then r and s are interchanged for a given central charge c, and the bcc operator then lies at $(1, 2)$.

[25] Some physicists will recognize this as the reverse Fokker–Planck equation.

boundary conditions, can be expressed in terms of hypergeometric functions (and is known as Schramm's formula.)

Now let us give the CFT derivation. In terms of correlation functions, P can be expressed as

$$P = \frac{\langle \phi_{2,1}(0)\Phi(\zeta,\bar{\zeta})\phi_{2,1}(\infty)\rangle}{\langle \phi_{2,1}(0)\phi_{2,1}(\infty)\rangle} .$$

The denominator is just the partition function restricted to there being a cluster boundary from 0 to infinity. Φ is an "indicator operator" which takes the value 0 or 1 depending on whether the curve passes to the right or left, respectively, of ζ. Since P is a probability it is dimensionless, so Φ has zero conformal dimension and transforms trivially.

Now suppose we insert $\int_C (2T(z)/z)(dz/2\pi i) + \text{c.c.}$ into the correlation function in the numerator, where C is a semicircular contour surrounding the origin but not ζ. Using the OPE of T with $\phi_{2,1}$, this gives

$$2L_{-2}\phi_{2,1}(0) = \frac{2}{g}\partial_x^2\phi_{1,2}(x)|_{x=0} .$$

Using translation invariance, this derivative can be made to act equivalently on the x-coordinate of ζ. On the other hand, we can also distort the contour to wrap around ζ, where it simply shifts the argument of Φ. The result is that we get exactly the same PDE as in eqn (2.48), with the identification

$$g = \frac{4}{\kappa} .$$

Of course, this was just one example. Let us see how to proceed more generally. In radial quantization, the insertion of the bcc field $\phi_{2,1}(0)$ gives a state $|\phi_{2,1}\rangle$. Under the infinitesimal mapping f_{dt}, we get the state

$$\left(1 - (2\hat{L}_{-2}\,dt + \hat{L}_{-1}\sqrt{\kappa}\,dB_t)\right)|\phi_{2,1}\rangle ,$$

or, over a finite time, a time-ordered exponential

$$\mathbf{T}\exp\left(-\int_0^t (\hat{L}_{-2}\,dt' + \hat{L}_{-1}\sqrt{\kappa}\,dB_{t'})\right)|\phi_{2,1}\rangle . \tag{2.49}$$

The conformal-invariance property of the measure on the curve then implies that, when averaged over $dB_{t'}$, this is in fact independent of t. Expanding to $O(t)$, we then again find eqn (2.47) with $g = 4/\kappa$. Since this is a property of the state, it implies an equivalence between the two approaches for all correlation functions involving $\phi_{2,1}(0)$, not just the one considered earlier. Moreover, if we replace $\sqrt{\kappa}\,dB_t$ by some more general random driving function dW_t, and expand eqn (2.49) to any finite order in t using the Virasoro algebra and the null state condition, we can determine all moments of W_t and conclude that it must indeed be a rescaled Brownian motion.

Of course, the steps we have used to arrive at this result in CFT are far less rigorous than the methods of SLE. However, CFT is more powerful in the sense that many other similar results can be conjectured which, at present, seem to be beyond the techniques of SLE. This is part of an ongoing symbiosis between the disciplines of theoretical physics and mathematics, which, one hopes, will continue.

2.8 Further reading

The basic reference for CFT is the "Big Yellow Book," *Conformal Field Theory* by P. di Francesco, P. Mathieu, and D. Senechal (Springer-Verlag, 1996.) See also volume 2 of *Statistical Field Theory* by C. Itzykson and J.-M. Drouffe (Cambridge University Press, 1989.) A gentler introduction is provided in the 1988 Les Houches lectures, by P. Ginsparg and J. Cardy in *Fields, Strings and Critical Phenomena*, eds. E. Brézin and J. Zinn-Justin (North-Holland, 1990.) Other specific pedagogical references are given below, preceded by the section numbers that they relate to.

2.2.1 J. Cardy, *Scaling and Renormalization in Statistical Physics*, Cambridge University Press, 1996.

2.3.2 P. Calabrese and J. Cardy, Entanglement entropy and quantum field theory—a non-technical introduction, *Int. J. Quantum Inf.* **4**, 429 (2006); arXiv:quant-ph/0505193.

2.6.1 V. Pasquier, *Nucl. Phys. B* **285**, 162 (1986); *J. Phys. A* **20**, L1229 (1987); I. Kostov, *Nucl. Phys. B* **376**, 539 (1992); arXiv:hep-th/9112059. The version in these lectures is discussed in J. Cardy, *J. Phys. A* **40**, 1427 (2007); arXiv:math-ph/0610030.

2.6.2 The basic reference is B. Nienhuis, in *Phase Transitions and Critical Phenomena*, Vol. 11, eds. C. Domb and J. L. Lebowitz, p. 1, Academic Press, 1987. A slightly different approach is explained in J. Kondev, *Phys. Rev. Lett.* **78**, 4320 (1997).

2.7.1, 2.7.2 For reviews, see V. B. Petkova and J.-B. Zuber, Conformal boundary conditions and what they teach us, lectures given at the Summer School and Conference on Nonperturbative Quantum Field Theoretic Methods and Their Applications, August 2000, Budapest, Hungary, arXiv:hep-th/0103007, and J. Cardy, Boundary conformal field theory, in *Encyclopedia of Mathematical Physics*, Elsevier, 2006, arXiv:hep-th/0411189.

2.7.3 For reviews on the connection between SLE and CFT, see J. Cardy, SLE for theoretical physicists, *Ann. Phys.* **318**, 81–118 (2005), arXiv:cond-mat/0503313, and M. Bauer and D. Bernard, 2D growth processes: SLE and Loewner chains, *Phys. Rep.* **432**, 115–221 (2006), arXiv:math-ph/0602049.

3
The quantum Hall effect

F. D. M. HALDANE

F. Duncan M. Haldane,
Princeton University,
327 Jadwin Hall,
Princeton, NJ 08544-0708,
USA.

4
Topological phases and quantum computation

A. KITAEV

Notes written by C. LAUMANN

Alexei Kitaev
California Institute of Technology,
Pasadena, CA 91125, USA.

Christopher Laumann,
Department of Physics, Princeton University,
Princeton, NJ 08544, USA.

4.1 Introduction: The quest for protected qubits

The basic building block of quantum computation is the qubit, a system with two (nearly) degenerate states that can be used to encode quantum information. Real systems typically have a full spectrum of excitations that are considered illegal from the point of view of a computation, and lead to decoherence if they couple too strongly into the qubit states during some process (see Fig. 4.1). The essential problem, then, is to preserve the quantum state of the qubit as long as possible to allow time for computations to take place.

Assuming that the gap Δ to the illegal states is reasonable, we can quite generally describe the dynamics of the qubit state by an effective Schrödinger equation

$$\frac{d}{dt}|\Psi\rangle = -iH_{\text{eff}}|\Psi\rangle\,, \tag{4.1}$$

where H_{eff} is the effective qubit Hamiltonian. In quantum optics, H_{eff} is often known with high precision. This is not so in condensed matter systems such as quantum dots. Even worse, H_{eff} may fluctuate or include interaction with the environment. This causes decoherence of the qubit state.

Ideally, we would like to arrange for H_{eff} to be zero (or $H_{\text{eff}} = \epsilon I$) for some good reason. Usually, we use a symmetry to protect degeneracies in quantum systems. For example, a quantum spin $\frac{1}{2}$ has a twofold degeneracy protected by the $\mathrm{SU}(2)$ symmetry, as do the $2s+1$ degeneracies of higher spins s. Indeed, any *non-Abelian* symmetry would work. Unfortunately, the $\mathrm{SU}(2)$ symmetry of a spin is lifted by magnetic fields and it's generally difficult to get rid of stray fields.

Rather than symmetry, in what follows we will look to topology to provide us with physically protected degeneracies in quantum systems. In particular, we will examine a number of exactly solvable models in one and two dimensions which exhibit topological phases—that is, gapped phases with a protected ground state degeneracy dependent on the topology of the manifold in which the quantum model is embedded. In Section 4.2, we warm up with a study of several quantum chains that exhibit Majorana edge modes and thus a twofold degeneracy on open chains. The topological phenomena available in two-dimensional models are much richer and will be the focus of the remaining three sections. We introduce and solve the toric code on the square lattice in Section 4.3, exhibiting its topological degeneracy and excitation

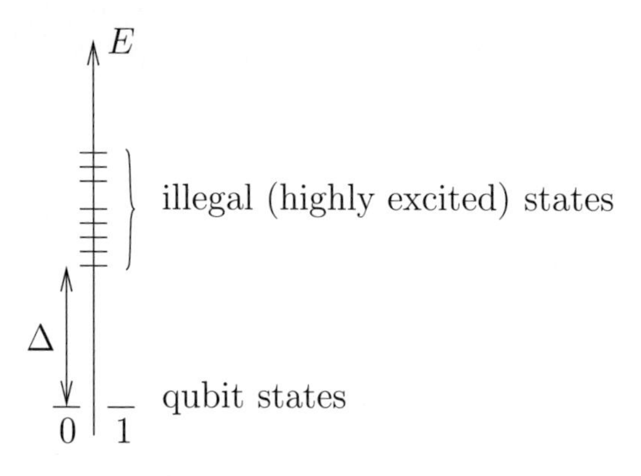

Fig. 4.1 Spectrum of a physical qubit system.

spectrum explicitly. The following section steps back to examine the general phenomenology of quasiparticle statistics braiding in two-dimensional models. Finally, in Section 4.5, we introduce the honeycomb lattice model, which exhibits several kinds of topological phases, including that of the simple toric code and, in the presence of time reversal symmetry breaking, a gapped phase with chiral edge modes protected by the topology of the Fermi surface.

4.2 Topological phenomena in 1D: Boundary modes in the Majorana chain

We will consider two examples of 1D models with $\mathbb{Z}_2$ symmetry and topological degeneracy: the *transverse-field Ising model* (TFIM) and the *spin-polarized superconductor* (SPSC). Although these models look rather different physically, we will find that they are mathematically equivalent and that they both exhibit a topological phase in which the ground state degeneracy is dependent on the boundary conditions of the chain. That is, the ground state on an open chain is twofold degenerate owing to the presence of boundary zero modes, whereas the ground state is unique on a closed loop. This topological degeneracy will be stable to small *local* perturbations that respect the $\mathbb{Z}_2$ symmetry. More details of these models may be found in Kitaev (2000).

1. The *transverse-field Ising model* is a spin-1/2 model with the Hamiltonian

$$H_S = -J \sum_{j=1}^{N-1} \sigma_j^x \sigma_{j+1}^x - h_z \sum_{j=1}^{N} \sigma_j^z \,. \tag{4.2}$$

Here J is the ferromagnetic exchange coupling in the x-direction and h_z is a uniform transverse (z) field. This model has a $\mathbb{Z}_2$ symmetry given by a global spin flip in the σ_x basis,

$$P_S = \prod_{j=1}^{N} \sigma_j^z \,. \tag{4.3}$$

2. The *spin-polarized 1D superconductor* is a fermionic system with the Hamiltonian

$$\begin{aligned} H_F = & \sum_{j=1}^{N-1} \left(-w(a_j^\dagger a_{j+1} + a_{j+1}^\dagger a_j) + \Delta a_j a_{j+1} + \Delta^* a_{j+1}^\dagger a_j^\dagger \right) \\ & - \mu \sum_{j=1}^{N} \left(a_j^\dagger a_j - \frac{1}{2} \right), \end{aligned} \tag{4.4}$$

where a_j and $a_j^\dagger$ are fermionic annihilation and creation operators, w is the hopping amplitude, Δ is the superconducting gap, and μ is the chemical potential. For simplicity, we will assume that $\Delta = \Delta^* = w$, so that

$$H_F = w \sum_{j=1}^{N-1} (a_j - a_j^\dagger)(a_{j+1} + a_{j+1}^\dagger) - \mu \sum_{j=1}^{N} \left(a_j^\dagger a_j - 1/2 \right). \tag{4.5}$$

This model has a $\mathbb{Z}_2$ symmetry given by the fermionic parity operator

$$P_F = (-1)^{\sum_j a_j^\dagger a_j} \,. \tag{4.6}$$

Although the two models are mathematically equivalent, as we will see in Section 4.2.2, they are clearly physically different. In particular, for the superconductor, the $\mathbb{Z}_2$ symmetry of fermionic parity cannot be lifted by any local physical operator, as such operators must contain an even number of fermion operators. Unfortunately, for the spin system the degeneracy is lifted by a simple longitudinal magnetic field $h_x \sum_j \sigma_j^x$ and thus the topological phase of the TFIM would be much harder to find in nature.

4.2.1 Nature of topological degeneracy (spin language)

Consider the transverse-field Ising model of eqn (4.2). With no applied field, there are a pair of Ising ground states ($h_z = 0$)

$$|\psi_\rightarrow\rangle = |\rightarrow\rightarrow\rightarrow\cdots\rightarrow\rangle,\ |\psi_\leftarrow\rangle = |\leftarrow\leftarrow\leftarrow\cdots\leftarrow\rangle. \tag{4.7}$$

The introduction of a small field h_z allows the spins to flip in the σ^x basis. In particular, tunneling between the two classical ground states arises via a soliton (domain wall) propagating from one side of the system to the other:

$$|\rightarrow\rightarrow\rightarrow\cdots\rightarrow\rangle \longrightarrow |\leftarrow:\rightarrow\rightarrow\cdots\rightarrow\rangle \longrightarrow |\leftarrow\leftarrow:\rightarrow\cdots\rightarrow\rangle \tag{4.8}$$

$$\longrightarrow |\leftarrow\leftarrow\leftarrow:\cdots\rightarrow\rangle \longrightarrow \cdots \longrightarrow |\leftarrow\leftarrow\leftarrow\cdots\leftarrow\rangle. \tag{4.9}$$

As usual, the tunneling amplitude t associated with this transition falls off exponentially in the distance the soliton must propagate, i.e.

$$t \sim e^{-N/\xi}, \tag{4.10}$$

where ξ is the correlation length of the model. The twofold degeneracy is therefore lifted by the effective Hamiltonian

$$H_{\text{eff}} = \begin{pmatrix} 0 & -t \\ -t & 0 \end{pmatrix}. \tag{4.11}$$

The splitting is exponentially small in the system size and the twofold degeneracy is recovered in the thermodynamic limit, as expected. Moreover, it is clear why introduction of a longitudinal field h_x will fully split the degeneracy.

4.2.2 Reduction of TFIM to SPSC by the Jordan–Wigner transformation

To show the equivalence of the one-dimensional models introduced above, we will use a standard Jordan–Wigner transformation to convert the spins of the Ising model into fermions. It is perhaps not surprising that a fermionic description exists for spin-1/2 systems—we simply identify the up and down state of each spin with the presence and absence of a fermion. The only difficulty arises in arranging the transformation so that the appropriate (anti)commutation relations hold in each description. The Jordan–Wigner transformation does this by introducing string-like fermion operators that work out quite nicely in 1D nearest-neighbor models.

To reduce H_S to H_F, we do the following:

1. Associate the projection of the spin onto the z-axis with the fermionic occupation number:

$$|\uparrow\rangle \leftrightarrow n = 0,\ \ |\downarrow\rangle \leftrightarrow n = 1. \tag{4.12}$$

   That is,

$$\sigma_j^z = (-1)^{a_j^\dagger a_j}. \tag{4.13}$$

2. Introduce the string-like annihilation and creation operators

$$a_j = \left(\prod_{k=1}^{j-1} \sigma_k^z\right) \sigma_j^+ ,$$
$$a_j^\dagger = \left(\prod_{k=1}^{j-1} \sigma_k^z\right) \sigma_j^- , \tag{4.14}$$

where σ^+ and σ^- are the usual spin-raising and lowering operators. At this stage, we can check that the usual fermionic anticommutation relations hold for the $a_j, a_j^\dagger$:

$$\left\{a_i, a_j^\dagger\right\} = \delta_{ij} . \tag{4.15}$$

3. Observe that

$$\sigma_j^x \sigma_{j+1}^x = -(a_j - a_j^\dagger)(a_{j+1} + a_{j+1}^\dagger), \tag{4.16}$$

so H_S (eqn (4.2)) reduces to H_F (eqn (4.5)) with

$$w = J, \quad \mu = -2h_z . \tag{4.17}$$

4.2.3 Majorana operators

Majorana operators provide a convenient alternative representation of Fermi systems when the number of particles is only conserved modulo 2, as in a superconductor. Given a set of N Dirac fermions with annihilation/creation operators $a_j, a_j^\dagger$, we can define a set of $2N$ real Majorana fermion operators as follows:

$$c_{2j-1} = a_j + a_j^\dagger ,$$
$$c_{2j} = \frac{a_j - a_j^\dagger}{i} . \tag{4.18}$$

These operators are Hermitian and satisfy a fermionic anticommutation relation

$$c_k^\dagger = c_k ,$$
$$c_k^2 = 1, \quad c_k c_l = -c_l c_k \quad (k \neq l) . \tag{4.19}$$

Or, more compactly,

$$\{c_k, c_l\} = 2\delta_{kl} . \tag{4.20}$$

From any pair of Majorana operators, we can construct an annihilation and a creation operator for a standard Dirac fermion ($a = (c_1 + ic_2)/2$ and h.c.), and thus the unique irreducible representation for the pair is a two-dimensional Hilbert space which is either occupied or unoccupied by the a fermion.

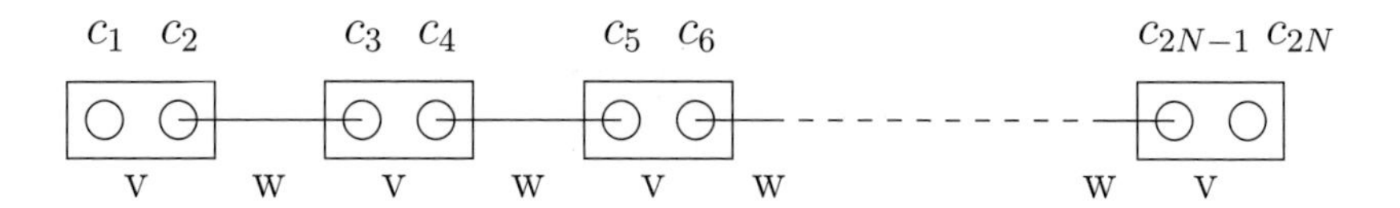

Fig. 4.2 Majorana chain representation of 1D superconductor. Each boxed pair of Majorana modes corresponds to one site of the original fermionic chain.

Both models H_S and H_F can be written as

$$H_{\text{maj}} = \frac{i}{2}\left(v\sum_{j=1}^{N} c_{2j-1}c_{2j} + w\sum_{j=1}^{N-1} c_{2j}c_{2j+1}\right), \tag{4.21}$$

where $v = h_z = -\mu/2$ and $w = J$. The $\mathbb{Z}_2$ symmetry of fermionic parity is given in the Majorana language by

$$P_{\text{maj}} = \prod_{k=1}^{N}(-ic_{2k-1}c_{2k}). \tag{4.22}$$

We can view this model graphically as a chain of coupled Majorana modes, two for each of the N sites of the original problem as in Fig. 4.2. If $v = 0$, then the Majorana modes at the ends of the chain are not coupled to anything. This immediately allows us to identify the twofold ground state degeneracy in H_{maj} as the tensor factor given by the two-dimensional representation of the boundary pair c_1, c_{2N}.

We will see in Section 4.2.4 that if $v \neq 0$ but $|v| < w$, the operators c_1 and c_{2N} are replaced by some *boundary mode operators* b_l, b_r. The effective Hamiltonian for this piece of the system is then

$$H_{\text{eff}} = \frac{i}{2}\epsilon b_l b_r = \epsilon\left(a^\dagger a - \frac{1}{2}\right), \tag{4.23}$$

where $\epsilon \sim e^{-N/\xi}$, and a, $a^\dagger$ are the Dirac fermion operators constructed from the boundary pair. Thus, the ground state degeneracy is lifted by only an exponentially small splitting in the system size.

4.2.4 General properties of quadratic fermionic Hamiltonians

We now step back and consider a generic quadratic fermionic Hamiltonian

$$H(A) = \frac{i}{4}\sum_{j,k} A_{jk}c_j c_k, \tag{4.24}$$

where A is a real, skew-symmetric matrix and the c_j are Majorana fermion operators. The normalization $i/4$ is convenient because it has the property that

$$[-iH(A), -iH(B)] = -iH([A, B]), \tag{4.25}$$

where $A, B \in \mathfrak{so}(2N)$, and $H(A), H(B)$ act on the Fock space $\mathfrak{F}_N = \mathbb{C}^{2^N}$. Thus $H(\cdot)$ provides a natural representation of $\mathfrak{so}(2N)$.

We now bring $H(A)$ to a canonical form,

$$H_{\text{canonical}} = \frac{i}{2}\sum_{k=1}^{m} \epsilon_k b'_k b''_k = \sum_{k=1}^{m} \epsilon_k \left(\tilde{a}_k^\dagger \tilde{a}_k - \frac{1}{2}\right), \tag{4.26}$$

where b'_k, b''_k are appropriate real linear combinations of the original c_j satisfying the same Majorana fermion commutation relations, and the $\tilde{a}_k, \tilde{a}_k^\dagger$ are the annihilation and creation operators associated with the b'_k, b''_k pair of Majorana modes. This form for H follows immediately from the standard block diagonalization of real skew-symmetric matrices

$$A = Q \begin{pmatrix} 0 & \epsilon_1 & & & \\ -\epsilon_1 & 0 & & & \\ & & 0 & \epsilon_2 & \\ & & -\epsilon_2 & 0 & \\ & & & & \cdots \end{pmatrix} Q^T, \quad Q \in O(2N), \epsilon_k \geq 0\,. \tag{4.27}$$

From this form, it is easy to check that the eigenvalues of A are $\pm i\epsilon_k$ and that the eigenvectors are the coefficients of c in $\tilde{a}_k$, $\tilde{a}_k^\dagger$.

If some of the ϵ_k vanish, then we refer to the associated fermions as *zero modes*. In particular, these will lead to ground state degeneracies, since occupation or nonoccupation of such modes does not affect the energy. For the Majorana chain of eqn (4.21), we have

$$A = \begin{pmatrix} 0 & v & & & & \\ -v & 0 & w & & & \\ & -w & 0 & v & & \\ & & -v & 0 & w & \\ & & & -w & 0 & \\ & & & & & \cdots \end{pmatrix}. \tag{4.28}$$

We can find a vector u such that $uA = 0$ by inspection:

$$u = \left(1, 0, \frac{v}{w}, 0, \left(\frac{v}{w}\right)^2, 0, \cdots\right). \tag{4.29}$$

This vector leads to a left boundary mode

$$b_l = \sum u_k c_k\,, \tag{4.30}$$

while an analogous calculation starting at the right end will find a right boundary mode b_r. These modes form a Majorana canonical pair, leading to a twofold degeneracy of the ground state of the chain. Clearly, $u_k \sim e^{-k/\xi}$ falls off exponentially from the edges of the chain with a correlation length $\xi^{-1} = \ln|w/v|$, as expected in Section 4.2.1.

4.2.5 Why are the boundary modes robust?

In the simple case of a quadratic fermion Hamiltonian, we know that the modes correspond to eigenvalues of a skew-symmetric real matrix. These come in pairs $\pm i\epsilon$, in general, and the case $\epsilon = 0$ is special. In particular, if the pair of Majorana modes corresponding to a zero mode are physically well separated, we expect perturbations to have trouble lifting the boundary degeneracy.

More generally, for interacting fermions, we can extend the symmetry group $\mathbb{Z}_2$, generated by $P = P_{\text{maj}}$, to a *noncommuting* algebra acting on the ground state space $\mathfrak{L}$. First, in the noninteracting limit, at $v = 0$, we define

$$X = Y \prod_{k=1}^{j} (-ic_{2k-1}c_{2k}) , \tag{4.31}$$

where $Y = c_{2j+1}$ is a local Majorana operator at site $2j + 1$. A straightforward calculation shows that

$$XP = -PX \tag{4.32}$$

and that $[H, X] = 0$, so that the algebra generated by X, P acts on $\mathfrak{L}$ nontrivially. We now allow Y to vary as we adiabatically turn on interactions and, so long as an energy gap is maintained, we expect Y to remain a local operator near $2j$, which we can separate from the boundary by a suitably large choice of j. That is, to find Y, one needs to know the ground state or at least the structure of the ground state near $2j$. This is a nontrivial operation; see Hastings and Wen (2005) for more details.

4.3 The two-dimensional toric code

The toric code is an exactly solvable spin-1/2 model on a square lattice. It exhibits a ground state degeneracy of 4^g when embedded in a surface of genus g, and a quasiparticle spectrum with both bosonic and fermionic sectors. Although we will not introduce it as such, the model can be viewed as an Ising gauge theory at a particularly simple point in parameter space (see Section 4.4.5). Many of the topological features of the toric-code model were essentially understood by Read and Chakraborty (1989), but those authors did not propose an exactly solved model. A more detailed exposition of the toric code may be found in Kitaev (2003).

We consider a square lattice, possibly embedded into a nontrivial surface such as a torus, and place spins on the edges, as in Fig. 4.3. The Hamiltonian is given by

$$H_T = -J_e \sum_s A_s - J_m \sum_p B_p , \tag{4.33}$$

where s runs over the vertices (stars) of the lattice and p runs over the plaquettes. The star operator acts on the four spins surrounding a vertex s,

$$A_s = \prod_{j \in \text{star}(s)} \sigma_j^x , \tag{4.34}$$

while the plaquette operator acts on the four spins surrounding a plaquette,

$$B_p = \prod_{j \in \partial p} \sigma_j^z . \tag{4.35}$$

Clearly, the A_s all commute with one another, as do the B_p. Slightly less trivially,

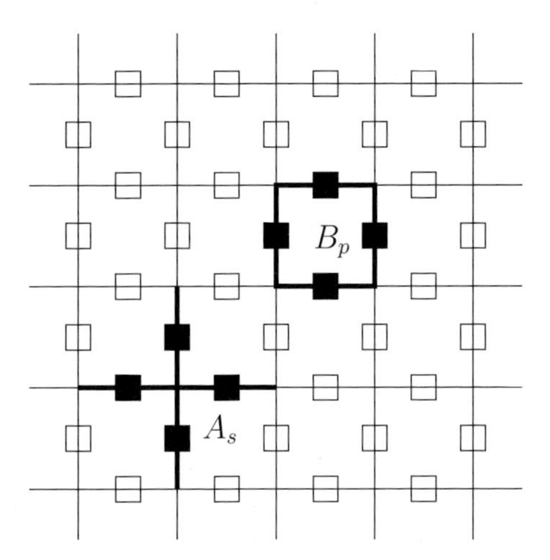

Fig. 4.3 A piece of the toric code. The spins live on the edges of the square lattice. The spins adjacent to a star operator A_s and a plaquette operator B_p are shown.

$$A_s B_p = B_p A_s \tag{4.36}$$

because any given star and plaquette share an even number of edges (either none or two) and therefore the minus signs arising from the commutation of σ^x and σ^z on those edges cancel. Since all of the terms of H_T commute, we expect to be able to solve it term by term.

In particular, we will solve H_T working in the σ^z basis. We define classical variables $s_j = \pm 1$ to label the σ^z basis states. For each classical spin configuration $\{s\}$, we can define the plaquette flux

$$w_p(s) = \prod_{j \in \partial p} s_j \,. \tag{4.37}$$

If $w_p = -1$, we say that there is a *vortex* on plaquette p.

4.3.1 Ground states

To find the ground states $|\Psi\rangle$ of H_T, we need to minimize the energy, which means to maximize the energy of each of the A_s and B_p terms. The plaquette terms provide the condition

$$B_p|\Psi\rangle = |\Psi\rangle \,, \tag{4.38}$$

which holds if and only if

$$|\Psi\rangle = \sum_{\{s: w_p(s)=1 \ \forall p\}} c_s |s\rangle \,. \tag{4.39}$$

That is, the ground state contains no vortices. The group of star operators acts on the configurations s by flipping spins. Thus, the star conditions

$$A_s|\Psi\rangle = |\Psi\rangle \tag{4.40}$$

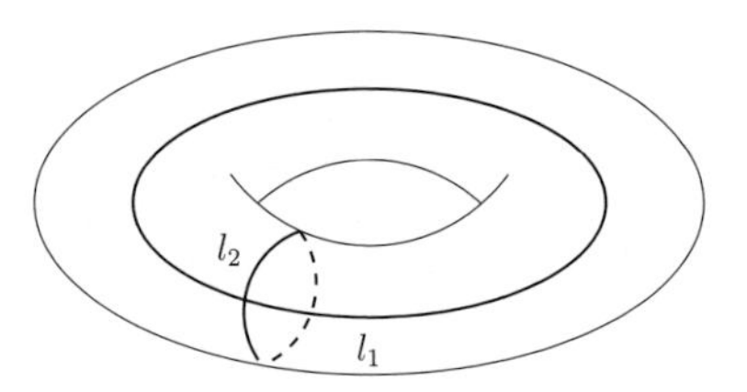

Fig. 4.4 Large cycles on the torus.

hold if and only if *all of the c_s are equal for each orbit of the action of star operators*. In particular, if the spin flips of A_s are ergodic, as they are on the plane, all c_s must be equal and the ground state is uniquely determined.

On the torus, the star operators preserve the *cohomology class* of a vortex-free spin configuration. In more physical terms, we can define conserved numbers given by the Wilson loop-like functions

$$w_l(s) = \prod_{j \in l} s_j, \quad l = l_1, l_2 \,, \tag{4.41}$$

where l_1 and l_2 are two independent nontrivial cycles on the square lattice wrapping the torus (Fig. 4.4). Any given star will overlap with a loop l on either zero or two edges, and therefore A_s preserves w_l. Since there are two independent loops on the torus, each of which can have $w_l = \pm 1$, there is a fourfold degenerate ground state

$$|\Psi\rangle = \sum_{\{s: w_p(s)=1 \ \forall p\}} c_{w_{l_1} w_{l_2}} |s\rangle \,. \tag{4.42}$$

4.3.2 Excitations

The excitations of the toric code come in two varieties: the *electric charges* and *magnetic vortices* of a $\mathbb{Z}_2$ gauge theory. We will see this connection more explicitly later. In the following, we restrict our attention to the planar system for simplicity.

To find the electric charges, let us define the electric path operator

$$W_l^{(e)} = \prod_{j \in l} \sigma_j^z \,, \tag{4.43}$$

where l is a path in the lattice going from s_1 to s_2 (see Fig. 4.5). This operator clearly commutes with the plaquette operators B_p and with all of the star operators A_s except at the endpoints s_1 and s_2, where only one edge overlaps between the star and the path and we have

$$W_l^{(e)} A_{s_1} = -A_{s_1} W_l^{(e)} \,. \tag{4.44}$$

Therefore, the state

$$|\Psi_{s_1,s_2}\rangle = W_l^{(e)} |\Psi_0\rangle \,, \tag{4.45}$$

where $|\Psi_0\rangle$ is the planar ground state, is an eigenstate of the Hamiltonian with excitations (charges) at s_1 and s_2 that each cost energy $2J_e$ to create relative to the ground state.

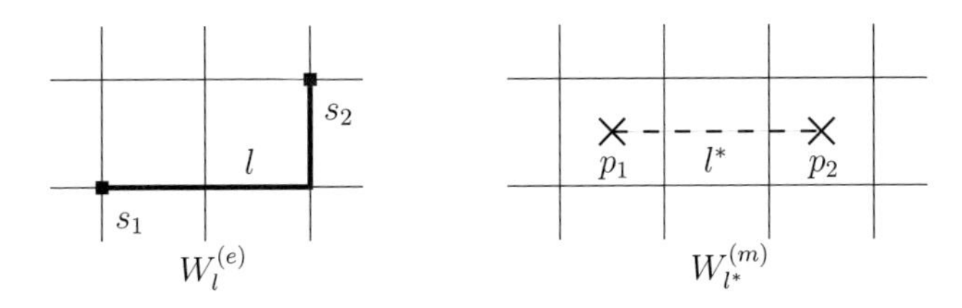

Fig. 4.5 Electric and magnetic path operators.

An analogous construction will find the magnetic vortices: we can define a dual path operator

$$W_{l^*}^{(m)} = \prod_{j \in l^*} \sigma_j^x , \tag{4.46}$$

where the path l^* lies in the dual lattice (see Fig. 4.5) and goes from p_1 to p_2. In this case, the stars A_s all commute with $W_{l^*}^{(m)}$, as do all of the plaquette operators B_p except the two at the endpoints of l^*, which anticommute. Thus, the $W_{l^*}^{(m)}$ operator creates a pair of magnetic vortices on the plaquettes p_1 and p_2 with an energy of $2J_m$ each.

4.4 Abelian anyons and quasiparticle statistics

Let us discuss what can possibly happen if we exchange two particles in two dimensions. To ensure that the particle statistics is well defined, we assume that there is no long-range interaction and that the phase is gapped. If we drag two particles around one another adiabatically,

then we expect both a dynamical phase accumulation and a statistical effect due to the exchange. We are well acquainted with this effect for everyday bosons and fermions, for which we have

$$\begin{aligned} &\text{Bosons:} \ |\Psi\rangle \mapsto |\Psi\rangle , \\ &\text{Fermions:} \ |\Psi\rangle \mapsto -|\Psi\rangle , \end{aligned} \tag{4.47}$$

where we have dropped the dynamical phase so as to focus on the statistics. In both of these standard cases, a full rotation (two exchanges),

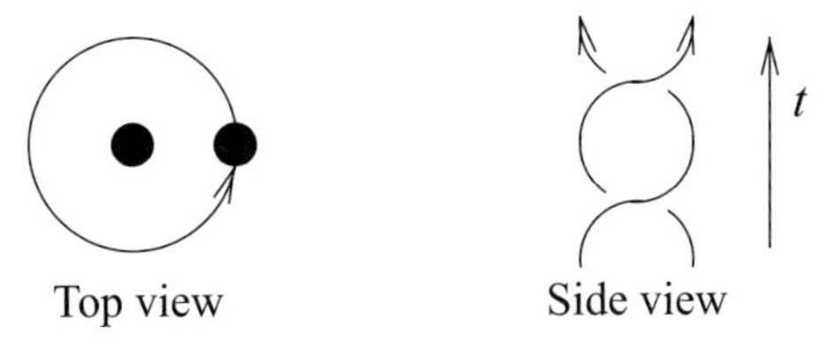

leaves $|\Psi\rangle$ unchanged.

In principle,

b a

$$R_{ab} =$$

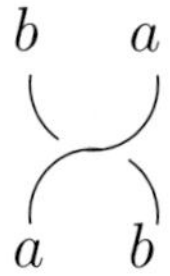

a b (4.48)

is an arbitrary phase factor or even an operator (a *braiding operator*). If the two particles are distinguishable ($a \neq b$), then R_{ab} does not have an invariant meaning, but the *mutual statistics*

a b

$$R_{ba} \cdot R_{ab} =$$

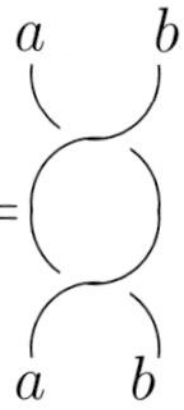

a b (4.49)

does.

Let us illustrate this in the toric-code model. In Section 4.3.2 we found two kinds of quasiparticle excitations in the toric code: electric charges (e) and magnetic vortices (m). Since path operators of the same type commute with one another, it is easy to show that both of these are bosons. However, they have nontrivial mutual statistics.

To calculate the mutual statistics, consider taking a charge e around a vortex m:

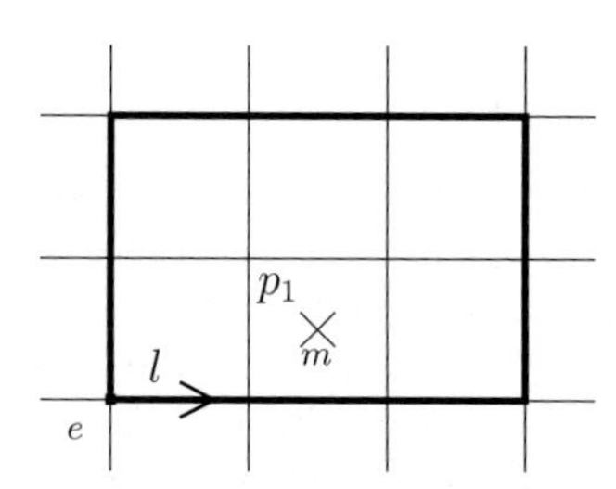

Let $|\xi\rangle$ be some state containing a magnetic vortex at p_1. Under the full braiding operation,

$$\begin{aligned} |\xi\rangle \mapsto & \left(\prod_{j \in l} \sigma_j^z\right) |\xi\rangle \\ = & \left(\prod_{p \text{ inside } l} B_p\right) |\xi\rangle , \end{aligned} \tag{4.50}$$

where the second line is a Stokes'-theorem-like result relating the product around a loop to the products for internal loops. Since

$$B_{p_1}|\xi\rangle = -|\xi\rangle \tag{4.51}$$

for the plaquette p_1 containing the vortex, we have that

$$|\xi\rangle \mapsto -|\xi\rangle\,, \tag{4.52}$$

or

e m = − e m

e m e m (4.53)

Using the bosonic self-statistics equations,

e e = e e m m = m m

e e e e m m m m (4.54)

we can derive the nontrivial corollary that composite e–m particles are fermions:

em em = e m e m = − e m e m

em em e m e m e m e m (4.55)

4.4.1 Superselection sectors and fusion rules

In the above, we initially described two kinds of bosonic excitations in the toric-code model (charges e and vortices m) in the solution of the Hamiltonian. After a bit of work, we discovered that a composite e–m object has a meaningful characterization within the model as well, at least in that it has fermionic statistics. This raises the question of how many particle types exist in the toric-code model and how we can identify them.

We take an algebraic definition of a particle type: each type corresponds to a *superselection sector*, which is a representation of the local operator algebra. In particular, we say that two particles (or composite objects) are *of the same type*, i.e.

$$a \sim b\,, \tag{4.56}$$

if a can be transformed to b by some operator acting in a finite region. For example, in the toric code, two e-particles are equivalent to having no particles at all,

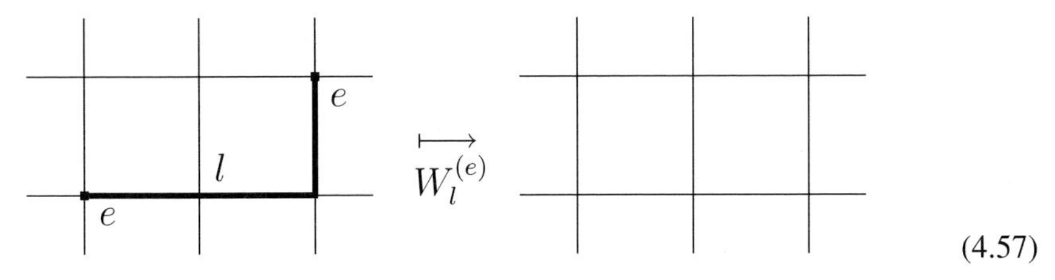

(4.57)

when we act with an appropriate geometrically bounded electric path operator $W_l^{(e)}$.

We introduce the notation

$$e \times e = 1 \tag{4.58}$$

to represent the *fusion rule* that two e-particles are equivalent to the vacuum sector 1. In the toric code, there are four superselection sectors,

$$1,\ e,\ m,\ \text{and}\ \epsilon = e \times m\,, \tag{4.59}$$

with the fusion rules

$$\begin{aligned} &e \times e = 1, \quad e \times m = \epsilon\,, \\ &m \times m = 1, e \times \epsilon = m\,, \\ &\epsilon \times \epsilon = 1, \quad m \times \epsilon = e\,. \end{aligned} \tag{4.60}$$

4.4.2 Mutual statistics implies degeneracy on the torus

This is an argument due to Einarsson (1990). Suppose that there are at least two particle types, e and m, with -1 mutual statistics. Let us define an operator Z acting on the ground state in an abstract fashion (not referring to the actual model) which creates an e pair, wraps one particle around the torus, and annihilates the pair. In the toric code, this will be the path operator $W_l^{(e)} = \prod_{j \in l} \sigma_j^z$ for a loop l winding around one of the nontrivial cycles on the torus, but we need not know that specifically.

We can define another operator X that creates a pair of the other type m and winds around the other nontrivial cycle on the torus. But now a bit of geometric inspection reveals the following fact for the combination $Z^{-1}X^{-1}ZX$:

$$Z^{-1}X^{-1}ZX = \quad (m) \quad (e) \quad = -1 \tag{4.61}$$

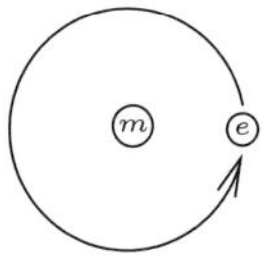

Thus, there are two noncommuting operators acting on the ground state space $\mathfrak{L}$, and we conclude that $\dim\ \mathfrak{L} > 1$. In fact, there are four such operators; each of the two particle types can be moved around each of the two nontrivial cycles. Working out the commutation relations of these operators implies that $\dim\ \mathfrak{L} = 4$.

4.4.3 The toric code in a field: Perturbation analysis

We now apply a magnetic field to the toric code that allows the quasiparticles to hop realistically and, unfortunately, destroys the exact solvability of the model (see Tupitsyn *et al.* (2008)). To wit:

$$H = -J_e \sum_s A_s - J_m \sum_p B_p - \sum_j \left(h_x \sigma_j^x + h_z \sigma_j^z\right). \tag{4.62}$$

For example, with $h_x = 0$ but $h_z \neq 0$, we can view the perturbation as an electric path operator of length 1 on each edge. Hence, it can cause charge pair creation and annihilation

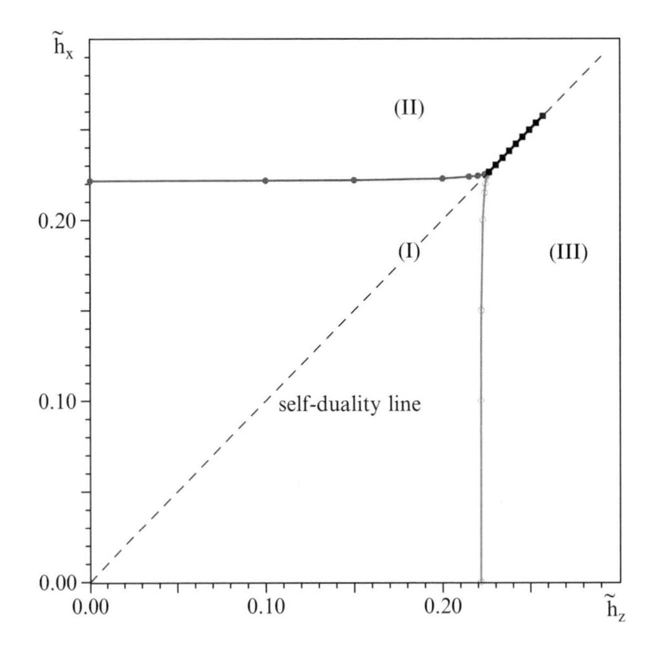

Fig. 4.6 Numerically determined phase diagram of the toric code in a field, from Tupitsyn *et al.* (2008). (I) labels the topological phase, and (II) and (III) the vortex and charge condensates (i.e. the paramagnetic phase). The numerics were done using discrete imaginary time with a rather large quantization step.

(at an energy cost $\sim 4J_e$) or cause existing charges to hop by one lattice displacement at no cost. For small h_z, this provides a nontrivial tight-binding dispersion to the charges,

$$\epsilon(q) \approx 2J_e - 2h_z(\cos q_x + \cos q_y)\,, \tag{4.63}$$

but does not close the gap or lead to a change in the topological degeneracy of the ground state in the thermodynamic limit.

At large $h_z \gg J_e, J_m$, the model should simply align with the applied field as a paramagnet. Clearly, in this limit the topological degeneracy has been destroyed and we have a unique spin-polarized ground state. The phase transition can be understood from the topological side as a Bose condensation of the charges, which proliferate as h_z increases.

The same argument is applicable if $h_x \gg J_e, J_m$. If h_x increases while $h_z = 0$, then vortices condense. However, the high-field phase is just a paramagnet, so one can continuously rotate the field between the x- and the z-direction without inducing a phase transition. Thus, the charge and vortex condensates are actually the same phase! This property was first discovered by Fradkin and Shenker (1979) for a 3D classical $\mathbb{Z}_2$ gauge Higgs model, where it appears rather mysterious. The full phase diagram is shown in Fig. 4.6.

4.4.4 Robustness of the topological degeneracy

The splitting of the ground state levels due to virtual quasiparticle tunneling is given by

$$\delta E \sim \Delta e^{-L/\xi}\,. \tag{4.64}$$

This follows from the effective Hamiltonian

$$H_{\text{eff}} = -(t_{1Z}Z_1 + t_{2Z}Z_2 + t_{1X}X_1 + t_{2X}X_2)\,, \tag{4.65}$$

where the operators Z_i, X_i are the winding-loop operators of Section 4.4.2. Physically, this is simply a statement of the fact that the only way to act upon the ground state is to wind quasiparticles around the torus. This is a process exponentially suppressed in the system size.

4.4.5 Emergent symmetry: Gauge formulation

There are two ways to introduce symmetry operators in the perturbed toric-code model:

1. One can define *loop operators* (e.g. Z_1, Z_2, X_1, and X_2), the definition of which depends on the actual ground state of the perturbed Hamiltonian. This is similar to the definition of the operator Y in the 1D case considered in Section 4.2.5, which also requires detailed knowledge of the ground state.
2. One can exploit *gauge invariance* by rewriting the model in a gauge-invariant form. This can be done for any spin model by introducing redundancy. In this case, the symmetry does not depend on the model but is only manifest in the topological phase.

We will take the second approach in order to avoid the difficulty of defining the appropriate loop operators and also to introduce the important gauge formulation of the model. To gauge the model, we proceed in steps:

1. We introduce one extra spin μ_v per vertex that always remains in the state

$$\frac{1}{\sqrt{2}}\left(|\uparrow\rangle + |\downarrow\rangle\right). \tag{4.66}$$

This state is characterized by the condition

$$\mu_v^x|\Psi\rangle = |\Psi\rangle\,, \tag{4.67}$$

where μ_v^x is the Pauli spin matrix for the spin μ at vertex v.

2. We change the spin operators from σ_{uv}, μ_v to $\tilde{\sigma}_{uv}, \tilde{\mu}_v$, where we represent the classical value of each old spin s_{uv} as $\tilde{m}_u\tilde{s}_{uv}\tilde{m}_v$. Here s_{uv} is the spin on the edge connecting u and v, and m_u, m_v are the classical values of the new spins (i.e. the labels in the μ^z basis).

Thus the complete transformation is given by

$$\begin{aligned}
\sigma_{uv}^z &= \tilde{\mu}_u^z\tilde{\sigma}_{uv}^z\tilde{\mu}_v^z\,, \\
\sigma_{uv}^x &= \tilde{\sigma}_{uv}^x\,, \\
\mu_u^z &= \tilde{\mu}_u^z\,, \\
\mu_u^x &= \tilde{\mu}_u^x \prod_{j\in\mathtt{star}(u)} \tilde{\sigma}_j^x = \tilde{\mu}_u^x\tilde{A}_u\,,
\end{aligned} \tag{4.68}$$

and the constraint in eqn (4.67) becomes the standard $\mathbb{Z}_2$ gauge constraint

$$\tilde{\mu}_u^x\tilde{A}_u|\Psi\rangle = |\Psi\rangle\,. \tag{4.69}$$

On states satisfying the gauge constraint of eqn (4.69), $A_u = \tilde{A}_u = \tilde{\mu}_u^x$. Therefore,

$$A_u|\Psi\rangle = \tilde{\mu}_u^x|\Psi\rangle \tag{4.70}$$

and we can rewrite the Hamiltonian as

$$H = -J_e \sum_v \tilde{\mu}_v^x - J_m \sum_p \tilde{B}_p - \sum_{\langle u,v\rangle} \left(h_x \tilde{\sigma}_{uv}^x + h_z \tilde{\mu}_u \tilde{\sigma}_{uv}^z \tilde{\mu}_v\right), \tag{4.71}$$

subject to the gauge constraint.

Viewed as a standard $\mathbb{Z}_2$ gauge theory, the protected topological degeneracy of the ground state is physically familiar as the protected degeneracy associated with the choice of flux threading the $2g$ holes of a genus-g surface.

4.5 The honeycomb lattice model

We now investigate the properties of another exactly solvable spin model in two dimensions, the *honeycomb lattice model*. This model exhibits a number of gapped phases that are perturbatively related to the toric code of the previous sections. Moreover, in the presence of time-reversal-symmetry-breaking terms, a new topological phase arises with different topological properties, including the nontrivial *spectral Chern number*. An extended treatment of the properties of this model, with much greater detail, can be found in Kitaev (2006).

In the honeycomb lattice model, the degrees of freedom are spins living on the vertices of a honeycomb lattice with nearest-neighbor interactions. The unusual feature of this model is that the interactions are link-orientation-dependent (see Fig. 4.7). The Hamiltonian is

$$H = -J_x \sum_{x \text{ links}} \sigma_j^x \sigma_k^x - J_y \sum_{y \text{ links}} \sigma_j^y \sigma_k^y - J_z \sum_{z \text{ links}} \sigma_j^z \sigma_k^z . \tag{4.72}$$

We might expect this model to be integrable because $[H, W_p] = 0$ for an extensive collection of plaquette operators

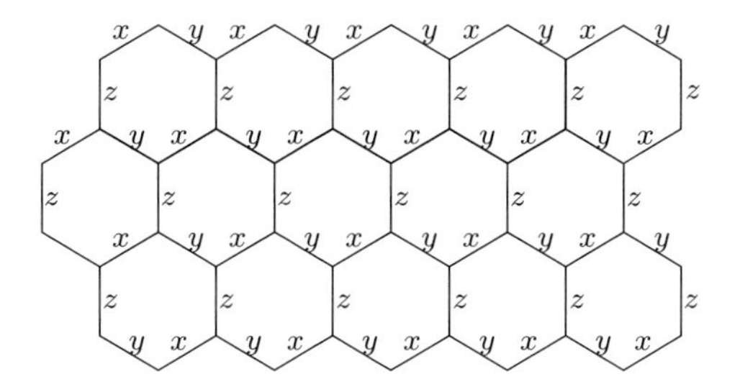

Fig. 4.7 The honeycomb model has spins living on the vertices of a honeycomb lattice with nearest-neighbor interactions that are link-orientation-dependent. x-links have $\sigma^x\sigma^x$ interactions, y-links have $\sigma^y\sigma^y$ interactions, and z-links have $\sigma^z\sigma^z$ interactions.

$$W_p = \sigma_1^x \sigma_2^y \sigma_3^z \sigma_4^x \sigma_5^y \sigma_6^z \tag{4.73}$$

where the spins and labels follow from the figure for each plaquette. Unfortunately, this is not quite enough: there are two spins but only one constraint per hexagon, so that half of each spin remains unconstrained. In fact, the remaining degrees of freedom are Majorana operators!

4.5.1 A (redundant) representation of a spin by four Majorana operators

We consider a collection of four Majorana operators c, b^x, b^y, and b^z that act on the four-dimensional Fock space $\mathfrak{F}$. We define the following three operators:

$$\begin{aligned} \tilde{\sigma}^x &= ib^x c\,, \\ \tilde{\sigma}^y &= ib^y c\,, \\ \tilde{\sigma}^y &= ib^y c\,. \end{aligned} \tag{4.74}$$

These operators do not obey the spin algebra relations on the full Fock space, but we clearly have two extra dimensions of wiggle room. In fact, the physical state space is identified with a two-dimensional subspace $\mathfrak{L} \subset \mathfrak{F}$ given by the constraint

$$D|\Psi\rangle = |\Psi\rangle, \quad \text{where } D = b^x b^y b^z c\,. \tag{4.75}$$

Within $\mathfrak{L}$, the $\tilde{\sigma}^\alpha$ act as the σ^α act on the actual spin. Of course, the $\tilde{\sigma}^\alpha$ also act on $\mathfrak{L}^\perp$, but we can ignore these states by enforcing the constraint.

To be careful, we need to check two consistency conditions:

1. $\tilde{\sigma}^\alpha$ preserves the subspace $\mathfrak{L}$, which follows from $[\tilde{\sigma}^\alpha, D] = 0$.
2. The $\tilde{\sigma}^\alpha$ satisfy the correct algebraic relations when restricted to $\mathfrak{L}$. For example,

$$\tilde{\sigma}^x \tilde{\sigma}^y \tilde{\sigma}^z = (ib^x c)(ib^y c)(ib^z c) = i^3(-1) b^x b^y b^z c^3 = iD = i\,, \tag{4.76}$$

where the last equality holds only in the physical subspace $\mathfrak{L}$.

4.5.2 Solving the honeycomb model using Majorana modes

We now use the Majorana representation of spins just introduced to rewrite each spin of the entire honeycomb model as in Fig. 4.8. This will greatly expand the 2^N-dimensional Hilbert space to the Fock space $\mathfrak{F}$ of dimension 2^{2N}, but the physical space $\mathfrak{L} \subset \mathfrak{F}$ is fixed by the gauge condition

$$D_j|\Psi\rangle = |\Psi\rangle \quad \text{for all } j\,, \tag{4.77}$$

where $D_j = b_j^x b_j^y b_j^z c$. We define a projector onto $\mathfrak{L}$ by

$$\Pi_{\mathfrak{L}} = \prod_j \left(\frac{1 + D_j}{2} \right). \tag{4.78}$$

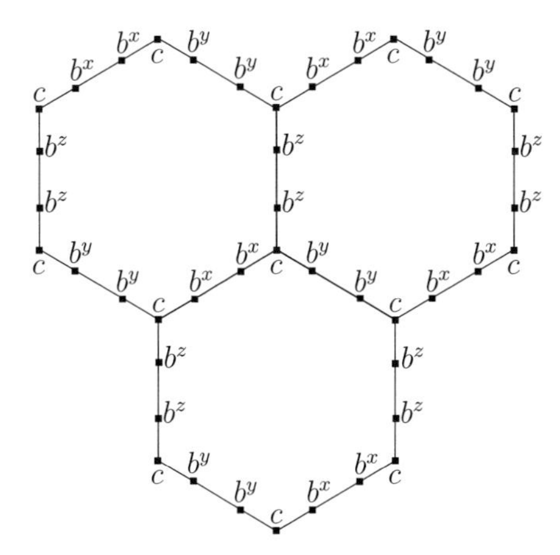

Fig. 4.8 Majorana representation of honeycomb model.

In the Majorana representation, the Hamiltonian (4.72) becomes

$$\begin{aligned}\tilde{H} &= \frac{i}{4} \sum_{\langle j,k \rangle} \hat{A}_{jk} c_j c_k \,, \\ \hat{A}_{jk} &= 2J_{\alpha(j,k)} \hat{u}_{jk} \,, \\ \hat{u}_{jk} &= i b_j^{\alpha(j,k)} b_k^{\alpha(j,k)} \,,\end{aligned} \tag{4.79}$$

where $\alpha(j,k) = x, y, z$ is the direction of the link between j and k.

We have suggestively written the Hamiltonian $\tilde{H}$ as if it were a simple quadratic fermion Hamiltonian as in Section 4.2.4, but of course $\hat{A}_{jk}$ is secretly an operator rather than a real skew-symmetric matrix. However, each operator b_j^α enters only one term of the Hamiltonian, and therefore the $\hat{u}_{jk}$ commute with each other and with $\tilde{H}$! Thus, we can fix $u_{jk} = \pm 1$, defining an orthogonal decomposition of the full Fock space

$$\mathfrak{F} = \bigoplus_u \mathfrak{F}_u, \quad \text{where } |\Psi\rangle \in \mathfrak{F}_u \text{ iff } \hat{u}_{jk}|\Psi\rangle = u_{jk}|\Psi\rangle \ \forall\, j,k\,. \tag{4.80}$$

Within each subspace $\mathfrak{F}_u$, we need to solve the quadratic Hamiltonian

$$\begin{aligned}\tilde{H}_u &= \frac{i}{4} \sum_{\langle j,k \rangle} A_{jk} c_j c_k \,, \\ A_{jk} &= 2J_{\alpha(j,k)} u_{jk} \,,\end{aligned} \tag{4.81}$$

which we know how to do in principle. On the other hand, the integrals of motion W_p (the hexagon operators) define a decomposition of the physical subspace $\mathfrak{L}$ labeled by the eigenvalues $w_p = \pm 1$,

$$\mathfrak{L} = \bigoplus_w \mathfrak{L}_w, \quad \text{where } |\Psi\rangle \in \mathfrak{L}_w \text{ iff } W_p|\Psi\rangle = w_p|\Psi\rangle \ \forall\, p\,. \tag{4.82}$$

We can relate these two decompositions by expressing W_p in the Majorana representation and noting that within the physical subspace,

$$\tilde{W}_p = \prod_{\langle j,k\rangle \in \partial p} \hat{u}_{jk} \,. \tag{4.83}$$

Thus, we find

$$\mathfrak{L}_w = \Pi_{\mathfrak{L}} \mathfrak{F}_u \,, \tag{4.84}$$

where $w_p = \prod_{(j,k)\in\partial p} u_{jk}$.

So we have a procedure for finding the ground state of the honeycomb model:

1. Fix $w_p = \pm 1$ for all p.
2. Find u_{jk} satisfying

$$w_p = \prod_{(j,k)\in\partial p} u_{jk} \,. \tag{4.85}$$

   There is a small subtlety here in that $u_{jk} = -u_{kj}$, so we must be careful about ordering. We can consistently take j in the even sublattice of the honeycomb and k in the odd sublattice in eqn (4.85).
3. Solve for the ground state of the quadratic Hamiltonian (4.81), finding the energy $E(w)$.
4. Project the state found onto the physical subspace (i.e. symmetrize over gauge transformations).
5. Repeat for all w; pick the w that minimizes the energy.

If there were no further structure to $E(w)$, this would be an intractable search problem in the space of w_p. Fortunately, owing to a theorem by Lieb (1994), the ground state has no vortices. That is,

$$E(w) = \min \text{ if } w_p = 1 \,\forall\, p \,. \tag{4.86}$$

Using this choice of w_p, it is easy to solve the model and produce the phase diagram shown in Fig. 4.9. The gapless phase has two Dirac points in the fermionic spectrum.

4.5.3 Fermionic spectrum in the honeycomb lattice model

We just need to diagonalize the Hamiltonian

$$\begin{aligned} \tilde{H}_u &= \frac{i}{4} \sum_{\langle j,k\rangle} A_{jk} c_j c_k \,, \\ A_{jk} &= 2J_{\alpha(j,k)} u_{jk} \,, \\ u_{jk} &= \begin{cases} +1 & \text{if } j \in \text{even sublattice}\,, \\ -1 & \text{otherwise}\,. \end{cases} \end{aligned} \tag{4.87}$$

This is equivalent to finding the eigenvalues and eigenvectors of the matrix iA. Since the honeycomb lattice has two sites per unit cell, by applying a Fourier transform we get a 2×2 matrix $A(\vec{q})$:

$$\begin{aligned} iA(\vec{q}) &= \begin{pmatrix} 0 & if(\vec{q}) \\ -if(\vec{q}) & 0 \end{pmatrix}, \\ \epsilon(\vec{q}) &= \pm |f(\vec{q})| \,, \end{aligned} \tag{4.88}$$

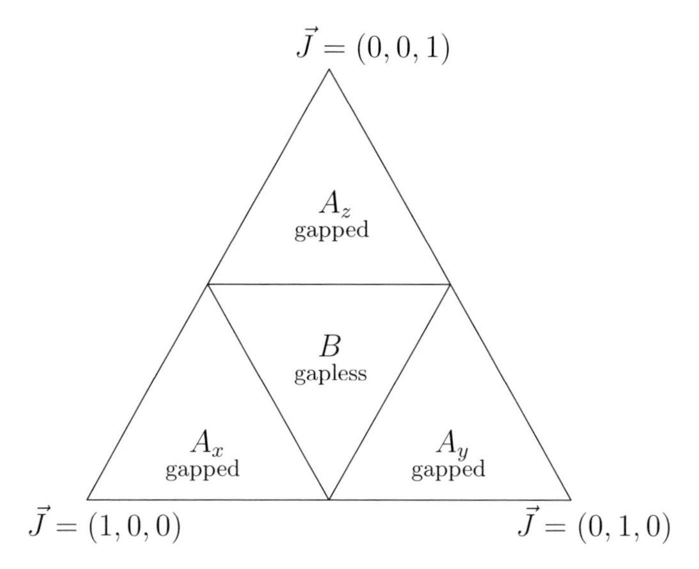

Fig. 4.9 Phase diagram of honeycomb model. This is a slice through the positive octant in the $\vec{J}$ coupling space along the $J_x + J_y + J_z = 1$ plane. The other octants are analogous.

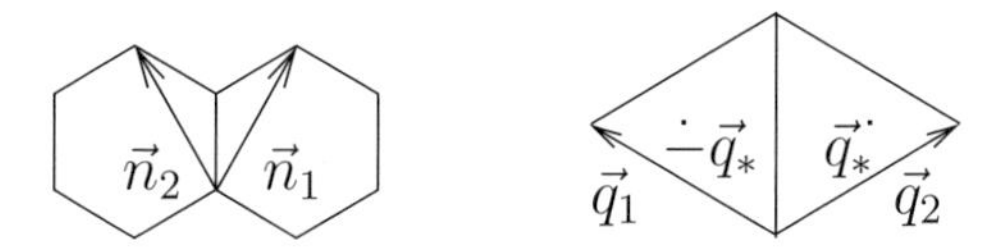

Fig. 4.10 Direct and reciprocal lattices of the honeycomb. The points $\pm\vec{q}_*$ are the two Dirac points of the gapless phase B.

where $f(\vec{q})$ is some complex function that depends on the couplings J_x, J_y, J_z. In the gapless phase (phase B in Fig. 4.9), $f(\vec{q})$ has two zeros, which correspond to Dirac points (see Fig. 4.10). At the transition to phase A, the Dirac points merge and disappear.

4.5.4 Quasiparticle statistics in the gapped phase

It appears that there are two particle types: fermions and vortices (hexagons with $w_p = -1$). The vortices are associated with a $\mathbb{Z}_2$ gauge field, where u_{jk} plays the role of the vector potential. Taking a fermion around a vortex results in multiplication of the state by -1 (compared with the no-vortex case). However, the details such as the fusion rules are not obvious.

Let us look at the model from a different perspective. If $J_x = J_y = 0, J_z > 0$, the system is just a set of dimers (see Fig. 4.11). Each dimer can be in two states: $\uparrow\uparrow$ and $\downarrow\downarrow$. The other two states have an energy that is higher by $2J_z$. Thus, the ground state is highly degenerate.

If $J_x, J_y \ll J_z$, we can use perturbation theory relative to the noninteracting-dimer point. Let us characterize each dimer by an effective spin

$$|\Uparrow\rangle = |\uparrow\uparrow\rangle, \quad |\Downarrow\rangle = |\downarrow\downarrow\rangle. \tag{4.89}$$

At the fourth order of perturbation theory, we get

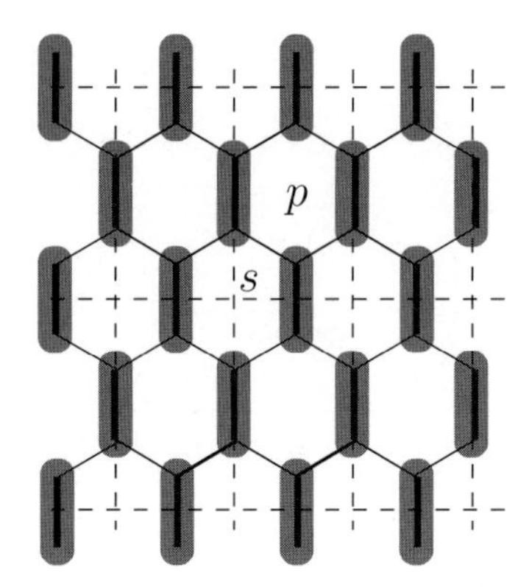

Fig. 4.11 The vertical dimers on the honeycomb lattice themselves form the edges of a square lattice (dashed lines). The plaquettes of alternate rows of the hexagonal lattice correspond to the stars and plaquettes of the square lattice. This is *weak breaking of translational symmetry.*

$$H_{\text{eff}}^{(4)} = \text{const} - \frac{J_x^2 J_y^2}{16 J_z^3} \sum_p Q_p \,, \tag{4.90}$$

where p runs over the square plaquettes of the dimer lattice (see Fig. 4.11) and

$$Q_p = \sigma_{p_1}^y \sigma_{p_2}^x \sigma_{p_3}^y \sigma_{p_4}^x \tag{4.91}$$

is a plaquette operator on the effective spin space $| \Uparrow \rangle, | \Downarrow \rangle$. By adjusting the unit cell and rotating the spins, we can reduce this Hamiltonian to the toric code!

The vertices and plaquettes of the new lattice correspond to alternating rows of hexagons. Thus, vortices on even rows belong to one superselection sector and vortices on odd rows to the other. It is impossible to move a vortex from an even row to an odd row by a local operator without producing other particles (e.g. fermions). The fermions and e–m pairs belong to the same superselection sector, ϵ, though these are different physical states.

4.5.5 Non-Abelian phase

In the gapless phase B, the vortex statistics is not well defined. However, a gap can be opened by applying a perturbation that breaks the time reversal symmetry, such as a magnetic field. Unfortunately, the honeycomb model in a field is not exactly solvable. Yao and Kivelson (2007) studied an exactly solvable spin model where the time reversal symmetry is spontaneously broken, but we will satisfy ourselves by introducing a T-breaking next-nearest-neighbor interaction on the fermionic level (which can be represented by a three-spin interaction in the original spin language).

We consider the following Hamiltonian, written in terms of Majorana fermions;

$$H = \frac{i}{4} \sum_{\langle j,k \rangle} A_{jk} c_j c_k \,, \tag{4.92}$$

where A_{jk} now has chiral terms connecting Majorana fermions beyond nearest neighbors in the honeycomb lattice (see Fig. 4.12). After Fourier transforming, we find

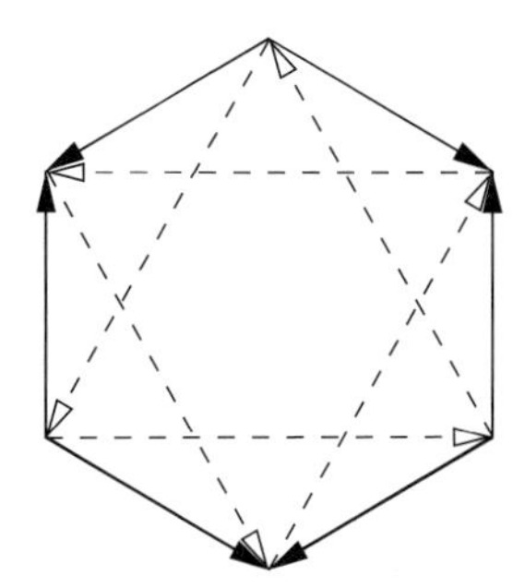

Fig. 4.12 Picture of the chiral interaction matrix A_{jk}. Forward arrows correspond to positive entries in the skew-symmetric real matrix A_{jk}. The solid arrows represent the interactions of the original honeycomb model; the dashed arrows represent the time-reversal symmetry-breaking perturbation.

$$iA(\vec{q}) = \begin{pmatrix} \Delta(\vec{q}) & if(\vec{q}) \\ -if(\vec{q}) & -\Delta(\vec{q}) \end{pmatrix}, \tag{4.93}$$

with the massive dispersion relation

$$\epsilon(\vec{q}) = \pm\sqrt{f(\vec{q})^2 + \Delta(\vec{q})^2}\,. \tag{4.94}$$

Within this massive phase, we will find nontrivial topological invariants of the quasiparticle spectrum. Let $iA(\vec{q})$ be a nondegenerate Hermitian matrix that depends continuously on $\vec{q}$. In our case, A acts in $\mathbb{C}^2$, but in general it can act in $\mathbb{C}^n$ for any n. Let us keep track of the "negative eigenspace" of $iA(\vec{q})$: the subspace $\mathfrak{L}(\vec{q}) \subseteq \mathbb{C}^n$ spanned by eigenvectors corresponding to negative eigenvalues. For the matrix (4.93), $\dim\, \mathfrak{L}(\vec{q}) = 1$. This defines a map F from momentum space (the torus) to the set of m-dimensional subspaces in $\mathbb{C}^n$. More formally,

$$F:\, \mathbb{T}^2 \longrightarrow U(n)/U(m) \times U(n-m)\,. \tag{4.95}$$

This map F may have nontrivial topology.

In the honeycomb model with T-breaking, we have $n = 2$, $m = 1$, and $U(2)/U(1) \times U(1) = \mathbb{C}P^1 = S^2$ is the unit sphere. Thus, $F : \mathbb{T}^2 \longrightarrow S^2$ and, for the matrix $iA(\vec{q})$ of eqn (4.93), F has degree 1. That is, the torus wraps around the sphere once. More abstractly, $\mathfrak{L}(\vec{q})$ defines a complex vector bundle over the momentum space $\mathbb{T}^2$. This has an invariant Chern number ν, which in this case is equal to 1.

What is the significance of the spectral Chern number? It is known to characterize the integer quantum Hall effect, where it is known as the "TKNN invariant." For a Majorana system, there is no Hall effect, since particles are not conserved. Rather, the spectral Chern number determines the number of chiral modes at the edge:

$$\nu = (\text{\# of left-movers}) - (\text{\# of right-movers})\,. \tag{4.96}$$

4.5.6 Robustness of chiral modes

A chiral edge mode may be described by its Hamiltonian

$$H_{\text{edge}} = \frac{iv}{4}\int \hat{\eta}(x)\partial_x\hat{\eta}(x)\,dx\,, \tag{4.97}$$

where $\hat{\eta}(x)$ is a real fermionic field. That is,

$$\hat{\eta}(x)\hat{\eta}(y) + \hat{\eta}(y)\hat{\eta}(x) = 2\delta(x-y)\,. \tag{4.98}$$

At temperature T, each mode carries an energy current

$$I_1 = \frac{\pi}{24}T^2\,. \tag{4.99}$$

The easiest explanation of this is a straightforward 1D Fermi gas calculation:

$$\begin{aligned} I_1 &= v\int_0^\infty n(q)\epsilon(q)\frac{dq}{4\pi} \\ &= \frac{1}{2\pi}\int_0^\infty \frac{\epsilon\, d\epsilon}{1+e^{\epsilon/T}}\,. \\ &= \frac{\pi}{24}T^2\,. \end{aligned} \tag{4.100}$$

However, it is useful to reexamine this current using conformal field theory (CFT), in order to understand better why the chiral modes are robust. We consider a disk of B phase extended into imaginary time at temperature T. That is, we have a solid cylinder with top and bottom identified:

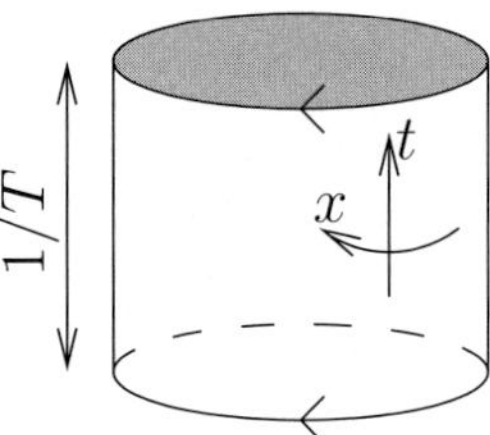

We have obtained a solid torus whose surface is a torus of the usual kind. The partition function is mostly determined by the surface.

Let the spatial dimensions be much greater than $1/T$. From this point of view, the cylinder looks more like

According to the usual CFT arguments, we have

$$\begin{aligned} Z &\sim q^{c/24}\bar{q}^{\bar{c}/24}\,, \\ q &= e^{2\pi i\tau}\,, \\ \tau &= i\frac{LT}{v} + \text{twist}\,. \end{aligned} \tag{4.101}$$

Twisting the torus changes the partition function by

$$\begin{aligned}\tau &\mapsto \tau + 1\,,\\ Z &\mapsto Ze^{2\pi i(c-\bar{c})/24}\,.\end{aligned} \tag{4.102}$$

On the other hand, the twist parameter ($\mathrm{Re}\,\tau$) couples to some component of the energy–momentum tensor, namely T_{xt}, which corresponds to the energy flow. This relation implies that

$$I = \frac{\pi}{12}(c - \bar{c})T^2\,. \tag{4.103}$$

The chiral central charge, $c - \bar{c}$, does not depend on the boundary conditions. Indeed, the energy current on the edge cannot change, because the energy cannot go into the bulk.

References

T. Einarsson (1990). Fractional statistics on a torus. *Phys. Rev. Lett.* **64**(17), 1995–1998.

E. Fradkin and S. H. Shenker (1979). Phase diagrams of lattice gauge theories with Higgs fields. *Phys. Rev. D* **19**(12), 3682–3697.

M. B. Hastings and X. G. Wen (2005). Quasiadiabatic continuation of quantum states: The stability of topological ground-state degeneracy and emergent gauge invariance. *Phys. Rev. B* **72**(4), 045141.

A. Kitaev (2000). Unpaired Majorana fermions in quantum wires. arXiv:cond-mat/0010440v2.

A. Kitaev (2003). Fault-tolerant quantum computation by anyons. *Ann. Phys.* **303**(1), 2–30.

A. Kitaev (2006). Anyons in an exactly solved model and beyond. *Ann. Phys.* **321**(1), 2–111.

E. Lieb (1994). Flux phase of the half-filled band. *Phys. Rev. Lett.* **73**(16), 2158–2161.

N. Read and B. Chakraborty (1989). Statistics of the excitations of the resonating-valence-bond state. *Phys. Rev. B.* **40**(10), 7133–7140.

I. S. Tupitsyn, A. Kitaev, N. V. Prokof'ev, and P. C. E. Stamp (2008). Topological multicritical point in the toric code and 3d gauge Higgs models. arXiv:0804.3175v1.

H. Yao and S. A. Kivelson (2007). Exact chiral spin liquid with non-abelian anyons. *Phys. Rev. Lett.* **99**(24), 247203.

5

Four lectures on computational statistical physics

W. KRAUTH

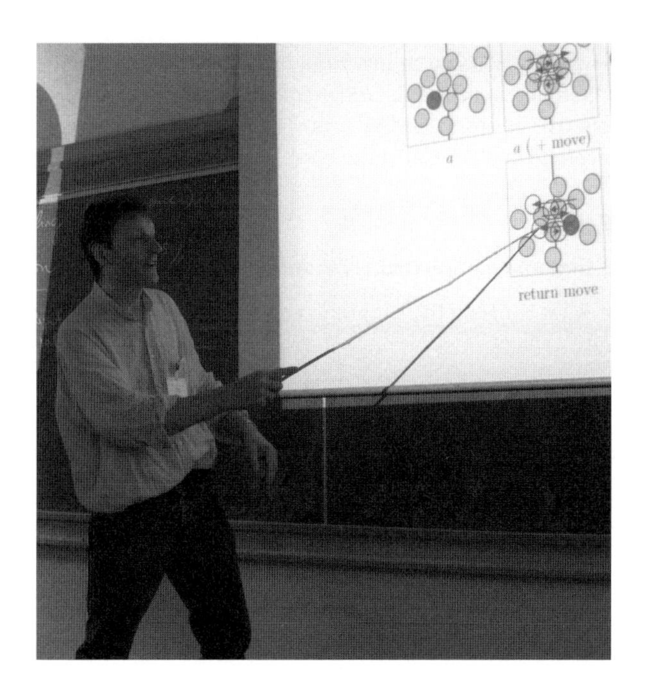

Werner Krauth,
Laboratoire de Physique Statistique,
CNRS-Ecole Normale Supérieure,
24 rue Lhomond,
75231 Paris Cedex 05.

In my lectures at this Les Houches Summer School in 2008, I discussed some central concepts of computational statistical physics, which I felt would be accessible to the very cross-cultural audience at the school.

I started with a discussion of sampling, which lies at the heart of the Monte Carlo approach. I especially emphasized the concept of perfect sampling, which offers a synthesis of the traditional direct and Markov chain sampling approaches. The second lecture concerned classical hard-sphere systems, which illuminate the foundations of statistical mechanics, but also illustrate the curious difficulties that beset even the most recent simulations. I then moved on, in the third lecture, to quantum Monte Carlo methods, which underlie much of the modern work on bosonic systems. Quantum Monte Carlo is an intricate subject. Yet one can discuss it in simplified settings (the single-particle free propagator and ideal bosons) and write direct-sampling algorithms for the two cases in two or three dozen lines of code only. These idealized algorithms illustrate many of the crucial ideas in the field. The fourth lecture attempted to illustrate some aspects of the unity of physics as realized in the Ising model simulations of recent years.

More details of what I discussed in Les Houches, and have written up (and somewhat rearranged) here, can be found in my book *Statistical Mechanics: Algorithms and Computations* (Krauth 2006) (referred to below as SMAC), as well as in recent papers. Computer programs are available for download and perusal at the book's website, www.smac.lps.ens.fr.

5.1 Sampling

5.1.1 Direct sampling and sample transformation

As an illustration of what is meant by sampling, and how it relates to integration, we consider the Gaussian integral

$$\int_{-\infty}^{\infty} \frac{\mathrm{d}x}{\sqrt{2\pi}} \exp\left(\frac{-x^2}{2}\right) = 1 . \tag{5.1}$$

This integral can be computed by taking its square,

$$\left[\int_{-\infty}^{\infty} \frac{\mathrm{d}x}{\sqrt{2\pi}} \exp\left(\frac{-x^2}{2}\right)\right]^2 = \int_{-\infty}^{\infty} \frac{\mathrm{d}x}{\sqrt{2\pi}} \mathrm{e}^{-x^2/2} \int_{-\infty}^{\infty} \frac{\mathrm{d}y}{\sqrt{2\pi}} \mathrm{e}^{-y^2/2} \tag{5.2}$$

$$\ldots = \int_{-\infty}^{\infty} \frac{\mathrm{d}x\,\mathrm{d}y}{2\pi} \exp\left[-(x^2+y^2)/2\right] , \tag{5.3}$$

and then switching to polar coordinates ($\mathrm{d}x\,\mathrm{d}y = r\,\mathrm{d}r\,\mathrm{d}\phi$),

$$\ldots = \int_0^{2\pi} \frac{\mathrm{d}\phi}{2\pi} \int_0^{\infty} r\,\mathrm{d}r\, \exp\left(\frac{-r^2}{2}\right) , \tag{5.4}$$

and performing the substitutions $r^2/2 = \Upsilon$ ($r\,\mathrm{d}r = \mathrm{d}\Upsilon$) and $\exp(-\Upsilon) = \Psi$,

$$\ldots = \int_0^{2\pi} \frac{\mathrm{d}\phi}{2\pi} \int_0^{\infty} \mathrm{d}\Upsilon\, \exp(-\Upsilon) \tag{5.5}$$

$$\ldots = \int_0^{2\pi} \frac{\mathrm{d}\phi}{2\pi} \int_0^{1} \mathrm{d}\Psi = 1 . \tag{5.6}$$

procedure gauss
$\phi \leftarrow \texttt{ran}\,(0, 2\pi)$
$\Psi \leftarrow \texttt{ran}\,(0, 1)$
$\Upsilon \leftarrow -\texttt{log}\,\Psi$
$r \leftarrow \sqrt{2\Upsilon}$
$x \leftarrow r\texttt{cos}\,\phi$
$y \leftarrow r\texttt{sin}\,\phi$
output $\{x, y\}$

Fig. 5.1 The algorithm gauss. SMAC pseudocode program for transforming two uniform random numbers ϕ, Ψ into two independent uniform Gaussian random numbers x, y. Except for typography, SMAC pseudocode is close to the Python programming language.

In our context, it is less important that we can do the integrals in eqn (5.6) analytically than that ϕ and Ψ can be sampled as uniform random variables in the interval $[0, 2\pi]$ (for ϕ) and in $[0, 1]$ (for Ψ). Samples $\phi = \texttt{ran}\,(0, 2\pi)$ and $\Psi = \texttt{ran}\,(0, 1)$ are readily obtained from the random number generator $\texttt{ran}\,(a, b)$ lingering on any computer. We can plug these random numbers into the substitution formulas which took us from eqn (5.2) to eqn (5.6), and that take us now from two uniform random numbers ϕ and Ψ to Gaussian random numbers x and y. We may thus apply the integral transformations in the above equation to the samples or, in other words, perform a "sample transformation." This is a practical procedure for generating Gaussian random numbers from uniform random numbers, and we can best discuss it as what it is, namely an algorithm (Fig. 5.1).

SMAC pseudocode can be implemented in many computer languages.[1] Of particular interest is the computer language Python, which resembles pseudocode, but is executable on a computer exactly as written. In the remainder of these lectures we use and show Python code, as in Algorithm 5.1.

Direct-sampling algorithms exist for arbitrary one-dimensional distributions (see SMAC, Section 1.2.4). Furthermore, arbitrary discrete distributions $\{\pi_1, \ldots, \pi_K\}$ can be directly sampled, after an initial effort of about K operations, with $\sim \log_2 K$ operations by "tower sampling" (see SMAC, Section 1.2.3). This means that sampling a distribution made up of, say, one billion terms takes only about 30 steps. We will use this fact in Section 5.3.3, for a direct-sampling algorithm for ideal bosons. Many trivial multidimensional distributions (for example for noninteracting particles) can also be sampled.

Direct-sampling algorithms also solve much less trivial problems, for example the free path integral, ideal bosons, and the two-dimensional Ising model; in fact, many problems which possess an exact analytic solution. These direct-sampling algorithms are often the racing-car engines inside general-purpose Markov chain algorithms for complicated interacting problems.

Let us discuss the computation of integrals derived from sampled ones $Z = \int \mathrm{d}x\ \pi(x)$ (in our example, the distribution $\pi(x)$ is the Gaussian):

[1]The SMAC Web page www.smac.lps.ens.fr provides programs in languages ranging from Python, Fortran, C, and Mathematica to TI Basic, the language of some pocket calculators.

Algorithm 5.1 gausstest.py

```
from random import uniform as ran , gauss
from math import sin , cos , sqrt , log , pi
def gausstest(sigma):
  phi = ran(0,2*pi)
  Upsilon = -log(ran(0.,1.))
  r = sigma*sqrt(2*Upsilon)
  x = r*cos(phi)
  y = r*sin(phi)
  return x,y
print gausstest(1.)
```

$$\frac{\int \mathrm{d}x \; \mathcal{O}(x)\pi(x)}{\int \mathrm{d}x \; \pi(x)} \simeq \frac{1}{N}\sum_{i=1}^{N} \mathcal{O}(x_i) \, ,$$

where the points x_i are sampled from the distribution $\pi(x)$. This approach, as shown, allows us to compute mean values of observables for a given distribution π. One can also bias the distribution function π using Markov chain approaches. This gives access to a large class of very nontrivial distributions.

5.1.2 Markov chain sampling

Before taking up the discussion of complicated systems (hard spheres, bosons, and spin glasses), we first concentrate on a single particle in a finite one-dimensional lattice $k = 1, \ldots, N$ (see Fig. 5.2). This case is even simpler than the aforementioned Gaussian, because the space is discrete rather than continuous, and because the site-occupation probabilities $\{\pi_1, \ldots, \pi_N\}$ are all the same,

$$\pi_k = \frac{1}{N} \quad \forall k \, . \tag{5.7}$$

We can sample this trivial distribution by picking k as a random integer between 1 and N. Let us nevertheless study a Markov chain algorithm, whose diffusive dynamics converges towards the probability distribution of eqn (5.7). For concreteness, we consider an algorithm where at

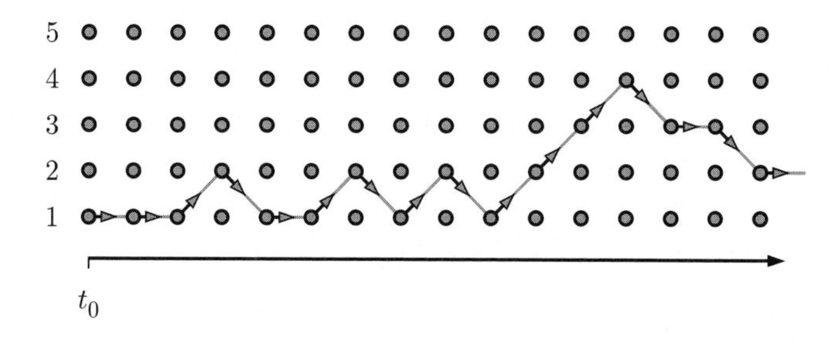

Fig. 5.2 Markov chain algorithm on a five-site lattice. A single particle hops towards a site's neighbor with probability $\frac{1}{3}$, and remains on its site with probability $\frac{1}{3}$ (probability $\frac{2}{3}$ on the boundary).

all integer times $t \in]-\infty, \infty[$, the particle hops with probability $\frac{1}{3}$ from one site to each of its neighbors:

$$p_{k \to k+1} = p_{k \to k-1} = \frac{1}{3} \quad \text{(if possible)}. \tag{5.8}$$

The probabilities to remain on the site are $\frac{1}{3}$ in the interior and, at the boundaries, $p_{1 \to 1} = p_{N \to N} = \frac{2}{3}$.

These transition probabilities satisfy the notorious detailed-balance condition

$$\pi_k p_{k \to l} = \pi_l p_{l \to k} \tag{5.9}$$

for the constant probability distribution π_k. Together with the ergodicity of the algorithm, this guarantees that in the infinite-time limit, the probability π_k to find the particle at site k is indeed independent of k. With appropriate transition probabilities, the Metropolis algorithm would allow us to sample any generic distribution π_k or $\pi(x)$ in one or more dimensions (see SMAC, Section 1.1.5).

Markov chain methods are more general than direct approaches, but the price to pay is that they converge to the target distribution π only in the infinite-time limit. The nature of this convergence can be analyzed by considering the transfer matrix $T^{1,1}$ (the matrix of transition probabilities, in our case a 5×5 matrix):

$$T^{1,1} = \{p_{k \to l}\} = \frac{1}{3} \begin{pmatrix} 2 & 1 & 0 & 0 & 0 \\ 1 & 1 & 1 & 0 & 0 \\ 0 & 1 & 1 & 1 & 0 \\ 0 & 0 & 1 & 1 & 1 \\ 0 & 0 & 0 & 1 & 2 \end{pmatrix}. \tag{5.10}$$

The eigenvalues of the transfer matrix $T^{1,1}$ are $(1, \frac{1}{2} \pm \sqrt{5}/6, \frac{1}{6} \pm \sqrt{5}/6)$. The largest eigenvalue, equal to one, expresses the conservation of the sum of all probabilities $\sum(\pi_1 + \cdots + \pi_5) = 1$. Its corresponding eigenvector is $| \bullet \circ \circ \circ\circ\rangle + \cdots + | \circ \circ \circ \circ\bullet\rangle$, by construction, because of the detailed-balance condition in eqn (5.9). The second-largest eigenvalue, $\lambda_2^{1,1} = \frac{1}{2} + \sqrt{5}/6 = 0.8727$, governs the decay of correlation functions at large times. This is easily seen by computing the probability vector $\boldsymbol{\pi}(t_0 + \tau) = \{\pi_1(t), \ldots, \pi_5(t)\}$, which can be written in terms of the eigenvalues and eigenvectors.

Let us suppose that the simulations always start[2] on site $k = 1$, and let us decompose this initial configuration onto the eigenvectors $\boldsymbol{\pi}(0) = \alpha_1 \boldsymbol{\pi}_1^e + \cdots + \alpha_N \boldsymbol{\pi}_N^e$:

$$\boldsymbol{\pi}(t) = \left(T^{1,1}\right)^t \boldsymbol{\pi}(0) = \boldsymbol{\pi}_1^e + \alpha_2 \lambda_2^t \boldsymbol{\pi}_2^e + \ldots .$$

At large times, corrections to the equilibrium state vanish as λ_2^τ, so that the site probabilities approach the equilibrium value as $\exp(\tau/\tau_{\text{corr}})$ with a time constant

$$\tau_{\text{corr}} = \frac{1}{|\log \lambda_2|} \simeq 7.874 \tag{5.11}$$

(see SMAC, Section 1.1.4, pp. 19f).

[2]We must assume that the initial configuration is different from the stationary solution, because otherwise all the coefficients $\alpha_2, \ldots, \alpha_N$ would be zero.

We retain the result that the exponential convergence of a Monte Carlo algorithm is characterized by a scale, the convergence time τ_{corr}. We also retain the result that convergence takes place after a small multiple of τ_{corr}, in our example say $3 \times 7.87 \sim 25$ iterations (there is absolutely no need to simulate for an infinite time). If our Markov chains always start at $t = 0$ on site 1, it is clear that for all times $t' < \infty$, the occupation probability $\pi_1(t')$ of site 1 is larger than, say, the probability $\pi_5(t')$ to be on site 5. This might make us believe that Markov chain simulations never completely decorrelate from the initial condition, and are always somehow less good than direct sampling. This belief is wrong, as we shall discuss in the next section.

5.1.3 Perfect sampling

We need to simulate for no more than a small multiple of τ_{corr}, but we must not stop our calculation short of this time. This is the critical problem in many real-life situations, where we cannot compute the correlation time as reliably as in our five-site problem: τ_{corr} may be much larger than we suspect, because the empirical approaches for determining correlation times may have failed (see SMAC, Section 1.3.5). In contrast, the problem of the exponentially small tail for $\tau \gg \tau_{\text{corr}}$ is totally irrelevant.

Let us take up again our five-site problem, with the goal of obtaining rigorous information about convergence from within the simulation. As illustrated in Fig. 5.3, the Markov chain algorithm can be formulated in terms of time sequences of random maps: instead of prescribing one move per time step, as we did in Fig. 5.2, we now sample moves independently for all sites k and each time t. At time t_0, for example, the particle should move horizontally from sites 1, 2, and 5, and down from sites 3 and 4, etc. Evidently, for a single particle, there is no difference between the two formulations, and detailed balance is satisfied either way. With the formulation in terms of random maps, we can verify that from time $t_0 + \tau_{\text{coup}}$ on, all initial conditions generate the same output. This so-called "coupling" is of great interest because after the coupling time τ_{coup}, the influence of the initial condition has completely disappeared. In the rest of this lecture, we consider extended Monte Carlo simulations like the one in Fig. 5.3, with arrows drawn for each site and time.

The coupling time τ_{coup} is a random variable ($\tau_{\text{coup}} = 11$ in Fig. 5.3) whose distribution $\pi(\tau_{\text{coup}})$ vanishes exponentially in the limit $\tau_{\text{coup}} \to \infty$ because the random maps at different times are independent.

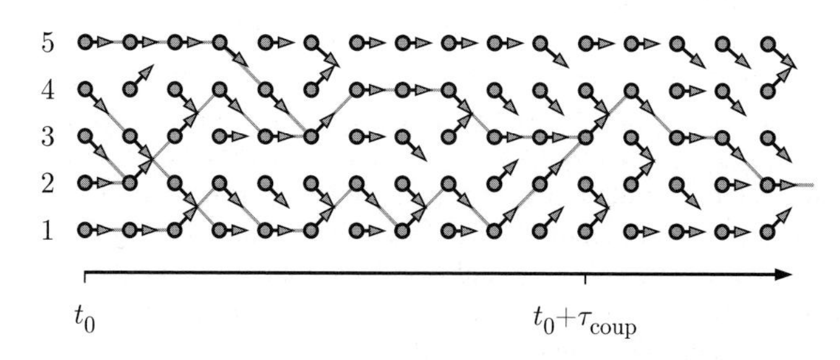

Fig. 5.3 Extended Monte Carlo simulation on $N = 5$ sites. In this example, the coupling time is $\tau_{\text{coup}} = 11$.

The extended Monte Carlo dynamics describes a physical system with from 1 to N particles, and transition probabilities from eqn (5.8), for example

$$\begin{aligned}
p^{\text{fw}}(|\circ\circ\bullet\circ\bullet\rangle &\to |\circ\bullet\circ\bullet\circ\rangle) = 1/9\,,\\
p^{\text{fw}}(|\circ\circ\bullet\circ\bullet\rangle &\to |\circ\circ\bullet\circ\bullet\rangle) = 2/9\,,\\
p^{\text{fw}}(|\circ\circ\bullet\circ\bullet\rangle &\to |\circ\circ\circ\bullet\circ\rangle) = 1/9\,,\\
p^{\text{fw}}(|\circ\circ\bullet\bullet\bullet\rangle &\to |\circ\circ\circ\bullet\circ\rangle) = 1/27\,,
\end{aligned} \tag{5.12}$$

etc. We may analyze the convergence of this system through its transfer matrix T^{fw} (in our case, a 31×31 matrix, because of the 31 nonempty states on five sites):

$$T^{\text{fw}} = \begin{pmatrix} T^{1,1} & T^{2,1} & \dots & \dots \\ 0 & T^{2,2} & T^{3,2} & \dots \\ 0 & 0 & T^{3,3} & \dots \\ \dots & & & \\ 0 & 0 & & T^{N,N} \end{pmatrix},$$

where the block $T^{k,l}$ connects states with k particles at time t to states with $l \leq k$ particles at time $t+1$. The matrix T^{fw} has an eigenvector with eigenvalue 1 which describes the equilibrium, and a second-largest eigenvalue which describes the approach to equilibrium and yields the coupling time τ_{coup}. Because of the block-triangular structure of T^{fw}, the eigenvalues of $T^{1,1}$ constitute the single-particle sector of T^{fw}, including the largest eigenvalue $\lambda_1^{\text{fw}} = 1$, with a corresponding left eigenvector $|\bullet\circ\circ\circ\circ\rangle + \dots + |\circ\circ\circ\circ\bullet\rangle$. The second-largest eigenvalue of T^{fw} belongs to the $(2,2)$ sector. It is given by $\lambda_2^{\text{fw}} = 0.89720 > \lambda_2^{\text{MC}}$, and describes the behavior of the coupling probability $\pi(\tau_{\text{coup}})$ for large times.

Markov chains shake off all memory of their initial conditions at time τ_{coup}, but this time changes from simulation to simulation: it is a random variable. Ensemble averages over (extended) simulations starting at t_0 and ending at $t_0 + \tau$ thus contain mixed averages over coupled and "not yet coupled" runs, and only the latter carry correlations with the initial condition. To reach pure ensemble averages over coupled simulations only, one has to start the simulations at time $t = -\infty$, and go up to $t = 0$. This procedure, termed "coupling from the past" (Propp and Wilson 1996), is familiar to theoretical physicists because in dynamical calculations, the initial condition is often put to $-\infty$. Let us now see how this trick can be put to use in practical calculations.

An extended simulation "from the past" is shown in Fig. 5.4. It has run for an infinite time so that all eigenvalues of the transfer matrix but $\lambda_1 = 1$ have died away and only the equilibrium state has survived. The configuration at $t = 0$ is thus a perfect sample, and each value k ($k \in \{1, \dots, 5\}$) is equally likely, if we average over all extended Monte Carlo simulations. For the specific simulation shown (the choice of arrows in Fig. 5.4), we know that the simulation must pass through one of the five points at time $t = t_0 = -10$. However, the Markov chain couples between $t = t_0$ and $t = 0$, so that we know that $k(t = 0) = 2$. If it did not couple at $t = t_0$, we would have to provide more information (draw more arrows, for $t = t_0 - 1, t_0 - 2, \dots$) until the chain coupled.

Up to now, we have considered the forward transfer matrix, but there is also a backward matrix T^{bw}, which describes the propagation of configurations from $t = 0$ back to $t = -1$,

$t = -2$, etc. A^{bw} is similar to the matrix A^{fw} (the two matrices have identical eigenvalues) because the probability distribution for coupling between times $t = -|t_0|$ and $t = 0$ equals the probability of coupling between times $t = 0$ and $t = +|t_0|$. This implies that the distributions of coupling times in the forward and backward directions are identical.

We have also seen that the eigenvalue corresponding to the coupling time is larger than that for the correlation time. For our one-dimensional diffusion problem, this can be proven exactly. More generally, this is because connected correlation functions decay as

$$\langle \mathcal{O}(0)\mathcal{O}(t)\rangle_c \sim \mathrm{e}^{-t/\tau_{\text{corr}}} .$$

Only the noncoupled (extended) simulations contribute to this correlation function, and their proportion is equal to $\exp\left(-t/\tau_{\text{coup}}\right)$. We arrive at

$$\exp\left[-t\left(\frac{1}{\tau_{\text{coup}}} - \frac{1}{\tau_{\text{corr}}}\right)\right] \sim \langle \mathcal{O}(0)\mathcal{O}(t)\rangle_c^{\text{noncoup}} ,$$

and $\tau_{\text{coup}} > \tau_{\text{corr}}$ because even the noncoupling correlation functions should decay with time.

Finally, in Figs. 5.3 and 5.4, we have computed the coupling time by following all possible initial conditions. This approach is impractical for more complicated problems, such as the Ising model on N sites with its 2^N configurations. Let us show, in the case of the five-site model, how this problem can sometimes be overcome. It suffices to define a new Monte Carlo

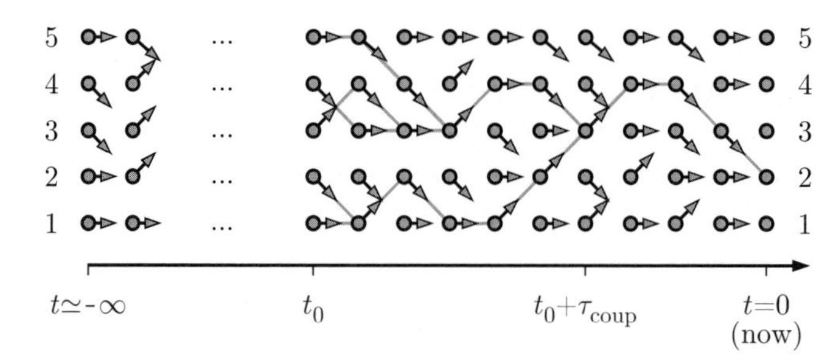

Fig. 5.4 Extended simulation on $N = 5$ sites. The outcome of this infinite simulation, from $t = -\infty$ to $t = 0$, is $k = 2$.

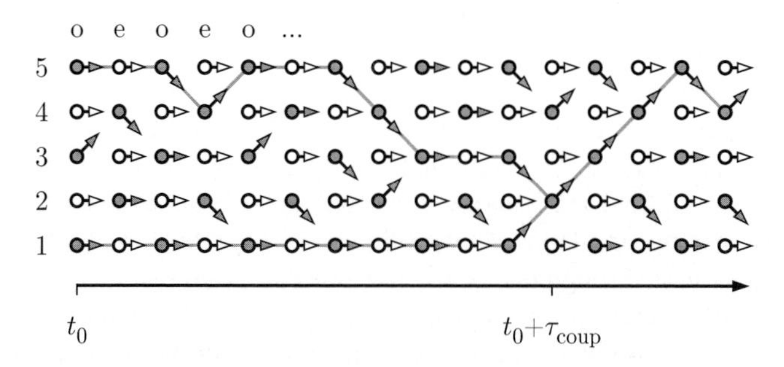

Fig. 5.5 Extended Monte Carlo simulation with odd and even times. Trajectories cannot cross, and the coupling of the two extremal configurations (with $k(t_0) = 1$ and $k(t_0 = 5)$) determines the coupling time.

dynamics in which trajectories cannot cross, by updating, say, even lattice sites ($k = 2, 4$) at every other time $t_0, t_0 + 2, \ldots$ and odd lattice sites ($k = 1, 3, 5$) at times $t_0 + 1, t_0 + 3, \ldots$ (Fig. 5.5). The coupling of the trajectories starting at sites 1 and 5 obviously determines τ_{coup}.

In the Ising model, the ordering relation of Fig. 5.5 survives in the form of a "half-order" between configurations (see SMAC, Section 5.2.2), but in the truly interesting models, such as spin glasses, no such trick is likely to exist. One really must supervise the 2^N configurations at $t = t_0$. This nontrivial task has been studied extensively (see Chanal and Krauth 2008).

5.2 Classical hard-sphere systems

5.2.1 Molecular dynamics

Before statistical mechanics, not so long ago, there was only classical mechanics. The junction between the two has fascinated generations of mathematicians and physicists, and nowhere can it be studied better than in the hard-sphere system: N particles in a box, moving about like billiard balls, with no interaction other than the hard-sphere exclusion (without friction or angular momentum). For more than a century, the hard-sphere model has been a prime example of statistical mechanics and a parade ground for rigorous mathematical physics. For more than fifty years, it has served as a test bed for computational physics, and it continues to play this role.

For concreteness, we first consider four disks (two-dimensional spheres) in a square box with walls (no periodic boundary conditions), as in Fig. 5.6. From an initial configuration, such as the one at $t = 0$, particles fly apart on straight trajectories either until one of them hits a wall or until two disks undergo an elastic collision. The time for this next "event" can be computed exactly by going over all disks (taken by themselves, in the box, to compute the time for the next wall collision) and all pairs of disks (isolated from the rest of the system and from the walls, to determine the next pair collision), and then treating the event closest in time (see SMAC, Section 2.1).

The event-chain algorithm can be implemented in a few dozen lines, just a few too many for a free afternoon in Les Houches (program listings are available on the SMAC website). It implements the entire dynamics of the N-particle system without time discretization, because there is no differential equation to be solved. The only error committed stems from the finite-precision arithmetic implemented on a computer.

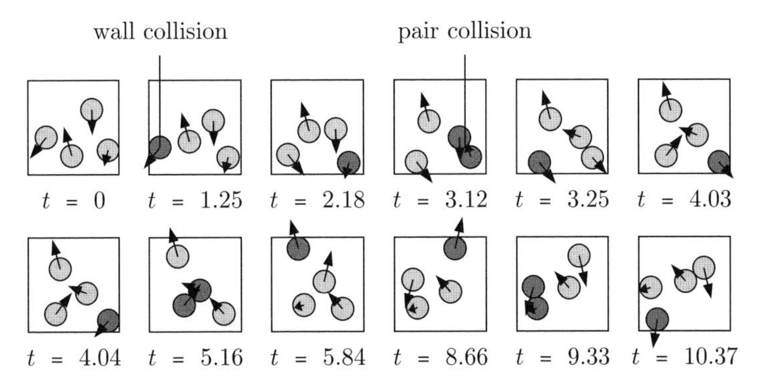

Fig. 5.6 Event-driven molecular dynamics simulation with four hard disks in a square box.

This error manifests itself surprisingly quickly, even in our simulation of four disks in a square: typically, calculations done with 64 bit precision (15 significant digits) get out of step with other calculations starting from identical initial conditions with 32 bit precision after a few dozen pair collisions. This is the manifestation of chaos in the hard-sphere system. The appearance of numerical uncertainties for given initial conditions can be delayed by using even higher-precision arithmetic, but it is out of the question to control a calculation that has run for a few minutes on our laptop and gone through a few billion collisions. Chaos in the hard-sphere model has its origin in the negative curvature of the surfaces of the spheres, which magnifies tiny differences in the trajectory at each pair collision and causes serious roundoff errors in computations.[3]

The mathematically rigorous analysis of chaos in the hard-disk system has met with resounding success: Sinai (1970) (for two disks) and Simanyi (2003, 2004) (for general hard-disk and hard-sphere systems) were able to mathematically prove the foundations of statistical physics for the system at hand, namely the equiprobability principle for hard spheres. This means that during an infinitely long molecular dynamics simulation, the probability density satisfies the following:

$$\left\{ \begin{array}{c} \text{probability of configuration with} \\ [\mathbf{x}_1, \mathbf{x}_1 + \mathrm{d}\mathbf{x}_1], \ldots, [\mathbf{x}_N, \mathbf{x}_N + \mathrm{d}\mathbf{x}_N] \end{array} \right\} \propto \pi(\mathbf{x}_1, \ldots, \mathbf{x}_N)\, \mathrm{d}\mathbf{x}_1, \ldots, \mathrm{d}\mathbf{x}_N\,,$$

where

$$\pi(\mathbf{x}_1, \ldots, \mathbf{x}_N) = \begin{cases} 1 & \text{if configuration legal } (|\mathbf{x}_k - \mathbf{x}_l| > 2\sigma \text{ for } k \neq l)\,, \\ 0 & \text{otherwise}\,. \end{cases} \tag{5.13}$$

An analogous property has been proven for the velocities:

$$\pi(\mathbf{v}_1, \ldots, \mathbf{v}_N) = \begin{cases} 1 & \text{if velocities legal } (\sum \mathbf{v}_k^2 = E_{\text{kin}}/(2m))\,, \\ 0 & \text{otherwise}\,. \end{cases} \tag{5.14}$$

The velocities are legal if they add up to the correct value of the conserved kinetic energy, and their distribution is constant on the surface of the $2N$-dimensional hypersphere of radius $\sqrt{E_{\text{kin}}/(2m)}$ (for disks).

The above two equations contain all of equilibrium statistical physics in a nutshell. The equalprobability principle of eqn (5.13) relates to the principle that two configurations of the same energy have the same statistical weight. The sampling problem for velocities in eqn (5.14) can be solved with Gaussians, as discussed in SMAC, Section 1.2.6. The distribution of velocities reduces to the Maxwell distribution for large N (see SMAC, Section 2.2.4) and implies the Boltzmann distribution (see SMAC, Section 2.3.2).

5.2.2 Direct sampling and Markov chain sampling

We now move on from molecular dynamics simulations to Monte Carlo sampling. To sample N disks with the constant probability distribution of eqn (5.13), we uniformly throw a set of N particle positions $\{\mathbf{x}_1, \ldots, \mathbf{x}_N\}$ into the square. Each of these sets of N positions is generated

[3] Just as it causes humiliating experiences at the billiard table.

Algorithm 5.2 direct-disks.py

```
1  from random import uniform as ran
2  sigma=0.20
3  condition = False
4  while condition == False:
5    L = [(ran(sigma,1-sigma),ran(sigma,1-sigma))]
6    for k in range(1,4): # 4 particles considered
7      b = (ran(sigma,1-sigma),ran(sigma,1-sigma))
8      min_dist = min((b[0]-x[0])**2+(b[1]-x[1])**2 for x in L)
9      if min_dist < 4*sigma**2:
10       condition = False
11       break
12     else:
13       L.append(b)
14       condition = True
15 print L
```

with equal probability. We then take out all those sets that are not legal hard-sphere configurations. The remaining ones (the gray configurations in Fig. 5.7) still have equal probabilities, exactly as called for in eqn (5.13). In the Python programming language, we can implement this algorithm in a few lines (see Algorithm 5.2): we place up to N particles at random positions (see line 7 of Algorithm 5.2). If two disks overlap, we break out of this construction and restart with an empty configuration. The acceptance rate $p_{\text{accept}}(N, \sigma)$ of this algorithm (the probability of generating legal (gray) configurations in Fig. 5.7),

$$p_{\text{accept}} = \frac{\text{number of valid configurations with radius } \sigma}{\text{number of valid configurations with radius } 0} = \frac{Z_\sigma}{Z_0}$$

is exponentially small both in the particle number N and in the density of particles; this is for physical reasons (see the discussion in SMAC, Section 2.2.2)).

For the hard-sphere system, Markov chain methods are much more widely applicable than the direct-sampling algorithm. In order to satisfy the detailed-balance condition of eqn (5.9), we must impose the condition that the probability to move from a legal configuration a to another, b, must be the same as the probability to move from b back to a (see Fig. 5.8). This is realized most easily by picking a random disk and moving it inside a small square around its original position, as implemented in Algorithm 5.3.

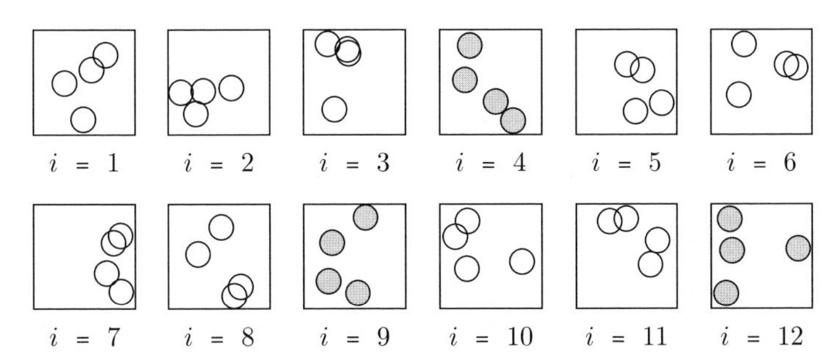

Fig. 5.7 Direct-sampling Monte Carlo algorithm for hard disks in a box without periodic boundary conditions (see Algorithm 5.2).

Algorithm 5.3 markov-disks.py

```
1  from random import uniform as ran , choice
2  L=[(0.25,0.25),(0.75,0.25),(0.25,0.75),(0.75,0.75)]
3  sigma, delta=0.20,0.15
4  for iter in range(1000000):
5    a = choice(L)
6    L.remove(a)
7    b = (a[0] + ran(-delta, delta),a[1] + ran(-delta, delta))
8    min_dist = min((b[0]-x[0])**2 + (b[1]-x[1])**2 for x in L)
9    box_cond = min(b[0],b[1]) < sigma or max(b[0],b[1]) >1-sigma
10   if box_cond or min_dist < 4*sigma**2:
11    L.append(a)
12   else:
13    L.append(b)
14 print L
```

The Monte Carlo dynamics of the Markov chain algorithm shown in Fig. 5.8 superficially resembles the molecular dynamics simulation in Fig. 5.6. Let us look at the essential differences between the two. The Monte Carlo dynamics is diffusive: it can be described in terms of diffusion constants and transfer matrices, and the convergence towards the equilibrium distribution of eqn (5.13) is exponential. In contrast, the dynamics of the event-chain algorithm is hydrodynamic (it contains eddies, turbulence, etc., and their characteristic timescales). Although the event-chain algorithm converges to the same equilibrium state, as discussed before, this convergence is algebraic. This has dramatic consequences, especially in two dimensions (see SMAC, Section 2.2.5), even though the long-time dynamics in a finite box is more or less equivalent in the two cases. This is the fascinating subject of long-time tails, discovered by Alder and Wainwright (1970), which, for lack of time could not be covered in my lectures (see SMAC, Section 2.2.5).

The local Markov chain Monte Carlo algorithm runs into serious trouble at high density, where the Markov chain of configurations effectively gets stuck for long times (although it remains ergodic). The cleanest illustration of this fact is obtained by starting the run with an easily recognizable initial configuration, such as the one shown in Fig. 5.9, which is slightly tilted with respect to the x-axis. We see in this example that $\tau_{\text{corr}} \gg 2.56 \times 10^9$ iterations, even though it remains finite and can be measured in a Monte Carlo calculation.

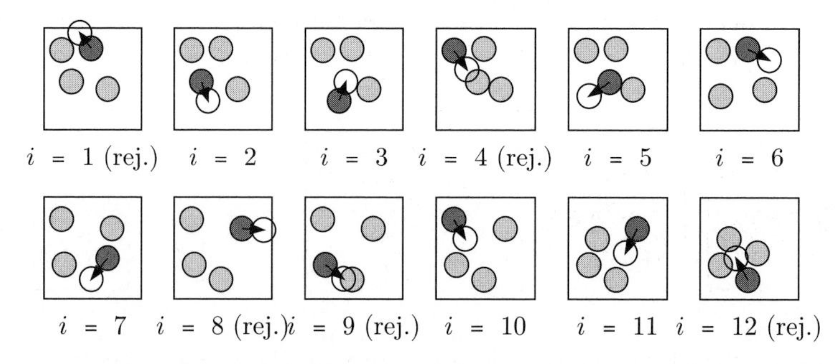

Fig. 5.8 Markov chain Monte Carlo algorithm for hard disks in a box without periodic boundary conditions.

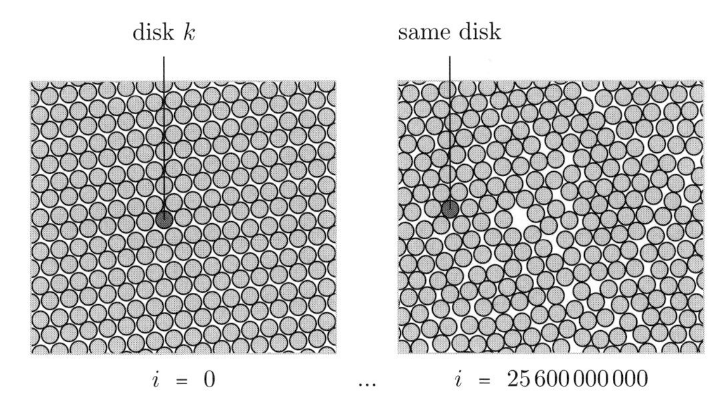

Fig. 5.9 Initial and final configurations for 256 disks at density $\eta = 0.72$ (much smaller than the close-packing density), for 25.6 billion iterations. The sample at iteration $i = 25.6 \times 10^9$ remembers the initial configuration.

Presently, no algorithm essentially faster than the local algorithm is known for uniform hard spheres, and many practical calculations (done with very large numbers of particles at high density) are clearly unconverged. Let us note that the slowdown of the Monte Carlo calculation at high density has a well-defined physical origin: the slowdown of single-particle diffusion at high density. However, this does not exclude the possibility of much faster algorithms, as we will discuss in the fourth lecture.

5.2.3 Cluster algorithms and birth-and-death processes

In the present section, we explore Monte Carlo algorithms that are not inspired by the physical process of single-particle diffusion underlying the local Markov chain Monte Carlo algorithm. Instead of moving one particle after another, cluster algorithms construct coordinated moves of several particles at a time, from one configuration, a, to a very different configuration, b in one deterministic step. The pivot cluster algorithm (Dress and Krauth 1995, Krauth and Moessner 2003) is the simplest representative of a whole class of algorithms.

In this algorithm, we use a "pocket," a stack of disks that eventually have to move. Initially, the pocket contains a random disk. As long as there are disks in the pocket, we take one of them out of it and move it. It gets permanently placed, and all particles that it overlaps with are added to the pocket (see Fig. 5.10; the "pocket particles" are colored dark). In order to satisfy detailed balance, the move $a \to b = T(a)$ must have a Z_2 symmetry $a = T^2(a)$ (such

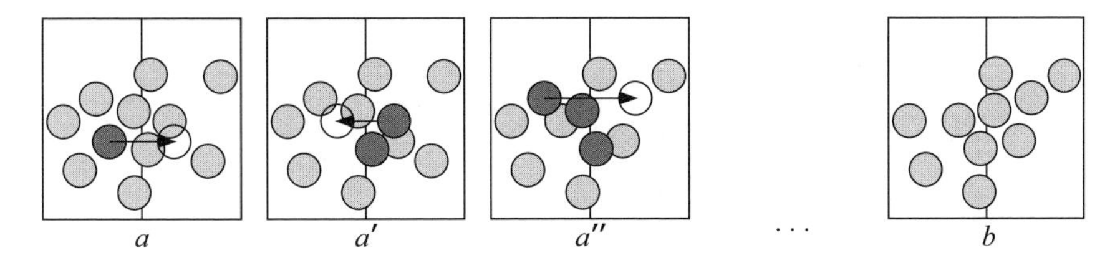

Fig. 5.10 A cluster move about a pivot axis involving five disks. The position and orientation (horizontal, vertical, or diagonal) of the pivot axis are chosen randomly. Periodic boundary conditions allow us to scroll the pivot into the center of the box.

as a reflection about a point, an axis, or a hyperplane), such that performing the move twice brings each particle back to its original position (see SMAC, Section 2.5.2). In addition, the transformation T must map the simulation box onto itself. For example, we can use a reflection about a diagonal in a square or cubic box, but not in a rectangle or cuboid. Periodic boundary conditions are an essential ingredient in this algorithm. In Python, the pocket algorithm can be implemented in a dozen lines of code (see Algorithm 5.4).

The pivot cluster algorithm fails at high density, for example under the conditions of Fig. 5.9, where the transformation T simply transforms all particles in the system. In that case the entire system is transformed without changing the relative particle positions. However, the algorithm is extremely efficient for simulations of monomer–dimer models and of binary mixtures, among other things. An example of this is given in Fig. 5.11.

To end this lecture, let us formulate the problem of hard-sphere systems with a grand canonical partition function and a fugacity λ:

Algorithm 5.4 pocket-disks.py in a periodic box of size 1×1

```
 1 from random import uniform as ran, choice
 2 import box_it, dist # see SMAC website for periodic distance
 3 Others=[(0.25,0.25),(0.25,0.75),(0.75,0.25),(0.75,0.75)]
 4 sigma_sq=0.15**2
 5 for iter in range(10000):
 6    a = choice(Others)
 7    Others.remove(a)
 8    Pocket = [a]
 9    Pivot=(ran(0,1),ran(0,1))
10    while Pocket != []:
11       a = Pocket.pop()
12       a = T(a,Pivot)
13       for b in Others[:]: # "Others[:]" is a copy of "Others"
14       if dist(a,b) < 4*sigma_sq:
15          Others.remove(b)
16          Pocket.append(b)
17       Others.append(a)
18 print Others
```

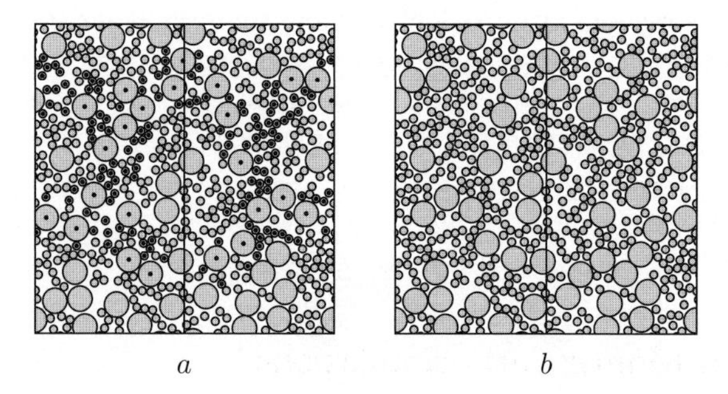

Fig. 5.11 A single iteration (from *a* to *b*) of the cluster algorithm for a binary mixture of hard spheres. The symmetry operation is a flip around the y-axis shown. Transformed disks are marked with a dot (see Buhot and Krauth, 1999).

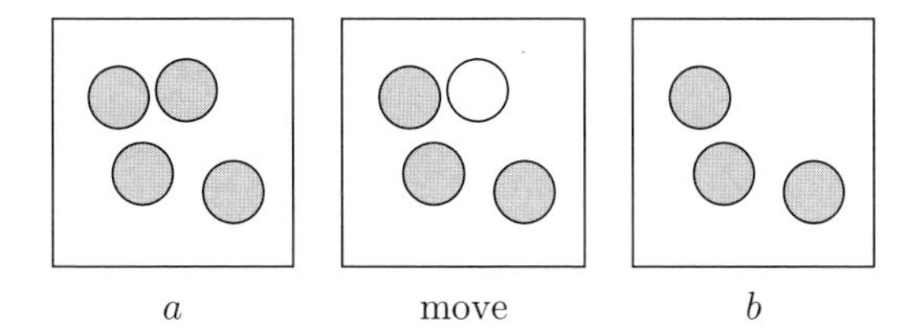

Fig. 5.12 Birth-and-death process for grand canonical hard disks.

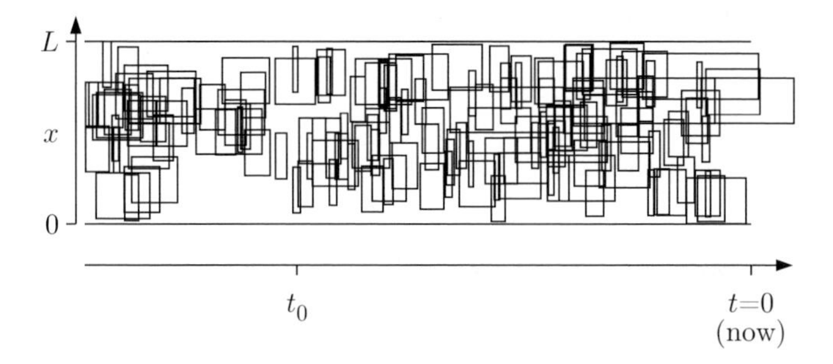

Fig. 5.13 One-dimensional time–space diagram of birth-and-death processes for hard disks in one dimension.

$$Z = \sum_{N=0}^{\infty} \lambda^N \pi(x_1, \ldots, x_N) . \tag{5.15}$$

We shall discuss the related Markov chain Monte Carlo algorithms in terms of birth-and-death processes (no existential connotation intended): between two connected configurations, the configuration can change only through the appearance ("birth") or disappearance ("death") of a disk (see Fig. 5.12). It follows from the detailed-balance condition in eqn (5.9) that the stationary probabilities satisfy $\pi(a) = \lambda\pi(b)$, so that $p(b \to a) = \lambda p(a \to b)$. This means that to sample eqn (5.15), we simply have to install one "death" probability (per time interval $\mathrm{d}t$) for any particle, and a birth probability for creating a new particle anywhere in the system (this particle is rejected if it generates an overlap).

As in the so-called "faster-than-the-clock" algorithms (see SMAC, Section 7.1), one does not discretize time $\tau \to \Delta_\tau$, but rather samples lifetimes and birth times from their exponential distributions. In Fig. 5.13, we show a time–space diagram of all of the events that can happen in one extended simulation (as in the first lecture), in the coupling-from-the-past framework, as used by Kendall and Moller (2000). We do not know the configuration of disks but, on closer examination, we can deduce it, starting from the time t_0 indicated.

5.3 Quantum Monte Carlo simulations

5.3.1 Density matrices and naive quantum simulations

After the connection between classical mechanics and (classical) statistical mechanics, we now investigate the junction between statistical mechanics and quantum physics. We first consider a single particle of mass m in a harmonic potential $V(x) = \frac{1}{2}\omega x^2$, for which the wave functions and energy eigenvalues are all known (see Fig. 5.14; we use units such that $\hbar = m = \omega = 1$).

As before, we are interested in the probability $\pi(x)$ for a particle to be at position x (compare with eqn (5.13)). This probability can be assembled from the statistical weight of level k, $\pi(E_k) \propto \exp(-\beta E_k)$, and the quantum mechanical probability $\psi_k(x)\psi_k^*(x)$ to be at position x while in level k:

$$\pi(x) = \frac{1}{Z(\beta)} \underbrace{\sum_k \exp(-\beta E_k)\, \Psi_k(x)\Psi_k^*(x)}_{\rho(x,x,\beta)} . \tag{5.16}$$

This probability involves a diagonal element of the density matrix given by $\rho(x, x', \beta) = \sum_k \exp(-\beta E_k)\, \Psi_k(x)\Psi_k^*(x)$, whose trace is the partition function

$$Z(\beta) = \int \mathrm{d}x\; \rho(x, x, \beta) = \sum_k \exp(-\beta E_k)$$

(see SMAC, Section 3.1.1). For the free particle and the harmonic oscillator, we can indeed compute the density matrix exactly (see eqn (5.20) later) but in general, eigenstates and energies are out of reach for more complicated Hamiltonians. The path integral approach obtains the density matrix without knowing the spectrum of the Hamiltonian, by using the convolution property

$$\rho^{\text{any}}(x, x', \beta) = \int \mathrm{d}x''\; \rho^{\text{any}}(x, x'', \beta')\, \rho^{\text{any}}(x'', x', \beta - \beta') \quad \text{(convolution)}, \tag{5.17}$$

which yields the density matrix at a given temperature through a product of density matrices at higher temperatures. In the high-temperature limit, the density matrix for the Hamiltonian $H = H^{\text{free}} + V$ is given by the Trotter formula

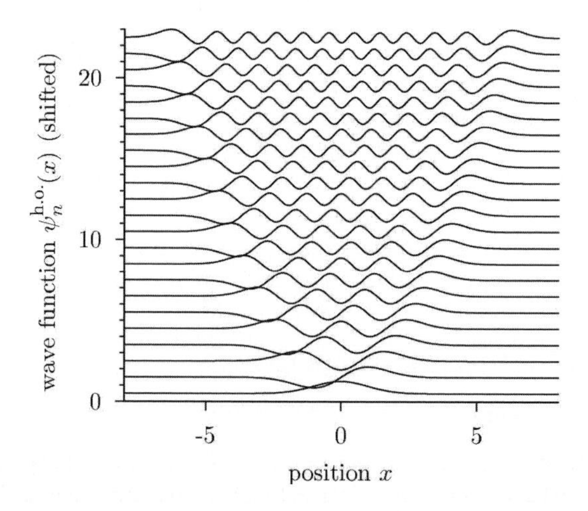

Fig. 5.14 Harmonic-oscillator wave functions ψ_n, shifted by the energy eigenvalue E_n.

$$\rho^{\text{any}}(x, x', \beta) \xrightarrow[\beta \to 0]{} \mathrm{e}^{-(1/2)\beta V(x)} \rho^{\text{free}}(x, x', \beta)\ \mathrm{e}^{-(1/2)\beta V(x')} \quad \text{(Trotter formula)} \tag{5.18}$$

(compare with SMAC, Section 3.1.2). For concreteness, we shall continue our discussion with the harmonic potential $V = \frac{1}{2}x^2$, although the methods discussed are completely general. Using eqn (5.17) repeatedly, we arrive at the path integral representation of the partition function in terms of high-temperature density matrices:

$$\begin{aligned} Z(\beta) &= \int \mathrm{d}x\ \rho(x, x, \beta) \\ &= \int \mathrm{d}x_0 \ldots \int \mathrm{d}x_{N-1}\ \underbrace{\rho(x_0, x_1, \Delta_\tau) \ldots \rho(x_{N-1}, x_0, \Delta_\tau)}_{\pi(x_0, \ldots, x_{N-1})}, \end{aligned} \tag{5.19}$$

where $\Delta_\tau = \beta/N$. In the remainder of this lecture we will focus on this multiple integral, but we are again more interested in sampling (that is, in generating "paths" $\{x_0, \ldots, x_{N-1}\}$ with probability $\pi(x_0, \ldots, x_{N-1})$) than in actually computing it. A naive Monte Carlo sampling algorithm can be set up in a few lines of code (see the algorithm `naive-harmonic-path.py` on the SMAC website). In the integrand of eqn (5.19), we choose a random point $k \in [0, N-1]$ and a uniform random displacement $\delta_x \in [-\delta, \delta]$ (see Fig. 5.15). The acceptance probability of the move depends on the weights $\pi(x_0, \ldots, x_{N-1})$, and thus both on the free density matrix part and on the interaction potential.

The naive algorithm is extremely slow because it moves only a single "bead" k out of N, and not very far from its neighbors $k-1$ and $k+1$. At the same time, displacements of several beads at the same time would be accepted only rarely. In order to go faster through configuration space, our proposed moves must learn about quantum mechanics. This is what we will teach them in the following section.

5.3.2 Direct sampling of a quantum particle: Lévy construction

The problem with naive path sampling is that the algorithm lacks insight: the probability distribution of the proposed moves contains no information about quantum mechanics. However, this problem can be solved completely for a free quantum particle and also for a particle in a harmonic potential $V(x) = \frac{1}{2}x^2$, because in both cases the exact density matrix for a particle of mass $m = 1$, with $\hbar = 1$, is a Gaussian:

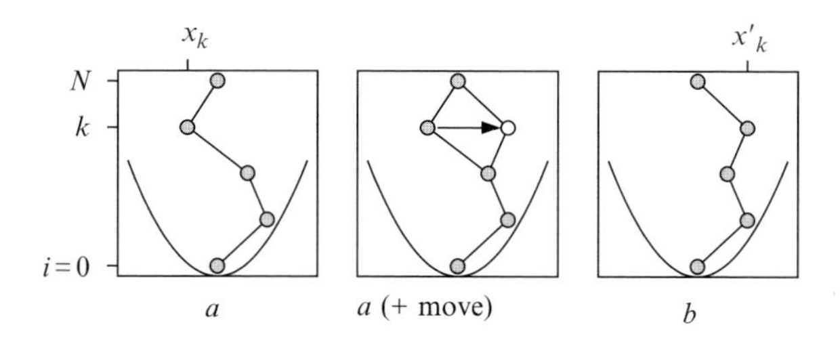

Fig. 5.15 Move of the path $\{x_0, \ldots, x_{N-1}\}$ via element $k \in [0, N-1]$, sampling the partition function in eqn (5.19). For $k = 0$, both 0 and N are displaced in the same direction. The external potential is indicated.

$$\rho(x,x',\beta) = \begin{cases} \frac{1}{\sqrt{2\pi\beta}} \exp\left[-\frac{(x-x')^2}{2\beta}\right] & \text{(free particle)}, \\ \frac{1}{\sqrt{2\pi \sinh\beta}} \exp\left[-\frac{(x+x')^2}{4}\tanh\frac{\beta}{2} - \frac{(x-x')^2}{4}\coth\frac{\beta}{2}\right] & \text{(osc.)}. \end{cases} \tag{5.20}$$

The distribution of an intermediate point, for example that in the left panel of Fig. 5.16, is given by

$$\pi(x_1|x_0,x_N) = \underbrace{\rho\left(x_0,x_1,\frac{\beta}{N}\right)}_{\text{Gaussian, see eqn (5.20)}} \underbrace{\rho\left(x_1,x_N,\frac{(N-1)\beta}{N}\right)}_{\text{Gaussian, see eqn (5.20)}}.$$

As a product of two Gaussians, this is again a Gaussian, and it can be sampled directly. After sampling x_1, one can go on to x_2, etc., until the whole path is constructed, and without rejections (Lévy 1940).

The Lévy construction is exact for a free particle in a harmonic potential and, of course, also for free particles (for which in the Python code is shown in Algorithm 5.5). Direct-sampling algorithms can be tailored to specific situations, such as periodic boundary conditions or hard walls (see SMAC, Sections 3.3.2 and 3.3.3). Moreover, even for generic interactions, the Lévy construction allows us to pull out, treat exactly, and sample without rejections the free-particle Hamiltonian. In the generic Hamiltonian $H = H_0 + V$, the Metropolis rejection then only takes care of the interaction term V. Much larger moves than before become possible and the phase space is run through more efficiently than in the naive algorithm, where the Metropolis rejection concerns the entire H. This makes possible nontrivial simulations with a very large number of interacting particles (Krauth 1996, Holzmann and Krauth 2008). We will illustrate this point in the following section, showing that N-body simulations of ideal (noninteracting) bosons (of arbitrary size) can be done by direct sampling without any rejection.

We note that the density matrices in eqn (5.20), for a single particle in a d-dimensional harmonic oscillator, yield the partition functions

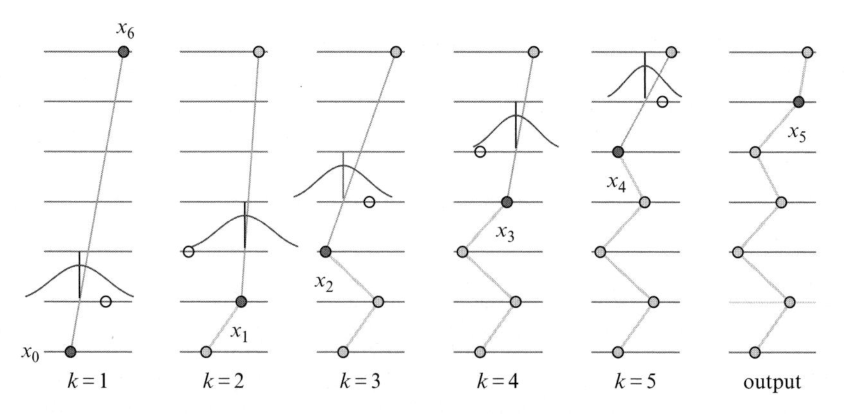

Fig. 5.16 Lévy construction of a path contributing to the free-particle density matrix $\rho(x_0, x_6, \beta)$. The intermediate distributions, all Gaussians, can be sampled directly.

Algorithm 5.5 levy-free-path.py

```
from math import pi, exp, sqrt
from random import gauss
N = 10000
beta = 1.
Del_tau = beta/N
x = [0 for k in range(N+1)]
y = range(N+1)
for k in range(1,N):
    Del_p = (N-k)*Del_tau
    x_mean = (Del_p*x[k-1] + Del_tau*x[N])/(Del_tau + Del_p)
    sigma = sqrt(1./(1./Del_tau + 1./Del_p))
    x[k] = gauss(x_mean, sigma)
print x
```

$$z^{\text{h.o.}}(\beta) = \int \mathrm{d}x\, \rho(x, x, \beta) = [1 - \exp(-\beta)]^{-d} \quad \text{(with } E_0 = 0\text{)}, \tag{5.21}$$

where we have used the lowercase symbol $z(\beta)$ in order to differentiate the one-particle partition function from its N-body counterpart $Z(\beta)$, which we will need in the next section.

5.3.3 Ideal bosons: Landsberg recursion and direct sampling

The density matrix for N distinguishable particles $\rho^{\text{dist}}(\{x_1, \ldots, x_N\}, \{x'_1, \ldots, x'_N\}, \beta)$ is assembled from normalized N-particle wave functions and their associated energies, as in eqn (5.16). The density matrix for indistinguishable particles is then obtained by symmetrizing the density matrix for N distinguishable particles. This can be done either by using symmetrized (and normalized) wave functions to construct the density matrix or, equivalently, by averaging the distinguishable-particle density matrix over all $N!$ permutations (Feynman 1972):

$$Z = \frac{1}{N!} \sum_P \int \mathrm{d}^N x\, \rho^{\text{dist}}(\{x_1, \ldots, x_N\}, \{x_{P(1)}, \ldots, x_{P(N)}\}, \beta) \,. \tag{5.22}$$

This equation is illustrated in Fig. 5.17 for four particles. In each of the diagrams in the figure we arrange the points x_1, x_2, x_3, x_4 from left to right and indicate the permutation by lines. The final permutation (at the bottom right of Fig. 5.17) corresponds, for example, to the permutation $P(1, 2, 3, 4) = (4, 3, 2, 1)$. It consists of two cycles of length 2, because $P^2(1) = P(4) = 1$ and $P^2(2) = P(3) = 2$.

To illustrate eqn (5.22), let us compute the contribution to the partition function stemming from this permutation for free particles:

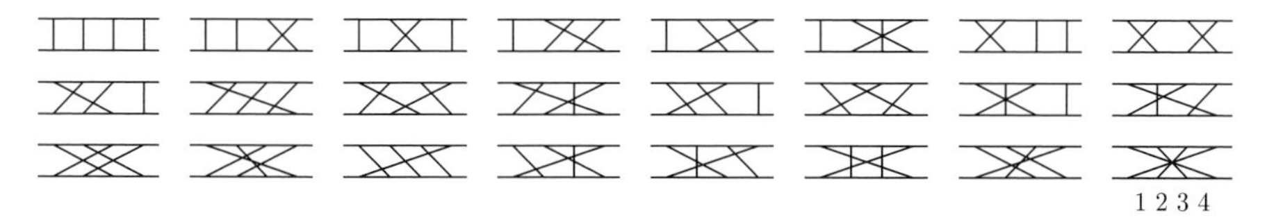

Fig. 5.17 The 24 permutations contributing to the partition function for four ideal bosons.

$$Z_{4321} = \int \mathrm{d}x_1\, \mathrm{d}x_2\, \mathrm{d}x_3\, \mathrm{d}x_4\, \rho^{\text{free}}(\{x_1, x_2, x_3, x_4\}, \{x_4, x_3, x_2, x_1\}, \beta) =$$

$$\underbrace{\left[\int \mathrm{d}x_1\, \mathrm{d}x_4\, \rho^{\text{free}}(x_1, x_4, \beta)\, \rho^{\text{free}}(x_4, x_1, \beta)\right]}_{\int \mathrm{d}x_1\, \rho^{\text{free}}(x_1, x_1, 2\beta) = z(2\beta) \quad \text{see eqn (5.17)}} \left[\int \mathrm{d}x_2\, \mathrm{d}x_3\, \rho^{\text{free}}(x_2, x_3, \beta)\, \rho^{\text{free}}(x_3, x_2, \beta)\right].$$

This partial partition function of a four-particle system can be written as a product of one-particle partition functions. The number of terms corresponds to the number of cycles, and the lengths of the cycles determine the effective inverse temperatures in the system (Feynman 1972).

The partition function for four bosons is given by a sum over 24 terms, five bosons involve 120 terms, and the number of terms in the partition function for a million bosons has more than 5 million digits. Nevertheless, this sum over permutations can be computed exactly through an ingenious recursive procedure with N steps due to Landsberg (1961) (see Fig. 5.18).

Fig. 5.18 contains the same permutations as Fig. 5.17, but they have been rearranged and the last-element cycle (the one containing particle 4) has been pulled out. The pulled-out terms outside the brackets make up the partition function of a single particle at temperature β (in the first row), at inverse temperature 2β (in the second row), etc., as already computed in eqn (5.21). The diagrams within the brackets in Fig. 5.18 contain the $3! = 6$ three-boson partition functions making up Z_3 (in the first row of the figure). The second row contains three times the diagrams making up Z_2, etc. All these terms yield together the terms in eqn (5.23):

$$Z_N = \frac{1}{N}\left(z_1 Z_{N-1} + \cdots + z_k Z_{N-k} + \cdots + z_N Z_0\right) \quad \text{(ideal Bose gas)} \tag{5.23}$$

(with $Z_0 = 1$; see SMAC, Section 4.2.3, for a proper derivation). Z_0 and the single-particle partition functions are known from eqn (5.21), so that we can determine first Z_1, then Z_2, and so on (see the algorithm `harmonic-recursion.py` in SMAC; the SMAC website contains a version with graphical output).

The term in the Landsberg relation of eqn (5.23) $\propto z_k Z_{N-k}$ can be interpreted as a cycle weight, the statistical weight of all permutations in which the particle N is in a cycle of length k:

$$\pi_k = \frac{1}{N Z_N} z_k Z_{N-k} \quad \text{(cycle weight)}.$$

From the Landsberg recursion, we can explicitly compute cycle weights for arbitrarily large ideal-boson systems at any temperature (see Fig. 5.19).

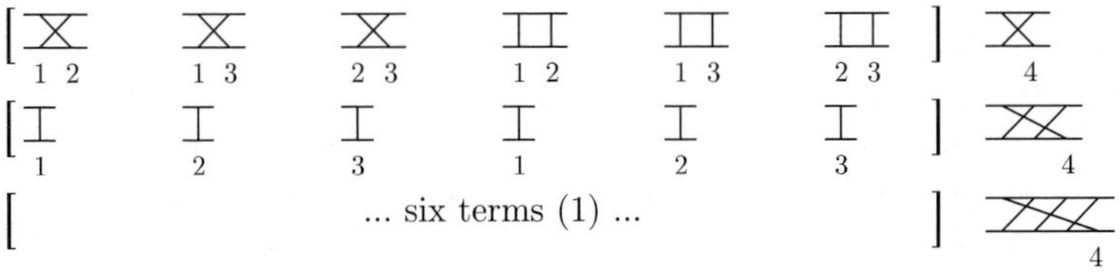

Fig. 5.18 The 24 permutations shown in Fig. 5.17, with the last-element cycle pulled out of the diagram.

Algorithm 5.6 canonic-recursion.py

```
1  import math
2  def z(beta,k):
3      sum = 1/(1- math.exp(-k*beta))**3
4  return sum
5  N=1000
6  beta=1./5
7  Z=[1.]
8  for M in range(1,N+1):
9      Z.append(sum(Z[k] * z(beta,M-k) for k in range(M))/M)
10 pi_list=[z(beta,k)*Z[N-k]/Z[N]/N for k  in range(1,N+1)]
```

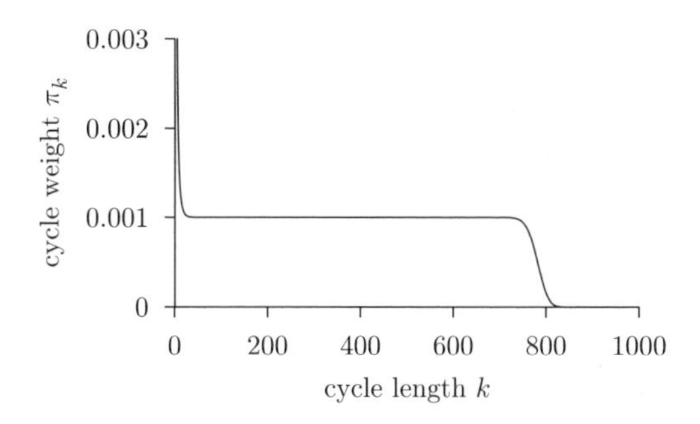

Fig. 5.19 Cycle weight distribution for 1000 ideal bosons in a three-dimensional harmonic trap, obtained from Algorithm 5.6 exactly as written.

In Fig. 5.19, obtained with Algorithm 5.6, we notice that the cycle weight distribution π_k is flat for most values of k before it drops to zero around $k \sim 780$. Curiously, the derivative of this function yields the distribution of the condensate fraction (Holzmann and Krauth 1999, Chevallier and Krauth 2007), so that we see that at the temperature chosen, there are about 780 particles in the ground state ($N_0/N \simeq 0.78$). At higher temperatures, the distribution of cycle weights is narrower. This means that there are no long cycles.

We now turn eqn (5.23) around, as we did first for the Gaussian integral: rather than computing $Z(\beta)$ from the cycle weights, we sample the cycle distribution from its weights; that is, we pick one of the π_k with

$$\pi_1, \ldots, \pi_k, \ldots, \pi_N .$$

(we pick a cycle length 1 with probability π_1, 2 with probability π_2, and so on; it is best to use the tower-sampling algorithm that we mentioned in the first lecture). Suppose we have sampled a cycle length k. We then know that our Bose gas contains a cycle of length k, and we can sample the three-dimensional positions of the k particles on this cycle from the three-dimensional version of the algorithm `levy-free-path.py` (Algorithm 5.5) adapted to the harmonic potential. We must consider k particles instead of N, at inverse temperature $k\beta$. Thereafter, we sample the next cycle length from the Landsberg recursion relation with $N - k$ particles instead of N, and so on, until all particles are used up (see the appendix to this chapter for a complete program in 44 lines). Some output from this program is shown in Fig. 5.20, projected onto two dimensions. As in a real experiment, the three-dimensional

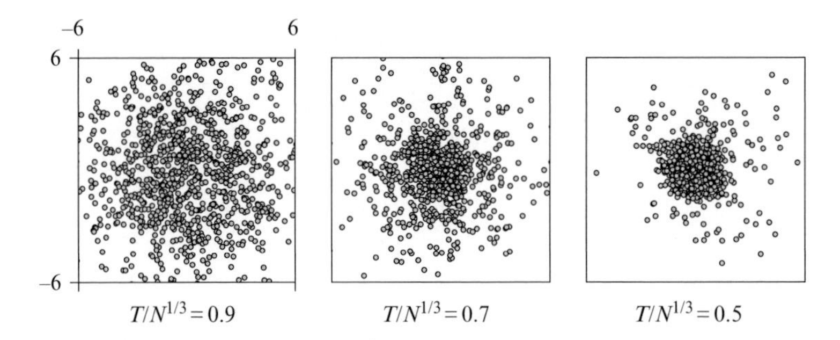

Fig. 5.20 Two-dimensional snapshots of 1000 ideal bosons in a three-dimensional harmonic trap (obtained with the direct-sampling algorithm).

harmonic potential confines the particles but, as we pass through the Bose–Einstein transition temperature (roughly at the temperature of the left panel of Fig. 5.20), they start to move to the center and to produce the landmark peak in the spatial density. At the temperature of the right panel, roughly 80% of particles are condensed into the ground state.

The power of the path integral approach resides in the facility with which interactions can be included. This goes beyond what can be treated in an introductory lecture (see SMAC, Section 3.4.2, for an in-depth discussion). For the time being we should take pride in our rudimentary sampling algorithm for ideal bosons, a true quantum Monte Carlo program in a nutshell.

5.4 Spin systems: Samples and exact solutions

5.4.1 Ising Markov chains: Local moves and cluster moves

In this final lecture, we study models of discrete spins $\{\sigma_1, \ldots, \sigma_N\}$ with $\sigma_k = \pm 1$ on a lattice with N sites, with energy

$$E = -\sum_{\langle k,l \rangle} J_{kl} \sigma_k \sigma_l \,. \tag{5.24}$$

Each pair of neighboring sites k and l is counted only once. In eqn (5.24), we may choose all of the J_{kl} equal to $+1$. We then have the ferromagnetic Ising model. If we choose random values $J_{kl} = J_{lk} = \pm 1$, we speak of the Edwards–Anderson spin glass model. Together with the hard-sphere systems, these models belong to the hall of fame of statistical mechanics, and have been the crystallization points for many developments in computational physics. Our goal in this lecture will be twofold. We shall illustrate several algorithms for simulating Ising models and Ising spin glasses in a concrete setting. We shall also explore the relationship between the Monte Carlo sampling approach and analytic solutions, in this case the analytic solution for the two-dimensional Ising model initiated by Onsager (1944), Kac and Ward (1952), and Kaufman (1949).

The simplest Monte Carlo algorithm for sampling the partition function

$$Z = \sum_{\text{confs } \boldsymbol{\sigma}} \exp\left[-\beta E(\boldsymbol{\sigma})\right]$$

picks a site at random, for example the central site of the square lattice in configuration a in Fig. 5.21. Flipping this spin would produce the configuration b. To satisfy detailed balance,

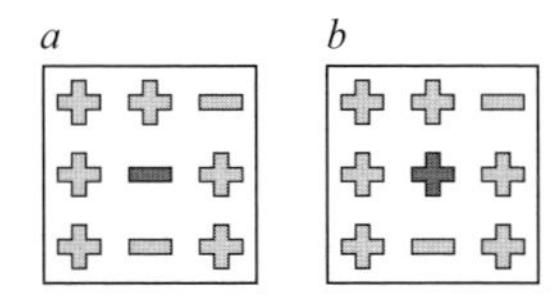

Fig. 5.21 A spin flip in the Ising model. Configuration a has a central field $h = 2$ (three "+" neighbors and one "−" neighbor), and configuration b has $h = -2$. The statistical weights satisfy $\pi_b/\pi_a = \exp(-4\beta)$.

eqn (5.9), we must accept the move $a \to b$ with probability $\min(1, \pi_b/\pi_a)$, as implemented in Algorithm 5.7 (see SMAC, Section 5.2.1). This algorithm is very slow, especially near the phase transition between the low-temperature ferromagnetic phase and the high-temperature paramagnet. This slowdown is due to the fact that, close to the transition, configurations with very different values of the total magnetization $M = \sum_k \sigma_k$ contribute with considerable weight to the partition function (in two dimensions, the distribution ranges practically from $-N$ to N). One step of the local algorithm changes the total magnetization by at most a tiny amount, ± 2, and it thus takes approximately N^2 steps (as in a random walk) to go from one configuration to an independent one. This, in a nutshell, is the phenomenon of critical slowing down.

One can overcome critical slowing down by flipping a whole cluster of spins simultaneously, using moves that know about statistical mechanics (Wolff 1989). It is best to start from a random site, and then to repeatedly add to the cluster, with probability p, the neighbors of sites already present if the spins all have the same sign. At the end of the construction, the whole cluster is flipped. We now compute the value of p for which this algorithm has no rejections (see SMAC, Section 5.2.3).

The cluster construction stops at the configuration a of Fig. 5.22 with probability $p(a \to b) = \text{const}(1-p)^{14}$, one factor of $(1-p)$ for every link "++" across the cluster boundary. Likewise, the cluster construction stops at the configuration b if the 18 neighbors "−−" have

Algorithm 5.7 markov-ising.py

```
1  from random import uniform as ran, randint, choice
2  from math import exp
3  import square_neighbors # defines the neighbors of a site
4  L = 32
5  N = L*L
6  S = [choice([-1,1]) for k in range(N)]
7  beta = 0.42
8  nbr = square_neighbors(L)
9  # nbr[k] = (right, up, left, down)
10 for i_sweep in range(100):
11     for iter in range(N):
12        k = randint(0,N-1)
13        h = sum(S[nbr[k][j]] for j in range(4))
14        Delta_E = 2.*h*S[k]
15        Upsilon = exp(-beta*Delta_E)
16        if ran(0.,1.) < Upsilon: S[k] = -S[k]
17 print S
```

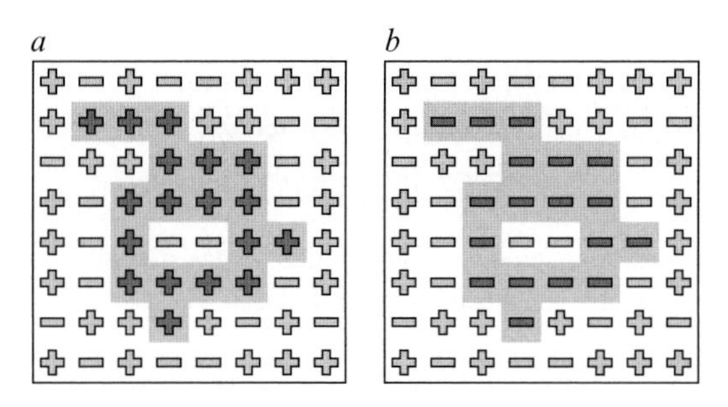

Fig. 5.22 A large cluster of like spins in the Ising model. The construction stops at the gray cluster in a with probability $(1-p)^{14}$, corresponding to the 14 "++" links across the boundary. The corresponding probability for b is $(1-p)^{18}$.

been considered without success (this happens with probability $p(b \to a) = \text{const}(1-p)^{18}$). The detailed-balance condition relates the construction probabilities to the Boltzmann weights of the configurations

$$\pi_a = \text{const}' \exp\left[-\beta(-14+18)\right] ,$$
$$\pi_b = \text{const}' \exp\left[-\beta(-18+14)\right] ,$$

where the "const$'$" describes all contributions to the energy not coming from the boundary. We may enter stopping probabilities and Boltzmann weights into the detailed-balance condition $\pi_a p(a \to b) = \pi_b \pi(b \to a)$, and find

$$\exp(14\beta)\exp(-18\beta)(1-p)^{14} = \exp(-14\beta)\exp(18\beta)(1-p)^{18} . \tag{5.25}$$

This is true for $p = 1 - \exp(-2\beta)$, and is independent of our example, with its 14 "++" and 18 "−−" links between neighbors (see SMAC, Section 4.2.3).

The cluster algorithm can be implemented in a few lines (see Algorithm 5.8) using the pocket approach of Section 5.2.3. Let the "pocket" comprise those sites of the cluster whose neighbors have not already been scrutinized. The algorithm starts by putting a random site into both the cluster and the pocket. It then takes a site out of the pocket, and adds neighboring sites (to the cluster and to the pocket) with probability p if their spins are the same, and if they are not already in the cluster. The construction ends when the pocket is empty. We then flip the cluster.

Cluster methods play a crucial role in computational statistical physics because, unlike local algorithms and unlike experiments, they do not suffer from critical slowing down. These methods have spread from the Ising model to many other fields of statistical physics.

Today, the nonintuitive rules for the cluster construction are well understood, and the algorithms are very simple. In addition, modern metalanguages are so powerful that a rainy Les Houches afternoon provides ample time to implement the method, even for a complete nonexpert in the field.

5.4.2 Perfect sampling: Semiorder and patches

The local Metropolis algorithm picks a site at random and flips it with a probability $\min(1, \pi_b/\pi_a)$ (as in Section 5.4.1). An alternative local Monte Carlo scheme is the heat bath algorithm, where the spin is equilibrated in its local environment (see Fig. 5.24). This means that

Algorithm 5.8 cluster-ising.py

```
from random import uniform as ran, randint, choice
from math import exp
import square_neighbors # defines the neighbors of a site
L = 32
N = L*L
S = [choice([-1,1]) for k in range(N)]
beta = 0.4407
p = 1-exp(-2*beta)
nbr = square_neighbors(L) # nbr[k]= (right, up, left, down)
for iter in range(100):
    k = randint(0,N-1)
    Pocket = [k]
    Cluster = [k]
    while Pocket != []:
        k = choice(Pocket)
        for l in nbr[k]:
            if S[l] == S[k] and l not in Cluster and ran(0,1) < p:
                Pocket.append(l)
                Cluster.append(l)
        Pocket.remove(k)
    for k in Cluster: S[k] = -S[k]
print S
```

Fig. 5.23 A large cluster with 1548 spins in a 64×64 Ising model with periodic boundary conditions. All the spins in the cluster flip together from "+" to "−."

in the presence of a molecular field h at site k, the spin points up and down with probabilities π_h^+ and π_h^-, respectively, where

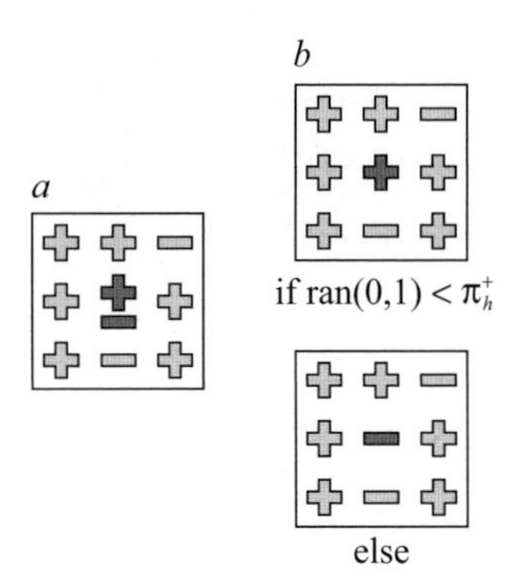

Fig. 5.24 Heat bath algorithm for the Ising model. The new position of the spin k (in configuration b) is independent of its original position (in a).

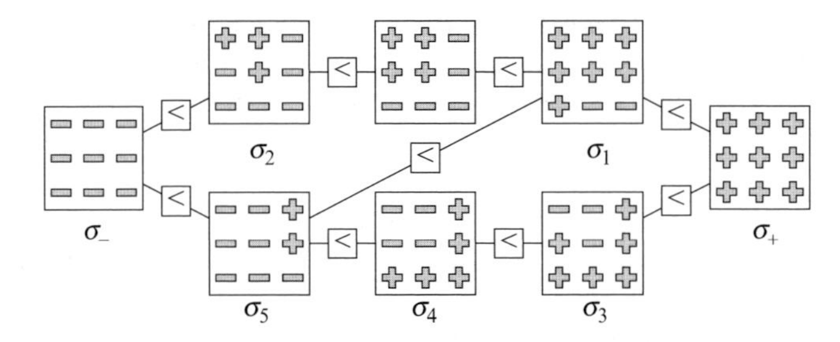

Fig. 5.25 Half-order in spin models: the configuration $\boldsymbol{\sigma}_-$ is smaller than and $\boldsymbol{\sigma}_+$ is larger than all other configurations, which are not all related to each other in this way.

$$\begin{aligned}\pi_h^+ &= \frac{\mathrm{e}^{-\beta E^+}}{\mathrm{e}^{-\beta E^+} + \mathrm{e}^{-\beta E^-}} = \frac{1}{1 + \mathrm{e}^{-2\beta h}}, \\ \pi_h^- &= \frac{\mathrm{e}^{-\beta E^-}}{\mathrm{e}^{-\beta E^+} + \mathrm{e}^{-\beta E^-}} = \frac{1}{1 + \mathrm{e}^{+2\beta h}}.\end{aligned} \tag{5.26}$$

The heat bath algorithm (which is again much slower than the cluster algorithm, especially near the critical point) couples just like our trivial simulation of the five-site model in Section 5.1.2. At each step, we pick a random site k and a random number $\Upsilon = \texttt{ran}\,(0, 1)$, and apply the Monte Carlo update of Fig. 5.24 with the same k and the same Υ to all configurations of the Ising model or spin glass. After a time τ_{coup}, all input configurations yield identical output (Propp and Wilson 1996).

For the Ising model (but not for the spin glass), it is very easy to compute the coupling time, because the half-order among spin configurations is preserved by the heat bath dynamics (see Fig. 5.25): we say that a configuration $\boldsymbol{\sigma} = \{\sigma_1, \ldots, \sigma_N\}$ is smaller than another configuration $\boldsymbol{\sigma}' = \{\sigma'_1, \ldots, \sigma'_N\}$ if, for all k, we have $\sigma_k \leq \sigma'_k$.[4] For the Ising model, the heat bath algorithm preserves the half-order between configurations upon update because a configuration which is smaller than another one has a smaller field on all sites, thus a smaller value of π_h^+. We just have to start the simulation from the all-plus-polarized and the all-minus-polarized

[4]This is a half-order because not all configurations can be compared with each other.

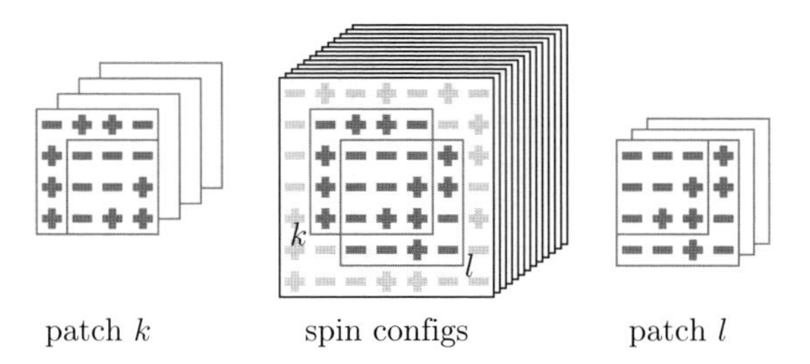

Fig. 5.26 The very large configuration space of an Ising spin glass on N sites viewed from the vantage point of N patches of smaller size M.

configurations and wait until these two extremal configurations couple. This determines the coupling time for all 2^N configurations, exactly as in the earlier trivial example of Fig. 5.5.

The Ising spin glass does not allow such simple tricks to be played, and we must explicitly check the coupling for all 2^N initial configurations. This enormous surveying task can be simplified considerably by breaking up the configurations on the entire lattice into smaller "patches" (see Fig. 5.26). For a given patch size M ($M = 16$ in Fig. 5.26), the total number of configurations on N patches is bounded by $N2^M \ll 2^N$. We can now follow the heat bath simulation on all individual patches and, at the end, assemble configurations on the whole lattice in the same way as we assemble an entire jigsaw puzzle from the individual pieces. The procedure can be made practical, and programmed very easily in modern metalanguages such as Python.

From a more fundamental point of view, the concept of coupling in spin glass models is of interest because the physical understanding of these systems has suffered seriously from long-standing doubts about the quality of Monte Carlo calculations and thus, at the most basic level, about the correctness of calculations of the correlation time.

5.4.3 Direct sampling and the Onsager solution

The two-dimensional Ising model is exactly solvable, as shown by Onsager (1944): we can compute its critical temperature, and many critical exponents. To be more precise, the partition function of the two-dimensional Ising model or a spin glass[5] on a planar lattice with N sites can be expressed as the square root of the determinant of a $4N \times 4N$ matrix (see SMAC, Section 5.1.4, for a practical algorithm). Periodic boundary conditions can also be handled; this gives four $4N \times 4N$ matrices. This was used very successfully by Saul and Kardar (1993). For the Ising model on a finite square lattice, these matrices can be diagonalized analytically. This gives the famous analytic expression for $Z(\beta)$ of Kaufman (1949) (see SMAC, Exercise 5.9, p. 265).

The solution of the Ising model amounts to summing its high-temperature series, and this solves an enumeration problem, as evidenced in the combinatorial solution by Kac and Ward (1952). Indeed, the density of states of the energy $\mathcal{N}(E)$ can be extracted from the analytic solution (see Beale (1996)); this means that we can obtain the (integer) number of states for any energy E for two-dimensional Ising spin glasses of very large sizes. The exact solution

[5] We speak here of one single "sample" of the spin glass, that is, a given choice of the couplings $\{J_{kl}\}$. To compute the average over all couplings is another matter.

of the Ising model thus performs an enumeration, but it only counts configurations; it cannot list them. The final point that we shall elaborate on in these lectures is that, although we are unable to list configurations, we can still sample them.

The subtle difference between counting and listing of sets implies that while we can access without any trouble the density of states $\mathcal{N}(E)$, we cannot obtain the distribution (the histogram) of magnetizations, and we know very little about the joint distribution of energies and magnetizations, $\mathcal{N}(M, E)$. Obtaining these distributions would in fact allow us to solve the Ising model in a magnetic field, which is not possible.

Let us now outline a sampling algorithm (Chanal and Krauth 2009), which uses the analytic solution of the Ising model to compute $\mathcal{N}(M, E)$ (with only statistical, and no systematic errors) even though this is impossible to do exactly. This algorithm constructs the sample one site after another. Let us suppose that the gray spins in the left panel of Fig. 5.27 are already fixed, as shown. We can now set a fictitious coupling J^*_{ll} either to $-\infty$ or to $+\infty$ and recalculate the partition function $Z_\pm$ with these two values. The statistical weight of all configurations in the original partition function with spin "+" is then given by

$$\pi_+ = \frac{Z_+ \exp(\beta J_{kl})}{Z_+ \exp(\beta J_{kl}) + Z_- \exp(-\beta J_{kl})}, \tag{5.27}$$

and this two-valued distribution can be sampled with one random number. Equation (5.27) resembles the heat bath algorithm of eqn (5.26), but it is not part of a Markov chain. After obtaining the value of the spin on site k, we keep the fictitious coupling, and add more sites. By going over all sites, we can generate direct samples of the partition function of eqn (5.24), at any temperature, and with a fixed, temperature-independent effort, for both the two-dimensional Ising model and the Ising spin glass. This allows us to obtain arbitrary correlation functions, the generalized density of states $\mathcal{N}(M, E)$, etc. We can also use this solution to determine the behavior of the Ising model in a magnetic field, which we cannot obtain from the analytic solution, although we use it in the algorithm.

The correlation-free direct-sampling algorithm is primarily of theoretical interest, as it is restricted to the two-dimensional Ising model and the Ising spin glass, which are already well understood. For example, it is known that the two-dimensional spin glass does not have a finite transition temperature. Nevertheless, it is intriguing that one can obtain, from the analytical solution, exactly and with a performance guarantee, quantities that the analytical solution

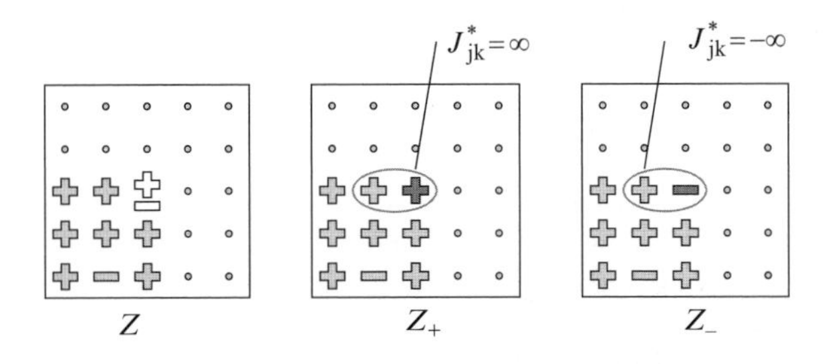

Fig. 5.27 One iteration in the direct-sampling algorithm for the two-dimensional Ising model. The probabilities $\pi(\sigma_k = +1)$ and $\pi(\sigma_k = -1)$ (with k the central spin) are obtained from the exact solution of the Ising model with fictitious couplings $J^*_{jk} = \pm\infty$.

cannot give. The "sampling" (even the very well-controlled, full-performance-guarantee direct sampling used here) does not have the limitations of complete enumerations.

Appendix

Algorithm 5.9 is an entire quantum Monte Carlo program (in Python 2.5) for ideal bosons in a harmonic trap (see the SMAC website for a program with graphical output).

Algorithm 5.9 direct-harmonic-bosons.py

```
from math import sqrt, sinh, tanh, exp
from random import uniform as ran, gauss
def z(beta,k):
    sum = 1/(1-exp(-k*beta))**3
    return sum
def canonic_recursion(beta,N):
    Z=[1.]
    for M in range(1,N+1):
        Z.append(sum(Z[k]*z(beta,M-k) for k in range(M))/M)
    return Z
def pi_list_make(Z,M):
    pi_list=[0]+[z(beta,k)*Z[M-k]/Z[M]/M for k in range(1,M+1)]
    pi_sum=[0]
    for k in range(1,M+1):
        pi_sum.append(pi_sum[k-1]+pi_list[k])
    return pi_sum
def tower_sample(data,Upsilon): #naive, cf. SMAC Sect. 1.2.3
    for k in range(len(data)):
        if Upsilon<data[k]: break
    return k
def levy_harmonic_path(Del_tau,N): #
    beta=N*Del_tau
    xN=gauss(0.,1./sqrt(2*tanh(beta/2.)))
    x=[xN]
    for k in range(1,N):
        Upsilon_1 = 1./tanh(Del_tau)+1./tanh((N-k)*Del_tau)
        Upsilon_2 = x[k-1]/sinh(Del_tau)+xN/sinh((N-k)*Del_tau)
        x_mean=Upsilon_2/Upsilon_1
        sigma=1./sqrt(Upsilon_1)
        x.append(gauss(x_mean,sigma))
    return x
N=512
beta=1./2.4
Z=canonic_recursion(beta,N)
M=N
x_config=[]
y_config=[]
while M > 0:
    pi_sum=pi_list_make(Z,M)
    Upsilon=ran(0,1)
    k=tower_sample(pi_sum,Upsilon)
    x_config+=levy_harmonic_path(beta,k)
    y_config+=levy_harmonic_path(beta,k)
    M -= k
```

Acknowledgments

I would like to thank the organizers of this school for giving me the opportunity to lecture about some of my favorite subjects. Thanks are due to C. Laumann for introducing me to Python programming and for pointing out the similarities to SMAC pseudocode, and to C. Chanal and M. Chevallier for a careful reading of the manuscript.

References

Alder, B. J. and Wainwright, T. E. (1970). Decay of the velocity autocorrelation function. *Physical Review A* **1**, 18–21.

Beale, P. D. (1996). Exact distribution of energies in the two-dimensional Ising model. *Physical Review Letters* **76**, 78–81.

Buhot, A. and Krauth, W. (1999). Phase separation in two-dimensional additive mixtures. *Physical Review E* **59**, 2939–2941.

Chanal, C. and Krauth, W. (2008). Renormalization group approach to exact sampling. *Physical Review Letters* **100**, 060601.

Chanal, C. and Krauth, W. (2010). Convergence and coupling for spin glasses and hard spheres. *Physical Review E*. In preparation.

Chevallier, M. and Krauth, W. (2007). Off-diagonal long-range order, cycle probabilities, and condensate fraction in the ideal Bose gas. *Physical Review E* **76**, 051109.

Dress, C. and Krauth, W. (1995). Cluster algorithm for hard spheres and related systems. *Journal of Physics A* **28**, L597–L601.

Feynman, R. P. (1972). *Statistical Mechanics: A Set of Lectures.* Benjamin/Cummings, Reading, MA.

Holzmann, M. and Krauth, W. (1999). Transition temperature of the homogeneous, weakly interacting Bose gas. *Physical Review Letters* **83**, 2687.

Holzmann, M. and Krauth, W. (2008). Kosterlitz–Thouless transition of the quasi two-dimensional trapped Bose gas. *Physical Review Letters* **100**, 190402.

Kac, M. and Ward, J. C. (1952). A combinatorial solution of the two-dimensional Ising model. *Physical Review* **88**, 1332–1337.

Kaufman, B. (1949). Crystal statistics. II. Partition function evaluated by spinor analysis. *Physical Review* **76**, 1232–1243.

Kendall, W. S. and Moller, J. (2000). Perfect simulation using dominating processes on ordered spaces, with application to locally stable point processes. *Advances in Applied Probability* **32**, 844–865.

Krauth, W. (1996). Quantum Monte Carlo calculations for a large number of bosons in a harmonic trap. *Physical Review Letters* **77**, 3695–3699.

Krauth, W. (2006). *Statistical Mechanics: Algorithms and Computations.* Oxford University Press, Oxford.

Krauth, W. and Moessner, R. (2003). Pocket Monte Carlo algorithm for classical doped dimer models. *Physical Review B* **67**, 064503.

Landsberg, P. T. (1961). *Thermodynamics, with Quantum Statistical Illustrations.* Interscience, New York.

Lévy, P. (1940). Sur certains processus stochastiques homogènes. *Composition Mathematica* **7**, 283–339.

Onsager, L. (1944). Crystal statistics. I. A two-dimensional model with an order–disorder transition. *Physical Review* **65**, 117–149.

Propp, J. G. and Wilson, D. B. (1996). Exact sampling with coupled Markov chains and applications to statistical mechanics. *Random Structures & Algorithms* **9**, 223–252.

Saul, L. and Kardar, M. (1993). Exact integer algorithm for the two-dimensional $\pm J$ Ising spin glass. *Physical Review E* **48**, R3221–R3224.

Simanyi, N. (2003). Proof of the Boltzmann–Sinai ergodic hypothesis for typical hard disk systems. *Inventiones Mathematicae* **154**, 123–178.

Simanyi, N. (2004). Proof of the ergodic hypothesis for typical hard ball systems. *Annales de l'Institut Henri Poincaré* **5**, 203–233.

Sinai, Y. G. (1970). Dynamical systems with elastic reflections. *Russian Mathematical Surveys* **25**, 137–189.

Wolff, U. (1989). Collective Monte-Carlo updating for spin systems. *Physical Review Letters* **62**, 361–364.

6
Loop models

B. NIENHUIS

Bernard Nienhuis,
Institute for Theoretical Physics,
Valckenierstraat 65, 1018 XE Amsterdam,
The Netherlands.

6.1 Historical perspective

Loop models have come up in the theory of phase transitions and (other) phenomena involving spatial scale invariance. One may say that the best understanding of phase transitions has come from a succession of theories that evolved in the 20th century. A phase transition can be defined as a locus in parameter space across which matter has very different properties. This implies that at a phase transition thermodynamic functions are singular, i.e. nonanalytic, as functions of the parameters. When the two different states of matter can coexist, the transition is called first-order; otherwise, it is continuous. A continuous phase transition is also called a critical point. At and near critical points, physical observables are related via power laws. The powers appearing in these relations, the critical exponents, became the focus of attention. The first breakthrough was mean-field (MF) theory, which was capable of predicting phase transitions starting from a microscopic description. In this approximate approach, the critical exponents can take only a few values, typically independent of the dimensions of the system. Over time it was observed that indeed critical exponents are highly universal, but they do depend on the dimension of the system, and on the symmetry involved in the phase transition. The exponents are insensitive to the details of the interaction. MF remained the dominant theory until renormalization group (RG) theory took over around 1970. Renormalization theories implement explicitly the concept of spatial scale invariance. Two approaches were developed. In one, called real-space RG [1, 2], a scale transformation was expressed in variables labeled by spatial coordinates. The other, momentum space RG [3], used variables labeled by coordinates in the Brillouin zone. Real-space RG was mainly done in two dimensions (2D), as this is more accessible to explicit numerical calculation. Momentum space RG focused on an expansion in powers of ε, where the dimension of space is $4 - \varepsilon$.

It was discovered that many two-dimensional models popular for the study of phase transitions could be formulated in terms of electric charges and magnetic monopoles interacting via the 2D version of electromagnetism [4, 5]. Combined with RG, this *Coulomb gas* (CG) method proved to be an effective way to calculate universal quantities. In this approach, loop models form the vocabulary of an intermediate language between, on one hand, lattice models with local interactions and, on the other hand, Coulomb gases. Loop models are ensembles of paths on a lattice with a statistical weight determined by the configuration of the steps in the paths and the total number of closed paths. Because the paths may not terminate in the interior of the lattice, they either form closed loops or run between two points on the boundary.

A natural extension of scale invariance is its extension from a global to a local symmetry, invariance under conformal transformations of space. The notion that conformal invariance is both plausible and powerful led to the study of conformal field theory (CFT) as the scaling limit of critical systems. It proved very effective in predicting many properties of the scaling limit. Finally, around 2000, stochastic Loewner evolution (SLE) allowed the mathematicians to prove what physicists had been so busy calculating all these years. While CFT and (for rigorous arguments) SLE are currently more powerful than CG, the latter approach has some clear advantages. In the CG approach, a lattice model and its operators are translated directly into a gas of interacting particles with specific properties, while in SLE and CFT the connection is indirect and is based on the correspondence of emerging properties.

Simultaneously with this development, a sequence of specific lattice models was solved exactly. The solution of the 2D Ising model had already shed doubt on the validity of MF theory in 1944 [6]. The six-vertex model [7], the eight-vertex model [8], the hard square model

[9], and the hard hexagon model [10] followed much later. In particular, the formulation of the Yang–Baxter equation (YBE) [11] led to an explosion of solvable models. These solutions were instrumental in the development of the theory of phase transitions, and served as anchor points for approximate or speculative approaches. It is noteworthy that, unlike MF theory, all predictions of RG theory are consistent with the existing exact solutions.

6.2 Brief summary of renormalization theory

The ideas that led to the application of renormalization to condensed matter systems can be formulated as follows. A many-body system typically has a number of relevant spatial scales. The microscopic scale might be, for example, the size of the molecules or the lattice constant (if the lattice is considered as given). The emerging length scales are correlation lengths, screening lengths, and the like. In generic systems, these various length scales are of the same order of magnitude. However, when a continuous phase transition (or critical point) is approached, the correlation lengths grow, and can become macroscopic. This implies that macroscopic properties involving space have a characteristic scale. However, when the critical point is reached, this spatial scale diverges. As a consequence the characteristic scale disappears, and the observables become scale invariant (that is, on scales sufficiently removed from the microscopic scales). If this train of thought is taken seriously, then it is plausible that one can formulate a scale transformation for which the critical system is invariant and the generic system covariant. The mere existence of such a transformation has many important consequences that can be verified. If the transformation can be calculated or approximated, even more can be deduced.

To make this explicit, consider a large but finite statistical system with N degrees of freedom s, associated with the sites of a lattice. It has a Hamiltonian $H_N(g, s)$, where g stands for a number of coupling constants. The partition sum (or integral) is given by

$$Z_N(g) = \sum_s \mathrm{e}^{-H_N(g,s)} \tag{6.1}$$

(the factor kT is absorbed into the coupling constants). To formulate a renormalization transformation (RT), we assume that we can coarse-grain the variables to a smaller number, say N'. This means that we split the degrees of freedom s into two sets s' and σ that describe the long- and short-wavelength fluctuations, respectively. Then a summation of the Boltzmann weights over the small-wavelength variables is performed, so that the sum over the remaining variables would result in the original partition sum. The summand after the partial summation is written as an effective Boltzmann weight again, so that

$$Z_N(g) = \sum_s \mathrm{e}^{-H_N(g,s)} = \sum_{s'} \sum_{\sigma} \mathrm{e}^{-H_N[g,s(s',\sigma)]} = \sum_{s'} \mathrm{e}^{-H_{N'}(g',s')} = Z_{N'}(g'), \tag{6.2}$$

where s represents the original variables, and s' and σ represent only the long- and short-wavelength fluctuations, respectively. The notation reflects the hypothesis that the partially summed Boltzmann weight can be written as an exponential function of an effective Hamiltonian of a smaller system of the same form as the original Hamiltonian but with altered values of the coupling constants. We assume further that the function $g'(g)$ is analytic or at least differentiable even at a phase transition. There is, at this point, no need to assume that we know how calculate $g'(g)$, only that in principle it exists.

Let $N = \ell^d N'$, with d the dimension of space, so that the linear change of scale is ℓ. In the limit in which N is very large, the free energy is proportional to N, so that the free energy per lattice site satisfies

$$f[g'(g)] = \ell^d f[g]\,. \tag{6.3}$$

Going back to the correlation lengths, we introduce the symbol ξ for one of the correlation lengths, in units of the lattice constant. This must satisfy

$$\xi[g] = \ell\xi[g'(g)]\,. \tag{6.4}$$

Since criticality corresponds to the divergence of ξ, eqn (6.4) shows that if the parameters g are a critical point, $g'(g)$ is also a critical point. In other words, critical manifolds are invariant under the map $g'(g)$. We may then assume that a critical manifold contains fixed points

$$g'(g^*) = g^*\,, \tag{6.5}$$

which are going to play a leading role.

Exercise 6.1 What possible values can $\xi[g^*]$ have if g^* is defined by eqn (6.5)?

Exercise 6.2 Equation (6.3) ignores the fact that integration (or summation) of a part of the degrees of freedom typically leads to an additional term in the Hamiltonian, independent of the remaining s'. Taking this into account leads to an inhomogeneous version of eqn (6.3),

$$f[g'(g)] = \ell^d f[g] + h[g]\,, \tag{6.6}$$

in which both $g'(g)$ and $h[g]$ are analytic functions. Show that one can write the free energy as the sum of an analytic part and a singular part, $f[g] = f_{\rm a}[g] + f_{\rm s}[g]$, such that $f_{\rm s}$ satisfies eqn (6.3).

If the microscopic scale is not given by a lattice, but by a (UV) integration cutoff, it is convenient to take ℓ as a variable, and analyze

$$\beta(g) \equiv \lim_{\ell\to 1} \frac{\partial g'(g)}{\partial \ell}\,, \tag{6.7}$$

called the β-function. In this case, the equation for the fixed point is $\beta(g^*) = 0$.

What can we say about the behavior of the system in the vicinity of a fixed point? Sufficiently near g^* we can linearize the transformation, and diagonalize the derivative matrix $\partial g'(g)/\partial g$. We denote the principal variations of g near g^* by u_j, and we can write the free energy and correlation length in terms of these u_j. The significance of a specific u_j depends on the corresponding eigenvalue. If an eigenvalue (assumed to be real) is larger than unity, the corresponding u_j grows, and if it is less than 1, u_j decreases.

Since the eigenvalues depend on ℓ, it is useful to make this dependence explicit. Since successive RTs with rescaling factors ℓ and ℓ' correspond to a rescaling by a factor $\ell\,\ell'$, and also the eigenvalues are multiplied, we write these eigenvalues in the form ℓ^y. To illustrate the distinction between eigenvalues greater than and less than one or, equivalently, positive and negative exponents y, we first discuss a case with two coupling constants and $y_2 < 0 < y_1$.

There is a nonlinear coordinate transformation $u(g)$ from g to u, with $u' = u(g')$, so that in terms of u the RG transformation has the simple form $u_i' = \ell^{y_i} u_i$. Consideration of the "flow lines" (i.e. the curves in the parameter space traced out by progressive renormalizations) in Fig. 6.1 shows that the line $u_1 = 0$ is a watershed between different limits of the flows. This implies that the physical behavior on either side of the curve $u_1 = 0$ is very different. If, for example, the effective variables at a very coarse scale are strongly coupled, the original variables will have long-range correlations, and if the effective variables are weakly coupled, the original variables will have only short-range correlations. Clearly, the locus $u_1 = 0$ is that of a phase transition.

Exercise 6.3 Argue why the u_1-axis is generally not a phase transition. Try to think of an extension of the RG flow lines outside of the borders of Fig. 6.1 which would form an exception.

When we write the free energy and correlation length as functions of u, the functions $\xi(u_1, u_2)$ and $f(u_1, u_2)$ satisfy

$$\begin{aligned} \xi(u_1, u_2) &= \ell \xi\left(u_1 \ell^{y_1}, u_2 \ell^{y_2}\right), \\ f(u_1, u_2) &= \ell^{-d} f\left(u_1 \ell^{y_1}, u_2 \ell^{y_2}\right). \end{aligned} \tag{6.8}$$

The solutions of these equation are given by

$$\begin{aligned} \xi(u_1, u_2) &= |u_1|^{-1/y_1} X_\pm\left(u_2 |u_1|^{-y_2/y_1}\right), \\ f(u_1, u_2) &= |u_1|^{d/y_1} S_\pm\left(u_2 |u_1|^{-y_2/y_1}\right), \end{aligned} \tag{6.9}$$

where X and S are unknown functions, and their subscripts ($+$ or $-$) are the sign of u_1.

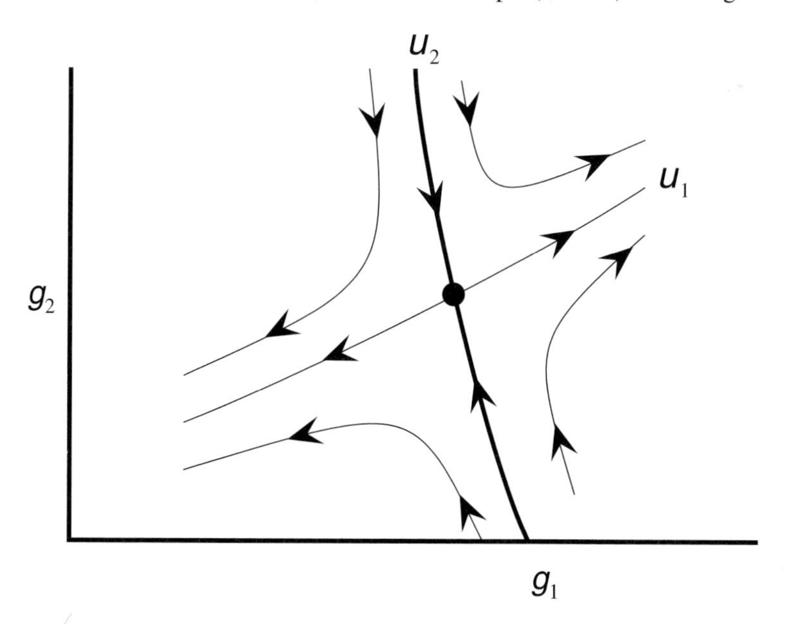

Fig. 6.1 The renormalization flows near a fixed point with one relevant and one irrelevant variable.

In general, the locus $u_2 = 0$ is not a phase transition. This implies that the thermodynamic functions are *analytic* at $u_2 = 0$ and the functions X and S can be expanded in powers of their argument:

$$\xi(u_1, u_2) = \sum_{k=0}^{\infty} X_{k,\pm}\, u_2^k\, |u_1|^{(-1-ky_2)/y_1} ,$$

$$f(u_1, u_2) = \sum_{k=0}^{\infty} S_{k,\pm}\, u_2^k\, |u_1|^{(d-ky_2)/y_1} . \tag{6.10}$$

Even though we have assumed only the existence of an RG transformation, and not its precise form, this result for the thermodynamic functions is very informative. We see here that the singular behaviors of the free energy and the correlation length are a sum of power-law terms, with increasing exponent. The noninteger powers concern the u_1 dependence. The leading term is determined by y_1. None of the exponents depends on the value of u_2. Because the leading critical behavior is not affected by u_2, this parameter is called *irrelevant*. In contrast, the parameters with a positive exponent y are called *relevant*. When $y = 0$ the corresponding parameter (or operator) is called marginal. Relevant and irrelevant parameters increase and decrease, respectively, under repeated RG transformation.

Exercise 6.4 Argue that a marginal operator can be marginally relevant or marginally irrelevant, that this behavior will generally be different on either side of the fixed point but need not be, and that one can also have strictly marginal operators.

The amplitudes of the nondominant terms depend explicitly on the value of the irrelevant parameter. When the nonlinear terms of $g'(g)$ are taken into account, the coefficient of the leading term will also depend on u_2. Finally, note that the exponents of the free energy and of the correlation length are not independent.

Now consider a case with two relevant variables, $y_1 > y_2 > 0$, omitting possible irrelevant variables. To make contact with a real-life case, think of u_1 as a magnetic field, and of u_2 as the temperature variation in the neighborhood of the critical temperature. Equation (6.9) for the free energy can be used to calculate various thermodynamic functions: the magnetization M and susceptibility χ as the first and second derivative of f with respect to the field h, and the heat capacity C as the second derivative of f with respect to the temperature T. The well-known critical exponents (at zero field, $C \propto |T - T_c|^{-\alpha}$, $M \propto (T_c - T)^{\beta}$, $\chi \propto |T - T_c|^{-\gamma}$, and $\xi \propto |T - T_c|^{-\nu}$, and at $T = T_c$, $h \propto M^{\delta}$) can now readily be calculated. We get

$$\alpha = 2 - \frac{d}{y_2}, \quad \beta = \frac{d - y_1}{y_2}, \quad \gamma = \frac{2y_1 - d}{y_2}, \quad \delta = \frac{y_1}{d - y_1}, \quad \nu = \frac{1}{y_2}. \tag{6.11}$$

Exercise 6.5 Verify these values of the critical exponents.

Evidently, these exponents are not independent but can expressed in terms of the two eigenvalues y_1 and y_2. The generalization to more than two parameters, some relevant and some irrelevant, does not introduce essential difficulties.

Also, the overall scaling of critical correlation functions can be calculated within the RG framework. Let the (extensive) operator conjugate to one of the parameters u_j be Q_j. Now consider the fluctuations of Q_j in a finite system with linear size L at a point in parameter space which is separated from the fixed point only in the u_j-direction,

$$G(u_j, L) \equiv \langle Q_j Q_j \rangle - \langle Q_j \rangle^2 = L^d \frac{\partial^2 f_L(u_j)}{\partial u_j^2} . \tag{6.12}$$

Analogously to eqn (6.8), this scales as

$$G(u_j, L) = \ell^{2y_j} \, G(\ell^{y_j} \, u_j, L/\ell) , \tag{6.13}$$

so that

$$G(0, L) \propto L^{2y_j} . \tag{6.14}$$

Let Q_j be written as a sum (or integral) over local operators

$$Q_j = \int_{L^d} \mathrm{d}^d r \; q_j(r) . \tag{6.15}$$

Then G can be written as an integral over the two-point function of $q_j(r)$:

$$G(0, L) = \int_{L^d} \mathrm{d}^d r \int_{L^d} \mathrm{d}^d r' \; \langle \, q_j(r) \, q_j(r') \, \rangle \; - \; \langle \, q_j(r) \, \rangle \, \langle \, q_j(r') \rangle . \tag{6.16}$$

Since the total integral behaves as a power of L (6.14), the integrand is expected to behave as a power of the distance:

$$\langle \, q_j(0) \, q_j(r) \, \rangle \; - \; \langle \, q_j(0) \, \rangle \, \langle \, q_j(r) \, \rangle \; \propto \; |r|^{2(y_j - d)} . \tag{6.17}$$

It is convenient to introduce $x_j = d - y_j$, because these are the exponents that appear in correlation functions. We thus see that both relevant and irrelevant operators have a power-law decay for their two-point function. If the decay is faster than $|r|^{-2d}$, the operator is irrelevant, and if the decay is slower, the operator is relevant.

6.2.1 Summary

The existence of an analytic RT is a strong and unproven assumption. It has, among others, the following consequences:

- *Relevant* and *irrelevant* parameters ($y \gtrless 0$) increase and decrease, respectively, under (repeated) renormalization.
- The critical exponents and scaling functions do not depend on irrelevant parameters, but can depend only on invariants of the RT, for example, symmetry and spatial dimension. In particular, they do not depend on the lattice (unless the symmetry of the lattice is broken on one side of the transition). This claim is known as *universality*.
- Two-point functions decay with a power of the distance equal to $2x_j = 2(d - y_j)$, where y_j is the exponent of the coupling parameter conjugate to the operators correlated.
- All correlation lengths diverge with the same exponent when a critical point is approached.

- The critical exponents are the same on either side of the transition, but the corresponding amplitudes may differ.
- The critical amplitudes of the dominant singularities and of corrections to scaling may depend on the irrelevant parameters.
- The ratio of the critical amplitudes of the same observable on either side of the transition is universal.
- The exponents for the various thermodynamic quantities are not independent, but satisfy a number of scaling relations.
- When integer linear combinations of the exponents y add up to d, the corresponding integer power laws are modified by a logarithm. This is not derived in the text above, but left as an exercise.
- RG theory admits the existence of a line of fixed points, along which critical exponents may vary.
- At a point where an irrelevant parameter becomes relevant, there are logarithmic corrections to the power-law behavior.

In all of the many cases where exact solutions of model systems or accurate experimental or numerical data are available, these consequences of the RG theory have been corroborated. For physics, this is the ultimate test of a theory. And, for this reason, the RG theory of phase transitions has the same status as any of the established theories in physics.

6.3 Loop models

Loop models are (lattice) gases of closed paths (loops). In the cases discussed here, the loops do not intersect themselves or each other, but other cases have been considered. Paths may be infinite or terminate on the boundary of the system, in which case they are not strictly loops. The models arose in the theory of phase transitions, but have been used also in other condensed matter contexts.

The possible configurations of loop segments around a single vertex are shown in Fig. 6.2. We consider here in particular the square and the hexagonal lattice, but the generalization to other lattices, possibly in higher dimensions, is immediate.

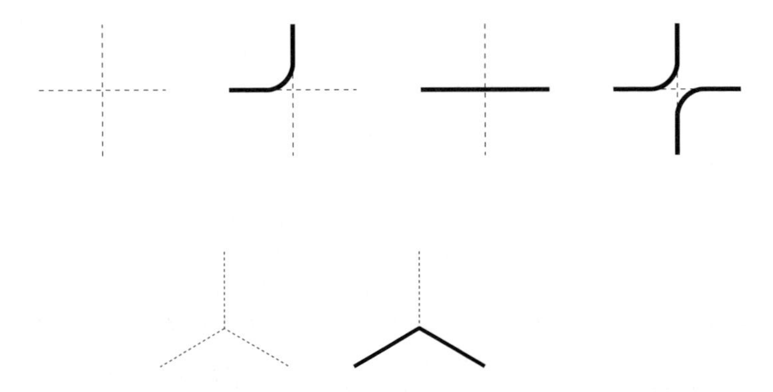

Fig. 6.2 The local configurations of paths around a single vertex, for the square lattice (top) and the hexagonal lattice (bottom). The vertex types are shown modulo the symmetry group of the lattice.

The different local configurations have different weights, which may depend on orientation, and each closed loop has weight n. The partition sum is thus given by

$$Z_{\text{loop}} = \sum_{\text{config.}} n^N \prod_v W_{\text{local}}(v) \,. \tag{6.18}$$

The sum is over all loop configurations, and N is the number of loops, so that each loop has weight n. The product is over the vertices v, where $W_{\text{local}}(v)$ is the weight of the configuration at vertex v, and depends only on which one of the elements of Fig. 6.2 is present, and possibly its orientation.

For a loop model on the hexagonal lattice with the full symmetry of the lattice, the partition sum simplifies to

$$Z_{\text{hex}} = \sum_{\text{config.}} n^N x^L \,, \tag{6.19}$$

where L is the total length of all loops. Here, the weights of the two local vertices in Fig. 6.2 have been chosen as 1 and x, respectively.

The $n = 0$ loop model denotes not just the $n \to 0$ limit, which is trivial, but the derivative of the partition sum with respect to n, which generates the configurations of a single loop. Higher derivatives describe a fixed number of such loops.

Exercise 6.6 Locality of a statistical model can be defined as the property that the change in the Boltzmann weight due to a local change in the configuration can be determined by inspection in the neighborhood of the locus of the change. Verify that the loop model for generic n is not local in this sense. For what value(s) of n should one make an exception?

Exercise 6.7 Show that the hexagonal version can be recovered from the square version by allowing the weights to be anisotropic, and requiring that the weight on the square factorizes into two triangles.

Similar models on the triangular lattice and on more general semiregular or even random lattices can be formulated analogously. Also, intersections can be introduced, and additional variables decorating the loops [12, 13]. These also have various applications to physical systems, and these models have been studied for this reason. Here we consider only the simplest versions.

In a particular model, the configurations shown in Fig. 6.2 need not all have nonzero weight. If the first three vertex configurations for the square lattice have zero weight, every edge is covered. This model will be referred to as the completely packed loop (CPL) model, with the defining property that all *edges* are covered. If only the first and the last edge are omitted, every *vertex* is visited once: this is called the fully packed loop (FPL) model. The CPL model can be defined also on the triangular lattice, but not on the hexagonal lattice. The FPL model exists for all three.

When all vertices have nonzero weight (on the square or hexagonal lattice), one can control the density of loops by means of the weight of each vertex configuration. The model has a noncritical phase when the vertex weights are dominated by the empty vertex, and a continuous phase transition into a critical phase for larger weights of the other vertex types. At the phase transition, the model is sometimes called the dilute loop model, and in the large-step-fugacity phase the dense loop model. This is distinct from the FPL and CPL models because the density is fluctuating.

The critical behavior of these models depends continuously on n, but is typically independent of the local weights. These models have a universal critical behavior (i.e. depending on n only) when the loop weight is $-2 \leq n \leq 2$. They also have other critical points, but these are induced by the specific lattice structure, and are therefore not universal in the traditional sense.

6.3.1 ADE models

Loop models can be used to represent lattice models with local variables and local interaction. One large class of such models is known as the ADE models. In these models, the faces of the lattice assume discrete values, represented by the nodes on a graph $\mathcal{G}$. Adjacent faces can either take the same value or take different values connected by a link in $\mathcal{G}$. The name ADE comes from the classification of graphs; the classes called A, D, and E play a special role (see Fig. 6.3). The graph $\mathcal{G}$ induces an adjacency matrix A with elements $A_{i,j}$ equal to 1 if the nodes i and j are connected, and 0 otherwise. Now we are ready to define the ADE model [14]. We start with the configurations of a loop model. The loops are interpreted as domain walls between domains of faces in the same state. The values on either side of a domain wall are adjacent on $\mathcal{G}$. The weight of the ADE model is completely local, and written as a product over the vertices:

$$W_{\mathrm{ADE}} = \prod_{\text{vert } v} W(v) \prod_{\text{turns}} A_{i,j} \left(\frac{S_j}{S_i} \right)^{\gamma_{\mathrm{b}}/2\pi} . \tag{6.20}$$

There is a factor $W(v)$ for each vertex that depends only on the local configuration of domain walls, equal to the weight of the corresponding vertex type of the loop model. And there is a factor for each turn of the domain wall, which depends on the states of the faces j and i on

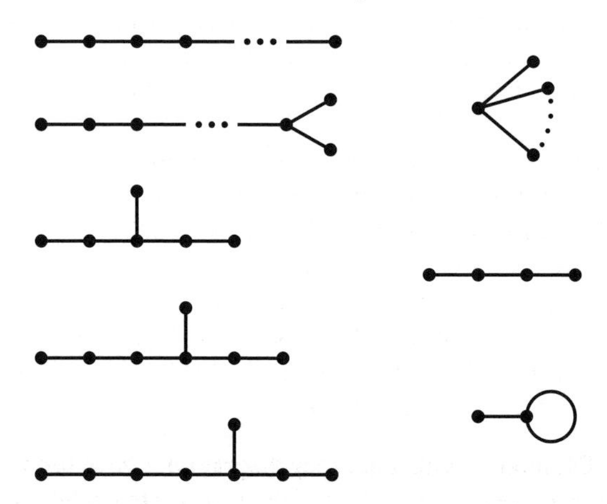

Fig. 6.3 The adjacency diagrams. In the left column, the top line is the A_m diagram, the second line is the D_m diagram, and the last three lines E_6, E_7, and E_8. The suffix refers to the number of nodes. In the right column, the adjacency diagram of the Potts model, the diagram A_4, and its folded version for the hard hexagon and hard square model are shown.

either side of the domain wall, and on the angle through which the domain wall bends at the vertex. The bending angles γ_b in eqn (6.20) are counted as positive when the state j is on the inside of the bend (and i on the outside). We choose S_j to be the j-th element of an eigenvector of A and call the eigenvalue n: $\sum_j A_{i,j} S_j = n S_i$. If we choose the Perron–Frobenius eigenvector, with only positive elements, then the local weights are real and positive. Since the bending angle summed over a closed loop is $\pm 2\pi$, the corresponding weight of a closed domain wall is

$$W_{\text{loop}} = A_{i,j} \frac{S_j}{S_i} \,. \tag{6.21}$$

When, for a fixed configuration of domain walls, the sum over compatible state configurations is performed, each closed domain wall thus contributes a factor equal to the eigenvalue of A for the eigenvector S, i.e. n. To see this, we start by summing over the states of the innermost domains, keeping the values in the other domains fixed. This results only in a factor n. Subsequently, we sum recursively over the states of the domains surrounding the ones just summed out. This works perfectly when all domain walls are contractible, for example, in a bounded flat space in which the boundary faces are all in the same state, and results in the partition sum eqn (6.19). With other boundary conditions, the paths may terminate on the boundary, and these paths may have a weight different from n, depending on the specific boundary condition.

On a torus, however, there may be noncontractible loops, which do not have the weight $A_{j,k} S_k / S_j$, but simply $A_{j,k}$. The domains which are separated by these winding loops form a closed chain. Once all domains surrounded by contractible walls have been summed over, the only sum that remains is over the noncontractible domains. This sum is a trace of a power of the matrix A. As a result, the ADE partition sum *on the torus* therefore reads

$$Z_{\text{ADE,T}} = \sum_j \sum_{\text{cfg}} \prod_{\text{vertices } k} W_k \, n^{N_c} \, \Lambda_j^{N_w} \,, \tag{6.22}$$

where the first sum is over the eigenvalues Λ_j of A, the second is over all loop configurations, N_c is the number of contractible loops, and N_w is the number of winding loops. For those familiar with CFT, each term in the sum over j corresponds to a character in the modular partition sum.

The other eigenvalues of A can also be used as possible values for n. In this case the ADE model may have complex weights, but this is not necessarily the case. The model will correspond to a CFT with negative conformal weights, which can be a blessing or a curse. This situation may be avoided by use of a selection rule that excludes the negative-weight sectors. These correspond to terms in the sum over j in eqn (6.22) in which $\Lambda_j > n$.

This equivalence between loop models and discrete lattice models was first shown by Pasquier [14] for the CPL model, and later generalized to general loop models [15]. We will refer to this translation as the ADE construction.

ABF models. The ADE models with adjacency diagram A_m have been studied by Andrews, Baxter, and Forrester, and are called ABF models or restricted solid-on-solid models. They can be viewed as a spin-$(m-1)/2$ Ising model. The critical point obtained from the ADE construction based on the CPL model is the Ising critical point for $m = 3$, and for larger values of m the model gives a sequence of multicritical points, each one selecting the point in parameter space where the transition corresponding to the previous point becomes first-order.

Transitions in the same universality class as in the CPL model are found for loop models in the dense phase. In the dilute loop models, the same sequence of universality classes is found, with m one less.

Exercise 6.8 Calculate the loop weights n for the ABF models as a function of m.

Height models. The A_m models in the limit $m \to \infty$ will play an important role in these lectures, since they are the basis for the Coulomb gas theory. The position along the diagram is sometimes referred to as the height, and the models as height models. The adjacency diagram has a continuous spectrum of eigenvalues $-2 \leq \Lambda \leq 2$, which makes these models very flexible. Also, they have running-wave modes as eigenstates, so that the elements of the eigenvector are all the same in magnitude, and are different only in phase. We will use these models mainly as a calculational tool: they admit any possible loop weight in the interval $[-2, 2]$, and the operators in other models can usually be translated into the A_∞ language.

The height model with $n = 2$ may be used to describe the behavior of a crystal surface, each height representing a stack of atoms. The noncritical phase is called the smooth phase, referring to the fact that all of the heights globally are concentrated around one given value (breaking of the $\mathbb{Z}$ symmetry). The regions of other heights form small islands or lakes and, occasionally, islands within islands.

The dense phase represents the rough phase of the crystal surface, because here the heights fluctuate wildly, the local averages of heights take continuous values, and the system (on a large scale) no longer has a preference for integer values. The transition between the smooth and the rough phase is called the roughening transition. In real crystals (and in sufficiently rich models), the roughening temperature depends on the direction in which the height is measured. The roughening temperature is lowest in the directions of high symmetry. As a result, a crystal in equilibrium with a melt, a vapor, or a solution has smooth facets separated by rounded (rough) edges and corners. Experimental findings are in good agreement with the theory. A more open question is the shape of a crystal under various conditions of growth outside of equilibrium.

If the CPL model is used as the basis of the A_∞ model, the weights are real and positive irrespective of the value of n, in spite of the fact that the eigenvector is complex. This is because, in the local loop configurations, two bends always come together. Each bend carries a complex weight, say z if the height on the inside of the bend is greater and z^* when the height is less than that on the outside. When the four heights around a single vertex have an overall gradient, the factor z for one bend is multiplied by z^* for the other. When the four heights form a saddle configuration, the weight z is multiplied by itself, but this height configuration is compatible with both vertex types, and if one type has weight z^2, the other has weight z^{*2}. Thus the sum is real and, in fact, non-negative.

The Potts model. Since symmetry is an important determinant of a universality class, it is useful to construct adjacency diagrams with special symmetries. An extensively investigated model is the Potts model [16], with S_q, the permutation group of q elements, as its symmetry. Its critical point can be found with an ADE construction. The appropriate state diagram is shown in the second column of Fig. 6.3. It has one central node, and a general number q legs. We refer to the q nodes as $1, 2, \cdots, q$, and to the neutral or central node as 0.

Exercise 6.9 Verify that the largest eigenvalue is $\sqrt{q}$ and its eigenstate is $(\sqrt{q}, 1, 1, \ldots)$.

It follows that the loop weight of the Potts model is $\sqrt{q}$. In the (homogeneous and isotropic) CPL model, since the square lattice is bipartite, one sublattice is completely frozen in the neutral state, while the q equivalent states, called spins, live on the other sublattice. We may ignore the frozen sublattice and deal with the effective interactions on the dynamic sublattice only. The Perron–Frobenius eigenvector has weight 1 for each of the q equivalent nodes, and $\sqrt{q}$ on the neutral node (up to normalization).

Exercise 6.10 Verify that in each square of four faces of the lattice, the weight of two unequal states across the diagonal of a square is 1, while the weight of two equal states is $1+\sqrt{q}$.

Another type of Potts model is found if, instead of the largest eigenvalue $\sqrt{q}$, we choose its conjugate $-\sqrt{q}$. The critical point for this model is in the antiferromagnetic regime, i.e. where the spins are preferentially different from their neighbors.

In principle, we can use any loop model in this construction, but then all $q+1$ states can sit anywhere in the lattice. Such models are known as *dilute* Potts models, referring to the neutral state as a vacancy or an unoccupied site.

Percolation, USTs, SAWs, and θ SAWs. Percolation is a description of transport behavior in a random mixture of permeable and nonpermeable materials (or conducting and insulating materials). One prototypical model is site percolation on the triangular lattice. This is a perfect model for a mixture of spherical beads of copper and glass on a tray, settled in a closed-packed array. Only when more than half of the beads are copper will a large array conduct. This model, at the transition, is well described by a hexagonal loop model with $n = 1$ and $x = 1$ (the loops being the boundary between the conducting and nonconducting beads). Another prototypical model is bond percolation on the square lattice. In bond percolation, neighboring sites are connected (by a conductor) with probability p, independently of each other. Again, precisely at $p = 1/2$ the model undergoes a transition between a conductor and an insulator. This model, at the transition, is described by the CPL loop model with $n = 1$. This fact makes it evident that percolation is the $q \to 1$ limit of the Potts model.

The uniform spanning trees (USTs) are the set of all tree-like subgraphs of a lattice, chosen with uniform weight. The CPL loop model with $n = 0$ is a description of this ensemble. The USTs are *the* (or rather *a* [17]) $q \to 0$ limit of the critical Potts model.

Exercise 6.11 Convince yourself of the relations between the two percolation models and the UST model on the one hand, and the corresponding loop models on the other.

Self-avoiding walks (SAWs) may be used to describe linear polymers in solution. The first derivative of eqn (6.18) or (6.19) with respect to n at $n = 0$ gives a partition sum of the closed SAW, and is grand canonical in the steps (or monomers). A small step fugacity describes a small polymer, and a critical (dilute) step fugacity a solution of a very long polymer.

In a poor solvent, a polymer has a tendency to demix or phase-separate. The phase transition between a well-dissolved polymer and a coagulated polymer is called the θ-point, and is described by the hexagonal loop model with $n = 1$ and $x = 1$ [18]. Of course, this model itself has many loops. But at the point $n = x = 1$ one can, for a fixed configuration of one open or closed walk, sum over all compatible configurations of the other loops. This sum results in an explicit attractive interaction between segments of the original walk which pass through the same hexagon but are distant along the chain.

Exercise 6.12 What is the precise nature of the attractive interaction?

Hard hexagon and hard square models. In the ABF model, not all m nodes of the diagram A_m need to be distinguished. Since the lattice is bipartite, one sublattice can take only even nodes as its states, and the other only odd. This admits a simplification by identifying even and odd nodes as the same state. In A_4, for instance, we may identify the nodes 1 and 4 and the nodes 2 and 3. The resulting diagram is shown in Fig. 6.3. The model can be interpreted as a lattice gas: nodes 1 and 4 can represent an occupied site, and 2 and 3 an empty one. In this model, every face can be occupied by at most one particle and two adjacent sites cannot be simultaneously occupied. On the square lattice, this model is called the hard square model [9], as the particles can be seen as tiles of size $\sqrt{2} \times \sqrt{2}$ placed in a diagonal orientation with respect to the lattice. The neighbor exclusion is equivalent to the restriction that the tiles are not permitted to overlap. The critical point resulting from the ADE construction is in the universality class of the tricritical Ising model.

Exercise 6.13 In the hard square model that follows from this construction, two particles occupying second-neighbor sites have an attractive interaction; calculate the strength of this interaction.

If we allow the weights of the two local loop configurations in the CPL model to be different, another simple model can be constructed, known as the hard hexagon model [10]. The meaning of the nodes on the A_4 diagram is the same: 1 and 4 signify occupied, and 2 and 3 empty. By a judicious choice of anisotropic loop weights in the CPL model, one may achieve the result that the Boltzmann weight of the square can be factorized into the weights of two triangles with full triangular symmetry. The resulting model can be seen as a lattice gas on the triangular lattice, in which simultaneous occupation of nearest neighbors is forbidden. This is the model known as the hard hexagon model, at its critical point.

Exercise 6.14 Verify that such a model can indeed be obtained by the ADE construction using A_4 and the anisotropic CPL model. One must permit negative local loop weights, but in the resulting hard hexagon model each Boltzmann weight is positive.

Exercise 6.15 Argue that the hard hexagon model must have the same critical exponents as the three-state Potts model.

6.3.2 Operators and correlation functions

Let the function $P(k)$ be the probability that an arbitrary face in the ADE model is in the state k. We will refer to this as the one-point distribution (1PD). Consider a loop well inside a large lattice. We assume that the 1PD is unaffected by the presence of the loop (or any other loop). In other words, we assume that the 1PD conditional on the presence of a loop is the same as the unconditional 1PD. That this is plausible follows from the calculation of the partition sum above: the contribution to the partition sum of a particular domain is independent of the domains it is contained in, and it is independent of all the domains once the states of these domains have been summed over. A stronger argument will result from the construction below.

The conditional probability $P(k|j)$ that the interior domain of a loop is in state k, provided the exterior domain is in a given state j, is determined by eqn (6.21) as $P(k|j) = A_{k,j}S_k/(nS_j)$. Thus, with the definition of the 1PD, we find the joint probability $P(k,j)$ that the outside of a loop is in state j *and* its inside is in state k as

$$P(k,j) = P(j)P(k|j) = P(j)A_{k,j}\,\frac{S_k}{nS_j}\,. \tag{6.23}$$

Summation over j now yields the probability that the interior domain is in state k, which should be equal to $P(k)$:

$$\sum_j P(k,j) = P(k)\,. \tag{6.24}$$

Equations (6.23) and (6.24) form a consistency condition on $P(k)$. Equation (6.24) has the form of an eigenvalue equation for the matrix A, which immediately gives the unique solution

$$P(k) = (S_k)^2\,. \tag{6.25}$$

Here and below, we assume that the eigenvectors are normalized, as $\sum_j (S_j)^2 = 1$.

We have thus calculated the 1PD for a spin well away from the boundary. However, there is a special boundary condition which makes eqn (6.25) exact for all spins. For this, we consider a bounded lattice of arbitrary size, with the faces on the boundary all in the same state, *but with eqn (6.25) as the probability distribution for that state*. Then, by induction, the same distribution holds for the domains separated from the boundary by a single domain wall, and so on recursively to the innermost domains. It is then assumed that in the thermodynamic limit the boundary condition should not matter, well away from the boundary.

Instead of considering the 1PD, we can take expectation values of any function of the state of one or more given faces. Let S^μ be the μ-th eigenvector of the adjacency matrix, with eigenvalue Λ_μ, and let S be the eigenvector with eigenvalue n. Then consider the function

$$w(k) = \frac{S_k^\mu}{S_k}\,. \tag{6.26}$$

If this function is an operator insertion in a correlation function $\langle \cdots S_k^\mu/S_k \cdots \rangle$, where k is the state of a given face, it effectively changes the weight of the loops *surrounding that face*. This is easily seen in eqn (6.21): the factor S_k^μ/S_k replaces the numerator by S_k^μ, so that the weight of the loop becomes that of the corresponding eigenvalue, as long as the loop does not surround other operator insertions. We will call these functions weight-changing operators of weight Λ_μ.

An interesting result comes from the two-point function of $w(k)$,

$$\left\langle \frac{S^\mu_j}{S_j} \frac{S^\nu_k}{S_k} \right\rangle ,$$

j and k being the states of two arbitrary faces. The weights of the loops surrounding either one of these faces but not the other are changed into the respective eigenvalues Λ_μ and Λ_ν, corresponding to the eigenvectors S^μ and S^ν. Now consider the innermost domain that surrounds both faces, and let j be its state. After the states of all of the domains nested inside it are summed over, the weight governing the state j of the final domain is

$$\frac{S^\mu_j S^\nu_j}{(S_j)^2} ,$$

aside from the weight of the domain wall surrounding it. This can be expanded as a linear combination of the other weight-changing operators:

$$S^\mu_j S^\nu_j = S_j \sum_\kappa C^\kappa_{\mu\,\nu} S^\kappa_j , \tag{6.27}$$

where (again provided the eigenvectors are normalized)

$$C^\kappa_{\mu\,\nu} = \sum_j \frac{S^\mu_j S^\nu_j S^\kappa_j}{S_j} . \tag{6.28}$$

The combination of two operators labeled μ and ν appears from a distance like a linear combination of operators κ.

These structure constants of the operator product expansion, or fusion rules, may of course be readily calculated explicitly for any adjacency diagram, but here we note only that (i) they are symmetric in μ, ν, and κ; (ii) they take integer values for the adjacency diagram of A_m; and (iii) they vanish if one of the indices corresponds to the largest eigenvalue while the other two are not the same.

Exercise 6.16 Verify (i) and (iii) of these properties and, for the A_m diagram, also (ii) (suggestion: do this numerically for some examples).

Property (iii) implies that the two-point correlation function of two different weight-changing operators vanishes in the thermodynamic limit. Obviously, these structure constants may be used just as well in correlation functions of more than two operators.

6.3.3 The O(n) model

Another representation of the loop model on the hexagonal model is the O(n) model, with

$$Z_{O(n)} = \int \mathcal{D}s \prod_{\langle j,k \rangle} (1 + x s_j \cdot s_k) . \tag{6.29}$$

The n-component spins s sit on the vertices j of the hexagonal lattice, and the interaction factors are between nearest neighbors. The name "O(n) model" refers to the global O(n)

symmetry under which the measure is invariant. The measure $\mathcal{D}s$ denotes integration over all spins and is normalized such that $\int \mathcal{D}s = 1$. The (mean) spin length is set by $\int \mathcal{D}s\ s \cdot s = n$. The loop model is obtained as an expansion of $Z_{\mathrm{O}(n)}$ in powers of x: the product under the integral is expanded, and the second term in the binomial is indicated by a bond on the corresponding lattice link. The only terms in the expansion that survive the integration over the spins have an even number of bonds incident on each of the lattice sites. The spin integration involves a sum over spin components, which results in a factor n per loop.

Exercise 6.17 Show that the square loop model can be obtained by placing O(n) spins on the vertices of the 4–8 lattice (a lattice with coordination number 3 where two octagonal faces and one square face meet at each vertex).

Ordinarily, one is interested in the critical behavior of a standard Heisenberg model with three spin components. This can be naturally generalized to n-component spins, but the natural partition sum is still

$$\int \mathcal{D}s \exp\left(\sum_{\langle j,k\rangle} K s_j \cdot s_k\right) \tag{6.30}$$

on any lattice. Universality would, in principle, assert that this model, having the same O(n) symmetry, has the same critical behavior as eqn (6.29), but let us look critically at the difference between the two models.

By expanding the exponent in eqn (6.30), we get multiply occupied bonds (powers of $K s_j \cdot s_k$). We then have to sum over all ways to connect the elementary steps of the diagrams pairwise in each vertex (see Exercise 6.18). Every closed polygon acquires a weight n in the summation over the indices. As a result of the multiple occupancy of the bonds, the polygons may cross each other as well as themselves. Thus the main differences between eqns (6.29) and (6.30) in the high-temperature (or weak-coupling) limit are (i) multiply occupied bonds and (ii) intersections. What remains intact is the fact that the diagrams consist of closed loops, with weight n for each loop.

Let us assume that the model (6.29) is exceptional in its critical behavior. This implies that it is a fixed point in a larger parameter space, and that deviations from it are relevant. This means that any such positive deviation will grow, but if it is zero it remains so.

When one looks at the graphs of the two models (6.29) and (6.30) from a large distance, multiply occupied bonds cannot be distinguished from neighboring parallel edges, each of which is singly occupied. This suggests that under RT, multiply occupied bonds will be generated. This is inconsistent with the assumption that these multiply occupied bonds contribute to a relevant perturbation which would differentiate between the models (6.30) and (6.29).

Thus loops seen from a large distance may begin to touch each other. However, even on a large scale such loops will not seem to intersect if they did not intersect to begin with. This suggests that intersections cannot be generated by an RT, and therefore a model without intersections may well be a fixed point in the space of models where intersections are allowed. The nonintersecting loop model will be generic if intersections are irrelevant, and exceptional if intersections are relevant. This will have to be verified.

Exercise 6.18 Consider the O(n) spin model (6.29) on the square lattice, and find the set of possible diagrams in the formal expansion of Z in powers of x, and the prescription for the weight of these diagrams.

Models with cubic symmetry. Cubic symmetry, in this context, denotes the symmetry group of the n-dimensional hypercube. The spins can be discrete and point, say, to the vertices of the cube or to the hyperfaces of the cube, but they can also be continuous, and have a weight function with cubic rather than O(n) symmetry, i.e. depend as a symmetric function on the components of the spin vector rather than on only its length. As long as the partition sum is defined as in eqn (6.29) (with the spin measure adjusted), the small-x expansion is again the loop model to all orders in the expansion.

Another possibility is to place spins on the faces of the lattice: let each of the n spin components be an Ising spin, and let the nearest neighbors be different in at most one of the components, at a weight cost x. Then the expansion of the partition sum in powers of x (in this case a low-temperature expansion) is again a loop model, the loops signifying domain walls between domains of equal spin. Since two neighboring domains can differ in one of their n components, the loop weight is again n. In fact, this is a form of an ADE model, in which the graph $\mathcal{G}$ is the hypercube, with vertices (corners) as nodes and edges as links.

Spin correlation function. In the case of the O(n) model, correlation functions such as $\langle s_0.s_r\rangle$ will, in the small-x expansion, correspond to a different set of diagrams than the partition sum, rather than to just a change in their weights. Indeed, in the expansion of $\langle s_0.s_r\rangle$ in powers of x, it is clear that the number of bonds incident on 0 and r should be odd rather than even. An example of such a diagram is shown in Fig. 6.7 later. The diagrams will consist of a path from 0 to r in combination with any number of closed loops. Generalizations to correlation functions involving more spins or higher powers of them are immediate. Later we will see the same diagrams for what is called the disorder operator in the Potts model.

6.4 The Coulomb gas

The CG approach is based on the ADE model with state diagram A_∞. The states are called heights and traditionally take values in $\pi\mathbb{Z}$, rather than $\mathbb{Z}$. We imagine an RG transformation in which the heights are not rescaled. Let us first consider the case $n = 2$, so that the loops can be considered as domain walls surrounding a hill or a valley, both with equal weight.

On a large scale, the heights will take continuous values with possibly a preference for integer multiples of π. Since the renormalized Hamiltonian can depend only on differences or gradients of h, it should be expandable in powers of the gradient:

$$H = \int \mathrm{d}^2r \, \frac{g}{4\pi}(\partial h)^2 + g_2(\partial h)^4 + \cdots - z_2 \cos 2h - z_4 \cos 4h - \cdots . \tag{6.31}$$

We consider even powers only, because the Hamiltonian has reflection symmetry in space. Because, in general, the Hamiltonian contains terms periodic in h, we define a renormalization such that h is invariant. This implies that the free field Hamiltonian, with only the first term, is unchanged under the RG, which follows from simple power counting. By the same argument, the higher-order powers of ∂h will be irrelevant (but may renormalize the leading term). To inspect the relevance of the parameters z_j, we can calculate the correlation function

$$\langle \cos(eh_0) \cos(eh_r) \rangle = \frac{1}{2} \langle \exp(\mathrm{i}eh_0 - \mathrm{i}eh_r) \rangle , \tag{6.32}$$

with some prefactor e. To calculate the r dependence, we need to do the integral

$$\int \mathcal{D}h \, \exp\left(\int \mathrm{d}^2 r \left[-\frac{g}{4\pi} (\partial h)^2 + \mathrm{i}eh_0 - \mathrm{i}eh_r \right] \right) , \tag{6.33}$$

a Gaussian integral. Since $\partial \cdot \partial \log h = 2\pi\delta(h)$, the result is

$$\langle \cos(eh_0) \cos(eh_r) \rangle \propto \exp\left(\frac{-e^2}{g} \log |r| \right) = |r|^{-e^2/g} . \tag{6.34}$$

The middle expression explains the notation e for the prefactor: the operators behave as particles with electric charge e (not to be confused with the base of the natural logarithm e). The $\cos(2h)$ term is relevant if the correlation function decays more slowly than $|r|^{-4}$, i.e. when $g > 1$. The higher harmonics will be less relevant, so we need only consider them if, for some special reason, z_2 vanishes.

To get a physical picture of what this all means, we should note that positive values of z_2 imply a positive density of charges: hence the name "Coulomb gas." If g is large, the coupling of the charges is weak, and they may effectively screen all Coulombic interaction between test charges. However, when g is small, the charges are strongly coupled, so that they may form charge-neutral complexes. This is what causes the crossover between the smooth and the rough crystal phase. In the language of charges, these phases can be called a plasma and a molecular phase, respectively. It may sound counterintuitive that the dynamic rough phase corresponds to the boring molecular phase.

The value of g resulting from the original loop model is not a priori known. It is evident that g will be a decreasing function of the original step fugacity. When the step fugacity is small, almost all heights will be the same: we have the smooth phase. If the domain walls are sufficiently numerous, the height fluctuations become more probable and it is reasonable that eventually g will fall below the threshold value 1. We can, however, conclude that everywhere in the rough phase, the exponent x_e of the correlation function

$$\langle \cos(eh_0) \cos(eh_r) \rangle \propto |r|^{-2x_e} \tag{6.35}$$

is proportional to e^2, and that at the roughening transition the prefactor is 1/2.

It should be noted that the presence of charges, even when they are irrelevant, will still weaken the electric interaction, and thus increase the value of g, by partial screening. We need not calculate the precise strength of this effect, but the sign is of importance.

6.4.1 Vortices

In some of our applications, we need the behavior of vortices or screw dislocations in the height model. Although in the A_∞ model they exist only as insertions, these operators have a natural abundance in systems in which the height is defined on a periodic interval. In order to calculate the exponent of the vortices, it is more convenient to consider two vortex insertions, in the absence of other vortices or periodic fields. A vortex can be seen as a source or sink of domain walls.

Consider thus the integral

$$\int \mathcal{D}h \, \exp\left(-\int \frac{g}{4\pi} \, \mathrm{d}^2 r \, (\partial h - 2\pi k)^2 \right) . \tag{6.36}$$

Here we take k to be a fixed vector field, zero almost everywhere, and with a value m on a line that connects two vortex insertions, in the direction orthogonal to the line. We use the continuum space notation because it is more compact, but it is convenient to think of the ∂'s as lattice differences, and the spatial integral as a sum over the lattice. However, we may "sum by parts" just as we may integrate by parts: $\int \mathrm{d}^2 r \, A \, \partial B = - \int \mathrm{d}^2 r \, B \, \partial A$, provided AB vanishes on the boundary of the integration

We have a Gaussian integral with terms in the exponent that are quadratic or linear in h. This integral can be done quickly after the introduction of the notation G for the inverse of the Laplace operator $\partial \cdot \partial$. This differs from the continuum version $\log(r - r')/2\pi$ only by a constant term and terms of order $\mathcal{O}(r^{-2})$. The action is then transformed successively as follows:

$$\left[-\frac{g}{4\pi} \, (\partial h - 2\pi k)^2 \right] \quad \rightarrow$$

(doing the spatial integration (summation) by parts)

$$\left(\frac{g}{4\pi} h \partial \cdot \partial h - g h \partial \cdot k - \pi g k \cdot k \right) \quad \rightarrow$$

(Gaussian integration over the height field)

$$\left(-\frac{g}{2} (\partial \cdot k) G (\partial \cdot k) - \pi g k \cdot k \right) \quad \rightarrow$$

(insertion of the spatial identity kernel $\partial \cdot \partial G$)

$$\left(-\frac{g}{2} (\partial \cdot k) G (\partial \cdot k) - \frac{g}{2} k \cdot (\partial \cdot \partial) G \, k \right) \quad \rightarrow$$

(spatial integration by parts)

$$\left(-\frac{g}{2} (\partial \cdot k) G (\partial \cdot k) - \frac{g}{2} k \cdot (\partial \cdot G \, \partial k) \right) \quad \rightarrow$$

(double integration by parts in the first term, followed by (backwards) rewriting of the product of two antisymmetric tensors)

$$\frac{g}{2} (\partial \times k) G (\partial \times k) \, .$$

After all this, noting that $(\partial \times k)$ vanishes everywhere except at the vortices, where it takes the values m and $-m$, respectively, we conclude that

$$\int \mathcal{D}h \, \exp\left(\int \mathrm{d}^2 r \; - \frac{g}{4\pi} \, (\partial h - 2\pi k)^2 \right) \propto |r|^{-g m^2} . \tag{6.37}$$

This result is very much like the correlation of two cosine functions of h, but here the exponent is proportional to g. The vortices are viewed as magnetic rather than electric charges.

The interaction between electric and magnetic charges is fairly obvious: when an electric charge is taken around a vortex, it must gain a phase factor equal to $\exp(2\pi iem)$, so that the interaction will look like

$$\exp[iem \arg(z)] \tag{6.38}$$

in terms of the complex number $z = r_x + \mathrm{i}r_y$, where r is the spatial vector between the two vortices.

6.4.2 Choosing the loop weight

When we want to consider loop models with general n, we can still make use of the A_∞ model, but select a subdominant eigenvalue, say, $n = 2\cos\pi e_0 \leq 2$. The corresponding eigenvector is $\exp(e_0 \mathrm{i}h)$. With this information, we can follow the protocol of Section 6.3.1 and construct the weights of the height model. Each bend of the loop with an angle γ_b, with the domain containing the higher h on the inside, has a weight $\exp\left[e_0 \mathrm{i}\gamma_\mathrm{b}/2\pi\right]$. When the lesser height is on the inside, the weight is conjugated.

As an aside, it is an interesting game (or a difficult exercise) to note that this addition in the continuum contributes

$$\frac{\mathrm{i}e_0}{2\pi} \frac{\sum_{\mu,\nu}(\partial_\mu h)\,(\partial_\mu\partial_\nu h)\,(\partial_\nu h)}{\sum_\lambda(\partial_\lambda h)\,(\partial_\lambda h)} - \frac{\mathrm{i}e_0}{2\pi}\sum_\mu \partial_\mu\partial_\mu h \tag{6.39}$$

to the action. This expression measures the curvature density of the contour lines (level sets) of h. Integrated over the entire configuration, this takes the form of the sum of the values of h at the extrema, minus the sum of h at the saddle points minus twice h at infinity. This is fairly obvious when we take into account where the expression comes from, because this counting of extrema and saddle points counts precisely the domain walls, with a sign for their orientation. On a curved surface, the expression should be incremented by $\mathrm{i}e_0 Rh/2\pi$, where R is the local curvature.

Now consider the model on a bounded domain, and take the heights at the boundary all equal to 0. This boundary condition precisely reproduces the loop model, with the condition that the loops are not permitted to penetrate the boundary. Now we introduce the weight-changing operator (6.26), based on the eigenvector $\exp(-e_0 \mathrm{i}h)$, with eigenvalue n, and name it $\mathcal{B}$. When we insert $\mathcal{B}$ at some position in the lattice, it has no effect on the weights of the loops in the configuration, because the changed weight is the same as the original. This observation is independent of the size and shape of the domain or the locus of the insertion. The operator $\mathcal{B}$, however, is not the identity operator. In particular, it must not be inserted twice (in the same or in different places), since then the weight of the loops that surround both insertions does change. When expressed directly in the height variables, the operator $\mathcal{B}$ has the form

$$\mathcal{B} = \exp\left(-2\mathrm{i}e_0 h\right), \tag{6.40}$$

an electric charge of strength $-2e_0$. It is sometimes called the background charge, or charge at infinity. This can be understood as follows. The loops surrounding a hill have weight $\exp\left(e_0 \mathrm{i}\pi\right)$, while the loops that surround a valley have weight $\exp\left(-e_0 \mathrm{i}\pi\right)$. For the loops that surround the operator insertion, these two weights are interchanged. This can be taken to mean that for those loops, the inside and outside are interchanged. In other words, the operator insertion plays the role of the point at infinity.

In summary, we note that there are two alternative ways to calculate the partition sum: with and without the insertion of the background charge. The same applies to correlation functions. This will play an important role in the sequel.

6.4.3 Weight-changing operators

Aside from the fact that the weights of all of the loops can be chosen from a continuum of values, the same applies to the weights of loops that change as the result of an operator insertion. If we want to construct a weight-changing operator with weight Λ, we select the eigenvector of the adjacency diagram A_∞ with that eigenvalue. This is $\exp(ieh)$ or, equivalently, $\exp(-ieh)$ with $\Lambda = 2\cos(\pi e)$. Note that the value of e is determined modulo 2. In the corresponding two-point function

$$\langle \exp(ieh_0 - ie_0 h_0) \exp(-ieh_r - ie_0 h_r) \rangle \,, \tag{6.41}$$

we combine the complex conjugate pair, and as in eqn (6.26), divide each by the eigenvector with eigenvalue n.

6.4.4 Application to the Potts model

In this section, we will show that the CG predicts the critical exponents of the Potts model. First we remember that the loop weight for the q-state Potts model is $\sqrt{q}$, and that the background charge follows from $\sqrt{q} = 2\cos(e_0\pi)$ (up to a sign and additive multiples of 2). We will first concentrate on the thermal, or energy, exponent.

The energy operator. We have seen that the Potts model has weight $1+\sqrt{q}$ for two neighboring equal spins and weight 1 for unequal spins. Changing the temperature of the Potts model would result in a change of the ratio of these weights. The operator conjugate to the (inverse) temperature is the energy operator $\mathcal{E}$.

To see what $\mathcal{E}$ looks like in the loop language, we note that the Potts spins live on only one of the two sublattices of the faces of the CPL model. We place the Potts variables on the sublattice where the heights are an even multiple of π, and the neutral state corresponds to the odd multiples of π. Changing the temperature results in a change in the relative weight of the two diagrams 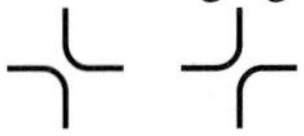 . Thus the energy operator gives opposite signs to the weights of these two diagrams. What does this do to the weight of the total configuration? Note that the Potts sites are always surrounded by an even number of loops. If we concentrate on the center of the diagram, this too must be surrounded by an even number of loops when the two Potts variables are "connected," and an odd number otherwise. In the latter case, the diagram has a negative weight. As a result, we can identify $\mathcal{E}$ with a weight-changing operator of weight $-n = -\sqrt{q}$.

In retrospect, it is not unexpected that $\mathcal{E}$ is this weight-changing operator, as only the eigenvalues $\pm\sqrt{q}$ have an eigenvector which is symmetric under permutation of the q states. This makes this operator the natural candidate.

We will now use this to see what $\mathcal{E}$ looks like when expressed in terms of the heights. Eigenvectors with eigenvalue $-\sqrt{q}$ are $\exp(ih - ie_0 h)$, $\exp(-ih + ie_0 h)$, $\exp(ih + ie_0 h)$, and $\exp(-ih - ie_0 h)$. In the two-point function, a complex conjugate pair must be chosen so that the total charge vanishes, and the loops surrounding both operator insertions have the appropriate weight $\sqrt{q}$. The ones with the smallest charge would be the natural choice, so,

Table 6.1 The weights of local configurations. The first line shows the heights around a single vertex, leaving out a factor π. The second line shows the average heights of these four faces. The third line is the Boltzmann weight in the partition, and the fourth line the weight when an energy operator is inserted at the position of the configuration.

Configuration	0 1 1 0	0 1 1 2	2 1 1 2	2 3 1 2	2 3 3 2
Height	$\frac{1}{2}$	1	$\frac{3}{2}$	2	$\frac{5}{2}$
Boltzmann weight	$2\cos(e_0\pi/2)$	1	$2\cos(e_0\pi/2)$	1	$2\cos(e_0\pi/2)$
$\mathcal{E}$	$2\mathrm{i}\sin(e_0\pi/2)$	-1	$-2\mathrm{i}\sin(e_0\pi/2)$	1	$2\mathrm{i}\sin(e_0\pi/2)$

with the convention that e_0 is positive, this would be the first two. However, in Table 6.1, we list a number of possible height configurations around a single vertex. While each individual face has integer heights (up to factor π), the mean height at a vertex can be a half-integer. This permits us to differentiate between the various eigenvectors. The third line in Table 6.1 shows the Boltzmann weight of the corresponding configuration, and the fourth line the factor by which it must be multiplied when an energy operator is inserted. This factor must be one of the eigenvectors given above, divided by the original $\exp(\mathrm{i}e_0 h)$, which generates the loop weight. Then we see that the sign of the imaginary terms in the third line of the table does not agree with the first candidate eigenvector, but does with the third. In a two-point function, it should be combined with the fourth eigenvector, so that the weight of loops surrounding both $\mathcal{E}$ insertions is $\sqrt{q}$ again. In terms of total charge, the operator $\mathcal{E}$ thus corresponds to a charge 1 or $-1-2e_0$, i.e.

$$\mathcal{E} = \exp(\mathrm{i}h) + \exp(-\mathrm{i}h - 2\mathrm{i}e_0 h)\,. \tag{6.42}$$

Having specified the nature of $\mathcal{E}$, we are now in the position to calculate its two-point function $\langle \mathcal{E}_0 \mathcal{E}_r \rangle$. Substituting eqn (6.42), there are at first sight three candidates for the two-point function, i.e. the following charge pairs: (1,1), $(-1-2e_0, -1-2e_0)$, and $(1, -1-2e_0)$. To verify if these charge pairs give the appropriate weights to the diagrams, note that the weight of each loop is given by $2\cos(e\pi + e_0\pi)$, where e is the total charge enclosed. The first candidate correctly gives the weight $2\cos(\pi + e_0\pi) = -\sqrt{q}$ for loops that surround one of the two $\mathcal{E}$ insertions, and $2\cos(2\pi + e_0\pi) = \sqrt{q}$ to loops that surround both. The second candidate gives a weight $2\cos(-e\pi - e_0\pi) = -\sqrt{q}$ to the loops that surround a single insertion, but the loops that surround both will have weight $2\cos(3\pi e_0)$, which is incorrect. The last option works fine: both $2\cos(\pi + e_0\pi)$ and $2\cos(-\pi + e_0\pi)$ are equal to $-\sqrt{q}$, while the loops surrounding both insertions get $2\cos(-e_0\pi) = \sqrt{q}$.

Note that in the case of the last charge pair the total charge is $2e_0$, which is that of the $\mathcal{B}$ operator, and should thus be considered as neutral. As a result, the corresponding two-point function is simply the Boltzmann weight of the two attracting charges:

$$\langle \mathcal{E}(0)\; \mathcal{E}(r) \rangle \;\propto\; |r|^{(-1-2e_0)/g}\,, \tag{6.43}$$

analogously to eqn (6.34). This must give the correct decay of the energy two-point function. The coefficient, however, cannot be obtained from these simple arguments.

The third candidate for the two-point function works equally well for the loop weights. However, in the CG language it has the difficulty that it is not charge neutral. We have learned to accept configurations of a total charge $-2e_0$, but not 2. This problem, however, can be corrected by noting that a charge of magnitude -2, inserted at any position, does not affect the weight of the loop diagram. Thus it may be used to correct the total charge. Effectively, the interaction between the two positive unit charges can be screened by a free-floating double negative charge. Thus we get

$$\langle \mathcal{E}(0)\, \mathcal{E}(r) \rangle \propto \int \mathrm{d}^2 r' \, |r|^{1/g} \, |r - r'|^{-2/g} \, |r'|^{-2/g} \propto |r|^{2-3/g} \,, \tag{6.44}$$

the result being obtained from simple power counting. Depending on the value of g, as yet unspecified, one may have to regularize the UV singularities of the integral. If, in the scaling limit, $\mathcal{E}$ consists of positive charges only, they must be prevented from forming a unit negative charge by picking up a -2.

Since these two calculations of the two-point function, eqns (6.43) and (6.44), represent exactly the same loop weights in the corresponding loop correlation function, they must be equal. This consistency condition gives a relation between g and e_0:

$$\frac{-1 - 2e_0}{g} = 2 - \frac{3}{g} \quad \text{so that} \quad e_0 = 1 - g \,. \tag{6.45}$$

This argument resolves the uncertainty in the value of g.

An independent argument [19, 20] to relate g and e_0 is based on the role of the screening charges just introduced in (6.44). If these operators were relevant, the ubiquitous presence of these charges would drive the system away from the dilute CG. However, if they were irrelevant, they would be ineffective at a sufficiently large spatial scale. Therefore, for the CG description to be applicable, they must be marginal. Remember that the two-point function of a marginal operator decays as r^{-4}. The exponent is calculated in the same way as in eqn (6.43) for the energy operator, and thus we have the exponent $(-2)(2-2e_0)/g = 4(e_0-1)/g$. The requirement that this be -4 leads to the same relation (6.45).

Exercise 6.19 A third way to calculate the two-point function is to take the integral (6.44), and insert $\mathcal{B}$ at an arbitrary fixed position. Show that this integral again leads to the same result, now assuming eqn (6.45) from the outset. Hint: note that the integrand behaves in a conveniently simple way under Moebius transformations, i.e. $z \to (a + bz)/(c + dz)$ with $ad - bc = 1$, in complex coordinates. Then, judiciously choose a Moebius transformation that simplifies the integral.

In conclusion, we have determined the thermal exponent of the q-state Potts model as

$$x_{\mathcal{E}} = 1 - \frac{3}{2g}, \quad \text{where} \quad \sqrt{q} = -2\cos \pi g \quad \text{and} \quad g \leq 1 \,. \tag{6.46}$$

The last condition goes back to the requirement that the charges ± 2 should be irrelevant, and the notation using $\sqrt{q}$ emphasizes that we need the positive root.

Exercise 6.20 Calculate the exponents α and the ν of the three-state Potts model.

Critical, tricritical, and antiferromagnetic points. The Potts model is defined for any positive integer q. Its low- and high-temperature series expansions are even for all q. It has been known for a long time that the transition is continuous for $q \leq 4$ and first-order for $q > 4$ [21]. The transition point is an analytic curve for all $q \neq 0$, but the analytic continuation of the thermal exponent beyond $q = 4$ folds back to smaller q (see eqn (6.46)). This suggests that it may describe critical behavior of another kind in a model with the same S_q symmetry. Continuity of the RG picture indicates that the other kind of critical behavior is a tricritical point [22], i.e. a special critical point where the transition changes from continuous to first-order.

When we constructed the Potts model by the ADE construction, we noted that in the case of the dense and dilute loop models, the central node of the diagram plays the role of a vacancy or unoccupied site. In principle, the introduction of a vacancy allows one to drive the transition to become discontinuous: if the vacancy has a large Boltzmann weight, and the Potts variables on the other hand have a strong interaction, there must be a first-order transition between the S_q symmetric dilute phase and an S_{q-1} symmetric ordered phase. Thus, in a dilute Potts model one may expect a critical and a tricritical RG fixed point. The observations above suggest that the critical and tricritical fixed points are the analytic continuation of each other, meeting at $q = 4$. This gives us an interpretation for another branch of eqn (6.46) for $g \geq 1$.

On this branch, the condition that double charges are irrelevant is no longer necessary, because in order to reach the tricritical point one has to tune another variable, which precisely suppresses the double charges. However, from the asymmetry between positive and negative charges noted before, $+2$ and -2 charges cannot both be suppressed by tuning a single parameter. The positive ones are suppressed at the tricritical point. This leads to a crossover exponent, which generates the RG flow from the tricritical to the critical point. The operator associated with this flow must correspond to the presence of positive double charges. As in the case of the leading thermal exponent, this crossover exponent can be calculated from two constructions of the two-point function. One is by letting two double positive charges be screened by two negative ones, and the other is by the simple correlation of $+2$ with $-2 - 2e_0$. As in the case of the leading thermal exponent, the result is the same,

$$x_{\varepsilon,2} = 4 - 2g \quad \text{with} \quad g \leq \frac{3}{2}. \tag{6.47}$$

Exercise 6.21 Show that both constructions result in the same exponent.

This is the exponent that governs the RG flow on the transition manifold from the tricritical to the critical regime, and then to the critical fixed point (where it has a smaller value g and is irrelevant).

The critical Boltzmann weight of the Potts model can be analytically continued beyond $q = 0$, as seen in Section 6.3.1. The model is in the antiferromagnetic domain, where unequal spins have higher weight than equal spins. Having calculated the leading thermal exponent, we see that in this antiferromagnetic regime, the temperature is an irrelevant parameter, and there should be a whole critical interval (in T), attracted to what is the analytic continuation of the ferromagnetic critical point. This critical phase is known as the Berker–Kadanoff phase [29].

The energy four-point function. The critical two-point functions are simple power laws, and thus contain no information beyond the value of the exponent. The three-point functions, given that they are conformally invariant, are also completely determined by their exponents, see e.g. [30]. The four-point functions, however, are nontrivial. Here we sketch a way to calculate them, mainly to demonstrate the power of the CG approach. We use complex numbers to denote positions in the plane.

When we considered the two-point function, we noted several ways to calculate it. Here too, we can take several different approaches. One is to take four positive unit charges to represent the operators $\mathcal{E}$. They should be screened by two double negative charges. The resulting fourfold integral,

$$\langle \mathcal{E}(r_1)\mathcal{E}(r_2)\mathcal{E}(r_3)\mathcal{E}(r_4)\rangle = \prod_{1\le j<k\le 4} |r_j - r_k|^{1/g} \int \mathrm{d}^2 w_1\, \mathrm{d}^2 w_2\; |w_1 - w_2|^{4/g} \prod_{j=1}^{2}\prod_{k=1}^{4} |w_j - r_k|^{-2/g} , \tag{6.48}$$

is not very attractive to even attempt. It becomes simpler if one of the $\mathcal{E}$'s is combined with a $\mathcal{B}$, so that quasi-neutrality is achieved with a single screening charge. The result is

$$\begin{aligned}\langle \mathcal{E}(r_1)\mathcal{E}(r_2)\mathcal{E}(r_3)\mathcal{E}(r_4)\rangle = &\prod_{1\le j<k\le 3} |r_j - r_k|^{1/g} \prod_{k=1}^{3} |r_k - r_4|^{2-3/g} \\ &\times \int \mathrm{d}^2 w\; |w - r_4|^{6/g-4} \prod_{k=1}^{3} |w - r_k|^{-2/g} .\end{aligned} \tag{6.49}$$

One may verify that the integrand is invariant for Moebius transformations (see Exercise 6.19), which can be used to simplify the expression. The area integral can reduced to a contour integral, making use of the following theorem, using complex notation for the plane. Let $p(u,v) = (\partial/\partial v)P(u,v)$ be a function analytic in domains D and $\bar{D}$ for the two arguments, respectively; then

$$\int_D \mathrm{d}^2 z\; p(z,\bar{z}) = \frac{1}{2\mathrm{i}} \oint_{\partial D} \mathrm{d}z\; P(z,\bar{z}) . \tag{6.50}$$

To apply this to eqn (6.49), we take D to be the complex plane with appropriate cuts, to exclude nonanalyticities due to the fractional powers. The calculation requires some bookkeeping [23], and the result is as follows, expressed in the complex variables $u_i = (r_i - r_4)(r_j - r_k)$, in which the indices i, j, k are a cyclic permutation of $1, 2, 3$:

$$\langle \mathcal{E}(r_1)\mathcal{E}(r_2)\mathcal{E}(r_3)\mathcal{E}(r_4)\rangle = -\frac{1}{2}\tan\frac{\pi}{g}\left\{\left(1 + 2\cos\frac{2\pi}{g}\right)|G|^2 + |H|^2\right\}, \tag{6.51}$$

where the contributions G and H are expressed in terms of the hypergeometric function $\mathrm{F}_{2,1}$ and the beta function B:

$$\begin{aligned} G &= \mathrm{B}\left(\frac{3}{g} - 1,\, 1 - \frac{1}{g}\right)\left(\frac{u_1}{u_2}\right)^{1/2g} u_3^{1-3/2g}\, \mathrm{F}_{2,1}\left(\frac{1}{g},\, 1 - \frac{1}{g};\, \frac{2}{g};\, -\frac{u_1}{u_2}\right), \\ H &= \mathrm{B}\left(1 - \frac{1}{g},\, 1 - \frac{1}{g}\right)\left(\frac{u_1 u_3}{u_2}\right)^{1-3/2g} \mathrm{F}_{2,1}\left(2 - \frac{3}{g},\, 2 - \frac{1}{g};\, 2 - \frac{2}{g};\, -\frac{u_1}{u_2}\right). \end{aligned} \tag{6.52}$$

It goes without saying that the four-point function found here can be confirmed. In particular, it can be verified that this four-point function is conformally invariant. We stress here that this property is a result, and not a property that has been inserted during the procedure.

An interesting result of this calculation, irrespective of the explicit outcome, is the prediction that the integrals (6.48) and (6.49) are the same up to an overall factor (which may actually be calculable).

The magnetic exponent, spin, and disorder operators. The adjacency matrix of the Potts model has eigenvalues $\sqrt{q}$, $-\sqrt{q}$, and 0, the latter $(q-1)$-fold degenerate. The first two are symmetric under the group S_q of permutations, but the last $q-1$ break that symmetry. The weight-changing operator associated with the second eigenvalue is the energy, as noted above. An example of an eigenvector in this case is $\exp(2\pi i s/q)$ for $s \in \{1, 2, \cdots, q\}$ and 0 in the symmetric state $s = 0$. The corresponding weight-changing operator represents the spin operator of the Potts model $\mathcal{S}$. In its two-point function, it gives weight zero to all loops that surround one of the insertions and not the other. Only those domain wall configurations contribute in which the two insertions are in the same domain. Can we find a specific charge operator that does precisely that?

We have already seen that when we fix all the heights at the boundary of the system to zero, each loop has a weight $2\cos(e\pi + e_0\pi)$, where e is the total charge enclosed. Since this weight must be zero for the operator $\mathcal{S}$, we choose $e_\pm = -e_0 \pm 1/2$. The exponent of the corresponding two-point function is [24]

$$x_s = \frac{e_+ \, e_-}{2g} = \frac{1/4 - (1-g)^2}{2g} \,. \tag{6.53}$$

Exercise 6.22 Verify this result and argue that a next-to-leading magnetic exponent is obtained by replacing the $1/4$ by $9/4$.

In the Potts model, we may also introduce what is called the disorder operator $\mathcal{D}$, which is the terminal of a cut across which the Potts spins s are identified with Ps, for a nontrivial permutation $P \in S_q$. The disorder operator is dual to the spin operator $\mathcal{S}$. That this is so can be seen as follows. Above, we argued that the $\mathcal{S}$ is the weight-changing operator with weight 0. Thus the only contributions to its two-point functions come from the domain wall configurations such that the two $\mathcal{S}$-insertions are in the same domain.

Now consider the insertion of two disorder operators into two different faces. For consistency, their state must be the central node of the adjacency diagram, i.e. the neutral state. There is then a cut between the two faces, across which s is identified with Ps. Let us for the moment assume that there is no state for which $Ps = s$, aside from the neutral state. Then any spin domain that surrounds either of the two $\mathcal{D}$-insertions and not the other has zero weight. If there is no such cluster, the two faces are in the same neutral cluster, and then the weight is not affected by the operator insertions. Clearly, the correlation function is the same as that of two spin operators, with an interchange between neutral and the spin clusters. The restriction that there is no state for which $Ps = s$ can be lifted when the unconnected part of the correlation is explicitly deducted.

Anyons. The combination of a spin operator and a disorder operator on adjacent faces turns out to form an anyon operator, i.e. excitations which pick up a phase factor if they circle one another [31].

Consider a disorder operator $\mathcal{D}$ associated with the symmetry operator $P \in \mathrm{S}_q$, and a spin operator $\mathcal{S}$ chosen to be a unitary representation of P, so that $P\mathcal{S} = \mathrm{e}^{2\pi\mathrm{i}\sigma}\mathcal{S}$, with σ a rational number with denominator $\leq q$. We consider the operator that results from the insertion of $\mathcal{S}$ and $\mathcal{D}$ on adjacent faces. We will show that this combination behaves as an anyon.

To inspect the properties of this operator, we consider the two-point function of two complementary anyons. Their disorder components are associated with the permutations P and P^{-1}, and their spin insertions are each others' complex conjugates. The spin operators are placed in a domain with spin states, and the disorder operators in a domain in the neutral state. The disorder operators are connected by a cut, indicated by the dashed line in Fig. 6.4, across which the state s on one side is identified with Ps on the other. The positions of the anyons are indicated by the small diamonds.

Since, as argued above, the two spin operators must sit in the same domain, and likewise the two disorder operators, their combined positions must be on the same domain wall. The domain wall that connects the two anyons is indicated by the bold loop in Fig. 6.4. The spins in the domain which it encloses should all be in the same state, as far as that state lives on one side of the cut. However, where it crosses the cut, this spin state s is replaced by Ps. Each time the domain crosses the cut in the same direction, on its way between the two $\mathcal{S}$-insertions, the spin operators generate a phase factor $\exp(2\pi\mathrm{i}\sigma)$. Each passage in the opposite direction gives the inverse phase factor. This gives the selection rule and the weight of the domain configurations contributing to the two-anyon correlator.

Can we accommodate these rules in the height representation of the model [25]? The restriction that the two operators must sit on the same domain wall can be represented by screw dislocations or vortices of strength 1 and -1. The domain wall that joins the two anyons reverses its signature as it passes one of the anyons. This causes the curvatures of the two halves of the domain wall to count in opposite ways. The total bending angle no longer adds up to 2π, but counts the number of times the domain walls wind around the anyons. In the figure, the domain wall winds twice around the right-hand anyon before reaching it. The weight of the

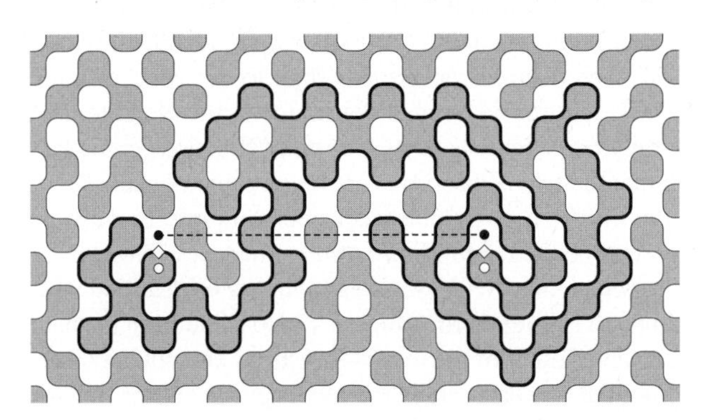

Fig. 6.4 A configuration contributing to a two-anyon correlation function. The gray domains contain the Potts spins, and the white domains the neutral state. The black dots are the disorder operators, and the open circles the spin operators. The dashed line is the cut associated with the disorder operators.

domain wall is therefore not equal to $\exp(\pm i\pi e_0)$; it has a factor $\exp(2i\pi e_0)$ for each time it winds around the insertions. This phase factor can be canceled by placing a factor $\exp(-ie_0 h)$ in each of the anyons, since the effective height at the anyon position is increased or decreased by such winding.

So far, the screw dislocations have selected the proper diagrams, and the charges $-e_0$ have neutralized the phase factor that would otherwise be associated with the net curvature of the domain walls. Above, we have seen that the diagrams pick up a phase factor $\exp(\pm 2\pi i\sigma)$ each time the cluster passes through the cut from above or from below, respectively. This phase factor can be accommodated by placing a charge $e = \pm\sigma$ on each of the anyons.

As a result, the exponent that governs the spatial decay of the two-point function is given by

$$x_{\mathrm{a}} = \frac{g}{2} m_1 m_2 + \frac{1}{2g} e_1 e_2 = \frac{g}{2} + \frac{\sigma^2 - e_0^2}{2g} = 1 + \frac{\sigma^2 - 1}{2g} . \tag{6.54}$$

In addition, these anyons have spin, i.e. they pick up a phase factor when they are taken around each other. This is caused by the interaction between the magnetic and electric charges

$$s_{\mathrm{a}} = \frac{1}{2}(m_1 e_2 + e_1 m_2) = \sigma . \tag{6.55}$$

Perspective. Before we continue, we wish to take stock of the qualitative assumptions that are ingredients of these calculations.

1. The RG is taken as valid.
2. It is assumed that the critical point of the Potts model approaches the Gaussian free-field model under progressive renormalization. While one can justify the assertion that the infinity of operators that separate the Potts models from the Gaussian free field (with screening charges and background charge) are irrelevant, *in the Gaussian model itself*, this says nothing about the possible occurrence of other fixed points which prevent the critical Potts model from renormalizing to the Gaussian model.
3. The value of e_0 follows directly from the value of q. But the coupling constant g could be determined only by indirect consistency arguments.
4. We proposed that the energy operator of the Potts model is a particular selection of weight-changing operators, which corresponds to the insertion of only positive unit charges. This implies that the temperature parameter of the Potts model traces out the special line where only positive unit charges are present. This conclusion will play a role in the next section. In principle, it would be possible to allow a small admixture of the other candidate weight-changing operators. This would spoil the argument that determines g, leaving it without a consistent solution.
5. As in the discussion of the RG, the proof of the pudding is in the eating. The many healthy results of this approach prove it to be correct.

6.4.5 The O(n) spin model

In the case of the Potts model, the critical point was known beforehand (by duality), which was a significant advantage. In the O(n) spin model eqn (6.29), no such prior result existed when the CG approach was brought to bear on the problem. In this section, we will deal with precisely that problem. By using eqn (6.29) as a starting point, we have already resolved a few difficulties. Every link can be occupied by at most one bond, and because the lattice is three-coordinated, intersections are not possible. The intention of this section is to make use of the knowledge we have acquired about the Potts model, in order to apply it to the O(n) model.

As a starting point, we make the following observations. Above, we argued that the Potts model at a noncritical temperature admits positive unit charges, but not negative ones. When negative unit charges are introduced, this affects the renormalization of g. If the product of fugacities $z_+ z_- > 0$, the coupling constant g will grow, thus weakening the electric interaction by screening. However, if $z_+ z_- < 0$, the effect is opposite and g will decrease. At first, $z_\pm$ will both grow because they are relevant. This will strengthen the effect on the value of g, which will decrease faster. However, this makes it unavoidable that g passes through the value where the $z_{\pm 1}$ are marginal, and become irrelevant. From there on, $z_{\pm 1}$ will both diminish, and the model returns to the Gaussian model, but with a much smaller value of g. A sketch of this RG flow in terms of g and the product $z_+ \, z_-$ is given in Fig. 6.5. If we assume that the fugacity of the positive unit charges z_+ is positive, we see that for large enough g the final direction of the RG flow depends critically on the sign of z_-. If that too is positive, it will grow indefinitely under the RG, but if it is negative, it will only grow initially, and will eventually diminish. This observation shows that the locus $z_- = 0$ is critical.

All this adds up to the conclusion that the Potts model (parameterized by its temperature) is a critical locus in a larger parameter space. A height model which embeds the height version

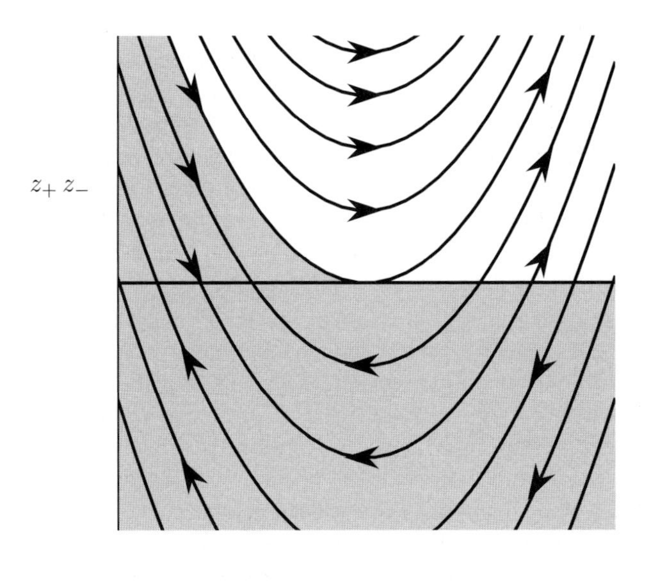

Fig. 6.5 The renormalization flows in terms of the Gaussian coupling constant g and the product of fugacities for positive and negative unit charges $z_+ \, z_-$. At large g, this product is relevant, but when it starts to become negative, it suppresses the value of g, so that eventually it becomes irrelevant. As a result, in the shaded region, the product $z_+ \, z_-$ eventually renormalizes to zero.

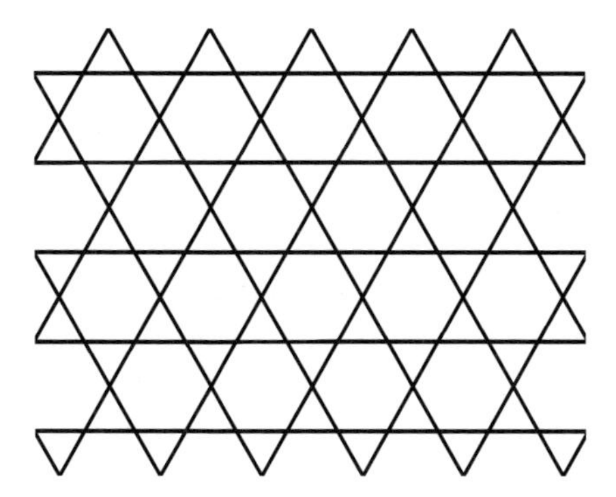

Fig. 6.6 A section of a kagome lattice.

of the Potts model will generally contain negative unit charges also. Thus, we propose that if a model $M(t)$, parameterized by t in a more general parameter space, intersects with the Potts model, it is to be expected that that intersection is the critical point of M, because at that point the negative unit charges disappear.

We will now proceed to show that the O(n) model defined by eqn (6.29) intersects with a Potts model at some value of the O(n) coupling constant. Consider a CPL model on a kagome lattice (see Fig. 6.6). We can use the ADE procedure to turn it into a Potts model or a height model. We choose the Potts variables to live in the hexagons, and the neutral state in the triangles. This corresponds to a Potts model on the triangular lattice. Likewise, in the height model the heights in the hexagons are even multiples of π and those in the triangles are odd multiples of π.

A sequence of local height configurations around a vertex of the kagome lattice is listed in Table 6.2. Equivalent configurations can be obtained from these by a shift in height or by a lattice symmetry. The configurations on the other sublattices of the kagome vertices are obtained by rotating the lattice through $2\pi/3$ or $4\pi/3$, the weights being unchanged.

The weights for the CPL model and the equivalent Potts model are determined by the weight ratio of the acute and obtuse bends as in the diagrams . Let the weight of the obtuse bends be 1 and the weight of (two) acute bends be B. The local weight for the height model corresponding to the Potts model is given in the second line of Table 6.2. When there is a local gradient in the configuration, only one of the two loop diagrams is compatible, with weight 1 or B as appropriate. When the heights form a saddle-like configuration, both

Table 6.2 A height model on the kagome lattice. The height configurations are given in the top line, with the even heights placed in the hexagons. The Boltzmann weights can be chosen so that the model is equivalent to the Potts model (second line) or the O(n) model (last line).

Configuration	0 1 1 2	2 1 3 2	2 1 1 2	2 3 3 2
Potts	1	B	$p + B/p^2$	$1/p + Bp^2$
O(n)	1	CD	C^2	D^2

loop diagrams fit, and must be summed over. The parameter p is the cube root of the ratio of eigenvector elements (see eqn (6.20)). For the Potts model, $p^3 + p^{-3} = \sqrt{q}$. The parameter B corresponds to temperature variations in the Potts model. The critical point of this Potts model is known, but is unimportant in this discussion.

In the height configurations permitted in this model, all hexagonal faces have a height which differs by π from the neighboring triangular faces. This implies that two adjacent hexagons can either be equal or be different in height by 2π. If the triangular faces could be ignored, this would precisely be equivalent to the loop model on the hexagonal lattice. Now we propose to choose weights for the height model on the kagome lattice such that the triangular heights can be summed over, given the configuration of the hexagonal faces. When a triangle has neighboring hexagons of unequal height, its own height is fixed to sit in between those of the hexagons. When a triangle sits in between three hexagons of equal height, its own height can be above or below that of its neighbors. These possibilities can be summed, when they are independent of the other triangular faces.

In the configurations listed in Table 6.2, the first has hexagons of unequal height, which thus fix the height of the triangles. In the following three configurations, however, the hexagons have height 2, and the triangles are free to take a height 1 or 3. The choices of the left and right triangle are independent if the weight ratio for the two options of one triangle does not depend on the state of the other triangle. This is expressed in the third line of Table 6.2, which gives a factor C for each triangle that is below its hexagonal neighbors, and D when it is above. For this choice of weights, the summation over the triangular faces can be done trivially. This will result in a height model with hexagonal faces only, hopefully a height version of the loop model on the hexagonal lattice (6.19).

From the reasoning above, this latter model will be critical where it intersects with the Potts model. It is formula manipulation to see if the models intersect: the weights in the second and third lines in the table must correspond. But, since both models trace a two-dimensional manifold, and there are just three weights to match, we can expect a one-dimensional intersection. At that intersection, we then expect the O(n) model to have its phase transition.

The value of q follows simply from the ADE construction, $q = (p^3 + p^{-3})^2$, but the value of n is not immediately obvious. A summation over the heights in the triangles has to be performed first: a triangle above a neighboring hexagon gives a weight D, and a triangle below gives C. When a triangle sits between three hexagons of equal height, it can have a height one unit above or one unit below them, independently of the other triangles. The weight, after summation, is then $C^3 + D^3$. Other triangles have either two hexagons above them and one below or two below and one above, with weights C^2D or CD^2, respectively. We will interpret this as the height representation of the hexagonal loop model (6.19).

Exercise 6.23 Calculate the intersection of the models and show that at the intersection point, (i) the O(n) parameters x and n in eqns (6.19) and (6.29) are related by $x^2(2 \pm \sqrt{2-n}) = 1$, (ii) the Potts model is in the antiferromagnetic regime and at zero temperature, and (iii) the relation between n and the number q of states in the Potts model is given by $n = 2 - (2-q)^2$.

In summary, the two height models defined in Table 6.2 are equivalent to the Potts and the O(n) model, respectively. The point where they intersect is expected to be a (or the) critical point of the O(n) model. The fact that the intersection takes place in the antiferromagnetic regime of the Potts model means that the thermal parameter of the Potts model is irrelevant, so that both models are in fact critical at the intersection.

Assertion (iii) of Exercise 6.23, together with the relation between q and g in eqn (6.46), implies that $n = -2\cos(4\pi g)$. This is different from the accepted notation in the literature i.e. $n = -2\cos(\pi g)$. This is related to the fact that in the model just constructed, the steps across domain walls have amplitude 2π, since they correspond to two steps of the Potts model. To keep contact with most of the literature, we will now rescale the heights so that $h_{\mathrm{O}(n)} = h_{\mathrm{Potts}}/2$, and rescale the charges so that $e_{\mathrm{O}(n)} = 2e_{\mathrm{Potts}}$ and the coupling constant so that $g_{\mathrm{O}(n)} = 4g_{\mathrm{Potts}}$ accordingly. Below, the suffix O(n) will be dropped.

The O(n) thermal exponent. Now we remember that the energy operator of the Potts model was a positive unit charge, which now becomes a positive double charge. The energy operator of the O(n) model is then a negative charge of the same magnitude, which the Potts model had managed to avoid so carefully. This leads to the following (charge-neutral) construction of the two-point energy function of the O(n) model:

$$|r_1 - r_2|^{4/g} \int \mathrm{d}^2 r_3\, \mathrm{d}^2 r_4\, |r_3 - r_4|^{4/g} \prod_{i=1,2;j=3,4} |r_i - r_j|^{-4/g} = |r_1 - r_2|^{4-8/g}\,, \tag{6.56}$$

the result from simple power counting. This corresponds to the exponent

$$x_\varepsilon = \frac{4}{g} - 2 \qquad \text{with} \qquad n = -2\cos(\pi g)\,. \tag{6.57}$$

Again, as in the case of the Potts model, the exponent can be analytically continued beyond $n = 2$. The case $g \geq 1$ belongs to $x^{-2} = 2 + (2-n)^{1/2}$, and $g \leq 1$ to $x^{-2} = 2 - (2-n)^{1/2}$. But since the thermal exponent for the large-x branch is irrelevant, we conclude that the whole interval $x^{-2} < 2 + (2-n)^{1/2}$ is attracted to this fixed point. Apparently, the phase with relatively large x, the phase dense with loops, is also critical. This is precisely what one might expect from Fig. 6.5, in which the shaded regime flows eventually to the Gaussian model. The behavior is a perfect image of that at the transition manifold of the Potts model, including the value of the exponents. It turns out that we have now established two relations between the Potts and O(n) models, one in which $n = \sqrt{q}$ (the transition manifold of the dilute Potts model, on the whole (n,x) sheet of the O(n) model), and one in which $n = 2 - (2-q)^2$ (O(n) critical $\leftrightarrow$ Potts antiferromagnetic). This explains why the thermal exponent of the O(n) model has the same expression as the crossover exponent of the tricritical Potts model (6.47).

The value of the background charge has changed in a nontrivial way in the above equivalence. This is because an "O(n) loop" is not simply two Potts loops with the same sign. It may be a local collaboration of Potts loops that go side by side for a while, and separate later. The background charge is determined completely by the magnitude of the step represented by a domain wall (again π, after the rescaling above), and the weight of closed loops (6.57): $n = -2\cos(\pi g) = 2\cos(e_0)$. It is natural to choose the smallest value of e_0, but for the sign we have to examine not only the weight of the loop but, separately, the weight of the domain wall encircling a lower or a higher plateau. The result is

$$e_0 = g - 1\,, \tag{6.58}$$

the same expression as eqn (6.45), but with opposite sign.

Exercise 6.24 Verify that the thermal exponent in eqn (6.57) can be calculated also without screening charges, but in a configuration with total charge $-2e_0$

The O(n) magnetic exponent. There are many other operators that one can construct in the O(n) model; the most notable is probably the spin operator itself, conjugate to the magnetic field. In this section, we will calculate its exponent. As before, a way to approach this is first to inspect the two-point function $\langle s_0 \cdot s_r \rangle$, the scalar product between two distant spin vectors. The diagrams that contribute to this, in the expansion in powers of x, are different from the ones in the partition sum: now the numbers of bonds incident on 0 and r must be odd, whereas the numbers incident on all other vertices are even. In the height language, this has an obvious interpretation: at sites 0 and r, there is a screw dislocation of one step π, i.e. a magnetic charge of strength $m = \pm\frac{1}{2}$.

At first sight one might expect an exponent equal to $gm^2/2 = g/8$, but this is not to be. The domain walls that terminate on 0 or r gain a phase factor when they curve. For closed domain walls, the integrated curvature is always the same, but for these open domain walls, this is not the case. The domain walls can spiral around their terminals indefinitely, and accordingly pick up arbitrary phase factors. We will have to correct this.

Note that as the domain walls spiral about their terminals, they either drill down the height of their terminal or screw it up. Consider the diagram in Fig. 6.7, showing a domain wall spiraling around its terminal. If the height at the terminal is defined by counting the steps up and down along a fixed path from infinity, this height is increased by the number of windings of the path. Thus we can counteract the phase factor of the spiraling by an extra insertion of an electric charge, neutralizing the curvature effect. This is done precisely by a charge $-e_0$. It takes a little thought to conclude that the same charge is needed at both the negative and the positive vortex, at either terminal of the open diagram.

As a result, the spin–spin correlation function is that of a combination of a magnetic charge $m = \frac{1}{2}$ and an electric charge $e = -e_0$ on one side and $m = -1/2$ and $e = -e_0$ on the other. The resulting exponent is

$$x_s = \frac{g}{8} - \frac{(g-1)}{2g} \,. \tag{6.59}$$

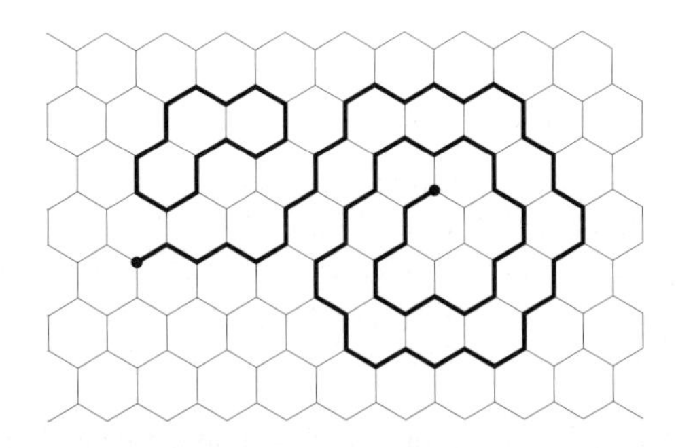

Fig. 6.7 Diagram contributing to the spin–spin correlation function of the O(n) model.

Exercise 6.25 It is not obvious that this electromagnetic charge has no spin, i.e. that there is no phase factor associated with rotating the position of s_r around that of s_0. Verify that this is the case.

Intersections. In this whole calculation, we have considered only O(n) models where the diagrams of the high-temperature phase have no intersections. In an O(n) spin model on an arbitrary (two-dimensional) lattice or with an arbitrary interaction, the paths in the diagrams do intersect (but still pick up a factor n for each closed loop). Therefore the special O(n) model defined by eqn (6.29) can be considered as generic only if the loops are irrelevant. Again, we will inspect this by considering the (connected) two-point function. This is the same as the two-point function of the joint terminals of four legs. This can be calculated in a manner completely analogous to that of the calculation of the magnetic exponent, and thus results in

$$x_\times = 2g - \frac{(g-1)^2}{2g}. \tag{6.60}$$

This is indeed irrelevant at the O(n) critical point.

Exercise 6.26 Verify that in this case also, the same charge $-e_0$ precisely cancels the total net curvature of the four legs. And verify that this exponent is irrelevant at the critical point of the O(n) model for $-2 \leq n \leq 2$.

Unfortunately, the operator is relevant in the dense phase (i.e. dense with loops), and for this reason this phase cannot serve as the low-temperature phase of the generic O(n) model. This argument is discussed in [26]. An intersecting loop model proposed there to describe the low-temperature phase of the O(n) spin model is described in [13].

Tricritical O(n) models. Just as in the Potts model, the transition of the O(n) model can be made discontinuous by altering the model by, for example, adding a four-spin interaction or thermally distributed vacancies. The description of this phase transition for a long time evaded description by the above methods, except for $n = 1$, the Ising model, because this model is identical to the $q = 2$ Potts model, and for $n = 0$ [18]. These cases could not be easily generalized to general n. Recently this problem was solved; for details, we refer to the original publication [27]. Unfortunately, the solution suffers from the same restriction as in the case of the dense phase mentioned above: it is applicable only to those O(n) spin models that can be represented by non-intersecting loops.

6.5 Summary and perspective

The applications of loop models are surprisingly numerous. Some methods to study critical systems, such as the CG and SLE, are based upon them. The language allows methods for and investigations of operators that are not accessible in completely local theories, but make physical sense (consider, for example, the probability that two spins in the Ising model sit in the same up cluster, or that two unequal spins sit on the same domain boundary as an unequal pair elsewhere.)

We have considered models that admit a single-component height representation. Several models with a more dimensional height lattice have been studied by these methods [12, 20,

28], with remarkable results. It appears that we have only scratched the surface of what is possible.

The assertion that alternative calculations of multipoint correlation functions give the same results leads to interesting conjectures for integral expressions. In the text, only one example was given, but the methods to calculate these correlation functions admit many more. A proof of these conjectures or, better, a mathematical framework in which they are obvious would be very interesting. At the moment, there is not even a numerical verification.

In these lectures we have, among the applications of loop models, focused on the Potts model and the O(n) model. A number of other applications, such as percolation, the self-avoiding walk, and spanning trees are special limits of these. For other applications, such as the hard hexagon and hard square models, these methods also give direct results for the critical exponents and other universal quantities.

The CG method provided the first analytic calculation of numerous critical exponents in two dimensions. The results have been confirmed by CFT and by exact solutions of integrable models. At the moment, the CFT approach is capable of giving more extensive results, and SLE more rigorous results. An advantage of the CG method that remains is that the results and the original models and operators are directly related. In SLE and CFT, this connection is usually made indirectly.

A challenging problem in these same models is the calculation of off-critical thermodynamic scaling functions, which are also universal objects. In general, these are thermodynamic quantities that are functions of (at least) two thermodynamic parameters infinitesimally close to the critical point. This problem has not been solved analytically, and it is conceivable that the CG approach is capable of doing this.

References

[1] L. P. Kadanoff, *Physics* **2**, 263 (1966).

[2] T. Niemeijer and J. M. J. van Leeuwen, *Physica* **71**, 17 (1974); in *Phase Transitions and Critical Phenomena*, Vol. 6, eds. C. Domb and M. S. Green, p. 425, Academic Press (1976).

[3] K. G. Wilson, *Phys. Rev. B* **4**, 3174 (1971).

[4] L. P. Kadanoff, *J. Phys A* **11**, 1399 (1978).

[5] B. Nienhuis, *J. Stat. Phys.* **34**, 731 (1984); in *Phase Transitions and Critical Phenomena*, eds. C. Domb and J. L. Lebowitz, p. 1, Academic Press, London (1987).

[6] L. Onsager, *Phys. Rev. B* **65**, 117 (1944).

[7] E. H. Lieb, *Phys. Rev. Lett.* **18**, 1046 (1967).

[8] R. J. Baxter, *Ann. Phys.* **70**, 193 (1972).

[9] R. J. Baxter and P. A. Pearce, *J. Phys. A* **16**, 2239 (1983).

[10] R. J. Baxter, *J. Phys. A* **13**, L61 (1980).

[11] R. J. Baxter, *Exactly Solved Models in Statistical Mechanics*, Academic Press, New York (1982).

[12] J. L. Jacobsen and J. Kondev, *Nucl. Phys. B* **532**, 635 (1998); J. Kondev and J. L. Jacobsen, *Phys. Rev. Lett.* **81**, 2922 (1998); J. L. Jacobsen and J. Kondev, *J. Stat. Phys.* **96**, 21 (1999).

[13] M. J. Martins, B. Nienhuis, and R. Rietman, *Phys. Rev. Lett.* **81**, 504 (1998).

[14] V. Pasquier, *J. Phys. A* **20**, 5707 (1987); *J. Phys. A* **20**, L1229 (1987).

[15] S. O. Warnaar, B. Nienhuis, and K. A. Seaton, *Phys. Rev. Lett.* **69**, 710 (1992).
[16] F. Y. Wu, *Rev. Mod. Phys.* **54**, 235 (1982).
[17] J. L. Jacobsen and H. Saleur, *Nucl. Phys. B* **716**, 439 (2005).
[18] B. Duplantier and H. Saleur, *Phys. Rev. Lett.* **59** (1987).
[19] O. Foda and B. Nienhuis, *Nucl. Phys. B* **324**, 643 (1989).
[20] J. Kondev, *Int. J. Mod. Phys. B* **11**, 153 (1997).
[21] R. J. Baxter, *J. Phys. A* **15**, 3329 (1982).
[22] B. Nienhuis, A. N. Berker, E. K. Riedel, and M. Schick, *Phys. Rev. Lett.* **43**, 737 (1979).
[23] B. Nienhuis, *Physica A* **251**, 104 (1998).
[24] M. Den Nijs, *Phys. Rev. B.* **27**, 1674 (1983).
[25] B. Nienhuis and H. J. F. Knops, *Phys. Rev. B* **32**, 1872 (1985).
[26] J. L. Jacobsen, N. Read, and H. Saleur, *Phys. Rev. Lett.* **90**, 090601 (2003).
[27] B. Nienhuis, W. A. Guo, and H. W. J. Blote, *Phys. Rev. B* **78**, 061104 (2008).
[28] J. Kondev, J. de Gier, and B. Nienhuis, *J. Phys.* **29**, 6489 (1996).
[29] A. N. Berker and L. P. Kadanoff, *J. Phys. A: Math. Gen.* **13**, L259 (1980).
[30] J. L. Cardy, in *Phase Transitions and Critical Phenomena*, eds. C. Domb and J. L. Lebowitz, Academic Press, London (1987).
[31] E. Fradkin and L. P. Kadanoff, *Nucl. Phys. B* **170**, 1 (1980).

7

Lectures on the integrability of the six-vertex model

N. Reshetikhin

Nicolai Reshetikhin,
Department of Mathematics,
University of California at Berkeley,
Berkeley, CA 94720-3840, USA,
and KDV Institute for Mathematics,
Universiteit van Amsterdam,
Plantage Muidergracht 24,
1018 TV, Amsterdam, The Netherlands.

7.1 Introduction

The goal of these notes is to outline the relation between solvable models in statistical mechanics, and classical and quantum integrable spin chains. The main examples are the six-vertex model in statistical mechanics, and spin chains related to the loop algebra Lsl_2.

The six-vertex model emerged as a version of the Pauling's ice model or, more generally, as a two-dimensional model of ferroelectricity. The free energy per site was computed exactly in the thermodynamical limit by Lieb [25]. The free energy as a function of an electric field was computed by Sutherland and Yang [39]. For details of and more references on earlier work on the six-vertex model, see [26]. The structure of the free energy as a function of the electric field was also studied in [6, 29].

Baxter discovered that the Boltzmann weights of the six-vertex model can be arranged into a matrix which satisfies what is now known as the Yang–Baxter equation. For more references on the six-vertex model and on the consequences of the Yang–Baxter equation for the weights of the six-vertex model, see [3].

The six-vertex model and similar "integrable" models in statistical mechanics became the subject of renewed research activity after the discovery by Sklyanin of the relation between the Yang–Baxter relation in the six-vertex model and the quantization of classical integrable systems [36]. This led to the discovery of many "hidden" algebraic structures of the six-vertex model and to the construction of quantizations of a number of important classical field theories [15, 16]. For more references see, for example [21]. Further development of this subject resulted in the development of quantum groups [13] and further understanding of the algebraic nature of integrability.

Classical spin chains related to the Lie group SL_2 are an important family of classical integrable systems related to the six-vertex model. The continuum version of this model is known as the Landau–Lifshitz model. One of its quantum counterparts, the Heisenberg spin chain, has a long history. Eigenvectors of the Hamiltonian of the Heisenberg model were constructed by Bethe in 1931 using a substitution which is now known as the Bethe ansatz. He expressed the eigenvalues of the Hamiltonian in terms of solutions to a system of algebraic equations now known as the Bethe equations. An algebraic version of this substitution was found in [16].

These lectures consist of three major parts. The first part is a survey of some basic facts about classical integrable spin chains. The second part is a survey of the corresponding quantum spin chains. The third part is focused on the six-vertex model and on the limit shape phenomenon. The appendix contains a number of random useful facts.

The author is happy to thank A. Okounkov, K. Palamarchuk, E. Sklyanin, and F. Smirnov for discussions, B. Sturmfels for an important remark about the solutions to the Bethe equations, and the organizers of the school for the opportunity to give these lectures.

This work was supported by the Danish National Research Foundation through the Niels Bohr initiative; the author is grateful to Aarhus University for hospitality. It was also supported by NSF grant DMS-0601912.

7.2 Classical integrable spin chains

The notion of a Poisson Lie group developed from the study of integrable systems and their relation to solvable models in statistical mechanics. It is a geometrical structure behind the

Poisson structures on the Lax operators, which emerged in the analysis of Hamiltonian structures in integrable partial differential equations. The first examples of these structures were related to taking the semiclassical limit of Baxter's R-matrix for the eight-vertex model. The notion of Poisson Lie groups and Lie bialgebras places these examples into the context of Lie theory and provides a natural version of such systems related to other simple Lie algebras.

7.2.1 Classical r-matrices and the construction of classical integrable spin chains

Here we recall the construction of classical integrable systems based on classical r-matrices.

A classical r-matrix with an (additive) spectral parameter is a holomorphic function on $\mathbb{C}$ with values in $End(V)^{\otimes 2}$ which satisfies the classical Yang–Baxter equation

$$[r_{12}(u), r_{13}(u+v)] + [r_{12}(u), r_{23}(v)] + [r_{13}(u+v), r_{23}(v)] = 0\,, \tag{7.1}$$

where $r_{ij}(u)$ act in $V^{\otimes 3}$, such that $r_{12}(u) = r(u) \otimes 1,\ \ r_{23}(u) = 1 \otimes r(u)$, etc.

Let $L(u)$ be a holomorphic function on $\mathbb{C}$ of a certain type (for example a polynomial). The matrix elements of the coefficients of this function satisfy quadratic r-matrix Poisson brackets if their generating function $L(u)$ has the Poisson brackets

$$\{L_1(u), L_2(v)\} = [r(u), L_1(u)L_2(v)]\,, \tag{7.2}$$

where $L_1(u) = L(u) \otimes 1$ and $L_2(u) = 1 \otimes L(u)$. The expression on the left is the collection of Poisson brackets $\{L_{ij}(u), L_{kl}(v)\}$.

Consider the product

$$T(u) = L^{(N)}(u - a_N) \dots L_1^{(1)}(u - a_1)\,.$$

The matrix elements of $T(u)$ are functions on $P_N \times \cdots \times P_1$. The factor $L^{(i)}$ is a function on P_i. The Poisson structure on P_i is as above.

The r-matrix Poisson brackets for $L^{(i)}(u)$ imply similar Poisson brackets for $T_{ij}(u)$:

$$\{T_1(u), T_2(v)\} = [r(u-v), T_1(u)T_2(v)]\,.$$

Taking the trace in this formula, we see that

$$\{t(u), t(v)\} = 0\,.$$

We fix symplectic leaves $S_i \subset P_i$ for each i.The restriction of the generating function $t(u)$ to the product of symplectic leaves $S = S_N \times \cdots \times S_1 \subset P_N \times \cdots \times P_1$ gives the generating function for commuting functions on this symplectic manifold.

Under the right circumstances, the generating function $t(u)$ will produce the necessary number of independent functions to produce a completely integrable system, i.e. $\dim(S)/2$.

One of the main conceptual questions in this construction is: Why do classical r-matrices exist? Indeed, if $n = \dim(V)$, eqn (7.1) is a system of n^6 functional equations for n^4 functions. The construction of the Drinfeld double of a Lie bialgebra provides an answer to this question [13].

7.2.2 Classical L-operators related to $\widehat{sl_2}$

In this section, we will focus on classical L-operators for the classical r-matrix corresponding to the standard Lie bialgebra structure on $\widehat{sl_2}$.

Such L-operators describe finite-dimensional Poisson submanifolds in the infinite-dimensional Poisson Lie group LSL_2. See the appendix to this chapter for some basic facts and references. Up to a scalar multiple, they are polynomials in the spectral variable z. One of the simplest such Poisson submanifolds corresponds to polynomials of first degree of the following form:

$$L(u) = \begin{pmatrix} a + a'z^2 & b' \\ z^2 b & c + c'z^2 \end{pmatrix}, \tag{7.3}$$

where $z = \exp(u)$. The r-matrix Poisson brackets on $L(u)$ with the r-matrix given by eqn (7.A.3) induce a Poisson algebra structure on the algebra of polynomials in $a, b, \ldots$. This Poisson algebra can be specialized further (by quotienting with respect to the corresponding Poisson ideal). As a result, we arrive at the following L-operator:

$$L(u) = \begin{pmatrix} zk - z^{-1}k^{-1} & z^{-1}f \\ zf & zk^{-1} - z^{-1}k \end{pmatrix}, \tag{7.4}$$

which satisfies the r-matrix Poisson brackets (7.2) with the following brackets on k, e, f:

$$\{k, e\} = \epsilon ke, \ \ \{k, f\} = -\epsilon kf\,, \tag{7.5}$$

$$\{e, f\} = 2\epsilon(k^2 - k^{-2})\,. \tag{7.6}$$

The function

$$c = ef + k^2 + k^{-2} \tag{7.7}$$

Poisson commutes with all other elements of this Poisson algebra, i.e. it is the Casimir function. It is easy to show that this is the only Casimir function on this Poisson manifold.

It is easy to check that

$$L(-u)^t = -D_z\sigma_2^y L(u)\sigma_2^y D_z^{-1}\,. \tag{7.8}$$

On the level surface of c parameterized as $c = t^2 + t^{-2}$, this matrix satisfies the extra identity

$$L(u)\theta(L(-u))^t = (zt - z^{-1}t^{-1})(tz^{-1} - t^{-1}z)I\,,$$

where I is the identity matrix and

$$\theta(e) = f, \ \ \theta(f) = e, \ \ \theta(k) = k$$

is the anti-Poisson involution: $\{\theta(a)\theta(b)\} = -\theta(\{a, b\})$, $\theta^2 = id$.

It is easy to find the determinant of $L(u)$:

$$\det(L(u)) = (zt - z^{-1}t^{-1})(t^{-1}z - tz^{-1}).$$

When $z = t, t^{-1}$, the matrix L degenerates to one-dimensional projectors

$$L(t) = \begin{pmatrix} \frac{tk - t^{-1}k^{-1}}{t^{-1}e} \\ 1 \end{pmatrix} \otimes \left(t^{-1}e, tk^{-1} - t^{-1}k\right), \tag{7.9}$$

$$L(t^{-1}) = \begin{pmatrix} \frac{t^{-1}k - tk^{-1}}{te} \\ 1 \end{pmatrix} \otimes \left(te, t^{-1}k^{-1} - tk\right). \tag{7.10}$$

7.2.3 Real forms

Here we will describe the Poisson manifold su_2^*. Let S_1, S_2, S_3 be the coordinates su_2^* corresponding to the usual orthonormal basis in su_2. The Poisson brackets between these coordinate functions are

$$\{S_1, S_2\} = 2S_3, \;\; \{S_2, S_3\} = 2S_1, \;\; \{S_3, S_1\} = 2S_2.$$

These coordinates are also known as classical spin coordinates. The center of this Poisson algebra is generated by $C = S_1^2 + S_2^2 + S_3^2$.

It is convenient to introduce $S^+ = (S_1 + iS^2)/2$, $S^- = (S_1 - iS^2)/2$. Since the S_i are real, $\overline{S^+} = S^-$. The Poisson brackets between these coordinates are

$$\{S^+, S^-\} = -iS_3, \;\; \{S_3, S^\pm\} = \mp 2iS^\pm.$$

The level surfaces of C are spheres. On the level surface with $C = l^2$, with two point $S_3 = \pm l$ removed, we have the following Darboux coordinates:

$$S^+ = e^{i\phi}\sqrt{pl - p^2}, \;\; S^- = e^{-i\phi}\sqrt{pl - p^2}, \;\; S_3 = l - 2p,$$

where $0 < p < l$ and $0 < \phi \leq 2\pi$ and $\{p, \phi\} = 1$.

The Poisson algebra

$$\{k, e\} = \epsilon ke, \;\; \{k, f\} = -2\epsilon kf,$$
$$\{e, f\} = 2\epsilon(k^2 - k^{-2})$$

has two real forms which are important for spin chains with compact phase spaces:

- the real form where $\epsilon = 1$, $|k| = 1$, $e = \overline{f}$,
- and the real form where $\epsilon = i$, $e = \overline{f}$, and $\overline{k} = k$.

In the first case, the level surface of eqn (7.7) with $c = 2\cos 2R$ and without the two points $e = f = 0$ has the following Darboux coordinates:

$$e = 2e^{i\phi}\sqrt{\sin(p)\sin(2R - p)}, \;\; f = 2e^{-i\phi}\sqrt{\sin(p)\sin(2R - p)}, \;\; k = e^{i(R-p)}$$

with $\{p, \phi\} = 1$, and $0 < p < l$ and $0 < \phi \leq 2\pi$.

Similarly, in the second case, the level surface $c = 2\cosh R$ without the two points $e = f = 0$ has Darboux coordinates

$$e = 2e^{i\phi}\sqrt{\sinh(p)\sinh(2R-p)}, \;\; f = 2e^{-i\phi}\sqrt{\sinh(p)\sinh(2R-p)}, \;\; k = e^{R-p}$$

where $\phi \in \mathbb{R}$, $0 < p < R$, and $\{p, \phi\} = 1$.

We will denote these level surfaces by $S^{(R)}$. In the compact case, $S^{(R)}$ is diffeomorphic to a sphere. It can be realized as a unit sphere with a symplectic form dependent on R.

7.2.4 Integrable classical local SU_2 spin chains

Because $S^{(R)}$ is a symplectic leaf of the Poisson Lie group LGL_2, the Cartesian product

$$\mathcal{M}^{R_1,\dots,R_N} = S^{(R_1)} \times \dots \times S^{(R_N)}$$

is a Poisson submanifold in LGL_2, which means that the matrix elements of the monodromy matrix

$$T(z) = L^{(R_1)}(za_1)L^{(R_2)}(za_2)\dots L^{(R_N)}(za_N) \tag{7.11}$$

satisfy the r-matrix Poisson brackets.

In order to obtain local Hamiltonians in the case of a homogeneous classical spin chain $a_1 = \dots = a_N = 1$, $R_1 = \dots = R_N = R$, we can use the degenerations (7.9) and (7.10). Combining eqns (7.9), (7.10), and (7.8), we obtain the following identities:

$$L(t) = \alpha \otimes \beta^t, \;\; L(t^{-1}) = -\sigma^y D\beta \otimes \alpha^t D^{-1}\sigma^y\,,$$

where the column vector α and row vector β^t are given in eqn (7.9). These identities imply

$$\mathrm{tr}(T(t)) = \prod_{n=1}^{N}(\beta_n, \alpha_{n+1}), \;\; \mathrm{tr}(T(t^{-1})) = (-1)^N \prod_{n=1}^{N}(\alpha_n, \beta_{n+1})\,.$$

Here we assume a periodicity $\alpha_{N+1} = \alpha_1$ and $\beta_{N+1} = \beta_1$. Now notice that

$$\mathrm{tr}(L_n(t)L_{n+1}(t)) = (\alpha_n, \beta_{n+1})(\beta_n, \alpha_{n+1})\,.$$

On the other hand, this trace can be computed explicitly:

$$\begin{aligned}\mathrm{tr}(L_n(t)L_{n+1}(t)) = e_n f_{n+1} + f_n e_{n+1} &+ (tk_n - t^{-1}k_n^{-1})(tk_{n+1} - t^{-1}k_{n+1}^{-1}) \\ &+ (tk_n^{-1} - t^{-1}k_n)(tk_{n+1}^{-1} - t^{-1}k_{n+1})\,.\end{aligned} \tag{7.12}$$

This gives the first local Hamiltonian

$$\begin{aligned}H = \log(\mathrm{tr}(T(t))\mathrm{tr}(T(t^{-1}))) = \sum_{n=1}^{N} \log(e_n f_{n+1} + f_n e_{n+1} \\ + (tk_n - t^{-1}k_n^{-1})(tk_{n+1} - t^{-1}k_{n+1}^{-1}) + (tk_n^{-1} - t^{-1}k_n)(tk_{n+1}^{-1} - t^{-1}k_{n+1}))\,.\end{aligned} \tag{7.13}$$

Other local spin Hamiltonians can be chosen as logarithmic derivatives of $\mathrm{tr}(T(z))$ at $z = t^{\pm 1}$; for details see [16] and references therein.

When N is even and the inhomogeneities are alternating such that $a_1 = a, a_2 = a^{-1}, a_3 = a, \ldots, a_N = a^{-1}$, there is a similar construction of local Hamiltonians also based on the degenerations (7.9) and (7.10). Again, all logarithmic derivatives of $T(z)$ at points $z = at^{\pm 1}, a^{-1}t^{\pm 1}$ are local spin Hamiltonians.

In the continuum limit, the Hamiltonian dynamics generated by these Hamiltonians converges to the Landau–Lifshitz equation; see [16] and references therein.

7.3 Quantization of local integrable spin chains

7.3.1 Quantum integrable spin chains

In the appendix, there is a short discussion of the integrable quantization of a classical integrable system.

A quantization of a local classical integrable spin chin is an integrable quantization of a classical local integrable spin chain such that quantized Hamiltonians remain local. That is, the collection of the following data:

- A choice of the quantization of the algebra of observables of the classical system (in the sense of Section 7.A.4), i.e. a family of associative algebras with a $*$-involution which deform the classical Poisson algebra of observables.
- A choice of a maximal commutative subalgebra in the algebra which is a quantization of the Poisson commuting algebra of classical integrals.
- In addition, the locality of the quantization means that the quantized algebra of observables is the tensor product of local algebras (one for each site of our one-dimensional lattice) $A_h = \otimes_{n=1}^{N} B_h^{(n)}$, and that the quantum Hamiltonian has the same local structure as the classical Hamiltonian (7.13):

$$H = \sum_n H_n$$

where $H_n = 1 \otimes \cdots \otimes H^{(k)} \otimes \cdots \otimes 1$ and $H^{(k)} \in B_h^{(n)} \otimes B_h^{(n+1)} \otimes \cdots \otimes B_h^{(n+k)}$.
- A $*$-representation of the algebra of observables (the space of pure states of the system).

7.3.2 The Yang–Baxter equation and quantization

Here we will describe the approach to the quantization of classical spin chains with an r-matrix Poisson bracket for polynomial Lax matrices based on the construction of the corresponding quantum R-matrices and quantum Lax matrices. In modern language, the construction of a quantum R-matrix means the construction of the corresponding quantum group, and the construction of the quantum L-matrix means the construction of the corresponding representation of the quantum group.

Suppose we have a classical integrable system with commuting integrals obtained as coefficients of the generating function $\tau(u)$ described in Section 7.2.1. The R-matrix quantization means the following:

- We find a family $R(u, h)$ of invertible linear operators acting in $V \otimes V$ such that for each h, they satisfy the quantum Yang–Baxter equation

$$R_{12}(u, h)R_{13}(u + v, h)R_{23}(v, h) = R_{23}(v, h)R_{13}(u + v, h)R_{12}(u, h)$$

and, when $h \to 0$,

$$R(u,h) = 1 + hr(u) + O(h^2)\,,$$

where $r(u)$ is the classical r-matrix.

- Let the classical Lax matrix $L(u)$ be a matrix-valued function of u of certain type (for example a polynomial of fixed degree), with matrix elements generating a Poisson algebra with Poisson brackets (7.2). For a given $R(u,h)$, we define the quantization of this Poisson algebra as the associative algebra generated by matrix elements of the matrix $\mathcal{L}(u)$ of the same type as $L(u)$ (for example a polynomial of the same degree), with defining relations

$$R(u,h)\mathcal{L}(u+v) \otimes \mathcal{L}(v) = (1 \otimes \mathcal{L}(v))(\mathcal{L}(u+v) \otimes 1)R(u,h)\,. \tag{7.14}$$

We denote such an algebra by B_h.

- Consider the generating function

$$T(u) = \mathcal{L}_1(u-w_1) \otimes \ldots \mathcal{L}_N(u-w_N) \tag{7.15}$$

acting in $End(V) \otimes B_h^{(1)} \otimes \cdots \otimes B_h^{(N)}$. It is easy to see that the commutation relations (7.14) imply

$$R(u,h)T(u+v) \otimes T(v) = (1 \otimes T(v))(T(u+v) \otimes 1)R(u,h)\,. \tag{7.16}$$

The invertibility of $R(u,h)$ together with the relations (7.16) implies that $t(u) = \mathrm{tr}_V(T(u)) \in B_h^{(N)} \otimes \cdots \otimes B_h^{(1)}$ is a generating function for a commutative subalgebra in $B_h^{(1)} \otimes \cdots \otimes B_h^{(N)}$:

$$[t(u), t(v)] = 0\,.$$

Under the right circumstances, this commutative subalgebra is maximal and defines an integrable quantization of the corresponding classical integrable spin chain.

The R-matrix was found by Baxter. Sklyanin discovered that when $h \to 0$, the classical R-matrix defines a Poisson structure on LGL_2 defined by eqn (7.A.5) and that it implies the Poisson commutativity of traces.

There is an algebraic way to derive Baxter's R-matrix from the universal R-matrix for $U_q(\widehat{gl_2})$. It is outlined in the appendix.

7.3.3 Quantum Lax operators and representation theory: Quantum LSL_2

Here is the formal definition of $C_q(\widehat{SL_2})$ in terms of generators and relations. Let q be a nonzero complex number. The algebra $C_q(\widehat{SL_2})$ is a complex algebra generated by the elements $T_{ij}^{(k)}$, where $i,j = 1,2$ and $k \in \mathbb{Z}$. Consider the matrix $\mathcal{T}(z)$ which is the generating function for the elements $T_{ij}^{(k)}$,

$$\mathcal{T}(z) = \sum_{k=1}^{\infty} T^{(k)} z^{2k} + \begin{pmatrix} T_{11}^{(0)} & T_{12}^{(0)} \\ 0 & T_{22}^{(0)} \end{pmatrix}. \tag{7.17}$$

The determining relations in $C_q(\widehat{SL}_2)$ can be written as the following matrix identity with entries in $C_q(\widehat{SL}_2)$:

$$R(z)\mathcal{T}(zw) \otimes \mathcal{T}(w) = (1 \otimes \mathcal{T}(w))(\mathcal{T}(zw) \otimes 1)R(z)\,, \qquad (7.18)$$
$$\mathcal{T}(qz)_{11}\mathcal{T}(z)_{22} - \mathcal{T}(qz)_{12}\mathcal{T}(z)_{21} = 1\,,$$

where the tensor product is a tensor product of matrices. The matrix elements in this formula are multiplied as elements of $C_q(\widehat{GL}_2)$ in the order in which they appear.

The matrix $R(z)$ acts in $\mathbb{C}^2 \otimes \mathbb{C}^2$ and has the following structure in the tensor product basis:

$$R(z) = \begin{pmatrix} 1 & 0 & 0 & 0 \\ 0 & f(z) & z^{-1}g(z) & 0 \\ 0 & zg(z) & f(z) & 0 \\ 0 & 0 & 0 & 1 \end{pmatrix}, \qquad (7.19)$$

where

$$f(z) = \frac{z - z^{-1}}{zq - z^{-1}q^{-1}}, \quad g(z) = \frac{(q - q^{-1})}{zq - z^{-1}q^{-1}}.$$

It satisfies the Yang–Baxter equation.

Remark 7.1 *There is an important function $s(z)$,*

$$s(z) = q^{-1/2}\frac{(z^2q^2; q^4)^2_\infty}{(z^2; q^4)_\infty (z^2q^4; q^4)_\infty},$$

where

$$(x; p)_\infty = \prod_{k=1}^{\infty}(1 - xp^n)\,.$$

The matrix

$$\mathcal{R}(z) = s(z)R(z)$$

satisfies what is known in physics as unitarity and the crossing symmetry:

$$\mathcal{R}(z)\mathcal{R}(z^{-1})^t = 1, \;\; \mathcal{R}(z^{-1})^{t_2} = C_2\mathcal{R}(zq^{-1})C_2\,,$$

where t is the transposition with respect to the standard scalar product in $\mathbb{C}^{2^\otimes}$, t_2 is the transposition with respect to the second factor in the tensor product, and $C_2 = 1 \otimes C$, where

$$C = \begin{pmatrix} 0 & -i \\ i & 0 \end{pmatrix}.$$

The Hopf algebra structure on $C_q(\widehat{SL}_2)$ is determined by the following action of the antipode on the generators:

$$\Delta(\mathcal{T}_{ij}(z)) = \sum_{k=1,2} \mathcal{T}_{ik}(z) \otimes \mathcal{T}_{kj}(z)\,. \tag{7.20}$$

The right-hand side here, as well as in the second relation in eqn (7.18), is understood as a product of Laurent power series. The antipode is determined by the relation

$$\mathcal{T}(z)(S \otimes id\, \mathcal{T}(z)) = 1\,.$$

Let d be a nonzero complex number. The identity (7.18) implies that the power series

$$\tau_1(z; d) = d\, \mathcal{T}_{11}(z) + d^{-1}\mathcal{T}_{22}(z) \tag{7.21}$$

generates a commutative subalgebra in $C_q(\widehat{SL}_2)$.

We choose a linear basis (for example ordered monomials in $T_{ij}^{(k)}$). The relations between generators will give the multiplication rule for monomials, which will depend on q. This multiplication turns into the commutative multiplication of coordinate functions when $q = 1$. The commutator of two monomials, divided by $q - 1$, at $q = 1$ becomes the Poisson bracket. It is easy to check that this Poisson bracket is exactly the one defined by the classical r-matrix (7.A.3).

Remark 7.2 *Let D be a diagonal matrix. It is easy to see that $[D \otimes 1 + 1 \otimes D, R(x)] = 0$. It is easy to show that if*

$$R(z)\mathcal{T}_1(zw)\mathcal{T}_2(w) = \mathcal{T}_2(w)\mathcal{T}_1(zw)R(z)\,,$$

then

$$\tilde{R}(z) = (z^D \otimes 1)R(z)(z^{-D} \otimes 1), \;\; \tilde{\mathcal{T}}(z) = z^D\mathcal{T}(z)z^{-D}$$

satisfy the same relation

$$\tilde{R}(z)\tilde{\mathcal{T}}_1(zw)\tilde{\mathcal{T}}_2(w) = \tilde{\mathcal{T}}_2(w)\tilde{\mathcal{T}}_1(zw)\tilde{R}(z)\,.$$

In particular, $\tilde{R}(z)$ satisfies the Yang–Baxter equation. Choosing $D = \mathrm{diag}(-1/2, 1/2)$ gives the R-matrix (7.19) but with no factors $z^{\pm 1}$ off-diagonal. This symmetric version of the R-matrix is the matrix of Boltzmann weights in the six-vertex model.

Remark 7.3 *If A is an invertible diagonal matrix such that $(A \otimes A)R(z) = R(z)(A \otimes A)$ and $\mathcal{T}(z)$ is as above, then*

$$\mathcal{T}^A(z) = A\mathcal{T}(z)A^{-1}$$

also satisfies the relations (7.18).

7.3.4 Irreducible representations

It is easy to check that the following matrix satisfies the R-matrix commutation relations (7.18):

$$\mathcal{L}(z) = \begin{pmatrix} zkq^{1/2} - z^{-1}k^{-1}q^{-1/2} & z^{-1}q^{-1/2}f \\ zq^{1/2}e & zk^{-1}q^{1/2} - z^{-1}kq^{-1/2} \end{pmatrix} \tag{7.22}$$

if e, f, k commute as

$$ke = qek, \quad kf = q^{-1}fk\,,$$
$$ef - fe = (q - q^{-1})(k^2 - k^{-2})\,.$$

We denote this algebra by C_q.

The element

$$c = fe + k^2q + k^{-2}q^{-1} \tag{7.23}$$

generates the center of this algebra.

It is clear that this algebra quantizes the Poisson algebra (7.5) (7.6). Indeed, the algebra C_1 is the commutative algebra generated by $e, f, k^{\pm 1}$. Consider the monomial basis $e^n k^m f^l$ in C_q. Fix the isomorphism between C_q and C_1 identifying this basis. The associative multiplication in C_q is given in this basis by the following function of q:

$$e_i e_j = \sum_k m_{ij}^k(q) e_k\,.$$

It is clear that when $q = 1$, this multiplication is the usual multiplication in the commutative algebra generated by e, f, k, k^{-1}. The skew-symmetric part of the derivative of $m(q)$ at $q = 1$ is the Poisson structure.

The algebra C_q is a Hopf algebra, with the comultiplication acting on generators as

$$\Delta k = k \otimes k, \quad \Delta e = e \otimes k + k^{-1} \otimes e, \quad \Delta f = f \otimes k + k^{-1} \otimes f\,.$$

The algebra C_q is closely related to the quantized universal enveloping algebra for sl_2. Indeed, the elements $E = ek/(q - q^{-1}), F = k^{-1}f/(q - q^{-1}), K = k^2$ are generators for $U_q(sl_2)$:

$$KE = q^2EK, \quad KF = q^{-2}FK, \quad EF - FE = \frac{K - K^{-1}}{q - q^{-1}}$$

with

$$\Delta K = K \otimes K, \quad \Delta E = E \otimes K + 1 \otimes E, \quad \Delta F = F \otimes 1 + K^{-1} \otimes F\,.$$

We assume that q is generic. We denote by $V^{(m)}$ the irreducible $m+1$-dimensional representation of C_q, and by $v_0^{(m)}$ the highest-weight vector in this representation:

$$kv_0^{(m)} = q^{m/2}v_0^{(m)}, \quad ev_0^{(m)} = 0\,.$$

The weight basis in this representation can be obtained by acting with f on the highest-weight vector. Properly normalized, the action of C_q on the weight basis is

$$kv_n^{(m)} = q^{m/2-n}v_n^{(m)}, \quad fv_n^{(m)} = (q^{m-n} - q^{-m+n})v_{n+1}^{(m)}, \quad ev_n^{(m)} = (q^n - q^{-n})v_{n-1}^{(m)}\,.$$

The Casimir element c acts on $V(m)$ by multiplication on $q^{m+1} + q^{-m-1}$. Because the algebra C_q is a Hopf algebra, it acts naturally on the tensor product of representations.

We denote the matrix (7.22) evaluated in the irreducible representation $V^{(m)}$ by $\mathcal{L}^{(m)}(z)$. It is easy to check that it satisfies the following identities:

$$\mathcal{L}^{(m)}(z^{-1})^t = -D_z C \mathcal{L}^{(m)}(zq) C^{-1} D_z^{-1}\,,$$

$$\mathcal{L}^{(m)}(z)^T \mathcal{L}^{(m)}(z^{-1}) = (zq^{(m+1)/2} - z^{-1}q^{-(m+1)/2})(z^{-1}q^{(m+1)/2} - zq^{-(m+1)/2})I\,,$$

where D_z is a diagonal matrix. Here t is the transposition with respect to the standard scalar product in $\mathbb{C}^2$, and T is the transposition t combined with transposition in $V^{(m)}$ with respect to the scalar product $(v_n^{(m)}, v_{n'}^{(m)}) = \delta_{n,n'}$.

The matrix (7.22) defines a family of two-dimensional representations of $C_q(\widehat{SL_2})$. If a is a nonzero complex number, such a representation is

$$\mathcal{T}(z) \mapsto g_m(z)\mathcal{L}^{(m)}(za)\,,$$

where the factor $g_m(z)$ is important only if we want to satisfy the following relations, which play the role of quantum counterparts of the unimodularity (i.e. $\det = 1$) of $\mathcal{T}(z)$:

$$g_m(z) = -z\frac{(t^{-1}q^3z^2;q^4)_\infty (tq^5z^2;q^4)_\infty}{(t^{-1}qz^2;q^4)_\infty (tq^3z^2;q^4)_\infty}\,.$$

Comultiplication defines the tensor product of irreducible representations described above,

$$\mathcal{T}(z) \mapsto \prod_{i=1}^{N} g_{m_i}(z/a_i) T^{(m_1,\ldots,m_N)}(z|a_1,\ldots,a_N)\,, \tag{7.24}$$

where

$$T^{(m_1,\ldots,m_N)}(z|a_1,\ldots,a_N) = \mathcal{L}_1(z/a_1)\ldots\mathcal{L}_N(z/a_N)\,, \tag{7.25}$$

where we have taken the matrix products of the matrices $\mathcal{L}(z)$ as 2×2 matrices. The n-th factor in eqn (7.25) acts on the n-th factor of $V^{(m_1)} \otimes \cdots \otimes V^{(m_N)}$.

Notice that this is also a tensor product of representations of C_q.

Real forms. As in the classical case, there are two real forms of the algebra C_q which are important for finite-dimensional spin chains.

Recall that a $*$-involution of a complex associative algebra is an anti-involution of the algebra, i.e. $(ab)^* = b^*a^*$, which is complex antilinear: $(\lambda a)^* = \bar{\lambda}a^*$. A real form of a complex A corresponding to this involution is a real algebra which is the real subspace in A spanned by the $*$-invariant elements.

When $|q| = 1$, we will write $q = \exp(i\gamma)$, and the relevant real form of C_q is characterized by a $*$-involution which acts on generators as

$$e^* = f,\;\; k^* = k^{-1}\,.$$

When q is real and positive, we will write $q = \exp(\eta)$. In this case the relevant real form is characterized by a $*$-involution acting on generators as follows:

$$e^* = f,\;\; f^* = e,\;\; k^* = k\,.$$

As $\gamma \to 0$ or $\eta \to 0$, these real forms become real forms of the corresponding Poisson algebras described in Section 7.2.3, with $\epsilon = i$ and $\epsilon = 1$, respectively.

7.3.5 The fusion of R-matrices and the degeneration of tensor products of irreducibles

This section describes the analogue of the construction of quantum L-operators by taking the tensor product of two-dimensional representations.

Consider the following product of R-matrices acting in the tensor product of $n+m$ copies of $\mathbb{C}^2$:

$$\begin{array}{lllll} R_{1'2'\dots m',12\dots n}(z) = & R_{1'1}(z) & R_{1'2}(zq) & \dots & R_{1'n}(zq^{n-1}) \\ & R_{2'1}(zq) & R_{2'2}(zq^2) & \dots & R_{2'n}(zq^n) \\ & \dots & \dots & \dots & \dots \\ & R_{m'1}(zq^{m-1}) & R_{m'2}(zq^m) & \dots & R_{m'n}(zq^{n+m-2}) . \end{array}$$

The operator $R(z)$ satisfies the identities

$$R(z)R^t(z^{-1}) = (zq - z^{-1}q^{-1})(z^{-1}q - zq^{-1}),$$

$$PR(z)P = R^t(z),$$

where t is the transposition operation (with respect to the tensor product of the standard basis in $\mathbb{C}^2$), and

$$\det(R(z)) = (zq - z^{-1}q^{-1})^3(z^{-1}q - zq^{-1}).$$

From this, we conclude that the matrix $PR(z)$ degenerates at $z = q$ and $z = q^{-1}$ to matrices of rank 3 and 1, respectively.

We define

$$\begin{array}{rl} P^+_{12\dots n} = & \check{R}_{12}(q)\, \check{R}_{23}(q^2) \dots \check{R}_{1n}(q^{n-1}) \\ & \qquad \check{R}_{23}(q) \dots \check{R}_{2n}(q^{n-2}) \\ & \qquad\qquad \dots \\ & \qquad\qquad \check{R}_{n-1n}(q) , \end{array} \tag{7.26}$$

where $\check{R}(z) = PR(z)$ and P is the permutation matrix such that $P(x \otimes y) = y \otimes x$.

We consider $(\mathbb{C}^2)^{\otimes n}$ as a representation of C_q. Because all finite-dimensional representations of this algebra are completely reducible, it decomposes into a direct sum of irreducible components. The irreducible representation $V^{(n)}$ appears in this decomposition with multiplicity 1. One can show that the operator (7.26) is the orthogonal projector to $V^{(n)}$. The proof can be found in [24].

Also, it is not difficult to show that

$$\begin{aligned} &P^+_{12\dots n} R_{1',12\dots n}(zq^{-(n-1)/2}) P^+_{12\dots n} \\ &= (zq^{-(n-3)/2} - z^{-1}q^{(n-3)/2}) \dots (zq^{(n-1)/2} - z^{-1}q^{-(n-1)/2}) R^{(1,n)}_{1',[12\dots n]}(z) , \end{aligned} \tag{7.27}$$

where the linear operator $R^{1,n}(z)$ acts in $\mathbb{C}^2 \otimes V^{(n)}$ and the second factor appears as the q-symmetrized part of the tensor product $\mathbb{C}^{2^{\otimes N}}$. Moreover, it is easy to show that this operator is conjugate by a diagonal matrix to $\mathcal{L}(z)$:

$$R^{(1,n)}(z) \simeq \begin{pmatrix} zkq^{1/2} - z^{-1}k^{-1}q^{-1/2} & z^{-1}q^{-1/2}f \\ zq^{1/2}e & zk^{-1}q^{1/2} - z^{-1}kq^{-1/2} \end{pmatrix},$$

where e, f, k act in the $(n+1)$-dimensional irreducible representation as described in Section 7.3.4. In this realization of the irreducible representation, weight vectors appear as $v^{(n)}_k =$

$P^+_{1\ldots n} e_1 \otimes e_1 \otimes e_2 \otimes e_2$, where we have $n-k$ copies of e_1 and k copies of e_2 in this tensor product.

Similarly,

$$P^+_{1'2'\ldots m'} P^+_{12\ldots n} \begin{matrix} R_{1'1}(zq^{-(n+m-2)/2}) & \ldots & R_{1'n}(zq^{(n-m)/2}) \\ R_{2'1}(zq^{-(n+m-4)/2}) & \ldots & R_{2'n}(zq^{(n-m+2)/2}) \\ \ldots & \ldots & \ldots \end{matrix} \; P^+_{1'2'\ldots m'} \quad P^+_{12\ldots n}$$

$$= \begin{matrix} (zq^{-(n+m-4)/2} - z^{-1}q^{(n+m-4)/2}) & \ldots & (zq^{-(n-m)/2} - z^{-1}q^{-(n-m)/2}) \\ \ldots & \ldots & \ldots \\ (zq^{-(n-m-2)/2} - z^{-1}q^{-(n-m-2)/2}) & \ldots & (zq^{(n+m-2)/2} - z^{-1}q^{-(n+m-2)/2}) \end{matrix} \; R^{(m,n)}(z)\,.$$

Here we assume that $m < n$. The matrix elements of $R^{(m,n)}(z)$ are Laurent polynomials of the form $z^{-m}P(z^2)$, where $P(t)$ is a polynomial of degree m.

The matrix $R^{(n,m)}$ also can be expressed in terms $e, f, k^{\pm 1}$. The matrices $R^{(k,l)}$ satisfy the Yang–Baxter equation

$$R^{(l,m)}_{12}(z) R^{(l,n)}_{13}(zw) R^{(m,n)}_{23}(w) = R^{(m,n)}_{23}(w) R^{(l,n)}_{13}(zw) R^{(l,m)}_{12}(z)\,.$$

In addition to this, they satisfy identities

$$R^{(l,m)}(z) R^{(m,l)}(z^{-1})^T = s_{ml}(z) s_{ml}(z^{-1})$$

and

$$R^{(l,m)}_{12}(z)^{t_1} = (-1)^l (D_z C^{(m)} \otimes 1) R^{(l,m)}_{12}(zq)(D_z^{-1} C^{(m)^{-1}} \otimes 1)\,,$$

where $s_{ml}(z)$ is a Laurent polynomial in z which is easy to compute, $C^{(m)} = P^+_{12\ldots n} \otimes CP^+_{12\ldots n}$, and we assume that $l \leq m$.

7.3.6 Higher transfer matrices

We define $C_q(\widehat{SL}_2)$-valued matrices

$$\mathcal{T}^{(m)}(z) = P^+_{12\ldots n} \mathcal{T}_1(zq^{(m-1)/2}) \ldots \mathcal{T}_m(zq^{-(m-1)/2}) P^+_{12\ldots n}\,,$$

where $P^+_{12\ldots n}$ is as defined above and $\mathcal{T}$ is the matrix (7.17). They satisfy the relations

$$R^{(l,m)}(z)_{12} \mathcal{T}^{(l)}_1(zw) \mathcal{T}^{(m)}_2(w) = \mathcal{T}^{(m)}_2(w) \mathcal{T}^{(l)}_1(zw) R^{(l,m)}(z)_{12}\,.$$

For nonzero d, we define the following elements of $C_q(\widehat{SL}_2)$:

$$\tau_\ell(z) = (\mathrm{id} \otimes \mathrm{tr}_{V^{(\ell)}})(\mathcal{T}^{(\ell)}(z) d^{(l)})\,,$$

where $d^{(l)} = \mathrm{diag}(d^l, d^{l-2}, \ldots, d^{-l})$ and $d \neq 0$.

The fusion relations for $\mathcal{T}(z)$ imply the following recursive relations for $\tau_\ell(z)$:

$$\tau_1(z)\tau_\ell(zq) = \tau_{\ell+1}(z) + \tau_{\ell-1}(zq^2)\,,$$

which can be solved in terms of determinants [5]:

$$\tau_\ell(z) = \det \begin{pmatrix} \tau_1(z) & 1 & & 0 \\ 1 & \tau_1(zq) & \ddots & \\ & \ddots & \ddots & 1 \\ 0 & & 1 & \tau_1(zq^{\ell-1}) \end{pmatrix}.$$

The remarkable fact is that the elements $\{\tau_\ell(z)\}$ also satisfy another set of relations, which follow from the fusion relations

$$\tau_\ell(zq^{1/2})\tau_\ell(zq^{-1/2}) = \tau_{\ell+1}(z)\tau_{\ell-1}(z) + 1\,. \tag{7.28}$$

7.3.7 Local integrable quantum spin Hamiltonians

The transfer matrices

$$t_m(u) = \mathrm{tr}_a(R_{a1}^{m,m_1}(z/a_1)\dots R_{a1}^{m,m_1}(z/a_1)d_a^{(l)})$$

form a commuting family of operators in $V^{(m_1)}\otimes\cdots\otimes V^{(m_N)}$. They quantize the generating functions for classical spin chains and can be used to construct local quantum spin chains. Below, we outline two common constructions of local Hamiltonians from such transfer matrices.

Homogeneous $SU(2)$ spin chains. The homogeneous Heisenberg model of spin S corresponds to the choice $m_1 = \cdots = m_N = l$ and $a_N = \cdots = a_1 = 1$. We will denote the corresponding transfer matrices by $\tau_m^{(l)}(u)$.

In the case $m = 1$, the transfer matrix $t_1^{(1)}(z)$ is

$$t_1^{(1)}(z) = \mathrm{tr}_0(R_{0N}(z)\dots R_{01}(z))\,, \tag{7.29}$$

where $R(z)$ is the matrix (7.19).

The linear operator (7.29) is the transfer matrix of the six-vertex model [3, 25]. It is also a generating function for local spin Hamiltonians:

$$H_1 = \frac{d}{du}\log(t_1^{(1)}(u))|_{u=0} = \sum_{n=1}^{N}(\sigma_n^x\sigma_{n+1}^y + \sigma_n^y\sigma_{n+1}^y + \Delta\sigma_n^z\sigma_{n+1}^z)\,, \tag{7.30}$$

$$H_k = \left(\frac{d}{du}\right)^k \log t_1^{(1)}(u)|_{u=0} = \sum_{n=1}^{N} H^{(k)}(\sigma_n,\dots,\sigma_{n+k})\,. \tag{7.31}$$

Here $\sigma^x, \sigma^y, \sigma^z$ are the Pauli matrices and σ_n is the collection of Pauli matrices acting nontrivially on the n-th factor of the tensor product:

$$\sigma^x = \begin{pmatrix} 0 & 1 \\ 1 & 0 \end{pmatrix}, \qquad \sigma^y = \begin{pmatrix} 0 & i \\ -i & 0 \end{pmatrix}, \qquad \sigma^z = \begin{pmatrix} 1 & 0 \\ 0 & -1 \end{pmatrix}. \tag{7.32}$$

If $l > 1$, a similar analysis can be done for the transfer matrix

$$t_l^{(l)}(z) = \mathrm{tr}_a(R_{a1}^{(l,l)}(z) \dots R_{aN}^{(l,l)}(z)) .$$

Since $R^{(l,l)}(1) = P$, we have

$$t_l^{(l)}(1) = \mathrm{tr}_a(P_{a1} \dots P_{aN}) = \mathrm{tr}_a(P_{12}P_{13} \dots P_{1N}P_{a1}) = P_{12}P_{13} \dots P_{1N} .$$

Here we have used the identities $P_{a1}A_aP_{a1} = A_1$ and $\mathrm{tr}_a(P_{a1}) = I_1$.

The operator $T = t_l^{(l)}(1)$ is the translation matrix:

$$T(x_1 \otimes x_2 \cdots \otimes x_n) = x_N \otimes x_1 \otimes \dots x_{N-1} .$$

Differentiating $t_l^{(l)}(z)$ at $z = 1$, we have

$$t_l^{(l)}(1)' = \sum_{i=1}^{N} \mathrm{tr}_a(P_{a1} \dots P_{ai-1}R_{ai}^{(l,l)}(1)' P_{ai+1} \dots P_{aN}) = T \sum_{i=1}^{N} H_{ii+1}^{(l)} .$$

Similarly, by taking higher logarithmic derivatives of $t_l^{(l)}(z)$ at $z = 1$, we obtain higher local Hamiltonians acting in $(\mathbb{C}^{l+1})^{\otimes N}$:

$$H_k = \left(z\frac{d}{dz}\right)^k t_l^{(l)}(z)|_{z=1} = \sum_{n=1}^{N} H_{n,\dots,n+k}^{(k)} .$$

Here the matrix $H^{(k)}$ acts in $(\mathbb{C}^{l+1})^{\otimes k+1}$. The subindices show how this matrix acts in $(\mathbb{C}^{l+1})^{\otimes N}$.

One can show that these local quantum spin chain Hamiltonians, in the limit $q \to 1$ and $l \to \infty$, become the classical Hamiltonians described in Section 7.2.4, assuming that q^l is fixed.

Inhomogeneous $SU(2)$ spin chains: Construction using degeneration points. The construction of inhomogeneous local operators is easy to illustrate with a spin chain where the inhomogeneities alternate,

$$t_m(z) = \mathrm{tr}_a(R_{a1}^{(m,l_1)}(za^{-1})R_{a2}^{(m,l_2)}(za) \dots R_{a,2N-1}^{(m,l_1)}(za^{-1})R_{a,2N}^{(m,l_2)}(za)) .$$

Now we have two sublattices and two translation operators

$$T_{\mathrm{even}} = P_{24}P_{26} \dots P_{2,2N}, \quad T_{\mathrm{odd}} = P_{13}P_{15} \dots P_{1,2N-1} .$$

It is easy to find the following special values of the transfer matrix:

$$t_{l_1}(a) = T^{\mathrm{even}} R_{21}^{(l_2,l_1)}(a^{-2}) \dots R_{2N,2N-1}^{(l_2,l_1)}(a^{-2}) ,$$

$$t_{l_2}(a^{-1}) = R_{12}^{(l_1,l_2)}(a^2) \dots R_{2N-1,2N}^{(l_1,l_2)}(a^2) T^{\mathrm{odd}} .$$

These operators commute, and

$$t_{l_1}(a)t_{l_2}(a^{-1}) = T^{\mathrm{even}}T^{\mathrm{odd}},$$

$$t_{l_1}(a)t_{l_2}(a^{-1})^{-1} = T^{\mathrm{even}}(T^{\mathrm{odd}})^{-1}$$
$$\times R^{(l_2,l_1)}_{2,2N-1}(a^{-2})R^{(l_2,l_1)}_{4,1}(a^{-2})\dots R^{(l_2,l_1)}_{2N,2N-3}(a^{-2})R^{(l_1,l_2)}_{2N-1,2N}(a^{-2})\dots R^{(l_1,l_2)}_{1,2}(a^{-2}). \tag{7.33}$$

Taking logarithmic derivatives of $t_{l_1}(z)$ at $z = a$ and of $t_{l_2}(z)$ at $z = a^{-1}$, we again have local operators, for example

$$z\frac{d}{dz}\log t_{l_1}(z)|_{z=a} = \sum_{n=1}^{N} R^{-1}_{2n+1,2n}(a^2)R'_{2n+1,2n}(a^2)$$
$$+\sum_{n=1}^{N} R^{-1}_{2n+1,2n}(a^2)P_{2n+1,2n-1}R'_{2n+1,2n-1}(1)R_{2n+1,2n}(a^2). \tag{7.34}$$

These Hamiltonians, in the semiclassical limit, reproduce the inhomogeneous classical spin chains described earlier.

7.4 The spectrum of transfer matrices

7.4.1 Diagonalizability of transfer matrices

Assume that q is real. Let $t(u)^*$ be the Hermitian conjugate of $t(u)$ with respect to the standard Hermitian scalar product on $(\mathbb{C}^2)^{\otimes N}$. It is easy to prove, using the identities for $R(u)$, that

$$t(z|a_1,\dots,a_N)^* = (-1)^N t(\overline{z}^{-1}q^{-1}|\overline{a_1}^{-1}\dots\overline{a_N}^{-1}),$$

where $\overline{z}$ is the complex conjugate of z.

Because $t(z)$ is the commutative family of operators, the operators $t(z)$ are normal when $\overline{a_i} = a_i^{-1}$. Therefore, for these values of a_i it is diagonalizable. Since $t(z)$ is linear (up to a scalar factor) in a_i^2, this implies that $t(z)$ is diagonalizable for all generic complex values of a_i and, for the same reasons, for all generic complex values of q.

7.4.2 Bethe ansatz for sl_2

In this section, we will recall the algebraic Bethe ansatz for an inhomogeneous finite-dimensional spin chain.

The quantum monodromy matrix for such a spin chain is

$$T(z) = \mathcal{L}^{(m_1)}_1(z/a_1)\dots\mathcal{L}^{(m_N)}_N(z/a_N)D, \tag{7.35}$$

where

$$D = \begin{pmatrix} Z & 0 \\ 0 & Z^{-1}\end{pmatrix}.$$

It is convenient to write this as

$$T(z) = \begin{pmatrix} A(z) & B(z) \\ C(z) & D(z) \end{pmatrix}.$$

In the basis $e_1 \otimes e_1, e_1 \otimes e_2, e_2 \otimes e_1, e_2 \otimes e_2$ of the tensor product $\mathbb{C}^2 \otimes \mathbb{C}^2$, we have

$$T_1(zw)T_2(w) = \begin{pmatrix} A(zw)A(w) & A(zw)B(w) & B(zw)A(w) & B(zw)B(w) \\ A(zw)C(w) & A(zw)D(w) & B(zw)C(w) & B(zw)D(w) \\ C(zw)A(w) & C(zw)B(w) & D(zw)A(w) & D(zw)B(w) \\ C(zw)C(w) & C(zw)D(w) & D(zw)C(w) & D(zw)D(w) \end{pmatrix}.$$

If we write the R-matrix in the tensor product basis as in eqn (7.19),

$$R(z) = \begin{pmatrix} 1 & 0 & 0 & 0 \\ 0 & f(z) & g(z)z^{-1} & 0 \\ 0 & g(z)z & f(z) & 0 \\ 0 & 0 & 0 & 1 \end{pmatrix},$$

the commutation relations (7.16) produce the following relations between A and B and between D and B:

$$A(z)B(v) = \frac{1}{f(vz^{-1})}B(v)A(z) - \frac{g(vz^{-1})zv^{-1}}{f(vz^{-1})}B(z)A(v)\,,$$

$$D(z)B(v) = \frac{1}{f(zv^{-1})}B(v)D(z) - \frac{g(zv^{-1})zv^{-1}}{f(zv^{-1})}B(z)D(v)\,,$$

where $f(z) = (z - z^{-1})/(zq - z^{-1}q^{-1})$, $g(z) = (q - q^{-1})(zq - z^{-1}q^{-1})$.

The L-operators act on the vector

$$\Omega = v_0^{(m_1)} \otimes v_0^{(m_2)} \otimes \cdots \otimes v_0^{(m_N)}$$

in a special way:

$$\mathfrak{L}_i(z)\Omega = \begin{pmatrix} (zq^{(m_i+1)/2} - z^{-1}q^{-(m_i+1)/2})\Omega & * \\ 0 & (zq^{-(m_i-1)/2} - z^{-1}q^{(m_i-1)/2})\Omega \end{pmatrix}.$$

From this it is clear that Ω is an eigenvector for the operators A and D and that C annihilates it:

$$T(z)\Omega = \begin{pmatrix} \alpha(z)\Omega & * \\ 0 & \delta(z)\Omega \end{pmatrix},$$

where

$$\alpha(z) = Z\prod_{i=1}^{N}(za_i^{-1}q^{(m_i+1)/2} - z^{-1}a_iq^{-(m_i+1)/2})\,,$$

$$\delta(z) = Z^{-1}\prod_{i=1}^{N}(za_i^{-1}q^{-(m_i-1)/2} - z^{-1}a_iq^{(m_i-1)/2})\,.$$

The details of the proof of the subsequent construction of eigenvectors can be found in [16].

Theorem 7.4 *The following identity holds:*

$$(A(z) + D(z))B(v_1) \dots B(v_n)\Omega = \Lambda(z|\{v_i\})B(v_1) \dots B(v_n)\Omega\,,$$

where

$$\Lambda(z|\{v_i\}) = \alpha(z)\prod_{i=1}^{n} \frac{v_i z^{-1} q - v_i^{-1} z q^{-1}}{v_i z^{-1} - v_i^{-1} z} + \delta(z)\prod_{i=1}^{n} \frac{v_i^{-1} z q - v_i z^{-1} q^{-1}}{v_i^{-1} z - v_i z^{-1}}\,, \tag{7.36}$$

if the numbers v_i satisfy the Bethe equations

$$\prod_{\alpha=1}^{N} \frac{v_i a_\alpha^{-1} q^{(m_\alpha+1)/2} - z^{-1} a_\alpha q^{-(m_\alpha+1)/2}}{v_i a_\alpha^{-1} q^{-(m_\alpha-1)/2} - z^{-1} a_\alpha q^{(m_\alpha-1)/2}} = -Z^2 \prod_{j=1}^{n} \frac{v_i v_j^{-1} q - v_i^{-1} v_j q^{-1}}{v_i v_j^{-1} q^{-1} - v_i^{-1} v_j q}\,. \tag{7.37}$$

Note that the formula for the eigenvalues in terms of solutions to the Bethe equations is a rational function. The Bethe equations can be regarded as conditions

$$\mathrm{res}_{z=v_j} \Lambda(z|\{v_i\}) = 0\,.$$

This agrees with the fact that $t(z)$ is a commuting family of operators which has no poles at finite z.

7.4.3 The completeness of Bethe vectors

The next step is to establish whether the construction outlined above gives all eigenvectors. We will focus here on a spin chain of spin $1/2$.

Assume that q, e^{2H}, and the inhomogeneities a_i are generic. Let us demonstrate that the vectors

$$B(v_1) \dots B(v_n)\Omega\,, \tag{7.38}$$

where the v_i are solutions to the Bethe equations, give all 2^N eigenvectors of the transfer matrix.

Consider the limit of eqn (7.38) when $a_N \to \infty$. Assume that v is fixed and $a_N \to \infty$. From the definition of $B(v)$, we have

$$B(v) = a_N q^{-1/2} v^{-1} (\tilde{A}(v) \otimes f - \tilde{B}(v) \otimes K)(1 + o(1))\,,$$

where $\tilde{A}(v)$, $\tilde{B}(v)$ are elements of the quantum monodromy matrix (7.35) with only $N-1$ first factors. On the other hand, if $a_N \to \infty$ and $v \to \infty$ such that $v = w a_N$ and w is finite, the asymptotics are different:

$$B(v) = w^{-1} q^{-1/2} \prod_{n=1}^{N} (w a_N a_n^{-1} q^{1/2}) k \otimes \dots \otimes k \otimes f(1 + o(1)\,.$$

From this we obtain the asymptotics of the Bethe vectors when all v_i are fixed and $a_N \to \infty$:

$$B(v_1)\dots B(v_n)\Omega_N \to q^{((m_N+1)/2)n}(-a_N q^{1/2})^n \prod_{i=1}^{n} v_i^{-1}$$
$$\times \left(\tilde{B}(v_1)\dots\tilde{B}(v_n)\Omega_{N-1}\otimes v_0^{(m_N)} - \sum_{i=1}^{n} q^{-(m_N+1)/2-i+1}\right.$$
$$\left.\times\, \tilde{B}(v_1)\dots\tilde{A}(v_i)\dots\tilde{B}(v_n)\Omega_{N-1}\otimes f v_0^{(m_N)}\right)(1+o(1))\,. \quad (7.39)$$

Similarly, when $v_1,\dots,v_{n-1}$ are fixed and $a_N \to \infty$ such that $v_n = wa_N$, we have

$$B(v_1)\dots B(v_n)\Omega_N \to q^{((m_N+1)/2)(n-1)-n}(-a_N q^{1/2})^{n-1} \prod_{i=1}^{n-1} v_i^{-1} q^{m_i/2}$$
$$\times\, \tilde{B}(v_1)\dots\tilde{B}(v_{n-1})\Omega_{N-1}\otimes f v_0^{(m_N)}(1+o(1))\,. \quad (7.40)$$

The solutions to eqn (7.37) have the following possible asymptotics when $a_N \to \infty$:

1. For all $j = 1,\dots,N$, $\lim_{a_N\to\infty} v_j = v'_j$, where $\{v'_j\}$ is a solution to the Bethe system for a spin chain of length $N-1$ with inhomogeneities $a_1,\dots,a_{N-1}$ and Z.
2. For one of the v_j, say v_n, we have $v_n = a_N w + O(1)$, and for the others, $\lim_{a_N\to\infty} v_j = v'_j$, where $\{v'_j\}$ is a solution to the Bethe system for a spin chain of length $N-1$ with inhomogeneities $a_1,\dots,a_{N-1}$ and Zq^{-1}. From the Bethe equation for v_n, we have
$$w^2 = \frac{1 - Z^2 q^{-N+2n}}{q^2 - Zq^{-N+2n}}\,.$$
3. More than one of the v_i is proportional to a_N.

Using induction and the asymptotics of the Bethe vectors (7.39) and (7.40), it is easy to show that only the first two options describe the spectrum of the spin-$1/2$ transfer matrix. Similar arguments were used in [20] to prove the completeness of the Bethe vectors in an SL_n spin chain.

This implies immediately that there are $\binom{N}{n}$ Bethe vectors for each $0 \le n \le N$, and that the total number of Bethe vectors is

$$2^N = \sum_{n=0}^{N} \binom{N}{n}\,.$$

Other solutions to the Bethe equations describe eigenvectors in infinite-dimensional representations of a quantized affine algebra with the same weights. They do not correspond to any eigenvectors of the inhomogeneous spin-$1/2$ system.

For special values of a_α, the solutions to the Bethe equations may degenerate (a level crossing may occur in the spectrum of $t(z)$). In this case the Bethe ansatz should involve derivatives of the vectors (7.38).

7.5 The thermodynamic limit

The procedure of "filling Dirac seas" is a way to construct physical vacua and the eigenvalues of quantum integrals of motion for integrable spin chains solvable by the Bethe ansatz.

To be specific, consider a homogenous spin chain of spins $1/2$. Let $H_1, H_2, \dots$ be the quasi-local Hamiltonians described in Section 7.3.7 with $q = \exp(\eta)$ for some real η. We take the linear combination

$$H(\lambda) = \sum_k H_k \lambda_k \,. \tag{7.41}$$

This operator is bounded. Let $\Omega_N(\lambda)$ be its normalized ground state. As $N \to \infty$, the matrix elements $(\Omega_N, a\Omega_N)$ converge to the state ω_λ in the inductive limit of the algebra of observables. The action of local operators on ω_λ generates the Hilbert space $\mathcal{H}$.

Since the eigenvalues of the coefficients of $t(u)$ can be computed in terms of solutions to the Bethe equations, the spectrum of these operators in the large-N limit is determined by the large-N asymptotics of solutions to the Bethe equations.

The main assumption in the analysis of the Bethe equations in the limit $N \to \infty$ is that the numbers $\{v_\alpha^{(0)}\}$ [1] are distributed along the real line with some density $\rho(u)$. The intervals where $\rho(u) \neq 0$ are called Dirac seas. For the Hermitian Hamiltonian (7.41), there is strong evidence that the Dirac seas are a finite collection of intervals $(B_1^+, B_1^-), \dots, (B_n^+, B_n^-)$. Here the numbers $B_\alpha^\pm$ are the boundaries of Dirac seas. We assume they increase from left to right. The boundaries of the Dirac seas are uniquely determined by $\{\lambda_l\}$.

A solution to the Bethe equations is said to contain an m-string when, as $N \to \infty$, there is a subset of $\{v_i\}$ of the form

$$v^{(m)} + i\frac{\eta}{2}m, v^{(m)} + i\frac{\eta}{2}(m-2), \dots, v^{(m)} - i\frac{\eta}{2}(m-2), v^{(m)} - i\frac{\eta}{2}m$$

with some real $v^{(m)}$.

The excitations over the ground states can be of the following types:

- A hole in the Dirac sea (B_k^+, B_k^-) corresponds to a solution to the Bethe equations which has one fewer number v_i, and as $N \to \infty$, the remaining v_i "fill" the same Dirac seas with the densities deformed by the fact that one of the numbers $\{v_\alpha^{(0)}\}$ is missing and the others are "deformed" by the missing one. The number which is "missing" is a state with a hole, which is the "rapidity" $v \in (B_k^+, B_k^-)$ of the hole.
- Particles correspond to "adding" one real number to the collection $\{v_\alpha^{(0)}\}$.
- There are also m-strings, where $m > 1$ (corresponding to adding one m-string solution to the collection $\{v_\alpha^{(0)}\}$).

There are convincing arguments that the Fock space of the system with the Hamiltonian (7.41) has the following structure. It has a vacuum state $\emptyset_\lambda$ corresponding to the solution of the Bethe equations with the minimal eigenvalue of $H(\lambda)$. The excited states are eigenvectors of the Hamiltonian (and of all other integrals) which correspond to solutions of the Bethe equations with finitely many holes, particles, and m-strings. The space has the following structure:

[1] Solutions to the Bethe equations corresponding to the minimum eigenvalue of $H(\lambda)$.

$$\mathcal{H}(\{B_i^+, B_i^-\}_{i=1}^k) = \bigoplus_{N_h \geq 0, N_p \geq 0, N \geq 0} \bigoplus_{n_1 + \dots + n_k = N_h, N_p} \cdot \bigotimes_{j=1}^{k} L_2^{\text{symm}}(I_1^{+ \times n_1} \times \dots I_k^{+ \times n_k} \times I^{N_p} \times S^{1^N}) .$$

Here we have used the notation $I_l^+ = (B_l^+, B_l^-), I_l^- = (B_{l-1}^+, B_l^-)$, where $B_0^+ = -\pi$ and $B_{k+1}^- = \pi$; n_k is the number of holes in the Dirac sea (B_k^+, B_k^-), N_p is the number of particles, I is the complement of the Dirac seas, and N is the number of strings with $m \geq 2$. The symbol "symm" means a certain symmetrization procedure which we will not discuss here (see, for example, [19] for a discussion of the antiferromagnetic ground state).

By varying $\{\lambda_k\}$ or, equivalently, the positions $\{B_k^\pm\}$ of the Dirac seas, we obtain a "large" part of the space of states of the spin chain in the limit $N \to \infty$: $(\mathbb{C}^2)^{\otimes N}$ in the limit $N \to \infty$ becomes the direct integral of separable Hilbert spaces

$$(\mathbb{C}^2)^{\otimes N} \to \bigoplus_{k \geq 0} \int_{[-\pi,\pi]^{\times 2k}}^{\oplus} \mathcal{H}(\{B_i^+, B_i^-\}_{i=1}^k) . \tag{7.42}$$

7.6 The six-vertex model

7.6.1 The six-vertex configurations and boundary conditions

The six-vertex model is a model in statistical mechanics where the states are configurations of arrows on a square planar grid (see Fig. 7.1 for an example). Weights are assigned to the vertices of the grid; they depend on the arrows on the edges surrounding the vertex. Nonzero weights correspond to the configurations in Fig. 7.2, configurations where the number of incoming arrows is equal to the number of outgoing arrows. This is also known as the ice rule [26].

Each configuration of arrows on the lattice can be equivalently described as a configuration of "thin" and "thick" edges (or "empty" and "occupied" edges), as shown in Fig. 7.2 also. There must be an even number of thick edges at each vertex as a consequence of the ice rule.

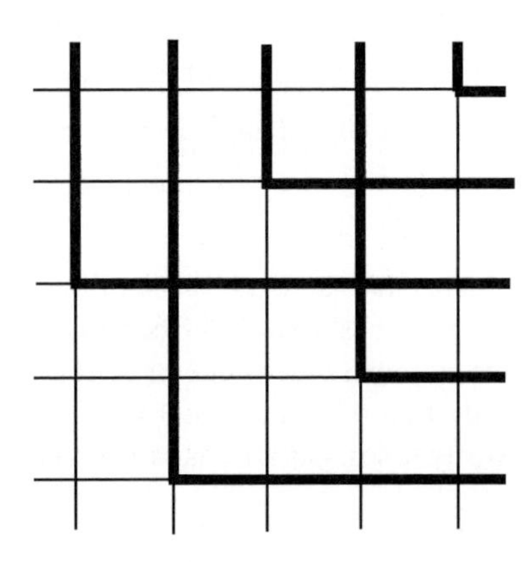

Fig. 7.1 A possible configuration of paths on a 5×5 square grid for domain wall boundary conditions.

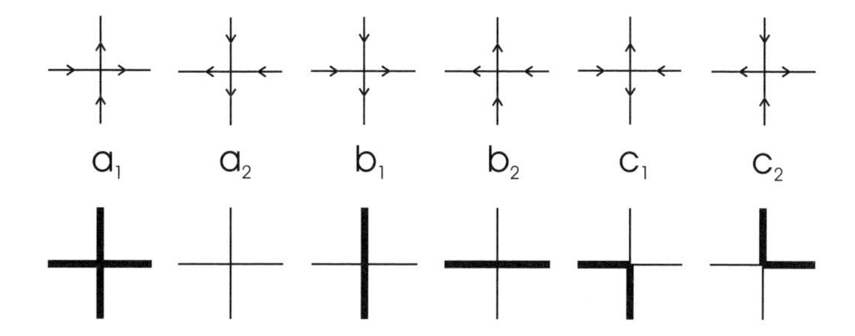

Fig. 7.2 The six types of vertices and the corresponding configurations of thin and thick edges.

The thick edges form paths going from northwest (NW) to southeast (SE). We assume that when four thick edges meet at a vertex, the corresponding paths meet at this point and then move apart. So, equivalently, the configurations of the six-vertex model can be regarded as configurations of paths going from NW to SE satisfying the rules in Fig. 7.2.[2]

7.6.2 Boundary conditions

It is natural to consider the six-vertex model on surface grids. If the surface is a domain on a plane, we will say that an edge is *outer* if it intersects the boundary of the domain. We assume that the boundary is chosen such that it intersects each edge at most once. Outer edges are attached to a 4-valent vertex on one side and to the boundary on the other side.

Fixed boundary conditions mean that the six-vertex configurations are fixed on outer edges. An example of fixed boundary conditions known as domain wall (DW) boundary conditions on a square domain is shown in Fig. 7.1.

We will be interested in three types of boundary conditions:

- A *domain* (simply connected on a plane) with *fixed boundary conditions* (see Fig. 7.3). We will also call these conditions Dirichlet boundary conditions.
- A *cylinder with fixed boundary conditions* (see Fig. 7.4). This case can be regarded as a domain with the states on the outer edges of two sides being identified and with fixed boundary conditions on the other sides.

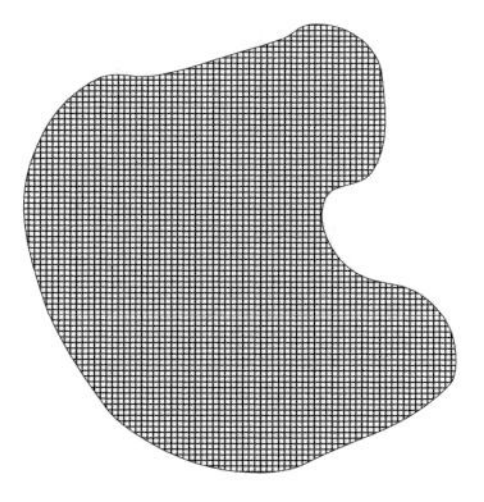

Fig. 7.3 A simply connected domain.

[2]One can consider such configurations on any 4-valent graph. But only for special graphs and special Boltzmann weights one can compute the partition function per site.

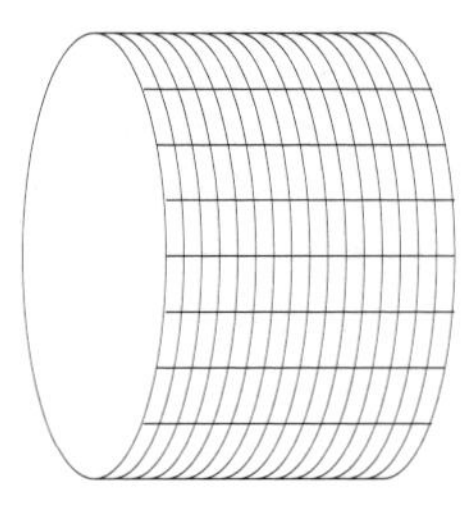

Fig. 7.4 Cylindrical boundary conditions.

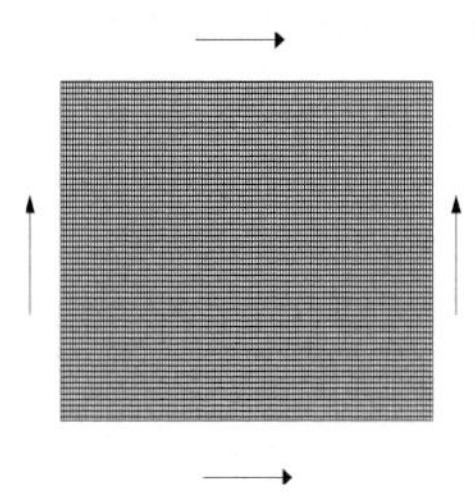

Fig. 7.5 Toric boundary conditions.

- Identification of the states on the outer edges of opposite sides of a rectangle gives the states for the six-vertex model on a torus (see Fig. 7.5). This model is also known as the six-vertex model with periodic boundary conditions.

7.6.3 The partition function and local correlation functions

To each configuration a_1, a_2, b_1, b_2, c_1, and c_2 in Fig. 7.2 we assign Boltzmann weights, which we denote by the same letters. The physical meaning of a Boltzmann weight is $\exp(-E/T)$, where E is the energy of a state and T is the temperature (in appropriate units), so all of the numbers a_1, a_2, b_1, b_2, c_1, and c_2 must be positive. The weight of a configuration is the product of the weights corresponding to vertices inside the domain, assigned to each vertex by the six-vertex rules.

A six-vertex model is called *homogeneous* if the weight assigned to a vertex depends only on the configuration of arrows on adjacent edges and not on the vertex itself. When the weights also depend on the vertex, the model is called *inhomogeneous*.

The partition function is the sum of the weights of all states of the model

$$Z = \sum_{\text{states}} \prod_{\text{vertices}} w(\text{vertex}),$$

where $w(\text{vertex})$ is the weight of a vertex assigned according to Fig. 7.2.

The weights of states define a probabilistic measure on the set of states of the six-vertex model. The probability of a state is given by the ratio of the weight of the state to the partition function of the model,

$$P(\text{state}) = \frac{\prod_{\text{vertices}} w(\text{vertex})}{Z}. \tag{7.43}$$

The *characteristic function* of an edge e is the function defined on the set of six-vertex states

$$\sigma_e(\text{state}) = \begin{cases} 1, & e \text{ is occupied by a path}, \\ 0, & \text{otherwise}. \end{cases}$$

A *local correlation function* is the expectation value of a product of such characteristic functions with respect to the measure (7.43):

$$\langle \sigma_{e_1} \sigma_{e_2} \dots \sigma_{e_n} \rangle = \sum_{\text{states}} P(\text{state}) \prod_{i=1}^{n} \sigma_{e_i}(\text{state}).$$

It is convenient to write the Boltzmann weights in exponential form

$$\begin{aligned} a_1 &= ae^{H+V}, & a_2 &= ae^{-H-V}, \\ b_1 &= be^{+H-V}, & b_2 &= be^{-H+V}, \\ c_1 &= ce^{-E}, & c_2 &= ce^{E}. \end{aligned}$$

From now on, we will assume that $E = 0$. If the weight are homogeneous (they do not depend on the vertex), the local correlation functions for a domain, cylinder, or torus do not depend on E. Also, the parameters H and V have a clear physical meaning; they can be regarded as horizontal and vertical electric fields. If $E = 0$ the weight of a state can be written as

$$\begin{aligned} w(S) = & \prod_{v \in \text{vertices}} w(v|S) \\ & \times \exp\left(\sum_{e \in E_h} \sigma_e(S)(H(e_+) + H(e_-))/2 + \sum_{e \in E_v} \sigma_e(S)(V(e_+) + V(e_-))/2 \right). \end{aligned} \quad (7.44)$$

Here S is a state of the model, E_h is the set of horizontal edges, E_v is the set of vertical edges, and $w(v|S)$ is the weight of the vertex v in the state S where $a_1 = a_2 = a$ and $b_1 = b_2 = b$. The symbol $\sigma_e(S)$ is the characteristic function of e: $\sigma_e(S) = 1$ if the arrow points up or to the left, and $\sigma_e(S) = -1$ if the arrow points down or to the right.

For a given *domain*, let us choose its boundary edge and enumerate all other boundary edges counterclockwise. The partition function for a domain for various fixed boundary conditions can be considered as a vector in $(\mathbb{C}^2)^{\otimes N}$ where the factors in the tensor product, counted from left to right, correspond to enumerated boundary vertices.

Similarly, the partition function for a (vertical) *cylinder* of size $N \times M$ (horizontal $\times$ vertical) can be regarded as a linear operator acting in $(\mathbb{C}^2)^{\otimes N}$, where the factors in the tensor product correspond to states on boundary edges.

The partition function for a *torus* of size $N \times M$ is a number and can be regarded as the trace of the partition function for a vertical cylinder of size $N \times M$ over the states on its horizontal sides, i.e. over $(\mathbb{C}^2)^{\otimes N}$. It can also be regarded as the trace of the partition function for a horizontal cylinder of size $M \times N$ over the states on its vertical sides, i.e. over $(\mathbb{C}^2)^{\otimes M}$. The result is an identity, which we will discuss later.

7.6.4 Transfer matrices

Let us write the matrix of Boltzmann weights for a vertex as a 4×4 matrix acting in the tensor product of the spaces of states on adjacent edges.

Let e_1 be a vector corresponding to an arrow pointing up on a vertical edge, and left on a horizontal edge. Let e_2 be a vector corresponding to an arrow pointing down on a vertical edge and right on a horizontal edge.

The matrix of Boltzmann weights with zero electric fields acts as

$$Re_1 \otimes e_1 = ae_1 \otimes e_1 \,, \tag{7.45}$$

$$Re_1 \otimes e_2 = be_1 \otimes e_2 + ce_2 \otimes e_1 \,, \tag{7.46}$$

$$Re_2 \otimes e_1 = be_2 \otimes e_1 + ce_1 \otimes e_2 \,, \tag{7.47}$$

$$Re_2 \otimes e_2 = ae_2 \otimes e_2 \,. \tag{7.48}$$

In the tensor product basis $e_1 \otimes e_1, e_1 \otimes e_2, e_2 \otimes e_1, e_2 \otimes e_2$, this matrix is the 4×4 matrix

$$\overline{R} = \begin{pmatrix} a\,0\,0\,0 \\ 0\; b\; c\; 0 \\ 0\; c\; b\; 0 \\ 0\,0\,0\,a \end{pmatrix} . \tag{7.49}$$

The six-vertex rules imply that the operator R commutes with the operator representing the total number of thick vertical edges, i.e.

$$[D^H \otimes D^H, \overline{R}] = 0 \,,$$

where

$$D^H = \begin{pmatrix} e^{H/2} & 0 \\ 0 & e^{-H/2} \end{pmatrix} .$$

The row-to-row transfer matrices with open boundary conditions, also known as the (quantum) *monodromy matrix*, is defined as

$$T_a = D_a^{H(a,1)} R_{a1} D_a^{H(a,1)} \dots D_a^{H(a,1)} R_{aN} \,.$$

It acts in the tensor product $\mathbb{C}_a^2 \otimes \mathbb{C}_1^2 \otimes \cdots \otimes \mathbb{C}_N^2$ of the spaces corresponding to horizontal edges and vertical edges. Each matrix R_{ai} is of the form (7.49); it acts trivially (as the identity matrix) in all factors of the tensor product except a and i. In the inhomogeneous case, the matrix elements of $\overline{R}_{ai}$ depend on i.

A matrix element of T is the partition function of the six-vertex model on a single row with fixed boundary conditions.

We define operators

$$D^{(a)} = D^{V(a,1)} \otimes \cdots \otimes D^{V(a,N)} \,,$$

$$D_a^H = 1 \otimes \cdots \otimes D^H \otimes \cdots \otimes 1 \,.$$

The row-to-row transfer matrix corresponding to a cylinder with a single horizontal row a and with an electric field $H(a, i)$ applied to the i-th horizontal edge of the a-th horizontal line is the following trace:

$$t_a = \mathrm{tr}_a(T_a) = \mathrm{tr}(D_a^{H(a,1)} R_{a1} \dots D_a^{H(a,N)} R_{aN}) \,.$$

This is an operator acting in $(\mathbb{C}^2)^{\otimes N}$. Its matrix elements are the partition functions of the six-vertex model on a cylinder with a single row with fixed boundary conditions on vertical edges.

The partition function of an inhomogeneous six-vertex model on a cylinder of height M with fixed boundary conditions on outgoing vertical edges is a matrix element of the linear operator

$$Z^{(C)} = D^{(M)} t_M \dots D^{(1)} t_1 ,$$

where

$$D^{(a)} = D^{V(a,1)} \otimes \dots \otimes D^{V(a,N)} .$$

The partition function for a torus with N columns and M rows is the trace of the partition function for the cylinder

$$Z^{(T)}_{N,M} = \mathrm{tr}_{(\mathbb{C}^2)^{\otimes N}} (D^{(M)} t_M \dots D^{(1)} t_1) . \tag{7.50}$$

Using the six-vertex rules, this trace can be transformed to

$$Z^{(T)}_{N,M} = \mathrm{tr}_{(\mathbb{C}^2)^{\otimes N}} (t_M^{(H_M)} \dots t_1^{(H_1)} (D_1^{V_1} \otimes \dots \otimes D_N^{V_N})) ,$$

where $H_a = \sum_{i=1}^N H_{ai}$ and $V_i = \sum_{a=1}^M V_{ai}$, and

$$t^{(H)} = \mathrm{tr}_a (R_{a1} \dots R_{aN} D_a^H) .$$

The partition function for a generic domain does not have a natural expression in terms of a transfer matrix. However, it is possible to give such an expression in a few exceptional cases, such as domain wall boundary conditions on a square domain; see for example [18] and references therein.

7.6.5 The commutativity of transfer matrices and positivity of weights

Commutativity of transfer matrices. Baxter discovered that matrices of the form (7.49) acting in the tensor product of two two-dimensional spaces satisfy the equation

$$R_{12} R'_{13} R''_{23} = R''_{23} R'_{13} R_{12} \tag{7.51}$$

if

$$\frac{a^2 + b^2 - c^2}{2ab} = \frac{a'^2 + b'^2 - c'^2}{2a'b'} = \frac{a''^2 + b''^2 - c''^2}{2a''b''} .$$

This parameter is denoted by Δ:

$$\Delta = \frac{a^2 + b^2 - c^2}{2ab} .$$

If each factor in the monodromy matrices $T'_a = R'_{a1} \dots R'_{aN}$ and $T''_b = R''_{a1} \dots R''_{aN}$ has the same value of Δ, eqn (7.51) implies that these monodromy matrices satisfy the relation

$$R_{ab} T'_a T''_b = T''_b T'_a R_{ab}$$

in $V_a \otimes V_b \otimes V_1 \otimes \dots \otimes V_N$. If R is invertible, this relation implies that row-to-row transfer matrices with periodic boundary conditions commute:

$$t' = \mathrm{tr}_a (T'_a D_a^H), \quad t'' = \mathrm{tr}_b (T''_b D_b^H), \quad [t', t''] = 0 .$$

It is easy to recognize that t is exactly the generating function for the commuting family of local Hamiltonians for spin chains constructed earlier. Thus, the problem of computing the partition function for periodic and cylindrical boundary conditions for the six-vertex model is closely related to finding the spectrum of Hamiltonians for integrable spin chains.

The parameterization. The set of positive triples of real numbers $a : b : c$ (up to a common multiplier) with fixed values of Δ has the following parameterization [3]:

1. When $\Delta > 1$, there are two cases. If $a > b + c$, the Boltzmann weights a, b, and c can be parameterized as

$$a = r\sinh(\lambda + \eta), \quad b = r\sinh(\lambda), \quad c = r\sinh(\eta)$$

with $\lambda, \eta > 0$. If $b > a + c$, the Boltzmann weights can be parameterized as

$$a = r\sinh(\lambda - \eta), \quad b = r\sinh(\lambda), \quad c = r\sinh(\eta)$$

with $0 < \eta < \lambda$. For both of these parameterizations of the weights, $\Delta = \cosh(\eta)$.

2. When $-1 < \Delta \leq 0$,

$$a = r\sin(\lambda - \gamma), \quad b = r\sin(\lambda), \quad c = r\sin(\gamma)\,,$$

where $0 < \gamma < \pi/2$, $\gamma < \lambda < \pi/2$, and $\Delta = -\cos\gamma$.

3. When $0 \leq \Delta < 1$,

$$a = r\sin(\gamma - \lambda), \quad b = r\sin(\lambda), \quad c = r\sin(\gamma)\,,$$

where $0 < \gamma < \pi/2$, $0 < \lambda < \gamma$, and $\Delta = \cos\gamma$.

4. When $\Delta < -1$, the parameterization is

$$a = r\sinh(\eta - \lambda), \quad b = r\sinh(\lambda), \quad c = r\sinh(\eta)\,,$$

where $0 < \lambda < \eta$ and $\Delta = -\cosh\eta$.

We will write $a = a(u), b = b(u), c = c(u)$ assuming these parameterizations.

Topological nature of the partition function of the six-vertex model. We fix a domain and a collection of simple, non-self-intersecting, oriented curves with simple (transverse) intersections. The result is a 4-valent graph embedded in the domain. We fix six-vertex states (arrows) at the boundary edges of this graph. Such a graph connects boundary points, and defines a perfect matching between boundary points.

The six-vertex rules define the partition function on such a graph. The Yang–Baxter equation implies invariance with respect to the two moves shown in Figs 7.6 and 7.7. Because of this, the partition function of the six-vertex model depends only on the connection pattern

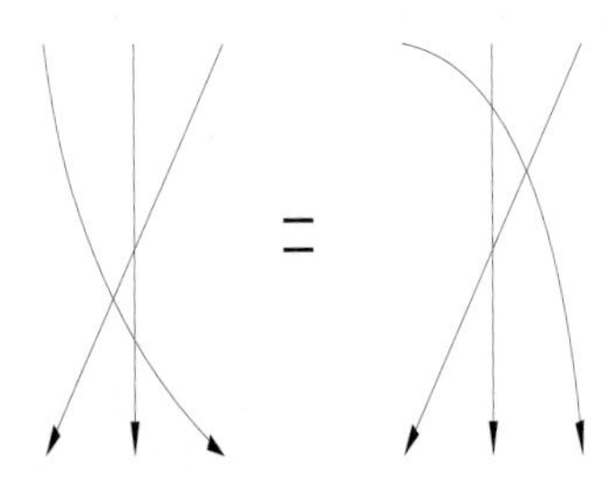

Fig. 7.6 The Yang–Baxter equation.

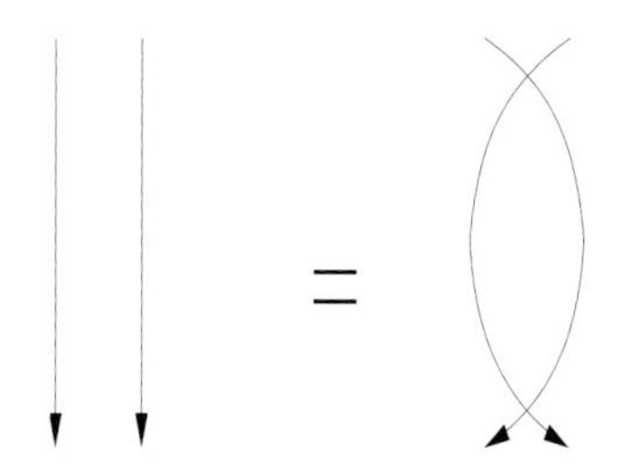

Fig. 7.7 Unitarity.

between boundary points, i.e. it depends only on the perfect matching on boundary points induced by the graph.

The "unitarity" relation involves the inverse of R, and therefore does not preserve the positivity of weights. But if it is used an "even" number of times it gives the equivalence of partition functions with positive weights. For example, it can be used to "permute" the rows in the case of cylindrical boundary conditions.

Inhomogeneous models with commuting transfer matrices. From now on, we will focus on the six-vertex model in a constant electric field. If the Boltzmann weights of the six-vertex model have the special inhomogeneity $a_{ij} = a(u_i - w_j), b_{ij} = b(u_i - w_j), c_{ij} = c(u_i - w_j)$, the partition function of the six-vertex model on a torus is

$$Z^{(T)}_{N,M}(\{u\}, \{w\}|H, V) = Tr_{\mathcal{H}_N}(t(u_1)\dots t(u_M)(D^{NV} \otimes \dots \otimes D^{NV})), \tag{7.52}$$

where $\mathcal{H}_N = (\mathbb{C}^2)^{\otimes N}$ and $t(u) = \mathrm{tr}_a(T_a(u))$, and where

$$T_a(u) = R_{a1}(u - w_1)\dots R_{aN}(u - w_N)D^{NH}. \tag{7.53}$$

Because the R-matrix satisfies the Yang–Baxter equation, we have

$$R_{ab}(u - v)T_a(u)T_b(v) = T_b(v)T_a(u)R_{ab}(u - v).$$

As a corollary, the traces of these matrices commute:

$$[t(u), t(v)] = 0;$$

and, as we have seen in previous sections, their spectrum can be described explicitly by the Bethe ansatz.

Notice that the positivity of weights restricts the possible values of inhomogeneities. For example, when $\Delta > 1$ and $a > b + c$ we must have $-u < w_i < u$.

For a diagonal matrix d such that $[d \otimes 1 + 1 \otimes d, R(u)] = 0$, we define

$$R^d(u) = (e^{ud} \otimes 1)R(u)(e^{-ud} \otimes 1).$$

It is clear that

$$\mathrm{tr}(R^d_{a1}(u - w_1)\dots R^d_{aN}(u - v_N)D^{NH}) = U\,\mathrm{tr}(R_{a1}(u - w_1)\dots R_{aN}(u - v_N)D^{NH})U^{-1},$$

where $U = \exp(-\sum_{i=1}^N w_i d_i)$. In particular, the partition functions for a torus with weights given by $R(u)$ and with weights given by $R^d(u)$ are the same.

Let $u_1, \ldots, u_N$ be the parameters of inhomogeneities along the horizontal direction, and let $w_1, \ldots, w_M$ be inhomogeneities along the vertical direction. The partition function (7.50) has the following properties:

- $Z^{(T)}_{N,M}(\{u\}, \{w\}|H, V)$ is a symmetric function of u and w.
- It has the form $\prod_{i=}^{N} e^{-Mu_i} \prod_{i=1}^{M} e^{-Nw_i} P(\{e^{2u}\}, \{e^{2w}\})$, where P is a polynomial of degree M in each e^{u_i} and of degree N in each e^{w_j}.
- It is a function of $u_i - w_j$.
- It satisfies the identity

$$Z^{(T)}_{N,M}(\{u\}, \{w\}|H, V) = Z^{(T)}_{M,N}(\{w\}, \{h - u\}|V, H), \tag{7.54}$$

where $h = \eta$ or γ depending on the regime. This identity corresponds to the "rotation" of the torus by 90 degrees and is known as the "crossing symmetry" or "modular identity".

7.7 The six-vertex model on a torus in the thermodynamic limit

7.7.1 The thermodynamic limit of the six-vertex model for periodic boundary conditions

By the thermodynamic limit we mean here the large-volume limit, when $N, M \to \infty$. The free energy per site in this limit is

$$f = -\lim_{N,M\to\infty} \frac{\log(Z_{N,M})}{NM},$$

where $Z_{N,M}$ is the partition function with periodic boundary conditions on a rectangular grid $L_{N,M}$. It is a function of the Boltzmann weights and magnetic fields.

For generic H and V, the six-vertex model in the thermodynamic limit has a unique translationally invariant Gibbs measure with a slope (h, v), where

$$h = -\frac{1}{2}\frac{\partial f}{\partial H} + \frac{1}{2}, \quad v = -\frac{1}{2}\frac{\partial f}{\partial V} + \frac{1}{2}. \tag{7.55}$$

This Gibbs measure defines local correlation functions in the thermodynamic limit. The parameter

$$\Delta = \frac{a^2 + b^2 - c^2}{2ab}$$

defines many characteristics of the six-vertex model in the thermodynamic limit.

7.7.2 The large-N limit of the eigenvalues of the transfer matrix

The row-to-row transfer matrix for the homogeneous six-vertex model on a lattice with periodic boundary conditions and with rows of length N is

$$t(u) = \mathrm{tr}_a(R_{a1}(u) \ldots R_{aN}(u)e^{H\sigma^z}) \exp\left(V \sum_{a=1}^{N} \sigma^z_a\right).$$

According to eqns (7.36) and (7.37), the eigenvalues of this linear operator are

$$\Lambda_{\{v_i\}} = a^N e^{NH} \prod_{i=1}^{n} \frac{v_i z^{-1} q - v_i^{-1} z q^{-1}}{v_i z^{-1} - v_i^{-1} z} + b^N e^{-NH} \prod_{i=1}^{n} \frac{v_i^{-1} z q - v_i z^{-1} q^{-1}}{v_i^{-1} z - v_i z^{-1}}, \tag{7.56}$$

where $0 \leq n \leq N$, and z, q parameterize a, b, c as $a = r(zq - zq^{-1}), b = r(z - z^{-1}), c = r(q - q^{-1})$. The numbers v_i are solutions to the Bethe equations

$$\left(\frac{v_i q - v_i^{-1} q^{-1}}{v_i - v_i^{-1}} \right)^N = -e^{-2H} \prod_{j=1}^{n} \frac{v_i v_j^{-1} q - v_i^{-1} v_j q^{-1}}{v_i v_j^{-1} q^{-1} - v_i^{-1} v_j q}. \tag{7.57}$$

As for the corresponding spin chains, it is expected that the numbers v_i corresponding to the largest eigenvalue concentrate, when $N \to \infty$, on a contour in the complex plane with a finite density. The Bethe equations provide a linear integral equation for this density. This conjecture is supported by numerical evidence and has been proven in some special cases, for example when $\Delta = 0$.

The partition function for the homogeneous six-vertex model on an $N \times M$ lattice with periodic boundary conditions is

$$Z_{N,M} = \sum_{\alpha} \Lambda_{\alpha}^{(N)\,M}, \tag{7.58}$$

where α parameterizes the eigenvalues of $t(u)$.

Let ω_N be the eigenvector of $t(u)$ corresponding to the maximal eigenvalue. The sequence of vectors $\{\Omega_N\}$ as $N \to \infty$ defines the Hilbert space of pure states for the infinite system. Let Λ_0 be the largest eigenvalue of $t(u)$. According to our main conjecture that the largest eigenvalue corresponds to numbers v_i filling a contour in the complex plane as $N \to \infty$, the largest eigenvalue has the following asymptotics:

$$\Lambda_0^{(N)} = \exp(-Nf(H,V) + O(1)). \tag{7.59}$$

The function $f(H, V)$, as we will see below, is the free energy of the system. It is computed in the next section.

The transfer matrix in this limit has the asymptotics

$$t(u) = \exp(-Nf(H,V))\widehat{t}(u),$$

where the operator $\widehat{t}(u)$ acts in the space H_∞ and its eigenvalues are determined by positions of "particles" and "holes," similarly to the structure of excitations in the large-N limit for spin chains.

7.7.3 Modularity

It is now easy to compute the asymptotics of the partition function in the thermodynamic limit. As $M \to \infty$, the leading term in the formula (7.58) is given by the largest eigenvalue:

$$Z_{N,M} = \Lambda_0^{(N)\,M}(1 + O(e^{-\alpha M}))$$

for some positive α.

Taking the limit $N \to \infty$ and taking into account the asymptotics of the largest eigenvalue (7.59), we identify the function $f(H,V)$ in eqn (7.59) with the free energy:

$$Z_{N,M} = e^{NMf(H,V)(1+o(1))} \,.$$

Note that we could have interchanged the roles of N and M by first taking the limit $N \to \infty$ and then $M \to \infty$. In this case we would have to compute first the asymptotics of

$$Z_{N,M} = \mathrm{tr}(t(u)^M)$$

as $N \to \infty$ and then take the limit $M \to \infty$.

The large-N limit of the trace can be computed by using the finite-temperature technique developed by Yang and Yang [41]. This was done by de Vega and Destry [12]. The leading term of the asymptotic expansion can be expressed in terms of the solution to a nonlinear integral equation.

This gives an alternative description for the largest eigenvalue. A similar description exists for all eigenvalues. This has been done by Zamolodchikov [42] for other integrable quantum field theories.

7.8 The six-vertex model at the free-fermionic point

When $\Delta = 0$, the partition function of the six-vertex model can be expressed in terms of the dimer model on a decorated square lattice. Because the dimer model can be regarded as a theory of free fermions, the six-vertex model is said to be free-fermionic when $\Delta = 0$.

At this point the row-to-row transfer matrix on N sites for a torus can be written in terms of the Clifford algebra of $\mathbb{C}^N$. The Jordan–Wiegner transform maps local spin operators to the elements of the Clifford algebra.

In the Bethe equations (7.37), the variables v_i disappear in the right-hand side, which becomes simply $(-1)^{n-1}e^{-2NH}$. After a change of variables, these equations can be interpreted as the periodic boundary conditions for a fermionic wave function.

7.8.1 Homogeneous case

At the free-fermionic point, $\Delta = 0$, i.e. $\gamma = \pi/2$ and the weights of the six-vertex model are parameterized as

$$a = \sin\left(\frac{\pi}{2} - u\right) = \cos(u), \;\; b = \sin(u), \;\; c = 1 \,.$$

Without losing generality, we may assume that $0 < u < \pi/4$.

The eigenvalues of the row-to-row transfer matrix are given by the Bethe ansatz formulas:

$$\Lambda(u) = ((\cos(u)^N(-1)^n e^{NH} + (\sin(u))^N e^{-NH})e^{NV} \prod_{i=1}^{n} (\cot(u - v_i)e^{-2V}) \,. \tag{7.60}$$

Here $0 \le n \le N$, and the v_i are distinct solutions to the Bethe equations:

$$\cot(v)^N = (-1)^{n-1}e^{-2NH} \,,$$

or

$$\cot(v) = \omega e^{-2H} \,,$$

where $\omega^N = (-1)^{n-1}$.

Using the identity

$$\cot(u-v) = \frac{\cot u \cot v + 1}{\cot v - \cot u} = \cot u - \frac{1-\cot^2 u}{\omega e^{-2H} - \cot u},$$

we can write the eigenvalues as

$$\Lambda(u) = \Big(\big(\cos(u)^N (-1)^n e^{NH} + (\sin(u))^N e^{-NH}\big) e^{(N-2n)V} \Big) \prod_{i=1}^{n} \left(\cot u - \frac{1-\cot^2 u}{\omega_i e^{-2H} - \cot u} \right). \tag{7.61}$$

To find the maximal eigenvalue in the limit $N \to \infty$, we must analyze the factors in the formula (7.60):

1. If

$$\max_{|\omega|=1} \cot(u-v) < e^{2V},$$

where $\cot(v) = \omega e^{-2H}$, all factors are less than one and the maximal eigenvalue

$$\Lambda_{\mathrm{ord}}(u) = (\cos(u)^N e^{NH} + (\sin(u))^N e^{-NH}) e^{NV}$$

is achieved when $n = 0$. As $N \to \infty$, the first term dominates when $\cot(u) < e^{2H}$. In this case the Gibbs state describing the six-vertex model in the thermodynamic limit is the ordered state A_1 shown in Fig. 7.8. When $\cot(u) > e^{2H}$, the second term dominates. In this case the Gibbs state is the ordered state B_2 shown in Fig. 7.8.

2. If $\min_{|\omega|=1} \cot(u-v) > e^{2V}$, all factors $\cot(u-v_j)e^{2V}$ are greater than one in absolute value. In this case the maximal eigenvalue is achieved when $n = N$. The corresponding ground state is ordered, and is A_2 in Fig. 7.8 when $\cot u > e^{-2H}$ and B_1 when $\cot(u) < e^{-2H}$.

3. If

$$\max_{|\omega|=1} \cot(u-v) < e^{2V},$$

then there exists an $\omega_0 = e^{iK}$ such that

$$|\cot(u-b)| = e^{2V},$$

where $\cot(b) = e^{\pm iK - 2H}$. It is easy to find K:

$$\cos(K) = \frac{(Ue^{2H} + U^{-1}e^{-2H})e^{4V} - (Ue^{-2H} + U^{-1}e^{2H})}{2(e^{4V}+1)}.$$

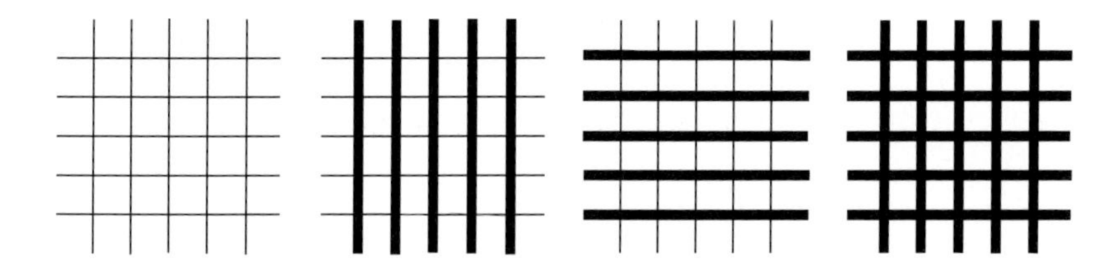

Fig. 7.8 Four frozen configurations (A_1, A_2, B_1, and B_2, from left to right) of the ferromagnetic phase.

In this case

$$|\cot(u-v)| > e^{2V}$$

when $\cot(v) = e^{i\alpha}e^{-2H}$ with $-K < \alpha < K$, and

$$|\cot(u-v)| < e^{2V}$$

when $-\pi \le \alpha < -K$ or $K < \alpha \le \pi$. The maximal eigenvalue in this case corresponds to the maximal n such that $\pi(n-1)/N < K$ and $\omega_j = \exp(i\pi(n+1-2j)/N)$, $j = 1, \ldots, n$, and is given by eqn (7.60).

As $N \to \infty$, the asymptotics of the largest eigenvalue are given by the following integral:

$$\log(\Lambda_{\text{disord}}(u)) = N(\log(\cos(u)e^H) + \frac{1}{2\pi i}\int_{-K}^{K} \log\left(\frac{\cot(u)\omega e^{-2H}+1}{\omega e^{-2H}-\cot(u)}\right)\frac{d\omega}{\omega}) + O(1)\,.$$

The six-vertex model in this regime is in the disordered phase.

The disordered and ordered phases are separated by the curve

$$\max_{|\omega|=1} |\cot(u-v)| = e^{-2V}, \quad \min_{|\omega|=1|} |\cot(u-v)| = e^{-2V},$$

or, more explicitly,

$$\left|\frac{Ue^{-2H} \mp 1}{U \pm e^{-2H}}\right| = e^{-2V}\,.$$

This curve is shown in Fig. 7.9.

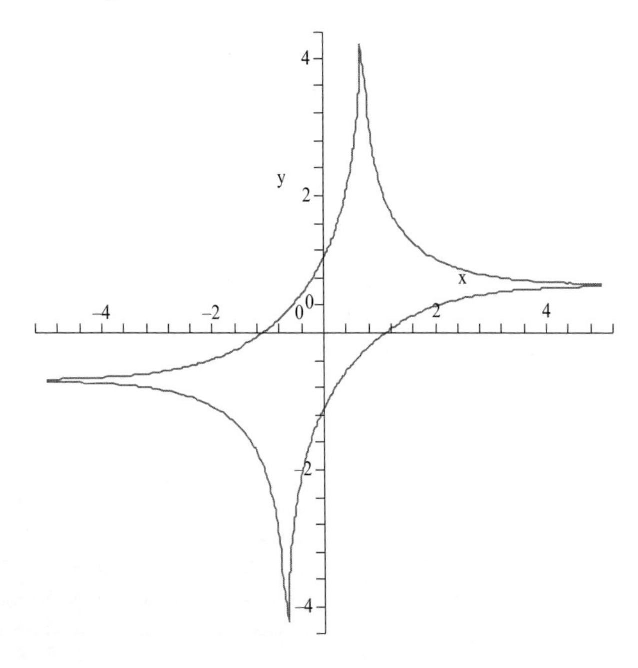

Fig. 7.9 The boundary between the ordered and disordered regions in the (H, V) plane for the homogeneous six-vertex model with $\Delta = 0$ and $U = \cot(u) = 2$.

7.8.2 Horizontally inhomogeneous case

The partition function for the six-vertex model at the free-fermionic point with inhomogeneous rows is

$$Z_{N,M}(u,a) = \sum_{n=0}^{N} \sum_{\omega_1,\dots,\omega_n} \Lambda(u+a|\omega)^M \Lambda(u-a|\omega)^M ,$$

where the $\omega_i \neq \omega_j$ are solutions to

$$\omega^N = (-1)^{n-1}$$

and $\Lambda(u|\omega)$ is given by eqn (7.61).

The boundary between the ordered and disordered phases is given by the equations

$$\max_{|\omega|=1} |\cot(u+a-v)\cot(u-a-v)| = e^{4V}, \quad \min_{|\omega|=1} |\cot(u+a-v)\cot(u-a-v)| = e^{4V} ,$$

or

$$\left|\frac{U_+e^{-2H} \pm 1}{U_+ \mp e^{-2H}}\right| \left|\frac{U_-e^{-2H} \pm 1}{U_- \mp e^{-2H}}\right| = e^{4V} . \tag{7.62}$$

This curve is shown in Fig. 7.10. The ordered phases A_1B_2 and A_2B_1 are shown in Fig. 7.11.

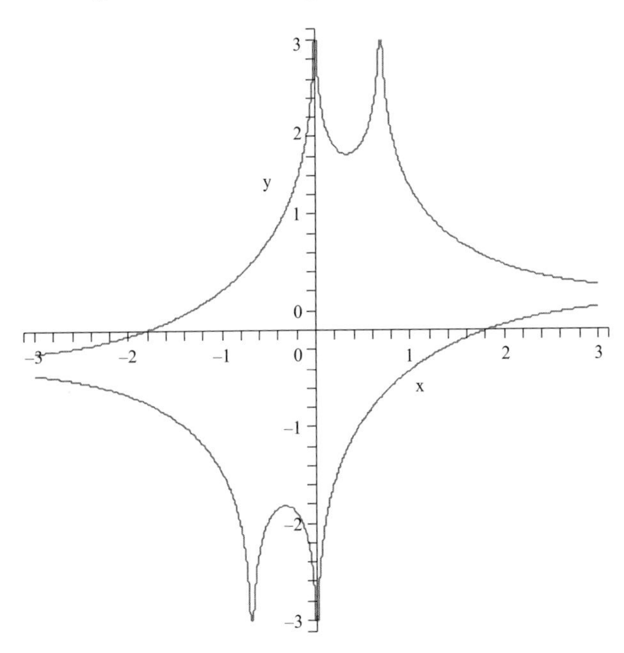

Fig. 7.10 The boundary between the ordered and disordered regions in the (H, V) plane for the inhomogeneous six-vertex model with $\Delta = 0$ and $U = \cot(u) = 3, T = \cot(a) = 1$.

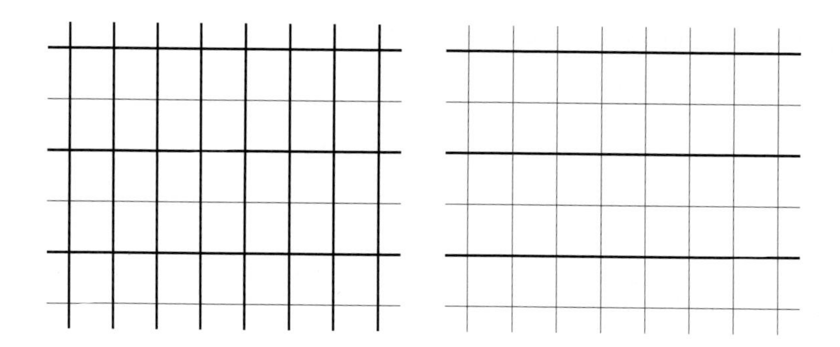

Fig. 7.11 The horizontally inhomogeneous ordered phases A_1B_2 (left) and A_2B_1 (right).

The free energy per site in the disordered region is given by

$$f(H,V) = \log\left(\cos(u+a)\cos(u-a)\right. + \frac{1}{2\pi i}\int_{-K_1}^{K_1}\log\left(\frac{\cot(u+a)\omega e^{-2H}+1}{\omega e^{-2H}-\cot(u+a)}\frac{\cot(u-a)\omega e^{-2H}+1}{\omega e^{-2H}-\cot(u-a)}\right)\frac{d\omega}{\omega} \left. + \frac{1}{2\pi i}\int_{-K_2}^{K_2}\log\left(\frac{\cot(u+a)\omega e^{-2H}+1}{\omega e^{-2H}-\cot(u+a)}\frac{\cot(u-a)\omega e^{-2H}+1}{\omega e^{-2H}-\cot(u-a)}\right)\frac{d\omega}{\omega}\right), \quad (7.63)$$

where $K_{1,2}$ are defined by the equation

$$|\cot(u+a-b)\cot(u-a-b)| = e^{2V},$$

with $\cot(b) = e^{\pm iK}e^{-2H}$.

7.8.3 Vertically inhomogeneous case

Assume that the weights are homogeneous in the vertical direction and alternate in the horizontal direction with parameters alternating as $\ldots, u+a, u-a, u+a, u-a, \ldots$. The positivity of weights in the region $0 \le u \le \pi/4$ requires $0 \le a \le u$.

The eigenvalues of the transfer matrix are given by the Bethe ansatz:

$$\Lambda(u) = \left((\cos(u+a)\cos(u-a))^N(-1)^n e^{NH} + (\sin(u+a)\sin(u-a))^N e^{-NH}\right) \times e^{(N-2n)V}\prod_{i=1}^{n}\cot(u-v_i), \quad (7.64)$$

where the v_i are distinct solutions to

$$(\cot(v-a)\cot(v+a))^N = (-1)^{n-1}e^{-2NH}.$$

The left side is the N-th power of

$$\frac{\cot(v)\cot(a)+1}{\cot(a)-\cot(v)}\frac{-\cot(v)\cot(a)+1}{-\cot(a)-\cot(v)} = \frac{\cot(v)^2\cot(a)^2-1}{\cot(a)^2-\cot(v)^2}.$$

From this it is easy to find a parameterization of the solutions by roots of unity:

$$\cot(v)^2 = \frac{\omega \cot(a)^2 e^{-2H} + 1}{\cot(a)^2 + \omega e^{-2H}}, \tag{7.65}$$

where

$$\omega^N = (-1)^{n-1}.$$

Largest eigenvalue. The analysis of the largest eigenvalue of the transfer matrix in the limit $N \to \infty$ is similar to the previous cases. When

$$\max_{|\omega|=1} |\cot(u-v)| \geq e^{2V}, \tag{7.66}$$

where v and ω are related as in eqn (7.65), the absolute values of all factors $|\cot(u-v)|e^{-2V}$ are greater than one in absolute value and the largest eigenvalue corresponds to $n = 0$. The rest of the analysis is similar.

Let us use the notation

$$T = \cot(a), \ \ x = e^{-H}, \ \ U = \cot(u), \ \ s = \cot(v).$$

The positivity of the weights implies

$$T \geq U > 0.$$

Proposition 7.5 *The curve separating the ordered phases from the disordered phase is*

$$X = \left|\frac{YU_+ - 1}{Y + U_+}\right| \left|\frac{YU_- - 1}{Y + U_-}\right|.$$

Proof In the notation defined above,

$$|\cot(u-v)|^2 = \left|\frac{\cot(u)\cot(v) + 1}{\cot(v) - \cot(u)}\right|^2 = \frac{Us+1}{s-U}\frac{U\bar{s}+1}{\bar{s}-U} = \frac{U^2|s|^2 + 2U\mathrm{Re}(s) + 1}{U^2 - 2U\mathrm{Re}(s) + |s|^2}.$$

The equation of the boundary of the disordered region is

$$\max_{|\omega|=1} |\cot(u-v)| = e^{2V}.$$

The equation defining v in terms of roots of unity ω can be solved explicitly for $s = \cot v$:

$$s^2 = \frac{T^2 x^2 \omega + 1}{T^2 + \omega x^2}.$$

We write $\omega = e^{i\alpha}$, and then

$$s^2 = \frac{(T^2 x^2 \cos(\alpha) + 1) + iT^2 x^2 \sin(\alpha)}{(T^2 + x^2 \cos(\alpha)) + ix^2 \sin(\alpha)},$$

$$\mathrm{Re}(s^2) = \frac{(T^2x^2\cos(\alpha)+1)(T^2+x^2\cos(\alpha)) + T^2x^2\sin(\alpha)x^2\sin(\alpha)}{(T^2+x^2\cos(\alpha))^2 + x^4\sin(\alpha)^2},$$

$$|s|^4 = \frac{T^4x^4 + 2T^2x^2\cos(\alpha)+1}{T^4+2T^2x^2\cos(\alpha)+x^4}.$$

Now, some simple algebra:

$$\mathrm{Re}(s) = \sqrt{\frac{\mathrm{Re}(s^2)+|s|^2}{2}}.$$

As we vary ω along the unit circle, the function $\cot(u-v)$ has a maximum at $\omega = 1$.

Taking this into account, the equation for the boundary curve becomes

$$\left|\frac{U\sqrt{1+x^2T^2} \pm \sqrt{T^2+x^2}}{U\sqrt{T^2+x^2} \pm \sqrt{1+x^2T^2}}\right| = e^{2V}.$$

Solving this equation for $X = x^2 = e^{-2H}$ in terms of $Y = e^{2V}$, we obtain

$$X = \left|\frac{YU_+ - 1}{Y+U_+}\right| \left|\frac{YU_- - 1}{Y+U_-}\right|,$$

which is, after changing the coordinates to X^{-1}, Y^{-1}, is the same curve as eqn (7.62). This is one of the implications of the modular symmetry. □

By modularity, i.e. by "rotating" the lattice with periodic boundary conditions by 90 degrees, all characteristics of the model with vertical inhomogeneities can be identified with the corresponding characteristics of the model with horizontal inhomogeneities.

7.9 The free energy of the six-vertex model

The computation of the free energy of the six-vertex model by taking the large-N, M limit of the partition function on a torus was outlined in Section 7.7.1. In this section, we will describe the free energy as a function of the electric fields (H, V) and its basic properties.

7.9.1 The phase diagram for $\Delta > 1$

The weights a, b, and c in this region satisfy one of the following two inequalities, either $a > b + c$ or $b > a + c$.

If $a > b + c$, the Boltzmann weights a, b, and c can be parameterized as

$$a = r\sinh(\lambda+\eta), b = r\sinh(\lambda), c = r\sinh(\eta) \tag{7.67}$$

with $\lambda, \eta > 0$.

If $a + c < b$, the Boltzmann weights can be parameterized as

$$a = r\sinh(\lambda-\eta), b = r\sinh(\lambda), c = r\sinh(\eta) \tag{7.68}$$

with $0 < \eta < \lambda$.

For both of these parameterizations of the weights, $\Delta = \cosh(\eta)$. The phase diagram of the model for $a > b + c$ (and, therefore, $a > b$) is shown in Fig. 7.12 and for $b > a + c$ (and, therefore, $a < b$) in Fig. 7.13.

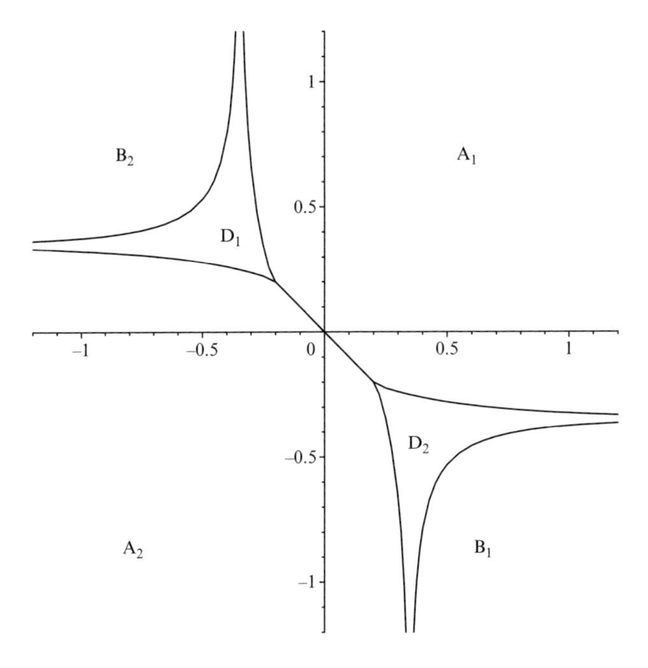

Fig. 7.12 The phase diagram in the (H, V) plane for $a = 2$, $b = 1$, and $c = 0.8$.

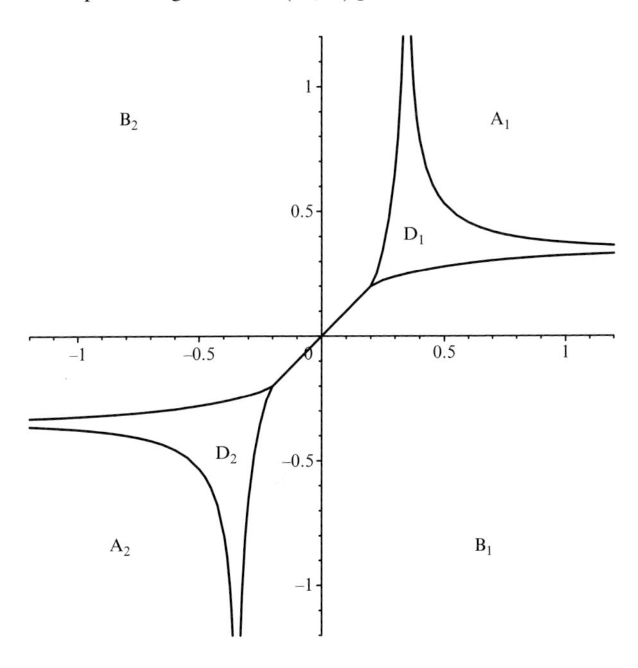

Fig. 7.13 The phase diagram in the (H, V) plane for $a = 1$, $b = 2$, and $c = 0.8$.

When the electric fields (H, V) are in one of the regions A_i, B_i of the phase diagram, the system in the thermodynamic limit is described by a translationally invariant Gibbs measure supported on the corresponding frozen (ordered) configurations. There are four frozen configurations A_1, A_2, B_1, and B_2, shown in Fig. 7.8. For a finite but large grid, the probability of any other state is of the order at most $\exp(-\alpha N)$ for some positive α.

The local correlation functions in a frozen state are products of expectation values of characteristic functions of edges,

$$\lim_{N\to\infty} \langle \sigma_{e_1} \dots \sigma_{e_n} \rangle_N = \sigma_{e_1}(S) \dots \sigma_{e_n}(S),$$

where S is the one of the ferromagnetic states A_i, B_i.

The boundary between the ordered phases in the (H, V) plane and the disordered phases, as in the free-fermionic case, is determined by the next to largest eigenvalue of the row-to-row transfer matrix.

Instead of going into the details of the computations as we did in the free-fermionic case, we will just present the answers:

- $a > b + c$ (see Fig. 7.12):

$$\begin{aligned}
&A_1 \text{ region:} \quad V + H \geq 0, \quad \cosh(2H) \leq \Delta,\\
&\quad (e^{2H} - b/a)(e^{2V} - b/a) \geq (c/a)^2, \quad e^{2H} > b/a, \quad \cosh(2H) > \Delta.\\
&A_2 \text{ region:} \quad V + H \leq 0, \quad \cosh(2H) \leq \Delta,\\
&\quad (e^{-2H} - b/a)(e^{-2V} - b/a) \geq (c/a)^2, \quad e^{-2H} > b/a, \quad \cosh(2H) > \Delta.\\
&B_1 \text{ region:} \quad (e^{2H} - a/b)(e^{-2V} - a/b) \geq (c/b)^2, \quad e^{2H} > a/b.\\
&B_2 \text{ region:} \quad (e^{-2H} - a/b)(e^{2V} - a/b) \geq (c/b)^2, \quad e^{-2H} > a/b.
\end{aligned}$$

- $b > a + c$ (see Fig. 7.13):

$$\begin{aligned}
&A_1 \text{ region:} \quad (e^{2H} - b/a)(e^{2V} - b/a) \geq (c/a)^2, \quad e^{2H} > b/a.\\
&A_2 \text{ region:} \quad (e^{-2H} - b/a)(e^{-2V} - b/a) \geq (c/a)^2, \quad e^{-2H} > b/a.\\
&B_1 \text{ region:} \quad V - H \geq 0, \quad \cosh(2H) \leq \Delta,\\
&\quad (e^{2H} - a/b)(e^{-2V} - a/b) \geq (c/b)^2, \quad e^{2H} > a/b, \quad \cosh(2H) > \Delta.\\
&B_2 \text{ region:} \quad V - H \leq 0, \quad \cosh(2H) \leq \Delta,\\
&\quad (e^{-2H} - a/b)(e^{2V} - a/b) \geq (c/b)^2, \quad e^{-2H} > a/b, \quad \cosh(2H) > \Delta.
\end{aligned}$$

The free energy is a linear function of H and V in the four frozen regions:

$$\begin{aligned}
f &= -\ln a - H - V \quad &&\text{in} \quad A_1,\\
f &= -\ln b + H - V \quad &&\text{in} \quad B_2,\\
f &= -\ln a + H + V \quad &&\text{in} \quad A_2,\\
f &= -\ln b - H + V \quad &&\text{in} \quad B_1.
\end{aligned} \tag{7.69}$$

The regions D_1 and D_2 are disordered phases. If (H, V) is in one of these regions, the local correlation functions are determined by a unique Gibbs measure with the polarization given by the gradient of the free energy. In these phases the system is disordered, which means that local correlation functions decay as a power of the distance $d(e_i, e_j)$ between e_i and e_j when $d(e_i, e_j) \to \infty$.

In the regions D_1 and D_2, the free energy is given by [39]

$$f(H,V) = \min\left(\min_\alpha\left(E_1 - H - (1-2\alpha)V - \frac{1}{2\pi i}\int_C \ln\left(\frac{b}{a} - \frac{c^2}{ab - a^2 z}\right)\rho(z)\,dz\right),\right.$$
$$\left.\min_\alpha\left(E_2 + H - (1-2\alpha)V - \frac{1}{2\pi i}\int_C \ln\left(\frac{a^2 - c^2}{ab} + \frac{c^2}{ab - a^2 z}\right)\rho(z)\,dz\right)\right), \quad (7.70)$$

where $\rho(z)$ can be found from the integral equation

$$\rho(z) = \frac{1}{z} + \frac{1}{2\pi i}\int_C \frac{\rho(w)}{z - z_2(w)}dw - \frac{1}{2\pi i}\int_C \frac{\rho(w)}{z - z_1(w)}dw\,, \quad (7.71)$$

in which

$$z_1(w) = \frac{1}{2\Delta - w}, \qquad z_2(w) = -\frac{1}{w} + 2\Delta\,.$$

$\rho(z)$ satisfies the following normalization condition:

$$\alpha = \frac{1}{2\pi i}\int_C \rho(z)\,dz\,.$$

The contour of integration C (in the complex z-plane) is symmetric with respect to the conjugation $z \to \bar{z}$, is dependent on H, and is defined by the condition that the form $\rho(z)\,dz$ has purely imaginary values on the vectors tangent to C:

$$\mathrm{Re}(\rho(z)\,dz)\Big|_{z\in C} = 0\,.$$

The formula (7.70) for the free energy follows from the Bethe ansatz diagonalization of the row-to-row transfer matrix. It relies on a number of conjectures that are supported by numerical and analytical evidence and are taken for granted in physics. However, there is no rigorous proof.

There are two points where three phases coexist (two frozen and one disordered phase). These points are called *tricritical*. The angle θ between the boundaries of D_1 (or D_2) at a tricritical point is given by

$$\cos(\theta) = \frac{c^2}{c^2 + 2\min(a,b)^2(\Delta^2 - 1)}\,.$$

The existence of such points makes the six-vertex model (and its degeneration known as the five-vertex model [17]) remarkably different from dimer models [22], where the generic singularities in the phase diagram are cusps. Physically, the existence of singular points where two curves meet at a finite angle manifests the presence of interaction in the six-vertex model.

Note that when $\Delta = 1$, the phase diagram of the model has a cusp at the point $H = V = 0$. This is the transitional point between the region $\Delta > 1$ and the region $|\Delta| < 1$, which is described below.

7.9.2 The phase diagram for $|\Delta| < 1$

In this case, the Boltzmann weights have a convenient parameterization by trigonometric functions. When $1 \geq \Delta \geq 1$,

$$a = r\sin(\lambda - \gamma), b = r\sin(\lambda), c = r\sin(\gamma)\,,$$

where $0 \leq \gamma \leq \pi/2$, $\gamma \leq \lambda \leq \pi$, and $\Delta = \cos\gamma$.

When $0 \geq \Delta \geq -1$,

$$a = r\sin(\gamma - \lambda), b = r\sin(\lambda), c = r\sin(\gamma)\,,$$

where $0 \leq \gamma \leq \pi/2$, $\pi - \gamma \leq \lambda \leq \pi$, and $\Delta = -\cos\gamma$.

The phase diagram of the six-vertex model with $|\Delta| < 1$ is shown in Fig. 7.14. The phases A_i, B_i are frozen and are identical to the frozen phases for $\Delta > 1$. The phase D is disordered. For magnetic fields (H, V), the Gibbs measure is translationally invariant with a slope $(h, v) = (\partial f(H,V)/\partial H, \partial f(H,V)/\partial V)$.

The frozen phases can be described by the following inequalities:

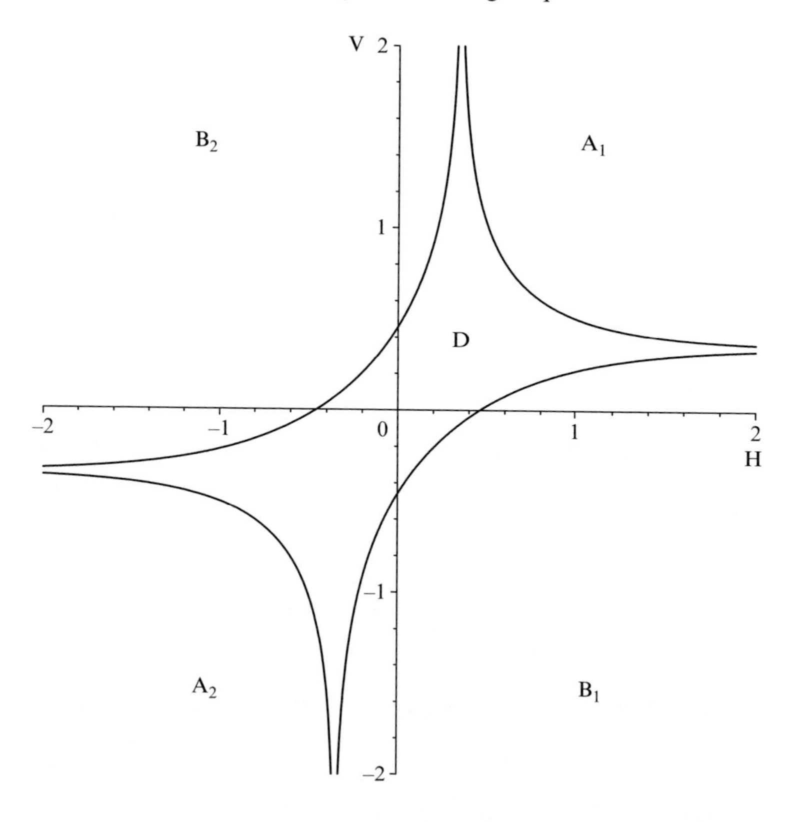

Fig. 7.14 The phase diagram in the (H, V) plane for $a = 1$, $b = 2$, and $c = 2$.

$$\begin{array}{llll} A_1 \text{ region:} & (e^{2H} - b/a)(e^{2V} - b/a) \geq (c/a)^2, & e^{2H} > b/a\,. & \\ A_2 \text{ region:} & (e^{-2H} - b/a)(e^{-2V} - b/a) \geq (c/a)^2, & e^{-2H} > b/a\,. & \\ B_1 \text{ region:} & (e^{2H} - a/b)(e^{-2V} - a/b) \geq (c/b)^2, & e^{2H} > a/b\,. & (7.72) \\ B_2 \text{ region:} & (e^{-2H} - a/b)(e^{2V} - a/b) \geq (c/b)^2, & e^{-2H} > a/b\,. & \end{array}$$

The free-energy function in the frozen regions is still given by the formulas (7.70). The first derivatives of the free energy are continuous at the boundaries of the frozen phases. The second derivative is continuous in the tangential direction at the boundaries of the frozen phases and is singular in the normal direction.

The free-energy function is smooth in the disordered region, where it is given by eqn (7.70), which, as in the case $\Delta > 1$, involves a solution to the integral equation (7.71). The contour of integration in eqn (7.71) is closed for zero magnetic field and, therefore, eqn (7.71) can be solved explicitly by Fourier transformation [3] .

The six-vertex Gibbs measure for zero magnetic field converges in the thermodynamic limit to a superposition of translationally invariant Gibbs measures with a slope $(1/2, 1/2)$. There are two such measures. They correspond to the double degeneracy of the largest eigenvalue of the row-to-row transfer matrix [3].

There is a very interesting relationship between the six-vertex model in zero magnetic field and the highest-weight representation theory of the corresponding quantum affine algebra. The double degeneracy of the Gibbs measure with a slope $(1/2, 1/2)$ corresponds to the fact that there are two integrable irreducible representations of $\widehat{sl_2}$ at level one. Correlation functions can be computed in this case using q-vertex operators [19]. For the latest developments, see [7].

7.9.3 The phase diagram for $\Delta < -1$

The phase diagram. The Boltzmann weights for these values of Δ can be conveniently parameterized as

$$a = r\sinh(\eta - \lambda), b = r\sinh(\lambda), c = r\sinh(\eta)\,, \tag{7.73}$$

where $0 < \lambda < \eta$ and $\Delta = -\cosh\eta$.

The Gibbs measure in the thermodynamic limit depends on the values of the magnetic fields. The phase diagram in this case is shown in Fig. 7.15 for $b/a > 1$. In the parameterization (7.73), this corresponds to $0 < \lambda < \eta/2$. When $\eta/2 < \lambda < \eta$, the four-tentacled "amoeba" is tilted in the opposite direction as in Fig. 7.12.

When (H, V) is in one of the regions A_i, B_i in the phase diagram, the Gibbs measure is supported on the corresponding frozen configuration (see Fig. 7.8).

The boundary between the ordered phases A_i, B_i and the disordered phase D is given by the inequalities (7.72). The free energy in these regions is linear in the electric fields and is given by (7.70).

If (H, V) is in the region D, the Gibbs measure is a translationally invariant measure with a polarization (h, v) determined by eqn (7.55). The free energy in this case is determined by the solution to the linear integral equation (7.71) and is given by the formula (7.70).

If (H, V) is in the region A, the Gibbs measure is a superposition of two Gibbs measures with a polarization $(1/2, 1/2)$. In the limit $\Delta \to -\infty$, these two measures degenerate to two measures supported on configurations C_1 and C_2, shown in Fig. 7.16. For a finite Δ, the

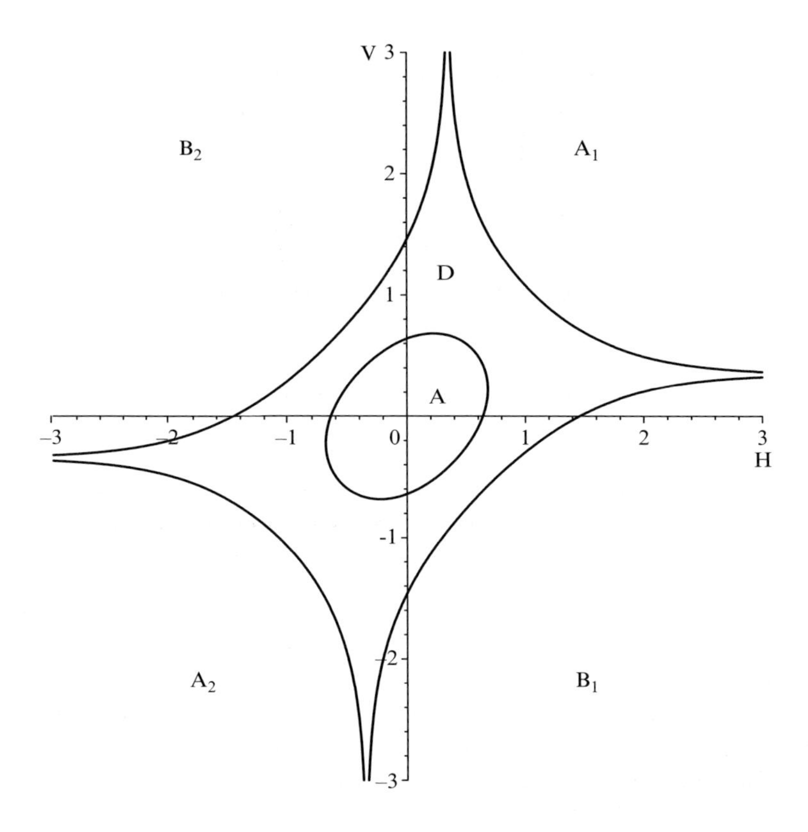

Fig. 7.15 The phase diagram in the (H, V) plane for $a = 1$, $b = 2$, and $c = 6$.

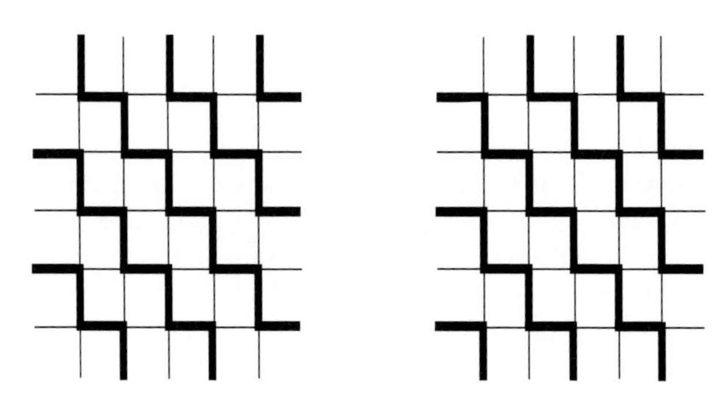

Fig. 7.16 The configurations C_1 and C_2.

support of these measures consists of configurations which differ from C_1 and C_2 in finitely many places on the lattice.

Remark 7.6 *Any two configurations lying in the support of each of these Gibbs measures can be obtained from C_1 or C_2 via flipping the path at a vertex "up" or "down" as shown in Fig. 7.17, finitely many times. It is also clear that it takes infinitely many flips to go from C_1 to C_2.*

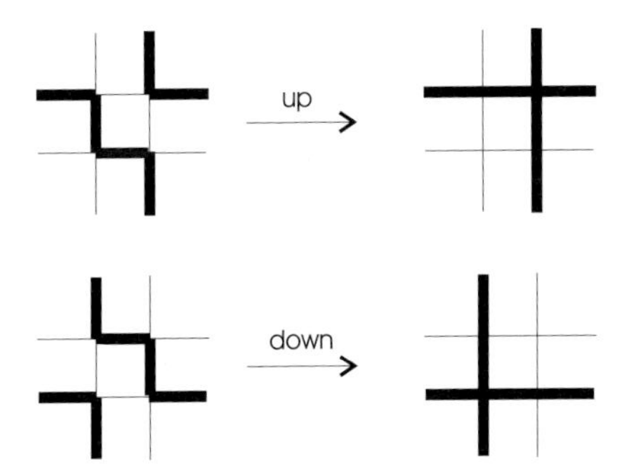

Fig. 7.17 The elementary up and down fluctuations in the antiferromagnetic phase.

The antiferromagnetic region. The six-vertex model in the phase A is disordered and is also noncritical. Here, the noncriticality means that the local correlation function $\langle \sigma_{e_i} \sigma_{e_j} \rangle$ decays as $\exp(-\alpha d(e_i, e_j))$ with some positive α as the distance $d(e_i, e_j)$ between e_i and e_j increases to infinity.

The free energy in the A-region can be explicitly computed by solving eqn (7.71). In this case the largest eigenvalue will correspond to $n = N/2$, the contour of integration in eqn (7.71) is closed, and the equation can be solved by a Fourier transform.[3]

The boundary between the antiferromagnetic region A and the disordered region D can be derived similarly to the boundaries of the ferromagnetic regions A_i and B_i by analyzing the next to largest eigenvalue of the row-to-row transfer matrix. This computation was done in [26, 39]. The result is a simple closed curve, which can be described parametrically as

$$H(s) = \Xi(s), \qquad V(s) = \Xi(\eta - \theta_0 + s),$$

where

$$\Xi(\varphi) = \cosh^{-1}\left(\frac{1}{\mathrm{dn}((K/\pi)\varphi|1-\nu)}\right),$$

$$|s| \leq 2\eta,$$

and

$$e^{\theta_0} = \frac{1 + \max(b/a, a/b)e^{\eta}}{\max(b/a, a/b) + e^{\eta}}.$$

The parameter ν is defined by the equation $\eta K(\nu) = \pi K'(\nu)$, where

$$K(\nu) = \int_0^{\pi/2} (1 - \nu \sin^2(\theta))^{-1/2} d\theta, \qquad K'(\nu) = \int_0^{\pi/2} (1 - (1-\nu)\sin^2(\theta))^{-1/2} d\theta.$$

The curve is invariant with respect to the reflections $(H, V) \to (-H, -V)$ and $(H, V) \to (V, H)$ since the function Ξ satisfies the identities

$$\Xi(\varphi) = -\Xi(-\varphi), \qquad \Xi(\eta - \varphi) = \Xi(\eta + \varphi).$$

This function is also 4η-periodic: $\Xi(4\eta + \varphi) = \Xi(\varphi)$.

[3] Strictly speaking, this is only a conjecture supported by numerical evidence.

As was shown in [31], this curve is algebraic in e^H and e^V and can be written as

$$\Big((1-\nu\cosh^2 V_0)\cosh^2 H + \sinh^2 V_0 - (1-\nu)\cosh V_0 \cosh H \cosh V\Big)^2$$
$$= (1-\nu\cosh^2 V_0)\sinh^2 V_0 \cosh^2 V \sinh^2 H(1-\nu\cosh^2 H)\,, \tag{7.74}$$

where V_0 is the positive value of V on the curve when $H=0$. Note that ν depends on the Boltzmann weights a, b, c only through η.

7.10 Some asymptotics of the free energy

7.10.1 The scaling near the boundary of the D-region

Assume that $\vec{H}_0 = (H_0, V_0)$ is a regular point at the boundary between the disordered region and the A_1 -region (see Fig. 7.14). Recall that this boundary is the curve defined by the equation

$$g(H,V)=0\,,$$

where

$$g(H,V) = \ln\left(b/a + \frac{c^2/a^2}{e^{2H}-b/a}\right) - 2V\,. \tag{7.75}$$

We denote the normal vector to the boundary of the D-region at $\vec{H}_0$ by $\vec{n}$, and the tangent vector pointing to the inside of the region D by $\vec{\tau}$.

We will study the asymptotics of the free energy along the curves $\vec{H}(r,s,t) = \vec{H}_0 + r^2 s\vec{n} + rt\vec{\tau}$, as $r \to 0$. It is clear that $\vec{H}(r,s,t)$ is in the D-region if $s \geq 0$.

Theorem 7.7 *Let $\vec{H}(r,s,t)$ be defined as above. The asymptotics of the free energy of the six-vertex model in the limit $r \to 0$ are given by*

$$f(\vec{H}(r,s,t)) = f_{\text{lin}}(\vec{H}(r,s,t)) + \eta(s,t)r^3 + O(r^5)\,, \tag{7.76}$$

where $f_{\text{lin}}(H,V) = -\ln(a) - H - V$ and

$$\eta(s,t) = -\kappa\left(\theta s + t^2\right)^{3/2}\,. \tag{7.77}$$

Here, the constants κ and θ depend on the Boltzmann weights of the model and on (H_0, V_0) and are given by

$$\kappa = \frac{16}{3\pi}\,\partial_H^2 g(H_0,V_0)$$

and

$$\theta = \frac{4 + (\partial_H g(H_0,V_0))^2}{2\partial_H^2 g(H_0,V_0)}\,,$$

where $g(H,V)$ is defined in eqn (7.75).

Moreover, $\partial_H^2 g(H_0,V_0) > 0$ and, therefore, $\theta > 0$.

We refer the reader to [31] for the details. This behavior is universal in the sense that the exponent $3/2$ is the same for all points at the boundary.

7.10.2 The scaling in the tentacle

Here, to be specific, we assume that $a > b$. The theorem below describes the asymptotics of the free-energy function when $H \to +\infty$ and

$$\frac{1}{2}\ln\left(\frac{b}{a}\right) - \frac{c^2}{2ab}e^{-2H} \leq V \leq \frac{1}{2}\ln\left(\frac{b}{a}\right) + \frac{c^2}{2ab}e^{-2H}, \qquad H \longrightarrow \infty\,. \tag{7.78}$$

These values of (H, V) describe points inside the right "tentacle" in Fig. 7.12.

Let us parameterize these values of V as

$$V = \frac{1}{2}\ln\left(\frac{b}{a}\right) + \beta\frac{c^2}{2ab}e^{-2H},$$

where $\beta \in [-1, 1]$.

Theorem 7.8 *When $H \to \infty$ and $\beta \in [-1, 1]$, the asymptotics of the free energy are given by the following formula:*

$$\begin{aligned} f(H,V) = &-\frac{1}{2}\ln(ab) - H \\ &- \frac{c^2}{2ab}e^{-2H}\Big(\beta + \frac{2}{\pi}\sqrt{1-\beta^2} - \frac{2}{\pi}\beta\arccos(\beta)\Big) + O(e^{-4H})\,. \end{aligned}$$

The proof is given in [31].

7.10.3 The five-vertex limit

The five-vertex model can be obtained as the limit of the six-vertex model when $\Delta \to \infty$. The magnetic fields in this limit behave as follows:

- $a > b + c$. In the parameterization (7.67), after the change of variables $H = \eta/2 + l$ and $V = -\eta/2 + m$, we take the limit $\eta \to \infty$, keeping λ fixed. The weights will converge (up to a common factor) to

$$a_1 : a_2 : b_1 : b_2 : c_1 : c_2 \to e^{\lambda+l+m} : e^{\lambda-l-m} : (e^{\lambda} - e^{-\lambda})e^{l-m} : 0 : 1 : 1\,.$$

- $a + c < b$. In the parameterization (7.68), after the change of variables $H = \eta/2 + l$ and $V = \eta/2 + m$, we take the limit $\eta \to \infty$, keeping $\xi = \lambda - \eta$ fixed. The weights will converge (up to a common factor) to

$$a_1 : a_2 : b_1 : b_2 : c_1 : c_2 \to (e^{\xi} - e^{-\xi})e^{l+m} : 0 : e^{\xi+l-m} : e^{\xi-l+m} : 1 : 1\,.$$

The two limits are related by inverting the horizontal arrows. From now on, we will focus on the five-vertex model obtained by taking the limit of the six-vertex one when $a > b + c$.

The phase diagram of the five-vertex model is easier than that for the six-vertex model but still sufficiently interesting. Perhaps the most interesting feature is the existence of the tricritical point in the phase diagram.

We will use the parameter

$$\gamma = e^{-2\lambda}.$$

Note that $\gamma < 1$.

The frozen regions on the phase diagram of the five-vertex model, denoted in Fig. 7.18 by A_1, A_2, and B_1, can be described by the following inequalities:

$$\begin{aligned}
&A_1 \text{ region:} && m \geq -l, \quad l \leq 0, \\
& && e^{2m} \geq 1 - \gamma(1 - e^{-2l}), \quad l > 1. \\
&A_2 \text{ region:} && m \leq -l, \quad l \leq 0, \\
& && e^{2m} \leq 1 - \frac{1}{\gamma}(1 - e^{-2l}), \quad l > 1. \\
&B_1 \text{ region:} && \left(e^{2l} - \frac{1}{1-\gamma}\right)\left(e^{-2m} - \frac{1}{1-\gamma}\right) \geq \frac{\gamma}{(1-\gamma)^2}, \quad e^{2l} > \frac{1}{1-\gamma}.
\end{aligned} \tag{7.79}$$

As follows from results presented in [17], taking the limit of the six-vertex model to obtain the five-vertex model commutes with the thermodynamic limit and, for the free energy of the five-vertex model, we can use the formula

$$f_5(l, m) = \lim_{\eta \to +\infty} \left(f(\eta/2 + l, -\eta/2 + m) - f(\eta/2, -\eta/2) \right), \tag{7.80}$$

where $f(H, V)$ is the free energy of the six-vertex model.

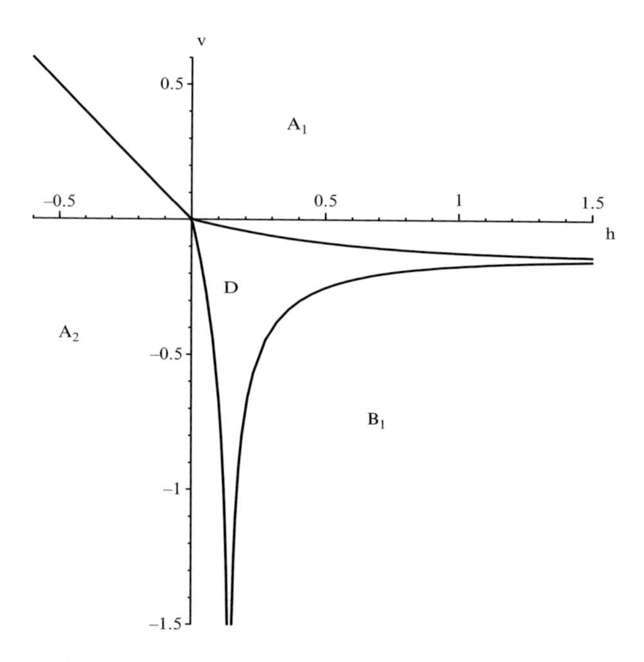

Fig. 7.18 The phase diagram of the five-vertex model with $\gamma = 1/4$ ($\beta = e^{-2h}$).

7.10.4 The asymptotics of the free energy near the tricritical point in the five-vertex model

The disordered region D near the tricritical point forms a corner

$$-\frac{1}{\gamma}\,l + O(l^2) \leq m \leq -\gamma\, l + O(l^2), \qquad h \to 0+\ .$$

The angle θ between the boundaries of the disordered region at this point is given by

$$\cos(\theta) = \frac{2\gamma}{1+\gamma^2}\,.$$

One can argue that the finiteness of the angle θ manifests the presence of interaction in the model. In comparison, translation-invariant dimer models most likely can only have cusps as such singularities.

Let $\gamma \leq k \leq 1/\gamma$ and

$$m = -kl\,.$$

As was shown in [6] for $m = -l$ and in [31] for $m = -kl$ with $\gamma \leq k \leq 1/\gamma$, the asymptotics of the free energy as $l \to +0$ are given by

$$f(l, -kl) = c_1(k,\gamma)l + c_2(k,\gamma)l^{5/3} + O(l^{7/3})\,, \tag{7.81}$$

where

$$c_1(k,\gamma) = \frac{1}{1-\gamma}\left(-(1+k)(1+\gamma) + 4\sqrt{k\gamma}\right) \tag{7.82}$$

and

$$c_2(k,\gamma) = (6\pi)^{2/3}\,\frac{2\gamma^{5/6}(1-\gamma)k^{3/2}(\sqrt{k} - 1/\sqrt{\gamma})^{4/3}}{5(\sqrt{k} - \sqrt{\gamma})^{4/3}}\,. \tag{7.83}$$

The scaling along any ray inside the corner near the tricritical point in the six-vertex model differ from this only by coefficients. The exponent $h^{5/3}$ is the same.

7.10.5 The limit $\Delta \to -1^-$

If $\Delta = -1$, the region A consists of one point located at the origin.

It is easy to find the asymptotics of the function $\Xi(\varphi)$ when $\eta \to 0+$, or $\Delta \to -1-$. In this limit $K' \to \pi/2$, $K \to \pi^2/2\eta$. Using the asymptotic approximations $1/\mathrm{dn}(u|1-m) \sim 1 + \frac{1}{2}(1-m)\sin^2(u)$ when $m \to 1-$ and $\cosh^{-1}(x) \sim \pm\sqrt{2(x-1)}$ when $x \to 1+$, we have

$$\theta_0 = \frac{|b-a|\eta}{a+b}$$

and [26]

$$\Xi(\varphi) \sim 4e^{-\pi^2/2\eta}\sin\left(\frac{\pi}{2\eta}\varphi\right).$$

The gap in the spectrum of elementary excitations vanishes in this limit at the same rate as $\Xi(\varphi)$. At lattice spacings of order $e^{\pi^2/2\eta}$, the theory has a scaling limit and becomes the relativistic $\mathrm{SU}(2)$ chiral Thirring model. The correlation functions of vertical and horizontal edges in the six-vertex model become correlation functions of the currents in the Thirring model.

7.10.6 The convexity of the free energy

The following identity holds in the region D [4, 28]:

$$f_{H,H}f_{V,V} - f_{H,V}^2 = \left(\frac{2}{\pi g}\right)^2 . \tag{7.84}$$

Here $g = 1/2D_0^2$. The constant D_0 does not vanish in the D-region, including its boundary. It is determined by the solution to the integral equation for the density $\rho(z)$.

Directly from the definition of the free energy, we have

$$f_{H,H} = \lim_{N,M\to\infty} \frac{\langle (n(L) - n(R))^2 \rangle}{NM} ,$$

where $n(L)$ and $n(R)$ are the number of arrows pointing to the left and the number of arrows pointing to the right, respectively. Therefore, the matrix $\partial_i \partial_j f$ of second derivatives with respect to H and V is positive definite.

As follows from the asymptotic behavior of the free energy near the boundary of the D-phase, despite the fact that the Hessian is nonzero and finite at the boundary of the interface, the second derivative of the free energy in the transverse direction at a generic point of the interface develops a singularity.

7.11 The Legendre transform of the free energy

The Legendre transform of the free energy

$$\sup_{H,V} \Big(xH + yV + f(H,V) \Big) ,$$

as a function of (x, y), is defined for $-1 \le x, y \le 1$.

The variables x and y are known as polarizations and are related to the slope of the Gibbs measure at $x = 2h - 1$ and $y = 2v - 1$. We will write the Legendre transform of the free energy as a function of (h, v),

$$\sigma(h,v) = \sup_{H,V} \Big((2h-1)H + (2v-1)V + f(H,V) \Big) . \tag{7.85}$$

$\sigma(h, v)$ is defined on $0 \le h, v \le 1$.

For periodic boundary conditions, the surface tension function has the following symmetries:

$$\sigma(x,y) = \sigma(y,x) = \sigma(-x,-y) = \sigma(-y,-x) .$$

The last two equalities follow from the fact that if all arrows are reversed, σ is the same, but the signs of x and y are changed. It follows that $\sigma_h(h, v) = \sigma_v(v, h)$ and $\sigma_v(h, v) = \sigma_h(v, h)$.

The function $f(H, V)$ is linear in the domains that correspond to conic and corner singularities of σ. Outside of these domains (in the disordered domain D), we have

$$\nabla\sigma \circ \nabla f = \mathrm{id}_D, \quad \nabla f \circ \nabla\sigma = \mathrm{id}_{\nabla f(D)} . \tag{7.86}$$

Here the gradient of a function is a mapping $\mathbb{R}^2 \to \mathbb{R}^2$.

When the six-vertex model is formulated in terms of a height function, the Legendre transform of the free energy can be regarded as a surface tension. The surface, in this terminology, is the graph of the height function.

Now let us describe some analytical properties of the function $\sigma(h, v)$ obtained as the Legendre transform of the free energy. The Legendre transform maps the regions where the free energy is linear with a slope $(\pm 1, \pm 1)$ to the corners of the unit square $\mathcal{D} = \{(h, v)| \quad 0 \leq h \leq 1, 0 \leq v \leq 1\}$. For example, the region A_1 is mapped to the corner $h = 1$ and $v = 1$, and the region B_1 is mapped to the corner $h = 1$ and $v = 0$. The Legendre transform maps the tentacles of the disordered region to regions adjacent to the boundary of the unit square. For example, the tentacle between the frozen regions A_1 and B_1 is mapped into a neighborhood of the $h = 1$ boundary of $\mathcal{D}$, i.e. $h \to 1$ and $0 < v < 1$.

Applying the Legendre transform to the asymptotics of the free energy in the tentacle between the frozen regions A_1 and B_1, we get

$$H(h, v) = -\frac{1}{2} \ln \left(\frac{\pi a b}{c^2} \frac{1-h}{\sin \pi(1-v)} \right), \qquad V(h, v) = \frac{1}{2} \ln \left(\frac{b}{a} \right) + \frac{\pi}{2}(1-h) \cot(\pi(1-v)),$$

and

$$\sigma(h, v) = (1-h) \ln \left(\frac{\pi a b}{c^2} \frac{1-h}{\sin(\pi(1-v))} \right) - (1-h) + v \ln \left(\frac{b}{a} \right) - \ln(b). \tag{7.87}$$

Here $h \to 1-$ and $0 < v < 1$. From eqn (7.87), we see that $\sigma(1, v) = v \ln(b/a) - \ln(b)$, i.e. σ is linear on the boundary $h = 1$ of $\mathcal{D}$. Therefore, its asymptotics near the boundary $h = 1$ are given by

$$\sigma(h, v) = v \ln \left(\frac{b}{a} \right) - \ln(b) + (1-h) \ln(1-h) + O(1-h),$$

as $h \to 1-$ and $0 < v < 1$. We note that this expansion is valid when $(1-h)/\sin(\pi(1-v)) \ll 1$.

Similarly, considering other tentacles of the region D, we conclude that the surface tension function is linear on the boundary of $\mathcal{D}$.

Next let us find the asymptotics of σ at the corners of $\mathcal{D}$ in the case where all points of the interfaces between the frozen and disordered regions are regular, i.e. when $\Delta < 1$. We use the asymptotics of the free energy near the interface between regions A_1 and D (7.76).

First let us fix the point (H_0, V_0) on the interface, and the scaling factor r in eqn (7.76). Then, from the Legendre transform, we get

$$1 - h = -\frac{3}{4} r \frac{\kappa(\theta s + t^2)^{1/2}}{(\partial_H g)^2 + 4} (\theta \partial_H g + 4rt)$$

and

$$1 - v = -\frac{3}{2} r \frac{\kappa(\theta s + t^2)^{1/2}}{(\partial_H g)^2 + 4} (-\theta + r \partial_H g t).$$

It follows that

$$\frac{1-h}{1-v} = \frac{\theta \partial_H g + 4rt}{2(-\theta + rt\partial_H g)}.$$

In the vicinity of the boundary, $r \to 0$ and, hence,

$$\frac{1-h}{1-v} = -\frac{\partial_H g}{2} = \frac{1-(b/a)\, e^{-2V_0}}{1-(b/a)\, e^{-2H_0}} \tag{7.88}$$

as $h, v \to 1$. Thus, under the Legendre transform, the slope of the line which approaches the corner $h = v = 1$ depends on the boundary point on the interface between the frozen and disordered regions.

It follows that the first terms of the asymptotics of σ at the corner $h = v = 1$ are given by

$$\sigma(h, v) = -\ln a - 2(1-h)H_0(h, v) - 2(1-v)V_0(h, v),$$

where $H_0(h, v)$ and $V_0(h, v)$ can be found from eqn (7.88) and $g(H_0, V_0) = 0$.

When $|\Delta| < 1$, the function σ is strictly convex and smooth for all $0 < h, v < 1$. It develops conical singularities near the boundary.

When $\Delta < -1$, in addition to the singularities on the boundary, σ has a conical singularity at the point $(1/2, 1/2)$. This corresponds to the "central flat part" of the free energy f (see Fig. 7.15).

When $\Delta > 1$, the function σ has corner singularities along the boundary as in the other cases. In addition to this, it has a corner singularity along the diagonal $v = h$ if $a > b$ and $v = 1 - h$ if $a < b$. We refer the reader to [6] for further details of the singularities of σ in the case when $\Delta > 1$.

When $\Delta = -1$, the function σ has a corner singularity at $(1/2, 1/2)$.

7.12 The limit shape phenomenon

7.12.1 The height function for the six-vertex model

Consider a square grid $L_\epsilon \subset \mathbb{R}^2$ with a step ϵ. Let $D \subset \mathbb{R}^2$ be a domain in $\mathbb{R}^2$. We denote by D_ϵ a domain in the square lattice which corresponds to the intersection $D \cap L_\epsilon$, assuming that the intersection is generic, i.e. the boundary of D does not intersect vertices of L_ϵ.

Faces of D_ϵ which do not intersect the boundary ∂D of D are called *inner faces*. Faces of D_ϵ which intersect the boundary of D are called *boundary faces*.

A height function h is an integer-valued function on the faces of D_ϵ of the grid L_N (including the outer faces), which has the following properties:

- it is nondecreasing when going up or to the right;
- if f_1 and f_2 are neighboring faces, then $h(f_1) - h(f_2) = -1, 0, 1$.

The *boundary value* of the height function is its restriction to the outer faces. Given a function $h^{(0)}$ on the set of boundary faces, we denote by $\mathcal{H}(h^{(0)})$ the space of all height functions with the boundary value $h^{(0)}$. We choose a marked face f_0 at the boundary. A height function is normalized at this face if $h(f_0) = 0$.

Proposition 7.9 *There is a bijection between the states of the six-vertex model with fixed boundary conditions and height functions with corresponding boundary values normalized at* f_0.

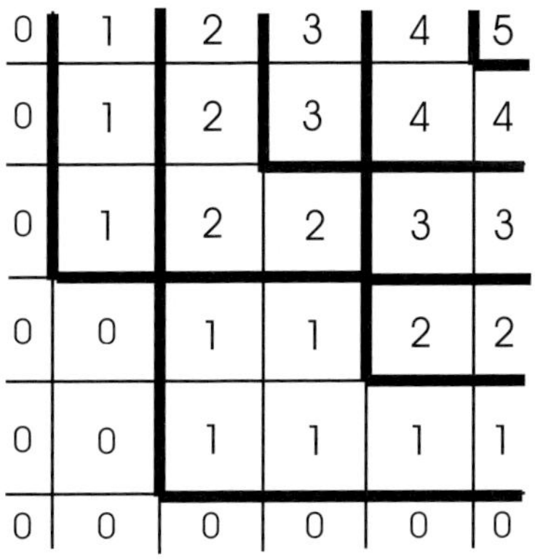

0	1	2	3	4	5
0	1	2	3	4	4
0	1	2	2	3	3
0	0	1	1	2	2
0	0	1	1	1	1
0	0	0	0	0	0

Fig. 7.19 The values of the height function for the configuration of paths given in Fig. 7.1.

Proof Given a height function, consider its "level curves," i.e. paths on D_ϵ where the height function changes its value by 1 (see Fig. 7.19). Clearly, this defines a state for the six-vertex model on D_ϵ with the boundary conditions determined by the boundary values of the height function.

On the other hand, given a state in the six-vertex model, consider the corresponding configuration of paths. There is a unique height function whose level curves are these paths and which satisfies the condition $h = 0$ at f_0.

It is clear that this correspondence is a bijection. □

There is a natural partial order on the set of height functions with given boundary values. One function is greater than another, $h_1 \leq h_2$, if it is entirely above the other, i.e. if $h_1(x) \leq h_2(x)$ for all x in the domain. There exist minimum and the maximum height functions $h_{\min}$ and $h_{\max}$, respectively, such that $h_{\min} \leq h \leq h_{\max}$ for all height functions h. Maximum and minimum height functions are shown in Fig. 7.20 for DW boundary conditions.

The characteristic function of an edge is related to the height function by

$$\sigma_e = h(f_e^+) - h(f_e^-), \tag{7.89}$$

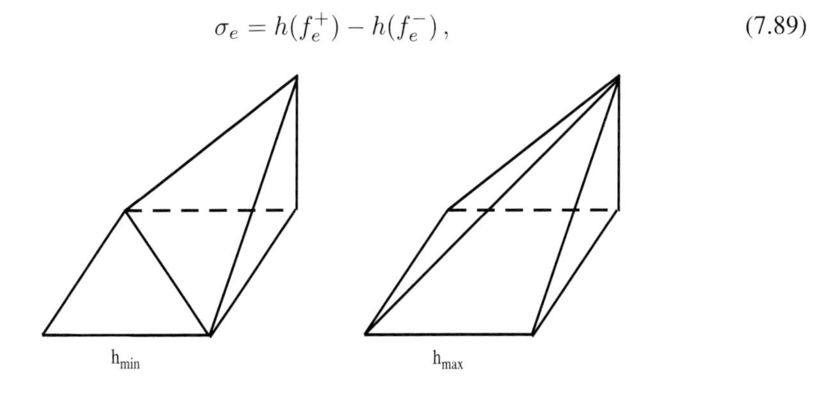

Fig. 7.20 Minimum and maximum height functions for DW boundary conditions.

where faces $f_e^{\pm}$ are adjacent to e, f_e^{+} is to the right of e for a vertical edge and above e for a horizontal edge. The face f_e^{-} is to the left for a vertical edge and below e for a horizontal edge.

Thus, we can consider the six-vertex model on a domain as a theory of fluctuating discrete surfaces constrained between $h_{\min}$ and $h_{\max}$. Each surface occurs with a probability given by the Boltzmann weights of the six-vertex model.

The height function does not exist when the region is not simply connected or when we have a square lattice on a surface with a nontrivial fundamental group. One can draw an analogy between states of the six-vertex model and 1-forms. States on a domain with a trivial fundamental group can be regarded as exact forms $\omega = dh$, where h is a height function.

7.12.2 Stabilizing fixed boundary conditions

Recall that the height function is a monotonic integer-valued function on the faces of the grid, which satisfies the Lipschitz condition (it changes at most by 1 on any two adjacent faces).

The normalized height function is a piecewise constant function on D_ϵ with the value

$$h^{\text{norm}}(x,y) = \epsilon\, h_\epsilon(n,m)\,,$$

where $h_\epsilon(n,m)$ is a height function on D_ϵ.

Normalized height functions on the boundary of D satisfy the inequality

$$|h(x,y) - h(x',y')| \leq |x - x'| + |y - y'|\,.$$

As for non-normalized height functions, there is a natural partial order on the set of all normalized height functions with given boundary values: $h_1 \geq h_2$ if $h_1(x) \geq h_2(x)$ for all $x \in D$. We define the operations

$$h_1 \vee h_2 = \min_{x\in D}(h_1(x), h_2(x)), \qquad h_1 \wedge h_2 = \max_{x\in D}(h_1(x), h_2(x))\,.$$

It is clear that

$$h_1 \vee h_2 \leq h_1, h_2 \leq h_1 \wedge h_2\,.$$

It is also clear that in this partial ordering there are unique minimum and maximum height functions, which we denote by $h^{\min}$ and $h^{\max}$, respectively.

The boundary value of the height function defines a piecewise constant function on the boundary faces of D_ϵ. Consider a sequence of domains D_ϵ with $\epsilon \to 0$. *Stabilizing fixed boundary conditions* for height functions are a sequence $\{h^{(\epsilon)}\}$ of functions on the boundary faces of D_ϵ such that $h^{(\epsilon)} \to h^{(0)}$, where $h^{(0)}$ is a continuous function of ∂D.

It is clear that in the limit $\epsilon \to 0$ we have at least two characteristic scales. On the *macroscopic* scale, we can "see" the region $D \subset \mathbb{R}^2$. The normalized height functions in the limit $\epsilon \to 0$ will be functions on D. As we will see in the next section, the height function in the six-vertex model develops a deterministic limit shape $h_0(x,y)$ on this scale. This is reflected in the structure of local correlation functions. Let e_i be edges with coordinates $(n_i, m_i) = (x_i/\epsilon, y_i/\epsilon)$. When $\epsilon \to 0$ and (x_i, y_i) are fixed,

$$\langle \sigma_{e_1} \ldots \sigma_{e_k} \rangle = \partial_{i_1} h_0(x_1, y_1) \ldots \partial_{i_k} h_0(x_k, y_k) + \ldots\,.$$

The randomness remains on smaller scales and, in particular, on the microscopic scale of the lattice. If the coordinates of the edges e_i are $(n_i, m_i) = (x/\epsilon + \Delta n_i, y/\epsilon + \Delta m_i)$, we expect that in the limit $\epsilon \to 0$, the correlation functions will have the limit

$$\langle \sigma_{e_1} \dots \sigma_{e_k} \rangle \to \langle \sigma_{e_1} \dots \sigma_{e_k} \rangle_{(\partial_x h_0(x,y), \partial_y h_0(x,y))} ,$$

where the correlation function on the right is taken with respect to a translation-invariant Gibbs measure with an average polarization $(\partial_x h_0(x,y), \partial_y h_0(x,y))$. The correlator on the right-hand side depends on the polarization and on $\Delta n_i - \Delta n_j, \Delta m_i - \Delta m_j$, i.e. it is translation invariant.

7.12.3 The variational principle

Here we will outline the derivation of the variational problem which determines the limit shape of the height function.

First, consider a sequence of rectangular domains of size M, M where $N, M \to \infty$ such that $a = N/M$ is finite and the boundary values of the height function stabilize to the function $\phi(x,y) = hy + vx, x \in [0,a], y \in [0,1]$. The partition function of such a system has the asymptotic behavior

$$Z_{N,M} \propto \exp(NM\sigma(h,v)) , \tag{7.90}$$

where $\sigma(h,v)$ is the Legendre transform of the free energy for the torus.

For a domain D_ϵ, we choose a subdivision of it into a collection of small rectangles. Taking into account eqn (7.90), the partition function of the six-vertex model with zero electric field and stabilizing boundary conditions can be written as

$$Z_{D_\epsilon} \simeq \sum_h e^{|D_\epsilon| \sum_i \sigma(\Delta h_{x_i}, \Delta h_{y_i}) \Delta_{x_i} \Delta_{y_i}} .$$

When $\epsilon \to 0$, the size of the region $|D_\epsilon|$ is increasing and the leading contribution to the sum comes from the height function which minimizes the functional

$$I(h) = \int_D \sigma(\nabla h)\, d^2x .$$

One can introduce an extra weight $q^{\mathrm{vol(h)}}$ into the partition function of the six-vertex model. This corresponds to an inhomogeneous electric field [31]. If $\epsilon \to 0$ and $q = \exp(-\lambda\epsilon)$, the leading contribution to the partition function comes from the height function which minimizes the functional

$$I_\lambda(h) = \int_D \sigma(\nabla h(x))\, d^2x + \lambda \int_D h(x)\, d^2x .$$

The variational principle. Thus, in order to find the limit shape in the thermodynamic limit of the six-vertex model on a disk with Dirichlet boundary conditions φ_0, we must minimize the functional

$$I_\lambda[\varphi] = \int_D \sigma(\nabla\varphi)\, d^2x + \lambda \int_D \varphi\, d^2x , \tag{7.91}$$

on the space $L(D, \varphi_0)$ of functions satisfying the Lipschitz condition

$$|\varphi(x,y) - \varphi(x',y')| \leq |x - x'| + |y - y'|$$

and satisfying boundary conditions that are monotonically increasing in the x and y directions

$$\varphi|_{\partial D} = \varphi_0 \, .$$

Since σ is convex, the minimizer is unique when it exists. Thus, we should expect that the variational problem (7.91) has a unique solution.

The large-deviation principle, when applied to this situation, should result in the convergence in probability, as $\epsilon \to 0$, of random normalized height functions $h(x,y)$ to the minimizer of eqn (7.91). See [9, 27] for the analysis of this phenomenon in dimer models.

If the vector $\nabla h(x,y)$ is not a singular point of σ, the minimizer h satisfies the Euler–Lagrange equation in a neighborhood of (x,y),

$$\operatorname{div}(\nabla\sigma \circ \nabla h) = \lambda \, . \tag{7.92}$$

We can also rewrite this equation in the form

$$\nabla\sigma(\nabla h(x,y)) = \frac{\lambda}{2}(x,y) + (-g_y(x,y), g_x(x,y)) \, , \tag{7.93}$$

where g is an unknown function such that $g_{xy}(x,y) = g_{yx}(x,y)$. This function is determined by the boundary conditions for h.

Applying eqn (7.86), it follows that

$$\nabla h(x,y) = \nabla f\left(\frac{\lambda}{2}x - g_y(x,y), \frac{\lambda}{2}y + g_x(x,y)\right) . \tag{7.94}$$

From the definition of the slope (see eqn (7.55)), it follows that $|f_H| \leq 1$ and $|f_V| \leq 1$. Thus, if the minimizer h is differentiable at (x,y), it satisfies the constrains $|h_x| \leq 1$ and $|h_y| \leq 1$. It is given that $h_x, h_y \geq 0$, and hence $0 \leq h_x \leq 1$ and $0 \leq h_y \leq 1$.

In particular, one can choose $g(x,y) = 0$. In this case the function (7.94) is the minimizer of the rate functional $I_\lambda[h]$ with very special boundary conditions. For an infinite region D (and finite λ), this height function reproduces the free energy as a function of the electric fields.

The limit shape of the height function is a real analytic function almost everywhere. One of the corollaries of eqn (7.94) is that curves on which this function is not analytic (interfaces of the limit shape) are images of boundaries between different phases in the phase diagram of $f(H,V)$. The ferroelectric and antiferroelectric regions, where f is linear, correspond to the regions where h is linear.

From eqn (7.94) and the asymptotics of the free energy near the boundaries between different phases, we can draw some general conclusions about the structure of the limit shapes. The first conclusion is that near regular pieces of the boundary between regions where the height function is real and analytic, the height function behaves as $h(x,y) - h(x_0,y_0) \simeq d^{3/2}$, where d is the distance from (x,y) to the closest point (x_0,y_0) on the boundary. The second obvious conclusion is that if (x,y) is inside the corner singularity in the boundary between the smooth parts, the height function behaves near the corner as $h(x,y) - h(x_0,y_0) \simeq d^{5/3}$, where d is the distance from (x,y) to the corner and (x_0,y_0) are the coordinates of the corner. Limit shapes for the six-vertex model with $\mathcal{DW}$ boundary conditions were studied numerically in [1]. For recent analytical results see [10].

7.12.4 Limit shapes for inhomogeneous models

So far, the analysis of limit shapes has been done in the homogeneous six-vertex model. The variational problem that determines the limit shapes involves the free energy as a function of the polarization. The extension of the analysis outlined above to an inhomogeneous periodically weighted case is straightforward. The variational principle is the same, but the function σ is determined from the Bethe ansatz for the inhomogeneous model. It is again the Legendre transform of the free energy as a function of the electric fields.

The computation of the free energy from the large-N asymptotics of the Bethe ansatz equations is similar and is based on a similar conjecture about the accumulation of solutions to the Bethe equations on a curve. This is illustrated in Section 7.8 for the free-fermionic case. If the size of the fundamental domain is $k \times m$, one should expect $2(k+m)$ cusps in the boundary between the ordered and disordered regions.

7.12.5 Higher-spin six-vertex model

The weights in the six-vertex model can be identified with matrix elements of the R-matrix of $U_q(\widehat{sl}_2)$ in the tensor product of two two-dimensional representations of this algebra. We have denoted such a matrix by $R^{(1,1)}(z)$.

In a similar fashion, one can consider matrix elements of $R^{(l_1,l_2)}(z)$ as weights of a higher-spin generalization of the six-vertex model. We will call this the higher-spin six-vertex model.

A state in such a model is an assignment of a weight of the irreducible representation $V^{(l_1)}$ to every horizontal edge, and of a weight of the irreducible representation of $V^{(l_2)}$ to every vertical edge. The natural local observables in such a model are sums of products of spin functions of an edge:

$$s_e(S) = \text{ the weight } s \text{ assigned to } e \text{ in the state } S\,.$$

The spin variables define the height function locally as

$$s_e = h(f_e^+) - h(f_e^-)\,,$$

where $f_e^{\pm}$ are faces adjacent to e. When the domain has a trivial fundamental group, the local height function extends to a global one. Otherwise, we must divide the region into simply connected pieces. The height function has the property

$$|h(n,m) - h(n',m')| \leq l_1|n-n'| + l_2|m-m'|\,.$$

Given a sequence of domains D_ϵ with $\epsilon \to 0$, we expect that the normalized height function $\epsilon h(\epsilon n, \epsilon m)$ will develop the limit shape $h_0(x,y)$ which minimizes the functional

$$I^{(l_1,l_2)}[h] = \int_D \sigma^{(l_1,l_2)}(\nabla h)\, d^2x$$

subject to the Dirichlet boundary conditions and the constraint

$$|h(x,y) - h(x',y')| \leq l_1|x-x'| + l_2|y-y'|\,.$$

The properties of this model and of its free energy as functions of the electric fields is an interesting problem which needs further research.

7.13 Semiclassical limits

7.13.1 Semiclassical limits in terms of Bethe states

The spectrum of the quantum spin chains described in Section 7.4 in terms of the Bethe equations has a natural limit when $m_i = R_i/h$, $q = e^h$, and $h \to 0$ with fixed R_i. In this limit, the quantum spin chain with the quantum monodromy matrix (7.35) becomes a classical spin chain with the monodromy matrix (7.11).

The semiclassical eigenvectors of quantum transfer matrices correspond to solutions to the Bethe equations which accumulate along contours representing branch cuts on the spectral curve of $T(z)$ (see [33] and, for more recent results, [38]).

7.13.2 Semiclassical limits in the higher-spin six-vertex model

Relatively little known about the semiclassical limit of the higher-spin six-vertex model. This is the limit when $l_1 = R_i/h$ and $h \to 0$. Here, depending on the values of Δ, $h = \eta$ or $h = \gamma$.

The first problem is to find the asymptotics of the conjugation action of the R-matrix in this limit. The mapping

$$x \to R^{(l_1,l_2)}(u) x R^{(l_1,l_2)}(u)^{-1}$$

is an automorphism of $End(V^{(l_1)} \otimes V^{(l_2)})$, which, as $h \to 0$, becomes the Poisson automorphism $\rho^{(R_1,R_2)}(u)$ of the Poisson algebra of functions on $S^{(R_1)} \otimes S^{(R_2)}$.

This automorphism was computed in [37] for the case $\Delta = 1$. It is easy to extend this result to $\Delta \neq 1$. For constant R-matrices, see [32].

The next problem is to find the semiclassical asymptotics for the R-matrix considered as an "evolution operator." For this purpose, we choose two Lagrangian submanifolds in $S^{(R_1)} \otimes S^{(R_2)}$, one corresponding to the initial data and the other corresponding to the target data. Points in these manifolds parameterize corresponding semiclassical states. Then the matrix elements of the R-matrix should have the asymptotic behavior

$$R^{(l_1,l_2)}(u)(\sigma_1, \sigma_2) = \text{const}\ \exp \frac{S^{(R_1,R_2)}(\sigma_1, \sigma_2)}{h} \sqrt{H(\sigma_1, \sigma_2)}(1 + O(h)),$$

where S is the generating function for the mapping ρ, and H is the Hessian of S. This asymptotic behavior of the R-matrix defines the semiclassical limit of the partition function and needs further investigation.

7.13.3 The large-N limit in a spin-$1/2$ spin chain as a semiclassical limit

The local spin operators in a spin-$1/2$ spin chain of length N are

$$S_n^a = 1 \otimes \dots \sigma_n^a \otimes \dots \otimes 1, \ \ n = 1, \dots, N.$$

They commute as

$$[S_n^a, S_m^b] = i \sum_c \epsilon_{abc} \delta_{nm} S_n^c.$$

As $N \to \infty$, the operators S_n^a converge to local continuous classical spin variables

$$S_n^a \to S^a\left(\frac{n}{N}\right),$$

where the $S^a(x)$ are local functionals on the classical phase space of the continuum spin system. The commutation relations become the following relations between Poisson brackets:

$$\{S^a(x), S^b(y)\} = \delta(x-y)\sum_c f^{ab}_{c} S^c(x)\,,$$

where the $S^a(x)$ are continuous local classical spin variables. This combination of classical and continuous limits looks more convincing in terms of the observables

$$S^a_N[f] = \frac{1}{N}\sum_{n=1}^{N} f\left(\frac{n}{N}\right) S^a_n\,.$$

As $N \to \infty$, such operators become $S^a[f] = \int_0^1 f(x) S^a(x)\, dx$, and the commutation relation

$$[S^a_N[f], S^b_N[g]] = \frac{i}{N}\sum_c \epsilon_{abc} S^c_N[fg]$$

becomes

$$\{S^a[f], S^b[g]\} = \sum_c \epsilon_{abc} S^c[fg]\,.$$

Here we have used the asymptotic approximation of commutators in the semiclassical limit $[a,b] = ih\{a,b\} + \dots$ and the fact that $1/N$ plays the role of the Planck constant.

Consider the limit $N \to \infty$ of the row-to-row transfer matrix in the six-vertex model. Assume that, at the same time, $\Delta \to 1$ such that $\Delta = 1 + \kappa^2/2N^2 + \dots$. To be specific, consider the case when $\Delta > 1$, so that $\kappa = N\eta$ is finite and real.

We can write the transfer matrix of the six-vertex model as

$$t(u) = (\sinh(u))^N \operatorname{tr}(T_N(u))\,,$$

where $T_N(u)$ is the solution to the difference equation

$$T_{n+1}(u) = \frac{R_n(u)}{\sinh(u)} T_n(u)$$

with the initial condition $T_0(u) = 1$. Here $R(u)$ is the R-matrix of the six-vertex model. For small η, we have

$$\frac{R_n(u)}{\sinh(u)} = 1 + \eta\left(\frac{1}{2}\coth(u)\sigma^3\sigma^3_n + \frac{\sigma^+\sigma^-_n + \sigma^-\sigma^+_n}{\sinh(u)}\right) + O(\eta^2)\,.$$

Taking this into account, we conclude that as $N \to \infty$, $T_n(u) \to T(u|x)$, where $T(u|x)$ is the solution to

$$\frac{\partial T(u|x)}{\partial x} = \left(\frac{1}{2}\coth(u)\sigma^3 S^3(x) + \frac{\sigma^+ S^-(x) + \sigma^- S^+(x)}{\sinh(u)}\right) T(u|x)$$

with the initial condition $T(u|0) = 1$. Here σ^z is the same as in eqn (7.32), $\sigma^\pm = \frac{1}{2}(\sigma^x \pm i\sigma^y)$, and $S^\pm = \frac{1}{2}(S^1 \pm iS^2)$.

The row-to-row transfer matrix of the six-vertex model has the following asymptotic behavior in this limit:

$$t(u) \to (\sinh(u))^N \tau(u)\,, \tag{7.95}$$

where $\tau(u) = \mathrm{tr}(T(u|1)$.

Now let us study the evolution of local spin operators in this limit.

Consider the partition function of the six-vertex model on a cylinder as a linear operator

$$Z^{(C)}_{M,N} = t(u)^M$$

acting in $\mathbb{C}^{2\otimes N}$. Consider the row-to-row transfer matrix as the evolution operator for one step and $Z^{(C)}_{M,N}$ as the evolution operator for M steps. In the Heisenberg picture, local spin operators evolve as

$$S^a_{n,m+1} = t(u) S^a_{n,m} t(u)^{-1}\,. \tag{7.96}$$

In the continuum semiclassical limit $N \to \infty$, the transfer matrix becomes a functional (7.95) on the phase space of continuous classical spins. Equation (7.96) can be written as

$$S^a_{n,m+1} - S^a_{n,m} = [t(u), S^a_{n,m}] t(u)^{-1}\,.$$

When $N \to \infty$ and $x = n/N$ and $t = m/N$ are fixed, this equation becomes the evolution equation for continuum spins:

$$\frac{\partial S^a(x,t)}{\partial t} = i\{H(u), S^a(x,t)\}\,,$$

where $H(u) = \log t(u)$.

The evolution with respect to the partition function on a cylinder of height M,

$$S^a_{n,m+M} = Z^{(C)}_{M,N} S^a_{n,m} {Z^{(C)}_{M,N}}^{-1}\,,$$

becomes an evolution in time $T = M/N$: $S^a(x,t) \mapsto S^a(x, t+T)$.

If we choose a Lagrangian submanifold in the phase space of a continuous spin system corresponding to the initial and target data (say, a version of the initial and target q-coordinates), the asymptotics of the partition function should be of the form

$$Z^{(C)}_{N,M}(\sigma_1,\sigma_2) = \mathrm{const}\, e^{-NS_T(\sigma_1,\sigma_2)} \sqrt{\mathrm{Hess}(\sigma_1,\sigma_2)}(1 + O(1/N))\,,$$

where S_T is the Hamilton–Jacobi action for the Hamiltonian $H(u)$, $\mathrm{Hess}(\sigma,\tau)$ is the Hessian of S_T, and on the left side we have the matrix element of $Z^{(C)}_{N,M}$ between semiclassical states corresponding to σ_1 and σ_2.

The Bethe equations and the spectrum of the Heisenberg Hamiltonian in this limit were studied in [11, 14].

7.14 The free-fermionic point and dimer models

We decorate a square grid by inserting a box with two faces into each vertex, as shown in Fig. 7.21. Recall that a dimer configuration on a graph is a perfect matching on a set of vertices

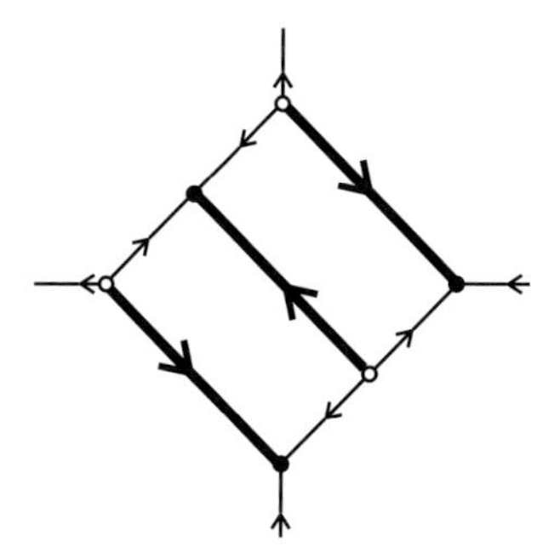

Fig. 7.21 Reference dimer configuration on $G^{(0)}_{nm}$.

connected by edges. In other words, it is a collection of "occupied edges" (occupied by dimers) such that two occupied edges never meet, and any vertex is an endpoint of an occupied edge.

Dimer configurations on the decorated square grid project to six-vertex configurations on a square grid, as shown in Fig. 7.22.

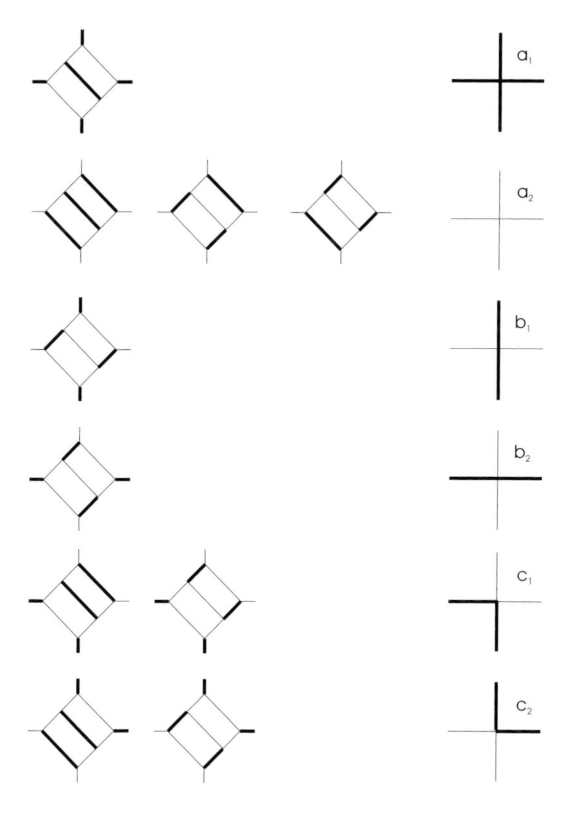

Fig. 7.22 Projection of dimer configurations on $G^{(0)}_{nm}$ onto the six different types of vertices at the (n, m) vertex.

It is easy to check that any edge weight system on the decorated lattice projects to six-vertex weights at the free-fermionic point, when $a^2 + b^2 - c^2 = 0$ at every vertex. Recall that the edge weights are a mapping w : Edges $\to \mathbb{R}_{\geq 0}$. The weight of a dimer configuration is

$$W(D) = \prod_{e \in D} w(e)\,.$$

The statement above means

$$\sum_{D \in \pi^{-1}(S)} W(D) = \prod_{v} w_v(S)\,,$$

where S is a six-vertex configuration on the square grid, π is the projection from the dimer configurations on the decorated square grid to the six-vertex configurations, and the six-vertex weights w_v are given by an explicit formula. The weights w_v satisfy the free-fermionic condition.

A pair of dimer configurations D, D_0 on bipartite graphs defines the height function. This height function agrees with the height function of the six-vertex model:

$$h_S(f) = h_{D,D_0}(f)\,,$$

where S is a six-vertex state on the square lattice, h_S is the corresponding height function, $D \in \pi^{-1}(S)$, D_0 is shown in Fig. 7.21, and f is a face of the decorated lattice which projects to a face of the square grid.

7.A Appendix

7.A.1 Symplectic and Poisson manifolds

Recall that an even-dimensional manifold equipped with a closed nondegenerate 2-form is called *symplectic.*

Let $(\mathcal{M}, \omega)$ be a $2n$-dimensional symplectic manifold. In local coordinates,

$$\omega = \sum_{ij=1}^{2n} \omega_{ij}\, dx^i \wedge dx^j, \;\; \det(\omega) \neq 0, \;\; d\omega = \sum_{k=1}^{2n} \sum_{ij=1}^{2n} \frac{\partial \omega_{ij}}{\partial x^k} dx^i \wedge dx^j = 0\,.$$

The latter identity is equivalent to the Jacobi identity for the bracket

$$\{f, g\} = \sum_{ij=1}^{2n} (\omega^{-1})^{ij} \frac{\partial f}{x^i} \frac{\partial g}{x^j}\,.$$

A smooth manifold M with a bivector field p (a section of the bundle $\wedge^2 TM$) such that the bracket between two smooth functions

$$\{f, g\} = p(df \wedge dg)$$

satisfies the Jacobi identity is called a Poisson manifold. In local coordinates, $x^1, \ldots, x^n$, $p(x) = \sum i, j = 1^n p^{ij}(x)(\partial/\partial x^i) \wedge (\partial/\partial x^i)$.

A Poisson tensor on a smooth manifold M defines a subspace mapping $p : T^*M \to TM$. The image is a system of subspaces $p(T^*M) \subset TM$ which is a distribution on M. Leaves of this distribution are spanned by curves which are flow lines of piecewise Hamiltonian vector fields. They are smooth submanifolds. Symplectic leaves of the Poisson manifold M are leaves of this distribution.

Well-known examples of symplectic leaves are co-adjoint orbits in the dual space to a Lie algebra.

7.A.2 Classical integrable systems and their quantization

Integrable systems in Hamiltonian mechanics. The notion of integrability is most natural in the Hamiltonian formulation of classical mechanics. For details, see [2, 35].

In the Hamiltonian formalism of classical mechanics, the dynamics takes place in the phase space and the equations of motion are of first order. When the system is a system of particles moving on a manifold M, the local coordinates are the positions and momenta of the particles. Globally, the phase space in this case is the cotangent bundle to the manifold M (see for example [2]).

The Hamiltonian formulation for spinning tops or other systems with more complicated constraints involves more complicated phase spaces. In all cases a phase space has the structure of a symplectic manifold; that is, it comes together with a nondegenerate closed 2-form on it.

Any symplectic manifold (and so any phase space of a Hamiltonian system) admits local coordinates (Darboux coordinates) in which the 2-form has the form

$$\omega = \sum_{i=1}^{n} dp_i \wedge dq^i \,.$$

These coordinates can be interpreted as momenta and positions, though in the case of spinning tops this interpretation does not have a lot of physical meaning, but Darboux coordinates are a convenient mathematical tool.

The dynamics in a Hamiltonian system is determined by the energy function H. The trajectories of such a system in local Darboux coordinates are solutions to the differential equations

$$\frac{dq^i}{dt} = \frac{\partial H}{\partial p_i}, \quad \frac{dp_i}{dt} = -\frac{\partial H}{\partial q^i} \,.$$

Geometrically, the trajectories are flow lines of the Hamiltonian vector field $v_H = \omega^{-1}(dH) \in \Gamma(\wedge^2 TM)$, where $\omega^{-1} : TM \to T^*M$ is the bundle isomorphism induced by the symplectic form ω.

Let $\mathcal{M}$ be a $2N$-dimensional symplectic manifold.

Definition 7.10 *An integrable system on $\mathcal{M}$ is a collection of N independent functions on $\mathcal{M}$ which commute with respect to the Poisson bracket.*

Recall that

- The "level surfaces"

$$\mathcal{M}(c_1, \ldots, c_n) = \{x \in \mathcal{M}, I_i(x) = c_i\} \tag{7.A.1}$$

are invariant with respect to the flow of any Hamiltonian $H = F(I_1, \ldots, I_N)$.

- For every such Hamiltonian and every level surface $\mathcal{M}(c_1, \dots, c_n)$, there exists an affine coordinate system $(p_1, \dots, p_n)$ in which the Hamiltonian flow generated by H is linear: $p_i = \text{constant}$.
- The coordinate system $(p_1, \dots, p_n)$ on $\mathcal{M}(c_1, \dots, c_n)$ can be completed to a canonical coordinate system (p_i, q_j) in every sufficiently small neighborhood of $\mathcal{M}(c_1, \dots, c_n)$. These coordinates are called action-angle variables, and in some cases these coordinates are global.

7.A.3 Poisson Lie groups

A Lie group G is called a Poisson Lie group if:

- G has a Poisson structure, i.e. it is given together with the Poisson tensor $p \in \Gamma(\wedge^2 TG)$, in local coordinates $p(x) = p^{ij}(x)(\partial/\partial x^i) \wedge (\partial/\partial x^j) \in \wedge^2 T_x G$. This tensor is the following Poisson bracket on smooth functions on G (the Lie bracket satisfying the Leibnitz rule with respect to pointwise multiplication):

$$\{f, g\}(x) = \sum_{ij} p^{ij}(x) \frac{\partial f}{\partial x^i} \frac{\partial g}{\partial x^j}(x) .$$

- The Poisson structure is compatible with the group multiplication. This means that the multiplication mapping $G \times G \to G$ brings the Poisson tensor on $G \times G$ to the Poisson tensor on G, or

$$\sum_{ij} p^{ij}(xy) \frac{\partial f}{\partial z^i} \frac{\partial g}{\partial z^j}(z)|_{z=xy} = \sum_{ij} p^{ij}(x) \frac{\partial f}{\partial x^i} \frac{\partial g}{\partial x^j}(xy) + \sum_{ij} p^{ij}(y) \frac{\partial f}{\partial y^i} \frac{\partial g}{\partial y^j}(xy)$$

  for any pair of functions f, g.

For more details of Poisson Lie groups and for numerous examples, see [8, 13, 23].

Factorizable Poisson Lie groups and classical r-matrices. The class of Poisson Lie groups relevant to integrable systems has so-called r-matrix Poisson brackets. Let $\mathfrak{g}$ be a Lie algebra corresponding to the Lie group G. A classical r-matrix for G is an element $r \in \mathfrak{g} \otimes \mathfrak{g}$ satisfying the bilinear identity

$$[r_{12}, r_{13}] + [r_{12}, r_{23}] + [r_{13}, r_{23}] = 0 . \tag{7.A.2}$$

The Poisson tensor

$$p(x) = r - Ad_x(r) \in \mathfrak{g} \wedge \mathfrak{g}$$

defines a Poisson Lie structure on G if r satisfies eqn (7.A.2) and $r + \sigma(r)$ is an invariant tensor.

Two remarks: we have identified $T_x G$ with $T_e G = \mathfrak{g}$ using left translations on G; and Ad_x is the diagonal adjoint action of G on $\mathfrak{g} \wedge \mathfrak{g}$, $u \wedge v \to xux^{-1} \otimes xvx^{-1}$, when G is a matrix Lie algebra.

Let $\pi^V : G \to GL(V)$ and $\pi^W : G \to GL(W)$ be two (finite-dimensional) representations of G and let π^V_{ij}, π^W_{ab} be matrix elements of G in a linear basis: $\pi^V_{ij} : g \in G \to \pi^V_{ij}(g) \in \mathbb{C}$. The r-matrix Poisson brackets between two such functions are given by

$$\{\pi^V_1, \pi^W_2\} = [(\pi^V \otimes \pi^W)(r), \pi^V_1, \pi^W_2] .$$

Taking the trace in this formula, we see that the characters of the finite-dimensional representations form a Poisson commutative algebra on G. If we restrict this subalgebra to a symplectic leaf of G, we will obtain an integrable system if the number of independent functions among χ_λ after this restriction is equal to half of the dimension of the symplectic leaf. The principal advantage of this approach is that it gives an algebraic way to construct classical r-matrices through the double construction of Lie bialgebras.

In the next section, we will see how r-matrices with spectral parameters appear naturally from Lie bialgebras on loop algebras.

Basic example $G = LSL_2$. Our basic example is an infinite-dimensional Poisson Lie group LGL_2 of mappings $S^1 = \{z \in \mathbb{C} | |z| = 1\} \to GL_2$ which are holomorphic inside the unit disk.

The Lie algebra Lgl_2 (we consider maps which are Laurent polynomials in t) of the Lie group LGL_2 has a linear basis $e_{ij}[n]$ with $e_{ij}[n](z) = e_{ij}t^{-n-1}$. It also has an invariant scalar product $(x[n], y[m]) = \delta_{n,m}$. The following element of the completed tensor product $Lgl_2 \otimes Lgl_2$,

$$r = \sum_{i \geq j; i,i=1,2} e_{ij}[0] \otimes e_{ji}[0] + \sum_{i,j=1,2; n \in \mathbb{Z}, n \geq 1} e_{ij}[n] \otimes e_{ji}[-n], \tag{7.A.3}$$

satisfies the classical Yang–Baxter equation (7.A.2). We will not explain why here, but this follows from the Drinfeld double construction for Lie bialgebras.

Let $\pi^V : GL_2 \to GL(V)$ be a representation of GL_2 and let $a \in \mathbb{C}^\vee$ be a nonzero complex number. The evaluation representation $\pi_{V,a}$ of Lgl_2 acts in V as

$$\pi_{V,a^2}(x[n]) = \pi^V(x)a^{-2n-2},$$

where a is a nonzero complex number.

Evaluating the element (7.A.3) in $\pi_{V,a} \otimes \pi_{V,b}$, where V is the two-dimensional representation, we get

$$\pi_{V,a} \otimes \pi_{V,b}(r) = r(a/b) + f(a/b)1,$$

where $f(z)$ is a scalar function and

$$r(z) = \frac{z + z^{-1}}{z - z^{-1}} \sigma^z \otimes \sigma^z + \frac{2}{z - z^{-1}} (\sigma^+ \otimes \sigma^- + \sigma^- \otimes \sigma^+). \tag{7.A.4}$$

Here $\sigma^z = e_{11} - e_{22}$, $\sigma^+ = e_{12}$, and $\sigma^- = e_{21}$ are Pauli matrices.

The Yang–Baxter equation for r implies

$$[r_{12}(z), r_{13}(zw)] + [r_{12}(z), r_{23}(w)] + [r_{13}(zw), r_{23}(w)] = 0.$$

The Poisson brackets between the coordinate functions $g_{ij}(z)$ are

$$\{g_1(z), g_2(w)\} = [r(z/w), g_1(z)g_2(w)]. \tag{7.A.5}$$

One of the important properties of such Poisson brackets is that characters of finite-dimensional representations define families of Poisson-commuting functions on LG_2:

$$t_V(z) = \mathrm{tr}_V(\pi^V(g(x)), \quad \{t_V(z), t_W(w)\} = 0.$$

The coefficients of these functions will produce Poisson-commuting integrals of motion for integrable systems, when restricted to symplectic leaves of LGL_2.

Symplectic leaves. It is a well-known, classical fact, which can be traced back to the work of Lie, that symplectic leaves of a Poisson manifold which is the dual space to a Lie algebra are co-adjoint orbits. Similarly, symplectic leaves of a Poisson Lie group G are orbits of the dressing action of the dual Poisson Lie group on G.

The structure of symplectic leaves of Poisson Lie groups is well known for finite-dimensional simple Lie algebras (see [23, 40]). For the construction of integrable spin chains related to SL_2, we need only some special symplectic leaves of LGL_2. These symplectic leaves are symplectic leaves of finite-dimensional Poisson submanifolds of polynomial maps of given degree.

7.A.4 Quantization

This section gives a brief outline of the quantization of Hamiltonian systems. There is also the very important point of view of path integral quantization, but we will not discuss that here.

Quantized algebra of observables. Let M be a symplectic manifold, which is the phase space of our mechanical system. We want to describe possible quantum mechanical systems which reproduce our system in the classical limit. This procedure is called quantization.

Let A be a Poisson algebra over $\mathbb{C}$, i.e. a complex vector space with a commutative multiplication ab and with a Lie bracket $\{a, b\}$ such that $\{a, bc\} = b\{a, c\} + \{a, b\}c$. Let $X \subset \mathbb{C}$ be a neighborhood of $0 \in \mathbb{C}$.

Definition 7.11 *A deformation quantization of A is a family of associative algebras A_h, parameterized by $h \in X$ together with two families of linear maps $\phi_h : A_h \to A$ and $\psi_h : A \to A_h$, such that*

- $\lim_{h\to 0} \phi_h \circ \psi_h \to id_A$,
- $\lim_{h\to 0} \phi_h(\psi_h(a)\psi_h(b)) = ab$,
- $\lim_{h\to 0} \phi_h(\psi_h(a)\psi_h(b) - \psi_h(a)\psi_h(b))/ih = \{a, b\}$.

We denote by $C(M)$ the classical algebra of observables, i.e. the algebra of real-valued functions on the phase space. If M is T^*N, this will be the algebra of functions on M which are polynomial in the cotangent direction and smooth on N. If M is an affine algebraic manifold, $C(M)$ will be an algebra of polynomial functions. If M is a smooth manifold, this algebra is $C^\infty M$, etc. The space $C(M)$ has a natural structure of a Poisson algebra, with the pointwise multiplication and the Poisson bracket determined by the symplectic form on M.

We denote its complexification by $C(M)_\mathbb{C}$. The real subalgebra $C(M) \subset C(M)_\mathbb{C}$ is the set of fixed points of complex conjugation, i.e. real-valued functions. We denote complex conjugation by σ, i.e. $\sigma(f)(x) = \overline{f(x)}$.

Deformation quantizations which are relevant to quantum mechanics are real deformation quantizations.

Definition 7.12 *A real deformation quantization of $C(M)$ is a pair $(A_h, \sigma_h, h \in X \subset \mathbb{R})$ where A_h is a deformation quantization (as above) of $C(M)_\mathbb{C}$, h is a real deformation parameter, and σ_h is a $\mathbb{C}$-antilinear anti-involution of A_h, i.e.*

$$\sigma_h^2 = 1, \;\; \sigma_h(ab) = \sigma_h(b)\sigma_h(a), \;\; \sigma_h(sa) = \bar{s}\sigma_h(a).$$

The subspace $C(M)_h \subset A_h$ of fixed points of σ_h is called the space of quantum observables. The linear maps ϕ_h and ψ_h in the definition of A_h must satisfy the extra condition

$$\lim_{h \to 0} \phi_h \circ \sigma_h \circ \psi_h = \sigma \,.$$

The space of fixed points of σ_h is called the *space of quantum observables*. Note that the multiplication does not preserve this space. However, it is closed with respect to operations $AB + BA$ and $i(AB - BA)$. Such structures are called Jordan algebras.

Of course, these definitions still need clarification. The vector space $C(M)$ is infinite-dimensional and we have to specify in which topology our linear isomorphism is continuous and in which sense we should take limits. The way to handle this is dictated by the nature of the specific problem, so we will discuss it later in relation to integrable spin chains, which are the main subject here.

Examples of family deformations. Take $M = \mathbb{R}^2$ and $A = \mathrm{Pol}_{\mathbb{C}}(\mathbb{R}^2) = \mathbb{C}[p, q]$, with the standard symplectic form $dp \wedge dq$ giving the bracket $\{p, q\} = 1$ (this determines the bracket). We have a natural monomial basis $p^n q^m$ on A. We define

$$A_h = \langle p, q | pq - qp = h \rangle \,.$$

It is clear that this is a family of associative algebras. To identify this family with a deformation quantization of A, we must find ϕ_h and ψ_h. For this purpose, we choose monomial bases $p^n q^m$ in A_h and A. We define linear maps ϕ_h and ψ_h as linear isomorphisms $A_h \simeq A$ identifying the monomial bases. It is clear that this choice makes A_h into a deformation quantization of A.

Let $\mathfrak{g}$ be a Lie algebra, and consider $\mathrm{Pol}(\mathfrak{g}^*) = \mathbb{C}[\mathfrak{g}]$. If $\{e_i\}$ is a basis for $\mathfrak{g}$, then we can think of the e_i as coordinate functions x_i on $\mathfrak{g}^*$. A theorem of Kostant states that $\{f, g\}(x) = \langle x, [df(x), dg(x)] \rangle$, $\{x_i, x_j\} = \sum_k c_{ij}^k x_k$. $\mathrm{Pol}(\mathfrak{g}^*)$ then gets a Poisson bracket.

We can now obtain a deformation quantization

$$A_h = \Big\langle x_1, \ldots, x_n | x_i x_j - x_j x_i = h \sum_k c_{ij}^k x_k \Big\rangle.$$

Note that $A_h \cong U\mathfrak{g}$ for any $h \neq 0$ (you just have to rescale the x's by h). On the other hand, we can choose the monomial basis $x_1^{a_1} \cdots x_n^{a_n}$ in $\mathbb{C}[x_1, \ldots, x_n]$ and the PBW basis in A_h. By identifying them, we get $A_h \cong \mathbb{C}[x_1, \ldots, x_n]$, which is how the PBW theorem is usually formulated:

$$U\mathfrak{g} \cong \mathrm{Pol}(\mathfrak{g}^*) \cong S(\mathfrak{g}) \,.$$

We get a linear isomorphism $\phi_h \colon A_h \cong A$. It is easy to check that this is a deformation quantization.

Quantization of integrable systems. As we have seen above the quantization of a classical Hamiltonian system on $\mathcal{M}$ with the Hamiltonian $H \in C(M)$ consists of the following:

- A family of associative algebras A_h quantizing the algebra of functions on $\mathcal{M}$.
- The choice of a quantum Hamiltonian $H_h \in A_h$ for each h, such that $\phi_h(H_h) \to H$ as $h \to 0$.

The quantization is integrable if, for each h, there is a maximal commutative subalgebra of $C_h(M)$ quantizing the subalgebra of classical integrals in $C(M)$ which contains H_h.

To be more precise, let $I(M) \in C(M)$ be the subalgebra generated by Poisson-commuting integrals in the classical algebra of observables, i.e. $\{H, F\} = 0$ for each $F \in I(M)$ and $\{G, F\} = 0$ for each $F, G \in I(M)$. Its integrable quantization consists of a commutative subalgebra $I_h \in A_h$ such that $FH = HF$ for each $F \in I_h$, such that $\lim_{h\to 0} \phi_h(I_h) = I$. For a precise definition in the case of formal deformation quantization, see [34].

When a quantized algebra of observables is represented in a Hilbert space, the important problem is the computation of the spectrum of commuting Hamiltonians.

References

[1] D. Allison and N. Reshetikhin, Numerical study of the 6-vertex model with domain wall boundary conditions, *Ann. Inst. Fourier (Grenoble)* **55** (2005), no. 6, 1847–1869.

[2] V. I. Arnold, *Mathematical Methods of Classical Mechanics*, 2nd edn, Springer, 1989.

[3] R. J. Baxter, *Exactly Solved Models in Statistical Mechanics*, Academic Press, San Diego, 1982.

[4] N. M. Bogolyubov, A. G. Izergin, and N. Yu. Reshetikhin, Finite-size effects and critical indices of 1D quantum models, *JETP Lett.* **44** (1986), no. 9, 521–523 (1987); translated from *Pisma Zh. Eksp. Teor. Fiz.* **44** (1986), no. 9, 405–407.

[5] V. V. Bazhanov and N. Reshetikhin, Restricted solid-on-solid models connected with simply laced algebras and conformal field theory, *J. Phys. A* **23** (1990), no. 9, 1477–1492.

[6] D. J. Bukman and J. D. Shore, The conical point in the ferroelectric six-vertex model, *J. Stat. Phys.* **78** (1995) 1277–1309.

[7] M. Jimbo, T. Miwa, and F. Smirnov, Hidden Grassmann structure in the XXZ model III: Introducing Matsubara direction, arXiv:0811.0439.

[8] V. Chari and A. Pressley, *A Guide to Quantum Groups*, Cambridge University Press, 1994.

[9] H. Cohn, R. Kenyon, and J. Propp, A variational principle for domino tilings, *J. Amer. Math. Soc.* **14** (2001), no. 2, 297–346.

[10] F. Colomo and A. Pronko, The arctic circle revisited, arXiv:0704.0362. The limit shape of large alternating sign matrices, arXiv:0803.2697.

[11] T. Bodineau, B. Derrida, V. Lecomte, and F. van Wijland, Long range correlations and phase transition in non-equilibrium diffusive systems, arXiv:0807.2394.

[12] H. J. de Vega and C. Destri, Unified approach to thermodynamic Bethe ansatz and finite size corrections for lattice models and field theories, *Nucl. Phys. B* **438** (1995) 413–454.

[13] V. G. Drinfeld, Quantum groups, in *Proc. Int. Congress of Mathematics* (Berkeley 1986), pp. 798–820, AMS, 1987.

[14] N. Gromov and V. Kazakov, Double scaling and finite size corrections in $sl(2)$ spin chain, *Nucl. Phys. B* **736** (2006) 199–224.

[15] E. K. Skljanin, L. A. Tahtadjan, and L. D. Faddeev, Quantum inverse problem method. I [in Russian], *Teoret. Mat. Fiz.* **40** (1979), no. 2, 194–220.

[16] L. A. Tahtadjan and L. D. Faddeev, The quantum method for the inverse problem and the XYZ Heisenberg model [in Russian], *Uspekhi Mat. Nauk* **34** (1979), no. 5(209), 13–63, 256.

[17] H. Y. Huang, F. Y. Wu, H. Kunz, and D. Kim, Interacting dimers on the honeycomb lattice: An exact solution of the five-vertex model, *Physica A* **228** (1996) 1–32.
[18] V. Korepin and P. Zinn-Justin, Thermodynamic limit of the six-vertex model with domain wall boundary conditions, *J. Phys. A* **33** (2000) 7053–7066; Inhomogeneous six-vertex model with domain wall boundary conditions and Bethe ansatz, *J. Math. Phys.* **43** (2002) 3261–3267.
[19] M. Jimbo and T. Miwa (eds.), *Algebraic Analysis of Solvable Lattice Models*, CBMS Regional Conference Series in Mathematics, Vol. 85, 1993.
[20] S. V. Kerov, A. N. Kirillov, and N. Yu. Reshetikhin, Combinatorics, the Bethe ansatz and representations of the symmetric group [in Russian], *Zap. Nauchn. Sem. Leningrad. Otdel. Mat. Inst. Steklov. (LOMI)* **155** (1986), *Differentsialnaya Geometriya, Gruppy Li i Mekh. VIII*, 50–64, 193; translation in *J. Sov. Math.* **41** (1988), no. 2, 916–924.
[21] V. E. Korepin, N. M. Bogolyubov, and A. G. Izergin, *Quantum Inverse Scattering Method and Correlation Functions*, Cambridge University Press, 1993.
[22] R. Kenyon, A. Okounkov, and S. Sheffield, Dimers and amoebae, arXiv:math-ph/0311005.
[23] L. I. Korogodski and Y. S. Soibelman, *Algebras of Functions on Quantum Groups*, Part I, Mathematical Surveys and Monographs, No. 56, American Mathematical Society, Providence, RI, 1998.
[24] P. P. Kulish, N. Yu. Reshetikhin, and E. K. Sklyanin, Yang–Baxter equations and representation theory. I. *Lett. Math. Phys.* **5** (1981), no. 5, 393–403.
[25] E. Lieb, *Phys. Rev.* **162** (1967) 162; E. H. Lieb, *Phys. Rev. Lett.* **18** (1967) 1046; *Phys. Rev. Lett.* **19** (1967) 108.
[26] E. H. Lieb and F. Y. Wu, Two dimensional ferroelectric models, in *Phase Transitions and Critical Phenomena*, Vol. 1, eds. C. Domb and M. S. Green, p. 321, Academic Press, London, 1972.
[27] B. Nienhuis, H. J. Hilhorst, and H. W. Bloete, Triangular SOS modles and cubic-crystal shapes, *J. Phys. A: Math. Gen.* **17** (1984) 3559–3581.
[28] J. D. Noh and D. Kim, Finite-size scaling and the toroidal partition function of the critical asymmetric six-vertex model, arXiv:cond-mat/9511001.
[29] I. M. Nolden, The asymmetric six-vertex model, *J. Stat. Phys.* **67** (1992) 155; Ph.D. thesis, University of Utrecht, 1990.
[30] K. Palamarchuk, Ph.D. thesis, University of California Berkeley, 2007.
[31] K. Palamarchuk and N. Reshetikhin, The six-vertex model with fixed boundary conditions, http://pos.sissa.it/archive/conferences/038/012/Solvay-012.pdf.
[32] N. Reshetikhin, Quasitriangularity of quantum groups at roots of 1. *Commun. Math. Phys.* 170 (1995), no. 1, 79–99.
[33] N. Yu. Reshetikhin and F. A. Smirnov, Quantum Floquet functions [in Russian], *Questions in Quantum Field Theory and Statistical Physics, 4. Zap. Nauchn. Sem. Leningrad. Otdel. Mat. Inst. Steklov. (LOMI)* **131** (1983) 128–141.
[34] N. Reshetikhin and M. Yakimov, Deformation quantization of Lagrangian fiber bundles, *Conférence Moshé Flato 1999*, Vol. II (Dijon), pp. 263–287, Mathematical Physics Studies 22, Kluwer Academic, Dordrecht, 2000.
[35] A. Reiman and M. Semenov-Tian-Shanskii, Group theoretical methods in integrable systems, in *Integrable systems VII*, eds. V. Arnold and S. Novikov, Encyclopaedia of Math-

ematical Sciences, Vol. 16, Springer, 1994.

[36] E. K. Skljanin, The method of the inverse scattering problem and the quantum nonlinear Schrödinger equation [in Russian], *Dokl. Akad. Nauk SSSR 244* (1979), no. 6, 1337–1341.

[37] E. K. Sklyanin, Classical limits of SU(2)-invariant solutions of the Yang–Baxter equation [in Russian], translated in *J. Sov. Math.* **40** (1988), no. 1, 93–107; *Differential Geometry, Lie Groups and Mechanics. VII. Zap. Nauchn. Sem. Leningrad. Otdel. Mat. Inst. Steklov. (LOMI)* **146** (1985), 119–136, 203, 206–207.

[38] F. A. Smirnov, Quasi-classical study of form factors in finite volume, in *L. D. Faddeev's Seminar on Mathematical Physics*, pp. 283–307, American Mathematical Society Translation Series 2, Vol. 201, American Mathematical Society, Providence, RI, 2000.

[39] B. Sutherland, C. N. Yang, and C. P. Yang, Exact solution of a model of two-dimensional ferroelectrics in an arbitrary external electric field, *Phys. Rev. Lett.* **19** (1967) 588.

[40] M. Yakimov, Symplectic leaves of complex reductive Poisson–Lie groups, *Duke Math. J.* **112** (2002), no. 3, 453–509.

[41] C. N. Yang and C. P. Yang, Thermodynamics of a one-dimensional system of bosons with repulsive delta-function interactions, *J. Math. Phys.* **10** (1967) 1115–1122.

[42] A. Zamolodchikov, Thermodynamic Bethe ansatz in relativistic models: Scaling 3-state-Potts and Lee–Yang models, *Nucl. Phys. B* **342** (1990) 695–720.

8 Mathematical aspects of 2D phase transitions

W. WERNER

Wendelin Werner,
Laboratoire de Mathémathiques,
Université Paris-Sud Orsay,
Bâtiment 425,
91405 Orsay Cedex, France.

Part II

Short lectures

9

Numerical simulations of quantum statistical mechanical models

F. ALET

Fabien Alet,
Université de Toulouse, UPS,
Laboratoire de Physique Théorique (IRSAMC),
F-31062 Toulouse, France,
and CNRS, LPT (IRSAMC),
F-31062 Toulouse, France.

9.1 Introduction

Quantum statistical physics models constitute an exciting world to study. Many models exist, with very rich physics. New phenomena and new concepts emerge from the introduction of quantum fluctuations. Quantum models are also closer to experiments, as any realistic description of materials certainly needs to correctly take quantum effects into account.

Unfortunately, quantum lattice models are very hard to treat. Analytical exact solutions are useful, but rare. In practice, they are limited to a few models, usually with a high symmetry and of low-dimensional nature (1D or 2D). What about numerical simulations of quantum lattice models? Various intertwined problems rapidly arise. First, quantum particles have their own rules (Bose or Fermi statistics, quantum spin algebra, etc.), and numerical simulations must follow them. Second, the Hamiltonian is usually composed of operators that do not commute: $H = \sum_i H_i$. The partition function therefore does not simply factorize: $Z = \mathrm{Tr}\, e^{-\beta H} \neq \mathrm{Tr} \prod_i e^{-\beta H_i}$. Third, the size of the Hilbert (configuration) space grows exponentially fast (e.g. as 2^N for N quantum spins $1/2$), seriously limiting methods based on exact enumeration of states.

Owing to these constraints, a straightforward adaptation of the numerical methods that are well suited for classical problems is just not possible. New techniques have to be developed. These notes present a gentle introduction to some aspects of the numerical simulation of quantum lattice models.

In Section 9.2, we first make a rapid tour of the most efficient methods to treat these models. There is, unfortunately, no best method in general, and the advantages and drawbacks of each technique are mentioned. The goal of this chapter is certainly not to provide a full review of each method, but simply to give the main ideas and to point towards more specialized references.

The rest of these lectures will focus on one family of methods, quantum Monte Carlo (QMC) methods. The first reason for such a specialization is the difficulty of knowing and commenting on all the details of each method, and the author of these notes is no exception to this. The second important reason is that the future of scientific computing probably resides in the use of massively parallel computers (see the various proposals to build a petaflop computer). Monte Carlo simulations are perfectly suited for that, as they are naively parallel.

We first give a rapid introduction in Section 9.3 to QMC path integral methods, which are in other respects well documented in the literature. This section also discusses the major drawback of QMC methods, namely the sign problem.

In these notes, we have deliberately chosen to put an emphasis on QMC methods that are not so well known in the field of quantum statistical physics. Section 9.4 presents details of a worm algorithm for a classical link-current model that emerges from the path integral description of a quantum bosonic model. This section also contains a general discussion of the use of worm algorithms for classically constrained models.

In Section 9.5, we switch to the description of another family of methods, namely projection Monte Carlo techniques. These methods are rarely used in quantum statistical physics, even though they are in principle able to treat all models. After an introduction to the general aspects common to all projection methods, we discuss in detail the reptation method, a particularly convenient variant.

Section 9.6 introduces a novel original projector method for quantum spin systems, which is formulated in the valence bond basis. This part also offers a quick tour of the unusual properties of this basis, which are at the heart of this method.

9.2 A rapid survey of methods

Numerical methods can be roughly divided into three classes: exact, approximate and "stochastic." Exact methods include techniques that give results which are bias-free, such as the full exact diagonalization (ED) method (up to machine precision) and the series expansion method (up to the order calculated). Some variational methods are usually included in this class (even though they are not, strictly speaking, exact), as they give, essentially, the correct results with very high precision. This is the case for the density matrix renormalization group (DMRG) method and iterative diagonalization methods. Approximate methods regroup a wide range of techniques which usually provide a lot of physical insight even though they are not exact. We can mention effective Hamiltonian methods (such as CORE), coupled-cluster methods, and dynamical mean-field theory (DMFT), even though the latter is not a numerical method in itself. Stochastic methods include all quantum Monte Carlo (QMC) algorithms, which is a generic name for a broad variety of algorithms. Some QMC techniques can be approximate, but the most recent algorithms are exact, at least in the statistical sense: there is no systematic error, and statistical errors can be reduced at will if one waits long enough. Finally, we also survey the ALPS project, a recent initiative to provide open-source codes for most of the aforementioned methods.

9.2.1 Exact diagonalization

Good recent reviews of the ED method for quantum lattice models can be found in [1, 2], so the discussion will be short. The idea of ED is simple: for a finite system, the Hamiltonian is just a big matrix that can diagonalized numerically.

Of course, the argument can be refined by using the symmetries of the model, as the Hamiltonian is block diagonal and each block can be diagonalized independently. These symmetries include Abelian continuous symmetries[1] related to conservation of the number of particles, and lattice symmetries (translation and point group symmetries such as rotations and reflections). The use of symmetries also allows one to obtain supplementary physical information from the quantum numbers (e.g. momentum and particle number) associated with the symmetries.

If the total spectrum of the Hamiltonian is needed, a *full diagonalization* of the matrices must be performed. This can be done using standard diagonalization routines (for example the LAPACK libraries [3]). In physics, one is often interested only in the low-energy spectrum and in that case, *iterative methods* allow one to focus only on the ground state and on a few excited states. Iterative methods are particularly interesting because they need to store only a few vectors in memory instead of full matrices. Amongst them, the Lanczos and Davidson algorithms are the most widely used.

The Lanczos algorithm is often preferable because it needs less memory (only three vectors), but spurious fake eigenstates can sometimes occur owing to round-off errors. These *ghosts* can be eliminated by a reorthogonalization procedure. The Lanczos algorithm works best when the matrices are sparse (lots of zeros), which is most often the case for Hamiltonians of quantum statistical physics models.

[1] Non-Abelian symmetries such as SU(2) are more difficult to handle.

The Davidson algorithm is more demanding in terms of memory but does not have the ghost problem. It is also preferable when a large number of excited states (typically more than 10) are required, as well as when the Hamiltonian has large diagonal elements.

ED is like pure gold: essentially, it allows one to simulate all models and to calculate every quantity (including quantities related to thermodynamics with full diagonalization and dynamics with the Lanczos algorithm). Unfortunately, like pure gold, which comes in small amounts, ED is limited to small systems by the exponential size of the Hilbert space. The matrices and vectors just do not fit into the CPU memory. To give a rough estimate of the sizes that can be reached with today's computers and algorithms, full diagonalization is limited to matrices of linear size of about 50 000 on a large-memory computer. The Lanczos algorithm can treat systems with up to roughly 200–300 million states on a lab computer and a few billion on supercomputers.[2] Taking quantum spin systems as an example and using all symmetries, this corresponds to obtaining the full spectrum of about 24 spins $1/2$ and the low-energy spectrum of 40 spins 1/2, respectively.

9.2.2 Density matrix renormalization group

The DMRG method was created by White in the early 1990s [5]. The idea of the DMRG method is to perform a real-space renormalization by increasing the lattice by a block at each step. Of course, this cannot be performed exactly, owing to the increasing size of the Hilbert space, and one has to select only a few states. One might naively expect that the lowest-lying states of the Hamiltonian would be the best choices, but it turns out that it is more efficient to consider the lowest-lying states of the *density matrix* reduced to one block (hence the name of the method). The Davidson algorithm is often used to obtain these states. A description of the DMRG algorithm requires some length and goes beyond the scope of this short survey. The original papers by White [5] and excellent existing reviews [1, 6] provide a nice overview of the method.

In practice, what type of calculations can the original DMRG algorithm perform? The algorithm allows one to compute, with a very high accuracy, the ground state properties (and sometimes those of excited states) of most quantum lattice models. Unfortunately, it is essentially limited in its original formulation to 1D or quasi-1D lattices. In the finite-size version of the algorithm (the most widely used), large systems of up to a few thousand sites can be studied. The use of open boundary conditions is often needed for the algorithm (again in its original formulation) to perform at its best.

Over the last five years, considerable breakthroughs have given the method new life. Relations to quantum information concepts have allowed the DMRG algorithm to be recast in a different framework. In particular, the finite-size variant can be understood as a variational algorithm over matrix product states (MPS). See [7] for a nice introduction to the DMRG algorithm from the MPS perspective. The quantum information insights allow us to understand how and when the DMRG algorithm converges or has difficulties. This has led to practical improvements: efficient simulations of systems with periodic boundary conditions at finite temperature, and of real time-evolution of quantum lattice models are now possible. These advances are partly reviewed in [6].

[2]The current record was established on the Earth Simulator with 159 billion states [4].

There is even more to come. Several recent publications have proposed ideas to construct algorithms for $d > 1$-dimensional systems using generalizations of MPS to higher dimensions. A nonexhaustive list includes projected entangled pair states [8] and related string states [9], the multiscale entanglement renormalization ansatz [10], and scale-renormalized states [11]. The field is evolving so rapidly that there is no review (to the best of the author's knowledge) yet that includes and summarizes these proposals.[3] It is also hard to decide at this stage if there is any superior method among those cited that gives unbiased results for a generic quantum statistical physics model in dimension $d > 1$.

9.2.3 Series expansion

There is an excellent recent review of all series methods in [12], and we will just give a few ideas here.

High-temperature series expansion works exactly as in the classical case [13]: we expand an extensive quantity as a function of a high-temperature parameter (typically β). Series expansion can also be performed at $T = 0$ in powers of a coupling constant λ of the Hamiltonian $H = H_0 + \lambda H_1$, as a systematic extension of perturbation theory in quantum mechanics [14].

Few orders are reachable analytically; a numerical approach is required to systematically compute higher-order terms. Computations can be performed on finite lattices, as the n-th order term concerns at most $n+1$ sites: this is the cluster expansion. The linked-cluster theorem [12] indicates that only connected clusters will contribute to observables in the thermodynamic limit. This allows computations of ground state properties in the case of the $T = 0$ expansion, but excited states can also be reached in some cases [15]. Numerically, efficient graph libraries (such as the Nauty library [16]) are needed to generate the clusters, find isomorphisms, and compute embeddings of finite clusters in an infinite lattice. One also needs to be able to compute the desired observables on the finite clusters.

Series expansion can provide results directly in the thermodynamic limit for all models in any dimension. Of course, it suffers from the same limitations as its classical counterpart: the order that can be reached is limited by CPU memory and time. In practice, one can go up to orders β^α or λ^α with $\alpha \sim 20$ (a bit more for some models). One then has to use extrapolation methods (such as Padé approximants) to check the convergence of the series. In some sense, finite-size scaling is traded off against series extrapolations. It is sometimes difficult to estimate the validity of an extrapolation scheme: this limits series expansion, in general, to a high-temperature (or small-coupling-constant) range, or at best an intermediate range. Also, the series expansion approach typically breaks down at a (quantum) phase transition.

9.2.4 Quantum Monte Carlo methods

The term "QMC" encompasses a large variety of methods which all have in common the fact that they are stochastic techniques to solve the Schrödinger equation. The goal is to replace the enumeration of all Hilbert space states by a sampling that selects only the most relevant ones.

QMC methods come in various flavors, depending on the temperature regime studied ($T = 0$ or finite temperature), the statistics of particles (fermions, bosons, spins etc.), the nature of

[3] See, however, the updated list of preprints related to the DMRG method maintained by T. Nishino at http://quattro.phys.sci.kobe-u.ac.jp/dmrg/condmat.html.

real space (continuum or lattice), the number of particles that one wishes to study (a finite number or the thermodynamic limit), etc. The related number of different schemes is large, resulting in a broad spectrum of QMC communities.

As far as quantum statistical mechanical models are concerned, two main families can be identified. In *path integral* (PI) methods, the QMC sampling is over the partition function at finite temperature. In *projection* methods, the ground state is sampled stochastically. Sections 9.3 and 9.5 describe these two families of methods in detail.

The most recent PI methods can be applied to a large variety of spin and bosonic models, with very high efficiency and precision. The system sizes that can be reached are much larger than those obtained with other methods: very large systems (of more than 10^6 particles) can be studied. Moreover, even though PI methods are formulated at finite T, the temperatures that can be reached are so low that only ground state properties are sampled. The precision becomes essentially as good as that of classical simulations, and sometimes even better! Reviews of the most efficient PI methods for lattice models can be found in [17–19]. A quick introduction will also be given in Section 9.3.

The precision and the range of lattice sizes obtained by projection methods are not that high, but are still superior to those for most other methods. The interesting aspect of projection methods is that they can be applied straightforwardly to any model (see, however, the main restriction below), whereas the PI formulation usually needs a Hamiltonian with higher symmetry and simple interactions. There is no review (to the best of the author's knowledge) available in the literature of projection QMC methods for quantum lattice models, and Section 9.5 constitutes an attempt to fill the gap.

All diagonal quantities (in the reference basis of the Hilbert space) are easily computed within QMC methods. Off-diagonal observables are more difficult to obtain, but such computations are possible. Time-dependent observables are also reachable; however, the PI and projection methods are formulated in *imaginary time*. One needs to perform an analytic continuation to obtain real-time and real-frequency observables, but this is an ill-defined problem as the imaginary-time data come with error bars. The usual solution is to use *maximum entropy* methods [20], recent improvements of which [21] have led to efficient practical schemes.

There is one major drawback of QMC methods: the *sign problem*. When off-diagonal elements of the Hamiltonian are positive, the stochastic principle underlying QMC methods breaks down. The sign problem can also arise from the statistics of particles, owing to the dephasing induced in the wave function when particles are exchanged. In such cases, QMC methods are totally inefficient. This severely restricts the range of models that can be simulated, and excludes, for instance, most interesting fermionic models, as well as frustrated quantum spin models.

In some models, clever tricks allow one to avoid the sign problem, or at least reduce it (see for instance the determinantal method [22] for half-filled or attractive fermionic models). However, a generic solution to the sign problem is most likely absent, as the sign problem possibly falls into the NP-hard complexity class [23]. If the sign problem were absent, the QMC method would certainly be the only numerical method in use for quantum lattice models. This also means that for models where the sign problem is absent, the QMC method is usually the method of choice (except maybe for 1D problems, where the performance of DMRG methods is comparable).

9.2.5 Approximate methods

This section includes a brief discussion of approximate numerical methods, such as the effective Hamiltonian CORE technique and the coupled-cluster method. We also mention the numerical aspects of dynamical mean-field theory.

Contractor renormalization (CORE) method. The CORE method was introduced in [24] and is essentially a real-space renormalization method, which allows one to build effective Hamiltonians that capture the low-energy physics of a given system. The effective Hamiltonian, which is formulated in a smaller Hilbert space, can then be studied analytically or numerically. In practice, the method is often associated with exact diagonalization, since the reduced size of the Hilbert space allows one to study larger samples than those of the original model.

Good reviews can be found in [25], and we give here only a brief summary of the method. Given an original Hamiltonian H, one first considers a small real-space cluster and selects a small number M of relevant states on this cluster—usually the lowest eigenstates of H on this cluster. The next step is to build a larger graph with r clusters and to obtain the low-energy spectrum $\{\epsilon_n\}$ of H on this graph; r denotes the range of the future cluster expansion to be performed.

By projecting the lowest-lying eigenstates into (usually) the tensor product space formed from M states kept for each cluster, one can build, after orthogonalization, the effective Hamiltonian on this graph $H_r^{\text{eff}} = \sum_{n=1}^{M^r} \epsilon_n |\psi_n\rangle\langle\psi_n|$, where the $|\psi_n\rangle$ form the orthogonal basis. The final CORE Hamiltonian is then deduced from a cluster expansion, where, as in the series expansion, the contribution from connected subclusters is subtracted so that $H^{\text{CORE}} = \sum_r H_r^{\text{eff,c}}$.

If $r \to \infty$, the lowest-lying spectrum of H is exactly reproduced by this approach. Of course, in practice, one cuts the expansion to a finite value of r and hopes that the low-energy physics is still well reproduced. As already stated, the study of this effective Hamiltonian is eased numerically by the reduced size (M^r for range r) of the Hilbert space. The CORE method can be applied in principle to all lattice models.

Coupled-cluster method. Another approximate technique, which is widely used in quantum chemistry, is the coupled-cluster method. The idea is to assume an exponential ansatz for the ground state of a many-body system $|\psi\rangle = e^S|\phi\rangle$, where $|\phi\rangle$ is a reference state and S is the *correlation operator*, expressed as an (unknown) linear combination of many-body creation/annihilation operators. The exponential is then usually expanded in a series of terms that concern only clusters of contiguous sites on the lattice, hence the name of the method. Minimization of the variational energy with this ansatz results in a set of equations, which are solved numerically. A recent review which focuses on the application of the coupled-cluster method to quantum spin models can be found in [26], which also includes further references.[4] The coupled-cluster method is variational in nature, works for most lattice models, and only concerns ground-state properties.

Dynamical mean-field theory. DMFT is not a numerical method, but rather a general theoretical framework for treating quantum many-body problems. Introductory reviews can be found

[4] A numerical package containing an implementation of the coupled-cluster method for lattice spin models can be found at http://www-e.uni-magdeburg.de/jschulen/ccm.

in [27], and again we will just state the main idea: DMFT maps a many-body problem onto a quantum impurity model located in a bath that obeys some self-consistent conditions. Originally, the effective problem consisted of a single impurity. Recent cluster methods [28] which treat a few impurities embedded in a self-consistent bath lead to a more faithful treatment of the original model.

The numerical aspects of DFMT come into play because the effective impurity problem is not solvable analytically in the general case. One needs to use a so-called impurity (or cluster) solver, which is a numerical method to solve the effective model. Such solvers can be of ED, DMRG, QMC, or some other type and are the crucial ingredients for a successfully converged self-consistent loop.

The DMFT approach works directly in the thermodynamic limit, and is exact in infinite dimensions but only approximate in finite dimensions. Nevertheless, it is certainly the best method to study fermionic lattice models (such as the Hubbard model) in $d > 1$, despite some inherent approximations. This statement becomes particularly true with the recent arrival of efficient cluster solvers [29] that allow one to simulate the effective impurity problem at lower temperatures. Note that DMFT can be nicely combined with band structure methods to give a coherent tool for understanding correlation effects in electronic materials [30]. The DMFT approach, owing to its formulation, is not easily applied to spin or bosonic lattice models (see, however, some recent proposals [31]).

ALPS project. In contrast to what happens in other fields (such as quantum chemistry or lattice gauge theories), there are no community codes in quantum statistical physics. Yet algorithms are sometimes well established, and there is an increasing demand from theoreticians (not specialized in numerics) and from experimentalists to have numerical results. Up to now, numerical simulations have been seen as a specialist's work. Typically, every Ph.D. student starting a thesis involving numerical simulations has started coding from scratch or from a previous code lent by a colleague.

Recently, specialists in numerical simulations of quantum lattice models have started the ALPS project, an initiative to provide open-source codes for most methods [32]. For instance, the source code of various QMC algorithms and of full-diagonalization and DMRG algorithms is provided. The webpage of the project[5] includes documentation of these codes and tutorials in their use.

The ALPS project also features libraries that are aimed at easing developers' work. The lattice, model, and Monte Carlo libraries help greatly in starting a new code without reinventing the wheel. For instance, coding a Monte Carlo method for a new model is particularly easy. The developer has just to concentrate on the essential part of the algorithm specific to the new model: Monte Carlo averages, treatment of error bars, I/O, checkpointing, parallelization, etc. are already taken care of. The libraries are written in C++ and can be downloaded from the project webpage. Contributions to this effort are of course possible and very much welcomed.

[5] http://alps.comp-phys.org.

9.3 Path integral and related methods

9.3.1 Path integral formalism

A simple stochastic evaluation of the partition function of a quantum system is not possible since, as already mentioned in the introduction, the different parts of the Hamiltonian do not commute. The basis of all PI methods is to cleverly split the Hamiltonian $H = \sum_i H_i$ into parts H_i such that the $\exp(H_i)$ are easily computed. Then the Trotter (or Trotter–Suzuki) formula [33],

$$\exp(-\epsilon H) = \prod_i \exp(-\epsilon H_i) + o(\epsilon^2)\,,$$

allows one to decouple these exponentials, up to some error varying as the square of the prefactor in front of the Hamiltonian. Usually there is no small prefactor in the partition function (β is arbitrary), but it can be created by splitting β into M parts with M large. One obtains a *path integral* representation of the partition function,

$$Z = \mathrm{Tr}\, e^{-\beta H} = \mathrm{Tr}(e^{-\Delta\tau \sum_i H_i})^M = \mathrm{Tr}\left(\prod_i e^{-\Delta\tau H_i}\right)^M + o(\Delta\tau^2)\,,$$

where $\Delta\tau = \beta/M$ is called the *time step*, as $\exp(-\beta H)$ is usually interpreted as the evolution operator $\exp(-itH)$ in *imaginary time* $\tau = it$, up to $\tau = \beta$.

One then usually injects representations of the identity operator in a given basis $\{|c\rangle\}$ to obtain

$$\begin{aligned} Z = \textstyle\sum_{c_1 \ldots c_{k.M}} & \langle c_1|e^{-\Delta\tau H_1}|c_2\rangle\langle c_2|e^{-\Delta\tau H_2}|c_3\rangle \ldots \\ & \times\langle c_{k.M-1}|e^{-\Delta\tau H_{k-1}}|c_{k.M}\rangle\langle c_{kM}|e^{-\Delta\tau H_k}|c_1\rangle + o(\Delta\tau^2)\,, \end{aligned} \tag{9.1}$$

where the Hamiltonian has been split into k terms. The transfer (or time evolution) matrices $e^{-\Delta\tau H_i}$ are chosen to be easily computable. Equation (9.1) is then seen as the partition function of a *classical* problem in $d+1$ dimensions, where the extra dimension is the imaginary time (going from $\tau = 0$ to β). The configuration space is the one formed by the basis states of the $k.M$ representations of identities, and is usually chosen to be easy to sample.

Powerful *cluster* methods known as loop or worm algorithms [34, 35] allow an efficient sampling of the partition function in the representation of eqn (9.1). They can be *directly* formulated [36] in the continuous-time limit $\Delta\tau \to 0$, avoiding the systematic error caused by the small but finite $\Delta\tau$ used in older algorithms. Multicanonical sampling of the partition function is also possible [37]. As excellent reviews of the PI methods and formalism can be found in [17–19], we will not expand further on this topic.

9.3.2 Stochastic series expansion

A different but related scheme performs a high-temperature expansion of the partition function

$$Z = \mathrm{Tr}\, e^{-\beta H} = \sum_{n=0}^{\infty} \frac{\beta^n}{n!} \mathrm{Tr}(-H)^n\,.$$

Expressing the various terms of the Hamiltonian and the trace, we have

$$
\begin{aligned}
Z &= \sum_{n=0}^{\infty} \frac{\beta^n}{n!} \sum_c \langle c|(-\sum_i H_i)^n|c\rangle \\
&= \sum_{n=0}^{\infty} \frac{\beta^n}{n!} \sum_c \sum_{(i_1 \ldots i_n)} \langle c| \prod_{p=1}^{n} (-H_{i_p})|c\rangle \,,
\end{aligned}
\tag{9.2}
$$

where the last sum is over all possible *sequences* of indices $(i_1 \ldots i_n)$ containing n terms. Of course, this summation is done via a Monte Carlo sampling and not explicitly.

The expression for the partition function in eqn (9.2) is the basis of the stochastic series expansion (SSE) QMC method created [38, 39] and developed [40] by Sandvik in the 1990s. The SSE formulation is in fact very similar to the PI form of eqn (9.1), which can be recovered via a time-dependent perturbation theory in imaginary time [35, 41]. In practice, the SSE representation is simpler to use, except when the Hamiltonian has large diagonal terms.

The sampling of eqn (9.2) can also be done with a cluster-type algorithm [40], which can be further improved with the use of "directed loops" [42, 43] (this notion will be reused later). This results in a powerful algorithm to study quantum lattice models of spins or bosons. Recent extensions allow one to develop the Hamiltonian into terms H_i that contain several lattice sites ("plaquette SSE"), improving the performance [44]. The SSE method is now mainstream and the original work of Sandvik [38–41] has been nicely covered in many articles, including articles on directed loops [42, 43] and on measurements of dynamical quantities [45].

9.3.3 Sign problem

QMC simulations of quantum lattice models are very efficient if they can be formulated in the PI or SSE formalism. This is not always possible for all models: for instance, models where the configuration space is *constrained* do not easily fit into the above formulations. Also, if complex many-body terms are present, algorithms slow down considerably or cannot be written at all.[6] In that case, projection methods are preferable (see Section 9.5).

There is a situation which is even worse: when the Hamiltonian possesses positive off-diagonal elements, some terms $\langle c|e^{-\Delta\tau H_i}|c'\rangle$ in eqn (9.1) or $\langle c|(-H_i)|c'\rangle$ in eqn (9.2) can be negative and render the full Monte Carlo weight negative: this is the minus sign problem, or simply *sign problem*. Note that for diagonal terms, there is no sign problem as a large negative constant can always be added to H without changing the physics. The sign problem also occurs with the same matrix elements for fermionic lattice models.[7]

As already stated, no simple solutions exist to the sign problem. For instance, sampling with the absolute value of the weight does not work, even if done properly. The sign problem usually leads to an *exponential* (in the system size and β) increase in the error bars in simulations. See [23] for a nice introduction to the sign problem. In practice, QMC simulations in the presence of the sign problem are barely competitive even for small samples or at high temperature: it is usually simpler and better to use exact diagonalization or series expansion techniques in these respective situations. Attempts to solve or reduce the sign problem for

[6] Note that some models with few-body but long-ranged interactions can be efficiently simulated using these techniques [47].

[7] For some 1D models, the fermionic sign problem can be avoided by the choice of open boundary conditions (such that two fermions never "cross").

specific models can work [46], but the reader should be warned that there have been more rewardless than successful tries.

9.4 Classical worm algorithm

QMC simulations can be performed quite systematically on a given Hamiltonian using the PI or SSE formalism. This leads to an effective $d+1$ classical problem which, if looked at from the statistical-physics point of view, is not remarkable: the equivalent problem is asymmetric and usually contains complex multibody interactions. This is precisely the opposite of the situation in statistical physics where we look for models that have simple ingredients but contain the essential physics.

It is sometimes possible to derive from a path integral approach $d+1$ classical models that are simple enough to merit their own appellation. The price to pay is, of course, an approximate treatment of the original quantum model in dimension d, but as far as universal behavior is concerned, the simplified $d+1$ classical model is enough. A further reason for interest in these classical models is that one can easily add extra interactions by hand to test a new behavior without having to search precisely for which exotic terms in the quantum model would allow this. In a word, these types of classical models are good, effective toy models for describing complex quantum behavior. A final argument in favor of these classical models is that they can often be simulated very efficiently with classical MC algorithms (which are much simpler to develop than their quantum counterparts).

In the following, we will discuss this approach in the context of deriving a *link-current* model from a quantum bosonic Hubbard model. The link-current model is a *constrained* classical model: the constraints are where the original quantum complexity hides. Simulations of this model can be performed with great efficiency with a *classical worm* MC algorithm, which will be described in detail. This worm algorithm has in fact nothing specific to do with the link-current model, and Section 9.4.3 describes how it can be generalized to many other constrained classical models.

9.4.1 The link-current model

We first sketch the derivation of a classical current model from a quantum bosonic model (the exact steps of the mapping can be found in [49]). As this derivation is not essential for the description of the link-current model and of the worm algorithm, the description of the derivation below can be skipped at first reading.

Derivation. We start from the soft-core bosonic Hubbard model

$$H = -t \sum_{\langle i,j \rangle} b_i^\dagger b_j + \text{h.c.} + \frac{U}{2} \sum_i n_i^2 - \mu \sum_i n_i \,, \tag{9.3}$$

where b_i and $b_i^\dagger$ are the boson annihilation and creation operators, respectively, and $n_i = b_i^\dagger b_i$ is the boson density at site i (many bosons can sit on the same site). t is the hopping amplitude on nearest-neighbor sites $\langle i,j \rangle$, U is the on-site repulsion, and μ is the chemical potential.

Let us decompose the boson operator in terms of its amplitude and phase, i.e. $b = |b| e^{i\phi}$. When the average number of bosons per site is large, i.e. $\langle n \rangle = n_0 \gg 1$, amplitude fluctuations around n_0 are very costly (owing to the large contribution of the on-site repulsion term).

It is sufficient to keep the phase degrees of freedom: this is the "phase only" approximation [48]. Integration of the amplitudes leads in this limit to the *quantum rotor model*,

$$H = -t \sum_{\langle i,j \rangle} \cos(\phi_i - \phi_j) + U \sum_i n_i^2 - \mu \sum_i n_i \,, \tag{9.4}$$

where n now represents the deviation of the density about n_0, and the phases ϕ and densities n are canonical conjugates, i.e. $[\phi_k, n_j] = i\delta_{kj}$. We now form the path integral by using eqn (9.1) (with H_1 the first term of H, and H_2 the two other terms) and inserting the resolution of identities in the occupation number basis $\{|n\rangle\}$. H_2 is diagonal in this basis but H_1 is not, and we now perform some approximations on this term to simplify calculations.

We first use the Villain approximation, where the exponential of a cosine is replaced by a sum of periodic Gaussians,

$$e^{-t\,\Delta\tau\,\cos(\phi)} \simeq \sum_m e^{\Delta\tau(\phi-2\pi m)^2/2} \,.$$

We then use Poisson summation,

$$\sum_m f(m) = \sum_J \int_{-\infty}^{\infty} dx\, e^{-2i\pi Jx} f(x) \,,$$

and perform the integration over x. We arrive at

$$e^{-t\,\Delta\tau\,\cos(\phi)} \simeq \sum_J e^{-J^2/2\,\Delta\tau\, t - iJ\phi} \,,$$

where the J's are integer numbers that live on the *links* (bonds) of the d-dimensional lattice. Essentially, they are bosonic *currents*: if $J = 1$ on a bond of the lattice, it means that one boson has hopped from one site at the end of this link to the other.

We reinject this form into the path integral, and remember that phase and density are conjugate: we have $\langle n|e^{-iJ\phi}|n'\rangle = \delta_{n+J,n'}$. Rearranging terms, we find that at every site $\nabla \cdot \mathbf{J}^{(d)} + (n_{\tau+1} - n_\tau) = 0$, where ∇ is the lattice divergence and $\mathbf{J}^{(d)}$ is the vector of currents J that start from a given site. n_τ is the boson occupation number of this site at time slice τ. This equation just means that the number of bosons is conserved. Now we can simply define $J^\tau \equiv n_\tau$ and have the constraint

$$\nabla \cdot \mathbf{J}^{(d+1)} = 0 \,,$$

with the current vector and lattice divergences now living in $d+1$ dimensions.

The classical model. Performing further approximations (a noteworthy one is a rescaling of the amplitude of terms in the $d+1$ direction—this should not affect universal behavior), we finally arrive at a classical link-current model

$$Z = \sum_{\substack{J^\alpha = 0, \pm 1, \pm 2 \ldots \\ \nabla \cdot \mathbf{J} = 0}} e^{-1/K \sum_{\mathbf{r}} (1/2)\mathbf{J}_{\mathbf{r}}^2 - \mu J_{\mathbf{r}}^\tau} \tag{9.5}$$

with $\mathbf{J} = (J^1, J^2, \ldots, J^d, J^\tau)$ integers living on the links of a $d+1$-dimensional lattice (at sites with coordinates $\mathbf{r}$). We use the notation J^τ for the current in the last dimension to keep

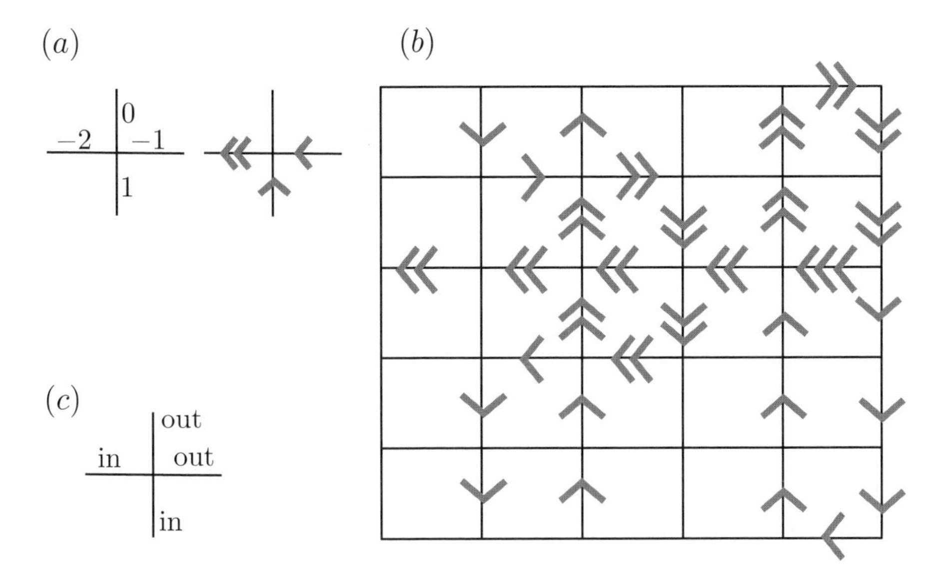

Fig. 9.1 Arrow representation of link currents: (a) definition, (b) example of a link-current configuration in (1+1) dimensions, (c) convention for incoming and outgoing links.

track of the original quantum problem, but this is not essential. Here $\mathbf{J_r}^2 = (J^1_\mathbf{r})^2 + \ldots + (J^d_\mathbf{r})^2 + (J^\tau_\mathbf{r})^2$. K plays the role of the temperature in the classical model and scales as t/U in the quantum problem.

If we stopped the description of the model now, this would be a trivial problem as all the variables J would be independent. The interactions come from the divergenceless condition $\nabla \cdot \mathbf{J} = 0$. To be consistent, we define

$$J^{-\alpha}_\mathbf{r} = -J^\alpha_{\mathbf{r}-\mathbf{e}^\alpha}\,, \tag{9.6}$$

where $\alpha = 1 \ldots \tau$ is a direction of the lattice, and $\mathbf{e}^\alpha$ is the unit vector in this direction. A configuration of link currents satisfying the divergenceless condition is nicely represented graphically by drawing $|J|$ arrows (oriented according to the sign of J) on each bond carrying a current J (see Fig. 9.1).

The divergence constraint makes for nonintegrability of the model, and numerical simulations are necessary. Recently, an efficient MC *worm algorithm* [50, 51] has been devised for the model of eqn (9.5).

9.4.2 Classical worm algorithm for the link-current model

The idea of the algorithm is the following: starting from a valid current configuration, we create a new configuration by the motion of a fictitious "worm" through the links of the lattice. Each time the worm passes through a link, the current carried by the link is modified. The configurations generated during the construction of the worm are not valid (the divergenceless condition is violated), but as the worm ends up forming a closed loop, the final configuration is valid.

To see what the exact action of a worm is when it passes through a link, we first have to choose a convention for the link orientation. Around each site $\mathbf{r}$ are located $2(d+1)$ links carrying currents $J_{\mathbf{r}}^{\sigma}$ with $\sigma = \pm 1, \ldots, \pm d, \pm\tau$ (see eqn (9.6)). We define as outgoing links those in the positive directions $1, \ldots, d, \tau$ and as incoming links those in the negative directions $-1, \ldots, -d, -\tau$. This convention is expressed in Fig. 9.1(c). Now the worm, on leaving the site at $\mathbf{r}$ by passing through an outgoing link $\sigma \in \{1, \ldots, d, \tau\}$, will cause the modification

$$J_{\mathbf{r}}^{\sigma} \to J_{\mathbf{r}}^{\sigma} + 1\,. \tag{9.7}$$

For an incoming link $\sigma \in \{-1, \ldots, -d, -\tau\}$,

$$J_{\mathbf{r}}^{\sigma} \to J_{\mathbf{r}}^{\sigma} - 1\,. \tag{9.8}$$

A different orientation convention can be chosen at the beginning of each MC step, but this is not necessary to ensure ergodicity.

We now describe the construction of the worm. First, we position the worm at an initial site $\mathbf{r}_0$ of the lattice. For its first move and only for its first,[8] we choose *randomly* one of the $2(d+1)$ directions and move the worm in this direction. The worm then continues its motion, but for all subsequent moves the next direction σ is chosen with a probability $P_{\mathbf{r}}^{\sigma}$ (which will given below) instead of randomly. Each time the worm passes through a link of the lattice, the corresponding current is modified according to the rules of eqns (9.7) and (9.8). Eventually, the worm reaches its initial point $\mathbf{r}_0$ (this is always true for a finite lattice): this is the end of the worm move. The final configuration of currents left by the worm is now entirely valid (see Fig. 9.2 for an illustration of a worm motion and the changes made to the original current configuration).

The only unknowns left are the values of $P_{\mathbf{r}}^{\sigma}$. One can show [51] that in order to respect the detailed-balance condition for the initial and final current configurations, it is sufficient to impose a *local detailed-balance* condition on $P_{\mathbf{r}}^{\sigma}$. Assume that the worm passes through the link σ at $\mathbf{r}$: this results in a change of the ratio of the global Boltzmann weights $(\Delta W)^{\sigma}$. This change is particularly easy to compute by noticing that the Boltzmann weight of the model in eqn (9.5) is *solely carried by the links*. The link "energy" is $E_{\mathbf{r}}^{\sigma} = \frac{1}{2}(J_{\mathbf{r}}^{\sigma})^2 - \mu\delta_{\sigma,\pm\tau}J_{\mathbf{r}}^{\sigma}$ *before* the worm passage and $E_{\mathbf{r}}^{\prime\sigma} = \frac{1}{2}(J_{\mathbf{r}}^{\sigma} \pm 1)^2 - \mu\delta_{\sigma,\pm\tau}(J_{\mathbf{r}}^{\sigma} \pm 1)$ *after* (the sign $\pm$ depends on whether σ is an outgoing or incoming link). We therefore have $(\Delta W)^{\sigma} = \exp(-\Delta E_{\mathbf{r}}^{\sigma}/K)$, where $\Delta E_{\mathbf{r}}^{\sigma} = E_{\mathbf{r}}^{\prime\sigma} - E_{\mathbf{r}}^{\sigma} = \pm J_{\mathbf{r}}^{\sigma} \mp \mu\delta_{\sigma,\pm\tau} + 1/2$. The local detailed-balance condition reads

$$\frac{P_{\mathbf{r}}^{\sigma}}{P_{\mathbf{r}+\mathbf{e}^{\sigma}}^{-\sigma}} = (\Delta W)^{\sigma} = \exp\left(-\frac{1}{K}\left[\pm J_{\mathbf{r}}^{\sigma} \mp \mu\delta_{\sigma,\pm\tau} + \frac{1}{2}\right]\right). \tag{9.9}$$

On the other hand, the worm must always move somewhere, which gives another condition:

$$\sum_{\sigma} P_{\mathbf{r}}^{\sigma} = 1\,. \tag{9.10}$$

A simple solution to the two constraints of eqns (9.9) and (9.10) is the "heat bath" solution

$$P_{\mathbf{r}}^{\sigma} = \frac{\exp(-(1/K)E_{\mathbf{r}}^{\prime\sigma})}{\sum_{\sigma}\exp(-(1/K)E_{\mathbf{r}}^{\prime\sigma})}\,, \tag{9.11}$$

which physically means that the worm follows a direction in proportion to the new Boltzmann weight that will be created by this move. $\Delta E_{\mathbf{r}}^{\sigma}$ and $P_{\mathbf{r}}^{\sigma}$ can be easily calculated on the fly

[8] This is a slight variation on the algorithms of [50, 51] which ensures a unit acceptance probability and simple computations of Green's functions.

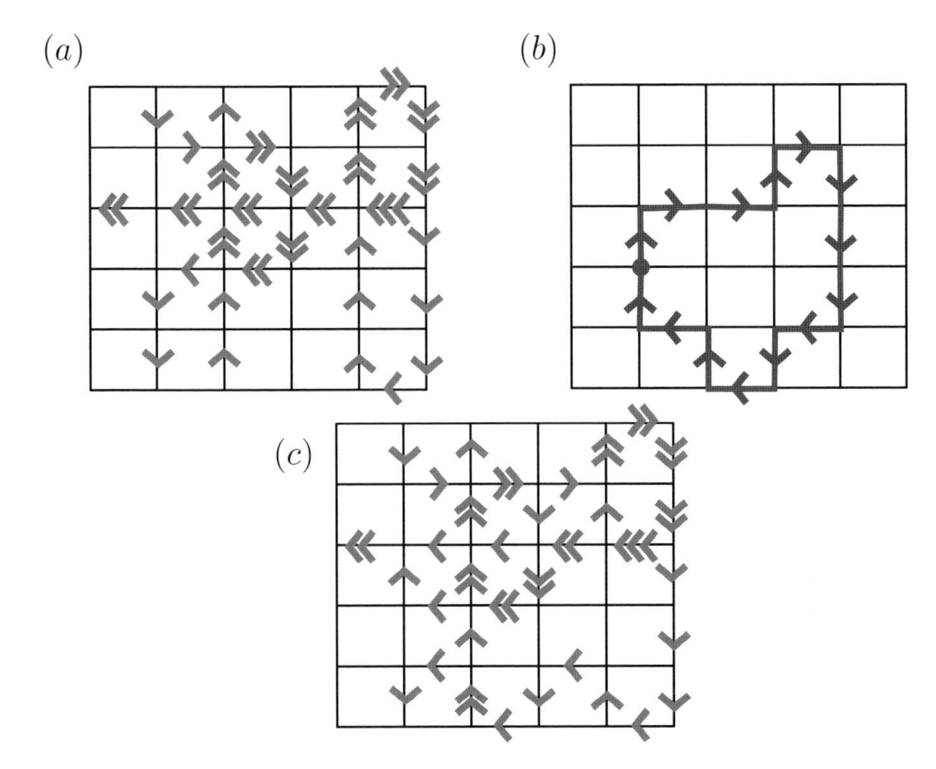

Fig. 9.2 An initial link-current configuration (a) is modified nonlocally by a worm (b) to give a different current configuration (c).

given the value of $J^{\sigma}_{\mathbf{r}}$. As the J's typically take only a few possible values, all the Boltzmann factors can be precalculated to speed up the calculations.

A close inspection of what happens locally to the worm reveals that we can do even better. Sometimes the worm goes in one direction and immediately comes back: this is known as a "bounce" process. Usually, bounces are bad because CPU time is lost in undoing some change that has just been performed. We thus introduce the notion of a "directed worm" algorithm [51], where bounce processes are avoided. The previous heat bath probabilities (9.11) do not allow us to do this, since they depend only on the direction of the worm. But there are, in general, more solutions to eqns (9.9) and (9.10).

The best way to proceed is to consider the conditional probability $p_{\mathbf{r}}(\sigma_i|\sigma_j)$ that the worm at site $\mathbf{r}$ goes in the direction σ_i, when we know that it came from direction σ_j. We form the $2(d+1)\times 2(d+1)$ scattering matrix $P_{\mathbf{r}}$ with matrix elements $P^{ij}_{\mathbf{r}} = p_{\mathbf{r}}(\sigma_i|\sigma_j)$, which give the probability that the worm scatters from direction σ_j to direction σ_i at site $\mathbf{r}$. What are the restrictions on this scattering matrix? First, $P_{\mathbf{r}}$ must be a stochastic matrix since $\sum_{i=1}^{2(d+1)} P^{ij}_{\mathbf{r}} = 1$ to satisfy the normalization of probabilities in eqn (9.10). Also, the off-diagonal elements should fulfill local detailed balance: $P^{ij}_{\mathbf{r}} \exp(-E'^{\sigma_i}_{\mathbf{r}}/K) = P^{ji}_{\mathbf{r}} \exp(-E'^{\sigma_j}_{\mathbf{r}}/K)$. Finally, we wish to minimize the bounce processes, that is, all *diagonal elements* of the matrix: for symmetry reasons, we try to minimize $\mathrm{Tr}(P_{\mathbf{r}})$.

All of the constraints and this last optimization goal define a simple *linear programming* problem. This can be solved analytically for small matrices, but in practice it is preferable to use standard numerical linear programming techniques [52] to precalculate the scattering matrices. There is a matrix P for all possible link configurations, and therefore it is also useful to precompute all these possible configurations. In practice, one is aided by the fact that the $|J|$'s never go beyond 3 or 4 for all cases of interest.[9] Also, the difference between all incoming and all outgoing currents is 1 (in absolute value) by construction, which further limits the number of configurations.

This small modification of the scattering probabilities leads to a very large gain in efficiency, as bounce processes are usually the major bottlenecks of worm algorithms. This strategy has also been shown to be very useful in the "directed loop" versions of SSE [42, 43], and therefore the investment of precalculating and optimizing these scattering matrices is really worthwhile (a comparison of the efficiency of all variants is given in [51]).

A nice aspect of the worm algorithm as described above is that all worms are accepted with a probability 1. This is a nontrivial feature for an MC method that generally leads to good algorithmic performance. This is indeed the case, as a gain of up to three orders of magnitude in the number of link variables has been obtained thanks to the worm algorithm [50, 51] as compared with previous local MC simulations of the same model [49]. Also, the possibility of the worm passing through the periodic boundary conditions allows an ergodic sampling of phase space (previous local MC algorithms lacked this possibility, resulting in biased estimates of observables).

Finally, a great advantage of the worm algorithm is that it allows calculations of observables that are typically difficult to obtain, namely Green's functions, and this is for free! Indeed, one can show [51] that the histogram $H(\mathbf{d})$ of the distance $\mathbf{d} = \mathbf{r} - \mathbf{r}_0$ between the current position of the worm and its original position is *exactly* proportional to the Green's function $G(\mathbf{d} = (\mathbf{s}, \tau)) = \langle \exp[i(\phi_{\mathbf{s}}(\tau) - \phi_{\mathbf{0}}(0))] \rangle$ of the original quantum rotor model. This is a particularly simple measurement to perform during the construction of the worm.

To finish, a remark about terminology: we have used the name "classical" worm algorithm, but the expert reader might wonder what the difference is with "quantum" worm algorithms, as used in path integral simulations of lattice models [35] and continuum models [53]. The distinction is admittedly quite loose, but one probably talks preferentially about a classical worm algorithm if the model obtained from the path integral is a simple effective classical model in the sense of the introduction to Section 9.4.

9.4.3 Worm algorithms in general

We have learned from an example how a worm algorithm actually works with a link-current model, but the idea behind worm algorithms can be easily generalized to many situations. This is particularly true for the large family of *locally constrained* (classical) models, for which the partition function can be written

$$Z = \sum_x{}' W(x)\,,$$

where the prime on the sum denotes a local constraint on the configurations x ($W(x)$ is the Boltzmann weight associated with x). Such local constraints can be directly built into the

[9] If, by a rare fluctuation, a few $|J|$'s were to be larger, scattering matrices for these rare events could always be computed on the fly.

model (for instance in dimer or vertex models). One can also perform transformations in order to express an unconstrained model in terms of constrained variables: this is the case for the dual transformation of the Ising model [54], and for the path integral formalism that leads from the soft-core bosonic Hubbard model to the link-current model. A more or less systematic way of performing these transformations for a generic model is through a high-temperature series expansion, which is the formalism in which the classical worm algorithm was originally developed [55].

Once the constrained model has been defined, the usual phenomenology of a classical worm algorithm is the following:

- Break the local constraint by introducing two defects (the head and tail of the worm).
- Propagate one of the defects, and modify the configuration on the fly: this leads to an invalid configuration at every propagation step. Measure path statistics (such as the histogram of the distance between defects) if needed.
- Propagation stops when the defect reaches its original position. Recombine the defects (the configuration is now valid).
- Quite generally, path statistics of the worm are related to physical quantities (typically Green's functions).

This strategy allows one to efficiently sample the constrained phase space. In particular, it usually allows one to sample all topological sectors (if there are any in the model) if the worm is allowed to move through the periodic boundary conditions of the sample. There is also a nice trick that in some situations allows one to simulate systems with periodic and antiperiodic boundary conditions at the same time. This results in an efficient calculation of the domain wall free energy (see [56] for an illustration using the Ising model).

One must be careful also to sample correctly the Boltzmann weight $W(x)$. The phenomenology here is to perform this sampling at the local level, that is to say, to implement a *local detailed-balance propagation* of the worm. Each elementary step of the worm must satisfy detailed balance, i.e. the worm must go in a given direction with a probability proportional to the Boltzmann weight resulting from a propagation in this direction.[10] In doing so, we ensure that the MC move is accepted *with a probability 1*. This is at odds with a naive algorithm, which would account for a change in the Boltzmann weight through a post-move acceptance rate. Such algorithms typically perform badly, as the acceptance rate vanishes exponentially with the worm size.

The simultaneous use of these two sampling strategies typically produces highly efficient nonlocal algorithms, leading to simulations of large systems with small autocorrelation times. The successful applications of the worm algorithm encompass a large variety of systems, including pure [55, 56] and disordered [57] Ising models, Potts and XY models [55], vertex models [58], dimer models [59, 60], link-current models [50, 51, 61], lattice gauge theories [62], charged Coulombian systems [63], and probably more to come.

One intriguing aspect of this family of algorithms is that their efficiency for a model in a given phase is intimately related to the physical behavior of defects in that phase: the more they are localized, the less efficient the algorithm will be (as the generated worms will be short). This is why we expect these algorithms to be rather efficient at second-order phase transitions, where the fluctuations of and correlations between defects are long-ranged. In a

[10] The meaning of this long sentence should be clear to the reader of Section 9.4.2.

word, worm algorithms are physical. An amusing consequence is that one can obtain a good guess of where a phase transition is located by looking at the behavior of internal algorithmic variables such as the average worm length.

9.5 Projection methods

9.5.1 Introduction

The idea of the second family of methods is simple: we apply an operator a large number of times to any wave function, and in most cases we end up with the eigenstate associated with the largest eigenvalue (in modulus) of the operator. This idea is at the heart of the success of exact iterative methods (such as the Lanczos and Davidson algorithms) and of the stochastic methods that we describe now. Choosing our notation, we have

$$\lim_{n\to\infty} K^n(H)|\psi_T\rangle \sim |\psi_0\rangle \,.$$

The operator K will of course be closely related to the Hamiltonian of the system, in order to obtain its ground state (GS) $|\psi_0\rangle$. Frequent choices are $K(H) = e^{-\tau(H-E_T)}$ and $K(H) = (1+\tau(H-E_T))^{-1}$ for systems in the continuum, and $K(H) = 1-\tau(H-E_T)$ for the lattice models that we are interested in. The first two choices are widely used in various fields of quantum physics (quantum chemistry, molecular physics, etc.) and are known under the respective names of the diffusion and Green's function methods. The latter choice is most often referred to as the power method. Even if it is not new [64], this method is not so often used for the study of quantum lattice statistical-physics models, and we now provide a gentle introduction to it. As for any method imported from other fields, it comes with a lot of variants and frightening jargon (indicated in italic). We will, however, see that the method is rather easy to understand and of great practical use. A pedagogical introduction is also given in [65].

In the expressions above, we have introduced a projection number n, a trial wave function $|\psi_T\rangle$, a time step τ (later considered infinitesimal), and a reference energy E_T.

The trial state $|\psi_T\rangle$ can be chosen randomly, as a result of a variational calculation, or according to physical intuition. It should not be orthogonal to the GS (otherwise the algorithm will converge to an excited state). A good choice is to consider a state that has all the known symmetries of the GS. Finally, it is important to emphasize that the closer $|\psi_T\rangle$ is to the GS, the faster the algorithm converges.

The projection number must be large in order to ensure proper convergence to the GS, but is of course finite. How large should it be? This depends on the value of the gap of the model (the total energy difference between the GS and the first excited state)—the larger the gap, the smaller n is—and on the quality of the trial and guiding (see later) wave functions (the closer to the GS, the smaller n is).

The role of the reference energy E_T is essentially to shift down the Hamiltonian such that the GS has the highest eigenvalue in modulus. If E_T is close to the GS energy, it also helps in improving the statistics.

9.5.2 Stochastic sampling

In MC projection methods, the application of $K(H)$ is simulated stochastically. This stochastic evolution can be rigorously formalized in a Markov chain process through the Feynman–Kac formula [66], but here we focus on a more intuitive and pragmatic approach. Ideally, one would like to set up a stochastic process (performed by a *walker*) in configuration space, where $\mathcal{P}_{c\to c'}(\tau) = \langle c|1 - \tau(H - E_T)|c'\rangle$ is the probability that the walker moves from configuration c to configuration c' during a time step. The first problem arises from the fact that H must have negative off-diagonal elements as probabilities would otherwise be negative: this is yet another instance of the sign problem. The projection MC community has put a lot of effort into circumventing this problem by using the *fixed-node approximation*, where one imposes the nodes of the GS wave function [67]. We will not take this option here, and will simply assume that H has no sign problem.

The second problem is that, strictly speaking, one cannot interpret $\mathcal{P}_{c\to c'}(\tau)$ as a probability, since it is not normalized: $\sum_{c'} \mathcal{P}_{c\to c'}(\tau) \neq 1$ in the general case. The cure is simple:

$$\mathcal{P}_{c\to c'}(\tau) = P_{c\to c'}(\tau).w_{c,c'}(\tau)$$

is decomposed into two terms, where

$$P_{c\to c'}(\tau) = \langle c|1 - \tau(H - E_L)|c'\rangle$$

is the probability to go from c to c', and

$$w_{c,c'}(\tau) = \frac{\langle c|1 - \tau(H - E_T)|c'\rangle}{\langle c|1 - \tau(H - E_L)|c'\rangle}$$

is the weight assigned to the walker performing this move. The expressions above have introduced the *local energy* $E_L(c) = \sum_{c'} \langle c|H|c'\rangle$, which is just the sum of the Hamiltonian matrix elements over a column. It is easy to check that $P_{c\to c'}(\tau)$ is now correctly normalized.

The two elementary processes in the random walk correspond to two different cases for the value of c': (i) $c' \neq c$, and the walker moves from c to c' with a probability $\mathcal{P}_{c\to c'}(\tau) = -\tau H_{cc'}$ and sees its weight unchanged, and (ii) $c' = c$, and the walker stays in configuration c with a probability $\mathcal{P}_{c\to c}(\tau) = 1 + \tau \sum_{c'\neq c} H_{cc'}$ and has its weight multiplied by $(1 - \tau(H_{cc} - E_T))/(1 - \tau(H_{cc} - E_L(c)))$. The physical interpretation is simple: if the diagonal elements of the Hamiltonian are large, the walker preferably stays for long periods of time in the same configuration, whereas if the off-diagonal elements dominate, it jumps frequently from one configuration to another.

9.5.3 First sampling schemes

Now the setup seems ready for a direct simulation (by the so-called *pure diffusion MC* method [66]). Instead of launching just one walker, it is preferable to use a larger number M of walkers to increase statistics. In this scheme, each walker follows an (imaginary-)time evolution according to the rules above ("change" or "stay" moves) with according modifications of its weight (first initialized to 1). After many time steps, one can perform measurements where the final configurations contribute according to their weights. This scheme works in principle, but in practice the weights will very likely over- or underflow. As a result, few walkers contribute actively to the averages and we are left with a statistical problem. To solve it, the next idea is of course to sample configurations according to their weights: this corresponds to *branching* techniques.

The idea of branching techniques is elementary: walkers with large weight are preferred. In consequence, those which have too small a weight are killed, while walkers with a large weight are duplicated. Thus, every now and then (typically not at every time step but after some control time adjusted empirically), each walker is duplicated into m copies (with the same current configurations c and weight $w_c(\tau)$). m is a non-negative integer, which should be proportional to the weight of the walker. A typical choice that fulfills detailed balance is $m_c(\tau) = \text{int}(w_c(\tau) + r)$, where int denotes the integer part and r is a random number uniformly distributed in $[0, 1]$. Other schemes (where, for instance, m is forced to be 0 or 2) also exist [68]. By the introduction of this branching process, the walker weight is correctly taken into account and all walkers will contribute equally to statistical averages. But yet another problem occurs: as the number of walkers fluctuates owing to the branching process, it might happen that this number actually explodes or vanishes—which is of no practical good. Successful solutions to this problem also exist and are known as *population control* methods. Among others, a convenient scheme for lattice models (*stochastic reconfiguration*) has been proposed in [69, 70], where the number of walkers is kept fixed without any bias in the simulation (see also [71]).

9.5.4 Measurements

Another potential worry for projection-type methods is the measurement part. All the schemes above generate configurations that are indeed representative of the GS $|\psi_0\rangle$, but we need two representatives of the GS (on the left- and right-hand sides) in order to form the expectation value $\langle \mathcal{O} \rangle = \langle \psi_0 | \mathcal{O} | \psi_0 \rangle / \langle \psi_0 | \psi_0 \rangle$ of any observable $\mathcal{O}$. Performing an average over the final configurations yields only *mixed estimates* of the form $\langle \psi_T | \mathcal{O} | \psi_0 \rangle / \langle \psi_T | \psi_0 \rangle$, which are in general different from $\langle \mathcal{O} \rangle$.

It turns out that this distinction is not relevant for observables that commute with the Hamiltonian. The GS is indeed an eigenvector of the observable, and it is easy to see that the expectation value and mixed estimate coincide. The most important observable having such a property is of course the Hamiltonian itself, yielding the GS energy E_0. It can easily be measured as an average of the local energy $E_0 = \langle E_L \rangle$ over configurations (there is no additional weight in the average once the configurations have been selected according to their weights as in branching methods).

For diagonal observables that do not commute with H, some extra effort must be performed, which goes under the name of *forward walking* [72]. We shall just give the idea here: starting from the final configurations on which the observable is measured, a few propagation steps are performed during a certain time (the *forward-walking time* t_{FW}). In this way, the second representative of the GS needed for the expectation value is created. This extra propagation generates another series of configurations with their own weights. Now these weights are used to reweight the previously performed measurements, in order to obtain the expectation value. If the forward-walking time is taken large enough, the expectation value converges to its correct value. This is the naive description of the forward-walking techniques and, of course, there are clever ways of reinterpreting the previous normal MC time steps as forward-walking steps in order to save CPU time and increase statistics. This involves, however, quite some bookkeeping of the various measurements.

9.5.5 Reptation Monte Carlo method

We will not go further into the details of population control and forward-walking techniques. Instead we will focus on another method, which turns out to be much simpler to describe and to use on both the population control and the measurement fronts. This method has been dubbed *reptation MC* [73, 74]. The idea of reptation MC is the same as that of the techniques described previously, i.e. to perform importance sampling on the walker weights, but it gives a simple and practical way of doing so that synthesizes all the good aspects of previous methods.

Description of configuration space. Reptation MC works with a single walker, which already offers a clear practical reason for interest. Starting from an initial configuration c_0, the walker evolves (with the standard "change" and "stay" rules) during a certain *control time* $\Delta\tau$, yielding a configuration c_1 and a weight w_1. This operation is performed n times in order to have a series of $n+1$ configurations $\{c_0, c_1, \ldots, c_n\}$ and n weights $\{w_1, \ldots, w_n\}$. The symbol n is by no coincidence the same as that for the projection number introduced in Section 9.5.1. The ensemble of the configurations is called a *reptile* $\mathcal{R}$ (for reasons that will become clear) and constitutes one MC configuration in the reptation method. The weight of the reptile $\mathcal{R}$ is the product of all weights $W = \prod_{i=1}^{n} w_i$.

Elementary move. An elementary move, which is called with no surprise a *reptation*, can now be performed. The idea is to generate a new reptile $\mathcal{R}'$ by removing one part (a *vertebra*) of $\mathcal{R}$ at either its head or its tail, and adding another one at the other end. In practice, a direction of propagation is chosen randomly: up (starting from c_n) or down (starting from c_0). Without loss of generality, let us choose up. c_n is evolved during a time $\Delta\tau$ in order to obtain c_{n+1}, with a corresponding weight w_{n+1}. Now the reptile $\mathcal{R}'$ is the ensemble of configurations $\{c_1, \ldots, c_n, c_{n+1}\}$ (note that c_0 has been removed from the list), and its weight is $W' = \prod_{i=2}^{n+1} w_i$. The probability to accept this reptation from $\mathcal{R}$ to $\mathcal{R}'$ can be chosen[11] to be of the Metropolis type $P(\mathcal{R} \to R') = \min(1, W'/W)$. Another practical advantage of reptation MC is that the ratio W'/W is readily evaluated as w_{n+1}/w_1, since all the other weights in the bulk of the reptile cancel. If the move is accepted, it is convenient to rename all configurations and weights from indices $1, \ldots, n+1$ to indices $0, \ldots, n$. This reptation move constitutes one MC step.

The reptation method is advantageous for various reasons. First, it solves the problem of the under- or overflow of weights (or of the number of walkers) without the need for population control methods. The implementation is simple, as there is only one walker and the transition rate from one reptile to another is easily computed. Any *vertebra* (configuration) inside the bulk of the reptile is representative of the GS, and it helps in performing measurements (see below). Finally, all tricks developed for other projection MC methods can be adapted without difficulty to reptation MC, as will be seen.

Measurements. Measurements can be performed between any successive reptation moves. Mixed estimates (for observables that commute with H) can be performed on any configuration in the reptile. To avoid autocorrelations, it is preferable to measure at the two ends. For instance, the GS energy is estimated as $E_0 = \frac{1}{2}\langle E_L(c_0) + E_L(c_n)\rangle$. As already stated, any configuration not too close to the ends of the reptile can be regarded as representative of the GS. This means that diagonal observables can be measured directly on the configurations in the middle of the reptile (say at vertebra $n/2$), without resorting to forward walking.

[11] There is a subtlety here, as this expression is valid only (see [74] for details) in the limit of continuous time, a notion that will described later. Since for all practical purposes continuous-time algorithms perform much better, we assume in the following that they are used.

Implicitly, a forward-walking measurement is performed at a forward time $t_{\rm FW} = n\,\Delta\tau/2$, but it comes for free. In practice, one usually measures diagonal observables at every vertebra and plots the measured value as a function of vertebra index. The curve should be symmetric around $n/2$. Convergence (which is ensured for large enough n) is readily checked if there is a plateau in such a curve close to the middle of the reptile. It is more tricky to measure off-diagonal and time-dependent observables, but this can also be done (see e.g. the discussion in [75, 76]).

Improvements to the reptation method. How can the efficiency of the method be further improved?

1. *Vertebrae.* The first trick that can be used is to add/remove M vertebrae (instead of one) in a reptation move. M and the control time $\Delta\tau$ are free parameters of the algorithm, and are usually chosen empirically such that the reptile acceptance rate is reasonable (say more than $1/3$).
2. *Continuous time.* The second trick is to perform the evolution of the configuration (the "change" and "stay" moves) directly in continuous time, that is to say to take the formal limit of vanishing time step $\tau \to 0$ directly in the algorithm. This idea is common to almost all projection methods for lattice models and can actually be found in various places in the MC literature (see e.g. [36]). The procedure is the following: we consider the limit of infinitesimal $\tau = d\tau$ in the "change" and "stay" probabilities, which become $\mathcal{P}_{c\to c'}(\tau) = -d\tau H_{cc'}$ and $\mathcal{P}_{c\to c}(\tau) \sim 1$, respectively (keeping the dominant term only). This means that most of the time, the configuration remains identical and CPU time is lost in proposing "change" moves that are accepted only infinitesimally. The situation is obviously reminiscent of radioactive decay. With this analogy in mind, the solution to this problem is clear: we can integrate all the useless "stays" by directly generating the next *decay time* t_d, distributed according to a Poisson distribution $P(t_d) = \lambda_c e^{-\lambda_c t_d}$, with "inverse lifetime" $\lambda_c = -\sum_{c'\neq c} H_{cc'}$. Consequently, the weight accumulated during all the "stay" processes that have been shortcut integrates out to $e^{-(E_L(c)-E_T)t_d}$. In continuous-time methods, a walker in configuration c changes to a configuration c' and the clock is directly advanced by an amount t_d. Up to this point, we have not decided what configuration c' really is, as all the possible decay channels have been added together in the inverse lifetime in order to speed up the simulation. c' can simply be chosen from among all possible decay channels with a simple heat-bath choice, that is to say with a probability $p(c') = H_{cc'}/\sum_{c'\neq c} H_{cc'}$. When this is adapted to the reptation method, one just has to check if the generated decay time t_d is smaller than the control time $\Delta\tau$. If no, then the configuration remains the same. If yes, the walker moves to another configuration c' and the procedure is repeated with the generation of new decay time(s) until $\Delta\tau$ is reached. The multiplicative weights have to be correctly computed according to the local energies of the configurations that have been visited during $\Delta\tau$. Continuous-time methods are clearly of great interest as they yield no systematic time-discretization error, are faster than discrete-time methods, and turn out to be simpler to implement.
3. *Guiding wave function.* Another improvement (the crucial nature of which does not allow us to use the word "trick") is to bias the simulation with a *guiding wave function* $|\psi_G\rangle$. The idea is to sample stochastically not the GS wave function $|\psi_0\rangle$ itself, but rather the product $|\psi_G\rangle.|\psi_0\rangle$ instead. When this is done, the walker will tend to move to configu-

rations where $|\psi_G\rangle$ is large and to avoid configurations where $|\psi_G\rangle$ is small. Why do we want to do this? Taking $|\psi_G\rangle$ random is clearly a bad choice. But if $|\psi_G\rangle$ is "close" to $|\psi_0\rangle$, the walker will be biased towards configurations where $|\psi_0\rangle$ is large, the statistics will be improved, and the simulation will converge much faster. In the ideal case where $|\psi_G\rangle = |\psi_0\rangle$, the local energy *is* the GS energy, all weights are equal to unity, and the reptile needs only one vertebra: in fact, we are just performing classical MC simulation! This is typical of situations where the exact GS wave function is known [77]. In the more realistic case where $|\psi_G\rangle$ is close but not equal to $|\psi_0\rangle$, this means in practice much smaller reptile lengths n than without a guiding wave function.

It is natural to expect that any physical input will help simulations but it is hard to imagine how much CPU and physical time one can gain by using a good guiding wave function without having tested this improvement. What is really appealing is that there is little modification to do: one just has to modify the off-diagonal matrix elements of the Hamiltonian $H_{cc'}$ to $H_{cc'}|\psi_G\rangle_{c'}/|\psi_G\rangle_c$, and correspondingly the local energy, which becomes $E_L(c) = \langle c|H|\psi_G\rangle/\langle c|\psi_G\rangle$. Otherwise, the rest of the scheme remains identical, including measurements.

In principle, the guiding wave function $|\psi_G\rangle$ may be different from the trial wave function $|\psi_T\rangle$ introduced earlier. However, for some methods such as the reptation Monte Carlo method, the distinction is meaningless and/or the two wave functions are chosen to coincide (which is not a bad idea if one already has a good guiding wave function).

If one can make any physical guess (even a small one) about what the GS wave function should look like, using it as a guiding wave function is really valuable. What if no guiding wave function is used? The simulation might be expected to be less efficient. In fact, even if not specified, there is *always* an implicit guiding wave function: it is $|\psi_G\rangle = \sum_c |c\rangle$ (i.e. all configurations are equiprobable). This is interesting for models that admit a so-called Rokhsar–Kivelson point (such as quantum dimer models), where $\sum_c |c\rangle$ *is* the exact GS wave function. Close to this point, the projection method is very efficient even without an explicit guiding function!

Conclusions on the reptation method. Is the projection method (especially without a guiding wave function) as efficient as other methods? In general, the answer is positive even if the typical lattice sizes that can be reached are usually smaller than those possible with the most recent efficient path integral methods (such as loop algorithms and SSE). However, there are many models for which it is not easy to write an efficient path integral algorithm. This is the case, for instance, for models with complex many-body interactions and for models with a constrained Hilbert space (such as the quantum dimer model). Also, there are situations where it is preferable to work in the canonical ensemble, which is generally used in the projection method, whereas efficient path integral methods work in the grand canonical ensemble instead.

9.6 Valence bond projection method

In this section, we describe a projection MC method for models of quantum spin systems based on an unconventional basis, namely the valence bond (VB) basis. The original properties of this basis induce several new aspects and variants of the projection method, which turn out to be very advantageous. Moreover, this method is quite recent and certainly contains several interesting things to discover. In particular, various quantities typical of quantum information

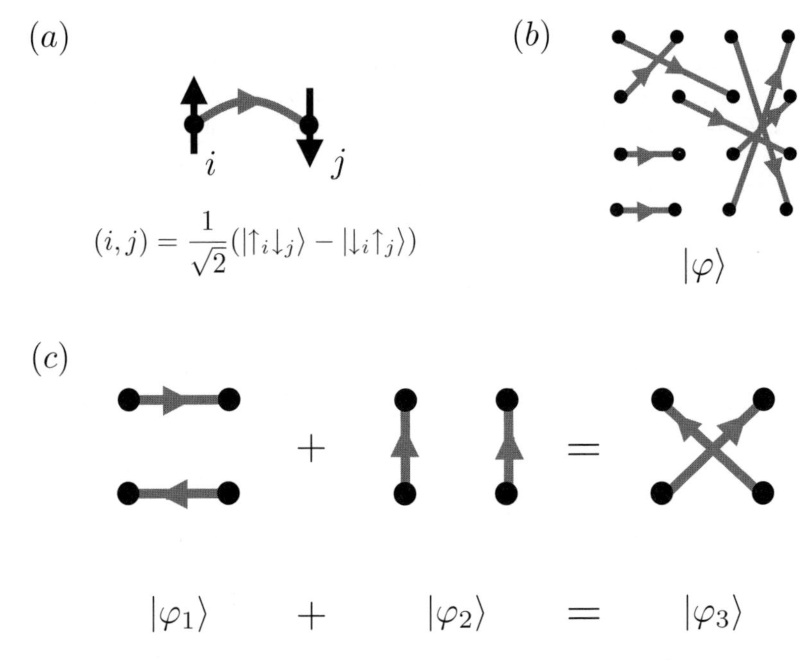

Fig. 9.3 (a) Graphical representation of a valence bond made of two spins $1/2$. (b) Example of a VB state. (c) Linear relation between VB states for four spins $1/2$.

theory are naturally formulated in the VB basis and can be computed efficiently. As the VB basis is seldom used in calculations, we first give a short tour of its properties.

9.6.1 A primer on the valence bond basis

Introduction to valence bonds. We consider two quantum spins $1/2$ labeled i and j, and form the unique state with a total spin $S = 0$, i.e. the *singlet* state $(i,j) = (1/\sqrt{2})(|\uparrow_i\downarrow_j\rangle - |\downarrow_i\uparrow_j\rangle)$. This two-site singlet is called a valence bond: the origin of the name comes from the early days of quantum mechanics and quantum chemistry, where this object was first introduced.[12] A VB (sometimes also called a "dimer") will be denoted graphically by a bond between the two spins. As the wave function is antisymmetric with respect to exchange of i and j, an arrow is drawn on the bond (see Fig. 9.3(a)).

Now consider N quantum spins $1/2$ (with N even from now on). How can states with a total spin $S = 0$ be described? This is not an easy task: indeed, in the standard S_z basis which considers the spin orientation $|\uparrow\rangle$ or $|\downarrow\rangle$ along the quantization axis z, it is easy to construct states with a fixed total $S_z = 0$, but not $S = 0$. Fortunately, the VBs come to our rescue, as all possible complete pairings of the N spins in two-site VBs constitute a basis of the singlet sector: this is the VB basis.

Before proceeding, let us mention the physical motivation for considering singlet states composed of N spins: almost all antiferromagnetic (finite-sized) quantum spin systems have

[12] Note that the notion of valence bond theory is still vividly active in quantum chemistry, as can be checked by an Internet search. What follows has little to do with this literature.

a singlet ground state. There is an exact theorem for this statement for nonfrustrated systems [78]. While there is no proof in the general case for frustrated systems, there is much compelling numerical evidence. Given the realm of experimental compounds that display antiferromagnetic interactions, this justifies the studies of singlet states.

Number of states. To continue, let us count the number of singlet states $\mathcal{N}_{S=0}(N)$ for N spins. For $N = 2$, we already know that there is a unique singlet $\mathcal{N}_{S=0}(N = 2) = 1$. When $N = 4$, two spins can form a singlet or a triplet, which can combine with the singlet or triplet formed by the two other spins. There are only two solutions (singlet + singlet or triplet + triplet) which lead to a total singlet, resulting in $\mathcal{N}_{S=0}(N = 4) = 2$. The same type of argument can be repeated for $N = 6$ to obtain $\mathcal{N}_{S=0}(N = 6) = 5$, but requires more work. It is clear, however, that this is not a good way to compute $\mathcal{N}_{S=0}(N)$ for arbitrary N. There is a simple argument that helps at this stage, which, ironically, is formulated in the S_z basis: all singlet $S = 0$ states have $S_z = 0$, all triplet $S = 1$ states have either $S_z = 1$, $S_z = -1$, or $S_z = 0$ in equal proportion, and, finally, all higher multiplets also have as many $S_z = 1$ states as $S_z = 0$. Consequently, $\mathcal{N}_{S=0}(N) = \mathcal{N}_{S_z=0}(N) - \mathcal{N}_{S_z=1}(N)$. The number of states with $S_z = 0$ or $S_z = 1$ is readily computed in the S_z basis by a simple combinatorial argument, and we have $\mathcal{N}_{S=0}(N) = N!/((N/2)!(N/2)!) - N!/((N/2-1)!(N/2+1)!) = N!/((N/2+1)!(N/2)!)$. The number of singlet states therefore scales as 2^N (i.e. as the total number of states).

Now let us consider the VB basis, formed from VB states. In a VB state $|\varphi\rangle$, all N spins are coupled pairwise in singlets:

$$|\varphi\rangle = 2^{-N/4} \prod_{(i,j)\in c} (|\uparrow_i \downarrow_j\rangle - |\downarrow_i \uparrow_j\rangle), \tag{9.12}$$

where c denotes a pair-covering of the system of N spins. An example of a VB state is given in Fig. 9.3(b).

How many VB states $\mathcal{N}_{\mathrm{VB}}(N)$ can be formed with N spins? Simple combinatorics again give the answer. To construct the VB state, let us pick the first pair. There are $\binom{N}{2}$ ways to do so. For the second pair, there are $\binom{N-2}{2}$ choices, and the argument can be repeated until a single possibility $\binom{2}{2}$ is left for the last singlet. Of course, the same states have been counted many times in this way, and one must account for the $(N/2)!$ corresponding permutations of the $N/2$ dimers. We finally have $\mathcal{N}_{\mathrm{VB}}(N) = (1/(N/2)!)\prod_{p=0}^{N/2-1}\binom{N-2p}{2} = N!/(2^{N/2}(N/2)!)$, a number which explodes as $N^{N/2}$.

VB basis. All VB states are of course singlets, and clearly there are many more VB states than singlets: $\mathcal{N}_{\mathrm{VB}}(N) \gg \mathcal{N}_{S=0}(N)$. This does not prove that the VB states form a basis of the singlet sector, as the VB states can (and actually are) linearly dependent. It turns out that the VB states do indeed form a basis for singlets, albeit overcomplete. The shortest proof of this statement that the author is aware of can be found in [79] (see also [80]), but is not particularly illuminating. An alternative proof will follow later, together with the notion of noncrossing valence bonds.

The overcompleteness of the VB basis implies a huge number of linear relations between VB states. For $N = 4$, there are three VB states for two singlets, and the unique linear relation is displayed in Fig. 9.3(c). This four-site example is instructive because it can already be seen that a state which has, graphically, a crossing between the VBs (the rightmost in Fig. 9.3(c)) can be "uncrossed" by expressing it as other states with no crossings. This will be useful in the next section.

Since VB states form a basis of the singlet sector, all singlet wave functions can be written as (nonunique) linear combinations of VB states: $|\psi\rangle = \sum_i c_i|\varphi_i\rangle$. The expectation values of observables read

$$\frac{\langle\psi|\mathcal{O}|\psi\rangle}{\langle\psi|\psi\rangle} = \frac{\sum_{i,j} c_i c_j \langle\varphi_i|\mathcal{O}|\varphi_j\rangle}{\sum_{i,j} c_i c_j \langle\varphi_i|\varphi_j\rangle}.$$

It is now sufficient to know how to compute *overlaps* $\langle\varphi_i|\varphi_j\rangle$ between VB states, and expectation values of the form $\langle\varphi_i|\mathcal{O}|\varphi_j\rangle$. This will be done later.

Other VB bases. The huge number of VB states is certainly very harmful for any numerical method based on full enumeration of the Hilbert space (such as exact diagonalization), but in fact this is not a problem for QMC calculations. Still, it might be helpful to consider smaller subsets of the VB basis.

The first possibility is to use the bipartite valence bond basis. Here, we divide the N spins into two equal sets A and B and only allow VB states where all VBs go from one set (say A) to the other (say B). An example of a bipartite VB state is displayed in Fig. 9.4(a). Bipartite VB states also form a basis of the singlet sector but are less numerous than unrestricted VB states. We simply have $(N/2)!$ bipartite VB states (still a huge number). For $N = 4$, there are only two bipartite VB states, which coincide with the number of singlets. The bipartite VB basis offers several simplifications for analytical calculations over the unrestricted VB basis.

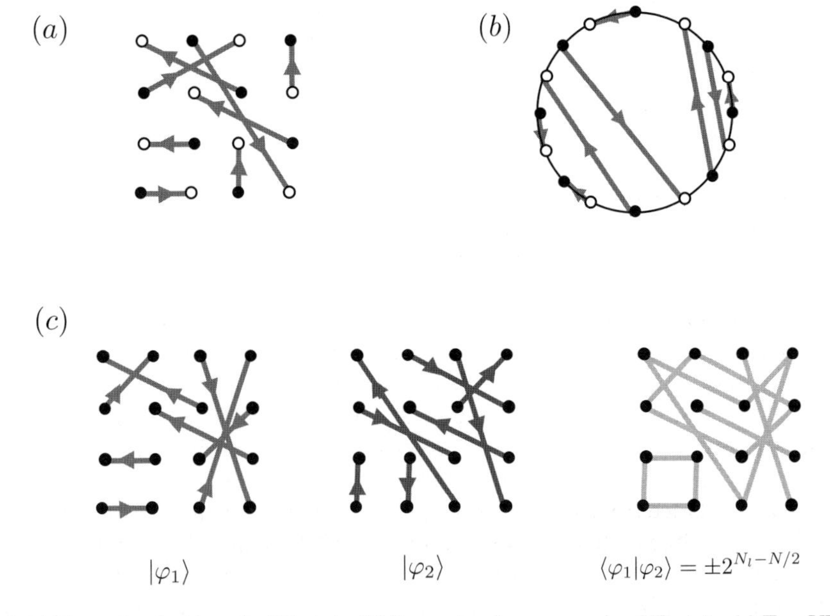

Fig. 9.4 (a) Example of a bipartite VB state. (b) Example of a noncrossing VB state. (c) Two VB states form closed loops when superposed. The overlap between these states is related to the number of loops in the superposition. Different colors have been used for the $N_l = 5$ loops in this example.

Most importantly, the projector MC method that will be described below uses the bipartite VB basis preferentially.

Finally, to complete the zoology of VB basis, it is important to mention the noncrossing basis. We align all spins on a circle and consider only VB states such that no crossing occurs in the graphical representation of the state (see Fig. 9.4(b)). It is easy to see that all noncrossing states are bipartite (for a partition where all A sites alternate with B sites along the circle). Moreover, it turns out that all VB states can be expressed as a linear combination of noncrossing states, a fact that was realized a long time ago [81]. An elegant proof of this can be found in [80], which is based on the fact that one can always uncross any crossing configuration by using the four-site relation of Fig. 9.3(c), and that when this is done the number of crossings always decreases. Counting the number of noncrossing VB states can be done by a combinatorial argument [82], and the result is $\mathcal{N}_{\mathrm{NC}} = N!/((N/2+1)!(N/2)!)$, exactly the number of singlet states. In fact, the noncrossing states form a complete basis of the singlet sector. Provided that the noncrossing states are linearly independent, this gives an alternative proof of the fact that unrestricted and bipartite VB states form a basis of singlets since all noncrossing VB states are regular or bipartite VB states, respectively. The linear independence of noncrossing states is nicely proven in the appendix of [82].

The noncrossing basis has been shown to be useful for quantum spin models in one dimension, as the noncrossedness is usually preserved by the Hamiltonian. This is particularly true for exact diagonalization in the singlet sector of spin chains [83], since the $S = 0$ sector is much smaller than the S_z sector. The noncrossing basis is also useful, but to a smaller extent, for QMC simulations of 1D systems.

Nonorthogonality of VB states. Along with the linear relations between states, the other unusual property of VB states is that they are all nonorthogonal: $\langle\varphi_1|\varphi_2\rangle \neq 0$ for all $|\varphi_1\rangle$ and $|\varphi_2\rangle$. The value of the overlap can be found very easily in a graphical representation where two VB states are superposed. This is typical of calculations in the VB basis, which often have a simple graphical interpretation.

To compute this overlap, let us use the representation of eqn (9.12) for both states, ignore the overall amplitude and sign for the moment and start from the first dimer wave function $(|\uparrow_{i_1}\downarrow_{j_1}\rangle - |\downarrow_{i_1}\uparrow_{j_1}\rangle)$. The first term $|\uparrow_{i_1}\downarrow_{j_1}\rangle$ is multiplied by the product of all dimer wave functions in the second VB state, but there is a nonzero contribution only from terms that have a down spin $|\downarrow_{j_1}\rangle$ at j_1. On the other hand, this state is associated in the second VB state with an up spin at another location, denoted i_2 as in $|\uparrow_{i_2}\downarrow_{j_1}\rangle$. In turn, this term will give a nonzero contribution only for terms in the first VB wave function that have an up spin at i_2, etc. Repeating the argument, the only way out is to come back to an up spin at the initial site i_1: by following the dimers alternately in the first and the second VB, we form a closed loop. Therefore, when the two VB configurations are superposed, only closed loops provide a contribution to the overlap: this is illustrated in Fig. 9.4(c). What is the value of the contribution? We started from the first term $|\uparrow_{i_1}\downarrow_{j_1}\rangle$, but the argument is identical for the other term $|\downarrow_{i_1}\uparrow_{j_1}\rangle$. Therefore each loop contributes a factor 2. Summing the amplitudes and signs, we obtain $\langle\varphi_1|\varphi_2\rangle = \pm 2^{N_l - N/2}$, where N_l is the total number of loops in the superposition.

The $\pm$ sign accounts for the exact sign encountered during the calculation of the contribution of each loop. The sign depends on the convention used to specify the direction for all possible VBs. It is not possible in the unrestricted VB basis to find a convention where all overlaps have the same sign: this causes a sign problem in subsequent QMC calculations.

In the bipartite VB basis, on the other hand, one can choose the convention where all bonds go from A sites to B sites: in that case, all overlaps are *positive*. This partially saves QMC simulations from the sign problem. For now, we restrict the discussion to the bipartite VB basis to simplify calculations.

In numerical methods, one usually prefers to work with orthogonal states. However, as we will see, the nonorthogonality of the VB basis will turn out to be useful in practice during projector QMC simulations.

Action of operators. We show here how simple spin operators act on VB states, and how expectation values can be computed. The discussion is simplified by using well-chosen SU(2) operators.

One of them is the *singlet projector* $P_{ij}^{S=0}$, which, acting on any VB state, will result in a state with a VB singlet between i and j. For two spins 1/2, it is easy to see that the singlet projector reads

$$P_{ij}^{S=0} = \frac{1}{4} - \mathbf{S}_i \cdot \mathbf{S}_j$$

in terms of spin operators.

There are basically three possibilities for $P_{ij}^{S=0}$ acting on a bipartite VB state $|\varphi\rangle$: either there is already a VB in $|\varphi\rangle$ between i and j or there is not, and it is useful to distinguish between the cases where i and j belong to the same subset or not. We naturally have the simple relation

$$P_{ij}^{S=0}(i,j) = (i,j) .$$

in the first case, which is depicted in Fig. 9.5.

When there is no VB between i and j belonging to different sublattices (say A for i and B for j), the projector will create a VB between i and j and accommodate the other VB in the following way (see Fig. 9.5 and note the factor 1/2):

$$P_{ij}^{S=0}(i,l)(k,j) = \frac{1}{2}(i,j)(k,l) .$$

When i and j belong to the same sublattice, it is not possible anymore to create a VB between these sites, since it is forbidden in the bipartite VB basis. More exactly, a state with a bond between i and j can be uncrossed thanks to the linear relation of Fig. 9.3(c) so that it can be expressed in terms of bipartite VB states. We have (see Fig. 9.5)

$$P_{ij}^{S=0}(i,l)(j,k) = \frac{1}{2}(i,l)(j,k) - \frac{1}{2}(i,k)(j,l) .$$

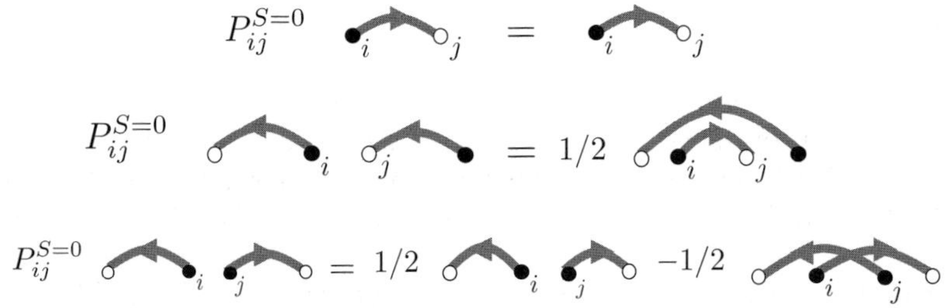

Fig. 9.5 Action of the singlet projection operator $P_{ij}^{S=0} = 1/4 - \mathbf{S}_i \cdot \mathbf{S}_j$ on VB states.

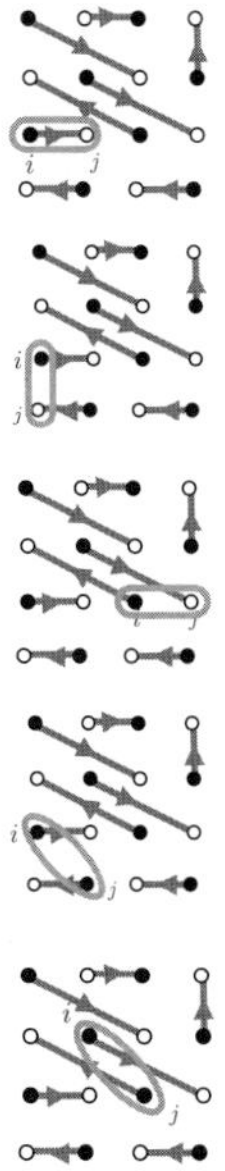

$\langle\varphi_1\vert P_{ij}^{S=0}\vert\varphi_2\rangle$	$\dfrac{\langle\varphi_1\vert\mathbf{S}_i\cdot\mathbf{S}_j\vert\varphi_2\rangle}{\langle\varphi_1\vert\varphi_2\rangle}$
$\langle\varphi_1\vert\varphi_2\rangle$	$-3/4$
$\langle\varphi_1\vert\varphi_2\rangle$	$-3/4$
$\frac{1}{4}\langle\varphi_1\vert\varphi_2\rangle$	0
$-\frac{1}{2}\langle\varphi_1\vert\varphi_2\rangle$	$3/4$
$\frac{1}{4}\langle\varphi_1\vert\varphi_2\rangle$	0

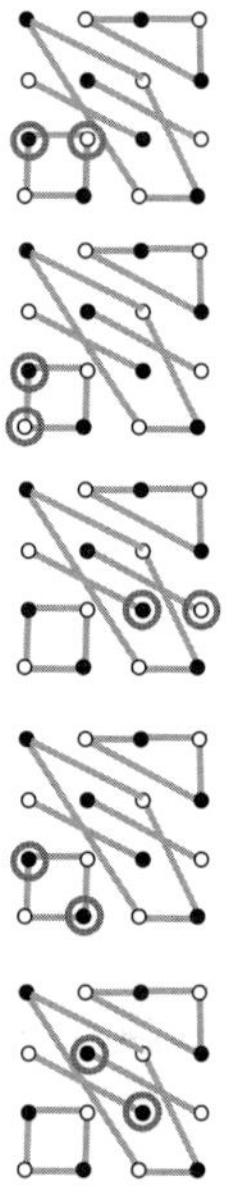

Fig. 9.6 Calculation of expectation values in the VB basis.

What is important to note in this case is that (i) $P^{S=0}$ no longer produces a single VB state but a linear combination (this situation is sometimes called "branching" but has nothing to do with branching in the projector method), and (ii) there is a minus sign in front of the last term. This can be again a source of a sign problem in QMC simulations.

As an exercise, the same type of rules can be derived for the *permutation operator* $\mathcal{P}_{ij} = 1/2 + 2\mathbf{S}_i \cdot \mathbf{S}_j$, which permutes the values of the spins at i and j.

Expectation values. We end this introduction to the VB basis by computing the expectation values of spin correlators. The idea is to take advantage of the action rules derived for $P_{ij}^{S=0}$ to compute its expectation value $\langle\varphi_1|P_{ij}^{S=0}|\varphi_2\rangle$. If $|\varphi_2\rangle$ has a VB between i and j then, trivially, $\langle\varphi_1|P_{ij}^{S=0}|\varphi_2\rangle = \langle\varphi_1|\varphi_2\rangle$. If not, one has to consider (for both cases) whether the action of $P_{ij}^{S=0}$ will *create* or *destroy* a loop in the loop representation of the overlap between $|\varphi_1\rangle$ and $|\varphi_2\rangle$. The five resulting situations are represented in Fig. 9.6, and it is left as an exercise to recover the values listed in the second column of the figure.

What is illuminating is to consider the spin correlator $\langle\varphi_1|\mathbf{S}_i \cdot \mathbf{S}_j|\varphi_2\rangle/\langle\varphi_1|\varphi_2\rangle$. The "loop rules" for this correlator are indeed particularly simple: it vanishes if i are j are located on different loops in the overlap diagram, and it takes the value $\pm 3/4$ if they are on the same loop ($+$ or $-$ for i and j on the same or different sublattices, respectively). This is illustrated in the last column of Fig. 9.6, where the positions of i and j in the original overlap diagram are highlighted.

We can finally summarize this calculation by writing the "loop estimator" [84]

$$\left.\frac{\langle \varphi_1 | \mathbf{S}_i \cdot \mathbf{S}_j | \varphi_2 \rangle}{\langle \varphi_1 | \varphi_2 \rangle}\right|_{\mathcal{L}} = \frac{3}{4} \epsilon_{i,j} \epsilon_{l_i,l_j} ,$$

where $\epsilon_{i,j} = 1$ if i and j are on the same sublattice and -1 if not, and $\epsilon_{l_i,l_j} = 1$ if i and j are on the same loop and 0 if not.

It turns out that almost all observables of interest do have a simple loop estimator $(\langle \varphi_1 | \mathcal{O} | \varphi_2 \rangle / \langle \varphi_1 | \varphi_2 \rangle)|_{\mathcal{L}}$, the value of which can be obtained from the overlap diagram of $\langle \varphi_1 | \varphi_2 \rangle$. Such loop estimators have been derived for many observables (and in a more elegant way) in [85]. These estimators also turn out to be very useful in QMC calculations.

9.6.2 Valence bond projector Monte Carlo method

All of the ingredients are now present for a projection MC algorithm in the VB basis. Such a projector VB algorithm was first proposed by Sandvik [86] in 2005.[13]

Heisenberg model. We concentrate the discussion on the $S = 1/2$ Heisenberg model

$$H = J \sum_{\langle i,j \rangle} \mathbf{S}_i \cdot \mathbf{S}_j ,$$

where $\langle \ldots \rangle$ denotes nearest-neighbor (n.n.) spins and $J = 1$ sets the energy scale. H can be expressed in terms of singlet projectors,

$$H = -J \sum_{\langle i,j \rangle} \left(P_{ij}^{S=0} - \frac{1}{4} \right) = -J \sum_{b=\langle i,j \rangle} P_b^{S=0} + E_T ,$$

by introducing the symbol b for n.n. bonds of the lattice. E_T is a constant energy shift which is such that all matrix elements of $(H - E_T)$ are negative.

If we consider a (nonfrustrated) bipartite lattice, great simplifications allow us to avoid the sign problem (this is also the case for other QMC methods). Indeed, in this case we can consider a bipartite VB basis where the *A*/*B* partition of sites is identical to the natural bipartition of the lattice. All overlaps are positive. The projection operator $P_{ij}^{S=0}$ never acts on sites i and j belonging to the same sublattice, avoiding the third rule of Fig. 9.5 (which is also the source of a sign problem). Moreover, the "branching" found in this third rule is also avoided, which eases the MC procedure described below.

Projector scheme. Now the setup is ready for a projection MC algorithm formulated in the VB basis:

$$\lim_{n \to \infty} (-H + E_T)^n | \varphi_T \rangle \sim | \psi_0 \rangle ,$$

with

$$(-H + E_T)^n = \left(\sum_b P_b^{S=0} \right)^n = \sum_{\mathcal{B}=\{b_1,b_2,\ldots,b_n\}} \prod_{i=1}^{n} P_{b_i}^{S=0} . \qquad (9.13)$$

An MC configuration is now given by a sequence of n.n. bonds $\{b_1, b_2, \ldots, b_n\}$ and a trial VB state $|\varphi_T\rangle$. It is convenient to represent such a configuration as in Fig. 9.7. There is

[13] Several ideas appearing in the algorithm can be traced back to older work by Liang [87].

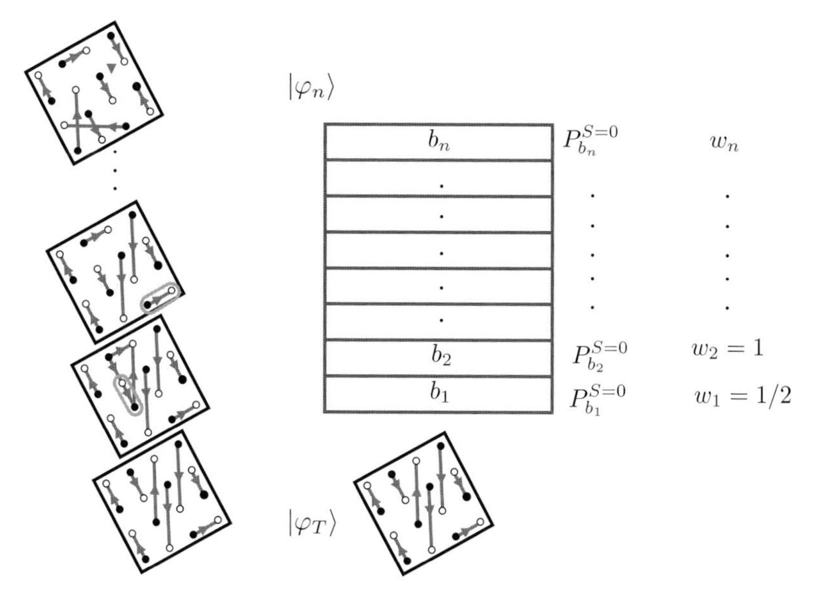

Fig. 9.7 Schematic representation of the valence bond projection method.

some resemblance to the "reptile" of the reptation Monte Carlo algorithm, but the methods are not similar, as will be discussed later. The sequence $\{b_1, b_2, \dots, b_n\}$ corresponds to a process where $|\varphi_T\rangle$ is propagated (acted on) by a sequence of singlet projectors to result in a propagated VB state $|\varphi_n\rangle$:

$$|\varphi_n\rangle = P^{S=0}_{b_n} \dots P^{S=0}_{b_2} P^{S=0}_{b_1} |\varphi_T\rangle \,.$$

The action of $P^{S=0}_{b_i}$ is very simple and corresponds to the rearrangements in the first two rules of Fig. 9.5, depending on whether there is a VB on bond b in the VB state. Owing to the prefactor in front of the VB state obtained, this action gives rise to a weight $w_i = 1$ for the first rule or $w_i = 1/2$ for the second rule. The total weight of the MC configuration is $W = \prod_{i=1}^n w_i$.

We understand now the importance of having *positive* numbers in the prefactors of Fig. 9.5, otherwise the weight can be negative and the sign problem occurs. Also, having no state that is annihilated by $P^{S=0}$ ensures that any sequence $\{b_1, b_2, \dots, b_n\}$ has a nonzero contribution (this would be the bottleneck if the same method were to be used in the S_z basis). Finally, the no-branching condition respected by the two first rules of Fig. 9.5 also helps as a single configuration is propagated step by step, instead of a linear combination.

Monte Carlo move. The simplest MC step that can be thought of is the following. To obtain a new MC configuration, a few n.n. bonds are proposed to be randomly changed in the sequence

$$\{b_1, b_2, \dots, b_n\} \to \{b'_1, b'_2, \dots, b'_n\} \,,$$

where only a few b'_i (typically between 1 and 5) are different from the original b_i. This will result in different propagation, final state $|\varphi'_n\rangle$, and weight $W' = \prod_{i=1}^n w'_i$. Such a move is accepted with a Metropolis rate $\min(1, W'/W)$.

This algorithm is extremely simple but turns out to be quite efficient. Indeed, changing only a few $b's$ might appear only as a small perturbation of the propagation. The very nature of the VB basis (where bonds can be generated on all length scales with only a few local operations) ensures that the propagated state $|\varphi'_n\rangle$ will differ substantially from $|\varphi_n\rangle$.

There are several improvements that can be made. First, instead of a single trial VB state $|\varphi_T\rangle$, one can use a linear combination of VB states. When the linear combination can be explicitly written down, a classical Monte Carlo simulation (for instance with the worm algorithm of Section 9.4) can be performed on the trial state. The weight of each VB state for this classical Monte Carlo simulation is just proportional to the amplitude of the state in the linear combination. One further improvement has been proposed in [88]: information about the VB length distribution in the propagated state $|\varphi_n\rangle$ can be reinjected into the trial state $|\varphi_T\rangle$. This is done again through a classical MC process, which samples the VB length distribution probability. Details of the procedure are given in [88]. The main consequence is that $|\varphi_T\rangle$ will be closer to the GS, which in turn reduces *considerably* the value of the projection number n and hence speeds up calculations. As in the case of guiding wave functions, the investment in implementing this improvement is really valuable. Note that the idea of reinjecting information about the propagated state into the trial (or guiding) wave function is seldom used in standard projection techniques (to the best of the author's knowledge), but might be a path to explore. Finally, even if it has never been done, the continuous-time trick of Section 9.5.5 should likely be applicable to the VB projector QMC method. Guiding wave functions could also be used.

Measurements. *Mixed estimates.* The GS energy of the Heisenberg Hamiltonian can be read from the final propagated state $|\varphi_n\rangle$ using the mixed estimate $\langle\psi_R|H|\varphi_n\rangle/\langle\psi_R|\varphi_n\rangle$, where the reference state $|\psi_R\rangle$ can be chosen at will. It is convenient to take again the equal-amplitude state $|\psi_R\rangle = \sum_i |\varphi_i\rangle$. In the bipartite VB basis, this is nothing but the Néel state $|\uparrow_A\downarrow_B\uparrow_A\downarrow_B\ \ldots\rangle$.[14] Noting that the Néel state has an equal overlap with all bipartite VB states and reinserting the energy shift E_T, we easily obtain

$$E_0 = -\frac{1}{2}\sum_b \langle n_b\rangle + 1\,,$$

where $n_b = 1$ if the n.n. bond b is occupied by a VB in the state $|\varphi_n\rangle$, and 0 otherwise.

Expectation values. Note that up to now, both for the propagation and for the measurement of mixed observables, the nonorthogonality of VB states is irrelevant. The latter arises in the calculation of expectation values of observables that do not commute with H. The idea here is simple: we perform two independent propagations to obtain two final states $|\varphi_n^L\rangle$ and $|\varphi_n^R\rangle$ (the indices stand for "left" and "right"), and then measure $\langle\varphi_n^L|\mathcal{O}|\varphi_n^R\rangle$ and $\langle\varphi_n^L|\varphi_n^R\rangle$ to form the estimate $\langle\varphi_n^L|\mathcal{O}|\varphi_n^R\rangle/\langle\varphi_n^L|\varphi_n^R\rangle$.

In an orthogonal basis, this simple method would fail because of a statistical problem: there would be a nonzero contribution to $\langle\psi_n^L|\psi_n^R\rangle$ if and only if the right and left states were identical, an event which has an almost vanishing probability.

On the other hand, the nonorthogonality of VB states is of great help: since all VB overlaps are nonzero, this leads to good statistics for the estimation of both $\langle\varphi_n^L|\mathcal{O}|\varphi_n^R\rangle$ and $\langle\varphi_n^L|\varphi_n^R\rangle$. We can go even further: instead of sampling independently the two propagations (i.e. with

[14] In the unrestricted VB basis, the equal-amplitude state is equal to the null vector, emphasizing the fact that this is not a good basis to perform a projection MC simulation.

a weight $W^L.W^R$), the weight $W^L.W^R.\langle\varphi_n^L|\varphi_n^R\rangle$ can be used. This not only increases the accuracy of calculations (since pairs of configurations with large overlap will be favored) but also allows us to use the loop estimators $(\langle\varphi_1|\mathcal{O}|\varphi_2\rangle/\langle\varphi_1|\varphi_2\rangle)|_{\mathcal{L}}$ derived in Section 9.6.1, based on the overlap diagram formed by $|\varphi_n^L\rangle$ and $|\varphi_n^R\rangle$. Measuring observables with a double-propagation scheme is unique to a nonorthogonal basis and avoids the forward-walking procedure typical of projector methods. Moreover, off-diagonal observables are simpler to measure as long as they have a loop estimator.

Spin gap. Finally, let us consider another interesting feature of the VB method: the calculation of the spin gap Δ_s. By definition, the spin gap is the difference between the total energy of the first triplet excited state and that of the singlet GS: $\Delta_s = E_0(S=1) - E_0(S=0)$. While computing the GS energy $E_0(S=0)$ is simple, it appears unlikely that a method exaggerately based on properties of the singlet sector can capture triplet states. This is, however, the case, and we do not even need to perform an extra simulation!

The trick is the following [86]: we inject one $S_z = 0$ triplet $[i,j] = (1/\sqrt{2})(|\uparrow_i\downarrow_j\rangle + |\downarrow_i\uparrow_j\rangle)$ inside the trial wave function and let it propagate. The action of $P_{ij}^{S=0}$ on a wave function containing a triplet is summarized in Fig. 9.8. If it is applied to the specific triplet bond, the triplet annihilates (as expected, since $P^{S=0}$ is a projector into a local singlet). If not, the propagation is exactly the same as for a singlet state (with the same prefactor $1/2$). This means that the full propagation of the singlet state $|\varphi_T\rangle$ can be *reinterpreted* as the propagation of a triplet state $|\varphi_T^{S=1}\rangle$, provided that the triplet can be killed during the propagation. In practice, one assumes that one of the VBs in $|\varphi_T\rangle$ is a triplet, and follows its evolution during the propagation. It can either die (if acted on by a $P^{S=0}$ operator), or survive up to the final state $|\varphi_n\rangle$. In the latter case, $|\varphi_n\rangle$ is reinterpreted as a triplet state where $E_0(S=1)$ can be measured by a mixed estimate. The nice aspect of this method is that, because of error cancelations between $E_0(S=1)$ and $E_0(S=0)$ (due to the fact that they emanate from the same propagation procedure), the value of the spin gap can be measured quite precisely. Of course, only triplets that have survived the propagation contribute to the expectation value of $E_0(S=1)$ and it might be that, owing to a large projection number n and/or a large spin gap, the triplet survival rate becomes too small. In that case, the statistics on the spin gap are not precise enough. A further trick can be used in this situation: an average over all possible initial positions of the triplet bond in the trial state $|\varphi_T\rangle$ can be performed, which greatly improves the statistics. Care must, however, be taken when performing the average, and the

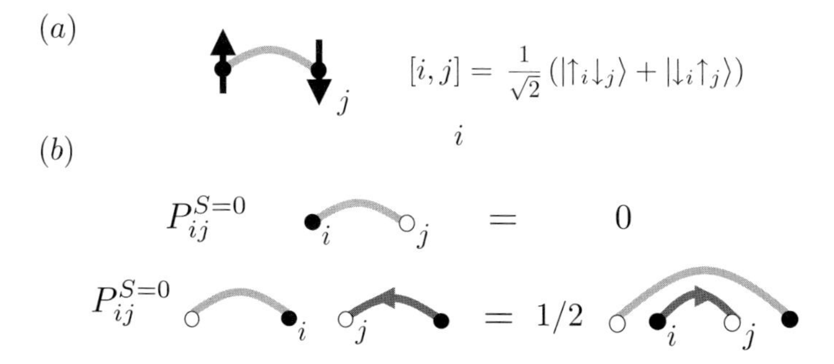

Fig. 9.8 (a) Graphical representation of a triplet. (b) Action of the singlet projection operator $P_{ij}^{S=0} = 1/4 - \mathbf{S}_i \cdot \mathbf{S}_j$ on VB states containing a triplet.

exact procedure is described in [85, 88].

The properties of other excited states can be computed. This is naturally the case for excited states which live in a different symmetry sector, for instance the energy of triplets at an arbitrary momentum $\mathbf{q}$ [88]. It remains to be tested that excited states in the *same* symmetry sector as the GS (i.e. singlets) can also be reached, using techniques known in other projection schemes [89]. It is likely that the nonorthogonality of the VB basis will again ease such a calculation.

Summary of the VB projection method. Let us conclude by comparing the VB projection method with other MC methods, and by providing some perspectives.

The relation to other projection MC methods is transparent. The simple VB projection method as described here is certainly much less elaborate than other schemes, but still quite efficient. Improvements are possible (some have already been made, and some are still to be discovered) and, in particular, it is tempting to try to adapt the reptation Monte Carlo method to the VB basis. The main problem to be faced is that the Hamiltonian is *non-Hermitian* in the VB basis, as can be clearly seen from the action of the operators in Fig. 9.5. Therefore, a direct application of the reptation procedure (as described in Section 9.5.5) is not possible, since there is a need to be able to reptate in both directions.

For the pure Heisenberg model, until very recently, VB projector simulations have not reached the large sample sizes and precision obtained with S_z-based recent path integral and SSE methods [18, 42]. The situation has just changed with the appearance of a new method [90], which can be seen as a hybrid SSE–VB projection method. The method uses the same decomposition of the Heisenberg Hamiltonian as in an SSE scheme (allowing an efficient loop update), and a VB representation for the initial boundary states (instead of the usual S_z basis). This results in an extremely powerful loop projector method for the Heisenberg model, which efficiency surpasses all other methods. This new method is, however, restricted to GS properties, whereas other conventional (PI or SSE) methods can also obtain finite-temperature properties.

The original VB projection method is also interesting because it offers direct access to VB physics (which is not easily expressed in the S_z basis): dimer–dimer correlators of the form $\langle(\mathbf{S}_i \cdot \mathbf{S}_j)(\mathbf{S}_k \cdot \mathbf{S}_l)\rangle$ are easily computed, and spinon degrees of freedom ($S = 1/2$ excitations) might also be reachable. Moreover, the spin gap can be efficiently computed, and some models are easier to simulate in the VB basis [91].

Let us now look at a few prospects: as with any new method, it is expected that many new aspects are to be discovered. One can imagine simulating other degrees of freedom than $S = 1/2$ spins. For instance, for some SU(N) models (with a specific representation of the generators of the SU(N) algebra), the update rules of Fig. 9.5 are *exactly* the same [92], except that the prefactor $1/2$ is replaced by $1/N$—resulting a trivial modification of a pre-existing algorithm.

Another interesting aspect is the relation to quantum information properties. As already seen, VBs are natural objects with which to characterize antiferromagnetic correlations. They are also the elementary building blocks of *entanglement*, as a VB is maximally entangled (entanglement is often measured in "singlet" or VB units). Recently, it has been shown that the VB basis (combined with the powerful VB projection method) could naturally help to characterize entanglement properties of quantum spin systems in any dimension [93]. For instance, the *entanglement entropy* shared by two subparts of a quantum system is in general very hard to compute in the S_z basis, but is fairly simple to calculate in a VB state (as a simple count of VBs shared by the subparts). Another quantity typical of quantum information that can be easily reached in the VB basis is the *fidelity* between two states (i.e. the modulus of their overlap). Fidelity is again a quantity hardly accessible to standard QMC methods. Since all VB states have a nonzero overlap, we expect fidelity measurements for quantum spin systems to be very precise using the VB projection method. This is of high interest, given the recent suggestions that fidelity may be able to automatically detect quantum phase transitions [94]. These two examples illustrate a situation typical of theoretical physics, where the complexity has been shifted from the problem (computing the entanglement entropy or fidelity) to the tool (using the VB basis and its unusual aspects).

Acknowledgments

I would like to thank the organizers of the School for creating such a nice, enjoyable atmosphere. I'm also taking this opportunity to thank all the colleagues with whom I had a chance to work on, develop, or learn the numerical techniques presented in these lecture notes. A nonexhaustive list includes K. Beach, M. Caffarel, S. Capponi, K. Harada, N. Kawashima, N. Laflorencie, A. Läuchli, M. Mambrini, I. McCulloch, O. Parcollet, D. Poilblanc, A. Sandvik, E. Sørensen, O. Syljuåsen, S. Todo, S. Trebst, M. Troyer, and S. Wessel. In particular, I would like to thank S. Capponi, M. Caffarel, M. Mambrini, and O. Syljuåsen for a careful reading of the manuscript.

References

[1] R. M. Noack and S. R. Manmana, *AIP Conf. Proc.* **789**, 93 (2005) [also available at arXiv:cond-mat/0510321].

[2] N. Laflorencie and D. Poilblanc, *Lect. Notes Phys.* **645**, 227 (2004) [also available at arXiv:cond-mat/0408363].

[3] LAPACK libraries, available at http://www.netlib.org/lapack.

[4] S. Yamada, T. Imamura, and M. Machida, in *Proceedings of International Conference for High Performance Computing, Networking and Storage (SC'05)*, 2005 [available at http://dx.doi.org/10.1109/SC.2005.1].

[5] S. R. White, *Phys. Rev. Lett.* **69**, 2863 (1992); Phys. Rev. B **48**, 345 (1993); see also S. R. White and R. M. Noack, *Phys. Rev. Lett.* **68**, 3487 (1992).

[6] U. Schollwöck, *Rev. Mod. Phys.* **77**, 259 (2005); K. Hallberg, *Adv. Phys.* **55**, 477 (2006) [also available at arXiv:cond-mat/0609039]; U. Schollwöck and S. R. White, *AIP Conf. Proc.* **816**, 155 (2006) [also available at arXiv:cond-mat/0606018].

[7] I. P. McCulloch, *J. Stat. Mech.* P10014 (2007); see also arXiv:0804.2509.

[8] J. Jordan *et al.*, *Phys. Rev. Lett.* **101**, 250602 (2008); F. Verstraete and J. I. Cirac, arXiv:cond-mat/0407066.

[9] N. Schuch *et al.*, *Phys. Rev. Lett.* **100**, 040501 (2008).

[10] G. Evenbly and G. Vidal, *Phys. Rev. B* **79**, 144108 (2009).

[11] A. W. Sandvik, *Phys. Rev. Lett.* **101**, 140603 (2008).

[12] J. Oitmaa, C. Hamer, and W. Zheng, *Series Expansion Methods for Strongly Interacting Lattice Models*, Cambridge University Press (2006).

[13] A. J. Guttmann, in *Phase Transitions and Critical Phenomena*, Vol. 13, eds. C. Domb and J. Lewovitz, p. 1, Academic Press (1989).

[14] M. P. Gelfand and R. R. P. Singh, *Adv. Phys.* **49**, 93 (2000).
[15] M. P. Gelfand, *Solid State Commun.* **98**, 11 (1996); S. Trebst *et al.*, *Phys. Rev. Lett.* **85**, 4373 (2000).
[16] Nauty library, available at http://cs.anu.edu.au/~bdm/nauty.
[17] M. Troyer, F. Alet, S. Trebst, and S. Wessel, *AIP Conf. Proc.* **690**, 156 (2003) [also available at arXiv:physics/0306128].
[18] N. Kawashima and K. Harada, *J. Phys. Soc. Japan* **73**, 1379 (2004).
[19] H.-G. Evertz, *Adv. Phys.* **52**, 1 (2003) [also available at arXiv:cond-mat/9707221].
[20] M. Jarell and J. E. Gubernatis, *Phys. Rep.* **269**, 133 (1996).
[21] K. S. D. Beach, arXiv:cond-mat/0403055.
[22] F. Assaad, in *Lecture Notes of the Winter School on Quantum Simulations of Complex Many-Body Systems: From Theory to Algorithms*, NIC Series, Vol. 10, John von Neumann Institute for Computing, Jülich, p. 99 (2002) [also available at http://www.fz-juelich.de/nic-series/volume10].
[23] M. Troyer and U.-J. Wiese, *Phys. Rev. Lett.* **94**, 170201 (2005).
[24] C. J. Morningstar and M. Weinstein, *Phys. Rev. Lett.* **73**, 1873 (1994).
[25] M. S. Siu and M. Weinstein, *Phys. Rev. B* **75**, 184403 (2007); A. Auerbach, *AIP Conf. Proc.* **816**, 1 (2006) [also available at arXiv:cond-mat/0510738]; S. Capponi, *AIP Conf. Proc.* **816**, 16 (2006) [also available at arXiv:cond-mat/0510785].
[26] J. Richter *et al.*, *Int. J. Mod. Phys. B* **21**, 2273 (2007) [also available at arXiv:cond-mat/0612146].
[27] A. Georges *et al.*, *Rev. Mod. Phys.* **68**, 13 (1996).
[28] T. Maier *et al.*, *Rev. Mod. Phys.* **77**, 1027 (2005); G. Biroli, O. Parcollet, and G. Kotliar, *Phys. Rev. B* **69**, 205108 (2004).
[29] P. Werner *et al.*, *Phys. Rev. Lett.* **97**, 076405 (2006); P. Werner and A. J. Millis, *Phys. Rev. B* **74**, 155107 (2006); K. Haule, *Phys. Rev. B* **75**, 155113 (2007); A. N. Rubtsov *et al.*, *Phys. Rev. B* **72**, 035122 (2005); E. Gull *et al.*, *Europhys. Lett.* **82**, 57003 (2008).
[30] G. Kotliar *et al.*, *Rev. Mod. Phys.* **78**, 865 (2006).
[31] K. Byczuk and D. Vollhardt, *Phys. Rev. B* **77**, 235106 (2008).
[32] A. F. Albuquerque *et al.*, *J. Magn. Magn. Mater.* **310**, 1187 (2007); F. Alet *et al.*, *J. Phys. Soc. Jpn. Suppl.* **74**, 30 (2005).
[33] H. F. Trotter, *Proc. Am. Math. Soc.* **10**, 545 (1959); M. Suzuki, *Prog. Theor. Phys.* **56**, 1454 (1976).
[34] H.-G. Evertz, G. Lana, and M. Marcu, *Phys. Rev. Lett.* **70**, 875 (1993).
[35] N. V. Prokof'ev, B. V. Svistunov, and I. S. Tupitsyn, *Sov. Phys. JETP* **87**, 310 (1998) [also available at arXiv:cond-mat/9703200].
[36] B. B. Beard and U.-J. Wiese, *Phys. Rev. Lett.* **77**, 5130 (1996).
[37] M. Troyer, S. Wessel, and F. Alet, *Phys. Rev. Lett.* **90**, 120201 (1996).
[38] A. W. Sandvik and J. Kurkijärvi, *Phys. Rev. B* **43**, 5950 (1991).
[39] A. W. Sandvik, *J. Phys. A* **25**, 3667 (1992).
[40] A. W. Sandvik, *Phys. Rev. B* **59**, R14157 (1999).
[41] A. W. Sandvik, R. R. P. Singh, and D. K. Campbell, *Phys. Rev. B* **56**, 14510 (1997).
[42] O. F. Syljuåsen and A. W. Sandvik, *Phys. Rev. E* **66**, 046701 (2002).
[43] F. Alet, S. Wessel, and M. Troyer, *Phys. Rev. E* **71**, 036706 (2005).
[44] K. Louis and C. Gros, *Phys. Rev. B* **70**, 100410 (2004).

[45] A. Dorneich and M. Troyer, *Phys. Rev. E* **64**, 066701 (2001).

[46] T. Nakamura, *Phys. Rev. B* **57**, R3197 (1998); S. Chandrasekharan and U.-J. Wiese, *Phys. Rev. Lett.* **83**, 3116 (1999); M. Nyfeler *et al.*, *Phys. Rev. Lett.* **100**, 247206 (2008).

[47] A. W. Sandvik, *Phys. Rev. E* **68**, 056701 (2003); K. Fukui and S. Todo, *J. Comp. Phys.* **228**, 2629 (2009).

[48] M. P. A. Fisher and G. Grinstein, *Phys. Rev. Lett.* **60**, 208 (1988); see also M. P. A. Fisher *et al.*, *Phys. Rev. B* **40**, 546 (1989).

[49] M. Wallin *et al.*, *Phys. Rev. B* **49**, 12115 (1994); see also E. S. Sørensen, Ph.D. thesis, University of California (1992).

[50] F. Alet and E. Sørensen, *Phys. Rev. E* **67**, 015701 (2003).

[51] F. Alet and E. Sørensen, *Phys. Rev. E* **68**, 026702 (2003).

[52] J. H. Wilkinson, *The Algebraic Eigenvalue Problem*, Oxford University Press (1988).

[53] M. Boninsegni, N. Prokof'ev, and B. Svistunov, *Phys. Rev. Lett.* **96**, 070601 (2006); *Phys. Rev. E* **74**, 036701 (2006).

[54] L. P. Kadanoff, *Statistical Physics, Statics, Dynamics and Renormalization*, World Scientific (2000).

[55] N. Prokof'ev and B. Svistunov, *Phys. Rev. Lett.* **87**, 160601 (2001).

[56] P. Hitchcock, E. Sørensen, and F. Alet, *Phys. Rev. E* **70**, 016702 (2004).

[57] J.-S. Wang, *Phys. Rev. E* **72**, 036706 (2005).

[58] O. F. Syljuåsen and M. B. Zvonarev, *Phys. Rev. E* **70**, 016118 (2004).

[59] A. W. Sandvik and R. Moessner, *Phys. Rev. B* **73**, 144504 (2006).

[60] F. Alet *et al.*, *Phys. Rev. Lett.* **94**, 235702 (2005); *Phys. Rev. E* **74**, 041124 (2006); *Phys. Rev. Lett.* **97**, 030403 (2006); G. Misguich, V. Pasquier, and F. Alet, *Phys. Rev. B* **78**, 100402(R) (2008).

[61] M.-C. Cha and J.-W. Lee, *Phys. Rev. Lett.* **98**, 266406 (2007); P. Hitchcock and E. S. Sørensen, *Phys. Rev. B* **73**, 174523 (2006); K. G. Balabanyan, *Phys. Rev. B* **75**, 144512 (2007).

[62] A. Vestergren, J. Lidmar, and T. H. Hansson, *Europhys. Lett.* **69**, 256 (2005); F. Alet, B. Lucini, and M. Vettorazzo, *Comput. Phys. Commun.* **169**, 370 (2005); M. G. Endres, *Phys. Rev. D* **75**, 065012 (2007).

[63] L. Levrel, F. Alet, J. Rottler, and A. C. Maggs, *Pramana* **64**, 1001 (2005).

[64] N. Trivedi and D. M. Ceperley, *Phys. Rev. B* **41**, 4552 (1990); W. Krauth, M. Caffarel, and J.-P. Bouchaud, *Phys. Rev. B* **45**, 3137 (1992).

[65] M. Caffarel and R. Assaraf, in *Mathematical Models and Methods for Ab Initio Quantum Chemistry*, eds. M. Defranceschi and C. Le Bris, Lecture Notes in Chemistry, Vol. 74, p. 45, Springer (2000).

[66] M. Caffarel and P. Claverie, *J. Chem. Phys.* **88**, 1088 (1988).

[67] H. J. M van Bemmel *et al.*, *Phys. Rev. Lett.* **72**, 2442 (1994); D. F. ten Haaf *et al.*, *Phys. Rev. B* **51**, 13039 (1995).

[68] O. F. Syljuåsen, *Phys. Rev. B* **71**, 020401(R) (2005).

[69] S. Sorella, *Phys. Rev. Lett.* **80**, 4558 (1998).

[70] M. Calandra Buonaura and S. Sorella, *Phys. Rev. B* **57**, 11446 (1998).

[71] R. Assaraf, M. Caffarel, and A. Khelif, *Phys. Rev. E* **61**, 4566 (2000).

[72] K. S. Liu, M. H. Kalos, and G. V. Chester, *Phys. Rev. A* **10**, 303 (1974).

[73] S. Baroni and S. Moroni, *Phys. Rev. Lett.* **82**, 4745 (1999).

[74] S. Baroni and S. Moroni, in *Quantum Monte Carlo Methods in Physics and Chemistry*, eds. M. P. Nightingale and C. J. Umrigar, p. 313, Kluwer (1999) [also available at arXiv:cond-mat/980821].
[75] O. F. Syljuåsen, *Int. J. Mod. Phys. B* **19**, 1973 (2005) [also available at arXiv:cond-mat/0504195].
[76] O. F. Syljuåsen, *Phys. Rev. B* **73**, 245105 (2006).
[77] C. L. Henley, *J. Stat. Phys.* **89**, 483 (1997).
[78] W. Marshall, *Proc. R. Soc. London Ser. A* **232**, 48 (1955); E. Lieb and D. C. Mattis, *J. Math. Phys.* **3**, 749 (1962).
[79] M. Karbach *et al.*, *Phys. Rev. B* **48**, 13666 (1993).
[80] R. Saito, *J. Phys. Soc. Jpn.* **59**, 482 (1990).
[81] G. Rumer, *Nachr. Gott., Math-physik. Klasse* **337** (1932); G. Rummer, E. Teller, and H. Weyl, *Nachr. Gott., Math-physik. Klasse* **499** (1932) [also available, in German, at http://www.digizeitschriften.de].
[82] H. N. V. Temperley and E. H. Lieb, *Proc. R. Soc. London Ser. A* **322**, 251 (1971).
[83] S. Ramasesha and Z. G. Soos, *Int. J. Quantum Chem.* **25**, 1003 (1984); K. Chang *et al.*, *J. Phys.: Cond. Matter* **1**, 153 (1989).
[84] B. Sutherland, *Phys. Rev. B* **37**, 3786 (1988).
[85] K. S. D. Beach and A. W. Sandvik, *Nucl. Phys. B* **750**, 142 (2006).
[86] A. W. Sandvik, *Phys. Rev. Lett.* **95**, 207203 (2005).
[87] S. Liang, *Phys. Rev. B* **42**, 6555 (1990).
[88] A. W. Sandvik and K. S. D. Beach, in *Computer Simulation Studies in Condensed Matter Physics XX*, eds. D. P. Landau, S. P. Lewis, and H.-B. Schüttler, Springer (2009) [also available at arXiv:0704.1469].
[89] D. M. Ceperley and B. Bernu, *J. Chem. Phys.* **89**, 6316 (1988).
[90] A. W. Sandvik and H. G. Evertz, arXiv:0807.0682.
[91] A. W. Sandvik, *Phys. Rev. Lett.* **98**, 227202 (2007).
[92] I. Affleck, *J. Phys. Cond. Matter* **2**, 405 (1990).
[93] F. Alet *et al.*, *Phys. Rev. Lett.* **99**, 117204 (2007).
[94] P. Zanardi and N. Paunkoviĉ, *Phys. Rev. E* **74**, 031123 (2006).

10
Rapidly rotating atomic Bose gases

N. R. Cooper

Nigel R. Cooper,
T.C.M. Group, Cavendish Laboratory,
University of Cambridge,
J.J. Thomson Avenue,
Cambridge CB3 0HE, United Kingdom.

10.1 Introduction

10.1.1 Scope of these lectures

One of the most remarkable characteristics of a Bose–Einstein condensate is its response to rotation. As was first understood in the context of superfluid helium-4 (Donnelly 1991), a Bose–Einstein condensate does not rotate in the manner of a conventional fluid, which undergoes rigid-body rotation. Rather, the rotation leads to the formation of an array of quantized vortex lines. This pattern is a dramatic manifestation of the existence of macroscopic phase coherence. The achievement of Bose–Einstein condensation in ultracold dilute atomic gases (Cornell and Wieman 2002, Ketterle 2002) and the creation of vortex arrays in stirred condensates (Madison *et al.* 2000) open up a wide range of new aspects of the physics of quantized vortices and vortex arrays. These systems allow access to parameter regimes unlike those of superfluid helium. As I will describe, this is predicted to lead to interesting strongly correlated phases in which the bosons are uncondensed. These phases can be understood as bosonic analogues of the strongly correlated phases that are responsible for the fractional quantum Hall effect (FQHE) of electrons in semiconductors (Prange and Girvin 1990, Das Sarma and Pinczuk 1997). In recent years, there have been advances in experimental capabilities and in the theoretical understanding of rotating ultracold atomic gases. The aim of these lectures is to explain some of the theoretical results for rapidly rotating atomic Bose gases. I shall focus on the regime of strong correlations and the connections to the FQHE. As these lectures are intended to be tutorial in nature, I will concentrate on the analytic statements that can be made. Details of the results of numerical studies are described in an extended review article (Cooper 2008). For a broader discussion of the physics of rotating atomic Bose gases, including experimental aspects, the reader is also referred to the excellent recent reviews of Bloch *et al.* (2008) and Fetter (2009).

10.1.2 Bose–Einstein condensation

Consider a noninteracting gas in three dimensions, consisting of identical particles of mass M with mean number density $\bar{n}$, and in equilibrium at a temperature T. The temperature sets the thermal de Broglie wavelength, λ_T, via $\hbar^2/M\lambda_T^2 \sim k_B T$, and the density sets the mean particle spacing $\bar{a} \sim \bar{n}^{-1/3}$. At low temperatures, when $\lambda_T \gtrsim \bar{a}$, the gas must be described by quantum theory, and its properties depend strongly on the statistics of the particles. For bosons, there is a phase transition when $\lambda_T \sim \bar{a}$, at a critical temperature

$$T_c = \frac{2\pi\hbar^2}{Mk_B}\left(\frac{\bar{n}}{\zeta(3/2)}\right)^{2/3}. \tag{10.1}$$

For $T < T_c$, the gas is a *Bose–Einstein condensate* (BEC), characterized by a finite fraction of the particles occupying the same quantum state.

Until recently there was only one experimental realization of an atomic BEC. The transition of helium-4 into a superfluid state below $T_c = 2.17\,K$ is known to be associated with Bose–Einstein condensation,[1] albeit in a system in which the interparticle interactions are relatively large. However, in recent years, Bose–Einstein condensation has been achieved in a

[1] Bose–Einstein condensation has been measured by neutron scattering, yielding a condensate fraction at low temperatures of about 9% (Sokol 1995).

wide variety of atomic species. These are prepared as metastable low-density gases, where $\bar{n} \sim 10^{12}$–$10^{15}\,\mathrm{cm}^{-3}$, confined in magnetic or optical traps. At such low densities, the BEC transition temperature is extremely small, $T_\mathrm{c} \simeq 100\,\mathrm{nK}$. Nevertheless, by a combination of laser and evaporative cooling, these low temperatures can be routinely achieved.

At the low temperatures involved, the thermal de Broglie wavelength, $\lambda_T \gtrsim \bar{a} \simeq 0.1$–$1\,\mu\mathrm{m}$, is much larger than the typical range of the interatomic potential. The two-particle scattering is therefore dominated by s-wave scattering, with a scattering length a_s that is typically of the order of a few nanometers (for ^{87}Rb, $a_\mathrm{s} \simeq 5\,\mathrm{nm}$). Thus, an atomic Bose gas is typically weakly interacting, in the sense that $\bar{n}a_\mathrm{s}^3 \ll 1$, so it is well described as an ideal Bose gas with very small condensate depletion (this is in contrast to superfluid helium, for which the strong interactions cause significant condensate depletion). That said, the interactions are nonzero and are important for many physical properties.

10.1.3 Quantized vortices

Quantized vortex line. It was noted by Onsager and Feynman (Donnelly 1991) that superfluid helium cannot rotate as a conventional fluid. A conventional fluid rotating at angular frequency $\mathbf{\Omega}$ has the velocity field of rigid-body rotation

$$\boldsymbol{v} = \mathbf{\Omega} \times \boldsymbol{r}\,, \tag{10.2}$$

for which the "vorticity" of the flow, $\boldsymbol{\nabla} \times \boldsymbol{v}$, is uniform,

$$\boldsymbol{\nabla} \times \boldsymbol{v} = 2\mathbf{\Omega}\,. \tag{10.3}$$

If, as is believed to be the case, the superfluid is described by a superfluid wave function $\psi_\mathrm{s} = \sqrt{n_\mathrm{s}}e^{i\phi(\boldsymbol{r})}$, then the superfluid velocity is

$$\boldsymbol{v}_\mathrm{s} = \frac{\hbar}{M}\boldsymbol{\nabla}\phi\,. \tag{10.4}$$

Hence, the fluid vorticity apparently vanishes:

$$\boldsymbol{\nabla} \times \boldsymbol{v}_\mathrm{s} = \frac{\hbar}{M}\boldsymbol{\nabla} \times \boldsymbol{\nabla}\phi = 0\,. \tag{10.5}$$

The last equality follows from the identity that the "curl of the gradient" of a smooth function vanishes. This overlooks the possibility that the phase ϕ might have line-like singularities (point-like in 2D) around which ϕ changes by an integer multiple of 2π. These are the *quantized vortex lines*. Integrating the vorticity over a 2D surface containing such a singularity gives

$$\int \boldsymbol{\nabla} \times \boldsymbol{v}_\mathrm{s} \cdot d\boldsymbol{S} = \oint \boldsymbol{v}_\mathrm{s} \cdot d\boldsymbol{l} = \frac{h}{M} \times \text{integer}\,, \tag{10.6}$$

indicating a delta-function contribution to the fluid vorticity on the vortex line. The "circulation" of the vortex, defined as

$$\kappa \equiv \oint \boldsymbol{v}_\mathrm{s} \cdot d\boldsymbol{l} = \frac{h}{M} \times \text{integer}\,, \tag{10.7}$$

is therefore quantized in units of h/M. This leads to a characteristic velocity profile, with an azimuthal velocity $|\boldsymbol{v}_\mathrm{s}| \sim 1/r$ that diverges as $r \to 0$ (see Fig. 10.1). In order to avoid

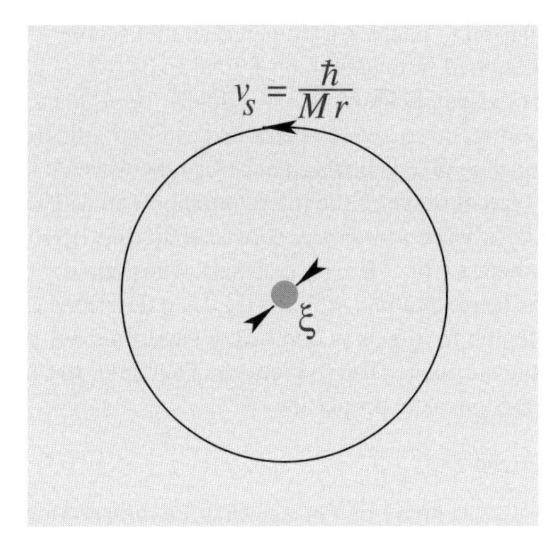

Fig. 10.1 Schematic diagram of the velocity field around a quantized vortex line with one quantum of circulation, $\kappa = h/M$. The superfluid density falls to zero within the vortex core, which has a scale set by the healing length ξ.

the infinite kinetic energy associated with this divergent velocity, in the core of the vortex the superfluid density n_{s} tends to zero, over a length scale of the *healing length*, ξ.

The healing length is an important characteristic of a superfluid. Within the Gross–Pitaevskii mean-field theory for an interacting Bose condensate, the energy is

$$E - \mu N = \int \left[\frac{\hbar^2}{2M} |\nabla\psi|^2 + \frac{1}{2} g |\psi|^4 - \mu |\psi|^2 \right] d^3\boldsymbol{r}\,, \tag{10.8}$$

where ψ is the condensate wave function. I have taken contact interactions, with a strength $g = 4\pi\hbar^2 a_{\mathrm{s}}/M$ chosen to reproduce the s-wave scattering length a_{s}. In many situations, eqn (10.8) is an accurate microscopic description of an atomic BEC, owing to the small condensate depletion $\bar{n}a_{\mathrm{s}}^3 \ll 1$. For (strongly interacting) superfluid helium, eqn (10.8) can be interpreted as the Ginzburg–Landau free energy of the superfluid order parameter, replacing ψ by ψ_{s} and interpreting the coefficients as phenomenological parameters. For a uniform fluid, minimization with respect to $|\psi|^2$ gives $\mu = g|\psi|^2$. There is therefore one characteristic length scale of the equilibrium fluid, set by

$$\frac{\hbar^2}{2M\xi^2} = g|\psi|^2 = \mu \quad \Rightarrow \quad \xi = \sqrt{\frac{\hbar^2}{2M\mu}} = \frac{1}{\sqrt{8\pi\bar{n}a_{\mathrm{s}}}}\,. \tag{10.9}$$

In the case of superfluid helium, for which interactions are strong, the healing length is very short; $\xi \sim 0.8\,\text{Å}$. Thus, the vortex cores have a size of the order of the interparticle spacing. For an atomic gas, the healing length is typically of the order of $\xi \simeq 0.5\,\mu\mathrm{m}$.

Vortex lattice. If superfluid helium is forced to rotate, for example by cooling liquid helium in a rotating vessel from above the superfluid transition (where the conventional fluid rotates as a rigid body) into the superfluid phase, the superfluid establishes an array of singly quantized vortex lines.

For dilute vortices, spaced by distances $a_{\rm v} \gg \xi$, the arrangement of vortices is dominated by the influence of the kinetic energy of the superfluid flow (Fetter 1965). This arrangement is conveniently analyzed by use of an analogy with 2D electrostatics, where we write

$$\boldsymbol{v}_{\rm s} = \frac{\hbar}{M}\boldsymbol{\nabla}\phi \equiv -\frac{\hbar}{M}\hat{\boldsymbol{z}} \times \boldsymbol{\nabla}\Phi\,, \tag{10.10}$$

which expresses the velocity field in terms of the function Φ, which is a smooth function of position away from the vortex cores. In this parameterization, the condition for steady-state flow $\boldsymbol{\nabla} \cdot \boldsymbol{v}_{\rm s} = 0$ is automatically satisfied, while the constraint of irrotational flow $\boldsymbol{\nabla} \times \boldsymbol{v}_{\rm s} = 0$ implies Laplace's equation $\nabla^2\Phi = 0$ away from the vortex cores. To complete the electrostatic analogy, note that a singly quantized vortex in the 2D fluid acts as a charge of strength 2π in the "electric field" $-\boldsymbol{\nabla}\Phi$, as may be seen from

$$\oint \boldsymbol{\nabla}\phi \cdot d\boldsymbol{l} = 2\pi \quad \Rightarrow \quad \oint -\hat{\boldsymbol{z}} \times \boldsymbol{\nabla}\Phi \cdot d\boldsymbol{l} = \oint [-\boldsymbol{\nabla}\Phi] \cdot (d\boldsymbol{l} \times \hat{\boldsymbol{z}}) = 2\pi\,, \tag{10.11}$$

i.e. the outward flux of the electric field $-\boldsymbol{\nabla}\Phi$ is 2π. Hence, in the presence of vortices of unit strength, the field Φ satisfies Poisson's equation

$$-\nabla^2\Phi = 2\pi \sum_i \delta^{(2)}(\boldsymbol{r} - \boldsymbol{R}_i)\,, \tag{10.12}$$

where the $\boldsymbol{R}_i$ are the positions of the vortices in the plane perpendicular to the rotation axis. Thus, Φ is the 2D electrostatic potential for a set of charges of strength 2π at the locations of the vortices.

For a superfluid forced to rotate with angular velocity $\boldsymbol{\Omega}$, one should find the flow that minimizes the kinetic energy (per unit length $\mathcal{L}$) *in the rotating frame*:

$$\begin{aligned}
\frac{K}{\mathcal{L}} &= \frac{1}{2}\rho_{\rm s} \int \left|\boldsymbol{v}_{\rm s} - \boldsymbol{\Omega} \times \boldsymbol{r}\right|^2 \, d^2\boldsymbol{r} && (10.13)\\
&= \frac{1}{2}\frac{\hbar^2}{M^2}\rho_{\rm s} \int \left|-\boldsymbol{z} \times \boldsymbol{\nabla}\Phi - \frac{M}{\hbar}\Omega\boldsymbol{z} \times \boldsymbol{r}\right|^2 \, d^2\boldsymbol{r} && (10.14)\\
&= \frac{1}{2}\frac{\hbar^2}{M^2}\rho_{\rm s} \int \left|-\boldsymbol{\nabla}\Phi - \frac{M\Omega}{\hbar}\boldsymbol{r}\right|^2 \, d^2\boldsymbol{r}\,, && (10.15)
\end{aligned}$$

where $\rho_{\rm s}$ is the mass density of the superfluid. This is the electrostatic field energy for the sum of the field $-\boldsymbol{\nabla}\Phi$ and a background field $-M\Omega\boldsymbol{r}/\hbar$.

From eqn (10.15), one can deduce some useful consequences:

(i) Since Φ is the potential for a set of discrete charges of size 2π, it cannot perfectly compensate the background charge density $\boldsymbol{\nabla}\cdot(-M\Omega\boldsymbol{r}/\hbar) = -2M\Omega/\hbar$. However, the condition of electrical neutrality on average fixes the mean vortex density to

$$n_{\rm v} = \frac{2M\Omega}{h}\,. \tag{10.16}$$

This is Feynman's result for the mean density of quantized vortices in a rotating superfluid (Donnelly 1991).

(ii) The vortices, being 2π point charges in 2D, experience logarithmic repulsion, leading to effective vortex–vortex interactions

$$\frac{\rho_s \hbar^2}{M^2} \sum_{i<j} -\frac{(2\pi)^2}{2\pi} \ln\left(\frac{|\boldsymbol{R}_i - \boldsymbol{R}_j|}{\xi}\right), \tag{10.17}$$

where the healing length cuts off the divergence in kinetic energy close to the vortex cores.[2] The ground state configuration of classical particles interacting with a logarithmic repulsion is a triangular lattice. One therefore expects the vortices to be arranged in a triangular lattice at low temperature. The observation of vortex lattice arrays in superfluid helium is hampered by the very small size of the vortex core $\xi \sim 0.8\,\text{Å}$. However, in remarkable experiments (Yarmchuk *et al.* 1979), images of small arrays of vortices have been obtained.

Vortex lattices in atomic Bose condensates. Vortex lattices may be generated in atomic BECs by confining the gas in a magnetic trap of cylindrical symmetry, and stirring the gas with a rotating deformation (Bloch *et al.* 2008).[3] Once the rotating deformation has spun the condensate up to a large angular momentum, the deformation can be removed; the gas then comes to equilibrium but now with a large angular momentum, which is conserved owing to the cylindrical symmetry. Images of the particle density, taken following expansion of the cloud, show large ordered arrays of vortex lines, which appear as regions in which the density is suppressed (see for example Madison *et al.* (2000) and Abo-Shaeer *et al.* (2001)).

There are many reasons why cold atomic gases are very interesting systems in which to study rotating BECs. The aspects that I shall focus on are:

- These atomic gases are typically weakly interacting. The healing length (10.9) is therefore relatively large; typically, $\xi \simeq 0.5\,\mu\text{m}$, which is comparable to the intervortex spacing $a_\text{v} \sim 1/\sqrt{n_\text{v}} \simeq 2\,\mu\text{m}$. As a result, one can envisage (Wilkin *et al.* 1998) entering a novel regime in which the rotation frequency is sufficiently high that the separation between the vortices a_v becomes less than the healing length ξ. In helium, where the zero-temperature healing length is $\xi \simeq 0.8\,\text{Å}$, achieving this regime would require $\Omega \gtrsim 10^{13}\,\text{rads}^{-1}$! It is of interest to understand the properties of a rotating Bose gas in the regime of high vortex density.
- In atomic gases, there is the possibility to tune the interparticle interactions. It is possible not only to change the strength of the contact interaction, but also to change the qualitative form of the interaction. As we shall see, the ground state of a rapidly rotating Bose gas is sensitive to the nature of the interaction.

10.2 Rapidly rotating atomic Bose gases

10.2.1 Rapid-rotation limit

In order to formulate the theoretical issues more precisely, I write down a Hamiltonian describing an interacting atomic gas,

[2] The total energy also contains single-vortex contributions, arising from the kinetic energy of the superfluid flow around individual vortices and their interaction with "image" vortices (Fetter 1965).

[3] The dynamics of vortex lattice formation is an interesting subject, involving surface wave instabilities and turbulent flow. I will discuss only the equilibrium states.

$$H = \sum_{i=1}^{N} \left[\frac{|\boldsymbol{p}_i|^2}{2M} + \frac{1}{2} M \omega_\perp^2 (x_i^2 + y_i^2) + \frac{1}{2} M \omega_\parallel^2 z_i^2 \right] + \sum_{i<j} V\left(\boldsymbol{r_i} - \boldsymbol{r_j}\right). \tag{10.18}$$

Most commonly, the interaction between ultracold atoms can be represented by a contact interaction

$$V(\boldsymbol{r}) = g\delta^{(3)}(\boldsymbol{r}), \tag{10.19}$$

with the interaction strength chosen as

$$g = \frac{4\pi\hbar^2 a_\mathrm{s}}{M} \tag{10.20}$$

to reproduce the s-wave scattering length. However, in later sections I shall also consider situations involving dipolar interactions or a Feshbach resonance in the interaction.

This Hamiltonian describes N identical particles which are confined in a harmonic trap, with natural frequencies $\omega_\perp$ and $\omega_\parallel$. The angular momentum about the z-axis is conserved. The question is: what is the lowest-energy state as a function of the angular momentum? That is, what is the ground state of a Bose gas that has been stirred so that it has picked up a large angular momentum?

I shall focus on the limit of weak interactions, for which the relevant single-particle states simplify. To motivate our choice of the form of the single-particle states, it is convenient to work not in terms of fixed angular momentum, but in terms of the conjugate variable, which is the rotation rate. To this end, let us consider the system in a frame of reference rotating about the z-axis with angular momentum $\boldsymbol{\Omega} = \Omega\hat{\boldsymbol{z}}$. In this frame, the Hamiltonian is (Landau and Lifshitz 1981)

$$H_\Omega = H - \boldsymbol{\Omega}\cdot\boldsymbol{L}. \tag{10.21}$$

The one-body terms can be written in a suggestive way,

$$H_\Omega^{(1)} = \frac{|\boldsymbol{p}|^2}{2M} + \frac{1}{2} M\omega_\perp^2(x^2+y^2) + \frac{1}{2} M\omega_\parallel^2 z^2 - \boldsymbol{\Omega}\cdot\boldsymbol{r}\times\boldsymbol{p} \tag{10.22}$$

$$= \frac{|\boldsymbol{p} - M\boldsymbol{\Omega}\times\boldsymbol{r}|^2}{2M} + \frac{1}{2} M(\omega_\perp^2 - \Omega^2)(x^2+y^2) + \frac{1}{2} M\omega_\parallel^2 z^2. \tag{10.23}$$

The kinetic term in this Hamiltonian is equivalent to that of a particle of charge q^* experiencing a magnetic field $\boldsymbol{B}^*$ with

$$q^*\boldsymbol{B}^* = 2M\boldsymbol{\Omega}. \tag{10.24}$$

This connection shows that the Coriolis force in the rotating frame plays the same role as the Lorentz force on a charged particle in a uniform magnetic field (Fröhlich *et al.* 1994). The radial confining potential is reduced, owing to the centrifugal forces. For weak interactions, the nontrivial ground states (those with L nonzero and finite) occur close to the point of instability of the confined gas, $\Omega = \omega_\perp$. Close to this value, the Hamiltonian (10.23) describes a quasi-2D system of particles in a uniform magnetic field.

I shall focus on situations in which the mean interaction energy $g\bar{n}$ is small compared with the single-particle level spacings

$$g\bar{n} \ll \hbar\omega_\parallel, 2\hbar\omega_\perp, \tag{10.25}$$

where the "2" in $2\hbar\omega_\perp$ is included to indicate that the effective cyclotron energy at $\Omega = \omega_\perp$ is $2\hbar\omega_\perp$. These conditions are equivalent to the conditions that the healing length is large compared with the intervortex spacing, i.e. $\xi \gg a_\perp$ (for a large vortex array, where $\Omega \simeq \omega_\perp$), and with the subband thickness in the z-direction, i.e. $\xi \gg a_\parallel$. Although eqn (10.25) is typically not satisfied for a nonrotating gas, for a rapidly rotating gas the centrifugal forces spread the cloud out, the density falls, and the system tends towards this weakly interacting regime.

Under the conditions (10.25), the single-particle states are restricted to (quasi)-2D and to the lowest Landau level (LLL) (Wilkin *et al.* 1998)

$$\psi_m(\boldsymbol{r}) \propto (x+iy)^m \, e^{-(x^2+y^2)/2a_\perp^2} \, e^{-z^2/2a_\parallel^2} \,, \tag{10.26}$$

where

$$a_{\perp,\parallel} \equiv \sqrt{\frac{\hbar}{M\omega_{\perp,\parallel}}} \tag{10.27}$$

are the trap lengths in the radial and axial directions. To make the connections to the fractional quantum Hall effect as clear as possible, I shall introduce the complex representation

$$\zeta \equiv \frac{x+iy}{\ell}\,, \qquad \ell \equiv \sqrt{\frac{\hbar}{2M\omega_\perp}} = \frac{a_\perp}{\sqrt{2}}\,, \tag{10.28}$$

where ℓ is the conventional magnetic length. The 2D LLL basis states are then

$$\psi_m(\zeta) = \frac{1}{\sqrt{2\pi 2^m m!}}\zeta^m \times \left[\frac{e^{-|\zeta|^2/4}}{\ell}\frac{1}{(\pi a_\parallel^2)^{1/4}}e^{-z^2/2a_\parallel^2}\right]. \tag{10.29}$$

In the following, for simplicity I shall omit the ubiquitous [bracketed] exponential terms in the wave functions, focusing only on the prefactors that are polynomial in ζ.

An alternative way to obtain this result is to work in the laboratory frame, and consider the total angular momentum as the control parameter. For weak interactions (10.25), to determine the ground state wave function one must first minimize the kinetic and potential energies. The single-particle energy is

$$E = \hbar\omega_\perp(2n_\perp + |m| + 1) + \hbar\omega_\parallel\left(n_\parallel + \frac{1}{2}\right), \tag{10.30}$$

where $m = 0, \pm 1, \pm 2, \ldots$ is the angular momentum (in units of $\hbar$) about the z-axis and $n_\perp, n_\parallel \geq 0$ are the radial and axial quantum numbers. The lowest-energy states have $n_\perp = n_\parallel = 0$. For fixed total angular momentum (in units of $\hbar$)

$$L = \sum_{i=1}^{N} m_i\,, \tag{10.31}$$

the lowest-energy states are also those for which $m_i \geq 0$ (Wilkin *et al.* 1998). The single-particle states with $n_\perp = n_\parallel = 0$ and $m_i \geq 0$ are the 2D LLL states (10.26). A collection of

N particles with total angular momentum (10.31) restricted to the 2D LLL has total single-particle energy

$$E = \sum_{i=1}^{N} \left[\hbar\omega_\perp \left(m_i + \frac{1}{2} \right) + \frac{\hbar\omega_\parallel}{2} \right] = \left(\hbar\omega_\perp + \frac{1}{2}\hbar\omega_\parallel \right) N + \hbar\omega_\perp L \,. \tag{10.32}$$

To see why one requires $m_i \geq 0$, consider moving particle $i = 1$ from a state $m_1 \geq 0$ to a state m_1'. To conserve the total angular momentum L, one also needs to add $m_1 - m_1'$ units of angular momentum to the other particles; keeping these particles in states with $m \geq 0$ leads to an overall change in energy

$$\Delta E = \hbar\omega_\perp \left[(-|m_1| + |m_1'|) + (m_1 - m_1') \right] = \hbar\omega_\perp \left[|m_1'| - m_1' \right],$$

which is an energy *increase* if $m_1' < 0$.

Determining the ground state of a rapidly rotating atomic gas poses a problem that is very closely related to that which appears in the FQHE. For a system of N particles with a given total angular momentum L, one must distribute the particles within the 2D LLL orbitals such that the angular momentum (10.31) is fixed. The total kinetic and potential energy of each of these many-particle states (10.32) is the same. Therefore, at the single-particle level (i.e. neglecting interactions), there is a very high degeneracy. The true ground state is determined from within this degenerate set of states by the action of the interactions. This is a fundamentally nonperturbative problem, as in the FQHE. The differences are firstly in the nature of the interparticle forces, and secondly, and most importantly, that here we are studying bosons, not fermions. We are interested in the nature of the phases that can emerge for rotating bosons, for the specific forms of interactions that are physically relevant in ultracold atomic gases. In the following, I shall describe the properties of rotating atomic gases in the weak-interaction limit (10.25). I shall refer to this regime either as the "2D LLL limit" or as the "rapid-rotation limit".

It is perhaps useful to pause to note that this is an apparently counterintuitive conclusion. In the *weak*-interaction limit (10.25), we have found that the ground state is strongly dependent on the nature of the interactions. This arises because the (large) kinetic and potential energies act to restrict the single-particle wave functions to the 2D LLL, but leave a very large residual degeneracy. It is then the interparticle interactions that must select the nature of the ground state. Note that we found a complementary result in the *strong*-interaction regime, $\xi \lesssim a_\mathrm{v}$. In that case, the interactions affect only the form of the vortex core. The nature of the large-scale structure of the vortex lattice is determined by minimizing the kinetic energy of the superfluid flow.

10.2.2 Gross–Pitaevskii mean-field theory

The Gross–Pitaevskii mean-field theory for an interacting Bose gas amounts to the assumption that the many-body state is a pure condensate, in which all particles occupy the same single-particle state $\psi(\boldsymbol{r})$:

$$\Psi(\{\boldsymbol{r}_i\}) = \prod_{i=1}^{N} \psi(\boldsymbol{r}_i) \,. \tag{10.33}$$

As an approximation to the ground state, the (normalized) condensate wave function ψ is chosen to minimize the expectation of the Hamiltonian. (I shall discuss the limits of applicability of this ansatz in Section 10.3.)

For a rapidly rotating Bose gas, the single-particle orbitals consist only of the states (10.26). Thus, the condensate wave function can be expanded in terms of these states:

$$\psi(\zeta, z) = \sum_{m\geq 0} c_m \psi_m \tag{10.34}$$

for $\sum_m |c_m|^2 = 1$. The mean-field ground state is obtained by choosing the coefficients c_m to minimize the expectation value of the interaction energy, which, for contact interactions (10.19), is

$$\frac{1}{2} g N^2 \int |\psi(\boldsymbol{r})|^4 \, d^3\boldsymbol{r} \, ,$$

with a constraint on the (average[4]) angular momentum,

$$L = N \sum_m m |c_m|^2 \, . \tag{10.35}$$

The mean-field ground state therefore depends only on L/N.

Noting that the condensate wave function (10.34) is a polynomial in ζ, we may express it in terms of its zeros as

$$\psi(\zeta, z) = A \prod_{\alpha=1}^{m_{\max}} (\zeta - \zeta_\alpha) \, , \tag{10.36}$$

where $m_{\max}$ is introduced as a cutoff in the degree of the polynomial (10.34), and A is a normalization factor. These zeros are the complex positions of the quantized vortices. Thus the LLL Gross–Pitaevskii wave function is fully described by the positions of the vortices. The process of choosing the $m_{\max} + 1$ complex coefficients c_m (with normalization) is equivalent to the choice of the locations of $m_{\max}$ vortices.

The minimization for contact interactions can be readily implemented numerically (Butts and Rokhsar 1999). A simple result applies at $L/N = 1$, for which the condensate wave function is found to be

$$\psi(\zeta, z) = A\zeta \, , \tag{10.37}$$

in which all particles are condensed in the $m = 1$ orbital. For other values of L/N, the system spontaneously breaks rotational symmetry (Cooper 2008). For large L/N, the number of vortices grows as $N_{\rm v} \simeq 3L/N$, and forms a triangular vortex lattice (weakly distorted by the confinement) (see Fig. 10.2).

The appearance of this triangular vortex lattice is very closely related to the appearance of a triangular lattice of flux lines in type II superconductors close to the upper critical field. In

[4] In general, eqn (10.34) is not an eigenstate of the angular momentum, whereas the Hamiltonian conserves total angular momentum. The fact that the condensate wave function does not preserve the rotational symmetry of the Hamiltonian should not be viewed as a deficiency. Rather, this wave function correctly captures the fact that the system spontaneously breaks rotational symmetry, in the limit $N \to \infty$ with L/N fixed.

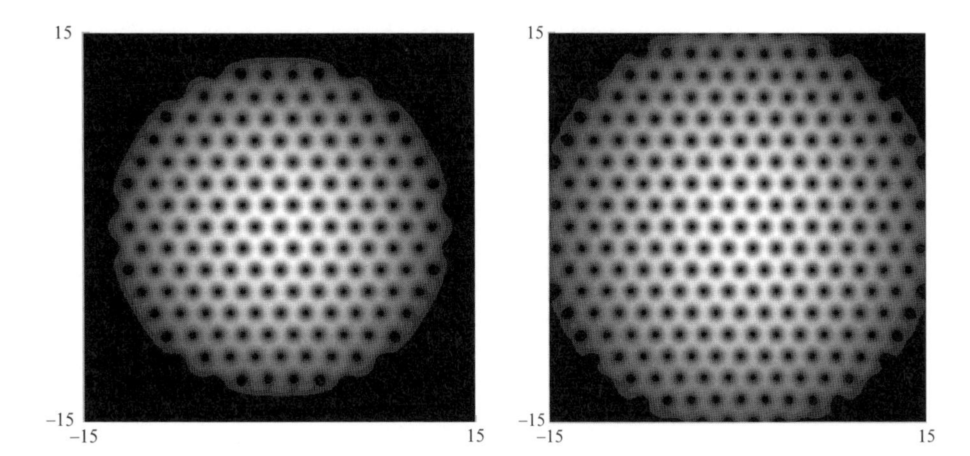

Fig. 10.2 Numerical results for the particle density of an atomic BEC in the weak-interaction limit (2D LLL), for two values of the angular momentum per particle $L/N = 60$ and 90. The lattice structure is determined only by the interactions, which are chosen here to be contact interactions (10.19). Close to the center of the trap, the vortices form a regular triangular lattice. Reproduced with permission from Cooper *et al.* (2004).

that case, the order parameter is determined by the Ginzburg–Landau energy functional (per unit length along the field direction)

$$\frac{F - F_n}{L} \sim \int \left[\frac{1}{2M} |(-i\hbar\boldsymbol{\nabla} + 2e\boldsymbol{A})\psi_{\rm s}|^2 + \alpha|\psi_{\rm s}|^2 + \frac{\beta}{2}|\psi_{\rm s}|^4 \right] d^2\boldsymbol{r}\,. \tag{10.38}$$

Close to H_{c2}, the order parameter is small, so to a first approximation one can neglect the quartic term and minimize the quadratic terms. These describe electron pairs (hence the charge $-2e$) in a uniform magnetic field: the lowest-energy states are the (degenerate) lowest-Landau-level orbitals. As explained by Abrikosov (1957), the ground state is selected as the linear combination of the lowest-Landau-level orbitals that minimizes the quartic term. This is mathematically equivalent to the problem described above, at least in the uniform case (i.e. close to the center of a trap containing many vortices); the ground state in both cases is a *triangular* vortex lattice.

For both $\xi \ll a_\perp$ (dilute vortices) and $\xi \gg a_\perp$ (dense vortices), the mean-field ground state is a triangular vortex lattice. As $\xi/a_\perp$ varies between these two limits, there is a smooth crossover (Fischer and Baym 2003). This crossover has been observed in experiments, which have achieved a chemical potential $\mu \sim g\bar{n}$ that is less than the effective cyclotron splitting $2\hbar\omega_\perp$. The apparent core size of the vortices shows the behavior expected at the crossover into the LLL regime (Coddington *et al.* 2004). In the regime of strong interactions $\xi \gtrsim a_{\rm v}$, the area of the density depletion close to a vortex core is found to be proportional to ξ^2. For the weak-interaction regime $\xi \lesssim a_{\rm v}$, this area is found to be proportional to $a_\perp^2$; this is consistent with the expectation that in this regime the single-particle states are restricted to the LLL wave functions, and only length scale is the mean vortex spacing $a_{\rm v} \simeq a_\perp$.

Given that very different physics controls the energetics in the two limits—for strong interactions $\xi \lesssim a_\perp$, the vortex lattice is determined by the kinetic energy, and for weak interactions $\xi \gtrsim a_\perp$, the vortex lattice is determined entirely by the interactions—one should view the smooth crossover as a coincidence. As I now show, changing the form of the interaction potential can lead to large changes in the vortex lattice structure for $\xi \gtrsim a_\perp$.

10.2.3 Dipolar interactions

It has been proposed that, under certain circumstances, the atoms (or molecules) in a cold trapped gas may carry an electric or magnetic dipole moment (Baranov *et al.* 2002). One should then add to the usual contact interaction the long-range dipole–dipole interaction. For two dipolar atoms with aligned dipole moments, the effective interaction is

$$V(\boldsymbol{r}) = \frac{4\pi\hbar^2 a_\mathrm{s}}{M}\,\delta^3(\boldsymbol{r}) + C_\mathrm{d}\frac{1-3\cos^2\theta}{r^3}\,, \tag{10.39}$$

where θ is the angle between the dipole moment and the line separating the atoms (see Fig. 10.3).

A BEC of atoms with significant dipolar interactions has been realized by condensing ^{52}Cr (Griesmaier *et al.* 2005), which is an atom with a very large magnetic dipole moment, $\mu = 6\mu_\mathrm{B}$, such that $C_\mathrm{d} = \mu_0\mu^2/4\pi$. The relative size of the dipolar and contact interactions is parameterized by the dimensionless ratio

$$\epsilon_\mathrm{dd} \equiv \frac{C_\mathrm{d} M}{3\hbar^2 a_\mathrm{s}}\,, \tag{10.40}$$

which is approximately 0.16 for native conditions, and has been further increased to approximately 1 by using a Feshbach resonance to reduce a_s (Lahaye *et al.* 2007).

Consider a rapidly rotating atomic gas (in the 2D LLL limit), and for simplicity assume that the dipole moments are directed parallel to the rotation axis. The mean-field theory requires one to minimize the expectation value of the interaction energy

$$\frac{1}{2}\int\!\!\int |\psi(\boldsymbol{r})|^2 V(\boldsymbol{r}-\boldsymbol{r}')|\psi(\boldsymbol{r}')|^2\, d^3\boldsymbol{r}\, d^3\boldsymbol{r}' \tag{10.41}$$

for $\psi(\boldsymbol{r})$ in the lowest Landau level, and the potential is that in eqn (10.39). Although this is a simple generalization of the Abrikosov problem, it is one that does not naturally appear in that context, where the microscopic physics determining the superconductivity acts on scales much less than the vortex lattice period.

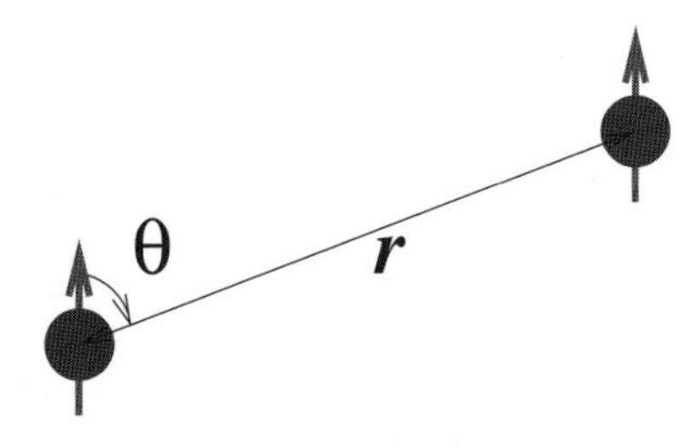

Fig. 10.3 Geometry of two atoms with aligned dipole moments, as discussed in the text.

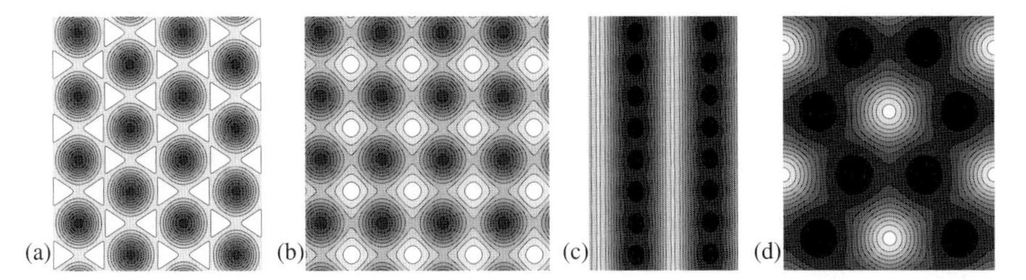

Fig. 10.4 Mean-field ground states of a rotating BEC in a 2D LLL, for particles interacting by both contact and dipolar interactions. The relative size of these is controlled by $\alpha = V_2/V_0$. The structure of the vortex lattice varies from (a) triangular ($0 \leq \alpha \leq 0.20$) to (b) square ($0.20 \leq \alpha \leq 0.24$) to (c) "stripe crystal", with a simple rectangular unit cell ($0.24 \leq \alpha \leq 0.60$), to "bubble crystal" phases ($\alpha \geq 0.60$), the simplest of which is shown in (d). Reproduced with permission from Cooper *et al.* (2005).

The mean-field ground states have been found numerically for the case of a translationally invariant vortex lattice (Cooper *et al.* 2005). The results are shown in Fig. 10.4, as a function of

$$\alpha \equiv \frac{V_2}{V_0}, \tag{10.42}$$

where

$$V_0 = \sqrt{\frac{2}{\pi}}\frac{\hbar^2 a_s}{M a_\perp^2 a_\parallel} + \sqrt{\frac{2}{\pi}}\frac{C_d}{a_\perp^2 a_\parallel} - \sqrt{\frac{\pi}{2}}\frac{C_d}{a_\perp^3}, \tag{10.43}$$

$$V_{m>0} = \sqrt{\frac{\pi}{2}}\frac{(2m-3)!!}{m!\,2^m}\frac{C_d}{a_\perp^3} \tag{10.44}$$

are the Haldane pseudopotentials for the interaction (10.39) in the limit $a_\parallel/a_\perp \ll 1$.[5]

The results show that the mean-field ground state is very sensitive to long-range repulsion, passing through a series of vortex lattice phases as α increases. The contact interaction aims to make $|\psi(\boldsymbol{r})|^2$ as uniform as possible, while the long-range repulsion causes $|\psi(\boldsymbol{r})|^2$ to cluster, leading to crystalline phases of clusters of particles at large α. These are referred to as "bubble crystal" phases, in analogy with the terminology used for structures of similar symmetry in 2D electron gases at high Landau levels (Koulakov *et al.* 1996, Moessner and Chalker 1996).

10.3 Strongly correlated phases

I now turn to describing the properties of rapidly rotating atomic BECs beyond the Gross–Pitaevskii mean-field approach. Much of our understanding is derived from numerical exact-diagonalization studies, and analytic results are limited.

[5] For the geometry considered, the net interaction (10.39) is rotationally symmetric about the z-axis. The two-body interaction is fully characterized by the Haldane pseudopotentials, which are the eigenvalues of the interaction for the two-particle states in the 2D LLL with relative angular momentum m. Situations in which the dipoles are tilted away from the rotation axis provide a natural way in which anisotropic two-body interactions can appear.

10.3.1 Low angular momentum

For small values of the total angular momentum L, the exact ground states of a rapidly rotating Bose gas are known. The following states have been shown to be exact eigenstates of the contact repulsion by both analytical (Smith and Wilkin 2000) and numerical (Bertsch and Papenbrock 1999) studies, and to be the ground states for a class of repulsive interactions (Hussein and Vorov 2002, Vorov *et al.* 2003) that includes the contact repulsion (10.19).

- For $L = 0$, there is only one state in the 2D LLL: the pure condensate in the $m = 0$ orbital

$$\Psi_{L=0}(\{\zeta_i\}) \propto 1 . \tag{10.45}$$

- For $L = 1$, the ground state is the center-of-mass excitation of the $L = 0$ state

$$\Psi_{L=1}(\{\zeta_i\}) \propto \zeta_c , \tag{10.46}$$

$$\zeta_c \equiv \frac{1}{N} \sum_{i=1}^{N} \zeta_i , \tag{10.47}$$

for which the interaction energy is the same as for the $L = 0$ ground state.

- For $2 \leq L \leq N$, the ground states are the elementary symmetric polynomials in the variables $(\zeta_i - \zeta_c)$

$$\Psi_L(\{\zeta_i\}) \propto \sum_{p_1<p_2<p_L} (\zeta_{p_1} - \zeta_c)(\zeta_{p_1} - \zeta_c) \ldots (\zeta_{p_L} - \zeta_c) . \tag{10.48}$$

Note, in particular, the case $L = N$, for which

$$\Psi_{L=N}(\{\zeta_i\}) \propto \prod_i (\zeta_i - \zeta_c) . \tag{10.49}$$

If the coordinate ζ_c were simply a number, this would be the wave function of a pure condensate with a vortex at complex position ζ_c, eqn (10.36). The fact that ζ_c is the center-of-mass coordinate (10.47), and therefore a function of the particle coordinates, means that this state is not fully condensed. Nevertheless, the fluctuations of the center of mass from its average value $\langle \zeta_c \rangle = 0$, computed for the state $\Psi_{L=N}$, are

$$\langle \zeta_c^2 \rangle \sim \frac{1}{N} . \tag{10.50}$$

Thus, in the limit $N \to \infty$,

$$\Psi_{L=N}(\{\zeta_i\}) \stackrel{N \to \infty}{\longrightarrow} \prod_i \zeta_i , \tag{10.51}$$

which is the fully condensed state with a single vortex at the origin (all particles are condensed in the $m = 1$ orbital).

10.3.2 The filling factor

In order to make sharp statements about *phases of matter* in rotating atomic Bose gases, one must consider the thermodynamic limit, $N \to \infty$. There are different ways in which to take this limit for a rotating gas, depending on how one chooses to scale the angular momentum.

The simplest situation to consider is $N \to \infty$ keeping the number of vortices fixed. Since $N_{\mathrm{v}} \sim L/N$, this requires $L \to \infty$ such that $L/N \to$ constant. The particle density then grows as $\bar{n} \propto N/(a_\perp^2 a_\parallel)$, so for the mean-field interaction $g\bar{n}$ to remain finite (for example, to stay in the 2D LLL regime), one should scale g such that $gN \to$ constant. With this set of scalings, the Gross–Pitaevskii theory is exact (Lieb and Seiringer 2006); the ground state is a pure condensate. I have given an explicit example of this limit for a special case, eqn (10.51).

Although this limit is appropriate for the description of atomic BECs in many experimental situations, from the perspective of many-body physics it is not the most interesting limit, leading only to fully condensed states. A much more interesting limit can be found by exploiting analogies with the FQHE. This regime may be of importance in future experiments at high vortex density.

The FQHE exhibits a series of strongly correlated quantum phases, which are characterized by the electron *filling factor*

$$\nu_{\mathrm{e}} \equiv n_{\mathrm{e}} \frac{h}{eB}, \tag{10.52}$$

where n_{e} is the 2D number density of electrons. From the above mapping of the rotating atomic gas (10.24), the analogous quantity is (Cooper *et al.* 2001)

$$\nu \equiv n_{2\mathrm{D}} \frac{h}{q^* B^*} = n_{2\mathrm{D}} \frac{h}{2M\Omega}, \tag{10.53}$$

which, from eqn (10.16), can be written in terms of the vortex density as

$$\nu = \frac{n_{2\mathrm{D}}}{n_{\mathrm{v}}}, \tag{10.54}$$

or as

$$\nu = \frac{N}{N_{\mathrm{v}}} \tag{10.55}$$

if one considers the particles to be uniformly distributed over an area containing N_{v} vortices.

It is interesting to study the nature of the ground state in the limit $N \to \infty$, $N_{\mathrm{v}} \to \infty$ such that $\nu \to$ constant (Cooper *et al.* 2001). What is the phase diagram of a rapidly rotating atomic gas as a function of the filling factor ν? This is the question that I shall address in the remainder of this section.

Evidence has been given (Cooper *et al.* 2001) that, for ν large (but finite), the mean-field theory is accurate—at least in correctly predicting a triangular vortex lattice (even though there may be quantitative corrections at finite ν). However, for small ν, quantum fluctuations are large, and can cause the mean-field theory to fail. There are several ways in which to understand why the filling factor determines the degree of quantum fluctuations. One approach, which I found very influential when first considering these issues, is to determine directly the quantum fluctuations of the vortices.

10.3.3 Quantum fluctuations of vortices

Consider the dynamics of a single vortex line in a 2D fluid (i.e. a straight vortex line). The classical dynamics of a 2D vortex, at a position X and Y in an external potential $V(X,Y)$, follows from the standard Magnus force dynamics of a classical fluid

$$-\rho_s \kappa \dot{Y} + F_X^{\text{ext}} = 0\,, \tag{10.56}$$
$$+\rho_s \kappa \dot{X} + F_Y^{\text{ext}} = 0\,, \tag{10.57}$$

where ρ_s is the mass density (per unit area) of the fluid, κ is the circulation of the vortex, and $(F_X^{\text{ext}}, F_Y^{\text{ext}})$ is an external applied force. The only amendment for a quantized vortex in a superfluid is that the circulation is quantized, i.e. $\kappa = h/M$, so one may write

$$\rho_s \kappa = (n_{2D} M)\frac{h}{M} = n_{2D} h\,. \tag{10.58}$$

A Lagrangian that reproduces this classical dynamics is[6]

$$L = n_{2D} h \dot{X} Y - V(X,Y) \qquad \left[\vec{F}^{\text{ext}} = -\vec{\nabla} V\right]. \tag{10.59}$$

Constructing the momentum conjugate to the particle coordinate X and applying canonical quantization leads to

$$\Pi_X \equiv \frac{\partial L}{\partial \dot{X}} = n_{2D} h Y\,, \tag{10.60}$$
$$[\hat{X}, \hat{\Pi}_X] = i\hbar \Rightarrow [\hat{X}, \hat{Y}] = \frac{i}{2\pi n_{2D}}\,. \tag{10.61}$$

The X and Y coordinates[7] are conjugate, and obey the generalized uncertainty relation

$$\Delta X\, \Delta Y \geq \frac{1}{4\pi n_{2D}}\,, \tag{10.62}$$

which implies

$$\Delta X^2 + \Delta Y^2 \geq \frac{1}{2\pi n_{2D}}\,. \tag{10.63}$$

The result (10.63) makes physical sense: one cannot locate the vortex line to a distance less than the mean 2D separation between the particles. It is interesting to note that this result has an entirely classical interpretation, but emerges from a *quantum* calculation owing to the cancelation of Planck's constant in the circulation with Planck's constant in the commutator. Furthermore, the result (10.63) is consistent with the calculation of the fluctuations of the vortex (10.50) for the exact one-vortex wave function (10.49). Equation (10.50) implies $\Delta X^2 + \Delta Y^2 \sim \ell^2/N \sim 1/n_{2D}$, where I take the typical particle density $n_{2D} \sim N/\ell$, noting that the N particles are within an area of $\sim \ell^2$ in this inhomogeneous state (10.49).

[6] This is easily checked by constructing the Euler–Lagrange equations.

[7] These are the guiding-center coordinates of a particle in a single Landau level.

The importance of the filling factor becomes clear if one applies a form of the Lindemann criterion and asserts that the vortex lattice will become unstable to quantum fluctuations if the r.m.s. fluctuation is larger than some multiple α_L of the vortex spacing:

$$\sqrt{\Delta X^2 + \Delta Y^2} = \frac{1}{\sqrt{2\pi n_{\rm 2D}}} \geq \alpha_L \times a_{\rm v} = \alpha_L \sqrt{\frac{2}{\sqrt{3} n_{\rm v}}}, \tag{10.64}$$

$$\nu \equiv \frac{n_{\rm 2D}}{n_{\rm v}} \geq \nu_{\rm c} = \frac{\sqrt{3}}{4\pi\alpha_L^2}. \tag{10.65}$$

For a typical value for the Lindemann parameter $\alpha_L^2 \simeq 0.02$ (Rozhkov and Stroud 1996), one finds $\nu_{\rm c} \simeq 7$.[8] Calculations of the r.m.s. vortex fluctuations that allow for the collective modes of a full vortex lattice (Sinova *et al.* 2002, Baym 2004) give very nearly the same results as this one-vortex result, and therefore the same $\nu_{\rm c}$ if the same Lindemann parameter is used.

While instructive, these considerations are hardly predictive, depending very sensitively on α_L, which, itself, is estimated from the thermal melting of 3D crystals! For this reason, it is useful to have a direct determination of the transition. This has been studied in large-scale exact-diagonalization studies (Cooper *et al.* 2001). The strategy is to work on a system with periodic boundary conditions (the torus geometry), which is consistent with the formation of a vortex lattice. The signal of crystallization is the collapse to very low energies (above the ground state energy) of a set of excitations at momenta that are reciprocal-lattice vectors of the vortex lattice. By looking for the emergence of broken translational symmetry, it was found that there is a transition to a triangular vortex lattice at $\nu_{\rm c} \simeq 6$.

10.3.4 Strongly correlated regime

For $\nu < \nu_{\rm c}$, the vortex lattice phase is unstable to quantum fluctuations and is replaced by a series of strongly correlated phases. These phases are best understood at small filling factors, far from the transition at $\nu_{\rm c}$. As ν approaches $\nu_{\rm c}$, our understanding becomes much poorer.

10.3.5 Laughlin state, $\nu = 1/2$

The simplest bosonic Laughlin state (Laughlin 1983) has the wave function

$$\Psi_L(\{\zeta_i\}) \propto \prod_{i<j}^{N} (\zeta_i - \zeta_j)^2. \tag{10.66}$$

This is the *exact* ground state for contact repulsion (10.19) at total angular momentum $L = N(N-1)$ (Wilkin *et al.* 1998). For $L > N(N-1)$, the ground state of the contact interaction becomes degenerate, with a subspace spanned by quasi-hole excitations of eqn (10.66).[9]

Although the particle density of this state is uniform, there are still vortices in the system. However, unlike the case for the vortex lattice phase (10.36), these vortices are not localized

[8]For a single vortex, the quantum fluctuations of the guiding-center coordinate and of the cyclotron motion give equal contributions. Combining the fluctuations of both these degrees of freedom, one finds that $\Delta x^2 + \Delta y^2 \geq 1/\pi n_{\rm 2D}$—that is, *twice* the value (10.63) from the guiding-center fluctuations alone. Using $\Delta x^2 + \Delta y^2$ in the Lindemann criterion leads to a critical value that is twice as large, $\nu_{\rm c} \simeq 14$.

[9]There are no incompressible states at $L > N(N-1)$ for the contact interaction. In particular, the other bosonic Laughlin states are compressible for this interaction.

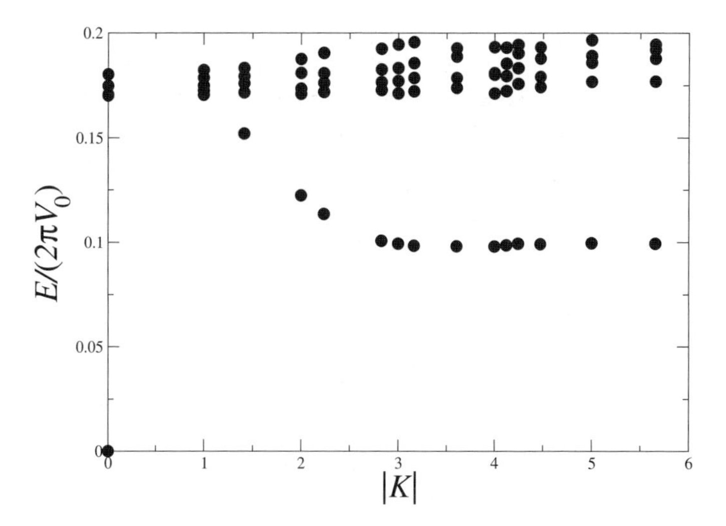

Fig. 10.5 Low-energy spectrum for the $\nu = 1/2$ Laughlin state of rotating bosons with contact interactions, obtained by exact-diagonalization studies on a torus with a square geometry for $N = 8$ and $N_\text{v} = 16$. The states are labeled by the magnitude of the conserved wave vector $\boldsymbol{K}$ in this periodic geometry. ($\boldsymbol{K}$ is measured in units of $2\pi/L$, where L is the side of the (square) torus. The characteristic energy scale is the $m = 0$ Haldane pseudopotential, $V_0 = \sqrt{2/\pi}\,(\hbar^2 a_\text{s}/M a_\parallel a_\perp^2)$.)

in space (translational symmetry is not broken). Rather, the vortices are *bound to the particles* (Girvin and MacDonald 1987, Read 1989). This can be seen by noting that in the state (10.66), the wave function for the particle i changes phase by $2 \times 2\pi$ around the position of every other particle $\zeta_{j\neq i}$. Each particle experiences two vortices bound to the position of every other particle, giving $N_\text{v} = 2(N-1)$ vortices in total. Thus, eqn (10.66) describes a phase that has filling factor $\nu = N/N_\text{v} = 1/2$ in the thermodynamic limit.[10]

The Laughlin state for bosons shows all of the characteristics familiar from the conventional FQHE (Prange and Girvin 1990): it is an incompressible fluid, with gapped collective excitations in the bulk (see Fig. 10.5) and gapless edge modes; the particle-like excitations have fractional particle number and fractional statistics (they are "Abelian anyons"). Incompressibility is evident from exact-diagonalization studies of the excitation spectrum, which show a gapped collective mode at all wave vectors (see Fig. 10.5 again).

10.3.6 Composite-fermion ansatz

At certain higher filling factors, i.e. $L < N(N-1)$, the strongly correlated states are accurately described by a "composite-fermion" ansatz. Here, since the underlying particles are bosons, one can form composite fermions by binding one vortex to the location of each particle. This may be achieved by the Jain construction (Jain 1989), writing the many-particle wave function as (Cooper and Wilkin 1999)

$$\Psi(\{\vec{r}_k\}) = \hat{P}_\text{LLL} \prod_{i<j} (\zeta_i - \zeta_j)\, \psi_\text{CF}(\{\vec{r}_k\})\,. \tag{10.67}$$

[10] The exact relation $N_\text{v} = 2N - 2$ should be interpreted as $N_\text{v} = N/\nu - \mathcal{S}$, where $\mathcal{S}$ is the "shift" of the state.

The Jastrow factor $\prod_{i<j}(\zeta_i - \zeta_j)$ causes any particle i to experience a single vortex at the location of each other particle $j \neq i$. Since this factor is completely antisymmetric under particle exchange, to obtain a bosonic wave function the function ψ_{CF} must also be antisymmetric: this is the wave function of the composite fermions. In general terms, one can appreciate why this is a useful variational state: the Jastrow factor suppresses the amplitude for two particles to approach each other, as does the composite-fermion wave function. Taken alone, these two factors would give a wave function that was a zero-energy eigenstate of the contact interactions. However, for $L < N(N-1)$, this construction requires ψ_{CF} to include basis states that are not in the lowest Landau level. One must project the single-particle states into the LLL, as represented by the operator $\hat{P}_{\mathrm{LLL}}$. For $L < N(N-1)$, the projected wave function does not vanish as two particles approach each other, so it is a state with a nonzero contact interaction energy. Nevertheless, it is found from numerical studies that the resulting trial states accurately describe the exact ground states of the two-body contact repulsion.

Treating the composite fermions as *noninteracting* particles and completely filling p Landau levels, one is led to the bosonic version of the "Jain" sequence

$$\nu = \frac{p}{p \pm 1}\,. \tag{10.68}$$

The states constructed in this way have large overlaps with exact ground states in the disk geometry (Cooper and Wilkin 1999) and account well for the bulk phases and their excitations in the spherical geometry at $\nu = 1/2, 2/3$, and (with less accuracy) $\nu = 3/4$ (Regnault and Jolicoeur 2003).

10.3.7 Moore–Read and Read–Rezayi states

One of the most interesting aspects of the physics of rapidly rotating Bose gases is the prediction of the appearance of *non-Abelian* phases: incompressible phases whose quasiparticle excitations obey non-Abelian exchange statistics.

Exact-diagonalization studies have shown very convincing evidence for the appearance of the Moore–Read state (Moore and Read 1991) and the Read–Rezayi states (Read and Rezayi 1999) in rotating Bose gases with realistic two-body interactions. The construction of these states may be viewed as a generalization of the Laughlin state, though the physics is very different (Abelian vs. non-Abelian quasiparticle excitations). The Laughlin state for bosons is the densest exact zero-energy eigenstate of the two-body contact interaction (10.19) within the 2D LLL. Similarly, the Moore–Read and Read–Rezayi states are densest exact zero-energy eigenstates of a $k+1$-body contact interaction

$$\sum_{i_1 < i_2 < \ldots i_{k+1} = 1}^{N} \delta(\boldsymbol{r}_{i_1} - \boldsymbol{r}_{i_2})\delta(\boldsymbol{r}_{i_2} - \boldsymbol{r}_{i_3})\ldots\delta(\boldsymbol{r}_{i_k} - \boldsymbol{r}_{i_{k+1}})\,. \tag{10.69}$$

For N divisible by k, the wave functions may be written in a simple way,

$$\Psi^{(k)}(\{\zeta_i\}) \propto \mathcal{S}\left[\prod_{i<j\in A}^{N/k}(\zeta_i - \zeta_j)^2 \prod_{l<m\in B}^{N/k}(\zeta_l - \zeta_m)^2 \ldots\right], \tag{10.70}$$

where $\mathcal{S}$ symmetrizes over all possible ways of dividing the N particles into k groups (A, $B, \ldots$) of N/k particles each (Cappelli *et al.* 2001). It is straightforward to convince oneself

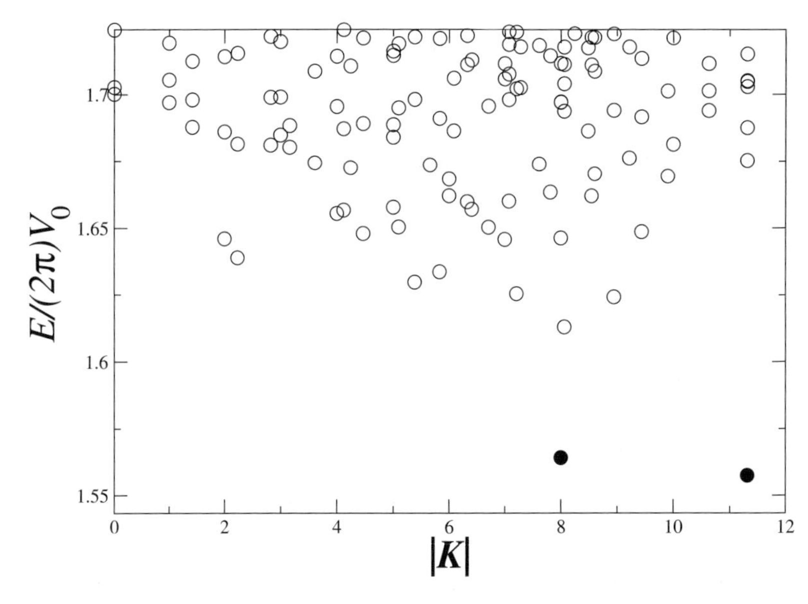

Fig. 10.6 Low-energy spectrum for the $\nu = 1$ state with contact interactions, obtained by exact-diagonalization studies on a torus with square geometry for $N = 16$ and $N_v = 16$. The states are labeled by the magnitude of the conserved wave vector in this periodic geometry. The appearance of low-energy states at $\boldsymbol{K} = (8, 8)$, $\boldsymbol{K} = (8, 0)$, and $\boldsymbol{K} = (0, 8)$ (indistinguishable in this plot, by symmetry) is consistent with the expected threefold degeneracy of the Moore–Read state. ($\boldsymbol{K}$ is measured in units of $2\pi/L$, where L is the side of the (square) torus.)

that eqn (10.70) vanishes when the positions of $k + 1$ particles coincide, as required. By counting the degree of the polynomial $L = N(N/k - 1)$, or the number of vortices $N_v = 2(N/k - 1)$, one sees that these states describe bosons at filling factor

$$\nu^{(k)} = \frac{k}{2}. \tag{10.71}$$

$k = 1$ is the Laughlin state; $k = 2$ is the Moore–Read state; $k \geq 3$ are the Read–Rezayi states.

Numerical evidence for the Moore–Read state ($k = 2$) for bosons interacting with contact interactions has been reported for disk (Wilkin and Gunn 2000), torus (Cooper *et al.* 2001), and spherical (Regnault and Jolicoeur 2003) geometries. Large overlaps of the exact wave functions with the model states (10.70) were found. While a large wave function overlap is encouraging, this is not necessarily the best way to characterize a phase of matter. In the thermodynamic limit the wave function overlap with a trial state will surely vanish, even if the two wave functions represent the same topological phase. A robust characterization of the topological phase is provided by the ground state degeneracy on a torus; this degeneracy is expected to survive (and even improve) in the thermodynamic limit, provided the wave function is in the same topological phase as the trial state (10.70). For the Moore–Read state, this degeneracy appears clearly in the spectrum even for relatively small system sizes (Fig. 10.6).

Evidence for the appearance of the Read–Rezayi states ($k \geq 3$) at $\nu = k/2$ for contact repulsion (10.19) has been found in the torus geometry (Cooper *et al.* 2001). While the wave

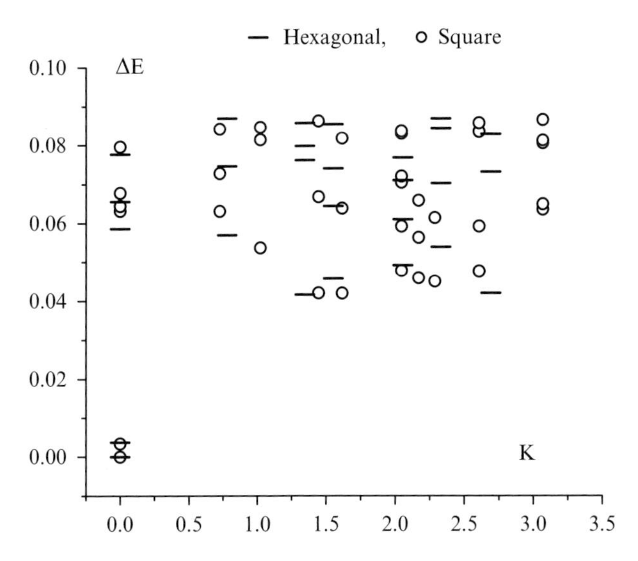

Fig. 10.7 Numerical evidence for the appearance of the $k = 3$ Read–Rezayi state for bosons at $\nu = 3/2$. (Excitation spectrum for $N = 18, N_{\mathrm{v}} = 12$ on a torus with two different unit cells, for contact interactions with a small amount of dipolar interaction, $\alpha = V_2/V_0 = 0.380$.) The spectrum depends weakly on the unit cell, so is representative of the thermodynamic limit. There is a clear twofold degeneracy of the ground state, as is expected for a topological phase described by the $k = 3$ Read–Rezayi state. (The overall fourfold degeneracy is recovered by considering the additional exact twofold center-of-mass degeneracy related to the half-integer filling factor.) Reproduced with permission from Rezayi *et al.* (2005).

function overlaps are large, at least for a k that is not too large, the expected ground state degeneracy is less clearly resolved than that for the Moore–Read state (Rezayi *et al.* 2005). However, the addition of a small nonlocal interaction leads to a very clear ground state degeneracy (Rezayi *et al.* 2005). These results (Fig. 10.7) provide by far the most convincing theoretical evidence for the existence of a Read–Rezayi topological phase in a system with realistic two-body interactions.

10.3.8 Smectic states, $2 \lesssim \nu \lesssim 6$

Exact-diagonalization studies on a torus at $\nu = 2$ show evidence for the appearance of a "smectic" state (Cooper and Rezayi 2007)—that is, a stripe state with broken translational order in only one direction. This can be viewed as a phase in which the vortex lattice is quantum melted owing to fluctuations of lines of vortices. While this state appears to describe the ground state in the system sizes that can be studied numerically ($N_{\mathrm{v}} \leq 12$), it is possible that finite-size effects may favor this over competing phases (vortex lattice and $k = 3$ Read–Rezayi phases). Estimates suggest that the finite-size effects that occur in experiments on cold gases (N_{v} is, of course, finite!) will mean that the smectic phases are likely to be important in practice.

10.3.9 Exact ground states close to a Feshbach resonance

The s-wave scattering length of two atoms passes through a "Feshbach resonance" when some external control parameter is used to tune the interatomic potential such that a new s-wave bound state appears (Bloch *et al.* 2008). For example, a magnetic field can be used to tune a bound state of two atoms in a "closed channel" through the "open channel" dissociation threshold, provided the atoms in these channels have differing magnetic moments. This is illustrated in Fig. 10.8, in which the energy of the bound state above the dissociation threshold is denoted by Δ. As Δ passes through 0, the scattering length $a_{\rm s}$ undergoes a resonance: $a_{\rm s}$ is large and negative for $\Delta > 0$, becoming large and positive on the appearance of the new bound state for $\Delta < 0$.

The physics of the resonance may be described by a "two-channel" model, in which a bosonic field for the closed-channel molecules $\hat{m}(\boldsymbol{r})$ is introduced in addition to the field operator for the bosonic atoms $\hat{a}(\boldsymbol{r})$, leading to an interaction Hamiltonian (Timmermans *et al.* 1999, Holland *et al.* 2001)

$$\hat{H}_F = \int \left\{ \Delta \hat{m}^\dagger_{\boldsymbol{r}} \hat{m}_{\boldsymbol{r}} + \frac{U_{aa}}{2} \hat{a}^\dagger_{\boldsymbol{r}} \hat{a}^\dagger_{\boldsymbol{r}} \hat{a}_{\boldsymbol{r}} \hat{a}_{\boldsymbol{r}} + \frac{g}{\sqrt{2}} \left[\hat{m}^\dagger_{\boldsymbol{r}} \hat{a}_{\boldsymbol{r}} \hat{a}_{\boldsymbol{r}} + \hat{m}_{\boldsymbol{r}} \hat{a}^\dagger_{\boldsymbol{r}} \hat{a}^\dagger_{\boldsymbol{r}} \right] \right\} d^3\boldsymbol{r} \,. \quad (10.72)$$

With an appropriate regularization of the contact terms, the parameters Δ and g can be chosen to reproduce the divergence of the scattering length at the Feshbach resonance.

What are the consequences of the Feshbach interaction for a rapidly rotating atomic Bose gas? Remarkably, there exist exact solutions for the ground state for a more general version of the model, in which molecule–atom and molecule–molecule scattering are also introduced (Cooper 2004).

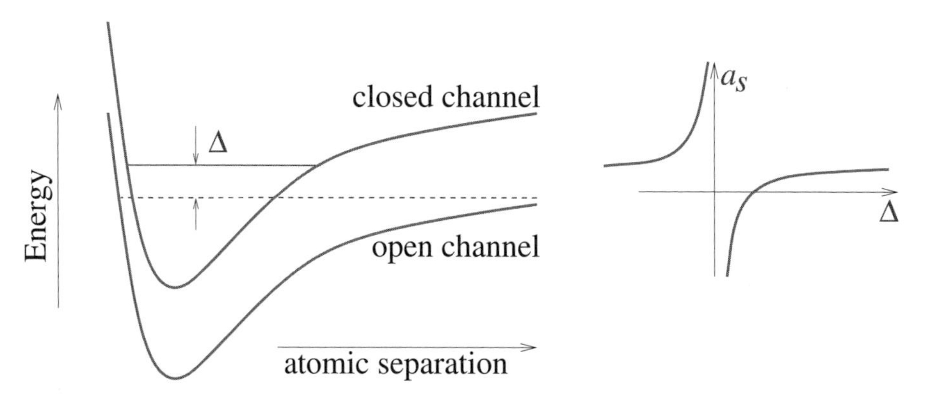

Fig. 10.8 Schematic diagram of the interatomic potentials for two atoms in the hyperfine states of the "open channel," and for two atoms in the hyperfine states of the "closed channel". These may be viewed as the potential surfaces obtained by adiabatically moving the atoms from wide separations, where the individual hyperfine levels of each atom are well defined. The closed channel has a high energy for widely separated atoms. However, a bound state of the closed-channel potential may pass through the dissociation threshold of the open channel, with the detuning Δ varying with external magnetic field if the two channels have different magnetic moments. Nonadiabatic processes mix the channels and lead to the appearance of a resonance in the scattering length of the open channel when the new bound state forms.

The Hamiltonian for the generalized model is written in the form

$$\hat{H} = \hat{H}_K + \hat{H}_F + \hat{H}_I + \left(\hbar\omega_\perp + \frac{\hbar\omega_\parallel}{2}\right)\hat{N} + \hbar\omega_\perp \hat{L}\,. \tag{10.73}$$

The strategy is to find parameter regimes for which $\hat{H}_K + \hat{H}_F + \hat{H}_I$ is positive semidefinite, and to determine exact zero-energy eigenstates of these operators for fixed N and L. These eigenstates are guaranteed to be ground states of the Hamiltonian (10.73) at fixed N and L.

The individual terms are:

- The kinetic and potential energy relative to the ground state energy,

$$\hat{H}_K \equiv \int \hat{a}^\dagger_{\boldsymbol{r}}\left[\hat{h}_a - \left(\hbar\omega_\perp + \frac{\hbar\omega_\parallel}{2}\right)\right]\hat{a}_{\boldsymbol{r}} + \hat{m}^\dagger_{\boldsymbol{r}}\left[\hat{h}_m - \left(\hbar\omega_\perp + \frac{\hbar\omega_\parallel}{2}\right)\right]\hat{m}_{\boldsymbol{r}}\, d^3\boldsymbol{r}\,. \tag{10.74}$$

  This is positive semidefinite by construction.

- Elastic atom–molecule and molecule–molecule interactions,

$$\hat{H}_I \equiv U_{am}\int \hat{m}^\dagger_{\boldsymbol{r}}\hat{a}^\dagger_{\boldsymbol{r}}\hat{a}_{\boldsymbol{r}}\hat{m}_{\boldsymbol{r}}\, d^3\boldsymbol{r} + \frac{U_{mm}}{2}\int \hat{m}^\dagger_{\boldsymbol{r}}\hat{m}^\dagger_{\boldsymbol{r}}\hat{m}_{\boldsymbol{r}}\hat{m}_{\boldsymbol{r}}\, d^3\boldsymbol{r}\,. \tag{10.75}$$

  This is positive semidefinite for $U_{mm}, U_{am} \geq 0$, which I now assume.

- Finally, $\hat{H}_F$ is the two-channel Feshbach interaction (10.72). In order to regularize the contact interactions appearing in eqn (10.72), I rewrite this as[11]

$$\hat{H}_F = \Delta\int \hat{m}^\dagger_{\boldsymbol{r}}\hat{m}_{\boldsymbol{r}}\, d^3\boldsymbol{r} + \frac{\tilde{U}_{aa}}{2}\int \hat{P}^\dagger_{\boldsymbol{r}}\hat{P}_{\boldsymbol{r}}\, d^3\boldsymbol{r} + \frac{\tilde{g}}{\sqrt{2}}\int \left[\hat{m}^\dagger_{\boldsymbol{r}}\hat{P}_{\boldsymbol{r}} + \hat{m}_{\boldsymbol{r}}\hat{P}^\dagger_{\boldsymbol{r}}\right] d^3\boldsymbol{r}\,, \tag{10.76}$$

  where I define

$$\hat{P}^{(\dagger)}_{\boldsymbol{r}} \equiv \int \phi(\boldsymbol{r}')\hat{a}^{(\dagger)}_{\boldsymbol{r}+\boldsymbol{r}'/2}\hat{a}^{(\dagger)}_{\boldsymbol{r}-\boldsymbol{r}'/2}\, d^3\boldsymbol{r}' \tag{10.77}$$

  in terms of a (closed-channel) molecular wave function, $\phi(\boldsymbol{r})$. This wave function is assumed to have a small length scale $\sigma \ll a_\perp, a_\parallel$, such that, to a good approximation,

$$\hat{P}_{\boldsymbol{r}} \simeq \hat{a}_{\boldsymbol{r}}\hat{a}_{\boldsymbol{r}}(\eta\sigma)^{3/2}\,, \tag{10.78}$$

  where the number η depends on the microscopic form of $\phi(\boldsymbol{r})$. As a result, the relation between the parameters is

$$\tilde{U}_{aa} = \frac{U_{aa}}{(\eta\sigma)^3}\,, \qquad \tilde{g} = \frac{g}{(\eta\sigma)^{3/2}}\,. \tag{10.79}$$

[11] Owing to the zero-point energies of atoms and molecules in the trap, there is a shift of the detuning between eqns (10.72) and (10.76) of $\hbar\omega_\perp + \hbar\omega_\parallel/2$.

The regularized Feshbach interaction (10.76) is a quadratic form, and is positive semidefinite for

$$\det\begin{pmatrix} \Delta & \tilde{g}/\sqrt{2} \\ \tilde{g}/\sqrt{2} & \tilde{U}_{aa}/2 \end{pmatrix} \geq 0\,. \tag{10.80}$$

This requires

$$\tilde{g}^2 \leq \Delta\tilde{U}_{aa} \quad \Rightarrow \quad g^2 \leq \Delta U_{aa}\,, \tag{10.81}$$

where I have used eqn (10.79). Expressing eqn (10.76) in terms of its eigenvectors, one finds

$$\begin{aligned} \hat{H}_F = \int \Bigg\{ & \frac{\lambda_-}{\alpha^2+\beta^2}\left(\alpha\hat{m}^\dagger_{\boldsymbol{r}} + \beta\hat{P}^\dagger_{\boldsymbol{r}}\right)\left(\alpha\hat{m}_{\boldsymbol{r}} + \beta\hat{P}_{\boldsymbol{r}}\right) \\ & + \frac{\lambda_+}{\alpha^2+\beta^2}\left(\beta\hat{m}^\dagger_{\boldsymbol{r}} - \alpha\hat{P}^\dagger_{\boldsymbol{r}}\right)\left(\beta\hat{m}_{\boldsymbol{r}} - \alpha\hat{P}_{\boldsymbol{r}}\right)\Bigg\}\, d^3\boldsymbol{r}\,, \end{aligned} \tag{10.82}$$

where, for eqn (10.81), the eigenvalues $\lambda_\pm$ are $\lambda \geq 0$ and $\lambda > 0$.

I will discuss two cases in turn.

(i) $g^2/\Delta > U_{aa}$. Both eigenvalues are positive, i.e. $\lambda_\pm > 0$. A zero-energy eigenstate $|\Psi\rangle$ of eqn (10.82) must satisfy

$$(\alpha\hat{m}_{\boldsymbol{r}} + \beta\hat{P}_{\boldsymbol{r}})|\Psi\rangle = (\beta\hat{m}_{\boldsymbol{r}} - \alpha\hat{P}_{\boldsymbol{r}})|\Psi\rangle = 0 \tag{10.83}$$

$$\Rightarrow \hat{m}_{\boldsymbol{r}}|\Psi\rangle = 0\,, \tag{10.84}$$

$$\hat{P}_{\boldsymbol{r}}|\Psi\rangle \rightarrow \hat{a}_{\boldsymbol{r}}\hat{a}_{\boldsymbol{r}}|\Psi\rangle = 0\,. \tag{10.85}$$

That is, there are no molecules in the state $|\Psi\rangle$, and vanishing amplitude for two atoms to be at the same position. Requiring, in addition, that $|\Psi\rangle$ is annihilated by $\hat{H}_K$, one finds that the only zero-energy eigenstates are the bosonic Laughlin state of atoms (with quasi-holes).[12] Therefore, exact zero-energy eigenstates exist only for $L \geq N(N-1)$, corresponding to $\nu \leq 1/2$. Since the Hamiltonian is positive semidefinite, these are exact ground states.

(ii) $g^2/\Delta = U_{aa}$. Now $\lambda_+ > 0$ but $\lambda_- = 0$. This is the situation where the two-body contact interaction *vanishes*. Since $\lambda_+ > 0$, a zero-energy eigenstate must obey

$$(\beta\hat{m}_{\boldsymbol{r}} - \alpha\hat{P}_{\boldsymbol{r}})|\Psi\rangle = 0\,. \tag{10.86}$$

One can satisfy this constraint by writing

$$|\Psi\rangle = \hat{R}|\Psi\rangle_a\,, \tag{10.87}$$

$$\hat{R} \equiv \exp\left(\frac{\alpha}{\beta}\int \hat{P}_{\boldsymbol{r}}\hat{m}^\dagger_{\boldsymbol{r}}\, d^3\boldsymbol{r}\right), \tag{10.88}$$

where $|\Psi\rangle_a$ is any state consisting only of atoms, so that $m_{\boldsymbol{r}}|\Psi\rangle_a = 0$.

$\hat{R}$ commutes with $\hat{N}$, $\hat{L}$. Furthermore, $\hat{R}$ preserves the 2D LLL condition: provided the state $|\Psi\rangle_a$ contains atoms only in the 2D LLL, then $|\Psi\rangle$ will contain atoms and molecules that

[12] Since there are no molecules, this state is also annihilated by $\hat{H}_I$.

are in the 2D LLL (and hence a zero-energy eigenstate of $\hat{H}_K$). This may be seen by noting that the 2D LLL wave functions for a molecule,

$$\langle \boldsymbol{r}|m\rangle_m \propto (x+iy)^m e^{-(x^2+y^2)/4\ell_m^2} e^{-z^2/2a_{\parallel m}^2} \tag{10.89}$$

with $m \geq 0$, provide a complete set of states for the product of any two 2D LLL atomic wave functions

$$\langle \boldsymbol{r}|m_1\rangle_a \langle \boldsymbol{r}|m_2\rangle_a \propto (x+iy)^{m_1+m_2} e^{-2(x^2+y^2)/4\ell_a^2} e^{-2z^2/2a_{\parallel a}^2}\,. \tag{10.90}$$

Finally, let us consider the interactions $\hat{H}_I$. This interaction, acting on states of the form (10.87), is equivalent to an effective interaction $\hat{R}^{-1}\hat{H}_I\hat{R}$ acting on the purely atomic states $|\Psi\rangle_a$. One may show (Cooper 2004) that this effective interaction has *three- and four-body contact interactions*. The origin of these effective many-particle interactions is illustrated in Fig. 10.9.

For $g^2/\Delta = U_{aa}$ and $U_{mm}, U_{am} > 0$, there are repulsive three- and four-body interactions. Therefore for $|\Psi\rangle$ to be a zero-energy eigenstate of $\hat{H}_I$, we require $|\Psi\rangle_a$ in eqn (10.87) to vanish when the positions of any three particles coincide. This can be achieved for $\nu = 1$ by choosing $|\Psi\rangle_a$ to be the bosonic Moore–Read state (or, for $\nu < 1$, by choosing the bosonic Moore–Read state with quasi-holes). Therefore, the *exact* ground state at $\nu = 1$ is a strongly correlated atom–molecule mixture formed from the Moore–Read state:

$$|\Psi\rangle = \hat{R}|\Psi_{\mathrm{MR}}\rangle_a\,. \tag{10.91}$$

For $g^2/\Delta = U_{aa}$ and $U_{mm} > 0$ but $U_{am} = 0$, the three-body forces vanish. Therefore for $|\Psi\rangle$ to be a zero-energy eigenstate of $\hat{H}_I$, we require $|\Psi\rangle_a$ in eqn (10.87) to vanish when the positions of any four particles coincide. By the same reasoning as above, we conclude that in this case the exact ground state at $\nu = 3/2$ is a strongly correlated atom–molecule mixture formed from the $k = 3$ Read–Rezayi state:

$$|\Psi\rangle = \hat{R}|\Psi_{k=3}\rangle_a\,. \tag{10.92}$$

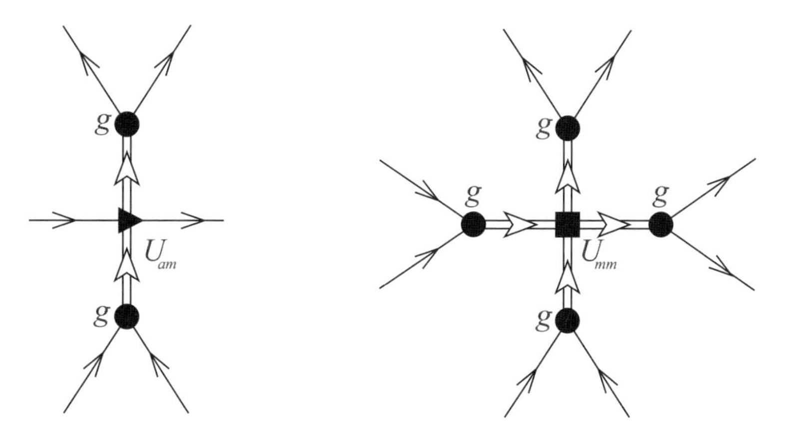

Fig. 10.9 The hybridization of pairs of atoms into molecules (via the Feshbach interaction) and subsequent molecule–molecule and atom–molecule scattering lead to effective three- and four-body interactions. Reproduced with permission from Cooper (2004).

Thus, we have established that the exact ground states of a rapidly rotating atomic gas with the Feshbach interaction (10.76) are the Moore–Read and Read–Rezayi states, up to the hybridization of atoms into molecules (10.87). It is remarkable that these strongly correlated phases—which were proposed for the sake of their interesting non-Abelian statistics—appear as the exact ground states of a model that arises very naturally in the context of cold atomic gases.

10.3.10 Experimental status

To date, the highest vortex densities in rotating atomic gases have been achieved by the Colorado group. In these experiments, the mean interaction energy $\mu \sim g\bar{n}$ is sufficiently small that the gases are in or close to the 2D LLL regime, $\mu \lesssim 2\hbar\omega_\perp, \hbar\omega_\parallel$. However, the smallest filling factor reported is $\nu \simeq 500$ (Schweikhard *et al.* 2004), deep in the regime $\nu > \nu_{\rm c}$, where one expects the ground state to be a vortex lattice. Clearly, further special efforts are required in order to reach the regime $\nu \lesssim 6$ where one expects quantum fluctuations of vortices to destabilize the lattice. Part of the difficulty in achieving this regime in experiments may be summarized by noting that one is trying to create a strongly correlated state in a weakly interacting gas. To add some meaning to this glib statement, note that, in the 2D LLL, the mean interaction energy at a filling factor $\nu \sim 1$ is

$$g\bar{n} \sim \frac{\hbar^2 a_{\rm s}}{M}\frac{1}{a_\perp^2 a_\parallel} \sim \frac{a_{\rm s}}{a_\parallel}\hbar\omega_\perp\,. \tag{10.93}$$

This energy scale sets the size of the incompressibility gap (up to a numerical factor), and therefore the temperature scale required to observe incompressibility. In typical experiments in magnetic traps, $\hbar\omega_\perp \sim 5\,{\rm nK}$, while $a_{\rm s}/a_\parallel \sim 0.01$, leading to gaps that are fractions of a nanokelvin, and much smaller than the temperatures that can currently be reached in such experiments ($\gtrsim 1\,{\rm nK}$). There exist realistic ways by which to increase the energy scale (10.93) and achieve the strongly correlated regime in experiments (Bloch *et al.* 2008, Fetter 2009, Cooper 2008). Future experiments on rotating ultracold atomic gases hold the promise of exploring this interesting regime.

10.4 Conclusions

Rapidly rotating atomic Bose gases offer the possibility to explore quantized vortices and vortex lattices in novel parameter regimes. There is strong theoretical evidence that novel strongly correlated phases of bosonic matter can appear in rapidly rotating Bose gases at small filling factors. These studies provide clear theoretical evidence that fascinating proposed non-Abelian phases can exist for physically realistic models. They also motivate and inform experimental efforts to progress to the regimes of interesting strong-correlation phenomena. There are many closely related issues that I have not addressed in these notes: spin degrees of freedom, rotating lattices, methods for experimental detection, edge excitations, and rotating Fermi gases. As in the case of the subjects discussed in these lectures, while there is some theoretical understanding of these issues, this remains limited. Many open questions remain to be addressed by future theory and experiment.

Acknowledgments

I would like to thank my collaborators in the work described in these notes: Mike Gunn, Stavros Komineas, Nick Read, Ed Rezayi, Steve Simon, and Nicola Wilkin.

References

Abo-Shaeer, J. R., Raman, C., Vogels, J. M., and Ketterle, W. (2001). Observation of vortex lattices in Bose–Einstein condensates. *Science* **292**, 476.

Abrikosov, A. (1957). On the magnetic properties of superconductors of the second group. *Zh. Eksp. Teor. Fiz.* **32**, 1442 [*Sov. Phys. JETP* **5**, 1174 (1957)].

Baranov, M., Dobrek, L., Goral, K., Santos, L., and Lewenstein, M. (2002). Ultracold dipolar gases—a challenge for experiments and theory. *Phys. Scripta* **T102**, 74.

Baym, G. (2004). Vortex lattices in rapidly rotating Bose–Einstein condensates: Modes and correlation functions. *Phys. Rev. A* **69**(4), 043618.

Bertsch, G. F. and Papenbrock, T. (1999). Yrast line for weakly interacting trapped bosons. *Phys. Rev. Lett.* **83**, 5412–5414.

Bloch, I., Dalibard, J., and Zwerger, W. (2008). Many-body physics with ultracold gases. *Rev. Mod. Phys.* **80**(3), 885.

Butts, D. A. and Rokhsar, D. S. (1999). Predicted signatures of rotating Bose–Einstein condensates. *Nature* **397**, 327–329.

Cappelli, A., Georgiev, L. S., and Todorov, I. T. (2001). Parafermion Hall states from coset projections of abelian conformal theories. *Nucl. Phys. B* **559**, 499.

Coddington, I., Haljan, P. C., Engels, P., Schweikhard, V., Tung, S., and Cornell, E. A. (2004). Experimental studies of equilibrium vortex properties in a Bose-condensed gas. *Phys. Rev. A* **70**(6), 063607.

Cooper, N. R. (2004). Exact ground states of rotating Bose gases close to a Feshbach resonance. *Phys. Rev. Lett.* **92**(22), 220405.

Cooper, N. R. (2008). Rapidly rotating atomic gases. *Adv. Phys.*, in press; arXiv:0810.4398.

Cooper, N. R. and Rezayi, E. H. (2007). Competing compressible and incompressible phases in rotating atomic Bose gases at filling factor $\nu = 2$. *Phys. Rev. A* **75**(1), 013627.

Cooper, N. R. and Wilkin, N. K. (1999). Composite fermion description of rotating Bose–Einstein condensates. *Phys. Rev. B* **60**, R16279–R16282.

Cooper, N. R., Wilkin, N. K., and Gunn, J. M. F. (2001). Quantum phases of vortices in rotating Bose–Einstein condensates. *Phys. Rev. Lett.* **87**, 120405.

Cooper, N. R., Komineas, S., and Read, N. (2004). Vortex lattices in the lowest Landau level for confined Bose–Einstein condensates. *Phys. Rev. A* **70**(3), 033604.

Cooper, N. R., Rezayi, E. H., and Simon, S. H. (2005). Vortex lattices in rotating atomic Bose gases with dipolar interactions. *Phys. Rev. Lett.* **95**(20), 200402.

Cornell, E. A. and Wieman, C. E. (2002). Nobel lecture: Bose–Einstein condensation in a dilute gas, the first 70 years and some recent experiments. *Rev. Mod. Phys.* **74**(3), 875–893.

Das Sarma, S. and Pinczuk, A. (eds.) (1997). *Perspectives in Quantum Hall Effects: Novel Quantum Liquids in Low-Dimensional Semiconductor Structures*. Wiley, New York.

Donnelly, R. J. (1991). *Quantized Vortices in Helium II*. Cambridge University Press, Cambridge.

Fetter, A. L. (1965). Vortices in an imperfect Bose gas. I. The condensate. *Phys. Rev.* **138**(2A), A429–A437.

Fetter, A. L. (2009). Rotating trapped Bose–Einstein condensates. *Rev. Mod. Phys.* **81**, 647.

Fischer, U. R. and Baym, G. (2003). Vortex states of rapidly rotating dilute Bose–Einstein condensates. *Phys. Rev. Lett.* **90**(14), 140402.

Fröhlich, J., Studer, U. M., and Thiran, E. (1994). Quantum theory of large systems of non-relativistic matter. In *Fluctuating Geometries in Statistical Mechanics and Field Theory* (eds. F. David, P. Ginsparg, and J. Zinn-Justin), pp. 771–912. Elsevier, Amsterdam.

Girvin, S. M. and MacDonald, A. H. (1987). Off-diagonal long-range order, oblique confinement, and the fractional quantum Hall effect. *Phys. Rev. Lett.* **58**(12), 1252–1255.

Griesmaier, A., Werner, J., Hensler, S., Stuhler, J, and Pfau, T. (2005). Bose–Einstein condensation of chromium. *Phys. Rev. Lett.* **94**, 160401.

Holland, M., Park, J., and Walser, R. (2001). Formation of pairing fields in resonantly coupled atomic and molecular Bose–Einstein condensates. *Phys. Rev. Lett.* **86**, 1915–1918.

Hussein, M. S. and Vorov, O. K. (2002). Generalized yrast states of a Bose–Einstein condensate in a harmonic trap for a universality class of interactions. *Phys. Rev. A* **65**, 035603.

Jain, J. K. (1989). Composite-fermion approach for the fractional quantum Hall effect. *Phys. Rev. Lett.* **63**(2), 199–202.

Ketterle, W. (2002). Nobel lecture: When atoms behave as waves: Bose–Einstein condensation and the atom laser. *Rev. Mod. Phys.* **74**(4), 1131–1151.

Koulakov, A. A., Fogler, M. M., and Shklovskii, B. I. (1996). Charge density wave in two-dimensional electron liquid in weak magnetic field. *Phys. Rev. Lett.* **76**(3), 499–502.

Lahaye, T., Koch, T., Fröhlich, B., Fattori, M., Metz, J., Griesmaier, A., Giovanazzi, S., and Pfau, T. (2007). Strong dipolar effects in a quantum ferrofluid. *Nature* **448**, 672.

Landau, L. D. and Lifshitz, E. M. (1981). *Statistical Physics*, Part 1, Volume 5. Butterworth Heinemann, Oxford.

Laughlin, R. B. (1983). Anomalous quantum Hall effect: An incompressible quantum fluid with fractionally charged excitations. *Phys. Rev. Lett.* **50**(18), 1395–1398.

Lieb, E. H and Seiringer, R. (2006). Derivation of the Gross–Pitaevskii equation for rotating Bose gases. *Commun. Math. Phys.* **264**, 505–537.

Madison, K.W., Chevy, F., Wohlleben, W., and Dalibard, J. (2000). Vortex formation in a stirred Bose–Einstein condensate. *Phys. Rev. Lett* **84**, 806–809.

Moessner, R. and Chalker, J. T. (1996). Exact results for interacting electrons in high Landau levels. *Phys. Rev. B* **54**(7), 5006–5015.

Moore, G. and Read, N. (1991). Nonabelions in the fractional quantum Hall-effect. *Nucl. Phys. B* **360**, 362–396.

Prange, R. E. and Girvin, S. M. (eds.) (1990). *The Quantum Hall Effect*, 2nd edn. Springer-Verlag, Berlin.

Read, N. (1989). Order parameter and Ginzburg–Landau theory for the fractional quantum Hall effect. *Phys. Rev. Lett.* **62**(1), 86–89.

Read, N. and Rezayi, E. H. (1999). Beyond paired quantum Hall states: Parafermions and incompressible states in the first excited Landau level. *Phys. Rev. B* **59**, 8084–8092.

Regnault, N. and Jolicoeur, T. (2003). Quantum Hall fractions in rotating Bose–Einstein condensates. *Phys. Rev. Lett.* **91**, 030402.

Rezayi, E. H., Read, N., and Cooper, N. R. (2005). Incompressible liquid state of rapidly rotating bosons at filling factor 3/2. *Phys. Rev. Lett.* **95**(16), 160404.

Rozhkov, A. and Stroud, D. (1996). Quantum melting of a two-dimensional vortex lattice at zero temperature. *Phys. Rev. B* **54**(18), R12697–R12700.

Schweikhard, V., Coddington, I., Engels, P., Mogendorff, V. P., and Cornell, E. A. (2004). Rapidly rotating Bose–Einstein condensates in and near the lowest Landau level. *Phys. Rev. Lett.* **92**(4), 040404.

Sinova, J., Hanna, C. B., and MacDonald, A. H. (2002). Quantum melting and absence of Bose–Einstein condensation in two-dimensional vortex matter. *Phys. Rev. Lett.* **89**, 030403.

Smith, R. A. and Wilkin, N. K. (2000). Exact eigenstates for repulsive bosons in two dimensions. *Phys. Rev. A* **62**, 061602.

Sokol, P. E. (1995). Bose–Einstein condensation in liquid helium. In *Bose–Einstein Condensation* (ed. A. Griffin, D. W. Snoke, and S. Stringari), p. 51. Cambridge University Press.

Timmermans, E., Tommasini, P., Cote, R., Hussein, M., and Kerman, A. (1999). Rarified liquid properties of hybrid atomic–molecular Bose–Einstein condensates. *Phys. Rev. Lett.* **83**, 2691–2694.

Vorov, O. K., Hussein, M. S., and Isacker, P. V. (2003). Rotating ground states of trapped atoms in a Bose–Einstein condensate with arbitrary two-body interactions. *Phys. Rev. Lett.* **90**(20), 200402.

Wilkin, N. K. and Gunn, J. M. F. (2000). Condensation of "composite bosons" in a rotating BEC. *Phys. Rev. Lett.* **84**, 6–9.

Wilkin, N. K., Gunn, J. M. F., and Smith, R. A. (1998). Do attractive bosons condense? *Phys. Rev. Lett.* **80**, 2265.

Yarmchuk, E. J., Gordon, M. J. V., and Packard, R. E. (1979). Observation of stationary vortex arrays in rotating superfluid helium. *Phys. Rev. Lett.* **43**(3), 214–217.

11
The quantum Hall effect

J. FRÖHLICH

Jürg Fröhlich,
Theoretische Phÿsik,
ETH Zürich,
CH-8093 Zürich,
Switzerland.

12
The dimer model

R. KENYON

Richard Kenyon,
Mathematics Department,
Brown University,
Providence, RI, USA.

12.1 Overview

Our goal is to study the planar dimer model and the associated random interface model. There are a number of references to the dimer model, including a few short survey articles, which can be found in the bibliography.

The dimer model is a classical statistical-mechanics model, first studied by Kasteleyn and by Temperley and Fisher in the 1960s, who computed the partition function. Later work by Fisher, Stephenson, Temperley, Blöte, Hilhorst, Percus, and many others contributed to understanding correlations, phase transitions, and other properties.

Essentially all of these authors considered the model on either $\mathbb{Z}^2$ or the hexagonal lattice. While these cases already contain much of interest, only when one generalizes to other lattices (with larger fundamental domains) does one see the "complete" picture. For example, gaseous phases and semifrozen phases (defined below) occur only in this more general setting.

Our goal here is to describe recent results on the planar, periodic, bipartite dimer model, obtained through joint work with Andrei Okounkov and Scott Sheffield (Kenyon *et al.* 2006, Kenyon and Okounkov 2006). We solve the dimer model in the general setting of a planar bipartite periodic lattice. We give a complete description of all possible phases, and we describe the phase diagram as a function of edge energies, in particular as a function of a constant external electric field (see Fig. 12.13).

The dimer model on a general periodic bipartite lattice has a rich behavior, with surprising connections to modern topics in real algebraic geometry. With a weighted, periodic, bipartite lattice we associate a two-variable polynomial $P(z, w)$, the *characteristic polynomial*. Most of the physical properties of the dimer model, i.e. the partition function, edge correlations, height fluctuations, and so on, are simple functions of P. Studying the set of P's which arise allows us to achieve a reasonably complete and satisfying theory for dimers in this setting: we have a good understanding of the set of translation-invariant Gibbs states, the influence of boundary conditions, and the scaling limits of fluctuations.

In Kenyon *et al.* (2006) it was proved that the Riemann surface $P(z, w) = 0$ is a *Harnack curve*; such curves were studied classically by Harnack and more recently by Mikhalkin and others (Mikhalkin and Rullgård 2001, Passare and Rullgård 2004). For example, Harnack curves have only the simplest kind of algebraic singularities (real nodes), and this fact has consequences for the possible behaviors of dimers. In Kenyon and Okounkov (2006) it was proved that every Harnack curve arises from a dimer model as above; in particular, the edge weights or edge energies give natural parameters for the set of Harnack curves.

Other statistical mechanical models such as the six-vertex model can be defined in similar generality but have not been solved, except for very special situations such as $\mathbb{Z}^2$ with periodic weights. One would expect that a detailed study of these models on general periodic lattices would also yield a wealth of new behavior, but at the moment the tools for this study are lacking.

Dimer models on nonbipartite lattices, of which the Ising model is an example, could also potentially be studied on general periodic lattices; Pfaffian methods exist for their solution. However, at the moment no one has undertaken such a study. I believe such a project would be very enlightening about the nature of the Ising model and, more generally, the nonbipartite dimer model.

12.2 Dimer definitions

A *dimer covering*, or *perfect matching*, of a graph is a subset of edges which covers every vertex exactly once, that is, every vertex is the endpoint of exactly one edge (see Fig. 12.1). Let $M(\mathcal{G})$ be the set of dimer covers of the graph $\mathcal{G}$.

In these lectures, we will deal only with *bipartite planar* graphs. A graph is bipartite when the vertices can be colored black and white in such a way that each edge connects vertices of different colors. Alternatively, this is equivalent to each cycle having an even length.

Let $\mathcal{E}$ be a function on the edges of a finite graph $\mathcal{G}$, defining the energy associated with having a dimer on that edge. With a dimer cover m we associate a total energy $\mathcal{E}(m)$, which is the sum of the energies of the edges covered by dimers. The partition function for a finite graph $\mathcal{G}$ is then

$$Z = Z(\mathcal{G}, \mathcal{E}) = \sum_{m \in M(\mathcal{G})} e^{-\mathcal{E}(m)} .$$

The corresponding Boltzmann measure is a probability measure that assigns a dimer cover m a probability $\mu(m) = (1/Z)e^{-\mathcal{E}(m)}$.

Note that if we add a constant energy $\mathcal{E}_0$ to each edge containing a given vertex, then the energy of any dimer cover changes by $\mathcal{E}_0$ and so the associated Boltzmann measure on $M(\mathcal{G})$ does not change. Therefore we define two energy functions $\mathcal{E}, \mathcal{E}'$ to be equivalent, $\mathcal{E} \sim \mathcal{E}'$, if one can be obtained from the other by a sequence of such operations. It is not hard to show that $\mathcal{E} \sim \mathcal{E}'$ if and only if the *alternating sums* along faces are equal: given a face with edges $e_1, e_2, \ldots, e_{2k}$ in cyclic order, the sums $\mathcal{E}_1 - \mathcal{E}_2 + \mathcal{E}_3 \cdots - \mathcal{E}_{2k}$ and

$$\mathcal{E}'_1 - \mathcal{E}'_2 + \mathcal{E}'_3 \cdots - \mathcal{E}'_{2k}$$

(which we call alternating sums) must be equal. It is possible to interpret these alternating sums as magnetic fluxes through the faces, but we will not take this point of view here.

Kasteleyn showed how to enumerate the dimer covers of any planar graph, and in fact how to compute the partition function for any $\mathcal{E}$: the partition function is a Pfaffian of an associated signed adjacency matrix. The random-interface interpretation that we will discuss is valid only for bipartite graphs, so we will only be concerned here with bipartite graphs. In this case one can replace the Pfaffian with a determinant. There are many open problems involving dimer coverings of nonbipartite planar graphs, which at present we do not have tools to attack.

Our prototypical examples are the dimer models on $\mathbb{Z}^2$ and the honeycomb graph. These are equivalent to, respectively, the domino tiling model (tilings with 2×1 rectangles) and the

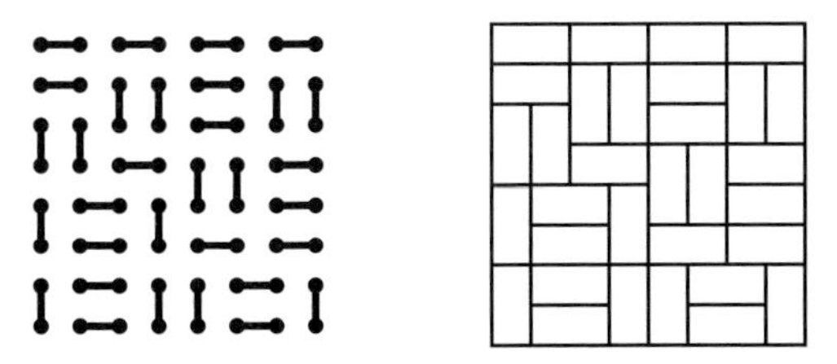

Fig. 12.1 The domino tiling model.

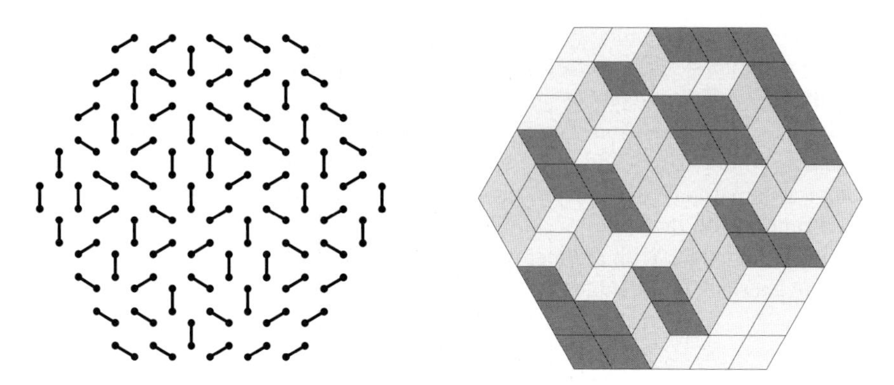

Fig. 12.2 The lozenge tiling model.

"lozenge tiling" model (tilings with 60° rhombi). The case when $\mathcal{E}$ is identically 0 on edges is already quite interesting (see Figs. 12.1 and 12.2). A less simple case (and one which is, in a sense, the most general) is the case of $\mathbb{Z}^2$ in which $\mathcal{E}$ is periodic with period n; that is, translations in $n\mathbb{Z}^2$ leave $\mathcal{E}$ invariant, but no larger group does.

12.2.1 Electric field

Let $\mathcal{G}$ be a planar, bipartite, periodic graph. This means that $\mathcal{G}$ is a planar bipartite weighted graph on which translations in $\mathbb{Z}^2$ (or some other rank-2 lattice) act by energy-preserving and color-preserving isomorphisms. Here, by *color-preserving* isomorphisms we mean isomorphisms which map white vertices to white and black vertices to black. Note, for example, that for the graph $\mathcal{G} = \mathbb{Z}^2$ with nearest-neighbor edges, the lattice generated by $(2,0)$ and $(1,1)$ acts by color-preserving isomorphisms, but $\mathbb{Z}^2$ itself does not act by color-preserving isomorphisms. So the fundamental domain (in this case) contains two vertices, one white and one black.

For simplicity we will assume that our periodic graphs are embedded so that the lattice of energy- and color-preserving isomorphisms is $\mathbb{Z}^2$, so that we can describe a translation using a pair of integers.

Let $\vec{E} = (E_x, E_y)$ be a constant electric field in $\mathbb{R}^2$. If we regard each dimer as a dipole, for example if we assign white vertices a charge $+1$ and black vertices a charge -1, then each dimer gets an additional energy due to its interaction with $\vec{E}$; this energy is just $\vec{E} \cdot (\mathrm{b} - \mathrm{w})$ for a dimer with vertices w and b.

On a finite graph $\mathcal{G}$, the Boltzmann measure on $M(\mathcal{G})$ is unaffected by the presence of $\vec{E}$, because for every dimer cover the energy due to $\vec{E}$ is

$$\sum_{m} \vec{E} \cdot (\mathrm{b} - \mathrm{w}) = \sum_{\mathrm{b}} \vec{E} \cdot \mathrm{b} - \sum_{\mathrm{w}} \vec{E} \cdot \mathrm{w}\,,$$

where the sums on the right are over all the black and all the white vertices, respectively. On (nonplanar) graphs with periodic boundary conditions, or on infinite graphs, $\vec{E}$ will influence the measure in a nontrivial way.

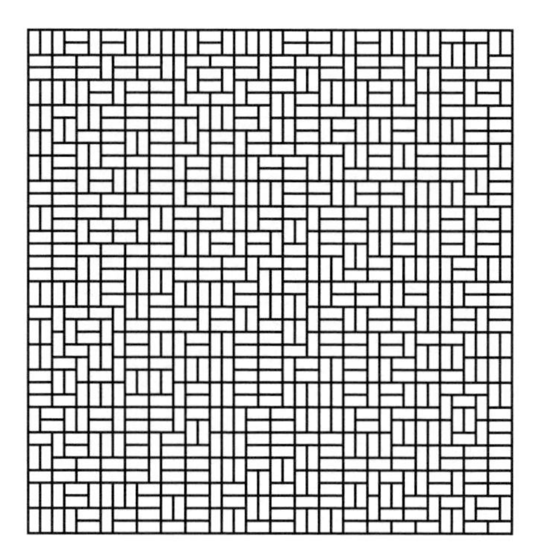

Fig. 12.3 Uniform domino tiling of a large square.

12.2.2 Random tilings

Look at the larger domino picture and lozenge picture in Figs. 12.3 and 12.4. These are both uniform random tilings of the corresponding regions; that is, they are chosen from the distribution in which all tilings are equally weighted. In the first case, there are about e^{455} possible domino tilings, and in the second, about e^{1255} lozenge tilings.[1] These two pictures clearly display some very different behavior. The first picture appears homogeneous, while in the second, the densities of the tiles of a given orientation vary throughout the region. The reason for this behavior is the boundary height function.

A lozenge tiling of a simply connected region is the projection along the $(1, 1, 1)$ direction of a piecewise linear surface in $\mathbb{R}^3$. This surface has pieces which are integer translates of the sides of the unit cube. Such a surface which projects injectively in the $(1, 1, 1)$ direction is called a *stepped surface* or *skyscraper surface*. Random lozenge tilings are therefore random stepped surfaces. A random lozenge tiling of a fixed region as in Fig. 12.4 is a random stepped surface spanning a fixed boundary curve in $\mathbb{R}^3$.

Domino tilings can also be interpreted as random surfaces. Here the third coordinate is harder to visualize, but see Fig. 12.5 for a definition. It is not hard to see that dimers on any bipartite graph can be interpreted as random surfaces.

When thought of as random surfaces, these models have a *limit shape phenomenon*. This says that, for a fixed boundary curve in $\mathbb{R}^3$ or sequence of converging boundary curves in $\mathbb{R}^3$, if we take a random stepped surface on finer and finer lattices, then with probability tending to 1 the random surface will lie close to a fixed nonrandom surface, the so-called *limit shape*. So this limit shape surface is not just the average surface but the only surface you will see if you take an extremely fine mesh... the measure is concentrating as the mesh size tends to

[1]How do you pick a random sample from such a large space? This is an important and difficult question, about which much is known. We will not discuss it here.

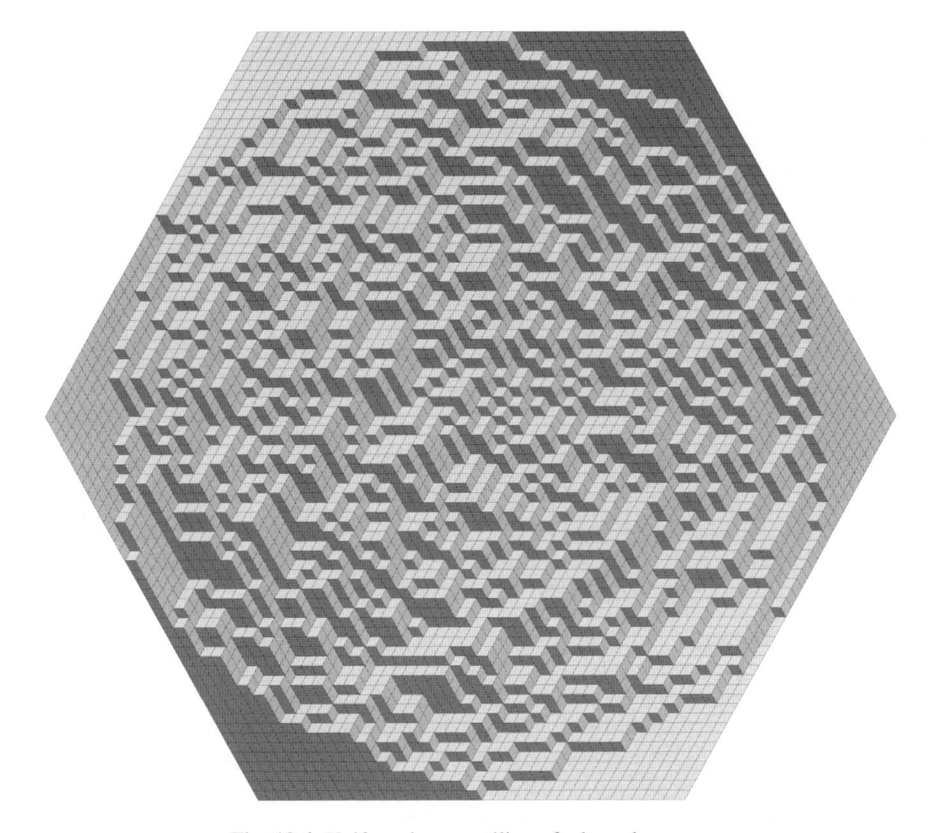

Fig. 12.4 Uniform lozenge tiling of a large hexagon.

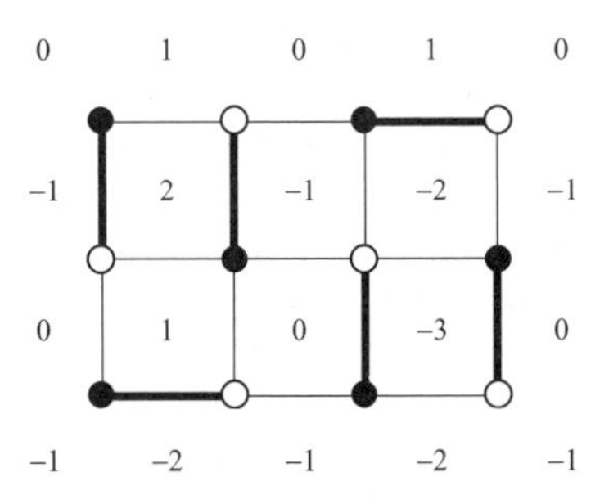

Fig. 12.5 The height is integer-valued on the faces and changes by ± 1 across an unoccupied edge; the change is $+1$ when crossing so that a white vertex is on the left.

zero on the delta-measure at this surface. This property is just a consequence of the law of large numbers, or, in the weighted case, a variational principle (principle of least action) (see Cohn *et al.* 2001). The limit shape surface is the surface which minimizes a "surface tension" integral, where the local surface tension is a function of the slope of the interface and is, by definition, the free energy for dimers constrained to have that slope.

In Fig. 12.3, the height function along the boundary is (with small variation) linear, and a random surface corresponding to a random domino tiling of the square is flat (with fluctuations of smaller order). For Fig. 12.4, however, the boundary height function is not flat: it consists of six sides of a large cube. Consequently, the stepped surface spanning it is forced to bend; this bending is responsible for the varying densities of tiles throughout the figure (indeed, the tile densities determine the slope of the surface locally).

12.2.3 Facets

One thing to notice about lozenge tiling in Fig. 12.4 is the presence of regions near the vertices of the hexagon where the lozenges are aligned. This phenomenon persists in the limit of small mesh size and, in fact, in the limit shape surface there is a *facet* near each corner, where the limit shape is planar. This is a phenomenon which does not occur in one dimension. The limit shapes for dimers generally contain facets and smooth (in fact analytic) curved regions separating these facets. It is remarkable that one can solve through analytic means, for reasonably general boundary conditions, for the entire limit shape, including the locations of the facets. We will not discuss this computation here, but refer the reader to Kenyon and Okounkov (2007).

12.2.4 Measures

What do we see if we zoom in to a point in Fig. 12.4? That is, consider a sequence of such figures with the same fixed boundary but decreasing mesh size. Pick a point in the hexagon and consider the configuration restricted to a small window around that point, a window which gets smaller as the mesh size goes to zero. One can imagine, for example, a window of side $\sqrt{\epsilon}$ when the mesh size is ϵ. This gives a sequence of random tilings of (relative to the mesh size) larger and larger domains, and in the limit (assuming that a limit of these "local measures" exists) we will get a random tiling of the plane.

We will see different types of behaviors, depending on which point we zoom in on. If we zoom in to a point in a facet, we will see in the limit a boring figure in which all tiles are aligned. This is an example of a measure on lozenge tilings of the plane which consists of a delta measure at a single tiling. This measure is an (uninteresting) example of an ergodic Gibbs measure (see the definition below). If we zoom in to a point in the nonfrozen region, one can again ask what limiting measure on tilings of the plane is obtained. One of the important open problems is to understand this limiting measure, in particular to prove that the limit exists. Conjecturally, it exists and depends only on the slope of the average surface at that point and not on any other property of the boundary conditions. For each possible slope (s, t), we will define below a measure $\mu_{s,t}$, and the *local statistics conjecture* states that, for any fixed boundary, $\mu_{s,t}$ is the measure which occurs in the limit at any point where the limit shape has slope (s, t). For certain boundary conditions, this has been proved (Kenyon 2008).

12.2.5 Phases

Gibbs measures on $M(\mathcal{G})$ come in three types, or *phases*, depending on the fluctuations of a typical (for that measure) surface. Suppose we fix the height at a face near the origin to be zero. A measure is said to be a *frozen phase* if the height fluctuations are finite almost surely; that is, the fluctuation of the surface away from its mean value is almost surely bounded, no matter how far away from the origin you are. A measure is said to be a *liquid phase* if the fluctuations have a variance increasing with increasing distance; that is, the variance of the

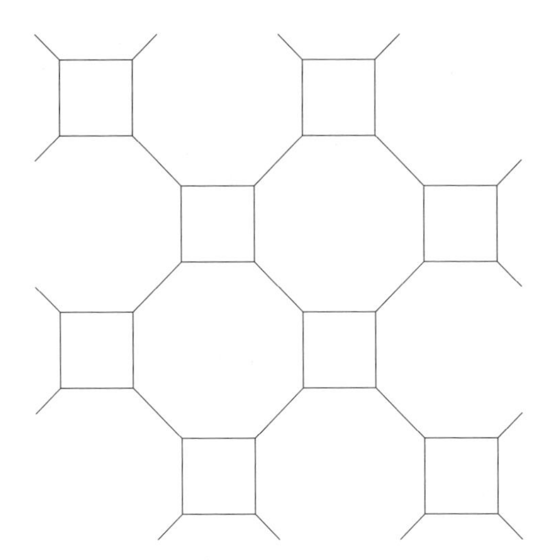

Fig. 12.6 The square–octagon lattice.

height at a point tends to infinity almost surely for points farther and farther from the origin. Finally, a measure is said to be a *gaseous phase* if the height fluctuations are unbounded but the variance of the height at a point is bounded independently of the distance from the origin.

We'll see that for uniform lozenge or domino tilings we can have both liquid and frozen phases but not gaseous phases. An example of a graph for which we have all three phases is the square–octagon dimer model (see Fig. 12.6). In general, the classification of phases depends on algebraic properties of the underlying graph and edge weights, as we'll see.

12.3 Gibbs measures

12.3.1 Definition

Let $X = X(\mathcal{G})$ be the set of dimer coverings of a graph $\mathcal{G}$, possibly infinite, with edge weight function w. Recall the definition of the Boltzmann probability measure on $X(\mathcal{G})$ when $\mathcal{G}$ is finite: a dimer covering has probability proportional to the product of its edge weights. When $\mathcal{G}$ is infinite, this definition will of course not work. For an infinite graph $\mathcal{G}$, a probability measure on X is a *Gibbs measure* if it is a weak limit of Boltzmann measures on a sequence of finite subgraphs of $\mathcal{G}$ filling out $\mathcal{G}$. By this we mean that, for any finite subset of $\mathcal{G}$, the probability of any particular configuration occurring on this finite set converges. That is, the probabilities of *cylinder sets* converge. Here a cylinder set is a subset of $X(\mathcal{G})$ consisting of all coverings containing a given finite subset of edges.

For a sequence of Boltzmann measures on increasing graphs, the limiting measure may not exist, but subsequential limits will always exist. The limit may not be unique, however; that is, it may depend on the approximating sequence of finite graphs. This will be the case for dimers, and it is this nonuniqueness which makes the dimer model interesting.

The important property of Gibbs measures is the following. Let A be a cylinder set defined by the presence of a finite set of edges. Let B be another cylinder set defined by a different set of edges but using the same set of vertices. Then, for the approximating Boltzmann measures, the ratio of the measures of these two cylinder sets is equal to the ratio of the product of the edge weights in A and the product of the edge weights in B. This is true along the finite growing sequence of graphs and so, in particular, the same is true for the limiting Gibbs measures. In fact this property *characterizes* Gibbs measures: given a finite set of vertices, the measure on this set of vertices conditioned on the exterior (that is, integrating over the configurations in the exterior) is just the Boltzmann measure on this finite set.

12.3.2 Ergodic Gibbs measures

For a periodic graph $\mathcal{G}$, a *translation-invariant measure* on $X(\mathcal{G})$ is simply one for which the measure of a subset of $X(\mathcal{G})$ is invariant under the translation-isomorphism action.

The *slope* (s, t) of a translation-invariant measure is the expected height change in the $(1, 0)$ and $(0, 1)$ directions; that is, s is the expected height change between a face and its translate by $(1, 0)$ and t is the expected height change between a face and its translate by $(0, 1)$ (recall that we have embedded our graph so that translations by $(1, 0)$ and $(0, 1)$ are color- and weight-preserving isomorphisms).

An *ergodic* Gibbs measure (EGM) is one in which translation-invariant sets have measure 0 or 1. A typical example of a translation-invariant set is the set of coverings which contain a translate of a particular pattern.

Theorem 12.1. (Sheffield 2005.) *For the dimer model on a periodic planar bipartite periodically edge-weighted graph, for each slope (s, t) for which there exists a translation-invariant measure, there exists a unique EGM $\mu_{s,t}$.*

In particular, we can classify EGMs by their slopes. The existence is not hard to establish by taking limits of Boltzmann measures on larger and larger tori where the height changes (h_x, h_y) are restricted; see below. The uniqueness is much harder; we won't discuss this here.

12.4 Kasteleyn theory

We shall show how to compute the number of dimer coverings of any bipartite planar graph using the KTF (Kasteleyn–Temperley–Fisher) technique. While this technique extends to non-bipartite planar graphs, we will have no use for this generality here.

12.4.1 Kasteleyn weighting

A *Kasteleyn weighting* of a planar bipartite graph is a choice of sign for each undirected edge with the property that each face with $0 \bmod 4$ edges has an odd number of minus signs and each face with $2 \bmod 4$ edges has an even number of minus signs.

In certain circumstances, it will be convenient to use complex numbers of modulus 1 rather than signs ± 1. In this case the condition is that the alternating product of edge weights (the first, divided by the second, times the third, and so on) around a face is negative real or positive real depending on whether the face has 0 or $2 \bmod 4$ edges.

This condition appears mysterious at first, but we'll see why it is important below.[2] It is not hard to see (using a homological argument) that any two Kasteleyn weightings are gauge

[2]The condition might appear more natural if we note that the alternating product is required to be $e^{\pi i N/2}$, where N is the number of triangles in a triangulation of the face.

equivalent: they can be obtained one from the other by a sequence of operations consisting of multiplying all edges at a vertex by a constant.

The existence of a Kasteleyn weighting is also easily established using spanning trees (put signs $+1$ on the edges of a spanning tree; the remaining edges have a determined sign). We leave this fact to the reader, as well as the proof of the following (easily proved by induction).

Lemma 12.2 *Given a cycle of length $2k$ enclosing ℓ points, the alternating product of signs around this cycle is $(-1)^{1+k+\ell}$.*

Note, finally, that for the (edge-weighted) honeycomb graph, all faces have $2 \bmod 4$ edges and so no signs are necessary in the Kasteleyn weighting.

12.4.2 Kasteleyn matrix

A *Kasteleyn matrix* is a weighted, signed adjacency matrix of the graph $\mathcal{G}$. Given a Kasteleyn weighting of $\mathcal{G}$, we define a $|B| \times |W|$ matrix K by $K(\mathrm{b}, \mathrm{w}) = 0$ if there is no edge from w to b, otherwise $K(\mathrm{b}, \mathrm{w})$ is the Kasteleyn weighting times the edge weight $w(\mathrm{bw}) = e^{-\mathcal{E}(\mathrm{b},\mathrm{w})}$.

For the graph in Fig. 12.7 with the Kasteleyn weighting indicated, the Kasteleyn matrix is

$$\begin{pmatrix} a & 1 & 0 \\ 1 & -b & 1 \\ 0 & 1 & c \end{pmatrix}.$$

Note that gauge transformation corresponds to pre- or post-multiplication of K by a diagonal matrix.

Theorem 12.3. (Kasteleyn 1961, Temperley and Fisher 1961.) $Z = |\det K|$.

In the above example, the determinant is $-a - c - abc$.

Here is the proof. If K is not square, the determinant is zero and there are no dimer coverings (each dimer covers one white and one black vertex). If K is a square $n \times n$ matrix, we expand

$$\det K = \sum_{\sigma \in S_n} \mathrm{sgn}(\sigma) K(\mathrm{b}_1, \mathrm{w}_{\sigma(1)}) \ldots K(\mathrm{b}_n, \mathrm{w}_{\sigma(1)}). \tag{12.1}$$

Each term is zero unless it pairs each black vertex with a unique neighboring white vertex. So there is one term for each dimer covering, and the modulus of this term is the product of its edge weights. We need only check that the signs of the nonzero terms are all equal.

Let us compare the signs of two different nonzero terms. Given two dimer coverings, we can draw them simultaneously on $\mathcal{G}$. We get a set of doubled edges and loops. To convert one dimer covering into the other, we can take a loop and move every second dimer (that is, dimers

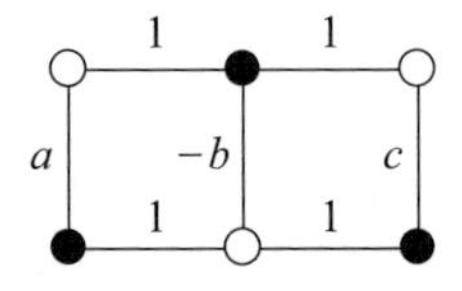

Fig. 12.7 Example Kasteleyn weighting.

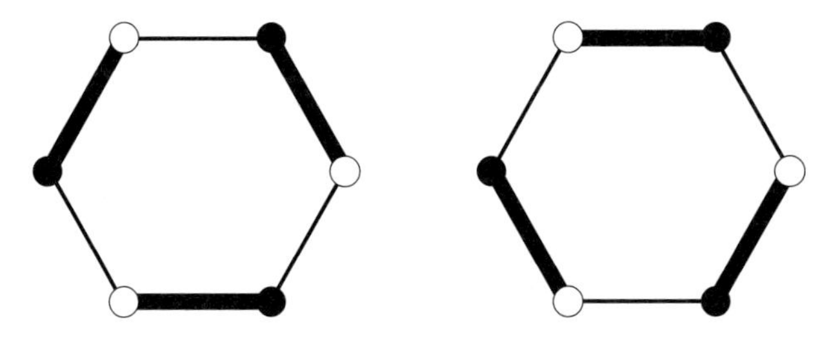

Fig. 12.8 Hexagon flip.

from the first covering) cyclically around by one edge so that they match the dimers from the second covering. When we do this operation for a single loop of length $2k$, we are changing the permutation σ by a k-cycle. Note that, by Lemma 12.2, the sign change of the edge weights in the corresponding term in eqn (12.1) is ± 1 depending on whether $2k$ is $2 \bmod 4$ or $0 \bmod 4$ (since ℓ is even there), exactly the same sign change as occurs in $\mathrm{sgn}(\sigma)$. These two sign changes cancel, showing that these two coverings (and hence any two coverings) have the same sign.

An alternate proof of this theorem for honeycomb graphs—which avoids using Lemma 12.2—goes as follows: if two dimer coverings differ only on a single face, that is, an operation of the type in Fig. 12.8 converts one covering into the other, then these coverings have the same sign in the expansion of the determinant, because the hexagon flip changes σ by a 3-cycle, which is an even permutation. Thus it suffices to notice that any two dimer coverings can be obtained from one another by a sequence of hexagon flips. This can be seen using the lozenge tiling picture, since applying a hexagon flip is equivalent to adding or subtracting a cube to or from the stepped surface. Any two surfaces with the same connected boundary can be obtained from one another by adding and/or subtracting cubes.

While there is a version of Theorem 12.3 (using Pfaffians) for nonbipartite planar graphs, there is no corresponding sign trick for nonplanar graphs in general (the exact condition is that a graph has a Kasteleyn weighting if and only if it does not have $K_{3,3}$ as a minor (Lovasz and Plummer 1986).

12.4.3 Local statistics

There is an important corollary to Theorem 12.3:

Theorem 12.4. (Kenyon 1997.) *Given a set of edges $X = \{\mathrm{w}_1\mathrm{b}_1, \ldots, \mathrm{w}_k\mathrm{b}_k\}$, the probability that all edges in X occur in a dimer cover is*

$$\left(\prod_{i=1}^{k} K(\mathrm{b}_i, \mathrm{w}_i)\right) \det(K^{-1}(\mathrm{w}_i, \mathrm{b}_j))_{1 \le i,j \le k} \,.$$

The proof uses the Jacobi Lemma, which says that a minor of a matrix A is $\det A$ times the complementary minor of A^{-1}.

The advantage of this result is that the probability of a set of k edges being present is only a $k \times k$ determinant, independently of the size of the graph. One needs only to be able to compute K^{-1}. In fact, the corollary is valid even for infinite graphs, once K^{-1} has been appropriately defined.

12.5 Partition function

12.5.1 Rectangle

Here is the simplest example. Assume mn is even, and let $\mathcal{G}_{m,n}$ be the $m \times n$ square grid. Its vertices are $V = \{1, 2, \ldots, m\} \times \{1, 2, \ldots, n\}$, and edges connect nearest neighbors. Let $Z_{m,n}$ be the partition function for dimers with edge weights 1. This is just the number of dimer coverings of $\mathcal{G}_{m,n}$.

A Kasteleyn weighting is obtained by putting weight 1 on horizontal edges and $i = \sqrt{-1}$ on vertical edges. Since each face has four edges, the condition in Section 12.4.1 is satisfied.[3]

The corresponding Kasteleyn matrix K is an $mn/2 \times mn/2$ matrix (recall that K is a $|W| \times |B|$ matrix). The eigenvalues of the matrix $\tilde{K} = \begin{pmatrix} 0 & K \\ K^t & 0 \end{pmatrix}$ are in fact simpler to compute. Let $z = e^{i\pi j/(m+1)}$ and $w = e^{i\pi k/(n+1)}$. Then the function

$$f_{j,k}(x, y) = (z^x - z^{-x})(w^y - w^{-y}) = -4 \sin\left(\frac{\pi j x}{m+1}\right) \sin\left(\frac{\pi k y}{n+1}\right)$$

is an eigenvector of $\tilde{K}$ with eigenvalue $z + 1/z + i(w + 1/w)$. To see this, we can check that

$$\lambda f(x, y) = f(x + 1, y) + f(x - 1, y) + if(x, y + 1) + if(x, y - 1)$$

when (x, y) is not on the boundary of $\mathcal{G}$, and it is also true when f is on the boundary, assuming we extend f to be zero just outside the boundary, i.e. when $x = 0$ or $y = 0$ or $x = m + 1$ or $y = n + 1$.

As j, k vary in $(1, m) \times (1, n)$, the eigenfunctions $f_{j,k}$ are independent (a well-known fact from Fourier series). Therefore we have a complete diagonalization of the matrix $\tilde{K}$, leading to

$$Z_{m,n} = \left(\prod_{j=1}^{m} \prod_{k=1}^{n} 2\cos\frac{\pi j}{m+1} + 2i \cos\frac{\pi k}{n+1} \right)^{1/2}. \tag{12.2}$$

Here the square root comes from the fact that $\det \tilde{K} = (\det K)^2$.

Note that if m, n are both odd then this expression is zero because of the term where $j = (m + 1)/2$ and $k = (n + 1)/2$ in eqn (12.2).

For example, $Z_{8,8} = 12\,988\,816$. For large m, n, we have

$$\lim_{m,n\to\infty} \frac{1}{mn} \log Z_{m,n} = \frac{1}{2\pi^2} \int_0^\pi \int_0^\pi \log(2\cos\theta + 2i\cos\phi)\, d\theta\, d\phi\,,$$

which can be shown to be equal to G/π, where G is Catalan's constant $G = 1 - 1/3^2 + 1/5^2 - \ldots$.

We can also write an explicit expression for the inverse Kasteleyn matrix and its limit. See Theorem 12.6 below.

[3] A weighting gauge equivalent to this one, and using only weights ± 1, is one that weights alternate columns of vertical edges by -1 and all other edges by $+1$. This was the weighting originally used by Kasteleyn (1961); our current weighting (introduced by Percus (1969)) is slightly easier for our purposes.

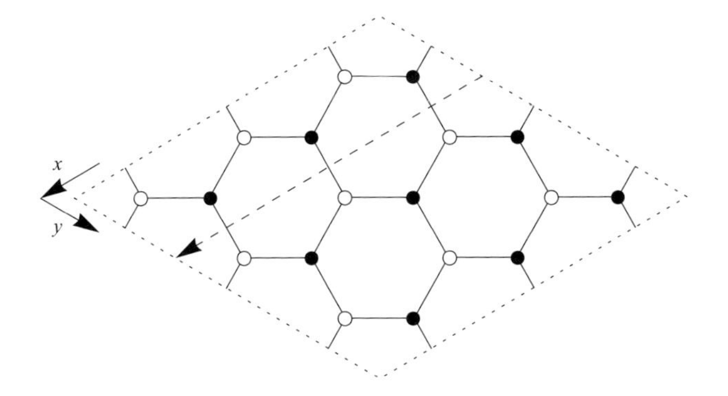

Fig. 12.9 Honeycomb graph on a torus.

12.5.2 Torus

A graph on a torus does not in general have a Kasteleyn weighting. However, we can still make a "local" Kasteleyn matrix whose determinant can be used to count dimer covers.

Rather that show this in general, let us work out a detailed example. Let H_n be the honeycomb lattice on a torus, as in Fig. 12.9, which shows H_3. It has n^2 black vertices and n^2 white vertices, and $3n^2$ edges. We weight edges with a, b, c according to direction: a for the horizontal edges, b for the NW–SE edges, and c for the NE–SW edges. Let $\hat{x}$ and $\hat{y}$ be the directions indicated. Given a dimer covering m of H_n, the numbers N_a, N_b, N_c of a-, b-, and c-type edges are multiples of n; for example, there is the same number of b-type edges crossing any NE–SW dashed line as in the figure. Let $h_x(m)$ and $h_y(m)$ be $1/n$ times the numbers of b- and c-type edges, respectively. That is, $h_x = N_b/n$ and $h_y = N_c/n$. These quantities measure the height change of m on a path winding around the torus; that is, if we think of a dimer covering as a stepped surface, there is a "locally defined" height function which changes by an additive constant on a path which winds around the torus; this additive constant is h_x in the $\hat{x}$ direction and h_y in the $\hat{y}$ direction.

Let K_n be the weighted adjacency matrix of H_n. That is, $K(\mathrm{b}, \mathrm{w}) = 0$ if there is no edge from b to w, and otherwise $K(\mathrm{b}, \mathrm{w}) = a, b,$ or c according to direction.

From the proof of Theorem 12.3, we can see that $\det K$ is a weighted, signed sum of dimer coverings. Our next goal is to determine the signs.

Lemma 12.5 *The sign of a dimer covering in* $\det K$ *depends only on its height change* (h_x, h_y) *modulo* 2. *Three of the four parity classes gives the same sign, and the fourth has the opposite sign.*

To see this, let N_b and N_c be the number of b- and c-type edges in a covering. If we take the union of a covering with the covering consisting of all a-type edges, we can compute the sign of the covering by taking the product of the sign changes when we shift along each loop. The number of loops is $q = \mathrm{GCF}(h_x, h_y)$ and each of these has homology class $(h_y/q, h_x/q)$ (note that the b edges contribute to h_x but to the $(0, 1)$ homology class). The length of each loop is $(2n/q)(h_x + h_y)$ and so each loop contributes sign $(-1)^{1+(n/q)(h_x+h_y)}$, for a total of $(-1)^{q+n(h_x+h_y)}$. Note that q is even if and only if h_x and h_y are both even. So if n is even then the sign is -1 unless $(h_x, h_y) \equiv (0, 0) \bmod 2$. If n is odd, the sign is $+1$ unless $(h_x, h_y) \equiv (1, 1) \bmod 2$. This completes the proof.

In particular, $\det K$, when expanded as a polynomial in a, b, and c, has coefficients which count coverings with particular height changes. For example, for n odd we can write

$$\det K = \sum (-1)^{h_x h_y} C_{h_x, h_y} a^{n(n-h_x-h_y)} b^{nh_x} c^{nh_y},$$

since $h_x h_y$ is odd exactly when h_x, h_y are both odd.

We define $Z_{00} = Z_{00}(a, b, c)$ to be this expression. We define

$$Z_{10}(a, b, c) = Z_{00}(a, be^{\pi i/n}, c),$$

$$Z_{01}(a, b, c) = Z_{00}(a, b, ce^{\pi i/n}),$$

and

$$Z_{11}(a, b, c) = Z_{00}(a, be^{\pi i/n}, ce^{\pi i/n}).$$

Then we can verify that, when n is odd,

$$Z = \frac{1}{2}(Z_{00} + Z_{10} + Z_{01} - Z_{11}), \tag{12.3}$$

which is equivalent to Kasteleyn's expression (Kasteleyn 1961) for the partition function. The case n even is similar and left to the reader.

12.5.3 Partition function

Dealing with a torus makes computing the determinant much easier, since the graph now has many translational symmetries. The matrix K commutes with translation operators and so can be diagonalized simultaneously with them. The simultaneous eigenfunctions of the horizontal and vertical translations (and K) are the exponential functions (also called Bloch–Floquet eigenfunctions) $f_{z,w}(x, y) = z^{-x} w^{-y}$, where $z^n = 1 = w^n$. There are n choices for z and n for w, leading to a complete set of eigenfunctions. The corresponding eigenvalue for K is $a + bz + cw$.

In particular,

$$\det K = Z_{00} = \prod_{z^n=1} \prod_{w^n=1} a + bz + cw.$$

This leads to

$$Z_{10} = \prod_{z^n=-1} \prod_{w^n=1} a + bz + cw,$$

$$Z_{01} = \prod_{z^n=1} \prod_{w^n=-1} a + bz + cw,$$

$$Z_{11} = \prod_{z^n=-1} \prod_{w^n=-1} a + bz + cw.$$

12.6 General graphs

As one might expect, this can be generalized to other planar, periodic, bipartite graphs. Representative examples are the square grid and the square–octagon grid (Fig. 12.6).

For simplicity, we're going to deal mainly with honeycomb dimers with periodic energy functions. Since it is possible, after simple modifications, to embed any other periodic bipartite planar graph in a honeycomb graph (possibly increasing the size of the period), we're actually not losing any generality. We'll also illustrate our calculations in an example in Section 12.6.4.

So let's start with the honeycomb with a periodic weight function ν on the edges, periodic with period ℓ in directions $\hat{x}$ and $\hat{y}$. As in Section 12.5.2, for any n we are led to an expression for the partition function for the $n\ell \times n\ell$ torus,

$$Z(H_{n\ell}) = \frac{1}{2}(Z_{00} + Z_{10} + Z_{01} - Z_{11}),$$

where

$$Z_{\tau_1,\tau_2} = \prod_{z^n=(-1)^{\tau_1}} \prod_{w^n=(-1)^{\tau_2}} P(z,w),$$

and where $P(z,w)$ is a polynomial, the *characteristic polynomial*, with coefficients depending on the energy function $\mathcal{E}$. The polynomial $P(z,w)$ is the determinant of $K(z,w)$, the Kasteleyn matrix for the $\ell \times \ell$ torus (with appropriate extra weights z and w on edges crossing fundamental domains); $K(z,w)$ is just the restriction of K to the space of Bloch–Floquet eigenfunctions (eigenfunctions for the operators of translation by $\ell\hat{x}, \ell\hat{y}$) with multipliers z, w. We can also consider $P(z,w)$ to be the signed weighted sum of dimer covers of the $\ell \times \ell$ torus consisting of a single $\ell \times \ell$ fundamental domain (with the appropriate extra weight $(-1)^{h_x h_y} z^{h_x} w^{h_y}$).

The algebraic curve $P(z,w) = 0$ is called the *spectral curve* of the dimer model, since it describes the spectrum of the K operator on the whole weighted honeycomb graph.

Many of the physical properties of the dimer model are encoded in the polynomial P.

Theorem 12.6 *The partition function per fundamental domain Z satisfies*

$$\log Z := \lim_{n\to\infty} \frac{1}{n^2} \log Z(H_{n\ell}) = \frac{1}{(2\pi i)^2} \int_{|z|=|w|=1} \log P(z,w) \frac{dz}{z}\frac{dw}{w}.$$

The inverse Kasteleyn matrix limit satisfies

$$K^{-1}(\mathrm{b},\mathrm{w}) = \frac{1}{(2\pi i)^2} \int_{|z|=|w|=1} \frac{z^x w^y \mathbf{Q}(z,w)_{b,w}}{P(z,w)} \frac{dz}{z}\frac{dw}{w},$$

where $\mathbf{Q}(z,w)/P(z,w) = K^{-1}(z,w)$, (x,y) is the translation between the fundamental domain containing b *to that containing* w, *and* $(b,w) \equiv (\mathrm{b},\mathrm{w}) \mod \mathbb{Z}^2$.

Note that the values of K^{-1} in the limit are linear combinations of Fourier coefficients of $1/P$ (since $\mathbf{Q}$ is a matrix of polynomials in z, w).

Recall the definition of h_x, h_y for a dimer cover of H_n. Any translation-invariant Gibbs state has an *expected slope*, which is the expected amount that the height function changes under translation by one fundamental domain in the $\hat{x}$- or $\hat{y}$-direction. It is a theorem of Sheffield that there is a unique translation-invariant ergodic Gibbs measure on dimer covers for each possible slope (s,t). The set of possible slopes for translation-invariant Gibbs measures can be naturally identified with the Newton polygon $\mathcal{N}(P)$ of P, that is, the convex hull in $\mathbb{R}^2$ of the set of integer exponents:

$$\mathcal{N}(P) = \text{cvxhull}(\{(i,j) : z^i w^i \text{ is a monomial of } P\}).$$

Note that there is a *periodic* dimer cover with slope $(h_x, h_y)/n$ for every $(h_x, h_y) \in N(p) \cap \mathbb{Z}^2$.

12.6.1 The amoeba of P

The *amoeba* of an algebraic curve $P(z,w) = 0$ is the set

$$\mathbb{A}(P) = \{(\log|z|, \log|w|) \in \mathbb{R}^2 \ : \ P(z,w) = 0\}.$$

See Fig. 12.13 later for an example. In other words, it is a projection to $\mathbb{R}^2$ of the zero set of P in $\mathbb{C}^2$, sending (z,w) to $(\log|z|, \log|w|)$. Note that for each point $(X,Y) \in \mathbb{R}^2$, the amoeba contains (X,Y) if and only if the torus $\{(z,w) \in \mathbb{C}^2 \ : \ |z| = e^X, |w| = e^Y\}$ intersects $P(z,w) = 0$.

The amoeba has "tentacles," which are regions where $z \to 0, \infty$, or $w \to 0, \infty$. Each tentacle is asymptotic to a line $\alpha \log|z| + \beta \log|w| + \gamma = 0$. These tentacles divide the complement of the amoeba into a certain number of unbounded complementary components. There may be bounded complementary components as well.

The *Ronkin function* of P is the function on $\mathbb{R}^2$ given by

$$R(X,Y) = \frac{1}{(2\pi i)^2} \int_{|z|=e^X} \int_{|w|=e^Y} \log P(z,w) \frac{dz}{z} \frac{dw}{w}.$$

The following facts are standard; see Mikhalkin and Rullgård (2001). The Ronkin function of P is convex in $\mathbb{R}^2$, and linear on each component of the complement of $\mathbb{A}(P)$.[4]

The gradient $\nabla R(X,Y)$ takes values in $\mathcal{N}(P)$. The Ronkin function is in fact Legendre dual to a function $\sigma(s,t)$ defined on $\mathcal{N}(P)$. That is, we define

$$\sigma(s,t) = \max_{X,Y}(-R(X,Y) + sX + tY).$$

This σ is the *surface tension* function for the dimer model.

This duality $(X,Y) \mapsto (s,t)$ maps each component of the complement of $\mathbb{A}(P)$ to a single point of the Newton polygon $\mathcal{N}(P)$. This is a point with integer coordinates. Unbounded complementary components correspond to integer points on the boundary of $\mathcal{N}(P)$; bounded complementary components correspond to integer points in the interior of $\mathcal{N}(P)$.[5]

See Figs. 12.10 and 12.11 for plots of σ and R in the case of uniform honeycomb dimers $(P(z,w) = 1 + z + w)$. The solid lines in Fig. 12.11 project downwards to the boundary of the amoeba of $1 + z + w$.

[4] Thus the complementary components of $\mathbb{A}(P)$ are convex.

[5] Not every integer point in $\mathcal{N}(P)$ may correspond to a complementary component of $\mathbb{A}$.

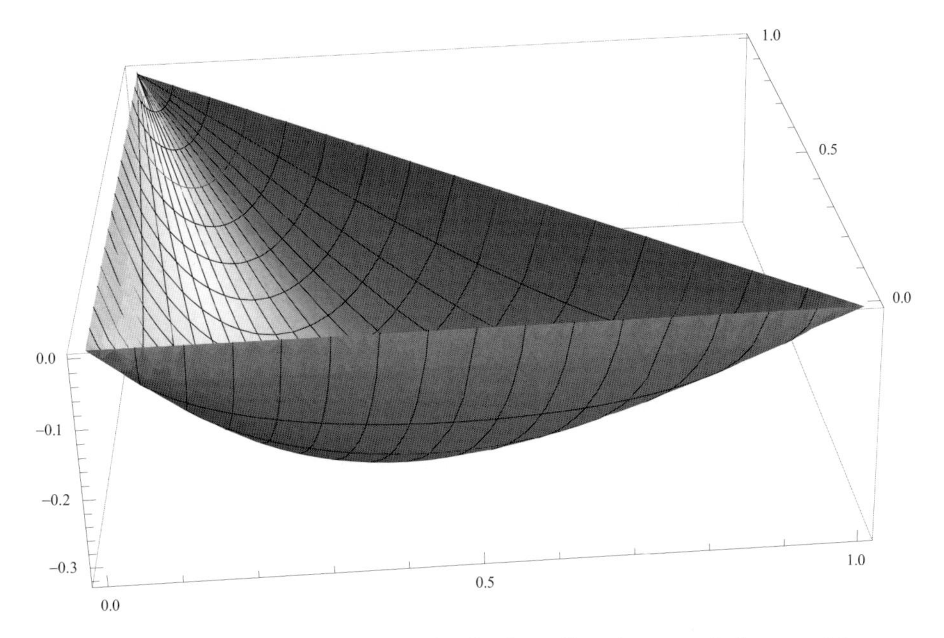

Fig. 12.10 Graph of the surface tension for the uniform honeycomb dimer model.

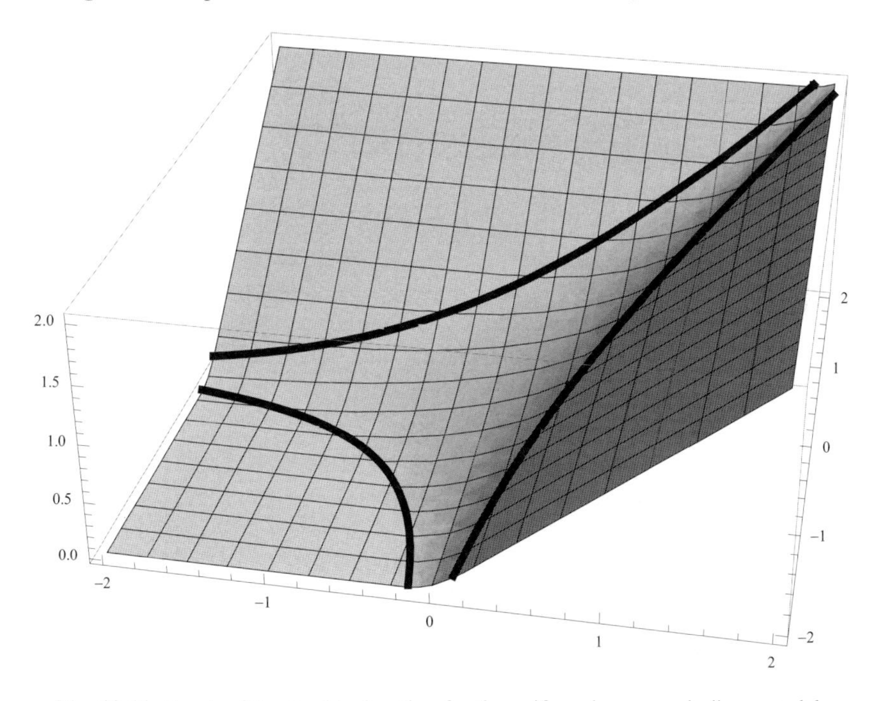

Fig. 12.11 Graph of the Ronkin function for the uniform honeycomb dimer model.

12.6.2 Phases of EGMs

Note that the Ronkin function can also be written

$$R(X,Y) = \frac{1}{(2\pi i)^2}\int_{|z|=1}\int_{|w|=1} \log P(e^X z, e^Y w)\frac{dz}{z}\frac{dw}{w}\,.$$

As such, the Ronkin function is the negative of the free energy as a function of the electric field:

$$\log Z(E_x, E_y) = -R(E_x, E_y)\,.$$

In particular, the amoeba of P can be thought of as the phase diagram for the dimer model as a function of the electric field $\vec{E} = (E_x, E_y)$. The amoeba boundaries are places where the partition function is not analytic as a function of $\vec{E}$. It remains to see what the different phases "mean."

To study the different phases, we study the decay of correlations for a fixed measure; that is, for a given $\mu_{s,t}$, how does the probability of two edges in a dimer cover compare with the product of their probabilities? The covariance of two edges is, by definition,

$$\mathbf{Cov}(e_1, e_2) = \mathbf{Pr}(e_1 \& e_2) - \mathbf{Pr}(e_1)\mathbf{Pr}(e_2)\,.$$

This can be written in terms of the inverse Kasteleyn matrix as (see Section 12.4.3)

$$\begin{aligned}\mathbf{Cov}(e_1, e_2) &= K_{\mathrm{w}_1\mathrm{b}_1}K_{\mathrm{w}_2\mathrm{b}_2}\det\begin{pmatrix}(K^{-1})_{\mathrm{b}_1\mathrm{w}_1} & (K^{-1})_{\mathrm{b}_1,\mathrm{w}_2}\\ (K^{-1})_{\mathrm{b}_2,\mathrm{w}_1} & (K^{-1})_{\mathrm{b}_2,\mathrm{w}_2}\end{pmatrix}\\ &\quad - K_{\mathrm{w}_1\mathrm{b}_1}K_{\mathrm{w}_2\mathrm{b}_2}(K^{-1})_{\mathrm{b}_1,\mathrm{w}_1}(K^{-1})_{\mathrm{b}_2,\mathrm{w}_2}\\ &= w(\mathrm{w}_1\mathrm{b}_1)w(\mathrm{w}_2\mathrm{b}_2)|K^{-1}(\mathrm{b}_1,\mathrm{w}_2)K^{-1}(\mathrm{b}_2,\mathrm{w}_1)|\,.\end{aligned}$$

The local statistics for a measure $\mu_{s,t}$ are determined by the inverse Kasteleyn matrix $K_{\vec{E}}^{-1}$, where $\vec{E} = (E_x, E_y)$ is related to (s,t) via Legendre duality, i.e. $\nabla R(E_x, E_y) = (s,t)$. As discussed in Theorem 12.6, values of $K_{\vec{E}}^{-1}$ are (linear combinations of) Fourier coefficients of $1/P(e^{E_x}z, e^{E_y}w)$. In particular, if $P(e^{E_x}z, e^{E_y}w)$ has no zeros on the unit torus $\{|z| = |w| = 1\}$, then $1/P(e^{E_x}z, e^{E_y}w)$ is analytic and so its Fourier coefficients decay exponentially fast. The corresponding covariance will decay exponentially fast in the separation between the edges. On the other hand, if $P(e^{E_x}z, e^{E_y}w)$ has simple zeros on the unit torus, its Fourier coefficients decay linearly, and the covariance of two edges will decay quadratically in the separation.

This, then, is the condition which separates the different phases of the dimer model. If a slope (s,t) is chosen so that (E_x, E_y) is in (the closure of) an unbounded component of the complement of the amoeba, then certain Fourier coefficients of $1/P$ (those contained in the appropriate dual cone) will vanish. This is enough to ensure that $\mu_{s,t}$ is in a frozen phase: the covariances of edges more than one fundamental domain apart are identically zero (this requires some argument, which we are not going to give here). For slopes (s,t) for which (E_x, E_y) is in (the closure of) a bounded component of the complement of the amoeba, the edge–edge covariances decay exponentially fast (in all directions). This is enough to show that the height fluctuations have bounded variance, and we are in a gaseous (but not frozen, since the correlations are nonzero) phase.

In the remaining case, (E_x, E_y) is in the interior of the amoeba, and P has zeros on a torus. It is a beautiful and deep fact that the spectral curves arising in the dimer model are special in that P has either two zeros, both simple, or a single node[6] over each point in the interior of $\mathbb{A}(P)$. As a consequence,[7] in this case the edge–edge covariances decay quadratically (quadratically in generic directions—there may be directions where the decay is faster). It is not hard to show that this implies that the height variance between distant points is unbounded, and we are in a liquid phase.

12.6.3 Harnack curves

Plane curves with the property described above, that they have at most two zeros (both simple) or a single node on each torus $|z| = \text{constant}$, $|w| = \text{constant}$, are called *Harnack curves*, or simple Harnack curves. They were studied classically by Harnack and more recently by Mikhalkin and others (Mikhalkin and Rullgård 2001).

The simplest definition is that a Harnack curve is a curve $P(z, w) = 0$ with the property that the map from the zero set to the amoeba $\mathbb{A}(P)$ is at most 2 to 1 over $\mathbb{A}(P)$. It will be 2 to 1 with a finite number of possible exceptions (the integer points of $\mathcal{N}(P)$), on which the map may be 1 to 1.

Theorem 12.7. (Kenyon and Okounkov 2006, Kenyon *et al.* 2006) *The spectral curve of a dimer model is a Harnack curve. Conversely, every Harnack curve arises as the spectral curve of some periodic bipartite weighted dimer model.*

In Kenyon and Okounkov (2006) it was also shown, using dimer techniques, that the areas of complementary components of $\mathbb{A}(P)$ and distances between tentacles are global coordinates for the space of Harnack curves with a given Newton polygon.

12.6.4 Example

Let's work out a detailed example illustrating the above theory. We'll take dimers on the square grid with a 3×2 fundamental domain (invariant under the lattice generated by $(0, 2)$ and $(3, 1)$). We take the fundamental domain with vertices labeled as in Fig. 12.12—we chose those weights to give us enough parameters (5) to describe all possible gauge equivalence classes of weights on the 3×2 fundamental domain. Letting z be the eigenvalue of translation in the direction $(3, 1)$ and w be the eigenvalue of translation by $(0, 2)$, the Kasteleyn matrix (where white vertices correspond to rows and black to columns) is

$$K = \begin{pmatrix} -1 + 1/w & 1 & e/z \\ c & a - w & d \\ z/W & 1 & -b + 1/w \end{pmatrix}.$$

We have

$$P(z, w) = \det K(z, w) = 1 + b + ab + bc + d + e \\ - \frac{1 + a + ab + c + d + ae}{w} + \frac{a}{w^2} - bw + \frac{ce}{z} + d\frac{z}{w}.$$

[6] A node is a point where $P = 0$ looks locally like the product of two lines, e.g. $P(x, y) = x^2 - y^2 + O(x, y)^3$ near $(0, 0)$.

[7] We have already shown what happens in the case of a simple pole. The case of a node is fairly hard.

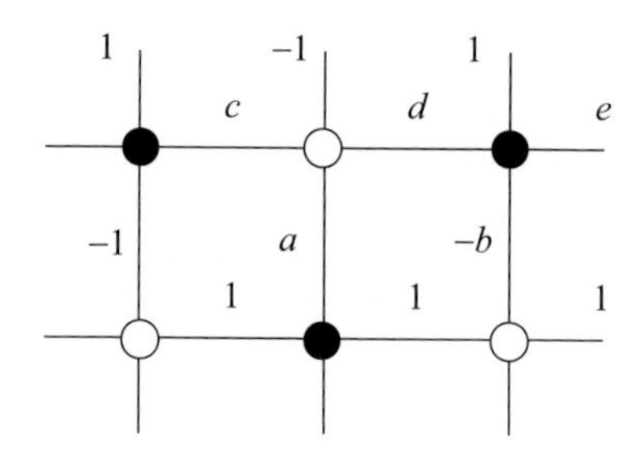

Fig. 12.12 Example Kasteleyn weighting.

This can of course be obtained by just counting dimer covers of $\mathbb{Z}^2/\{(0,2),(3,1)\}$ with these weights, and with an appropriate factor $(-1)^{ij+j}z^iw^j$ when there are edges going across fundamental domains. Let's specialize to $b = 2$ and all other edges of weight 1. Then,

$$P(z,w) = 9 - 2w + \frac{1}{w^2} - \frac{7}{w} + \frac{1}{z} + \frac{z}{w}.$$

The amoeba is shown in Fig. 12.13. There are two gaseous components, corresponding to EGMs with slopes $(0,0)$ and $(0,-1)$. The four frozen EGMs correspond to slopes $(1,-1)$, $(0,1)$, $(0,-2)$, and $(-1,0)$. All other slopes correspond to liquid phases.

If we take $c = 2$, $b = \frac{1}{2}(3 \pm \sqrt{3})$, and all other weights 1 then there remains only one gaseous phase; the other gas "bubble" in the amoeba shrinks to a point and becomes a node in $P = 0$. If we take all edges 1 then both gaseous phases disappear; we just have the uniform measure on dominos again.

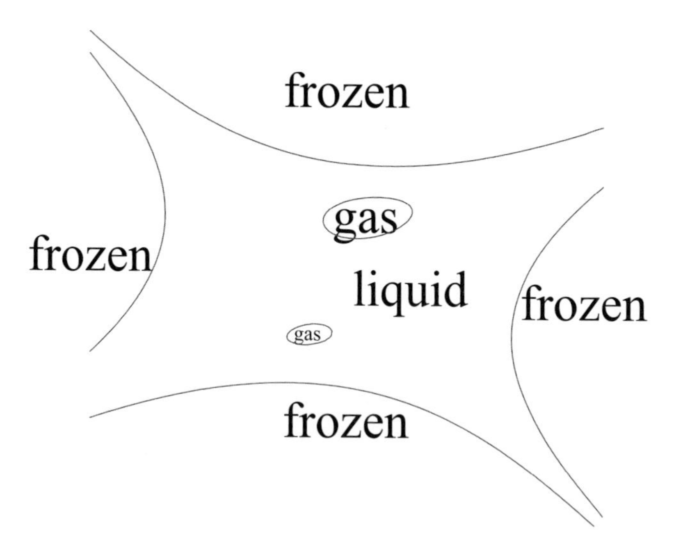

Fig. 12.13 Amoeba for the example discussed in the text.

References

Cohn, H., Kenyon, R., and Propp, J. (2001). A variational principle for domino tilings. *J. Am. Math. Soc.* **14**, 297–346.

Kasteleyn, P. (1961). The statistics of dimers on a lattice: i. The number of dimer arrangements on a quadratic lattice. *Physica* **27**(12), 1209–1225.

Kasteleyn, P. (1967). Graph theory and crystal physics. In *Graph Theory and Theoretical Physics* (ed. F. Harary), pp. 43–110. Academic Press, London.

Kenyon, R. (1997). Local statistics of lattice dimers. *Ann. Inst. H. Poincaré, Probabilités* **33**, 591–618.

Kenyon, R. (2004). An introduction to the dimer model. *School and Workshop on Probability, ICTP Lecture Notes* (ed. G. Lawler), math.CO/0310326.

Kenyon, R. (2008). Height fluctuations in the honeycomb dimer model. *Commun. Math. Phys.* **281**, 675–709.

Kenyon, R. and Okounkov, A. (2006). Planar dimers and Harnack curves. *Duke Math. J.* **131**(3), 499–524.

Kenyon, R., Okounkov, A., and Sheffield, S. (2006). Dimers and amoebae. *Ann. Math. (2)* **163**(3), 1019–1056.

Lovasz, L. and Plummer, M. (1986). *Matching Theory*, North-Holland Mathematics Studies 121. Annals of Discrete Mathematics, 29. North-Holland, Amsterdam.

Mikhalkin, G. and Rullgård, H. (2001). Amoebas of maximal area. *Int. Math. Res. Not.* No. 9, 441–451.

Passare, M. and Rullgård, H. (2004). Amoebas, Monge-Ampère measures, and triangulations of the Newton polytope. *Duke Math. J.* **121**(3), 481–507.

Percus, J. (1969). One more technique for the dimer problem. *J. Math. Phys.* **10**, 1881–1888.

Sheffield, S. (2005). Random surfaces. *Astérisque* No. 304.

Temperley, W. and Fisher, M. (1961). Dimer problem in statistical mechanics—an exact result. *Philos. Mag. (8)* **6**, 1061–1063.

13
Boundary loop models and 2D quantum gravity

I. KOSTOV

Ivan Kostov,
Institute de Physique Théorique,
CNRS URA 2306, C.E.A. Saclay,
F-91191 Gif-sur-Yvette, France.

13.1 Introduction

The solvable two-dimensional statistical models that have a geometrical description in terms of self-avoiding and mutually avoiding clusters, such as the Ising, $O(n)$, and Potts models, can also be formulated and solved on a dynamical lattice. In this case we say that the model is coupled to quantum gravity. The methods of solution are, in most cases, based on an equivalence with large-N matrix models. Two classical reviews of 2D gravity are Di Francesco *et al.* (1995) and Ginsparg and Moore (1993).

For systems coupled to gravity, it is possible to solve analytically problems whose exact solution is inaccessible on a flat lattice. But the solution of a spin model on a dynamical lattice allows us to reconstruct the phase diagram and the critical exponents of the same spin model on the flat lattice. Therefore the solvable models of 2D quantum gravity are a powerful tool for exploring the critical phenomena in two dimensions.

In these lectures, we give an introduction to 2D quantum gravity and its uses in studying geometrical critical phenomena in two dimensions, paying special attention to boundary phenomena. We will focus on the $O(n)$ model coupled to gravity, which is particularly useful in studying problems of quantum geometry because of its representation as a gas of self-avoiding and mutually avoiding loops. Our aim is to explain how the method of 2D gravity works in practice and to arm the student with the necessary tools to solve concrete problems. This is why we do not avoid calculations when they can be instructive. An understanding of these notes needs some basic notions of quantum field theory (QFT), statistical mechanics, and conformal field theory (CFT) (John Cardy's lecture notes are more than sufficient), as well as a rudimentary knowledge of complex analysis.

First we give a sketch of Liouville gravity and a derivation of the KPZ scaling relation between the critical exponents of a spin model on flat and dynamical lattices. In particular, we will explain the statistical meaning of the two branches of the KPZ relation, which is missing in the classical review papers. Then we consider the $O(n)$ vector model on random graphs, its loop gas expansion, and the correspondence with a large-N matrix model with $O(n)$ symmetry.

From this correspondence, we evaluate the partition function of the $O(n)$ model on a sphere and on a disk. The $O(n)$ model coupled to gravity can be solved exactly for any value of the temperature coupling, which is not the case on a flat lattice. The partition function on the sphere is computed using an equation for the susceptibility. This is a transcendental equation with two singular points, corresponding to the dilute and the dense phase of the loop gas.

Once the solution on the disk is known, we use the solution to compute the bulk and boundary correlation functions of the $O(n)$ field. We evaluate the boundary two-point function for Neumann and Dirichlet boundary conditions, as well as for a continuum of boundary conditions that break the $O(n)$ symmetry.

13.2 Continuous world-sheet description: Liouville gravity

13.2.1 Integral over world sheet geometries

The path integral of 2D quantum gravity includes an independent summation with respect to the fluctuations of the matter field and the fluctuations of the metric of the two-dimensional Universe.

Let $Z_{\rm matter}[g_{ab}]$ be the partition function of the matter field on a two-dimensional manifold $\mathcal{M}$ with Riemann metric

$$ds^2 = g_{ab}(\sigma)\, d\sigma^a d\sigma^b\,.$$

The path integral of 2D quantum gravity involves an additional integration with respect to the fluctuations of the Riemann metric defined in the above equation. We will refer to the manifold $\mathcal{M}$ as a world sheet. If the world sheet is homeomorphic to a sphere, the measure over Riemann metrics is controlled by a parameter μ coupled to its area

$$A = \int_{\mathcal{M}} d^2\sigma\, \sqrt{\det g_{ab}}\,. \tag{13.1}$$

The partition function on the sphere, which we denote by $F(\mu)$, is written symbolically as a functional integral over all Riemann metrics on the sphere,

$$F(\mu) = \int [\mathcal{D}g_{ab}]\; e^{-\mu A}\; Z_{\rm matter}^{\rm (sphere)}[g_{ab}]\,. \tag{13.2}$$

Since A is the volume of our two-dimensional Universe, the parameter μ is called the *(bulk) cosmological constant.*

For a world sheet $\mathcal{M}$ with a boundary $\partial\mathcal{M}$ we can introduce a second parameter, the *boundary cosmological constant* μ_B, coupled to the boundary length

$$\ell = \int_{\partial\mathcal{M}} \sqrt{g_{ab}\, d\sigma^a d\sigma^b}\,. \tag{13.3}$$

The partition function on the disk, which we denote by $U(\mu,\mu_B)$, is defined as a functional integral over all Riemann metrics on the disk,

$$U(\mu,\mu_B) = \int [\mathcal{D}g_{ab}]\; e^{-\mu A - \mu_B \ell}\; Z_{\rm matter}^{\rm (disk)}[g_{ab}]\,. \tag{13.4}$$

Here we have assumed that the matter field satisfies a certain boundary condition. We will also consider more complicated situations where the boundary is split into several segments with different boundary cosmological constants and different boundary conditions for the matter field.

In the conformal gauge

$$g_{ab}(\sigma) = e^{2\phi(\sigma)}\; \hat{g}_{ab}\,, \tag{13.5}$$

where $\hat{g}_{ab}$ is some fixed background metric, the fluctuations of the metric are described by the local scale factor $\phi(\sigma)$. At the critical points, where the matter QFT is conformally invariant, the scale factor ϕ couples to the matter field through a universal classical action induced by the conformal anomaly $c_{\rm matter}$. In addition, the conformal gauge necessarily introduces Faddeev–Popov reparameterization ghosts with central charge $c_{\rm ghosts} = -26$ (Polyakov 1981). The effective action of the field ϕ is called the *Liouville action* because the corresponding classical equation of motion is the equation for metrics with constant curvature, the Liouville equation.

Liouville gravity is a term for the two-dimensional quantum gravity whose action is induced by a conformally invariant matter field. The simplest correlation functions in Liouville gravity (up to three-point functions of the bulk fields and two-point functions of the boundary fields) factorize into matter and Liouville pieces. This factorization does not hold anymore away from the critical points, where the effective action of the scale field ϕ contains additional interactions.

13.2.2 Liouville gravity on a disk

Topologically, the world sheet is characterized by its Euler characteristics

$$\chi = 2 - 2\#(\text{handles}) - \#(\text{boundaries})\,. \tag{13.6}$$

We will be mainly interested in the topology of a disk ($\chi = 1$), but for the moment let us consider a world sheet with general topology. We denote by $\hat{R}$ the background curvature in the bulk and by $\hat{K}$ the geodesic background curvature at the boundary, normalized so that

$$\int_{\mathcal{M}} \hat{R}^{(2)} + 2\int_{\partial\mathcal{M}} \hat{K} = 4\pi\chi\,. \tag{13.7}$$

The effective action of the Liouville gravity consists of three pieces, associated with the matter, Liouville, and ghost fields. For our discussion, we will need only the explicit form of the Liouville piece $\mathcal{A}_{\mu,\mu_B}[\phi]$. This is the action of a Gaussian field with linear coupling to the curvature and exponential terms coupled to the bulk and boundary cosmological constants:

$$\begin{aligned}\mathcal{A}_{\mu,\mu_B}[\phi] &= \frac{1}{4\pi}\int\limits_{\mathcal{M}} \left(g\,(\nabla\phi)^2 + Q\phi\hat{R}\right) + \frac{1}{2\pi}\int\limits_{\partial\mathcal{M}} d\sigma\, Q\phi\hat{K} \\ &\quad + \mu A + \mu_B \ell\,.\end{aligned} \tag{13.8}$$

The coupling to the curvature is through the (imaginary) background charge iQ. The area and the length are defined by eqns (13.1) and (13.3) for the metric (13.1). We will assume that the coupling constant g of the Liouville field is larger than one: $g \geq 1$.

Boundary Liouville CFT. For a world sheet homeomorphic to a disk, it is always possible to choose $\hat{g}_{ab}(x) = \delta_{ab}$. In this gauge, all the curvature is concentrated on the boundary. By a conformal transformation, the disk can be mapped to the upper half-plane (UHP). After that, the action ceases to depend on the curvature,

$$\mathcal{A}_{\mu,\mu_B}[\phi] = \frac{g}{4\pi}\int\limits_{\text{UHP}} (\partial_a\phi)^2 d^2z + \mu\int\limits_{\text{UHP}} e^{2\phi(z,\bar{z})} d^2z + \mu_B\int\limits_{\mathbb{R}} e^{\phi(x)} dx\,, \tag{13.9}$$

but the curvature terms are replaced with the asymptotics at infinity

$$\phi(z,\bar{z}) \simeq -Q\log(z\bar{z})\,. \tag{13.10}$$

The boundary potential is equivalent to imposing a generalized Neumann boundary condition on the real axis, $i(\partial - \bar{\partial})\phi = 4\pi\mu_B e^{\phi}$.

The quantum theory based on this classical action (13.9) is called boundary Liouville theory. The theory is conformally invariant, with conformal anomaly

$$c_L = 1 + \frac{6Q^2}{g}\,. \tag{13.11}$$

The Liouville field has a continuous spectrum of boundary conditions labeled by the boundary cosmological constant μ_B.

The Liouville theory has a continuous spectrum of bulk and boundary primary states. The bulk primary fields with respect to the Liouville energy–momentum tensor,

$$T(z) = -g(\partial_z\phi)^2 + Q\partial_z^2\phi, \quad \bar{T}(\bar{z}) = -g(\partial_{\bar{z}}\phi)^2 + Q\partial_{\bar{z}}^2\phi\,, \tag{13.12}$$

are the exponential fields

$$V_\alpha \sim e^{2\alpha\phi(z,\bar{z})} \qquad (\alpha \in \mathbb{C})\,, \tag{13.13}$$

which have conformal weights

$$\Delta_\alpha^{\text{Liouv}} = \bar{\Delta}_\alpha^{\text{Liouv}} = \frac{\alpha(Q-\alpha)}{g}\,. \tag{13.14}$$

The charges of the normalizable primary states are of the form $\alpha = Q/2 + iP$, which reminds us of the spin of the continuous principal series of the representations of the noncompact group $SL(2,\mathbb{R})$. The states that correspond to local operators on the world sheet have real Liouville charge α and are not normalizable.

Similarly, the boundary primaries are represented by boundary exponential fields

$$B_\beta \sim e^{\beta\phi(x)}\,. \tag{13.15}$$

With this normalization, their conformal weights $\Delta_\beta^{\text{Liouv}} = \beta(Q-\beta)/g$ are given by the same expression as that for the bulk fields. The correlation functions of the boundary Liouville fields depend on the boundary conditions on both sides, which is a general feature of the boundary operators (Cardy 1984). For a boundary operator with left and right boundary conditions μ and μ', respectively, we will use the notation ${}^{\mu}[B_\beta]^{\mu'}$.

The Liouville theory possesses a reflection symmetry which relates the correlation functions of the exponential fields V_α and $V_{Q-\alpha}$. When restricted to normalizable states, this symmetry coincides with Hermitian conjugation.

Having a continuous spectrum of primary fields, the Liouville theory cannot be studied by the methods developed by Belavin *et al.* (1984). The conformal bootstrap methods were extended to the case of a "noncompact" CFT, such as Liouville theory, by Fateev, Teschner, and the Zamolodchikovs, following previous work by Dorn and Otto. The basic correlation functions in boundary Liouville theory were obtained by Fateev *et al.* (2000), Ponsot *et al.* (2002), and Hosomichi (2001).

Liouville gravity. The full theory is, by construction, invariant with respect to the diffeomorphisms of the world sheet manifold $\mathcal{M}$. 2D quantum gravity is a topological theory in the sense that the correlation functions cannot depend on the distances. It is more appropriate to call them instead "correlation numbers." The conformal symmetry, which is part of the general covariance, must be exact. Therefore, when considered in the context of 2D gravity, the background charge of the Liouville field must be tuned so that the total conformal anomaly vanishes:

$$c_{\text{tot}} \equiv c_\phi + c_{\text{matter}} + c_{\text{ghosts}} = \left(1 + \frac{6Q^2}{g}\right) + c_{\text{matter}} - 26 = 0\,. \tag{13.16}$$

The balance of the central charge gives

$$c_{\text{matter}} = 1 - \frac{6(g-1)^2}{g}\,. \tag{13.17}$$

This is the conformal anomaly of a Gaussian field with coupling constant g and real background charge $e_0 = g - 1$.

The observables in Liouville gravity are integrated local densities

$$O \sim \int_{\mathcal{M}} \Phi_\Delta\, e^{2\alpha\phi}\,, \tag{13.18}$$

where Φ_Δ represents a scalar ($\Delta = \bar{\Delta}$) matter field[1] and V_α is a Liouville "dressing" field, which completes the conformal weights of the matter field to $(1,1)$. The balance of the conformal dimensions, $\Delta + \Delta_\alpha^{\text{Liouv}} = 1$, gives a quadratic relation between the Liouville dressing charge and the conformal weight of the matter field:

$$\Delta + \frac{\alpha(Q-\alpha)}{g} = 1\,. \tag{13.19}$$

In particular, the Liouville interaction $e^{2\phi}$ dresses the identity operator. The equation for the balance of the dimensions (13.19) determines, for $\alpha = 1$ and $\Delta = 0$, the value of the background charge Q:

$$Q = g + 1\,. \tag{13.20}$$

Similarly, the boundary matter fields B_Δ with boundary dimensions Δ are dressed by boundary Liouville exponential fields,

$$B \sim \int_{\partial\mathcal{M}} B_\Delta\, e^{\beta\phi}\,, \tag{13.21}$$

where the Liouville exponent β is related to Δ^B as in eqn (13.19). The boundary operators in Liouville gravity are completely determined when the matter boundary conditions b, b' and Liouville boundary conditions μ_B, μ'_B on both sides are specified. In the expressions for the boundary correlation functions, we will use the notation ${}^{\mu_B}_{\;b}[B]^{\mu'_B}_{b'}$.

The physical distance ℓ_{12} between two boundary operators B_1 and B_2 is controlled by the boundary cosmological constant μ^B_{12} associated with the boundary segment between them.

[1] The fields cannot have spin, because any local rotation can be "unwound" by a coordinate transformation.

For example, we can study the operator product expansion (OPE) of such operators by taking the limit $\mu_{12} \to \infty$ of their correlation functions.

The quadratic relation (13.19) has two solutions, α and $\tilde{\alpha}$, such that $\alpha \leq Q/2$ and $\bar{\alpha} \geq Q/2$. The Liouville dressing fields corresponding to the two solutions are related by the Liouville reflection $\alpha \to \bar{\alpha} = Q - \alpha$. Generically, the scaling limit of the operators on the lattice is described by smaller of the two solutions of the quadratic constraint (13.19). The corresponding field V_α has a good quasi-classical limit. Such fields are said to satisfy the Seiberg bound $\alpha \leq Q/2$; see the review by Ginsparg and Moore (1993).[2]

13.2.3 Correlation functions in Liouville gravity and KPZ scaling relation

We will derive a scaling relation which allows us to extract the conformal weights of the matter operators from the scaling properties of their correlation functions after coupling to gravity. The unnormalized correlation functions for a world sheet with Euler characteristics χ are formally defined by the path integral

$$\langle O_1 O_2 \ldots \rangle_{\mu,\mu_B} = \int [\mathcal{D}\phi]\, e^{\mathcal{A}[\phi]} O_1 O_2 \ldots . \tag{13.22}$$

The scaling of the correlation function with the bulk cosmological constant μ follows from the covariance of the Liouville action (13.8) with respect to translations of the Liouville field:

$$\mathcal{A}_{\mu,\mu_B}\left[\phi - \frac{1}{2}\ln\mu\right] = -\frac{1}{2}Q\chi\ln\mu + \mathcal{A}_{1,\,\mu_B/\sqrt{\mu}}[\phi]. \tag{13.23}$$

Applying this to the correlation function of the bulk and boundary operators

$$O_j = \int_{\mathcal{M}} \Phi_{\Delta_j} e^{2\alpha_j\phi}, \qquad B_k = \int_{\partial\mathcal{M}} B_{\Delta_k} e^{\beta_k\phi}, \tag{13.24}$$

we obtain the scaling relation

$$\left\langle \prod_j O_j \prod_k B_k \right\rangle_{\mu,\mu_B} = \mu^{(1/2)\chi Q - \sum_j \alpha_j}\, \mu_B^{-\sum_k \beta_k} \left\langle \prod_j O_j \prod_k B_k \right\rangle_{1,\,\mu_B/\sqrt{\mu}} . \tag{13.25}$$

We see that for any choice of the local fields, the unnormalized correlation function depends on the global curvature of the world sheet through a universal exponent

$$\gamma_{\rm str} = 2 - Q = 1 - g, \tag{13.26}$$

which is called the *string susceptibility exponent*. The scaling of the bulk and boundary local fields are determined by their *gravitational anomalous dimensions* δ_i and δ_j^B, related to the exponents α_i and β_j by

$$\delta_i = 1 - \alpha_i, \qquad \delta_i^B = 1 - \beta_i. \tag{13.27}$$

[2] Originally it was believed that the second solution was unphysical, until it was realized that it describes a specially tuned measure over surfaces. We will return to this point later.

The scaling relation (13.25) generalizes in an obvious way if the boundary operators separate different Liouville boundary conditions:

$$B_i \;\to\; {}^{\mu_i^B}[B_{\Delta_i}]^{\mu_{i+1}^B} \,. \tag{13.28}$$

As a consequence, any bulk or boundary correlation function must be of the form

$$\left\langle \prod_j O_j \prod_k {}^{\mu_k^B}[B_k]^{\mu_{k+1}^B} \right\rangle_\mu = \mu^{(1/2)\chi(g+1)+\sum_j(\delta_j-1)} \prod_k (\mu_k^B)^{\delta_k^B-1} f \,, \tag{13.29}$$

where f is some scaling function of the dimensionless ratios $\mu_B^k/\sqrt{\mu}$.

The central charge c_{matter} and conformal weights Δ_i of the matter fields are related to the exponents δ_i and γ_{str}, which characterize the observables in 2D quantum gravity, by a quadratic relation

$$\Delta = \frac{\delta(\delta-\gamma_{\text{str}})}{1-\gamma_{\text{str}}}, \qquad c_{\text{matter}} = 1 - 6\frac{\gamma_{\text{str}}^2}{1-\gamma_{\text{str}}} \qquad (\gamma_{\text{str}} = 1-g)\,, \tag{13.30}$$

and a similar relation for the boundary fields. This relation, which follows from eqns (13.17), (13.19), and (13.26), is known as the KPZ correspondence between the flat and gravitational dimensions of the local fields (Knizhnik *et al.* 1988, David 1988, Distler and Kawai 1989).

The "physical" solution of the quadratic equation (13.19) can be parameterized by a pair of real numbers r and s, not necessarily integers:[3]

$$\Delta_{rs}(g) = \frac{(rg-s)^2-(g-1)^2}{4g}, \qquad \alpha_{rs}(g) = \frac{g+1-|rg-s|}{2}\,. \tag{13.31}$$

The corresponding gravitational anomalous dimension $\delta_{rs} = 1-\alpha_{rs}$ is

$$\delta_{rs}(g) = \frac{|rg-s|-(g-1)}{2}\,. \tag{13.32}$$

We recall that our conventions are such that $g > 1$. The solution (13.31) applies also for the boundary exponents, after the replacements $\Delta \to \Delta^B, \delta \to \delta^B, \alpha \to \beta$.

13.2.4 The "wrong" branch of the Liouville dressing of the identity operator

The quadratic condition for the balance of dimensions (13.19) has a second solution

$$\tilde{\alpha} = Q - \alpha\,, \tag{13.33}$$

and similarly for β. The second solution corresponds to the negative branch of the square root in the Kac parameterization (13.32). For the identity operator, $\Delta = 0$, and the two solutions are $\alpha = 1$ and $\tilde{\alpha} = g$. The bulk and boundary interaction terms in the Liouville action (13.9) correspond to the first solution, which is the one that makes sense in the quasi-classical limit $g \to 0$ or $c_\phi \to \infty$. Generically, this is the action that describes the continuum limit of statistical systems on random graphs. However, there are examples of microscopic theories

[3] When r and s are integers, Δ_{rs} are the conformal weights of the degenerate fields of the matter CFT.

whose continuum limit is described by the "wrong" dressing of the bulk or boundary identity operators. Such theories are defined by the action (13.9) with $e^{2\phi(z,\bar{z})} \to e^{2g\phi(z,\bar{z})}$ and/or $e^{\phi(x)} \to e^{g\phi(x)}$. The modified Liouville theory is related to the original one by a remarkable duality property (Fateev *et al.* 2000). Below, we will discuss the geometrical meaning of this duality.

Microscopic mechanism for the "wrong" bulk Liouville term. In the space of Riemann metrics, there are singular configurations that have nontrivial closed geodesics of vanishing length. The world sheet for such a metric looks like a tree-like cluster of spherical bubbles ("baby universes"), connected to each other by microscopic cylinders as shown in Fig. 13.1 (left). One can control the measure of these configurations by assigning a weight factor to each such "neck." As was shown by Klebanov (1995), the effect of the fine-tuning of this weight is to replace the Liouville potential $\mu e^{2\phi}$ by $\tilde{\mu} e^{2g\phi}$. We will say that a world sheet is *fractal* if it allows baby universes and *smooth* if it does not. Only a smooth world sheet has a good quasi-classical limit.

The intuitive reason for the change of the branch of the gravitational dressing of the identity operator is that the bubbles alter the ultraviolet behavior of the theory. After a gauge corresponding to a flat background metric is chosen, all the bubbles but one are mapped to points and the only effect they can produce is to change the definition of the local operators.

Let $\tilde{F}(\tilde{\mu})$ and $F(\mu)$ be the partition functions of the theories with and without bubbles, respectively, i.e. with fractal and smooth worldsheets. The two partition functions are related by a Legendre transformation (Klebanov and Hashimoto 1995):

$$\tilde{F}(\tilde{\mu}) = F(\mu) + \mu\tilde{\mu}\,. \tag{13.34}$$

Differentiating with respect to μ, one finds

Fig. 13.1 World sheets with baby universes in the bulk (left) and on the boundary (right). The effect of the baby universes is taken into account by the "wrong" bulk and/or boundary Liouville dressing of the identity operator.

$$\tilde{\mu} = -\partial_\mu F(\mu)\,. \tag{13.35}$$

These relations mean that the theory with bubbles is essentially equivalent to the theory without bubbles but with a different cosmological constant $\tilde{\mu}$, equal to the sum of the surfaces with one marked point in the theory without bubbles.

From the scaling of the free energy $F(\mu) \sim \mu^{2-\gamma_{\rm str}} = \mu^{1+g}$, we get $\tilde{\mu} \sim \mu^{g}$. This is indeed the scaling of the bulk cosmological constant for the "wrong" Liouville interaction $e^{2g\phi}$. The corresponding theory of 2D gravity is characterized by a positive string susceptibility exponent:

$$\tilde{F}(\tilde{\mu}) \sim \tilde{\mu}^{1+g^{-1}} = \tilde{\mu}^{2-\tilde{\gamma}_{\rm str}} \quad \Rightarrow \quad \tilde{\gamma}_{\rm str} = 1 - \frac{1}{g} > 0\,. \tag{13.36}$$

Equation (13.35) implies that the susceptibilities $u(\mu) = -\partial_\mu^2 F$ and $\tilde{u}(\tilde{\mu}) = -\partial_{\tilde{\mu}}^2 \tilde{F}$ are reciprocal:

$$\tilde{u}(\tilde{\mu}) = \frac{1}{u(\mu)}\,. \tag{13.37}$$

In a theory with a fractal world sheet, the relation between the Liouville dressing charge and the gravitational dimension is modified because of the anomalous dimension of the cosmological constant: $\alpha = (1-\tilde{\delta})/g$. Comparing this with the relation (13.27) for a theory with a smooth world sheet, we find

$$\tilde{\delta} - 1 = \frac{\delta - 1}{g}\,. \tag{13.38}$$

Extending the parameterization (13.32) to a theory with a fractal world sheet, we have that $\tilde{\delta}_{r,s}(g) = \delta_{s,r}(1/g)$. This is compatible with the symmetry $\Delta_{r,s}(g) = \Delta_{s,r}(1/g)$ of the conformal weights of the matter fields.

Microscopic models for the "wrong" boundary Liouville term. The other type of singularity is associated with the boundary. Such singularities lead to a world sheet that splits into a cluster of "baby universes" having the topology of a disk and connected by microscopic strips, as shown in Fig. 13.1 (right). The Riemann metric for such a world sheet has geodesics of vanishing length that connect pairs of distinct points at the boundary. We will say, in short, that a boundary with such disk baby universes is *fractal*, while a boundary without them is *smooth*. An example of a microscopic theory that can have a fractal boundary is the dense phase of the $O(n)$ loop model coupled to gravity.

Let us denote by $U(\mu_B)$ and $\tilde{U}(\tilde{\mu}_B)$ the disk partition functions with a smooth and a fractal boundary, respectively. The two disk partition functions are related by a Legendre transformation:

$$\tilde{U}(\tilde{\mu}_B) = U(\mu_B) + \mu_B \tilde{\mu}_B\,. \tag{13.39}$$

According to the general scaling relation (13.29) with $\chi = 1$, the disk partition function with a smooth boundary scales as $U(\mu, \mu_B) \sim \mu^{1-\gamma_{\rm str}/2} \sim \mu_B^{1+g}$. Differentiating (13.39) with respect to μ_B, we find that the boundary cosmological constant in the theory with a fractal boundary scales as

$$\tilde{\mu}_B = -\partial_{\mu_B} U \sim \mu_B^g \,. \tag{13.40}$$

This is exactly the scaling that corresponds to the "wrong" dressing $e^{g\phi}$ of the boundary identity operator. In a theory with a smooth bulk but a fractal boundary, the boundary length has an anomalous scaling $\ell \sim A^{g/2}$.

In the following, we will encounter boundary operators which separate smooth and fractal segments of the boundary. In this case, the left and right gravitational dimensions will be different. This is not unnatural, since the gravitational dimension characterizes the fluctuations of the geometry in the presence of a matter field and not directly the matter field itself. The left and right gravitational dimensions δ and $\tilde{\delta}$ of such an operator are related by (Duplantier 2004, 2005)

$$\tilde{\delta}^B - 1 = \frac{\delta^B - 1}{g} \,. \tag{13.41}$$

With $\tilde{\delta}$ defined as above, the KPZ formula (13.30) takes the form[4]

$$\Delta^B = \delta^B \, \tilde{\delta}^B \,. \tag{13.42}$$

In the Kac parameterization (13.32), the gravitational dimensions δ_{rs} and $\tilde{\delta}_{rs}$ are obtained from each other by the replacements $s \leftrightarrow r, \; g \leftrightarrow 1/g$:

$$\tilde{\delta}_{rs}(g) = \delta_{s,r}\left(\frac{1}{g}\right) . \tag{13.43}$$

The flat dimension of matter Δ_{rs} is invariant with respect to this operation, as it should be:

$$\Delta_{rs}(g) = \Delta_{s,r}\left(\frac{1}{g}\right) . \tag{13.44}$$

The duality $g \to 1/g$ appears also in the context of SLE curves (Duplantier 2000). It relates the scaling properties of a strongly fractal SLE curve with $\tilde{\kappa} = 4g$ to its external perimeter, which behaves as a smoother SLE curve with $\kappa = 4/g$.

13.3 Discrete models of 2D gravity

13.3.1 Discretization of the path integral over metrics

The statement that an integral over Riemann metrics can be discretized by a sum over a sufficiently large ensemble of planar graphs has no rigorous proof, but it has passed a multitude of quantitative tests. In order to build models of discrete 2D gravity, it is sufficient to consider the ensemble of trivalent planar graphs which are dual to triangulations, such as the one shown in Fig. 13.2. To avoid ambiguity, we represent the propagators by double lines ('t Hooft 1974). The faces of such a "fat" graph are bounded by the polygons formed by single lines.

[4] To our knowledge, this formula was first written down by B. Duplantier.

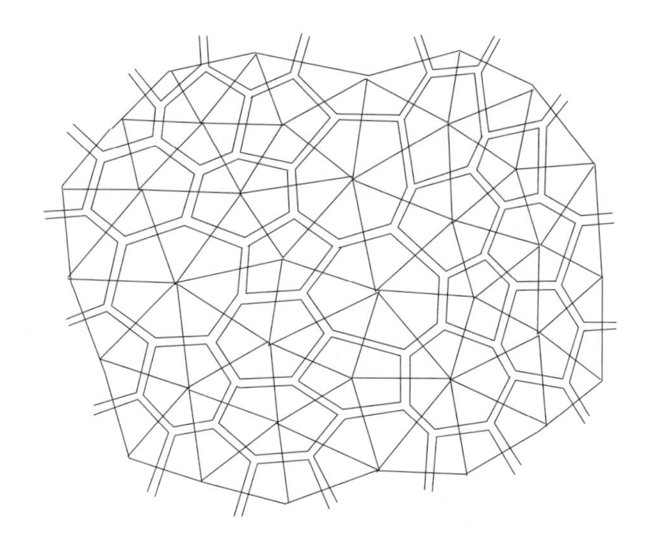

Fig. 13.2 A trivalent fat planar graph Γ and its dual triangulation.

Each fat graph Γ defines a triangulated surface with a natural metric for which all triangles are flat and equilateral and have edges of unit length. With this choice of the metric, the curvature is concentrated at the vertices r of the triangulation. The curvature R_r (or the boundary curvature K_r if the vertex is on the boundary) is expressed through the coordination number c_r, equal to the number of triangles meeting at the point r,

$$R_r = \pi \frac{6 - n_r}{3}, \qquad K_r = \pi \frac{3 - n_r}{3}. \tag{13.45}$$

The total curvature is equal to the Euler number $\chi(\Gamma) = \#\text{vertices} - \#\text{lines} + \#\text{faces}$:

$$\frac{1}{2\pi} \sum_{r \in \text{bulk}} R_r + \frac{1}{\pi} \sum_{r \in \text{boundary}} K_r = \chi(\Gamma). \tag{13.46}$$

For each fat graph Γ, we define

$$|\Gamma| \stackrel{\text{def}}{=} \#\,(\text{faces of }\Gamma)\,, \quad |\partial\Gamma| \stackrel{\text{def}}{=} \#\,(\text{external lines of }\Gamma)\,. \tag{13.47}$$

Then the area and the boundary length of the triangulated surface are given by

$$\begin{aligned} \text{Area} &= \frac{\sqrt{3}}{2} \#(\text{triangles}) = \sqrt{3}\,(|\Gamma| - \chi(\Gamma))\,, \\ \text{Length} &= \#(\text{boundary edges}) = |\partial\Gamma|\,. \end{aligned} \tag{13.48}$$

Assuming that the matter field is a spin model which can be defined on any trivalent fat graph, it is now obvious how to discretize path integrals of the type of eqn (13.2) or (13.4).

13.3.2 Example: Pure discrete gravity

In the simplest model of discrete 2D gravity, there are no matter fields. This model gives a discretization of pure Liouville gravity. Let us denote by $\{\text{Sphere}\}$ and $\{\text{Disk}\}$ the ensembles of the fat graphs dual to triangulations of the sphere and disk, respectively. The path integrals for the partition function for the sphere (13.2) and for the disk (13.4) are discretized as

$$\mathcal{F}(\bar{\mu}) = \sum_{\Gamma \in \{\text{Sphere}\}} \frac{1}{k(\Gamma)} \bar{\mu}^{-|\Gamma|}, \tag{13.49}$$

$$\mathcal{U}(\bar{\mu}, \bar{\mu}_B) = \sum_{\Gamma \in \{\text{Disk}\}} \frac{1}{|\partial\Gamma|} \bar{\mu}^{-|\Gamma|} \bar{\mu}_B^{-|\partial\Gamma|}. \tag{13.50}$$

Here $\bar{\mu}$ and $\bar{\mu}_B$ are the lattice bulk and boundary cosmological constants, respectively, and $k(\Gamma)$ is the volume of the symmetry group of the planar graph Γ. The symmetry factor of the fat graph discretizing the disk is equal to the number $|\partial\Gamma|$ of its external lines.

The matrix model. The evaluation of the partition functions boils down to the problem of counting planar graphs. This problem was solved in a seminal paper by Brezin *et al.* (1978) after being reformulated in terms of a large-N matrix integral,

$$\mathcal{Z}_N(\beta) = \int d\mathbf{M}\, e^{\beta\left(-(1/2)\operatorname{Tr}\mathbf{M}^2 + (1/3)\operatorname{Tr}\mathbf{M}^3\right)}, \tag{13.51}$$

where $d\mathbf{M}$ is the flat measure in the linear space of all Hermitian $N \times N$ matrices. This integral can be considered as the partition function of a zero-dimensional QFT with the Feynman rules given in Fig. 13.3. The integral is divergent but makes sense as a perturbative expansion around the Gaussian measure. The logarithm of the partition function, the "vacuum energy" of the matrix field, is equal to the sum of all connected diagrams.

Up to the symmetry factor $[k(\Gamma)]^{-1}$, the weight of each connected Feynman graph Γ is a product of weight factors associated with its vertices, lines, and faces,

$$\text{Weight}(\Gamma) = \frac{1}{k(\Gamma)} \beta^{\#\text{vertices}-\#\text{lines}} N^{\#\text{faces}} = \frac{1}{k(\Gamma)} \beta^{\chi(\Gamma)} \left(\frac{N}{\beta}\right)^{|\Gamma|}. \tag{13.52}$$

Therefore we can identify the ratio $\bar{\mu} = \beta/N$ with the lattice bulk cosmological constant. In the large-N limit with the ratio β/N fixed,

$$N \to \infty, \qquad \frac{\beta}{N} = \bar{\mu}, \tag{13.53}$$

i j — i j $1/\beta$ i k i j k j β

Fig. 13.3 Feynman rules for the matrix model for pure gravity. The two lines of the double-line propagator symbolize Kronecker symbols. Each closed "index line" contributes a factor of N.

only Feynman graphs dual to triangulations of the sphere ($\chi = 2$) contribute, and their weight will be $\beta^2 \bar{\mu}^{-|\Gamma|}$. As a result, the sphere partition function (13.49) of discrete 2D pure gravity is equal to the vacuum energy of the matrix model in the large-N, or planar, limit

$$\mathcal{F}(\bar{\mu}) = \lim_{N\to\infty} \frac{1}{\beta^2} \log \mathcal{Z}, \quad \bar{\mu} = \frac{\beta}{N}. \tag{13.54}$$

It is advantageous to consider the disk partition function (13.50) for any complex value of the boundary cosmological constant, which we will denote by the letter z. The disk partition function for $\bar{\mu}_B = z$ is given by the following normalized expectation value in the matrix model:

$$\mathcal{U}(\bar{\mu}, z) = -\frac{1}{\beta} \langle \operatorname{Tr} \log(z - \mathbf{M}) \rangle . \tag{13.55}$$

The derivative $\mathcal{W} = -\partial_z \mathcal{U}$, which is the partition function on a disk with a marked point on the boundary, is the expectation value of the resolvent of the random matrix:

$$\mathcal{W}(z) = \frac{1}{\beta} \left\langle \operatorname{Tr} \frac{1}{z - \mathbf{M}} \right\rangle . \tag{13.56}$$

Solution of the one-matrix model. The $U(N)$ invariance allows us to write the partition function (13.51) as the integral

$$\mathcal{Z}_N \sim \int_{\mathbb{R}} \prod_{i=1}^{N} d\lambda_i \, e^{-\beta V(\lambda_i)} \prod_{i<j}^{N} (\lambda_i - \lambda_j)^2, \qquad V(z) = \frac{z^2}{2} - \frac{z^3}{3}, \tag{13.57}$$

where $\lambda_1, \ldots, \lambda_N$ are the eigenvalues of the Hermitian matrix $\mathbf{M}$. In the thermodynamic limit (13.53), the integral is saturated by the equilibrium distribution of the eigenvalues in some interval $[a, b]$ on the real axis. The disk partition function

$$\mathcal{U}(\bar{\mu}, z) = -\frac{1}{\beta} \sum_{j=1}^{N} \log(z - \lambda_j) \tag{13.58}$$

is a classical observable with respect to this distribution. The condition that charges are in equilibrium means that the effective potential of a probe charge placed at the point $z \in \mathbb{C}$,

$$V_{\text{eff}}(z) = V(z) + 2\,\mathcal{U}(z), \tag{13.59}$$

is constant on the interval $[a, b]$. As a consequence, the function

$$\mathcal{G}(z) = \mathcal{W}(z) - \frac{1}{2} V'(z) \tag{13.60}$$

has a vanishing real part for $z \in [a, b]$ and therefore $\mathcal{G}^2(z)$ is analytic everywhere. This condition, together with the asymptotics $-\frac{1}{2}(z - z^2) + (\bar{\mu} z)^{-1}$ at infinity, determines the meromorphic function $\mathcal{G}(z)$ completely:

$$\begin{aligned} \mathcal{G}(z) &= (z + S - 1)\sqrt{(z - S)^2 - R^2}, \\ R^2 &= 2S(1 - S), \quad \frac{1}{2} S(1 - S)(1 - 2S) = \frac{1}{\bar{\mu}}, \end{aligned} \tag{13.61}$$

where we have written $a = S + R$ and $b = S - R$.

The continuum limit. We are interested in the universal critical properties of the ensemble of triangulations in the limit when the size of the typical triangulation diverges. This limit is achieved by tuning the bulk and boundary cosmological constants near their critical values,

$$\bar{\mu} = \bar{\mu}_* + \epsilon^2 \mu, \qquad z = \bar{\mu}_B^* + \epsilon x, \tag{13.62}$$

where ϵ is a small cutoff parameter with dimensions of length and x is proportional to the renormalized boundary cosmological constant μ_B. (We have assumed that the bulk and the boundary are smooth.) The renormalized bulk and boundary cosmological constants μ and x are coupled to the renormalized area A and length ℓ, respectively, of the triangulated surface, defined as

$$A = \epsilon^2 |\Gamma|, \quad \ell = \epsilon |\partial\Gamma|. \tag{13.63}$$

In the continuum limit $\epsilon \to 0$, the universal information is contained in the singular parts of the observables, which scale as fractional powers of ϵ. This is why the continuum limit is also called the scaling limit. In particular, for the partition functions (13.2) and (13.4), we have

$$\begin{aligned} \mathcal{F}(\bar{\mu}) &= \text{regular part} + \epsilon^{2(2-\gamma_{\text{str}})} F(\mu), \\ \mathcal{U}(\bar{\mu}, z) &= \text{regular part} + \epsilon^{2-\gamma_{\text{str}}} U(\mu, x). \end{aligned} \tag{13.64}$$

The powers of ϵ follow from the general scaling formula (13.29) with $\chi = 2$ for the sphere and $\chi = 1$ for the disk.

The partition functions for fixed length and/or area are the inverse Laplace images with respect to x and/or μ. For example,

$$W(\mu, x) = -\partial_x U(\mu, x) = \int_0^\infty d\ell\, e^{-\ell x}\, \tilde{W}(\mu, \ell). \tag{13.65}$$

Now let us take the continuum limit of the solution of the matrix model. First we have to determine the critical values of the lattice couplings. From the explicit form of the disk partition function, eqn (13.61), it is clear that the latter depends on μ only through the parameter $S = (a+b)/2$. The singularities in 2D gravity are typically of third order, and one can show that S is proportional to the second derivative of the free energy of the sphere. This means that the derivative $\partial_{\bar{\mu}} S$ diverges at $\bar{\mu} = \bar{\mu}^*$. Solving the equation $dS/d\bar{\mu} = 0$, we find

$$\bar{\mu}^* = 12\sqrt{3}. \tag{13.66}$$

Furthermore, the critical boundary cosmological constant coincides with the right end of the eigenvalue distribution at $\bar{\mu} = \bar{\mu}^*$:

$$\bar{\mu}_B^* = b(\bar{\mu}^*) = \frac{3+\sqrt{3}}{6}. \tag{13.67}$$

In the limit (13.62), the resolvent behaves as

$$\mathcal{W}(z)\frac{dz}{dx} = \text{regular part} \; + \; \epsilon^{5/2}\, W(x), \qquad z = b^* + \epsilon x, \tag{13.68}$$

with

$$W(x) = (x+M)^{3/2} - \frac{3}{2}M\,(x+M)^{1/2}, \quad M = \sqrt{2\mu}\,. \tag{13.69}$$

We have dropped the overall numerical factor, which does not have a universal meaning. With this normalization of W and μ, the derivative $\partial_\mu W$ is given by

$$\partial_\mu W(x) = -\frac{3}{4}\frac{1}{\sqrt{x+M}}\,. \tag{13.70}$$

The scaling exponent $5/2$ in eqn (13.68) corresponds to $\gamma_{\rm str} = 2 - 5/2 = -1/2$ and $c_{\rm matter} = 0$.

13.3.3 The $O(n)$ loop model coupled to gravity

Definition and loop expansion of the $O(n)$ model on dynamical triangulations. The $O(n)$ loop model (Domany *et al.* 1981) can be defined on any trivalent fat graph Γ, such as the one shown in Fig. 13.2. The local fluctuating variable associated with the vertices $r \in \Gamma$ is an $O(n)$ spin with components $S_1(r), \dots, S_n(r)$, normalized so that ${\rm tr}[S_a(r)S_b(r')] = \delta_{ab}\delta_{rr'}$. Here, "tr" means integration with respect to the positions of the classical spins. The partition function on the graph Γ depends on the temperature T and is defined as

$$\mathcal{Z}_{O(n)}(T;\Gamma) = {\rm tr}\left[\prod_{\langle rr'\rangle\in\Gamma}\left(1 + \frac{1}{T}\sum_a S_a(r)S_a(r')\right)\right], \tag{13.71}$$

where the product runs over all links $\langle rr'\rangle$ of the graph Γ. By expanding the trace as a sum of monomials, the partition function can be written as a sum over all configurations of self-avoiding, mutually avoiding loops that can be drawn on Γ, each counted with a factor of n:

$$Z_{O(n)}(T;\Gamma) = \sum_{\rm loops} T^{-[\text{total length of the loops}]}\, n^{\#\text{loops}}\,. \tag{13.72}$$

The temperature T controls the length of the loops. A sample of a loop configuration is shown in Fig. 13.4. Unlike the original formulation, the loop gas representation (13.72) makes sense also for noninteger n.

If the graph Γ has a boundary, the definition (13.71) corresponds to a *Neumann boundary condition* for the $O(n)$ field, which prescribes that the measure for the spins on the boundary is the same $O(n)$-invariant measure as in the bulk.[5] In terms of the loop gas, the loops in the bulk avoid the boundary just as they avoid the other loops and themselves.

The path integral of 2D gravity with an $O(n)$ matter field is defined by a sum discretized by performing the sum over all trivalent fat graphs of a given topology. It is important that the sum

[5] In the loop gas expansions of the SOS and ADE height models, the loops appear as domain walls separating two adjacent heights. A Neumann boundary condition for the $O(n)$ model corresponds to a Dirichlet (constant-height) boundary condition for a height model.

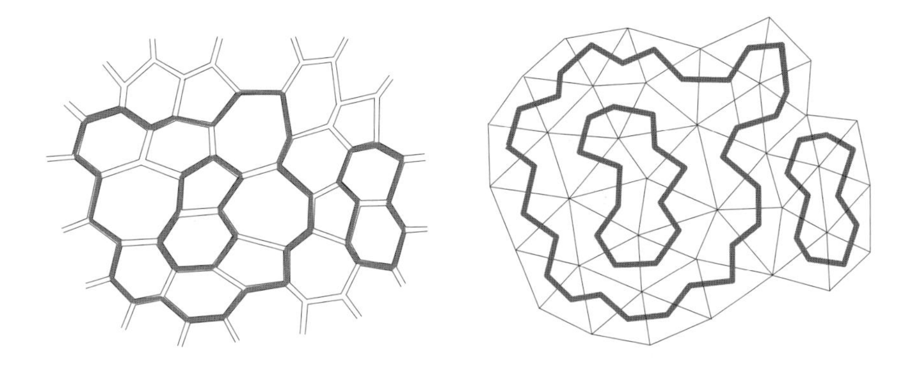

Fig. 13.4 Loops on a trivalent fat planar graph (left) and on its dual triangulation (right).

over the graphs and the integration over the spin configurations are performed independently. The partition functions on the sphere (13.2) and on the disk (13.4) are defined as

$$\mathcal{F}(T,\bar{\mu}) = \sum_{\Gamma\in\{\text{Sphere}\}} \frac{1}{k(\Gamma)}\,\bar{\mu}^{|\Gamma|}\, Z_{O(n)}(\Gamma;T)\,, \tag{13.73}$$

$$\mathcal{U}(T,\bar{\mu},\bar{\mu}_B) = \sum_{\Gamma\in\{\text{Disk}\}} \frac{1}{|\partial\Gamma|}\,\bar{\mu}^{|\Gamma|}\,\bar{\mu}_B^{|\partial\Gamma|}\, Z_{O(n)}(\Gamma;T)\,. \tag{13.74}$$

Continuum limit and critical thermodynamics (qualitative picture predicted by Liouville gravity). The $O(n)$ loop model on a regular hexagonal lattice has a continuous transition in two dimensions for $|n| \leq 2$ (Nienhuis 1982). In this interval, the number of flavors has the standard parameterization

$$n = -2\cos(\pi g)\,. \tag{13.75}$$

Assuming that the $O(n)$ field coupled to gravity exhibits qualitatively the same critical regimes as the $O(n)$ field on a flat lattice, we expect the following phase diagram.

The *dilute phase* of the loop gas appears at a critical temperature $T = T^*$, at which the typical length of the loops diverges. The dilute phase is expected to be described by a Liouville gravity with a central matter charge

$$c_{\text{critical}} = 1 - 6\frac{(g-1)^2}{g}, \qquad g \in [1,2]\,. \tag{13.76}$$

The deviation from the critical point is measured by the temperature coupling $t \sim T - T^*$, which can be positive or negative. For positive t, the loops become massive and the matter field has a finite correlation length. Therefore, for positive t, we have pure gravity with $c_{\text{matter}} = 0$. On the other hand, for negative t, the loops acquire negative mass and grow until they fill all the lattice. This is the *dense phase* of the loop gas, which is also critical and is described by a matter CFT with a lower central charge

$$c_{\text{dense}} = 1 - 6\frac{(\tilde{g}-1)^2}{\tilde{g}}, \qquad \tilde{g} \equiv 2 - g \in [0,1]. \tag{13.77}$$

On a flat lattice, the massless flow off the critical point is generated by the thermal operator $\Phi_{1,3}$ (Nienhuis 1984, Fendley *et al.* 1993) with conformal dimensions

$$\Delta_{1,3} = \bar{\Delta}_{1,3} = \frac{2-g}{g}. \tag{13.78}$$

According to this result, we expect that the continuum limit of the $O(n)$ model on a dynamical lattice is described for finite t by an action of the form

$$\mathcal{A} = \mathcal{A}_{\text{critical}} + tO_{1,3} \qquad \text{with} \qquad O_{1,3} = \int_{\mathcal{R}} \Phi_{1,3}\, e^{2(g-1)\phi}. \tag{13.79}$$

Then the disk partition function for finite t must be equal to the expectation value

$$U(t,\mu,\mu_B) = \left\langle e^{tO_{1,3}} \right\rangle_{\mu,\mu_B} = U(\mu,\mu_B) + t\,\langle O_{1,3}\rangle_{\mu,\mu_B} + \cdots. \tag{13.80}$$

The coefficients in the expansion in t are the multipoint correlation functions of the thermal operator $O_{1,3}$ evaluated at the critical point. The expected scaling behavior of $U(t,\mu,\mu_B)$ can be read off from the gravitational dressing exponent $\alpha_{1,3} = g-1$ of the perturbing operator:

$$U(t,\mu,\mu_B) = \mu^{(g+1)/2} f\left(\frac{\mu_B^2}{\mu}, \frac{t}{\mu^{g-1}}\right). \tag{13.81}$$

After these preliminaries, we can formulate the double scaling limit (in $\bar{\mu}$ and T) that we should take in order to explore the critical thermodynamics of the theory. Together with the renormalized bulk and boundary cosmological constants defined by eqn (13.62), we introduce the renormalized temperature coupling t by

$$T = T^* + \epsilon^{2\theta} t, \tag{13.82}$$

where T^* is the critical temperature of the loop gas on planar graphs, and the exponent θ is determined by the gravitational dimension of the perturbing operator as

$$\theta = 1 - \delta_{1,3} = g - 1. \tag{13.83}$$

To avoid confusion, let us emphasize that although we denote them by the same letter t, the thermal couplings defined by eqns (13.82) and (13.79) are equal only up to a numerical factor, which can be determined by comparing the two-point boundary correlation functions in the discretized and continuous theories.

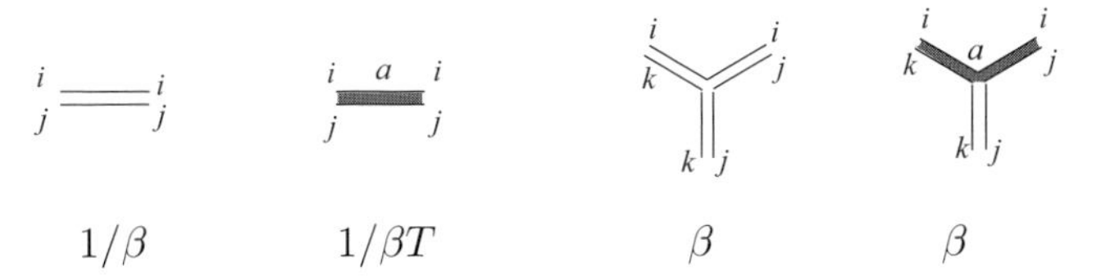

Fig. 13.5 Feynman rules for the $O(n)$ matrix model.

The $O(n)$ matrix model. The $O(n)$ matrix model (Kostov 1989, Gaudin and Kostov 1989) is obtained from the matrix model for pure 2D gravity by introducing n extra $N \times N$ Hermitian matrix variables $\mathbf{Y}_1, \ldots, \mathbf{Y}_n$:

$$\mathcal{Z}_N \sim \int d\mathbf{M} \prod_{a=1}^{n} d\mathbf{Y}_a \, e^{\beta\left(-(1/2)\mathrm{tr}\,\mathbf{M}^2 - (T/2)\sum_{a=1}^{n} \mathbf{Y}_a^2 + (1/3)M^3 + \sum_{a=1}^{n} \mathbf{M}\mathbf{Y}_a^2\right)} . \quad (13.84)$$

The Feynman rules for the $O(n)$ matrix model are shown in Fig. 13.5. Each connected Feynman diagram represents a trivalent fat graph with a set of loops on it. Apart from the factor (13.52), the loops contribute a factor $nT^{-\mathrm{length}}$. The disk partition function (13.74) is again given by eqn (13.55), with the expectation value evaluated in the ensemble (13.84).

It is convenient to shift the matrix variable according to[6]

$$\mathbf{M} = \frac{1}{2}T + \mathbf{X} . \quad (13.85)$$

After this linear change of the variables, the matrix model partition function becomes

$$\mathcal{Z}_N \sim \int d\mathbf{X} \prod_{a=1}^{n} d\mathbf{Y}_a \, e^{\beta\,\mathrm{tr}\,[-V(\mathbf{X}) + \sum_{a=1}^{n} \mathbf{X}\mathbf{Y}_a^2]} , \quad (13.86)$$

where $V(\mathbf{X})$ is a cubic potential with coefficients which depend only on T. We use the $U(N)$ invariance of the measure and the action to diagonalize the matrix $\mathbf{X}$:

$$\mathbf{g}\mathbf{X}\mathbf{g}^{-1} = \{X_1, \ldots, X_N\}, \quad d\mathbf{X} \to \prod_{i=1}^{N} dX_i \prod_{i<j} (X_i - X_j)^2 . \quad (13.87)$$

Then we do the Gaussian integration with respect to the Y-matrices, which leads to the following eigenvalue integral:

$$\mathcal{Z}_N \sim \int_{\mathbb{R}} \prod_{i=1}^{N} dX_i \, e^{-\beta V(X_i)} \prod_{i<j}^{N} (X_i - X_j)^2 \prod_{i,j=1}^{N} (X_i + X_j)^{-n/2} . \quad (13.88)$$

The integral (13.88) can be imagined, as in the case of the one-matrix model, as the partition function of a two-dimensional Coulomb gas of N equal charges constrained on the real

[6] To justify this change of variable, let us mention, anticipating the solution, that $\bar{\mu}_B^* = \frac{1}{2}T$ is the critical value of the boundary cosmological constant in eqn (13.73) at $T = T^*$.

axis and subjected to an external potential $V(X)$. The effect of the $O(n)$ matter field is that each charge is attracted, with force n, by the mirror images of the other charges.

In the limit $N, \beta \to \infty$ with $\beta/N = \bar{\mu}$, the integral (13.88) is saturated by the equilibrium distribution of the eigenvalues. A smooth distribution is possible only if β grows sufficiently fast with N, so that the potential retains the charges to the left of the origin. Therefore the charges must be distributed in some compact interval $[a, b]$ with $b < 0$. When the right edge b of the distribution hits the origin, the integral becomes singular and no sensible large-N limit exists. Therefore $b(\bar{\mu}, T) = 0$ gives a critical line in the space of the couplings $\bar{\mu}$ and T.

Another condition for having a smooth large-N limit is that $n \leq 2$, which justifies the parameterization (13.75). If $n < -2$, the charges are pushed to $-\infty$. If n is larger than 2, then the attraction prevails and all the charges collapse at the origin. In the case $n = 2$, the repulsion and the attraction nearly compensate each other, which leads to logarithmic critical behavior.

In the large-N limit, the disk partition function becomes a classical observable with respect to the equilibrium distribution of the charges,

$$\mathcal{U}(X) = -\frac{1}{\beta} \sum_{j=1}^{N} \log(X - X_j)\,. \tag{13.89}$$

The effective potential $V_{\text{eff}}(X)$ for a probe charge at the point $X \in \mathbb{C}$ is expressed through this observable as

$$V_{\text{eff}}(X) \equiv V(X) - \frac{2}{\beta} \sum_{i=1}^{N} \log|X - X_j| - \frac{n}{\beta} \sum_{i=1}^{N} \log(X + X_j) \tag{13.90}$$

$$= V(X) + 2\mathcal{U}(X) - n\mathcal{U}(-X)\,. \tag{13.91}$$

The condition of equilibrium means that $V_{\text{eff}}(X)$ must be constant on the support of the eigenvalue density:

$$V_{\text{eff}}(X) \equiv V(X) + 2\text{Re}\,\mathcal{U}(X) - n\mathcal{U}(-X) = \text{const}, \qquad X \in [a, b]\,. \tag{13.92}$$

This condition, which is valid only on the eigenvalue interval, can be reformulated as a bilinear functional identity valid on the whole complex plane, written for the meromorphic function

$$\mathcal{G}(X) = -\frac{\partial \mathcal{U}}{\partial X} - \frac{2V'(X) - nV'(-X)}{4 - n^2}\,, \tag{13.93}$$

which represents the singular part of the resolvent

$$\mathcal{W}(X) = -\frac{\partial \mathcal{U}}{\partial X} = \frac{1}{\beta} \left\langle \text{Tr} \frac{1}{X - \mathbf{X}} \right\rangle. \tag{13.94}$$

The bilinear identity in question reads (Kostov 1992)

$$\mathcal{G}^2(X) + \mathcal{G}^2(-X) + n\,\mathcal{G}(X)\mathcal{G}(-X) = A + BX^2 + CX^4\,. \tag{13.95}$$

Proof By its definition (13.93), the function $\mathcal{G}(X)$ has no singularities other than the cut $[a, b]$ on the real axis and behaves at infinity as a quadratic polynomial. According to eqn (13.92), its values on the two sides of the cut are related by

$$\mathcal{G}(X + i0) + \mathcal{G}(X - i0) + n\mathcal{G}(-X) = 0, \qquad X \in [a, b]\,. \tag{13.96}$$

We multiply both sides of eqn (13.96) by $\mathcal{G}(X + i0) - \mathcal{G}(X - i0)$ and symmetrize with respect to $X \to -X$. The left-hand side of the resulting equation coincides with the discontinuity in the left-hand side of eqn (13.95) across the real axis, which therefore vanishes in the intervals $[a, b]$ and $[-b, -a]$. Since the function in question is regular outside these intervals, it must be analytic everywhere. Since $\mathcal{G}(X)$ grows at infinity not faster than X^2, the right-hand side of eqn (13.95) must be quadratic polynomial in X^2. The coefficients A, B, C are evaluated from the large-X expansion

$$\mathcal{G}(X) = -\frac{2V'(X) - nV'(-X)}{4 - n^2} + \frac{\bar{\mu}^{-1}}{X} + \frac{\mathcal{W}_1(T, \bar{\mu})}{X^2} + o(X^{-3})\,. \tag{13.97}$$

□

In Section 13.A.1, we give an alternative derivation of eqn (13.95) from the Ward identities for the $O(n)$ matrix model.

Before we proceed with the solution in the scaling limit, let us make an important observation concerning the derivative

$$\mathcal{H}(X) = -\bar{\mu}^2 \partial_{\bar{\mu}} \mathcal{G}(X)\,. \tag{13.98}$$

The function $\mathcal{H}(X)$ satisfies the same boundary condition on the cut $[a, b]$ as does $\mathcal{G}(X)$, eqn (13.96), but has different asymptotics at infinity. Differentiating both sides of eqn (13.97), we find

$$\mathcal{H}(X) = \frac{1}{X} + o(X^{-2}) \qquad (X \to \infty)\,. \tag{13.99}$$

Therefore $\mathcal{H}(X)$ depends on the coupling constants T and $\bar{\mu}$ only through the positions a and b of the branch points.

Finally, let us remark, although this will not be used in these lectures, that the partition function of the $O(n)$ matrix model can be represented in terms of a chiral Gaussian field defined on the complex plane z. The expectation value of this field is the disk partition function of the $O(n)$ model. This representation allows one to apply some of the conformal-field-theory techniques developed for other matrix models (Kostov 2004).

Exercise 13.1 Represent the partition function $\mathcal{Z}_{N_c}$ of the $O(n)$ matrix model as an expectation value $\langle N_c|\mathcal{O}|0\rangle$ in the Fock space of a chiral Gaussian field

$$\varphi(z) = q - ip \ln z - i \sum_{n \neq 0} \frac{J_n}{n} z^{-n}, \quad [J_n, J_m] = n\delta_{n+m,0}, \quad [p, q] = 1\,,$$

with the left vacuum states (of charge N_c) and right vacuum states (of charge 0) defined by $J_n|0\rangle = 0$, $\langle N|J_{-n} = 0,\ n > 0,\ p|0\rangle = 0,\ \langle N|p = N\langle N|$.

Hint: Show that the vertex operator

$$E(z) =: e^{i\Gamma(z)} : , \qquad \Gamma(z) = e_+\varphi(z) + e_-\varphi(-z) ,$$

with $e_+^2 + e_-^2 = 2$, $2e_+e_- = -n$, satisfies the OPE

$$E(z)E(z') = (z - z')^2(z + z')^{-n} : E(z)E(z') : .$$

Solution for the disk partition function in the scaling limit. It is possible to evaluate the resolvent $\mathcal{W}(X)$ exactly (Eynard and Kristjansen 1995, 1996). The exact result, which is expressed in terms of Jacobi theta functions, is somewhat involved and contains nonuniversal information that we do not need. Below, we will evaluate directly the singular parts of the partition functions on the disk and on the sphere. They are given by eqn (13.64) with $\gamma_{\rm str} = 1 - g$.

We introduce the cutoff parameter ϵ and define the renormalized coupling constants

$$T - T^* \sim \epsilon^{2\theta} t , \qquad \frac{1}{\bar{\mu}^*} - \frac{1}{\bar{\mu}} \sim \epsilon^2 \mu , \qquad X \sim \epsilon x . \tag{13.100}$$

The value of T^* will be determined in a moment. The value of $\bar{\mu}^*$ cannot be determined without the information about the behavior of the resolvent at infinity which was lost in the scaling limit, but we will not actually need it, since it only fixes the normalization of the solution, which is not universal.

Now we write the bilinear equation (13.95) in terms of the scaling resolvent $W(x) = -\partial_x U(x)$, which is related to the function $\mathcal{G}(X)$ by

$$\mathcal{G}(X) \sim \epsilon^g W(x) . \tag{13.101}$$

The first simplification occurs because the third term on the right-hand side then contains a positive power of ϵ and therefore can be neglected.[7] Therefore the equation to be solved is

$$W^2(x) + W^2(-x) + nW(x)W(-x) = \epsilon^{-2g} A + \epsilon^{2-2g} Bx^2 . \tag{13.102}$$

Since the left-hand side does not depend on ϵ, the constants on the right-hand side should scale as $A \sim \epsilon^{2g}$ and $B \sim \epsilon^{2g-2}$. In particular, B must vanish at $T = T^*$, which determines the critical temperature

$$T^* = 1 + \sqrt{\frac{2-n}{6+n}} . \tag{13.103}$$

It follows from the explicit expression of B that B is proportional to $T - T^* \sim \epsilon^{2\theta} t$ with a positive coefficient. The matching of powers of ϵ gives the expected value $\theta = g - 1$ of the exponent in eqn (13.100). We choose the normalization of t so that

$$B = \epsilon^{2\theta} t . \tag{13.104}$$

[7] For this to be possible, we must assume that $1 < g < 2$. At $g = 2$, when this term becomes relevant, the resolvent exhibits logarithmic critical behavior (Kostov and Staudacher 1992).

The second simplification comes from the fact that in the limit $\epsilon \to 0$, the right edge of the eigenvalue distribution remains at a finite distance M from the origin, while the left edge is sent to $-\infty$:

$$[a, b] \to [-\infty, -M] \qquad (\epsilon \to 0)\,. \tag{13.105}$$

In this limit, the branch points of the Riemann surface of $W(x)$ can be resolved by a hyperbolic map

$$x = M \cosh \tau\,. \tag{13.106}$$

The two sides of the cut $[-\infty < x < -M]$ are parameterized by $\tau \pm i\pi,\ \tau > 0$.

Using the explicit form we have found for B, we write the bilinear equation (13.102) in the τ-parameterization:

$$W^2(\tau + i\pi) + W^2(\tau) + nW(\tau + i\pi)W(\tau) = \tilde{A} + tM^2 \cosh^2 \tau\,, \tag{13.107}$$

where the first term $\tilde{A} = \epsilon^{-2g} A$ on the right-hand side depends only on μ and t. The unique solution of this equation is, up to an irrelevant normalization factor,

$$W(x) = M^g \cosh g\tau \; + gtM^{2-g} \cosh(2 - g)\tau\,. \tag{13.108}$$

The value of M is determined by $(M^g + tgM^{2-g})^2 = 4g\tilde{A}$, but we cannot use this information because the constant A is unknown. Instead, we will find the function $M(\mu, t)$ using the following trick (Kostov 2006). Remember that the derivative $H = \partial_\mu W(x)$ depends on μ and τ only through the position of the branch point given by M. By imposing this constraint on the solution (13.108), we will obtain the information we need. So we take the derivative of the solution with respect to μ at fixed x,

$$\partial_\mu W|_x = \partial_\mu M \left(\partial_M - \frac{1}{M \tanh \tau} \partial_\tau \right) G(x) \tag{13.109}$$

$$= -\partial_\mu M \left(gM^{g-1} - gt(2 - g)M^{1-g} \right) \frac{\sinh(g - 1)\tau}{\sinh \tau}\,. \tag{13.110}$$

The factor in front of the hyperbolic function must be proportional to M^{g-2}. We choose the coefficient for later convenience as

$$\partial_\mu W|_x = -\frac{g}{2}\, M^{g-2}\, \frac{\sinh(g - 1)\tau}{\sinh \tau}\,. \tag{13.111}$$

In this way, the self-consistency of the solution leads to an equation for the derivative $\partial_\mu M$:

$$\partial_\mu M \left(M^{g-1} - t(2 - g)M^{1-g} \right) = \frac{1}{2} M^{g-2}\,. \tag{13.112}$$

Integrating with respect to μ, we find a transcendental identity that determines M as a function of μ and t:

$$\mu = M^2 - \; tM^{2(2-g)}\,. \tag{13.113}$$

To summarize, we have obtained the following parametric representation for the derivatives of the disk partition function in the continuum limit $U(x)$:

$$\begin{aligned} -\partial_x U|_\mu &= M^g \cosh(g\tau) + gtM^{2-g} \cosh(2-g)\tau\,, \\ -\partial_\mu U|_x &= \frac{g}{2(g-1)} M^{g-1} \cosh(g-1)\tau\,, \\ x &= M \cosh\tau\,, \end{aligned} \tag{13.114}$$

where the function $M(\mu, t)$ is determined from the transcendental equation (13.113). An expression for $\partial_\mu U$ was obtained by integrating eqn (13.111).

Equation of state and partition function of the loop gas on the sphere. The function $M(t,\mu)$ plays a central role in the solution. Its physical meaning can be revealed by taking the limit $x \to \infty$ of the observable $\partial_\mu U(x)$, which is the partition function for a disk with a puncture. Since x is coupled to the length of the boundary, in the limit of large x the boundary shrinks to a point and the result is the partition function of a sphere with two punctures, or the susceptibility $\partial_\mu^2 \mathcal{F}$. Expanding at $x \to \infty$, we find

$$-\partial_\mu U \sim x^{g-1} - M^{2g-2}\, x^{1-g} + \text{lower powers} \tag{13.115}$$

(the numerical coefficients have been omitted). We conclude that the susceptibility is given, up to a normalization, by

$$u \equiv -\partial_\mu^2 F = M^{2g-2}\,. \tag{13.116}$$

The normalization of u can be absorbed into the definition of the string coupling constant $g_s \sim 1/\beta$. Thus the transcendental equation (13.113) for M gives the equation of state of the loop gas on the sphere,

$$u^{1/(g-1)} - tu^{(2-g)/(g-1)} = \mu\,. \tag{13.117}$$

Using this equation, one can evaluate the partition function on the sphere as a Taylor series in t,

$$\begin{aligned} F(t,\mu) &= (g-1)\mu^{g+1} \sum_{n=0}^{\infty} \frac{\Gamma[g(n-1)-n-1]\left(-t\mu^{1-g}\right)^n}{\Gamma(n+1)\Gamma[(g-2)(n-1)]} \\ &= -\frac{1}{g(g+1)}\mu^{g+1} - \frac{g-1}{2} t\mu^2 + \frac{(g-1)}{2(g-3)} t^2 \mu^{3-g} + \cdots\,. \end{aligned} \tag{13.118}$$

We give an instructive calculation (Zamolodchikov 2005) in Section 13.A.2. This expansion, whose coefficients are the n-point correlation functions of the thermal operator $O_{1,3}$ in Liouville gravity, was analyzed by Belavin and Zamolodchikov (2005), who showed agreement with Liouville gravity up to $o(t^5)$.

The equation of state (13.117) confirms the expected phase diagram described earlier in this section. The three singular points of the equation of state reproduce the expressions for the susceptibility at the three conformal points, where the theory decouples into a Liouville and a matter CFT. For $t = 0$, the second term vanishes and we are in the dilute phase with

$$\gamma_{\rm str} = 1 - g \qquad \text{(dilute phase)}\,. \tag{13.119}$$

For $t \to -\infty$, one can neglect the first term, and $u \sim \mu^{(g-1)/(2-g)}$. The exponent $\gamma_{\rm str}$ is that for the dense phase,

$$\gamma_{\rm str} = -\frac{g-1}{2-g} - 1 - \frac{1}{(2-g)} \qquad \text{(dense phase)}\,. \tag{13.120}$$

In the dense phase, we have an effective Liouville gravity action with coupling constant

$$g' = \frac{1}{2-g} > 1\,. \tag{13.121}$$

Finally, for t positive there exists a critical value t_c where χ develops a square-root singularity. This is the critical point of pure gravity ($\gamma_{\rm str} = -1/2$).

The disk partition function for fixed length. It is useful to have the solution (13.114) written in the ℓ-representation, where we fix the lengths ℓ of the boundaries instead of the boundary cosmological constants x. The x- and ℓ-representations are related by

$$W(x) = \int_0^\infty d\ell\, e^{-\ell x}\, \tilde{W}(\ell)\,, \quad \tilde{W}(\ell) = \frac{1}{2\pi i}\int_{i\mathbb{R}} dz\, e^{\ell x}\, W(x)\,. \tag{13.122}$$

The disk partition function for fixed boundary length is

$$\tilde{W}(\ell) = \frac{\sin(\pi(g-1))}{\pi}\left(g\frac{K_g(M\ell)}{\ell} - (2-g)\,t\frac{K_{2-g}(M\ell)}{\ell}\right). \tag{13.123}$$

It is positive both in the dilute ($t = 0$) and in the dense ($t \to -\infty$) phase. The solution has power-like behavior in the range $\epsilon \ll \ell \ll 1/M$:

$$\tilde{W}(\ell) \sim \ell^{-g-1} + t\,\ell^{-(2-g)-1}\,. \tag{13.124}$$

For large ℓ, the disk partition function decays exponentially as

$$\tilde{W}(\ell) \sim e^{-M\ell} \qquad (M\ell \gg 1)\,. \tag{13.125}$$

From this, we see that the position of the right edge of the eigenvalue distribution, $M = M(t,\mu)$, has the significance of an effective "mass" of the loops induced by the fluctuations of the bulk theory. According to eqn (13.113), in the dilute phase, $M \sim \sqrt{\mu}$ and the length ℓ scales as the square root of the area,

$$\ell \sim A^{1/2} \qquad \text{(dilute phase)}\,. \tag{13.126}$$

In the world sheet theory, the boundary Liouville term is e^{ϕ} as in eqn (13.9).

In the dense phase, however, $M \sim \mu^{1/(2(2-g))}$, which means that the length has an anomalous dimension,

$$\ell \sim A^{1/(2(2-g))} \qquad \text{(dense phase)} . \tag{13.127}$$

Therefore the dense $O(n)$ model gives an example of a theory with a fractal boundary in the sense discussed in Section 13.2.4. The boundary Liouville term in the corresponding world sheet theory, $e^{g'\phi}$ with $g' = 1/(2-g)$, corresponds to the "wrong" dressing of the boundary identity operator in Liouville gravity with the coupling g'.

Note that the property of the boundary being smooth or fractal is related only to the measure over the world sheet metrics and not to the matter field. We will see later that the same (Neumann) boundary condition for the matter field can be realized with both smooth and fractal boundaries.

13.4 Boundary correlation functions

The basic local operators in the $O(n)$ model are the operators S_L obtained by insertion of L spins at a point r of the graph,

$$S_L(r) \sim S_{a_1}(r) S_{a_2}(r) \ldots S_{a_L}(r) . \tag{13.128}$$

We will assume that all the indices are different. The point r can be either in the bulk or on the boundary. In the second case, S_L is a boundary L-spin operator.

The correlation functions of the $O(n)$ spins have a nice geometrical interpretation in terms of the loop gas expansion. For example, the two-point function of the operator S_L may be evaluated by means of the partition function of the loop gas in the presence of L open lines connecting the points r and r'. The open lines are self-avoiding and mutually avoiding and are not allowed to intersect the loops. In general, the loop gas expansion of any correlation function of the operators (13.128) is obtained by assuming that the operator $S_L(r)$ creates L open lines starting at the point r and having flavor indices $a_1, \ldots, a_n$. This is why the operator S_L is sometimes called an L-leg operator. Unlike the two-point function, the correlation functions of more than two operators depend on the way the flavor indices are contracted. Each type of contraction leads to a loop gas in the presence of a planar polymer network with a given topology.

In the $O(n)$ matrix model, the bulk and boundary L-spin operators are constructed from the $N \times N$ matrices

$$\mathbf{S}_L \sim \mathbf{Y}_{a_1} \mathbf{Y}_{a_2} \ldots \mathbf{Y}_{a_L} . \tag{13.129}$$

The bulk operator S_L is represented by the trace $\mathrm{Tr}(\mathbf{S}_L)$, while the boundary operator S_L is represented by the matrix $\mathbf{S}_L$ with open indices.

13.4.1 Bulk L-spin operators

As a first demonstration of the power of the KPZ scaling formula, we will evaluate the dimensions of the bulk L-spin operators in the dilute and dense phases of the loop gas. Consider the (unnormalized) correlation function $\mathcal{G}_L$ of two bulk L-spin operators on a sphere. The only possible contraction of the flavor indices gives the partition function of the loop gas in the presence of L nonintersecting open lines, as shown in Fig. 13.6. In the matrix model, $\mathcal{G}_L$ is the correlation function of two traces,

$$\mathcal{G}_L \sim \frac{1}{\beta} \langle \operatorname{Tr}(\mathbf{S}_L)\operatorname{Tr}(\mathbf{S}_L) \rangle . \tag{13.130}$$

It follows from the Ward identities for the matrix model (Section 13.A.1) that the two-point function $\mathcal{G}_L$ can be expressed as a convolution-type integral of the product of L disk partition functions. We write this representation directly for the singular part G_L, defined by

$$\mathcal{G}_L(T, \bar{\mu}) = \text{regular part } + \epsilon^{2(1+g)-2(1-\delta_{S_L})} G_L(t, \mu) . \tag{13.131}$$

Let $W(\ell)$ be the inverse Laplace image of $W(x)$ given in the continuum limit by eqn (13.123). Then

$$G_L(\mu, t) = \int_0^\infty d\ell_1\, d\ell_2 \ldots d\ell_L\, W(\ell_1 + \ell_2) W(\ell_2 + \ell_3) \ldots W(\ell_L + \ell_1) . \tag{13.132}$$

This integral representation has the following geometrical interpretation (Duplantier and Kostov 1988, Kostov 1989, Duplantier and Kostov 1990): the right-hand side is obtained by cutting the world sheet along the L open lines and using the fact that the measures on the two sides of the cut are independent.

Let us evaluate the gravitational dimensions of the L-spin operators from the scaling of G_L in the dilute and the dense phase.

Dilute phase ($t = 0$). The dilute phase is described by Liouville gravity with a coupling g, so that $\gamma_{\text{str}} = 1 - g$. According to eqn (13.124), $W(\ell) \sim \ell^{-g-1}$ and the integral scales as

$$\mathcal{G}_L \sim M^{Lg} \sim \mu^{Lg/2} = \mu^{2-\gamma_{\text{str}}-2(1-\delta_L)} . \tag{13.133}$$

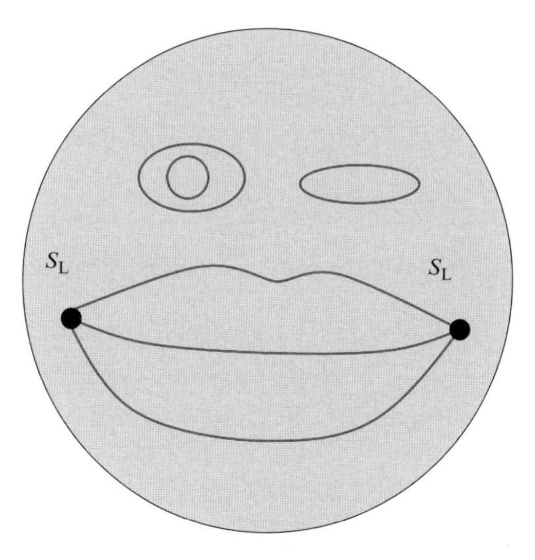

Fig. 13.6 Loop expansion for the bulk two-point function of the L-leg operator ($L = 3$).

Hence the dimension of the L-leg operator is

$$\delta_{S_L} = \frac{1}{4}Lg - \frac{1}{2}(g-1) = \delta_{L/2,0}(g) \qquad \text{(dilute phase)}. \tag{13.134}$$

Dense phase $(t \to -\infty)$. The dense phase is described by Liouville gravity with a coupling $1/\tilde{g}$ with $\tilde{g} = 2 - g$, so that $\gamma_{\rm str} = 1 - \tilde{g}^{-1}$. Here $W(\ell) \sim \ell^{-\tilde{g}-1}$, and the integral over the L lengths gives

$$\mathcal{G}_L \sim M^{L\tilde{g}} \sim \mu^{L/2} \sim \mu^{2-\gamma_{\rm str}-2(1-\delta_L)} = \mu^{\tilde{g}^{-1}-1+2\delta_L}, \tag{13.135}$$

$$\delta_{S_L} = \frac{1}{4}L - \frac{1}{2}\left(\tilde{g}^{-1} - 1\right) = \delta_{0,-L/2}\left(\frac{1}{\tilde{g}}\right) \qquad \text{(dense phase)}. \tag{13.136}$$

These expressions give, by the KPZ scaling formula (13.31), the conformal weights of the L-leg operators on a flat lattice (Saleur 1986, Duplantier and Saleur 1986):

$$\Delta_{S_L}(S_N) = \begin{cases} \Delta_{L/2,0} = \dfrac{L^2g^2 - 4(g-1)^2}{16g} & \text{dilute phase}, \\[2ex] \Delta_{0,L/2} = \dfrac{L^2 - 4(\tilde{g}-1)^2}{16\,\tilde{g}} & \text{dense phase}. \end{cases} \tag{13.137}$$

13.4.2 Boundary L-spin operators with Neumann/Neumann boundary conditions

The simplest boundary correlation functions in the $O(n)$ model are those of the L-spin operators with Neumann/Neumann (N/N) boundary conditions. The Liouville boundary conditions on the two sides are given by the values of the boundary cosmological constants $x_1 = M\cosh\tau_1$ and $x_2 = \cosh\tau_2$. We denote an operator S_L with such boundary conditions by ${}^{x_1}_{\rm N}[S^\dagger_L]^{x_2}_{\rm N}$. The correlation function of two such operators,

$$D_L(x_1, x_2) = \langle\, {}^{x_1}_{\rm N}[S_L]^{x_2}_{\rm N}[S_L]^{x_1}_{\rm N}\,\rangle, \tag{13.138}$$

is the partition function of a disk with L open lines connecting two points on the boundary, as shown in Fig. 13.7.

The function $D_L(x_1, x_2)$ is obtained as the scaling limit of the following expectation value in the matrix model:

$$\mathcal{D}_L(X_1, X_2) \stackrel{\rm def}{=} \frac{1}{\beta}\left\langle \mathrm{Tr}\left(\frac{1}{X_1 - \mathbf{X}}\mathbf{S}_L\frac{1}{X_2 - \mathbf{X}}\mathbf{S}^\dagger_L\right)\right\rangle. \tag{13.139}$$

The role of the operators $(X_1 - \mathbf{X})^{-1}$ and $(X_2 - \mathbf{X})^{-1}$ is to create two segments of the boundary with boundary cosmological constants X_1 and X_2, respectively. The two insertions $\mathbf{S}_L$ and $\mathbf{S}^\dagger_L$ generate L open lines at the points separating the two segments.

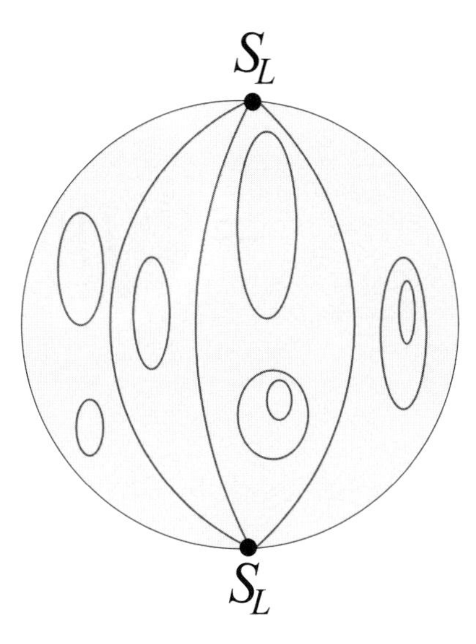

Fig. 13.7 Loop gas expansion for the boundary two-point function of the L-spin operator.

It is useful to extend this definition to the case $L = 0$, assuming that S_0 is the boundary identity operator. In this simplest case, the expectation value (13.139) is evaluated instantly as

$$\mathcal{D}_0(X_1, X_2) = \frac{1}{\beta}\left\langle \operatorname{Tr}\left(\frac{1}{X_1 - \mathbf{X}}\frac{1}{X_2 - \mathbf{X}}\right)\right\rangle = \frac{\mathcal{W}(X_2) - \mathcal{W}(X_1)}{X_1 - X_2}. \quad (13.140)$$

The scaling limit of this correlation function,

$$D_0(x_1, x_2) = \frac{W(x_2) - W(x_1)}{x_1 - x_2}, \quad (13.141)$$

is the double Laplace transform of the two-point function of the boundary identity operator in ℓ-space. The latter is given by the amplitude for a disk with two marked points, separating segments with lengths ℓ_1 and ℓ_2:

$$\mathcal{D}_0(x_1, x_2) = \int_0^\infty d\ell \, e^{-x_1\ell_1 - x_2\ell_2} \, \tilde{W}(\ell_1 + \ell_2). \quad (13.142)$$

The two-point function $\mathcal{D}_L$ for $L \geq 1$ can be computed from a recurrence relation between $\mathcal{D}_L$ and $\mathcal{D}_{L-1}$. Here we give an intuitive geometrical derivation of this relation, which can also be rigorously derived from the matrix model (Section 13.A.1). We write the recurrence relation directly in the continuum limit.

We assume that $L \geq 1$ and express $D_L(x_1, x_2)$ as the Laplace integral

$$D_L(x_1, x_2) = \int_0^\infty d\ell_1 \, d\ell_2 \, e^{-x_1\ell_1 - x_2\ell_2} \, \tilde{D}_L(\ell_1, \ell_2). \quad (13.143)$$

We cut the world sheet along the first of the L open lines and use the fact that, for a given length ℓ of the line, the measure factorizes to a product of $W(\ell_1 + \ell)$ and $D_{L-1}(\ell, \ell_2)$, as

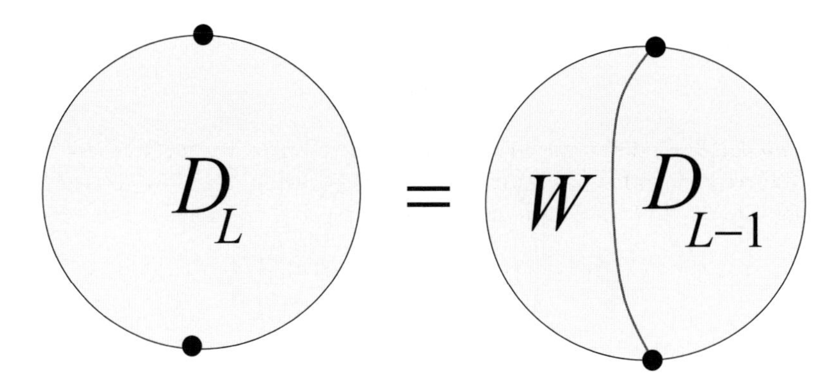

Fig. 13.8 The recurrence equation (13.148) for D_L.

shown in Fig. 13.8. Then $\tilde{D}_L(\ell_1, \ell_2)$ is equal to the integral of the product with respect to the length of the first line:

$$\tilde{D}_L(\ell_1, \ell_2) = \int_0^\infty d\ell\, \tilde{W}(\ell_1 + \ell)\, \tilde{\mathcal{D}}_{L-1}(\ell, \ell_2)\,. \tag{13.144}$$

After Laplace transformation, this relation takes the form

$$\tilde{D}_L(x_1, x_2) = \oint_{\mathcal{C}_-} \frac{dx'}{2\pi i} \frac{W(x') - W(x)}{x - x'} D_{L-1}(x')\,, \tag{13.145}$$

where the contour $\mathcal{C}_-$ goes along the imaginary axis and closes at $-\infty$. Since the only singularity of the integrand in the half-plane $\mathrm{Re}\, x < 0$ is the cut $[-\infty, -M]$, the contour C_- can be deformed so that the integral goes along this cut.[8]

Since such type of integrals will reappear many times, we introduce a compact notation for them. For any pair of meromorphic functions analytic in the right half-plane $\mathrm{Re}(x) \geq 0$ and vanishing at infinity, we define the star-product

$$[f \,\mathrm{star}\, g](x) \stackrel{\mathrm{def}}{=} -\oint_{\mathcal{C}_-} \frac{dx'}{2\pi i} \frac{f(x) - f(x')}{x - x'} g(-x')\,, \tag{13.146}$$

where the contour $\mathcal{C}_-$ encircles the left half-plane $\mathrm{Re}\, x < 0$. The star-product possesses the important property

$$[f \,\mathrm{star}\, g](x) + [g \,\mathrm{star}\, f](-x) = f(x) g(-x)\,, \tag{13.147}$$

which can be proved by deforming the contour of integration.

[8]The contour integral (13.145) is defined unambiguously only before taking the scaling limit, when the integrand vanishes sufficiently fast at infinity In the continuum limit, the amplitudes grow when $x \to \infty$ and the integral must be regularized. One regularization prescription which does not spoil the analytical properties of the integrand consists in taking the analytic continuation in g and evaluating the integral for the domain where it converges. In this way, we can also treat the singular behavior of the integral (13.145) at $\ell \to 0$.

The recurrence relation (13.145) or, in terms of the star-product,

$$D_L = W \,\mathrm{star}\, D_{L-1}\,, \tag{13.148}$$

can be reformulated as a difference equation for the uniformization variable τ defined by eqn (13.106). To derive the difference equation, we evaluate the discontinuity in the two sides of eqn (13.148) for $x < -M$ and then use the map $x = M\cosh\tau$. The result is

$$D_L(\tau_1 + i\pi, \tau_2) - D_L(\tau_1 - i\pi, \tau_2) = [W(\tau_1 + i\pi) - W(\tau_1 - i\pi)]D_{L-1}(\tau_1, \tau_2)\,, \tag{13.149}$$

or, in operator form,

$$D_L = \frac{1}{\sin \pi\partial_\tau} \cdot ((\sin \pi\partial_\tau W) \cdot D_{L-1})\,. \tag{13.150}$$

The right-hand side can be considered as the definition of the star-product in τ-space. It is defined up to a periodic function in τ, which represents an analytic function in the x-space. Since only the singular part of the integral carries information about the universal behavior, this ambiguity is irrelevant in the continuum limit. At the critical point $t = 0$, the difference equation (13.149) is identical to the one obtained for boundary Liouville theory by Fateev *et al.* (2000) using the operator product expansion with the lowest degenerate boundary Liouville field.

Let us now determine the boundary scaling dimension of the operator ${}^{x_1}_{\mathrm{N}}[S_L]^{x_2}_{\mathrm{N}}$, which is related by the KPZ formula to the boundary gravitational dimension δ^B_L. The exponent δ^B_L can be evaluated from the general scaling relation (13.29) once we know the scaling behavior of the two-point function. At the critical point,

$$D_L \sim x_1^{(L+1)g-1} \sim \mu^{(g+1)/2} x_1^{2\delta^B_L - 2} \sim x_1^{g-1+2\delta_L}\,, \tag{13.151}$$

$$\delta^B_L = \frac{1}{2} gL = \delta_{L+1,1}(g) \qquad (\mathrm{N/N,\ dilute\ phase})\,. \tag{13.152}$$

In the dense phase,

$$D_L \sim x_1^{(L+1)\tilde{g}-1} \sim \mu^{(\tilde{g}^{-1}+1)/2} x_1^{2\delta^B_L - 2} \sim x_1^{\tilde{g}-1+2\delta_L}\,, \tag{13.153}$$

$$\delta^B_L = \frac{1}{2} L\tilde{g} = \delta_{L+1,1}(\tilde{g}) = \delta_{1,-L-1}\left(\frac{1}{\tilde{g}}\right) \qquad (\mathrm{N/N,\ dense\ phase})\,. \tag{13.154}$$

By the KPZ correspondence, we find the conformal weight of the boundary L-spin operator:

$$\Delta^B_L(S_L) = \begin{cases} \Delta_{L+1,1}, & \text{N/N, dilute phase}\,, \\ \Delta_{1,L+1}, & \text{N/N, dense phase}\,. \end{cases} \tag{13.155}$$

Here is the place to make an important remark concerning the recurrence relation (13.148). The derivation of this relation uses only the fact that on one of the two boundaries, let us call it the left boundary, the $O(n)$ field satisfies a Neumann boundary condition. Therefore the recurrence equation (13.148) determines the boundary correlation function of the L-spin operator ${}^{x_1}_{N}[S_L]^{x_2}_{\text{any}}$ with a Neumann boundary condition on the left boundary and *any* boundary condition on the right boundary, provided we know the result for $L = 0$. In particular, the gravitational boundary dimension of S_L with a Neumann condition on the left boundary grows linearly with L:

$$\begin{aligned}
\delta_L^B &= \frac{1}{2}Lg + \delta_0^B \qquad \text{(dilute phase)}, \\
\delta_L^B &= \frac{1}{2}L\tilde{g} + \tilde{\delta}_0^B \qquad \text{(dense phase)}.
\end{aligned} \tag{13.156}$$

13.4.3 Boundary L-spin operators with Neumann/Dirichlet boundary conditions

Another obvious boundary condition is the *Dirichlet boundary condition*, which can be achieved by putting a strong magnetic field $\vec{H}$ on the boundary. To obtain the partition function on a disk with a Dirichlet boundary condition, we have to insert in the trace (13.71) the following product over the boundary sites:

$$\prod_{r\in\partial\{\text{Disk}\}} \vec{H}\cdot\vec{S}. \tag{13.157}$$

Without loss of generality, we can assume that $H_a = H\delta_{a,1}$. Then the loop gas expansion will involve, together with the "vacuum" loops, a set of open lines having their ends at the boundary sites, so that each boundary site is a source of an open line.

Let us first focus on the simplest boundary correlation function with Neumann/Dirichlet (N/D) boundary conditions, which is the analogue of D_0 in the previously considered N/N case. Microscopically, this two-point function represents the partition function of a disk whose boundary is split into two segments. On the first segment, we impose a Neumann boundary condition for the spins and $\mu_B = x$, while on the second segment we have a Dirichlet condition for the spins and $\mu_B = y$.

In the Ising model ($n = 1$), the operator which separates Neumann and Dirichlet boundary conditions is called the *boundary twist operator*. We will save this name for the case of general n. In the $O(n)$ matrix model, this correlation function is given by the continuum limit of the expectation value

$$\Omega(X,Y) = \frac{1}{\beta}\left\langle \operatorname{Tr}\left(\frac{1}{X-\mathbf{X}}\,\frac{1}{Y-\mathbf{Y}_1}\right)\right\rangle. \tag{13.158}$$

The loop gas expansion of the two-point function $\Omega(x, y)$ is a sum over closed loops and open lines having their ends at sites on the Neumann boundary as shown in Fig 13.9.

The correlation function satisfies a nonlinear equation which can be written, using the compact notation (13.146), as

$$Y\Omega = \Omega \,\text{star}\, \Omega + W. \tag{13.159}$$

We give the derivation in Section 13.A.1. This equation was originally derived by cutting the world sheet along the open line that starts from the first site of the Neumann boundary as

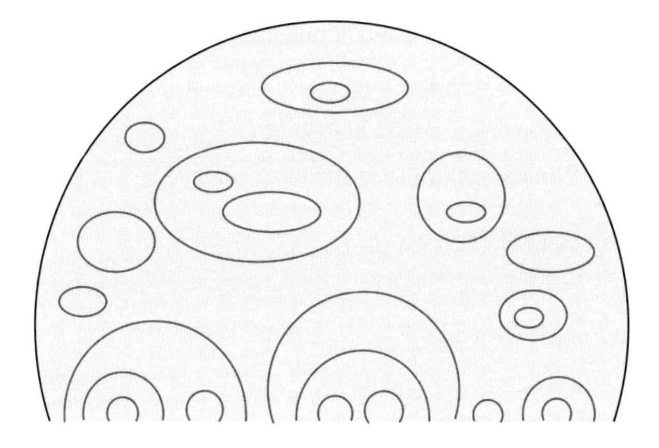

Fig. 13.9 The two-point function with mixed Neumann/Dirichlet boundary conditions.

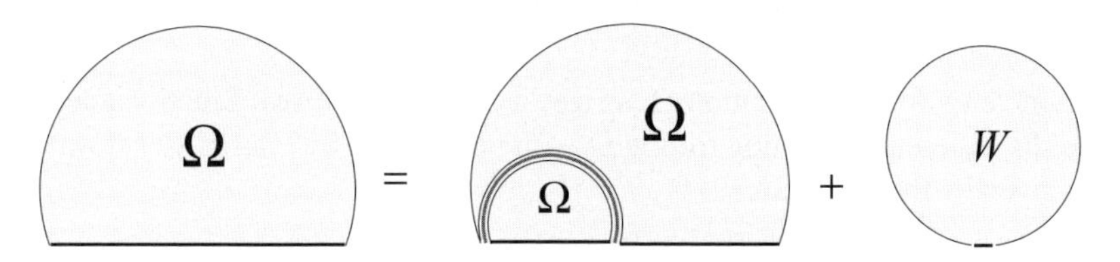

Fig. 13.10 Pictorial form of the loop equation (13.159).

shown in Fig. 13.10 and then using the factorization of the measure on both sides (Kazakov and Kostov 1992, Kostov *et al.* 2004). Using the property (13.147) of the star-product, we can write (13.159) as the functional equation

$$Y\Omega(X,Y) + Y\Omega(-X,Y) = \Omega(X,Y)\Omega(-X,Y) + \mathcal{W}(X) + \mathcal{W}(-X)\,. \quad (13.160)$$

Let us extract the scaling limit of this equation. For this purpose, we write $\mathcal{W}(X)$ in terms of the rescaled variables defined by eqn (13.100), retaining all regular terms that vanish more slowly than the singular term when $\epsilon \to 0$:

$$\mathcal{W}(X) = C_0 - C_1 x\epsilon + \epsilon^g W(x)\,, \quad X = \epsilon x\,. \quad (13.161)$$

Here C_0, C_1 are positive coefficients depending on T. For us, it is important that the constant C_0^2 is positive, where

$$C_0(T) = \frac{T(2-T)}{4(2+n)}\,. \quad (13.162)$$

To determine the critical value of Y, we replace $\mathcal{W}(X)$ by the right-hand side of eqn (13.161) and consider the functional equation at the point $X = 0$, where it becomes algebraic. Looking at the right-hand side of the solution

$$\Omega(0,Y) = Y - \sqrt{Y^2 - 2C_0 + \epsilon^g W(0)}\,, \quad (13.163)$$

we see that $Y^* = \sqrt{2C_0}$ and that we should define the singular part of $\Omega(X)$ as

$$\Omega(X,Y) - Y = \epsilon^{g/2} D_0(x,y)\,,$$
$$Y - \sqrt{2C_0} = 2\epsilon^g y\,. \tag{13.164}$$

The scaling of X and T means that the Dirichlet boundary is fractal, while the Neumann boundary is smooth.

Substituting this in eqn (13.159), we get

$$D_0(x,y)D_0(-x,y) + W(x) + W(-x) = 2y \tag{13.165}$$

or, in terms of the uniformization map $x = M\cosh\tau$,

$$D_0(\tau + i\pi/2)D_0(\tau - i\pi/2) + W(\tau + i\pi/2) + W(\tau - i\pi/2) = 2y\,. \tag{13.166}$$

This equation can be solved explicitly in the dense and dilute phases (Kazakov and Kostov 1992). Let us consider the dilute phase ($t = 0$), where $W(\tau) = M^g \cosh(g\tau)$. In this case it is convenient to parameterize the cosmological constant of the Dirichlet boundary by

$$y = \cosh(g/2)\, M^g \cosh(g\sigma)\,. \tag{13.167}$$

In this parameterization, the solution factorizes to a product of two functions depending on $\tau \pm \sigma$, and the difference equation for $\log D_0$ can be inverted by Fourier transformation. The solution is given by the integral

$$\log D_0(\tau,\sigma) = \frac{g}{4}\log\mu - \frac{1}{2}\int_{-\infty}^{\infty} \frac{dk}{k}\left[\frac{\cos(\tau k)\cos(\sigma k)}{\cosh(\pi k/2)\,\sinh(\pi k/g)} - \frac{g}{\pi k}\right]. \tag{13.168}$$

In the dense phase, D_0 can be computed in the same way, and the result is given by the expression on the right-hand side of (13.168) with the first term replaced by $\frac{1}{4}\log\mu$ and, in the second term, g replaced by $g' = 1/(2-g)$. In this phase, the Neumann boundary is fractal and the Dirichlet boundary is smooth.

The scaling of the two-point function, $D_0 \sim \mu^{g/4}$ (dilute phase) and $D_0 \sim \mu^{1/2}$ (dense phase), determines the gravitational dimension of the boundary twist operator:

$$\delta_0^B = \begin{cases} \delta_{1/2,0}, & \text{N/D, dilute phase}\,, \\ \delta_{0,1/2}, & \text{N/D, dense phase}\,. \end{cases} \tag{13.169}$$

As in the case of the N/N matter boundary conditions, we can insert L spins at the points separating the two boundaries. The operator ${}^{\tau_1}_{N}[S_L]^{\tau_2}_{\mathrm{D}}$ generates L nonintersecting lines at the point which separates the Dirichlet and Neumann boundaries. The two-point function of this operator,

$$D_L(x,y) = \langle {}^{x}_{N}[S_L]^{y}_{\mathrm{D}}[S_L]^{x}_{N}\rangle\,, \tag{13.170}$$

is obtained as the singular part of the normalized expectation value

$$\Omega_L(X,Y) = \frac{1}{\beta}\left\langle \mathrm{Tr}\left(\frac{1}{X-\mathbf{X}}\mathbf{S}_L\,\frac{1}{Y-\mathbf{Y}_1}\mathbf{S}_L^\dagger\right)\right\rangle, \tag{13.171}$$

with $\mathbf{S}_L$ defined by eqn (13.129). This correlation function satisfies the same recurrence equation (13.148), but with a different initial condition given by eqn (13.168). The gravitational dimensions δ_L^B of the operators ${}^{x}_{N}[\mathbf{S}_L]^{y}_{\mathrm{D}}$ are, according to eqn (13.156),

$$\delta_L^B = \begin{cases} \delta_{L+1/2,0}, & \text{N/D, dilute phase}, \\ \delta_{0,L+1/2}, & \text{N/D, dense phase}. \end{cases} \tag{13.172}$$

13.4.4 More general boundary conditions

Here we will consider more general boundary conditions, which were proposed for the dense phase of the loop gas by Jacobsen and Saleur (2008) and which we therefore call JS boundary conditions. The idea is to restrict the integration over the boundary spins to the first m components of $\vec{S}$, while the remaining $n - m$ components are kept fixed. This more general boundary condition breaks the $O(n)$ symmetry to $O(m) \times O(n-m)$. We will denote this boundary condition by the symbol JS_m. The JS_m boundary conditions contain as particular cases the Neumann ($m = n$) and Dirichlet ($m = 1$) boundary conditions.

The partition function on a disk with a JS_m boundary condition is defined by eqn (13.71) with the weight for the boundary spins replaced by[9]

$$\prod_{r \in \partial\Gamma} \left(1 + T_B^{-1} \sum_{k=1}^{m} S_k(r) S_k(r) \right). \tag{13.173}$$

The boundary temperature coupling T_B is to be tuned appropriately together with the bulk coupling T. In terms of the loop gas, the JS_m boundary condition prescribes that the loops that touch the boundary at least once are counted with a different fugacity m, while the loops that do not touch the boundary are counted with fugacity n. Once the observables are reformulated in terms of the loop gas, the boundary parameter m can be given any real value.

Since the JS boundary condition breaks the $O(n)$ symmetry to $O(m) \times O(n-m)$, the insertion of a spin S_a has different effects depending on whether $a \in \{1, \ldots, m\}$ or $a \in \{m+1, \ldots, n\}$. In the first case, it creates an open line that can pass through the sites of the JS_m boundary without being penalized by an extra weight factor. In the second case, such an insertion creates a line that avoids the sites on the JS_m boundary. Therefore the boundary spin operators (13.128) with N/JS_m boundary conditions split into two classes,

$$\begin{aligned} S_L^{||} &= S_{a_1} \ldots S_{a_L}, & a_L &\in \{1, \ldots, m\}, \\ S_L^{\perp} &= S_{a_1} \ldots S_{a_L}, & a_L &\in \{m+1, \ldots, n\}. \end{aligned} \tag{13.174}$$

Since the L lines are mutually avoiding, it is in fact sufficient to specify the flavor of only the rightmost one. We denote the corresponding boundary two-point functions by

$$D_L^{||} = \left\langle {}_{\text{N}}^{x}[S_L^{||}]_{\text{JS}}^{y}[S_L^{||}]_{\text{N}}^{x} \right\rangle, \quad D_L^{\perp} = \left\langle {}_{\text{N}}^{x}[S_L^{\perp}]_{\text{JS}}^{y}[S_L^{\perp}]_{\text{N}}^{x} \right\rangle. \tag{13.175}$$

The loop expansion of the two-point functions $\mathcal{D}_0$ and $\mathcal{D}_1^{||}$ is illustrated in Fig. 13.11.

In the matrix model, the derivative of the disk partition function with a JS_m boundary condition can be defined by the expectation value

$$\mathcal{H}_m(Y) = \frac{1}{\beta} \langle \operatorname{Tr} \mathbf{H}(Y) \rangle, \qquad \mathbf{H}_m(Y) \stackrel{\text{def}}{=} \frac{1}{Y - T_B \mathbf{X} - \sum_{a=1}^{m} \mathbf{Y}_a^2}. \tag{13.176}$$

[9] By "boundary sites" we understand the vertices of the fat graph which are also endpoints of external lines.

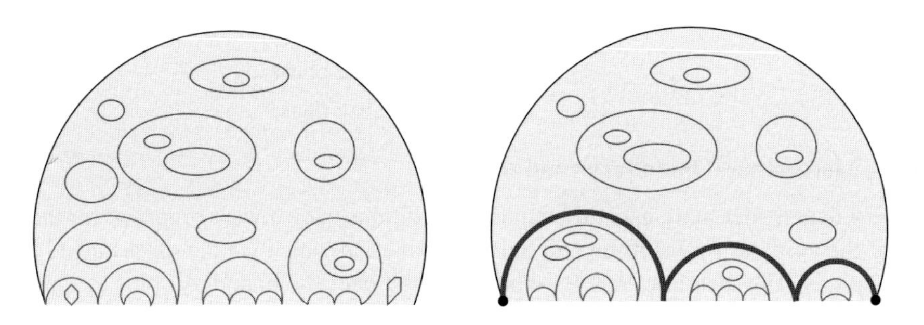

Fig. 13.11 The boundary correlation functions $\mathcal{D}_0$ (left) and $\mathcal{D}_1^{\|}$ (right).

In this way, there are two realizations of the Neumann boundary conditions for the $O(n)$ model coupled to gravity. The first one corresponds to the previously considered expectation value

$$\mathcal{W}(X) = \frac{1}{\beta}\langle \operatorname{Tr} \mathbf{W}(X)\rangle, \qquad \mathbf{W}(X) \stackrel{\text{def}}{=} \frac{1}{X - \mathbf{X}}. \tag{13.177}$$

The second realization of the Neumann boundary condition is given by eqn (13.176) with $m = n$ and $T_B = T$. Then the boundary weight (13.173) is exactly the same as in the bulk.

In the previous subsection, we derived a quadratic identity for the simplest two-point function on a disk with N/D boundary conditions. In the more general case of N/JS$_m$ boundary conditions, one obtains instead a pair of bilinear equations for the two-point functions $D_0(x, y)$ and $D_1^{\|}(x, y)$, which are the singular parts of the expectation values

$$\mathcal{D}_0(X, Y) = \frac{1}{\beta}\langle \operatorname{Tr}[\mathbf{W}(X)\,\mathbf{H}(Y)]\rangle, \tag{13.178}$$

$$\mathcal{D}_1^{\|}(X, Y) = \sum_{a=1}^{m} \frac{1}{\beta}\langle \operatorname{Tr}[\mathbf{W}(X)\mathbf{Y}_a\,\mathbf{H}(Y)\mathbf{Y}_a]\rangle \tag{13.179}$$

in the matrix model. These equations, first obtained by combinatorics (Kostov 2007) can also be derived from the Ward identities for the matrix model (Bourgine and Hosomichi 2009). In Section 13.A.1, we present the second, more elegant, derivation. The equations in question are written in terms of the star-product (13.146) with respect to the variable X, while the second variable Y enters as a parameter:

$$(Y - T_B X)\mathcal{D}_0 - \mathcal{W} + T_B\mathcal{H} = \mathcal{D}_0 \operatorname{star}\left(\mathcal{D}_1^{\|} + m\mathcal{W}\right), \tag{13.180}$$

$$\mathcal{D}_1^{\|}(X) = (\mathcal{D}_1^{\|} + m\mathcal{W}) \operatorname{star} \mathcal{D}_0. \tag{13.181}$$

The two equations have a simple geometrical interpretation, shown in Fig 13.12. Using the property (13.147) of the star-product, we combine these two equations into one functional identity

$$A(X, Y)B(-X, Y) = C(X, Y), \tag{13.182}$$

where

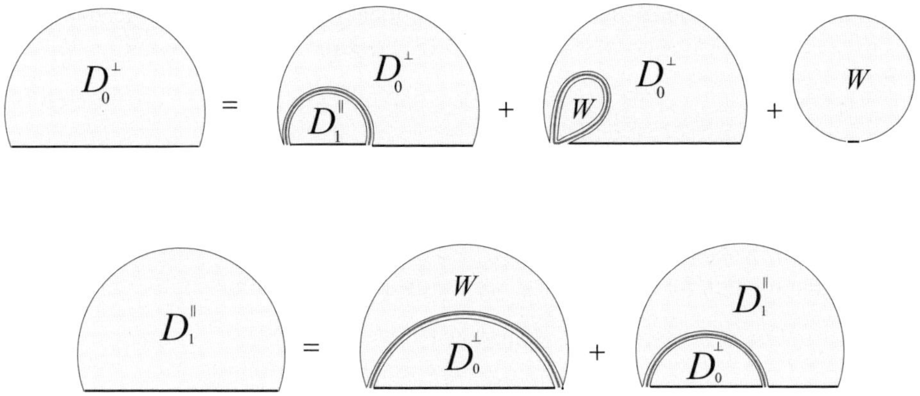

Fig. 13.12 Pictorial representation of eqn (13.180) (top) and eqn (13.181) (bottom).

$$\begin{aligned} A(X,Y) &= \mathcal{D}_0(X) - 1\,, \\ B(X,Y) &= \mathcal{D}_1^{\|}(X) + m\mathcal{W}(X) - Y - T_B X\,, \\ C(X,Y) &= Y + T_B \mathcal{H}(Y) - \mathcal{W}(X) - m\mathcal{W}(-X) - T_B X\,. \end{aligned} \tag{13.183}$$

Let us see if this equation admits a good continuum limit. We consider the dense phase, which is simpler to analyze. In this case we can take $T_B = 0$. In the dense phase, the scaling behavior of $\mathcal{W}$ is

$$\mathcal{W}(X) = C_0 - C_1 \epsilon x + (T^* - T)\, \epsilon^{\tilde{g}}\, W(x)\,, \quad X = \epsilon x\,. \tag{13.184}$$

Substituting this into the right-hand side of eqn (13.182), we see that the critical value of the JS boundary cosmological constant is $Y^* = (1+m)C_0$. We define the renormalized boundary cosmological constant y by

$$Y = Y^* + 2m\,(T^* - T)\,\epsilon^{\tilde{g}} y \tag{13.185}$$

and assume that the singular parts of $\mathcal{D}_0$ and $\mathcal{D}_1^{\|}$ scale as

$$\begin{aligned} A(X,Y) &\sim \epsilon^{\alpha_0} D_0(x)\,, \\ B(X,Y) &\sim m\epsilon^{\beta_1} D_1^{\|}(x)\,, \end{aligned} \tag{13.186}$$

where α_0 and β_1 are some scaling exponents such that $\alpha_0 + \beta_1 = \tilde{g}$. Then the functional equation (13.182) takes the form

$$D_0(x)\, D_1^{\|}(-x) + \frac{1}{m} W(x) + W(-x) = 2y\,. \tag{13.187}$$

The linear term in x has vanished because it is accompanied by a factor $\epsilon^{1-\tilde{g}}$.

We observe that, as in the case of N/D boundary conditions, the cosmological constants of the two boundaries have different scalings. The Neumann boundary is fractal, while the JS boundary is smooth.

In order to find α_0 and α_1, let us consider the critical point $\mu = y = 0$, where all observables must be given by powers of x and eqn (13.187) reduces to

$$C x^{\alpha_0}(-x)^{\beta_1} + x^{\tilde{g}} + m(-x)^{\tilde{g}} = 0\,. \tag{13.188}$$

Comparing the imaginary parts of eqn (13.187) for $x > 0$ and $x < 0$, we get

$$m = \frac{\sin(\pi\beta_1)}{\sin(\pi\alpha_0)} = \frac{\sin[\pi(\alpha_0 + 1 - \tilde{g})]}{\sin(\pi\alpha_0)}\,. \tag{13.189}$$

To find the solution for finite μ and y, we use the parameterization

$$\begin{aligned} x &= M\cosh(\tau)\,, \\ y &= (\mathrm{const}) M^{\tilde{g}} \cosh(\tilde{g}\sigma)\,, \\ W &= M^{\tilde{g}} \cosh(\tilde{g}\tau)\,. \end{aligned} \tag{13.190}$$

Then eqn (13.187) can be solved in the same way as in the case of $\mathrm{N/D}$ boundary conditions. The solution is given, up to a normalization, by the boundary two-point correlation function in Liouville theory (Fateev *et al.* 2000):

$$D_0(\tau,\sigma) \sim M^{\alpha_0} d_{\alpha_0}(\tau,\sigma), \quad D_1^{\parallel}(\tau,\sigma) \sim M^{\alpha_1} d_{\beta_1}(\tau,\sigma)\,, \tag{13.191}$$

where

$$\ln d_\alpha(\tau,\sigma) = -\int_{-\infty}^{\infty} \frac{d\omega}{\omega}\left(\frac{\cos(\omega\tau)\cos(\omega\sigma)\sinh(\pi\omega\alpha/\tilde{g})}{\sinh(\pi\omega)\sinh(\pi\omega/\tilde{g})} - \frac{\alpha}{\pi\omega}\right). \tag{13.192}$$

The function d_α satisfies the relation $d_{\alpha+1} \sim W \,\mathrm{star}\, d_\alpha$, which is equivalent to the difference equation obtained by Fateev *et al.* (2000). Therefore the correlation functions for ${}^{x}_{\mathrm{N}}[S_L^{\perp}]^{y}_{\mathrm{JS}_m}$ and ${}^{x}_{\mathrm{N}}[S_L^{\parallel}]^{y}_{\mathrm{JS}_m}$, which satisfy eqn (13.148), can again be expressed in terms of this function:

$$\begin{aligned} D_L^{\perp} &\sim M^{\alpha_L} d_{\alpha_L}, \quad \alpha_L = L + \alpha_0\,, \\ D_L^{\parallel} &\sim M^{\beta_L} d_{\beta_L}, \quad \beta_L = L + \beta_1\,. \end{aligned} \tag{13.193}$$

The gravitational dimensions are evaluated from

$$\alpha_L = \tilde{g} - 1 + 2\delta_L^{\perp}, \quad \beta_L = \tilde{g} - 1 + 2\delta_L^{\parallel}\,. \tag{13.194}$$

Let us pass to the Kac parameterization (13.32), where $\alpha_0 = r(1 - \tilde{g})$ and

$$m(r) = \frac{\sin\left[\pi(r+1)(1-\tilde{g})\right]}{\sin[\pi r(1-\tilde{g})]}\,. \tag{13.195}$$

Then the gravitational dimensions of the boundary operators ${}^{x}_{\mathrm{N}}[S_0^{\perp}]^{y}_{\mathrm{JS}_m}$ and ${}^{x}_{\mathrm{N}}[S_1^{\parallel}]^{y}_{\mathrm{JS}_m}$ are

$$\delta_0^{\perp} = \delta_{r,r-L}\,, \qquad \delta_1^{\parallel} = \delta_{r,r+L}\,. \tag{13.196}$$

They indeed match with the flat conformal dimensions $\Delta_L^{\perp} = \Delta_{r,r-L}$ and $\Delta_L^{\parallel} = \Delta_{r,r+L}$ conjectured by Jacobsen and Saleur (2008).

We can extend the recurrence equations (13.148) to the case $L = 1$ if we define the correlators

$$D_0^{\perp} = D_0 = M^{\alpha_0} d_{\alpha_0}, \qquad D_0^{\parallel} = -\frac{1}{D_0} = M^{-\alpha_0} d_{-\alpha_0} \,. \tag{13.197}$$

The correlator $D_0^{\parallel}$ corresponds to the wrong dressing of the two boundary operators. It is obtained from $D_0^{\perp}$ by Liouville reflection. These relations would follow naturally if the Neumann and JS boundaries were allowed to touch in $D_0^{\parallel}$ but not allowed to do so in $D_0^{\perp}$.

Our result for the two-point function implies the symmetry

$$S_L^{\parallel} \leftrightarrow S_L^{\perp}, \quad m \leftrightarrow \nu - m \,, \tag{13.198}$$

which looks quite natural given the definition of these operators and resembles the duality symmetry that exchanges Dirichlet and Neumann boundary conditions. In the parameterization (13.195), exchanging m and $n - m$ is equivalent to changing the sign of r, owing to the identity

$$m(r) + m(-r) = \nu \,. \tag{13.199}$$

In the loop gas formulation, the symmetry (13.198) can be spelled out as

$$\{\parallel, r\} \leftrightarrow \{\perp, -r\} \tag{13.200}$$

and is respected by the exponents (13.196).

Exercise 13.2 Find the scaling limit of the bilinear equation (13.182) and the conformal dimensions of the boundary operators ${}^{x}_{\mathrm{N}}[S_0^{\perp}]^{y}_{\mathrm{JS}_m}$ and ${}^{x}_{\mathrm{N}}[S_1^{\parallel}]^{y}_{\mathrm{JS}_m}$ in the dilute phase.

Hint: Tune the boundary temperature coupling T_B to cancel the linear term in x on the right-hand side and then proceed as in the dense phase.

13.A Appendices

There is a wealth of boundary correlation functions that can be evaluated using the Ward identities in the $O(n)$ matrix model. In general, the boundary of the disk can be split into several intervals with different JS boundary conditions associated with the different intervals. Let $I_j \subset \{1, 2, \ldots, n\}$ be the set of the components of the $O(n)$ field that are free on the j-th interval. As before, the loop expansion of the partition function with such a boundary condition prescribes special weights to be given to the loops that touch the boundary. However, the rules get more complicated. If a loop touches the boundary (once or several times) in a single interval, say the j-th one, then it is counted with a factor $m_j = \mathrm{card}(I_j)$. If the loop touches the boundary simultaneously in the j-th and the k-th interval, it must be counted with a weight $m_{jk} = \mathrm{card}(I_j \cap I_k)$, and so on. The case of a flat cylinder with two different JS boundary conditions was recently studied by Dubail *et al.* (2008).

The above picture reminds us of the constructions with intersecting branes in string theory. (In string theory, the boundary conditions are called branes.) Indeed, the $O(n)$ model coupled to 2D gravity can be viewed as a solvable model of bosonic string theory with a curved target space, representing the $(n-1)$-dimensional sphere. In the interval $1 < n < 2$, the curvature $R = (n-1)(n-2)$ of the target space is negative. This explains the continuous spectrum of boundary conditions.

13.A.1 Ward identities for the matrix model

The translational invariance of the matrix measure implies, for any matrix $\mathbf{F}$ made out of X and Y_a, the Ward identities

$$\frac{1}{\beta}\langle \mathrm{Tr}\, \partial_{\mathbf{X}} \mathbf{F} \rangle = \left\langle \mathrm{Tr} \left[\left(V'(\mathbf{X}) - \sum_a \mathbf{Y}_a^2 \right) \mathbf{F} \right] \right\rangle = 0\,, \tag{13.A.1}$$

$$\frac{1}{\beta}\langle \mathrm{Tr}(\partial_{\mathbf{Y}_a} \mathbf{F})\rangle + \langle \mathrm{Tr}(\mathbf{Y}_a \mathbf{X} + \mathbf{X}\mathbf{Y}_a)\mathbf{F}\rangle = 0\,. \tag{13.A.2}$$

A very useful form of the second equation was suggested by Bourgine and Hosomichi (2009). We take a function F of the form

$$\mathbf{F} = \int_0^\infty dL\, e^{L\mathbf{X}} \mathbf{G}\, e^{L\mathbf{X}}\,. \tag{13.A.3}$$

Then $\mathbf{G} = -(\mathbf{X}\mathbf{F} + \mathbf{F}\mathbf{X})$, and the second equation takes the form

$$\langle\, \mathrm{Tr}\,(\mathbf{Y}_a \mathbf{G})\,\rangle = \frac{1}{\beta}\int_0^\infty dL\, \langle\, \mathrm{Tr}\,(e^{L\mathbf{X}} \partial_{\mathbf{Y}_a} e^{L\mathbf{X}} \mathbf{G})\,\rangle\,. \tag{13.A.4}$$

The second Ward identity, written in the form (13.A.3), has a simple geometrical meaning. Whenever we have an expectation value of the type $\langle \mathrm{Tr} \mathbf{Y}_a \mathbf{G}\rangle$, where $\mathbf{G}$ can be expanded as a series in the $\mathbf{Y}$'s, it is equal to the sum over all possible Wick contractions of the matrix $\mathbf{Y}_a$ with the other matrices, with the "propagator"

$$\overline{[\mathbf{Y}_a]_i^j [\mathbf{Y}_b]_k^l} = \delta_{a,b} \int_0^\infty dL\, [e^{L\mathbf{X}}]_i^l [e^{L\mathbf{X}}]_k^j = \delta_{a,b} \left[(\mathbf{X} \otimes \mathbf{1} + \mathbf{1} \times \mathbf{X})^{-1} \right]_{ik}^{jl}. \tag{13.A.5}$$

Each such contraction splits the trace into two traces. After performing the contractions, we need to use the factorization property of the traces,

$$\left\langle \frac{1}{\beta}\mathrm{Tr}(\mathbf{A}) \frac{1}{\beta}\mathrm{Tr}(\mathbf{B}) \right\rangle = \left\langle \frac{1}{\beta}\mathrm{Tr}(\mathbf{A}) \right\rangle \left\langle \frac{1}{\beta}\mathrm{Tr}(\mathbf{B}) \right\rangle. \tag{13.A.6}$$

As a very simple example, take the two-point function of the L-spin operators

$$G_L = \langle\, \mathrm{Tr}(\mathbf{Y}_{a_1} \dots \mathbf{Y}_{a_N}) \mathrm{Tr}(\mathbf{Y}_{a_N} \dots \mathbf{Y}_{a_1})\,\rangle\,. \tag{13.A.7}$$

We apply the Ward identity (13.A.4) to the ensemble of the matrices in the left trace. The only way to perform the L Gaussian contractions produces the integral (13.132) for the product of L traces.

Equation for the resolvent. To derive an equation for the resolvent

$$\mathcal{W}(X) = \frac{1}{\beta}\langle \mathrm{Tr}\, \mathbf{W}(X)\rangle, \qquad \mathbf{W}(X) \equiv \frac{1}{X - \mathbf{X}}\,, \tag{13.A.8}$$

we take $\mathbf{F} = \mathbf{W}(X)$. The first Ward identity (13.A.1) then gives

$$\mathcal{W}^2(X) = \frac{1}{\beta}\langle \operatorname{Tr} V'(\mathbf{X})\mathbf{W}(X)\rangle - \sum_a \frac{1}{\beta}\left\langle \operatorname{Tr} \mathbf{Y}_a^2\, \mathbf{W}(X)\right\rangle. \tag{13.A.9}$$

The second Ward identity, in the form (13.A.4), applied to the last term, gives

$$\frac{1}{\beta}\langle \operatorname{Tr} \mathbf{Y}_a^2 \mathbf{W}(X)\rangle = \int_0^\infty dL \langle \operatorname{Tr} e^{L\mathbf{X}} \partial_{\mathbf{Y}_a} e^{L\mathbf{X}} \mathbf{Y}_a \mathbf{W}(X)\rangle\,. \tag{13.A.10}$$

The derivative splits the trace into a product of two traces and, using the factorization property of the traces, we write the right-hand side as

$$\begin{aligned} \int_0^\infty dL \langle \operatorname{Tr} e^{L\mathbf{X}} \mathbf{W}(X)\rangle \langle \operatorname{Tr} e^{L\mathbf{X}}\rangle &= \int_0^\infty dL\, dL'\, e^{-L'X} \tilde{\mathcal{W}}(L+L')\tilde{\mathcal{W}}(L) \\ &= \int_{i\mathbb{R}} \frac{dX'}{2\pi i} \frac{\mathcal{W}(X') - \mathcal{W}(X)}{X - X'} W(-X') = [\mathcal{W} \operatorname{star} \mathcal{W}](X)\,, \end{aligned} \tag{13.A.11}$$

where we have used the definition (13.146) of the star-product. Hence the identity

$$\mathcal{W}(X)^2 - V'(x)\mathcal{W}(X) + n\mathcal{W} \operatorname{star} \mathcal{W}(x) = \left\langle \operatorname{Tr}\left(\frac{V'(\mathbf{X}) - V'(X)}{X - \mathbf{X}}\right)\right\rangle. \tag{13.A.12}$$

For a cubic potential, the expectation value on the right-hand side is a polynomial of degree one. By applying the identity $\mathcal{W} \operatorname{star} \mathcal{W}(x) + \mathcal{W} \operatorname{star} \mathcal{W}(-x) = \mathcal{W}(x)\mathcal{W}(-x)$, we obtain eqn (13.95).

Equations for the boundary two-point functions.

- *Proof of the recurrence equation (13.148).* The correlators of the form

  $$D_L(X) = \langle \operatorname{Tr} \mathbf{W}(X)\mathbf{Y}_{a_1} \dots \mathbf{Y}_{a_L} \mathbf{F} \mathbf{Y}_{a_N} \dots \mathbf{Y}_{a_1}\rangle\,, \tag{13.A.13}$$

  where $\mathbf{F}$ is an arbitrary function made from $\mathbf{X}$ and $\mathbf{Y}$, satisfy a recurrence equation obtained by applying the Ward identity (13.A.4) to one of the two variables $\mathbf{Y}_{a_1}$. The contraction with the second matrix $\mathbf{Y}_{a_1}$ splits the trace into two and the result gives eqn (13.148),

  $$D_L = W \operatorname{star} D_{L-1}\,. \tag{13.A.14}$$

- *Proof of eqn (13.159).* We write the correlator (13.158) as

  $$Y\Omega(X) = \mathcal{W}(X) + \frac{1}{\beta}\left\langle \operatorname{Tr} \mathbf{W}(X)\, \frac{\mathbf{Y}_1}{y - \mathbf{Y}_1}\right\rangle \tag{13.A.15}$$

  and apply eqn (13.A.4) to the variable $\mathbf{Y}_1$ in the numerator. The result is

  $$Y\Omega = \mathcal{W} + \Omega \operatorname{star} \Omega\,.$$

- *Proof of eqn (13.180).* We represent eqn (13.178) as

  $$(Y - T_B X)\mathcal{D}_0(X,Y) = \mathcal{W}(X) - T_B \mathcal{H}(Y) + \sum_{a=1}^m \langle \operatorname{Tr}[\mathbf{W}(X)\mathbf{Y}_a \mathbf{Y}_a \mathbf{H}(Y)]\rangle$$

  and apply eqn (13.A.4) to the first variables $\mathbf{Y}_a$. The result is

  $$(Y - T_B X)\mathcal{D}_0 = \mathcal{W} - T_B \mathcal{H} + \left(m\mathcal{D}_0 \operatorname{star} \mathcal{W} + \mathcal{D}_0 \operatorname{star} \mathcal{D}_1^{\parallel}\right). \tag{13.A.16}$$

- *Proof of eqn (13.181).* We apply eqn (13.A.4) to one of the variables $\mathbf{Y}_a$ in

$$\mathcal{D}_1^{\|}(X,Y) = \sum_{a=1}^{m} \langle \operatorname{Tr} [\mathbf{W}(X)\mathbf{Y}_a\mathbf{H}(Y)\mathbf{Y}_a] \rangle . \tag{13.A.17}$$

The result is

$$\mathcal{D}_1^{\|} = \mathcal{D}_1^{\|} \operatorname{star} \mathcal{D}_0 + m\mathcal{W} \operatorname{star} \mathcal{D}_0 . \tag{13.A.18}$$

13.A.2 Partition function on the sphere

Here we give a solution to the transcendental equation (13.113),

$$\mu = M^2 - tM^{2(2-g)} , \tag{13.A.19}$$

following Zamolodchikov (2005). First we perform an inverse Laplace transformation of $M^{2\gamma}$ with respect to μ and express the result as a series in t:

$$\begin{aligned}
\int \frac{d\mu}{2\pi i} M^{2\gamma} e^{\mu A} &= -\frac{1}{2\pi i A} \int e^{\mu A} d[M^{2\gamma}] = -\frac{1}{2\pi i A} \int e^{A(M^2 - tM^{2(2-g)})} d[M^{2\gamma}] \\
&= -\frac{\gamma}{2\pi i A} \int \frac{v^{\gamma-1} dv}{A^\gamma} e^{v - tA^{g-1} v^{2-g}} \\
&= -\frac{\gamma}{A^{\gamma+1}} \sum_{n=0}^{\infty} \frac{(-tA^{g-1})^n}{n!} \int \frac{dv}{2\pi i} v^{\gamma-1+n(2-g)} e^v \\
&= -\frac{\gamma}{A^{\gamma+1}} \sum_{n=0}^{\infty} \frac{(-tA^{g-1})^n}{n!\, \Gamma[(g-2)n + 1 - \gamma]} .
\end{aligned} \tag{13.A.20}$$

Then we Laplace transform back to the variable μ:

$$\begin{aligned}
M^{2\gamma} &= -\gamma \sum_{n=0}^{\infty} \frac{(-t)^n}{n!\, \Gamma[(g-2)n + 1 - \gamma]} \int_0^{\infty} dA\, e^{-A\mu} A^{-1-\gamma+n(g-1)} \\
&= -\gamma \mu^\gamma \sum_{n=0}^{\infty} \frac{\Gamma(n(g-1) - \gamma)}{n!\, \Gamma[(g-2)n + 1 - \gamma]} \left(-\frac{t}{\mu^{g-1}} \right)^n .
\end{aligned} \tag{13.A.21}$$

Applying this expansion formula for $u = M^{2g-2}$, we get

$$\begin{aligned}
u &= -(g-1)\mu^{g-1} \sum_{n=0}^{\infty} \frac{\Gamma[(n-1)(g-1)]}{n!\, \Gamma[(n-1)(g-2)]} \left(-\frac{t}{\mu^{g-1}} \right)^n \\
&= \mu^{g-1} + (g-1)t - \frac{1}{2}(g-2)(g-1)t^2 \mu^{1-g} + \cdots .
\end{aligned} \tag{13.A.22}$$

Integrating twice with respect to μ, we obtain

$$\begin{aligned}
\mathcal{F}(t,\mu) &= (g-1)\mu^{g+1} \sum_{n=0}^{\infty} \frac{\Gamma[g(n-1) - n - 1] \left(-t\mu^{1-g} \right)^n}{\Gamma(n+1)\Gamma[(g-2)(n-1)]} \\
&= -\frac{1}{g(g+1)} \mu^{g+1} - \frac{g-1}{2} t\mu^2 + \frac{(g-1)}{2(g-3)} t^2 \mu^{3-g} - \frac{g-1}{6} t^3 \mu^{4-2g} + \cdots .
\end{aligned} \tag{13.A.23}$$

References

A. A. Belavin and A. B. Zamolodchikov (2005). Moduli integrals, ground ring and four-point function in minimal Liouville gravity. In arXiv:hep-th/0510214, p. 16.

A. Belavin, A. Polyakov, and A. Zamolodchikov (1984). Infinite conformal symmetry in two-dimensional quantum field theory. *Nucl. Phys. B* **241**, 333–380.

E. Brezin, C. Itzykson, G. Parisi, and J. B. Zuber (1978). Planar diagrams. *Commun. Math. Phys.* **59**, 35.

J. E. Bourgine and K. Hosomichi (2009). Boundary operators in the O(n) and RSOS matrix models. arXiv:0811.3252 [hep-th].

J. L. Cardy (1984). Conformal invariance and surface critical behavior. *Nucl. Phys. B* **240**, 514.

F. David (1988). Conformal field theories coupled to 20D gravity in the conformal gauge. *Mod. Phys. Lett. A* **3**, 1651.

P. Di Francesco, P. Ginsparg, and J. Zinn-Justin (1995). 2D gravity and random matrices. *Phys. Rep.* **254**, 1–133; arXiv:hep-th-9306153.

J. Distler and H. Kawai (1989). Conformal field theory and 2-D quantum gravity or who's afraid of Joseph Liouville? *Nucl. Phys. B* **321**, 509.

E. Domany, D. Mukamel, B. Nienhuis, and A. Schwimmer (1981). Duality relations and equivalences for models with O(N) and cubic symmetry. *Nucl. Phys. B* **190**, 279.

J. Dubail, J. L. Jacobsen, and H. Saleur (2008). Conformal two-boundary loop model on the annulus. arXiv:0812.2746.

B. Duplantier (2000). Conformally invariant fractals and potential theory. *Phys. Rev. Lett.* **84**, 1363.

B. Duplantier (2004). Conformal fractal geometry and boundary quantum gravity. *Proc. Symposia Pure Math.*, vol. 72, Part 2, pp. 365–482. AMS, Providence, RI; arXiv:math-ph/0303034.

B. Duplantier (2005). Random geometry. *Les Houches*, Session LXXXIII; arXiv:math-ph/0608053.

B. Duplantier and I. Kostov (1988). Conformal spectra of polymers on a random surface. *Phys. Rev. Lett.* **61**, 1433.

B. Duplantier and I. Kostov (1990). Geometrical critical phenomena on random surfaces with arbitrary genus. *Nucl. Phys. B* **340**, 491.

B. Duplantier and H. Saleur (1986). Exact surface and wedge exponents for polymers in two dimensions. *Phys. Rev. Lett.* **57**, 3179.

B. Eynard and C. Kristjansen (1995). Exact solution of the O(n) model on a random lattice. *Nucl. Phys. B* **455**, 577.

B. Eynard and C. Kristjansen (1996). More on the exact solution of the $O(n)$ model on a random lattice and an investigation of the case $|n| > 2$. *Nucl. Phys. B*, **466**, 463.

V. Fateev, A. Zamolodchikov, and A. Zamolodchikov (2000). Boundary Liouville field theory I. Boundary state and boundary two-point function. arXiv:hep-th/0001012.

P. Fendley, H. Saleur, and A. Zamolodchikov (1993). Massless flows I: The sine–Gordon and $O(n)$ models. *Int. J. Mod. Phys. A* **8**, 5751.

M. Gaudin and I. Kostov (1989). O (n) model on a fluctuating planar lattice. Some exact results. *Phys. Lett. B* **220**, 200.

P. Ginsparg and G. Moore (1992). Lectures on 2D gravity and 2D string theory (TASI 1992). arXiv:hep-th/9304011.

A. Hashimoto and I. R. Klebanov (1995). Matrix model approach to $D > 2$ noncritical superstrings. *Mod. Phys. Lett. A* **10**, 2639; arXiv:hep-th/9507062.

K. Hosomichi (2001). Bulk–boundary propagator in Liouville theory on a disc. *JHEP* **0111**, 044; arXiv:hep-th/0108093.

J. L. Jacobsen and H. Saleur (2008). Conformal boundary loop models. *Nucl. Phys. B* **788**, 137; arXiv:math-ph/0611078.

V. A. Kazakov and I. K. Kostov (1992). Loop gas model for open strings. *Nucl. Phys. B* **386**, 520; arXiv:hep-th/9205059.

I. R. Klebanov (1995). Touching random surfaces and Liouville gravity. *Phys. Rev. D* **51**, 1836; arXiv:hep-th/9407167.

I. R. Klebanov and A. Hashimoto (1995). Nonperturbative solution of matrix models modified by trace squared terms. *Nucl. Phys. B* **434**, 264; arXiv:hep-th/9409064.

V. Knizhnik, A. Polyakov, and A. Zamolodchikov (1988). Fractal structure of 2D quantum gravity. *Mod. Phys. Lett. A* **3**, 819.

I. K. Kostov (1989). $O(n)$ vector model on a planar random lattice: Spectrum of anomalous dimensions. *Mod. Phys. Lett. A* **4**, 217.

I. K. Kostov (1992). Strings with discrete target space. *Nucl. Phys. B* **376**, 539; arXiv:hep-th/9112059.

I. Kostov (2004). Matrix models as conformal field theories. *Lectures Given at the Summer School Applications of Random Matrices in Physics*, Les Houches, 6–25 June 2004.

I. Kostov (2006). Thermal flow in the gravitational $O(n)$ model. arXiv:hep-th/0602075.

I. Kostov (2007). Boundary loop models and 2D quantum gravity. *J. Stat. Mech.* **0708**, P08023; arXiv:hep-th/0703221.

I. Kostov and M. Staudacher (1992). Multicritical phases of the O(n) model on a random lattice. *Nucl. Phys. B* **384**, 459.

I. K. Kostov, B. Ponsot, and D. Serban (2004). Boundary Liouville theory and 2D quantum gravity. *Nucl. Phys. B* **683**, 309; arXiv:hep-th/0307189.

B. Nienhuis (1982). Exact critical point and critical exponents of O(N) models in two-dimensions. *Phys. Rev. Lett.* **49**, 1062.

B. Nienhuis (1984). Critical behavior of two-dimensional spin models and charge asymmetry in the Coulomb gas. *J. Stat. Phys.* **34**, 731.

A. Polyakov (1981). Quantum geometry of bosonic strings. *Phys. Lett. B* **103**, 207.

B. Ponsot and J. Teschner (2002). Boundary Liouville field theory: Boundary three point function. *Nucl. Phys. B* **622**, 309.

H. Saleur (1986). New exact exponents for two-dimensional self-avoiding walks. *J. Phys. A* **19**, L807.

H. Saleur (1987). Conformal invariance for polymers and percolation. *J. Phys. A* **20**, 455.

G. 't Hooft (1974). A planar diagram theory for strong interactions. *Nucl. Phys. B* **72**, 461.

A. B. Zamolodchikov (2005). On the three-point function in minimal Liouville gravity. *Theor. Mat. Phys.* **142**, 183; arXiv:hep-th/0505063.

14
Real-space condensation in stochastic mass transport models

S. N. MAJUMDAR

Satya Majumdar,
Laboratoire de Physique Théorique et Modèles Statistiques,
Université Paris-Sud, Bât. 100,
91405 Orsay Cedex, France.

14.1 Introduction

The phenomenon of Bose–Einstein condensation (BEC) in an ideal Bose gas is by now textbook material and has recently seen a huge revival of interest, driven mostly by new experiments. Consider an ideal gas of N bosons in a d-dimensional hybercubic box of volume $V = L^d$. In the thermodynamic limit $N \to \infty$, $V \to \infty$ but with the density $\rho = N/V$ fixed, as one reduces the temperature below a certain critical value $T_c(\rho)$ in $d > 2$, a macroscopically large number of particles ($\propto V$) condense into the ground state, i.e. the zero-momentum quantum state. Alternately, one encounters the same condensation transition upon fixing the temperature but increasing the density ρ beyond a critical value $\rho_c(T)$.

The traditional BEC happens in momentum (or, equivalently, energy) space. In contrast, over the last two decades it has been realized that a "similar" Bose–Einstein-type condensation can also occur in *real space* in the steady state of a variety of nonequilibrium systems, such as in the case of cluster aggregation and fragmentation [1], jamming in traffic and granular flow [2, 3], and granular clustering [4]. The common characteristic feature that these systems share is the stochastic transport of some conserved scalar quantity, which can simply be called mass. A condensation transition occurs in these systems when, above some critical mass density, a single "condensate" captures a finite fraction of the total mass of the system. The "condensate" corresponds to a dominant cluster in the context of granular clustering, or a single large jam in the context of traffic models. Another example of condensation is found in the phase separation dynamics of one-dimensional driven systems, where the condensation manifests itself in the emergence of a macroscopic domain of one phase [5]. Other examples of such real-space condensation can be found in socioeconomic contexts: for example, in the case of wealth condensation in macroeconomics, where a single individual or a company (the condensate) owns a finite fraction of the total wealth [6], and in the case of growing networks, where a single node or hub (such as Google) may capture a finite fraction of the links in the network [7].

This real-space condensation mentioned above has been studied theoretically in very simple stochastic mass transport models defined on lattices. These are typically nonequilibrium models without any Hamiltonian and are defined by microscopic dynamical rules that specify how some scalar quantities such as masses or a certain number of particles are transported from site to site of the lattice. These rules typically violate detailed balance [8]. Under these rules, the system evolves into a *stationary* or *steady* state, which is typically not a Gibbs–Boltzmann state as the system lacks a Hamiltonian [8]. For a certain class of transport rules, the system can reach a steady state where, upon increasing the density of mass or particles beyond a critical value, a macroscopically large mass (or number of particles) condenses onto a single site of the lattice, signaling the onset of "real-space" condensation.

In this article we will mostly focus on *homogeneous* systems, where the transport rules are independent of the sites, i.e. the system is translationally invariant. In the condensed phase in an infinite system, the condensate forms at a single site, which thus breaks the translational invariance spontaneously. In a finite system, a condensate at a given site has a finite lifetime, beyond which it dissolves and then becomes relocated at a different site, and the various timescales associated with the formation/relocation of the condensate diverge with increasing system size (see later). In *heterogeneous* systems, where the transport rules may differ from site to site, the condensate may form at a site with the lowest outgoing mass transport rate [9–11]. The mechanism of the condensation transition in such heterogeneous systems is exactly analogous to the traditional BEC in momentum space, and the site with the lowest outgoing mass transfer rate plays the role of the ground state in the quantum system of an ideal Bose gas. In contrast, the mechanism of condensation in homogeneous systems, the subject of our focus here, is rather different: the onset and formation of a condensate in an infinite system is associated with the spontaneous breaking of translational invariance. Also, unlike the case of the traditional equilibrium Bose gas in a box, this real-space condensation in nonequilibrium mass transport models can occur *even in one dimension*.

The purpose of these lectures is to understand the phenomenon of real-space condensation in homogeneous systems within the context of simple one-dimensional mass transport models. The main questions we will be addressing are threefold: (i) When does the condensation happen? That is, to find the criterion for condensation. (ii) How does the condensation happen? That is, to unfold the mathematical mechanism behind such a transition if it happens. (iii) What is the nature of the condensate? That is, for example, to compute the distribution of mass or the number of particles in the condensate.

The article is organized as follows. In Section 14.2, we will discuss three simple, well-studied lattice models of stochastic mass transport. In Section 14.3, we will consider a generalized mass transport model that includes the previous three models as special cases, and investigate its steady state. In particular, we will study in detail steady states that are factorizable. The necessary and sufficient conditions for such a factorizability property will be discussed. Thanks to this property, a detailed analytical study of the condensation is possible for such steady states, which will be illustrated in Section 14.4. In Section 14.5, we will illustrate how various results associated with the condensation transition in factorizable steady states can be simply understood in terms of sums and extremes of independent identically distributed (i.i.d) random variables. Finally, we will conclude in Section 14.6 with a summary and other possible generalizations/issues associated with real-space condensation.

14.2 Three simple mass transport models

14.2.1 Zero-range process

The zero-range process (ZRP), introduced by Spitzer [12], is perhaps one of the simplest analytically solvable models of mass/particle transport that exhibits a real-space condensation in a certain range of its parameters—for a review, see [13]. The ZRP is defined on a lattice with periodic boundary conditions. For simplicity, we will consider a 1D lattice with L sites; the generalization to higher dimensions is straightforward. On each site of the lattice at any instant of time rest a number of particles, say m_i at site i, where $m_i \geq 0$ is a non-negative integer. We can also think of each particle as carrying a unit mass, so that m_i represents the total mass at site i. A configuration of the system at any given instant is specified by the masses at all sites $\{m_1, m_2, \ldots, m_L\}$. The system starts from an arbitrary initial condition with a total mass $M = \sum_i m_i$. The subsequent dynamics conserves this total mass or total particle number, or, equivalently, the density $\rho = M/L$.

The system evolves via a continuous-time stochastic dynamics specified by the following rules:

- In a small time interval dt, a single particle from site i with a number of particles m_i is transported to its right neighbor $i+1$ with probability $U(m_i)\,dt$, provided $m_i \geq 1$. In terms of mass, this means that a single unit of mass is transferred from site i to site $i+1$ with rate $U(m_i)$.

- Nothing happens with probability $1 - U(m_i)\,dt$.

Here $U(m)$ is an arbitrary positive function, with the constraint that $U(0) = 0$, since there cannot be any transfer of unit mass if the site has no mass at all. Thus in the ZRP, the particle or mass transfer rate $U(m)$ depends only on the number of particles/mass m at the departure site prior to the transfer. One can of course easily generalize the ZRP to discrete-time dynamics, with symmetric transfer of particles to both neighbors, etc. [13]. But here we shall stick to the asymmetric continuous-time model for simplicity.

As the system evolves under this dynamics, the probability of a configuration $P(m_1, m_2, \ldots, m_L, t)$ evolves in time, and in the long-time limit $t \to \infty$, it approaches a time-independent stationary joint distribution of masses $P(m_1, m_2, \ldots, m_L)$. This is the basic quantity of interest, since the statistics of all other physical observables in the steady state can, in principle, be computed from this joint distribution. In many such nonequilibrium systems, computing the steady state $P(m_1, m_2, \ldots, m_L)$ is indeed the first big hurdle [8]. Fortunately, in the ZRP, this can be computed explicitly and has a rather simple factorized form [12, 13]

$$P(m_1, m_2, \ldots, m_L) = \frac{1}{Z_L(M)} f(m_1) f(m_2) \ldots f(m_L)\, \delta\left(\sum_i m_i - M\right), \tag{14.1}$$

where the weight function $f(m)$ is related to the transfer rate $U(m)$ via

$$\begin{aligned} f(m) &= \prod_{k=1}^{m} \frac{1}{U(k)} \quad \text{for} \quad m \geq 1 \\ &= 1 \quad \text{for} \quad m = 0\,. \end{aligned} \tag{14.2}$$

The delta function in eqn (14.1) specifies the conserved total mass M, and $Z_L(M)$ is just a normalization factor that ensures that the total probability is unity and satisfies a simple recursion relation

$$Z_L(M) = \sum_{m_i} \prod_{i=1}^{L} f(m_i)\, \delta\left(\sum_i m_i - M\right) = \sum_{m=0}^{M} f(m) Z_{L-1}(M-m)\,. \tag{14.3}$$

To prove the result in eqn (14.1), one simply writes down the master equation for the evolution of the probability in the configuration space and then verifies [13] that the stationary solution of this master equation is indeed given by eqn (14.1).

Finally, the single-site mass distribution $p(m)$, defined as the probability that any site has mass m in the steady state, is just the marginal obtained from the joint distribution

$$p(m) = \sum_{m_2, m_3, \ldots, m_L} P(m, m_2, m_3, \ldots, m_L) = f(m) \frac{Z_{L-1}(M-m)}{Z_L(m)}\,. \tag{14.4}$$

Note that $p(m)$ implicitly depends on L, but this L dependence has been suppressed for notational simplicity. This single-site mass distribution is important, as any signature of the existence of a condensate will definitely show up in the explicit form of $p(m)$.

Evidently, the steady-state $p(m)$ depends on the transfer rate $U(m)$ through the weight function $f(m)$ in eqn (14.2). Not all choices of $U(m)$ lead to a steady state with a condensation transition. Indeed, one may ask what choices of $U(m)$ may lead to a condensation transition. An example of such a choice is given by $U(m) \propto (1 + \gamma/m)$ for large m, which leads to, using eqn (14.2), a power-law weight function $f(m) \sim m^{-\gamma}$ for large m. In this case, it has been shown [2, 13] that for $\gamma > 2$, the system undergoes a condensation transition as the density ρ is increased through a critical value $\rho_c = 1/(\gamma - 2)$. The condensation transition shows up in $p(m)$ in the thermodynamic limit, which has different forms for $\rho < \rho_c$, $\rho = \rho_c$, and $\rho > \rho_c$ [13]:

$$\begin{aligned} p(m) &\sim \frac{1}{m^\gamma} \exp\left[-\frac{m}{m^*}\right] \quad \text{for} \quad \rho < \rho_c\,, \\ &\sim \frac{1}{m^\gamma} \quad \text{for} \quad \rho = \rho_c\,, \\ &\sim \frac{1}{m^\gamma} + \text{“condensate”} \quad \text{for} \quad \rho > \rho_c\,. \end{aligned} \tag{14.5}$$

Thus for $\rho < \rho_c$, the single-site mass distribution decays exponentially with a characteristic mass m^* that diverges as $\rho \to \rho_c$ from below, and has a power-law form exactly at $\rho = \rho_c$; for $\rho > \rho_c$, while the power-law form remains unchanged, all the additional mass $(\rho - \rho_c)L$ condenses onto a single site, which shows up as a bump in $p(m)$ at the tail of the power-law form (see Fig. 14.3 later). The term “condensate” in eqn (14.5) refers to this additional bump. Physically, this means that a single condensate coexists with a background critical fluid for $\rho > \rho_c$. This change of behavior of $p(m)$ as one increases ρ through ρ_c is a prototype signature of real-space condensation, and this behavior is found in various other stochastic mass transport models that will be discussed below. In addition, many details of the condensation phenomena in the ZRP also follow as special cases of the more general mass transport model defined in Section 14.3.

Before ending this subsection, it is useful to point out that there are several other issues/studies in relation to ZRP that are not covered here. Interested readers may consult the reviews [13, 14].

14.2.2 Symmetric chipping model

Here we discuss another simple one-dimensional mass transport model that also exhibits a condensation phase transition in its steady state. As in the ZRP, this model is also defined on a lattice with periodic boundary conditions where each site i carries an integer mass $m_i \geq 0$ [1]. A nonlattice mean-field version of the model was studied in [15]. The system evolves via the continuous-time dynamics defined by the following rules [1, 16]:

- *Diffusion and aggregation*: in a small time interval dt, the entire mass m_i from site i moves either to its right neighbor $(i + 1)$ with probability $dt/2$ or to its left neighbor $(i - 1)$ with probability $dt/2$.
- *Chipping*: in the same interval dt, only one unit of mass is chipped off site i, with mass m_i (provided $m_i \geq 1$), and moves to either its right neighbor with probability $w\,dt/2$ or its left neighbor with probability $w\,dt/2$.
- With probability $1 - (1 + w)\,dt$, nothing happens.

Once again, the total mass $M = \rho L$ is conserved by the dynamics. The model thus has two parameters ρ (the density) and w (the ratio of the chipping rate to the diffusion rate). At long times, the system evolves into a steady state where the single-site mass distribution $p(m)$, for large L, exhibits a condensation phase transition at a critical density [1] $\rho_c(w) = \sqrt{w+1} - 1$. Remarkably, this equation of state $\rho_c(w) = \sqrt{w+1} - 1$ turns out to be exact in all dimensions [17] and is thus "superuniversal." For $\rho < \rho_c(w)$, the mass is homogeneously distributed in the system with a mass distribution that has an exponential tail for large mass. At $\rho = \rho_c(w)$, the mass distribution decays as a power law and for $\rho > \rho_c$, a condensate forms on a single site, which carries the additional macroscopic mass $(\rho - \rho_c)L$ and coexists with a critical background fluid [1]:

$$
\begin{aligned}
p(m) &\sim \exp\left[-\frac{m}{m^*}\right] \quad \text{for} \quad \rho < \rho_c(w)\,, \\
&\sim \frac{1}{m^\tau} \quad \text{for} \quad \rho = \rho_c(w)\,, \\
&\sim \frac{1}{m^\tau} + \text{"condensate"} \quad \text{for} \quad \rho > \rho_c\,,
\end{aligned}
\tag{14.6}
$$

where the exponent τ is equal to $5/2$ within the mean-field theory [1] and is conjectured to have the same mean-field value even in one dimension [17].

Note that unlike the situation for the ZRP, the exact joint distribution of masses $P(m_1, m_2, \ldots, m_L)$ in the steady state is not known for the symmetric chipping model. In fact, it is believed [17] that $P(m_1, m_2, \ldots, m_L)$ does not have a simple product measure (factorizable) form as in the ZRP in eqn (14.1). Another important difference is that in the ZRP, the condensation transition happens for both asymmetric and symmetric transfer of masses to the neighbors, as long as the rate $U(m)$ is chosen appropriately. In contrast, for the chipping model, a true condensation transition happens in the thermodynamic limit only for a symmetric transfer of masses. For an asymmetric transfer of masses (say only to the right neighbor), the condensed phase disappears in the thermodynamic limit even though, for finite L, one does see a vestige of a condensation transition [18]. However, a generalization that includes both the chipping model and the ZRP as special cases does appear to have a condensation transition even with asymmetric hopping [19]. Finally, when the diffusion rate depends on the mass of the departure site in the chipping model in a certain manner, the condensation transition disappears [20, 21].

This simple chipping model with aggregation and fragmentation rules has been useful in various experimental contexts, such as the growth of palladium nanoparticles [22]. In addition, the possibility of such a condensation phase transition driven by an aggregation mechanism has been discussed in the case of a system of Au sputtered by swift heavy ions [23]. Finally, the chipping model and its various generalizations have also been studied in the context of traffic [24], finance [25], and networks [26].

14.2.3 Asymmetric random average process

Another simple mass transport model that has been studied extensively [27–30] is the asymmetric random average process (ARAP). As in the previous two models, the ARAP is defined on a one-dimensional lattice with periodic boundary conditions. However, in contrast to the ZRP and the chipping model, here the mass m_i at each site i is assumed to be a continuous positive variable. The model has been studied for both continuous-time and discrete-time dynamics [27, 28]. In the continuous-time version, the microscopic evolution rules are [27, 28]:

- In a small time interval dt, a random fraction $r_i m_i$ of the mass m_i at site i is transported to the right neighbor $(i+1)$ with probability dt, where $r_i \in [0,1]$ is a random number chosen, independently for each site i, from a uniform distribution over $[0,1]$.
- With probability $1-dt$, nothing happens.

The dynamics evidently conserves the total mass $M = \rho L$. At long times, the system reaches a steady state. Once again, the joint distribution of masses $P(m_1, m_2, \ldots, m_L)$ in the steady state does not have a factorized product measure form as in the ZRP.

What does the single-site mass distribution $p(m)$ look like in the large-L limit? The first important point that one notices here is that the density ρ obviously sets the overall mass scale in this model. In other words, the mass distribution $p(m,\rho)$ for any given ρ must have an exact scaling form

$$p(m,\rho) = \frac{1}{\rho} F\left(\frac{m}{\rho}\right), \tag{14.7}$$

where the scaling function $F(x)$ must satisfy the conditions

$$\int_0^\infty F(x)\, dx = 1, \qquad \int_0^\infty x F(x)\, dx = 1\,. \tag{14.8}$$

The first condition follows from the normalization, $\int p(m,\rho) dm = 1$, and the second from mass conservation, $\int m p(m,\rho)\, dm = \rho$. Since the dynamics involves transferring a uniform fraction of mass from one site to a neighbor, the scaling in eqn (14.7) is preserved by the dynamics. This is in contrast to the ZRP and the chipping model, where a single unit of mass is chipped off from one site to a neighbor and thereby the dynamics introduces a separate mass scale (the unit mass), in addition to the overall density ρ.

The scaling function $F(x)$ for the ARAP has been computed within the mean-field theory [27, 28] and is given by

$$F(x) = \frac{1}{\sqrt{2\pi x}}\, e^{-x/2}\,; \tag{14.9}$$

this mean-field result is remarkably close to the numerical results in one dimension, even though one can prove rigorously [28] that the joint distribution of masses does not factorize. In contrast, for the ARAP defined with a parallel discrete-time dynamics (where all sites are updated simultaneously), it has been proved [27, 28] that the joint distribution of masses factorizes as in the ZRP and the scaling function $F(x)$ for the single-site mass distribution can be computed exactly, $F(x) = 4xe^{-2x}$. The steady state of the discrete-time ARAP is also related to the steady state of the so-called q-model of force fluctuations in granular materials [31].

What about condensation? In the ARAP, one does not find a condensation transition. This is of course expected, since the density ρ just sets the mass scale and one does not expect to see a change of behavior in the mass distribution upon increasing ρ, apart from a trivial rescaling of the mass at all sites by a constant factor ρ. However, one can induce a condensation transition in the ARAP by inducing an additional mass scale, for example by imposing a maximum threshold on the amount of mass that may be transferred from a site to a neighbor [30].

14.3 A generalized mass transport model

Let us reflect for a moment on what we have learnt so far from the three models discussed above. It is clear that the dynamics of mass transport often, though not always, may lead to a steady state that exhibits real-space condensation. For example, the ZRP and the symmetric chipping model exhibit real-space condensation, but not the ARAP. Also, we note that some of these models, such as the ZRP, have a simple, factorizable steady state. But *factorizability of the steady state is clearly not a necessary condition for a system to exhibit real-space condensation*, as we have learnt, for example, from the study of the symmetric chipping model, where the steady state is not factorizable. The factorizability, if present, of course helps the mathematical analysis.

So, a natural question is: given a set of microscopic mass transport rules, what are the necessary and sufficient conditions that they may lead to a steady state that exhibits real-space condensation? For example, from the study of the three models above, it seems that in order to have a condensation one needs to introduce via the dynamics a different mass scale, in addition to the density, such as the chipping of a single unit of mass in the ZRP and the symmetric chipping model, or a maximum cap on the mass to be transferred in the ARAP. If there is only one overall mass scale (set by the density) that is preserved by the dynamics as in the usual ARAP, one does not expect to see a phase transition in the mass distribution as one changes the density.

This question about finding the conditions for real-space condensation in a generic mass transport model seems too general, and is perhaps difficult to answer. Instead, one useful strategy is to restrict ourselves to a special class of mass transport models that have a factorizable steady state and then ask about the criterion, mechanism, and nature of the real-space condensation phenomenon within this restricted class of mass transport models, which includes the ZRP as a special class. This strategy has been demonstrated to work rather successfully in a recent series of papers [32–36], and a fairly good understanding of the real-space condensation phenomenon has been developed within this restricted class of mass transport models. This is what we will briefly discuss in this section.

One can include all three models discussed in Section 14.2 in a more generalized mass transport model [32]. For simplicity, we define the model here on a one-dimensional ring of L sites with asymmetric transfer rules, but it can be generalized in a straightforward manner to arbitrary graphs and arbitrary transfer rules. Similar mass transport models with open boundaries and with dissipation at each site have also been studied [37], though here we restrict ourselves to periodic boundary conditions and nondissipative dynamics that preserve the total mass. On each site of the ring, there is a scalar continuous mass m_i. At any given time t, we choose a mass $0 \leq \mu_i \leq m_i$ independently at each site from a probability distribution $\phi(\mu_i|m_i)$, normalized such that $\int_0^m \phi(\mu|m)\, d\mu = 1$. In the time interval $[t, t+dt]$, the mass μ_i is transferred from site i to site $i+1$, simultaneously for all sites i (see Fig. 14.1). In a ring geometry, the site $(L+1)$ is identified with the site 1. Thus, after this transfer, the new masses at time $t + dt$ are given by [32]

$$m_i(t+dt) = m_i(t) - \mu_i(t) + \mu_{i-1}(t)\,, \tag{14.10}$$

where the second term on the right-hand side denotes the mass that has left site i and the third term denotes the mass that has come to site i from site $(i-1)$. The function $\phi(\mu|m)$ that specifies the distribution of the stochastic mass to be transferred from any given site will

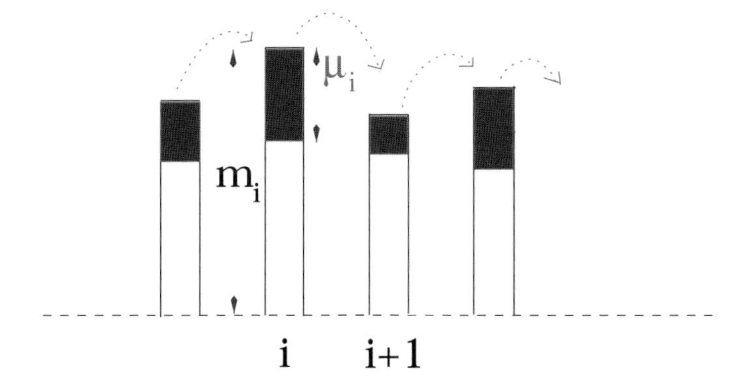

Fig. 14.1 Generalized mass transport model where a random mass μ_i is transferred from site i, with mass m_i, to site $i + 1$.

be called the "chipping kernel." Here we take a homogeneous chipping kernel $\phi(\mu|m)$ which does not depend on the site index. Note that the model above has been defined with parallel dynamics, where all sites are updated simultaneously. Of course, by choosing $dt \to 0$, one can recover a continuous-time random sequential dynamics where the probability that two sites will be updated simultaneously is very small, $\sim O((dt)^2)$. Thus the parallel dynamics includes the continuous-time (random sequential) dynamics as a special case. Note that for random sequential dynamics, $\phi(\mu|m)$ must generically be of the form

$$\phi(\mu|m) = \alpha(\mu|m)\, dt + \left[1 - dt \int_0^m \alpha(\mu'|m)\, d\mu'\right] \delta(\mu)\,, \tag{14.11}$$

where $\alpha(\mu|m)$ denotes the *rate* at which a mass μ leaves a site with mass m, and the second term denotes the probability that no mass leaves the site. The form of eqn (14.11) is designed so that it automatically satisfies the normalization condition $\int_0^m \phi(\mu|m)\, d\mu = 1$.

This model with a general chipping kernel $\phi(\mu|m)$ includes the previously discussed ZRP, chipping model and ARAP as special cases [32]. Since we introduced these models in the previous section in continuous time, we will consider here a generalized model with a chipping kernel of the form of eqn (14.11) with a general chipping rate $\alpha(\mu|m)$. But, of course, one can consider a more general $\phi(\mu|m)$ with a parallel dynamics that includes the continuous-time dynamics as a special case. Let us consider the three previous examples:

(i) As the first example, we see that the ZRP is recovered if, in eqn (14.11), for $0 \le \mu \le m$, we choose

$$\alpha(\mu|m) = U(m)\delta(\mu - 1)\,. \tag{14.12}$$

Note that $U(m)$ is zero if $m < 1$.

(ii) Similarly, the asymmetric chipping model is recovered if we choose

$$\alpha(\mu|m) = w\delta(\mu - 1) + \delta(\mu - m)\,, \tag{14.13}$$

where the first term refers to the event of a transfer of a single unit of mass with rate w, and the second term refers to the transfer of the full mass m with rate 1.

(iii) Finally, we recover the ARAP by choosing

$$\alpha(\mu|m) = \frac{1}{m} \tag{14.14}$$

for all $0 \leq \mu \leq m$, corresponding to the transfer of a fraction of the mass that is chosen uniformly in $[0, 1]$, leading to a uniform rate $\alpha(\mu|m)$ independent of μ.

By appropriately choosing the chipping kernel $\phi(\mu|m)$ or, equivalently, the rate $\alpha(\mu|m)$ for continuous-time dynamics, one can construct a whole class of mass transport models, thus justifying the name "generalized mass transport model."

Given a general chipping kernel $\phi(\mu|m)$ or, equivalently, the chipping rate $\alpha(\mu|m)$ for continuous-time dynamics, one can ask two important questions: (i) What is the steady-state joint mass distribution $P(m_1, m_2, \ldots, m_L)$? (ii) Which types of $\phi(\mu|m)$ or, equivalently, $\alpha(\mu|m)$ for continuous-time dynamics, lead to a real-space condensation transition in the steady state? As discussed earlier, answers to either of these questions are hard to provide for a general chipping kernel $\phi(\mu|m)$ (or chipping rate $\alpha(\mu|m)$). However, let us now restrict ourselves only to those chipping kernels $\phi(\mu|m)$ (or $\alpha(\mu|m)$) that lead to a factorized steady-state distribution of the form

$$P(m_1, m_2, \ldots, m_L) = \frac{1}{Z_L(M)} f(m_1)\, f(m_2) \ldots f(m_L)\, \delta\left(\sum_i m_i - M\right), \tag{14.15}$$

where the partition function $Z_L(M)$ is just a normalization constant. This then leads us to a restricted mass transport model with a factorizable steady state, and we will address our questions regarding real-space condensation within this restricted class. The answers to these questions turn out to be easier for this restricted class, since one can make use of the exact form of the steady-state joint mass distribution (14.15).

14.3.1 A restricted mass transport model

Here we restrict ourselves to those chipping kernels that lead to a factorizable steady state (14.15). Let us first investigate the following question: given $\phi(\mu|m)$ (or, equivalently, $\alpha(\mu|m)$ for continuous-time dynamics), what is the necessary and sufficient condition on $\phi(\mu|m)$ that leads to a factorizable steady state as in eqn (14.15), and if it happens, what is the exact form of the weight function $f(m)$ in terms of $\phi(\mu|m)$? Fortunately, answers to both questions can be obtained exactly, which we state below without giving the details of the proof (see ([32]) for details). The necessary and sufficient condition for factorizability is that $\phi(\mu|m)$ must be of the form [32]

$$\phi(\mu|m) = \frac{v(\mu)\, w(m-\mu)}{\int_0^m v(y)\, w(m-y)\, dy}, \tag{14.16}$$

where $v(x)$ and $w(x)$ are arbitrary positive functions and the denominator is chosen to ensure normalizability, $\int_0^m \phi(\mu|m)\, d\mu = 1$. In other words, if the chipping kernel $\phi(\mu|m)$ factorizes into a function of the mass that leaves the site and a function of the mass that stays on the site, then the steady state is guaranteed to be factorizable as in eqn (14.15), with a weight function whose exact form is given by the denominator in eqn (14.16),

$$f(m) = \int_0^m v(y)\, w(m-y)\, dy. \tag{14.17}$$

This is a sufficient condition. On the other hand, this condition can also be proved to be necessary, i.e. given that the steady state is of the form (14.15) with some $f(m)$, the chipping kernel has to be of the type (14.16), and $f(m)$ then must have the form (14.17).

An analogous condition can be found for continuous-time dynamics, where $\phi(\mu|m)$ has the form (14.11) with a chipping rate $\alpha(\mu|m)$. The necessary and sufficient condition stated above for $\phi(\mu|m)$ translates into the following condition on $\alpha(\mu|m)$ [32],

$$\alpha(\mu|m) = y(\mu)\,\frac{z(m-\mu)}{z(m)}\,, \tag{14.18}$$

where $y(x)$ and $z(x)$ are two arbitrary positive functions. Note that $\alpha(\mu|m)$ is a rate (and not a probability), and hence there is no normalization condition here to be satisfied. If the rate $\alpha(\mu|m)$ has the form (14.18) then we are guaranteed to reach a factorizable steady state (14.15) with a simple weight function [32]

$$f(m) = z(m)\,. \tag{14.19}$$

As an example, it is easy to verify that the chipping rate in the ZRP (14.12) can indeed be written in the form of eqn (14.18) by choosing $y(\mu) = \delta(\mu - 1)$ and $z(m) = \prod_{k=1}^{m}(1/U(k))$ for $m \geq 1$, $z(0) = 1$, and $z(m < 0) = 0$. Thus, the ZRP with sequential dynamics is guaranteed to have a factorizable steady state (14.15) with a weight function $f(m) = z(m) = \prod_{k=1}^{m}(1/U(k))$ for $m \geq 1$ and $f(0) = z(0) = 1$. In contrast, both for the asymmetric chipping model and for the ARAP, the chipping rates, given in eqns (14.13) and (14.14), respectively, cannot be written in the form (14.18) with some choice of non-negative functions $y(x)$ and $z(x)$, proving that neither of these two models has a factorizable steady state.

One can ask several other interesting related questions. For example, suppose we are given a chipping kernel $\phi(\mu|m)$ and we want to know if it has a factorizable steady state or not. This amounts to an explicit search for suitable non-negative functions $v(x)$ and $w(x)$ such that the kernel $\phi(\mu|m)$ can be written as in eqn (14.16). This is often laborious. Can one devise a simple test which will allow us to do this search quickly just by looking at the functional form of $\phi(\mu|m)$? It turns out that indeed there is such a simple test, which can be stated as follows [33]. Given $\phi(\mu|m)$, first set $m = \mu + \sigma$ and compute the following two derivatives:

$$q(\mu,\sigma) = \partial_\mu \partial_\sigma \log\left[\phi(\mu|\mu+\sigma)\right], \tag{14.20}$$

which, in general, is a function of both variables μ and σ. The test devised in [33] states that a given $\phi(\mu|m)$ will lead to a factorizable steady state if and only if this function $q(\mu,\sigma)$ is a function only of the single variable $\mu + \sigma$, i.e.

$$q(\mu,\sigma) = h(\mu+\sigma)\,, \tag{14.21}$$

and in that case the weight function $f(m)$ for the factorizable steady state (14.15) is given explicitly by

$$f(m) = \exp\left[-\int^{m} dx \int^{x} dy\, h(y)\right]. \tag{14.22}$$

In the discussion above, we have focused only on the ring geometry with an asymmetric transfer of mass. Some of these results can be partially generalized to higher dimensions and even to arbitrary graphs [36, 38], and also to transport models with more than one species [39] of scalar variables, such as mass and energy, for instance.

14.4 Condensation in mass transport models with a factorizable steady state

In this section, we discuss issues related to condensation within the restricted class of mass transport models that have a special steady state—namely, a factorizable joint distribution (14.15) with a suitable weight function $f(m)$. Some aspects of the condensation transition in such a factorizable steady state, notably properties in the fluid state, were first studied in the context of the backgammon model [40] without recourse to the dynamics that gives rise to such a factorizable steady state. A more complete analysis, including a study of the properties of the condensed phase, was undertaken in [34, 35], which will be summarized in this section.

There are three main issues here. (i) *Criterion*: what kind of weight functions $f(m)$ lead to a condensation transition? (ii) *Mechanism*: what is the mechanism of the condensation transition when there is one? (iii) *Nature*: what is the nature of the condensate, for example what are the statistics of the mass in the condensate in the condensed phase? All of these questions can be answered in detail for factorizable steady states. We briefly mention the main results here; the details can be found in [34, 35].

14.4.1 Criterion

The factorization property (14.15) allows us to find the criterion for a condensation transition rather easily by working in the grand canonical ensemble (GCE). Within the GCE framework, we introduce a fugacity $\exp[-sm]$, where s is the negative of the chemical potential associated with each site. This is just equivalent to taking the Laplace transform of eqn (14.15) with respect to the total mass M (with s being the Laplace variable), which replaces the delta function by $\exp[-s(m_1+m_2+\ldots m_L)]$. Then s is chosen such that the constraint $M=\sum m_i$ is satisfied on average. Given that each site now has a mass distribution $p(m)=f(m)\exp[-sm]$ (up to a normalization constant), the equation that fixes the value of s for a given $M=\rho L$ is simply

$$\rho=\rho(s)\equiv\frac{\int_0^\infty mf(m)e^{-sm}\,dm}{\int_0^\infty f(m)e^{-sm}\,dm}. \tag{14.23}$$

The criterion for condensation can be derived easily by analyzing the function $\rho(s)$ defined in eqn (14.23). If, for a given ρ, we find a solution to this equation $s=s^*$ such that the single-site mass distribution is normalizable, i.e. $\int p(m)\,dm=\int f(m)\,\exp[-s^*m]\,dm$ is finite, then there is no condensation, in the sense that for all values of ρ, the single-site mass distribution has an exponential tail and there is not one special site that needs to accommodate extra mass. On the other hand, it may be that for certain $f(m)$'s, as ρ is increased, there may be a critical value ρ_c below which a good solution s to eqn (14.23) is found, but such a solution ceases to exist for $\rho>\rho_c$. This will then signal the onset of a condensation because for $\rho>\rho_c$, the system needs to break up into two parts: (a) a critical background fluid part consisting of $(L-1)$ sites, at each of which the average density is the critical value ρ_c, and (b) a single condensate site which accommodates the additional mass $(\rho-\rho_c)L$.

As an example, let us consider an $f(m)$ that decays more slowly than an exponential, but faster than $1/m^2$ for large m. Since $f(m)$ decays more slowly than an exponential, in order that the single-site mass distribution $p(m)=f(m)\,e^{-s^*m}$ is normalizable (i.e. $\int p(m)\,dm=1$), the possible solution s^* of eqn (14.23) cannot be negative. Thus the lowest possible solution is $s^* = 0$. Now as $s \to 0$, the function $\rho(s)$ in eqn (14.23) approaches a critical value,

$$\rho_c = \rho(s \to 0) = \frac{\int_0^\infty m f(m)\, dm}{\int_0^\infty f(m)\, dm}, \tag{14.24}$$

which is finite since $f(m)$ decays faster than $1/m^2$ for large m. Thus, as long as $\rho < \rho_c$, by solving eqn (14.23) one will get a positive solution s^* and hence no condensation. As $\rho \to \rho_c$ from below, $s^* \to 0$ from above. But for $\rho > \rho_c$, there is no positive solution s^* to eqn (14.23), which signals the onset of a condensation transition.

A detailed analysis of eqn (14.23) shows [35] that in order to have condensation, the weight function $f(m)$ must have a large-m tail that lies above an exponential but below $1/m^2$, i.e. $\exp[-cm] < f(m) < 1/m^2$ for large m with some positive constant $c > 0$. A natural candidate satisfying this criterion is

$$f(m) \simeq A m^{-\gamma} \quad \text{with} \quad \gamma > 2 \tag{14.25}$$

for large m. Indeed, the ZRP discussed in the previous sections with the choice $U(m) \sim (1 + \gamma/m)$ for large m leads to a weight function $f(m)$ of the form of eqn (14.25), and then condensation happens only for $\gamma > 2$.

14.4.2 Mechanism and nature

Given an appropriate weight function $f(m)$ such as that in eqn (14.25) that leads to condensation, one can then ask about the mathematical mechanism that drives the condensation. Actually, there is a very simple way to understand this mechanism in terms of sums of random variables [35], which we will discuss in the next section. For now, we notice that in an infinite system, where the GCE is appropriate, the single-site mass distribution $p(m)$ has the form $p(m) = f(m)\exp[-sm]$, with an appropriate s which is the solution of eqn (14.23) as long as $\rho < \rho_c$. For $\rho > \rho_c$, there is no solution to eqn (14.23). In fact, for $\rho > \rho_c$, the value of s remains at its critical value s_c and the GCE framework is no longer valid. To understand how the condensation manifests itself in the single-site mass distribution, one has to study a system with a finite size L and work in the canonical ensemble with a strict delta function constraint as in eqn (14.15).

In a finite system of size L, the single-site mass distribution $p(m)$ can be obtained by integrating the joint distribution (14.15) over the masses at all sites except one, where the mass is fixed at m. It is easy to see from eqn (14.15) that

$$p(m) = \int P(m, m_1, m_2, \ldots, m_L)\, dm_2\, dm_3 \ldots dm_L = f(m)\frac{Z_{L-1}(M-m)}{Z_L(m)}, \tag{14.26}$$

where the partition function $Z_L(M)$ is given by

$$Z_L(M) = \int f(m_1) f(m_2) f(m_L) \delta\left(\sum_i m_i - M\right) dm_1\, dm_2 \ldots dm_L\,. \tag{14.27}$$

Taking the Laplace transform with respect to M, we obtain

$$\int_0^\infty Z_L(M)\, e^{-sM}\, dM = \left[\int_0^\infty f(m)\, e^{-sm}\, dm\right]^L, \tag{14.28}$$

which can be formally inverted using the Bromwich formula

$$Z_L(M) = \int_{s_0-i\infty}^{s_0+i\infty} \frac{ds}{2\pi i} \exp\left[L\left(\ln g(s) + \rho s\right)\right], \tag{14.29}$$

where we have used $M = \rho L$, the integral runs along the imaginary axis $(s_0 + iy)$ in the complex s plane to the right of all singularities of the integrand, and

$$g(s) \equiv \int_0^\infty f(m)\, e^{-sm}\, dm\,. \tag{14.30}$$

We can write a similar integral representation of the numerator $Z_L(M - m)$ in eqn (14.26). Next, we analyze $Z_L(M)$ and $Z_L(M-m)$ using the method of steepest descent in the large-L limit.

As long as there is a saddle point solution to eqn (14.29), say at $s = s^*$, we see immediately from eqn (14.26) that the single-site mass distribution for large L has the form $p(m) \sim f(m)\exp[-s^* m]$, i.e. we recover the GCE result. Thus the GCE approach is valid as long as there is a saddle point solution s^*. As ρ is increased, the saddle point s^* starts moving towards 0 in the complex s plane and when ρ hits ρ_c, $s^* \to 0$. For $\rho > \rho_c$, there is no saddle point and one has to analyze the Bromwich integrals by correctly choosing the contour to evaluate $p(m)$ for $\rho > \rho_c$. This was done in detail in [35]. We shall omit the details here and mention only the main results for $Z_L(M)$ in eqn (14.29) and subsequently for $p(m)$ in eqn (14.26), when the weight function $f(m)$ is chosen to be of the power-law form (14.25).

If we normalize $f(m)$ such that $\int_0^\infty f(m)\, dm = 1$, the partition function in eqn (14.29) can be interpreted as the probability that the sum of L i.i.d. variables, each drawn from a distribution $f(m)$, is M (see Section 14.5). By analyzing the Bromwich integral (14.29) for large L, using the small-s behavior of $g(s)$ in eqn (14.30), we find [35] that the asymptotic behavior of the distribution $Z_L(M)$ is different for $2 < \gamma \leq 3$ and for $\gamma > 3$.

- $2 < \gamma \leq 3$. In this regime, the following scaling behavior of $Z_L(M)$ is found:

$$Z_L(M) \simeq \frac{1}{L^{1/(\gamma-1)}}\, V_\gamma\left[\frac{\rho_c L - M}{L^{1/(\gamma-1)}}\right], \tag{14.31}$$

where $\rho_c = \mu_1 = \int_0^\infty m f(m)\, dm$ is the first moment, and the function $V_\gamma(z)$ is given explicitly by [35]

$$V_\gamma(z) = \frac{1}{\pi}\int_0^\infty dy\, e^{-c_3 y^{\gamma-1}} \cos\left[b\cos\left(\frac{\pi\gamma}{2}\right) y^{\gamma-1} + yz\right]. \tag{14.32}$$

Here $c_3 = -b\sin(\pi\gamma/2) > 0$ and $b = A\Gamma(1-\gamma)$ for $2 < \gamma < 3$, with A being the amplitude in eqn (14.25). The precise asymptotic tails of this scaling function can be computed [35]:

$$V_\gamma(z) \simeq A\,|z|^{-\gamma} \quad \text{as } z \to -\infty\,, \tag{14.33}$$

$$= c_0 \quad \text{at } z = 0\,, \tag{14.34}$$

$$\simeq c_1 z^{(3-\gamma)/2(\gamma-2)}\, e^{-c_2 z^{(\gamma-1)/(\gamma-2)}} \quad \text{as } z \to \infty\,, \tag{14.35}$$

where c_0, c_1, and c_2 are known constants [35]. Thus the function is manifestly non-Gaussian.

- $\gamma > 3$. In this regime, the partition function $Z_L(M)$ has a Gaussian peak

$$Z_L(M) \simeq \frac{1}{\sqrt{2\pi\Delta^2 L}} e^{-(M-\rho_c L)^2/2\Delta^2 L} \quad \text{for } |M - \rho_c L| \ll O(L^{2/3}), \tag{14.36}$$

where $\Delta^2 = \mu_2 - \mu_1^2$, with $\mu_k = \int_0^\infty m^k f(m)\, dm$ being the k-th moment. But far to the left of the peak, $Z_L(M)$ has a power-law decay [35].

So, what does the single-site mass distribution $p(m)$ in eqn (14.26) look like? We have to use the result for the partition function derived above in eqn (14.26). We find different behavior of $p(m)$ in different regions of the ρ–γ plane. For $\gamma > 2$, there is a critical curve $\rho_c(\gamma)$ in the ρ–γ plane that separates a fluid phase (for $\rho < \rho_c(\gamma)$) from a condensed phase (for $\rho > \rho_c(\gamma)$). In the fluid phase, the mass distribution decays exponentially for large m, i.e. $p(m) \sim \exp[-m/m^*]$, where the characteristic mass m^* increases with increasing density and diverges as the density approaches its critical value ρ_c from below. At $\rho = \rho_c$, the distribution decays with a power law, $p(m) \sim m^{-\gamma}$, for large m. For $\rho > \rho_c$, the distribution, in addition to the power-law-decaying part, develops an additional bump, representing the condensate, centered around the "excess" mass

$$M_{ex} \equiv M - \rho_c L\,. \tag{14.37}$$

Furthermore, by our analysis within the canonical ensemble, we can show that even inside the condensed phase ($\rho > \rho_c(\gamma)$), there are two types of behavior of the condensate depending on the value of γ. For $2 < \gamma < 3$, the condensate is characterized by anomalous non-Gaussian fluctuations, whereas for $\gamma > 3$, the condensate has Gaussian fluctuations. This leads to a rich phase diagram in the ρ–γ plane, a schematic picture of which is presented in Fig. 14.2.

The detailed forms of $p(m)$ for $\rho < \rho_c$ (fluid phase), $\rho > \rho_c$ (condensed phase), and $\rho = \rho_c$ (critical point) are summarized below (see also Fig. 14.3).

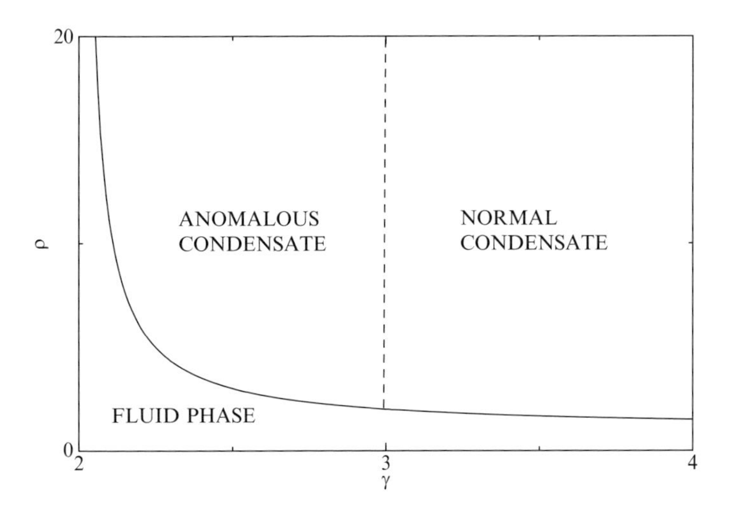

Fig. 14.2 Schematic phase diagram in the ρ–γ plane.

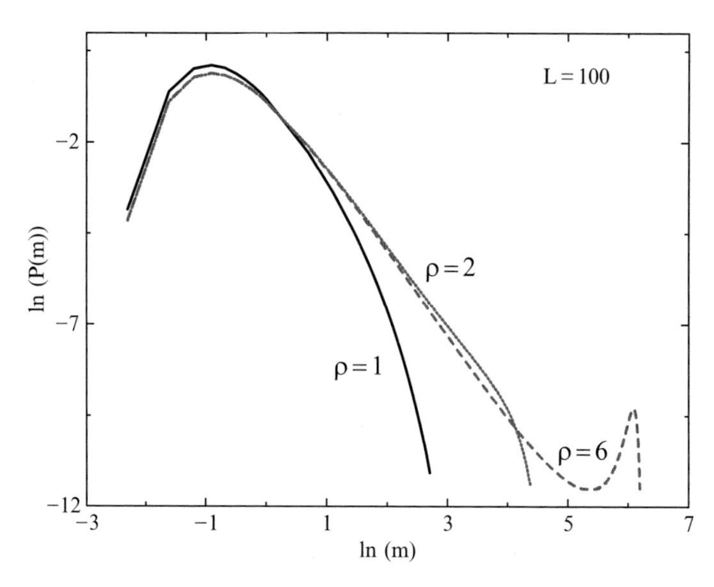

Fig. 14.3 The exact single-site mass distribution $p(m)$ plotted for a particular choice of $f(m)$ with $\gamma = 5/2$ and $\rho_c = 2$, and system size $L = 100$. Full line, $\rho = 1$ (subcritical, fluid phase); dotted line, $\rho = 2$ (critical); dashed line, $\rho = 6$ (supercritical, condensed phase). The condensate bump, $p_{\text{cond}}(m)$, is evident in the supercritical phase.

- *Fluid phase*, $\rho < \rho_c$. In this case, we find

$$p(m) \sim f(m)\, e^{-m/m^*} \quad \text{for} \quad 1 \ll m \ll M\,, \tag{14.38}$$

where the characteristic mass m^* diverges as ρ approaches ρ_c from below as $(\rho - \rho_c)^{-1}$ for $\gamma > 3$ and $(\rho - \rho_c)^{-1/(\gamma-2)}$ for $2 < \gamma < 3$.

- *Condensed phase*, $\rho > \rho_c$. In this case, we find

$$p(m) \simeq f(m) \quad \text{for} \quad 1 \ll m \ll O(L)\,, \tag{14.39}$$

$$p(m) \simeq f(m)\frac{1}{(1-x)^{\gamma}} \quad \text{for} \quad m = xM_{ex}, \quad \text{where} \quad 0 < x < 1\,, \tag{14.40}$$

$$p(m) \sim p_{\text{cond}}(m) \quad \text{for} \quad m \sim M_{ex}\,. \tag{14.41}$$

Here p_{cond} is the piece of $p(m)$ which describes the condensate bump (see Fig. 14.3): centered on the excess mass M_{ex} and with an integral equal to $1/L$, it takes on two distinct forms according to whether $\gamma < 3$ or $\gamma > 3$. For $2 < \gamma < 3$,

$$p_{\text{cond}}(m) \simeq L^{-\gamma/(\gamma-1)}\, V_{\gamma}\left[\frac{m - M_{ex}}{L^{1/(\gamma-1)}}\right], \tag{14.42}$$

where the function $V_{\gamma}(z)$ is given explicitly in eqn (14.32). Thus, clearly, the condensate bump has a non-Gaussian shape for $2 < \gamma < 3$, and we refer to this as an "anomalous" condensate. On the other hand, for $\gamma > 3$,

$$p_{\text{cond}}(m) \simeq \frac{1}{\sqrt{2\pi\Delta^2 L^3}}\, e^{-(m-M_{ex})^2/2\Delta^2 L} \quad \text{for } |m - M_{ex}| \ll O(L^{2/3})\,, \tag{14.43}$$

i.e. $p_{\mathrm{cond}}(m)$ is Gaussian on the scale $|m - M_{ex}| \ll O(L^{2/3})$, but, far to the left of the peak, $p(m)$ decays with a power law.

- *Critical density*, $\rho = \rho_c$. In this case, we find

$$p(m) \propto f(m) V_\gamma \left(m / L^{1/(\gamma-1)} \right) \quad \text{for} \quad 2 < \gamma < 3 \,, \tag{14.44}$$

$$p(m) \propto f(m)\, e^{-m^2/2\Delta^2 L} \quad \text{for} \quad \gamma > 3 \,, \tag{14.45}$$

where the scaling function $V_\gamma(z)$ is as before. Thus, at criticality, $p(m)$ decays with a power law $m^{-\gamma}$ for large m which is cut off by a finite-size scaling function, and the cutoff mass scales as

$$m_{\mathrm{cutoff}} \sim L^{1/(\gamma-1)} \qquad \text{for } 2 < \gamma < 3 \,, \tag{14.46}$$

$$\sim L^{1/2} \qquad \text{for } \gamma > 3 \,. \tag{14.47}$$

14.4.3 Physical picture

It is useful to summarize the main physical picture that emerges out of this mathematical analysis. We notice from the joint distribution (14.15) that the masses at each site are "almost" independent random variables, each with a power law distribution $f(m)$, except for the global constraint of mass conservation imposed by the delta function, which actually makes them "correlated." The system feels this correlation for $\rho < \rho_c$ and $\rho > \rho_c$ in different ways, and exactly at the critical point $\rho = \rho_c$, the effect of the constraint is actually the least. For $\rho < \rho_c$, the effective mass distribution at each site acquires an exponential tail, $p(m) \sim f(m) \exp[-s^* m]$, which is induced by the constraint. For $\rho = \rho_c$, $s^* \to 0$ and $p(m) \sim f(m)$, and thus the system does not feel the constraint at all and the masses behave as completely independent random variables, each distributed via $f(m)$. But for $\rho > \rho_c$, while $(L-1)$ sites behave as a critical fluid, i.e. as if the mass at each of these $(L-1)$ sites was distributed via $f(m)$, there is one single condensate site which acquires an additional mass $(\rho - \rho_c)L$.

For $\rho > \rho_c$, the resulting nonmonotonic shape of the single-site mass distribution (i.e. with an additional bump) in Fig. 14.3 can then be understood very easily from this physical picture. Basically, for $\rho > \rho_c$, the total mass M of the system splits into a critical fluid and a condensate part:

$$M = m_{\mathrm{cond}} + M_{\mathrm{fluid}} \,, \tag{14.48}$$

where m_{cond} denotes the mass in the condensate, and the critical background fluid mass

$$M_{\mathrm{fluid}} = \sum_{i=1}^{L-1} m_i \tag{14.49}$$

is a sum of $(L-1)$ independent random variables (masses) each distributed via $f(m)$. Thus the probability distribution M_{fluid} is given precisely by the partition function $Z_{L-1}(M_{\mathrm{fluid}})$ in eqn (14.27). This partition function can be computed explicitly, and the results are given in eqns (14.31) and (14.36). Knowing this partition function, the distribution of the condensate mass m_{cond} can be obtained using eqn (14.48), giving

$$\mathrm{Prob}(m_{\mathrm{cond}} = y) = Z_{L-1}(M - y) \,. \tag{14.50}$$

The overall single-site mass distribution for $\rho > \rho_c$ can then be computed as follows. If we choose a site at random, with probability $(L-1)/L$ it belongs to the background fluid and hence its mass distribution is $f(m)$, whereas with probability $1/L$ it will be the condensate site with the mass distribution given in eqn (14.50). Thus, for $\rho > \rho_c$,

$$p(m) \approx \frac{(L-1)}{L} f(m) + \frac{1}{L} Z_{L-1}(M-m). \tag{14.51}$$

The second term is what we referred to before as

$$p_{\text{cond}}(m) = \frac{1}{L} Z_{L-1}(M-m), \tag{14.52}$$

and it is this piece that describes the bump in Fig. 14.3 for $\rho > \rho_c$. Its precise asymptotic behavior is detailed in eqns (14.42) and (14.43) for $2 < \gamma \leq 3$ and $\gamma > 3$, respectively.

14.5 Interpretation as sums and extremes of random variables

There is a very nice and simple way [35], using sums of random variables, to understand the mechanism of the condensation transition for factorizable steady states (14.15) with a given weight function $f(m)$, say of the form (14.25). Let us consider a set of L positive i.i.d random variables $\{m_1, m_2, \ldots, m_L\}$ each drawn from a distribution $f(m)$ (we choose $f(m)$ such that it is normalized to unity). Let $M = \sum_{i=1}^{L} m_i$ be the sum. For instance, M can be interpreted as the position of a random walker after L independent steps of lengths $m_1, m_2, \ldots, m_L$. We then notice that the partition function $Z_L(M)$ in eqn (14.27) can be interpreted as the probability that the walker reaches M in L steps starting from the origin.

How does one interpret condensation in this random-walk language? Note from eqn (14.24) that if $f(m)$ is normalized to unity, the critical density

$$\rho_c = \mu_1 = \int_0^\infty m f(m)\, dm \tag{14.53}$$

is just the mean step length of the random walker's steps. Thus if the final position M is less than $\rho_c L = \mu_1 L$ (i.e. $\rho = M/L < \rho_c$), we expect that the typical configuration of the random walker's path will consist of steps each of which is of order $O(1)$. But, for $M > \mu_1 L$ (i.e. $\rho > \rho_c$), the ensemble will be dominated by configurations where $(L-1)$ steps are of order $O(1)$, but also by one single big step of order $(M - \mu_1(L-1)) \sim (\rho - \rho_c)L$ to compensate for the deficit in the distance. This single big step is precisely the condensate. Within this interpretation, it also becomes clear that for an $f(m)$ of the form (14.25), there are two possibilities, depending on whether the second moment of the step length distribution $\mu_2 = \int_0^\infty m^2 f(m)\, dm$ is divergent ($2 < \gamma \leq 3$) or finite ($\gamma > 3$). In the former case, the corresponding random walk is a Lévy flight with anomalously large fluctuations that leads to an anomalously large fluctuation in the condensate mass. In the latter case, by virtue of the central limit theorem, one recovers a Gaussian fluctuation, leading to a Gaussian distribution of the condensate mass. This explains the two types of condensate phases in the phase diagram in Fig. 14.2.

14.5.1 Condensation and extreme statistics

Another interesting issue intimately related to the condensation is the associated extreme value statistics—for example, what is the distribution of the largest mass $m_{\max}$ in the system, where

$$m_{\max} = \max(m_1, m_2, \ldots, m_L)\,? \tag{14.54}$$

This is particularly important in the condensed phase, where the largest mass is carried by the condensate, at least in models with simple factorized steady states where there is a single condensate. The theory of extreme value statistics is well developed in cases where the extremes (e.g. the maximum) of a set of i.i.d random variables are studied [41]. In our case, the factorized steady state in eqn (14.15) shows that the masses are not completely independent, but are correlated via the global mass conservation constraint explicitly manifest in the delta function in eqn (14.15). Without this delta function constraint, with $f(m) \sim m^{-\gamma}$ and $\gamma > 2$, the scaled distribution of the maximal mass would have been a Fréchet distribution [41]. However, the presence of the constraint induces important correlations that change the nature of the distribution of the maximal mass.

In [42], the authors of that publication studied rigorously how the typical value of the extremal mass scales with the system size in the ZRP with $\gamma > 3$. For more general mass transport models with factorized steady states as in eqn (14.15), the full distribution of the extremal mass was studied recently in [43]. In was found that in the fluid phase ($\rho < \rho_c$), where the single-site masses effectively become uncorrelated but with an additional factor $\exp[-s^* m]$ that comes from the conservation constraint, $p(m) \sim f(m)\exp[-s^* m]$ and hence the distribution of the maximal mass, in the scaling limit, becomes a Gumbel distribution [43]. At the critical point, where $s^* = 0$ (i.e. where the constraint is least effective), one recovers the Fréchet distribution. But for $\rho > \rho_c$, the maximal-mass distribution is the same as that of the condensate mass, i.e.

$$m_{\max} = m_{\text{cond}}\,. \tag{14.55}$$

However, as mentioned in the previous section, the distribution of $m_{\text{cond}} = M - M_{\text{fluid}}$ can be computed via computing the distribution of M_{fluid} as a sum of independent random variables each distributed via $f(m)$, and is given by eqn (14.50). Thus,

$$\text{Prob}\,(m_{\max} = y) \approx Z_{L-1}(M - y) = L\, p_{\text{cond}}(y)\,, \tag{14.56}$$

where $p_{\text{cond}}(m)$ is given in eqns (14.42) and (14.43) for $2 < \gamma < 3$ and $\gamma > 3$, respectively. Thus, in the condensed phase, one has a completely new type of extreme value distribution of correlated random variables which is exactly computable [43]. Moreover, it is interesting to note that for $\rho > \rho_c$, the computation of the distribution of an extreme of correlated variables reduces to a calculation of the distribution of a sum of independent random variables.

14.6 Conclusion

In this brief review, I have discussed recent developments in understanding the physics of real-space condensation in a class of mass transport models. Here, real-space condensation refers to the phenomenon when, upon increasing the density beyond a critical value, a macroscopically large mass settles onto a single site in real space in the steady state. The system discussed here is homogeneous in the sense that the transport rules do not depend on the sites or particles. Thus, in the limit of an infinite system size, the formation of a condensate at one single site actually breaks the translational symmetry in the system spontaneously. The criterion for and mechanism of the transition, as well as the detailed finite-size dependence of the distribution of the condensate mass in the condensed phase, were discussed within a restricted class of one-dimensional mass transport models that have a factorizable steady state.

There are several directions in which the questions addressed here can be extended, some of which are briefly mentioned below.

- *Pair-factorized steady states.* Here we have discussed only mass transport models that have a factorized steady state (14.15). A natural question is whether real-space condensation can happen in other types of steady states that are not simply factorizable as in eqn (14.15) and, if so, (i) are there natural *local* transport rules that lead to such steady states and (ii) does the nature of the condensate change fundamentally from the one with factorizable steady states? Recently, a generalization of eqn (14.15), called pair-factorized steady states (PFSSs), was introduced in [44],

$$P(m_1, m_2, \ldots, m_L)] = \frac{1}{Z_L(M)} \prod_{i=1}^{L} g(m_i, m_{i+1})\, \delta\left(\sum_{i=1}^{L} m_i - M\right). \tag{14.57}$$

  Thus there is one factor $g(m_i, m_{i+1})$ for each pair of neighboring sites on a ring of L sites. The transport rules, involving three neighboring sites, that lead to such steady states were also found [44]. Interestingly, the condensate in this PFSS, for a class of weight functions $g(m, n)$ that are short-ranged, was found to be spread over a relatively large number of sites $\sim O(L^{1/2})$ [44], in contrast to the condensate that forms on a single site in the usual factorized steady state (14.15). The average shape of this subextensive condensate for a class of weight functions $g(m, n)$ and the precise form of the "condensate bump" in the single-site mass distribution in the condensed phase were recently computed in a very nice paper [45]. In addition, the transport rules that lead to a PFSS on an arbitrary graph were also found recently [46], thus generalizing one-dimensional models with a PFSS.
- *Dynamics.* Here we have discussed only static properties associated with the condensation transition. Another interesting issue is the nature of the dynamics in the steady state and in the approach to the steady state, in particular in the condensed phase [47–51]. In a finite system, a condensate forms at a site, then survives there for a long time T_s, and then dissolves and forms at another site. For the ZRP with $f(m) \sim m^{-\gamma}$ with $\gamma > 2$, in the stationary state, it was found that while the condensate lifetime T_s was of order $(\rho - \rho_c)^{\gamma+1} L^{\gamma}$ for large L [49], there is another, shorter timescale associated with the relocation of the condensate, $T_r \sim (\rho - \rho_c)^2 L^2$ [50]. In addition, the current fluctuations in the steady state show a striking change of behavior [50] as one goes from the fluid phase ($\rho < \rho_c$) to the condensed phase ($\rho > \rho_c$). On the fluid side, the current fluctuations shows an interesting oscillating behavior due to the presence of kinematic waves [50], similar to the oscillatory behavior of the variance in the displacement of a tagged particle in 1D asymmetric exclusion processes [52]. Another interesting dynamical quantity is the power spectra associated with the time series depicting the evolution of the total number of particles over a fixed segment of a ring, studied recently in the context of the ZRP [53].

  It would be interesting to study the dynamics for mass transport models in higher dimensions or on arbitrary graphs, and also for more generalized steady states such as PFSSs.

Other interesting directions involve studying condensation phenomena in multispecies models [13, 39] and misanthropic processes [13]; instabilities of the condensed phase due to nonconserving rates in the chipping model [1, 54] and in the ZRP [55, 56], and also in the ZRP due to quench-disordered particle transfer rates [57]; a ZRP leading to multiple condensates [58]; condensation in polydisperse hard spheres [59]; and others. Most of these generalizations have been carried out so far in the context of the ZRP, but it would be interesting to study the general mass transport model discussed here with these additional generalizations. It would also be interesting to compute the distribution of maximal mass in PFSSs where the condensate is subextensive. In this case, the extremal mass is not the total mass carried by the full condensate, but rather the mass on the site inside the condensate that carries the largest mass.

Acknowledgments

It is a pleasure to thank my collaborators in this subject: M. Barma, M. R. Evans, C. Godrèche, S. Gupta, T. Hanney, S. Krishnamurthy, R. Rajesh, E. Trizac, and R. K. P. Zia. I also thank A. Comtet, D. Dhar, J. Krug, K. Mallick, H. Meyer-Ortmanns, D. Mukamel, G. Schütz, and C. Sire for many useful discussions. This article is based on a series of lectures that I first gave at the summer school "The Principles of the Dynamics of Non-Equilibrium Systems" held at the Isaac Newton Institute, Cambridge, in 2006, and later at the present summer school "Exact Methods in Low-Dimensional Statistical Physics and Quantum Computing" held at Les Houches in 2008. I thank the organizers of both of the schools for hospitality. Support from Grant No. 3404-2 of the Indo-French Center for the Promotion of Advanced Research (IFCPAR/CEFIPRA) is also gratefully acknowledged.

References

[1] S. N. Majumdar, S. Krishnamurthy, and M. Barma, *Phys. Rev. Lett.* **81**, 3691 (1998).
[2] O. J. O'Loan, M. R. Evans, and M. E. Cates, *Phys. Rev. E* **58**, 1404 (1998).
[3] D. Chowdhury, L. Santen, and A. Schadschneider, *Phys. Rep.* **329**, 199 (2000).
[4] D. van der Meer, K. van der Weele, and D. Lohse, *Phys. Rev. Lett.* **88**, 174302 (2002); J. Torok, *Physica A* **355**, 374 (2005); D. van der Meer *et al.*, *J. Stat. Mech.: Theor. Exp.* P07021 (2007).
[5] Y. Kafri, E. Levine, D. Mukamel, G. M. Schütz, and J. Török, *Phys. Rev. Lett.* **89**, 035702 (2002).
[6] Z. Burda *et al.*, *Phys. Rev. E* **65**, 026102 (2002).
[7] S. N. Dorogovstev and J. F. F. Mendes, *Evolution of Networks*, Oxford University Press, Oxford (2003).
[8] M. R. Evans, *Braz. J. Phys.* **30**, 42 (2000).
[9] M. R. Evans, *Europhys. Lett.* **36**, 13 (1996).
[10] J. Krug and P. A. Ferrari, *J. Phys. A: Math. Gen.* **29**, L465 (1996).
[11] A. G. Angel, M. R. Evans, and D. Mukamel, *J. Stat. Mech.: Theor. Exp.* P04001 (2004).
[12] F. Spitzer, *Adv. Math.* **5**, 246 (1970).
[13] M. R. Evans and T. Hanney, *J. Phys. A: Math. Gen* **38**, R195 (2005).
[14] C. Godrèche, Lecture notes for Luxembourg summer school (2005), arXiv:cond-mat/0604276.
[15] P. L. Krapivsky and S. Redner, *Phys. Rev. E* **54**, 3553 (1996).
[16] S. N. Majumdar, S. Krishnamurthy, and M. Barma, *J. Stat. Phys.* **99**, 1 (2000).

[17] R. Rajesh and S. N. Majumdar, *Phys. Rev. E* **63**, 036114 (2001).
[18] R. Rajesh and S. Krishnamurthy, *Phys. Rev. E* **66**, 046132 (2002).
[19] E. Levine, D. Mukamel, and G. Ziv, *J. Stat. Mech.: Theor. Exp.* P05001 (2004).
[20] R. Rajesh, D. Das, B. Chakraborty, and M. Barma, *Phys. Rev. E* **56**, 056104 (2002).
[21] D. J. Lee, S. Kwon, and Y. Kim, *J. Korean Phys. Soc.* **52**, S154 (2008).
[22] V. Agarwal *et al.*, *Phys. Rev. B* **74**, 035412 (2006).
[23] P. K. Kuiri *et al.*, *Phys. Rev. Lett.* **100**, 245501 (2008).
[24] E. Levine *et al.*, *J. Stat. Phys.* **117**, 819 (2004).
[25] H. Yamamoto *et al.*, *Japan. J. Ind. Appl. Math.* **24**, 211 (2007); A. Svorencik and F. Slanina, *Eur. Phys. J. B* **57**, 453 (2007).
[26] S. Kwon, S. Lee, and Y. Kim, *Phys. Rev. E* **73**, 056102 (2006); *Phys. Rev. E* **78**, 036113 (2008); M. Tang and Z. H. Liu, *Commun. Theor. Phys.* **49**, 252 (2008); D. Y. Hua, *Chinese Phys. Lett.* **26**, 018901 (2009).
[27] J. Krug and J. Garcia, *J. Stat. Phys.* **99**, 31 (2000).
[28] R. Rajesh and S. N. Majumdar, *J. Stat. Phys.* **99**, 943 (2000).
[29] R. Rajesh and S. N. Majumdar, *Phys. Rev. E* **64**, 036103 (2001).
[30] F. Zielen and A. Schadschneider, *Phys. Rev. Lett.* **89**, 090601 (2002).
[31] C.-H. Liu *et al.*, *Science* **269**, 513 (1995); S. N. Coppersmith *et al.*, *Phys. Rev. E* **53**, 4673 (1996).
[32] M. R. Evans, S. N. Majumdar, and R. K. P. Zia, *J. Phys. A: Math. Gen.* **37**, L275 (2004).
[33] R. K. P. Zia, M. R. Evans, and S. N. Majumdar, *J. Stat. Mech.: Theor. Exp.* L10001 (2004).
[34] S. N. Majumdar, M. R. Evans, and R. K. P. Zia, *Phys. Rev. Lett.* **94**, 180601 (2005).
[35] M. R. Evans, S. N. Majumdar, and R. K. P. Zia, *J. Stat. Phys.* **123**, 357 (2006).
[36] M. R. Evans, S. N. Majumdar, and R. K. P. Zia, *J. Phys. A: Math. Gen.* **39**, 4859 (2006).
[37] E. Bertin, *J. Phys. A: Math. Gen.* **39**, 1539 (2006).
[38] R. L. Greenblatt and J. L. Lebowitz, *J. Phys. A: Math. Gen.* **39**, 1565 (2006).
[39] T. Hanney, *J. Stat. Mech.: Theor. Exp.* P12006 (2006).
[40] P. Bialas, Z. Burda, and D. Johnston, *Nucl. Phys. B* **493**, 505 (1997).
[41] E. J. Gumbel, *Statistics of Extremes*, Columbia University, New York (1958).
[42] I. Jeon, P. March, and B. Pittel, *Ann. Probab.* **28**, 1162 (2000).
[43] M. R. Evans and S. N. Majumdar, *J. Stat. Mech.: Theor. Exp.* P05004 (2008).
[44] M. R. Evans, T. Hanney, and S. N. Majumdar, *Phys. Rev. Lett.* **97**, 010602 (2006).
[45] B. Waclaw, J. Sopik, W. Janke, and H. Meyer-Ortmanns, arXiv:0901.3664.
[46] B. Waclaw, J. Sopik, W. Janke, and H. Meyer-Ortmanns, arXiv:0904:0355.
[47] C. Godrèche, *J. Phys. A. Math. Gen.* **36**, 6313 (2003).
[48] S. Groskinsky, G. M. Schütz, and H. Spohn, *J. Stat. Phys.* **113**, 389 (2003).
[49] C. Godréche and J. M. Luck, *J. Phys. A* **38**, 7215 (2005).
[50] S. Gupta, M. Barma, and S. N. Majumdar, *Phys. Rev. E* **76**, 060101(R) (2007).
[51] G. M. Schütz and R. J. Harris, *J. Stat. Phys.* **127**, 419 (2007).
[52] S. Gupta, S. N. Majumdar, C. Godréche, and M. Barma, *Phys. Rev. E* **76**, 021112 (2007).
[53] A. G. Angel and R. K. P. Zia, *J. Stat. Mech.: Theor. Exp.* P03009 (2009).
[54] S. N. Majumdar, S. Krishnamurthy, and M. Barma, *Phys. Rev. E* **61**, 6337 (2000).
[55] A. G. Angel, M. R. Evans, E. Levine, and D. Mukamel, *Phys. Rev. E* **72**, 046132 (2005); *J. Stat. Mech.: Theor. Exp.* P08017 (2007).

[56] S. Grosskinsky and G. M. Schütz, *J. Stat. Phys.* **132**, 77 (2008).
[57] S. Grosskinsky, P. Chleboun, and G. M. Schütz, *Phys. Rev. E* **78**, 030101(R) (2008).
[58] Y. Schwarzkopf, M. R. Evans, and D. Mukamel, arXiv:0801:4501 (2008).
[59] M. R. Evans, S. N. Majumdar, I. Pagonabarraga, and E. Trizac, *J. Chem. Phys.* **132**, 014102 (2010).

15
Quantum spin liquids

G. Misguich

Grégoire Misguich,
Institut de Physique Théorique,
CEA, IPhT, CNRS, URA 2306,
F-91191 Gif-sur-Yvette, France.

15.1 Introduction: Band and Mott insulators

Depending on the context (experiments, theory, simulations, ...), "quantum spin liquid" is sometimes used with several rather different meanings. But let us start with a first simple definition: the ground state of a lattice quantum spin model is said to be a quantum spin liquid (QSL) if it spontaneously breaks *no* symmetry. According to this definition, a QSL is realized if the spins fail to develop any kind of long-range order at zero temperature ($T = 0$) (hence the word "liquid," as opposed to solids, which are ordered and break some symmetries). Of course, this first definition raises a number of questions: Does this define new distinct states of matter? Do QSLs have interesting properties? Are there any experimental examples? To answer these questions, it is useful to go back to the origin of magnetism in insulators.

Generally speaking, there are two kinds of insulators: *band* insulators and *Mott* insulators. The first kind can be qualitatively understood from the limit of noninteracting (or weakly interacting) electrons. Consider, for instance, a periodic lattice[1] with an *even* number n of sites per unit cell, with an average electron density of one electron per site (this is called half-filling). The Hamiltonian describing how the electrons hop from site to site looks like $H_K = -t\sum_{\langle i,j\rangle,\sigma=\uparrow,\downarrow}\left(c^\dagger_{i\sigma}c_{i\sigma} + \text{H.c.}\right)$, where only first-neighbor hopping is considered for simplicity. H can be diagonalized in Fourier space and gives n dispersive bands. The ground state is just the Fermi sea obtained by filling the lowest-energy states. Since the density is one electron per site, the $n/2$ lowest-energy bands are completely filled (one up and one down electron in each single-particle state). Assuming that the band $n/2+1$ is separated by a gap Δ in energy from the $n/2$ lower bands, all the excitations are gapped and, at temperatures smaller than the gap, there are no charge carriers to carry an electric current. This is the well-known picture of a band insulator: there are no low-energy charge degrees of freedom, no magnetic (spin) degrees of freedom, and the ground state (Fermi sea) is unique and breaks no symmetry. To get an interesting QSL, we should instead look at Mott insulators. There, the number of sites per unit cell is *odd* and the noninteracting limit is unable to give the correct insulating behavior (at least one band is partially filled, and hence has low-energy charge excitations). It is more useful to look at the system in the opposite limit of very large electron–electron repulsion, as in the large-U limit of the Hubbard model: $H = H_K + U\sum_i c^\dagger_{i\uparrow}c_{i\uparrow}c^\dagger_{i\downarrow}c_{i\downarrow}$. At $U = \infty$ and $t = 0$ (still at half-filling), the ground state is highly degenerate ($= 2^V$, where V is the total number of sites), since any state with one electron per site is a ground state, whatever the spin orientations. To describe how this degeneracy is lifted at weak but finite t/U, a second-order perturbation has to be computed.[2] The result is an effective Hamiltonian acting in the subspace of spin configurations, and takes the form of a quantum spin-$\frac{1}{2}$ Heisenberg model:

$$H = \frac{1}{2}\sum_{ij} J_{ij}\vec{S}_i \cdot \vec{S}_j\,, \tag{15.1}$$

[1] We use a tight-binding model where the solid is modeled by one state per site, neglecting (or, more precisely, integrating out) filled orbitals and high-energy empty states.

[2] To first order in t, a single electron hopping inevitably leads to a doubly occupied site.

where $J_{ij} = t_{ij}^2/U$ involves the hopping amplitude t_{ij} between sites i and j and measures the strength of the antiferromagnetic (AF) interaction between the (electron) spins $\vec{S}_i$ and $\vec{S}_j$.[3]

Although the model of eqn (15.1) is in general a complicated quantum many-body problem with very few exact results,[4] its ground state and low-energy properties are qualitatively well understood in many cases. In particular, the ground state can be *antiferromagnetically ordered* (also called a Néel state). Such a state can be approached from the semiclassical point of view described in Section 15.3: the spins point in well-defined directions and form a regular structure. Most of the Mott insulators studied experimentally belong to this family. The simplest example is the nearest-neighbor Heisenberg model on a bipartite lattice such as the square, cubic, or hexagonal lattice. There, on average, all the spins of sublattice A point in the direction $+\vec{S}_0$ (spontaneous symmetry breaking of the $\mathrm{SU}(2)$ rotation symmetry), and all the spins of sublattice B point in the direction $-\vec{S}_0$. The difference from a classical spin configuration is that the magnetization of one sublattice (this is the order parameter for a Néel state) is reduced by the quantum zero-point fluctuations of the spins, even at $T = 0$. Such ordered states are not QSLs (they might instead be called spin "solids"), since they break rotation symmetry.

The main question addressed in these notes is the fate of the ground state of eqn (15.1) when the lattice and the interactions J_{ij} are such that the spins *fail* to develop any such Néel ordered state. A state without any order is not necessarily interesting from a theoretical point of view. For instance, a spin system at very high temperature is completely disordered and does not have any rich structure. As we will see, the situation in Mott insulators at $T = 0$ is completely different. A first hint that Mott QSLs host some interesting topological properties will be discussed in Section 15.4 (on the Lieb–Schultz–Mattis–Hastings theorem [1, 2]). A concrete (but qualitative) picture of QSL wave functions is given in Section 15.5, in terms of short-range valence bond configurations and deconfined spinons (magnetic excitations carrying a spin $\frac{1}{2}$). Finally, Section 15.6 presents a formalism which puts some of the ideas above on firmer ground. It is based on a large-N generalization of the Heisenberg models ($\mathrm{SU}(2) \to \mathrm{Sp}(N)$) which allows us to describe some gapped QSLs and to establish a connection to topologically ordered states of matter, such as the ground state of Kitaev's toric code [3].

15.2 Some materials without magnetic order at $T = 0$

There are many magnetic insulators that do order at $T = 0$.[5] For instance, the magnetic, properties of many compounds are described by 1D spin chains of spin ladder Hamiltonians. Thanks to the Mermin–Wagner theorem and the reduced dimensionality, these systems cannot

[3] In real materials, there are often tens or hundreds of electrons per unit cell, several ions, and many atomic orbitals. Although the description of the magnetic properties in terms of lattice spin models is often very accurate, the spin–spin interactions are often more complicated than in this antiferromagnetic Heisenberg model. It is quite frequent that some interactions violate the $\mathrm{SU}(2)$ symmetry of the Heisenberg model, owing to spin–orbit couplings in a crystalline environment. In these notes, we focus on models with an $\mathrm{SU}(2)$ symmetry.

[4] In these notes, we focus here on dimension $D > 1$, but much more is known about one-dimensional (1D) spin chains.

[5] Some order at a temperature with is very small compared the typical energy scale of the Heisenberg spin–spin interactions. This is often due to perturbations that are not included in the simplest Heisenberg model description.

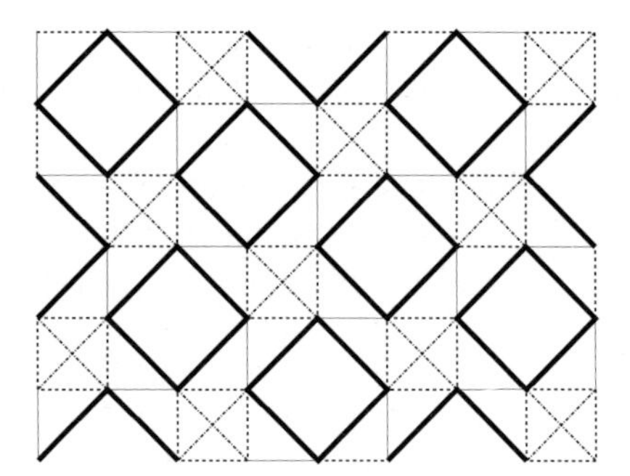

Fig. 15.1 Depleted square lattice model for the magnetic properties of CaV_4O_9. The different exchange energies are shown by different types of line. The strongest J corresponds to the thick lines forming the large, tilted square plaquettes.

develop long-range spin–spin correlations, even at $T = 0$.[6] They certainly deserve to be called QSLs and represent a very rich field of activity. In these notes, we will instead focus on QSLs in $D > 1$ systems, where our present understanding is less complete.

CaV_4O_9 is the first Heisenberg system in $D > 1$ where the magnetic excitations were experimentally shown to be gapped, in 1995, [4]. This compound can be modeled by an antiferromagnetic spin-$\frac{1}{2}$ Heisenberg model on a *depleted* square lattice where one site out of five is missing (Fig. 15.1). The remaining sites correspond to the locations of the vanadium ions, which carry the magnetically active electrons (spins). The magnetic interactions J_{ij} turn out to be significant not only between nearest neighbors, but also between second nearest neighbors (the electrons hop through oxygen orbitals, which have a complex geometry). Through magnetic susceptibility measurements, it was shown that the ground state is a rotationally invariant spin singlet, thus excluding any Néel ordering. This QSL behavior can be understood by taking a limit where only the strongest J_{ij} are kept, and the others are set to zero. It turns out that these strongest couplings are between second nearest neighbors, and form a set of *decoupled* four-site plaquettes (of area $\sqrt{2} \times \sqrt{2}$ and surrounding a missing site). Since the ground state of such a four-site Heisenberg cluster is a unique singlet $S = 0$ state, separated by a gap from other states, the model is trivially gapped and without any broken symmetry in this limit. But this is not the kind of QSL we want to focus on here, since it can be adiabatically transformed into a band insulator. Switching off the electron–electron interactions would make the system metallic, but one can proceed in a different way. Starting with realistic values of the J_{ij}, the interplaquette couplings are gradually turned off. In doing so, one can check (numerically, for instance) that the spin gap does not close and no (quantum) phase transition is encountered. Then, in this system of decoupled four-electron clusters, the Hubbard repulsion U can be switched to zero, without causing any phase transition. The final model is evidently a band insulator and smoothly connected to the initial Heisenberg model.

Since then, numerous 2D and 3D (Heisenberg) magnetic systems with an *even* number of spins $\frac{1}{2}$ per unit cell have been found to be gapped. To our knowledge, their ground state can be qualitatively understood from a limit of weakly coupled clusters in all cases and can

[6] Owing to residual 3D couplings, there can be a finite-temperature phase transition to an ordered state at very low temperature.

therefore be "classified" as band insulators (like CaV_4O_9 above). Some of them can be very interesting for various reasons,[7] but their ground states are not fundamentally new states of matter.

In recent years, experimentalists have also discovered a number of materials which are well described by 2D Heisenberg models with an *odd* number of spins $\frac{1}{2}$ per crystal unit cell,[8] and which do not develop any Néel order when $T \to 0$. Some examples are the minerals herbertsmithite ($ZnCu_3(OH)_6Cl_2$)[9] [7] and volborthite ($Cu_3V_2O_7(OH)_2 \cdot 2H_2O$) [8] (both with a kagome lattice geometry), triangle-based organic materials [9, 10], and triangular atomic layers of He^3 adsorbed onto graphite [11] (there, the spin is not electronic, but nuclear). It turns out that all these systems seem to have *gapless* magnetic excitations, and a complete theoretical understanding of these systems is still lacking. The present theories for gapless QSLs are rather elaborate [12], and many questions remain open (stability, nature of the excitations, correlation exponents, etc.). However, as we will see, *gapped* QSLs are simpler from a theoretical point of view. Intriguingly, to our knowledge, no gapped QSL has been discovered so far in nature, although many spin models do have gapped QSL ground states.

15.3 Spin wave theory, zero modes, and breakdown of the $1/S$ expansion

To understand why an AF Heisenberg spin model can *fail* to order at zero temperature, it is useful to briefly review the standard approach to Néel phases: the semiclassical $1/S$ spin-wave expansion [13]. This approach (i) starts from a classical spin configuration which minimizes the Heisenberg interaction, (ii) assumes that the quantum deviations from this ordered direction are *small*, and (iii) treats these deviations as a collection of harmonic oscillators (the leading term in a $1/S$ expansion). In this approximation, the Hamiltonian is written using boson creation and annihilation operators, is quadratic, and can be diagonalized by a Bogoliubov transformation. One can then check a posteriori if the spin deviations are indeed small. If this is not the case, we have a strong indication that the magnetic long-range order is in fact "destroyed" by the quantum fluctuations, thus opening a route for a QSL ground state.

15.3.1 Holstein–Primakoff representation

The starting point is the representation of the spin operators using Holstein–Primakoff [14] bosons

$$S_i^z = S - a_i^\dagger a_i\,, \quad S^+ = \sqrt{2S - a_i^\dagger a_i}\; a_i\,, \quad S^- = a_i^\dagger \sqrt{2S - a_i^\dagger a_i}\,, \tag{15.2}$$

from which one can check that the commutation relations $[S_i^\alpha, S_i^\beta] = i\epsilon^{\alpha\beta\delta} S_i^\delta$ and $\vec{S}_i^2 = S(S+1)$ are satisfied (using $[a_i, a_i^\dagger] = 1$).

Let $\{\vec{z}_i\}$ be a classical ground state of eqn (15.1), minimizing $E = \frac{1}{2}\sum_{ij} J_{ij}\vec{z}_i \cdot \vec{z}_j$ with $\vec{z}_i^2 = 1$. These directions can be used as local quantization axes: we use eqn (15.2) in a local (orthogonal) frame $(\vec{x}_i, \vec{y}_i, \vec{z}_i = \vec{x}_i \wedge \vec{y}_i)$ adapted to the classical ground state. Under the

[7] For instance, $TlCuCl_3$ [6] is a coupled dimer system with a Bose–Einstein condensation of magnetic excitations in the presence of an external magnetic field, and $SrCu_3(BO_3)_2$ [5] has a magnetization curve with quantized plateaus.

[8] In such cases, the absence of long-range order cannot be attributed to band insulator physics.

[9] Although the spin–spin interaction strength is of the order of $J \sim 200$ K, no order has be found down to 50 mK.

assumption that $\vec{S}_i$ shows small deviations from the classical vector $S\vec{z}_i$, the typical number $\langle a^\dagger a\rangle$ of Holstein–Primakoff bosons should be small compared with S. We can therefore simplify S^+ (and S^-) in eqn (15.2) by keeping only $\sqrt{2S}$ in the square roots, to obtain [13]

$$\vec{S}_i \simeq \left(\left(S+\frac{1}{2}\right) - \vec{\pi}_i^2\right)\vec{z}_i + \sqrt{2S}\vec{\pi}_i\,, \tag{15.3}$$

where

$$\vec{\pi}_i = \frac{1}{2}(a_i + a_i^\dagger)\vec{x}_i + \frac{1}{2i}(a_i - a_i^\dagger)\vec{y}_i\,, \tag{15.4}$$

$$\vec{\pi}_i^2 = a_i^\dagger a_i + \frac{1}{2}\,, \tag{15.5}$$

$$\text{and} \quad \vec{z}_i \cdot \vec{\pi}_i = 0\,. \tag{15.6}$$

Replacing eqn (15.3) in the Hamiltonian gives

$$H = \frac{1}{2}\left(S+\frac{1}{2}\right)^2 \sum_{ij} J_{ij}\,\vec{z}_i\cdot\vec{z}_j + S\sum_{ij} J_{ij}\,\vec{\pi}_i\cdot\vec{\pi}_j - \frac{1}{2}S\sum_{ij} J_{ij}\left(\vec{\pi}_i^2 + \vec{\pi}_j^2\right)\vec{z}_i\cdot\vec{z}_j + \mathcal{O}(S^0)\,. \tag{15.7}$$

The first term is a constant, proportional to the classical energy E_0. The two other terms, proportional to S, are quadratic in the boson operators and describe the spin fluctuations as a set of coupled harmonic oscillators.[10] The operators for the positions $q_i = (1/\sqrt{2})(a_i + a_i^\dagger)$ and momenta $p_i = (1/\sqrt{2}i)(a_i - a_i^\dagger)$ of these oscillators can be conveniently grouped into a column vector of size $2N$ (where N is the total number of spins),

$$\mathbf{V} = \begin{bmatrix} q_1 \\ \vdots \\ q_N \\ p_1 \\ \vdots \\ p_N \end{bmatrix}, \tag{15.8}$$

so that H becomes

$$H = \left(S+\frac{1}{2}\right)^2 E_0 + \frac{S}{2}\mathbf{V}^t\mathcal{M}\mathbf{V}\,, \tag{15.9}$$

where $\mathcal{M}$ is a $2N \times 2N$ matrix given by

$$\mathcal{M} = \begin{bmatrix} J^{xx} - J^{zz} & J^{xy} \\ (J^{xy})^t & J^{yy} - J^{zz} \end{bmatrix} \tag{15.10}$$

[10] Owing to the fact that $\{\vec{z}_i\}$ minimizes the classical energy, $\sum_j J_{ij}\vec{z}_j$ is perpendicular to $\vec{z}_i$ and thus orthogonal to $\vec{\pi}_i$, and there is no term *linear* in $\vec{\pi}$.

and the $N \times N$ matrices J^{xx}, J^{yy}, J^{xy}, and J^{zz} are defined by

$$J^{xx}_{ij} = J_{ij}\, \vec{x}_i \cdot \vec{x}_j\,, \quad J^{yy}_{ij} = J_{ij}\, \vec{y}_i \cdot \vec{y}_j\,, \quad J^{xy}_{ij} = J_{ij}\, \vec{x}_i \cdot \vec{y}_j\,, \tag{15.11}$$

$$\text{and} \quad J^{zz}_{ij} = \delta_{ij} \sum_k J_{ik} \vec{z}_i \cdot \vec{z}_k\,. \tag{15.12}$$

15.3.2 Bogoliubov transformation

Diagonalizing H amounts to finding bosonic creation operators $b^\dagger_\alpha$ and corresponding energies $\omega_\alpha \geq 0$ such that $H = \sum_\alpha \omega_\alpha \left(b^\dagger_\alpha b_\alpha + \frac{1}{2}\right)$ (up to a constant). A necessary condition is that the operators $b^\dagger_\alpha$ and b_α are "eigenoperators" of the commutation with H, with the eigenvalues ω_α and $-\omega_\alpha$, respectively: $\left[H, b^\dagger_\alpha\right] = \omega_\alpha b^\dagger_\alpha$ and $[H, b_\alpha] = -\omega_\alpha b_\alpha$. We thus seek the eigenvectors of the action of $[H, \bullet]$ in the space of linear combinations of q_i and p_j. The commutators of H (eqn (15.9)) with the operators q and p are simple to obtain using $[q_i, q_j] = [p_i, p_j] = 0$ and $[q_i, p_j] = i\delta_{ij}$. For an arbitrary linear combination of the q_i and p_i parameterized by the complex numbers $x_1, \cdots, x_{2N}$, the result is

$$\begin{aligned} &[H, x_1 q_1 + x_N q_N + x_{N+1} p_1 + \cdots x_{2N} p_N] \\ &= y_1 q_1 + y_N q_N + y_{N+1} p_1 + \cdots y_{2N} p_N\,, \end{aligned} \tag{15.13}$$

with the coefficients $y_1, \cdots, y_{2N}$ given by

$$\begin{bmatrix} y_1 \\ \vdots \\ y_{2N} \end{bmatrix} = iS\mathcal{M} \begin{bmatrix} 0 & \mathbf{1} \\ & \\ -\mathbf{1} & 0 \end{bmatrix} \begin{bmatrix} x_1 \\ \vdots \\ x_{2N} \end{bmatrix}, \tag{15.14}$$

where $\mathbf{1}$ is the $N \times N$ identity matrix. So, finding the operators $b^\dagger_\alpha$ (spin-wave creation operators) amounts to finding the eigenvectors of the "commutation matrix" $\mathcal{C} = i\mathcal{M} \begin{bmatrix} 0 & \mathbf{1} \\ -\mathbf{1} & 0 \end{bmatrix}$.

But $\mathcal{C}$ is not symmetric, and cannot always be fully diagonalized (in contrast to $\mathcal{M}$). It can be shown that if all the eigenvalues of $\mathcal{M}$ were *strictly* positive, $\mathcal{C}$ could be diagonalized, and its eigenvalues would be real and come in pairs $-\omega, \omega$.[11]

However, $\mathcal{M}$ in fact has some zero eigenvalues. The matrix $\mathcal{M}$ is not specific to the quantum spin problem. The quadratic form describing the classical energy variation for a small perturbation around the chosen classical ground state $\{\vec{z}_i\}$ is described by the same matrix $\mathcal{M}$.[12] In particular, if the classical ground state admits some zero-energy (infinitesimal) spin rotations, $\mathcal{M}$ possesses some eigenvector with an eigenvalue 0. Because global rotations should not change the energy, $\mathcal{M}$ has at least two zero eigenvalues. Nevertheless, these global rotations do not cause difficulties in diagonalizing the spin-wave Hamiltonian; they just correspond to some $\omega_\alpha = 0$ (the associated collective coordinate Q *and* conjugate momentum P simply do not appear in H).

[11] Let P be an orthogonal matrix which diagonalizes a symmetric matrix M: $M = P^{-1}\lambda P$, where λ is a diagonal matrix and $PP^t = 1$. If the eigenvalues of M are *strictly* positive, $K = P^{-1}\sqrt{\lambda}$ is invertible and $M = KK^t$. We write $C = iSKK^t\sigma$, where $\sigma = \begin{bmatrix} 0 & \mathbf{1} \\ -\mathbf{1} & 0 \end{bmatrix}$. Then, $\tilde{C} = K^{-1}CK = iSK^t\sigma K$ is Hermitian (since σ is real and antisymmetric, and K is real). $\tilde{C}$ can therefore be diagonalized and its spectrum is real. Since C and $\tilde{C}$ have the same spectrum, C can also be diagonalized and has real eigenvalues. Finally, we use $C^t = -C$. Since C and C^t should have the same spectrum, the eigenvalues of C come in pairs $-\omega, \omega$.

[12] Equations (15.3) and (15.7) also hold if $\vec{\pi}_i$ is a classical spin deviation of length $\vec{\pi}_i^2 \ll 1$.

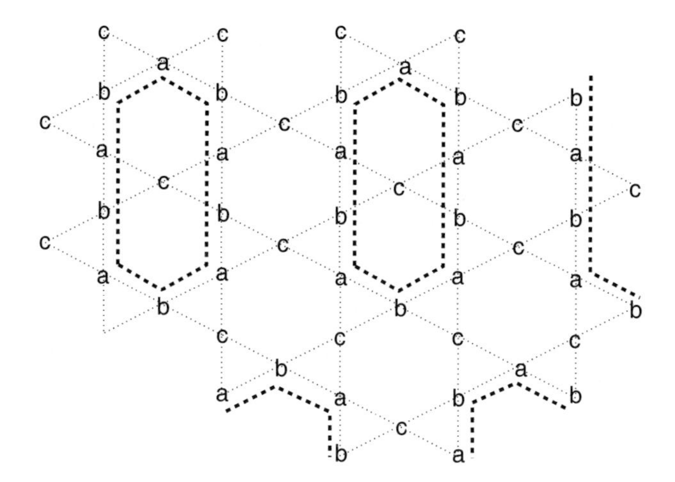

Fig. 15.2 Classical planar ground state on the kagome lattice. The loops where the spins alternate between the $\vec{a}$ and $\vec{b}$ directions are marked with dashed lines; they host independent zero modes (by rotation around the $\vec{c}$ axis).

15.3.3 Zero modes on the kagome lattice

However, some Heisenberg models admit classical zero modes (and hence zero eigenvalues in $\mathcal{M}$) which do not correspond to global rotations. As an example, consider the Heisenberg model on the kagome lattice [15] (for another classic example, the J_1–J_2 model on the square lattice, see [16]). Any classical spin configuration such that the sum $\vec{z}_i + \vec{z}_j + \vec{z}_k$ vanishes on each triangle (ijk) minimizes the classical energy. Among the numerous ways to achieve these conditions, there are the *planar* ground states, where all the spins lie in the same plane. In such a state, the spins take only three possible directions, $\vec{a}$, $\vec{b}$, and $\vec{c}$, at 120 degrees from each other. On the kagome lattice, there is an exponential number of ways to assign these three orientations such that the same letter is never found twice on the same triangle (the three-coloring problem, see Fig. 15.2). Now, we choose one of these "abc" states, and find a closed loop of the type $ababab\cdots$. Because of the three-coloring rule, the spins which are neighbors of this loop all point in the $\vec{c}$ direction. Now, we can rotate the spins of the loop about the $\vec{c}$ axis by any angle. This transforms the planar ground state into another (nonplanar) ground state, without any energy cost. So, for a generic planar ground state, we get as many zero modes (in $\mathcal{M}$) as closed loops with two alternating "colors". This number typically grows like the number of sites in the system.

What are the consequences of such classical zero modes for the quantum problem? As explained previously, the operators describing the two transverse directions along which the spins can deviate from the $\vec{z}_i$ axis obey the same commutation rules (to leading order in the $1/S$ expansion) as the position q and momentum p of a harmonic oscillator. In the case of the kagome loop modes discussed above, the energy is zero in one direction (rotation about the $\vec{c}$ direction), and quadratic in the other direction. Using the associated collective coordinates P and Q, we expect the Hamiltonian to be proportional to $H = \frac{1}{2}(P^2 + \omega^2 Q^2)$ with $\omega = 0$, since there is no classical energy cost for spin deviations in the direction Q. The corresponding commutation matrix is $C = i \begin{bmatrix} 0 & \omega^2 \\ -1 & 0 \end{bmatrix}$ and cannot be diagonalized when $\omega = 0$, as anticipated. In general, each such local zero mode will lead to an irreducible 2×2 Jordan block of this kind.[13] The ground state $|0\rangle$ of the oscillator is simple to obtain and corresponds to a zero-point motion of the coordinate Q which *diverges* when $\omega \to 0$ (no restoring force, as for a free particle): $\langle 0|Q^2|0\rangle = 1/2\omega$.

As long as the number of such zero modes is *finite* in the thermodynamic limit (this is the case when the classical ground state has no special degeneracy, beyond those implied by global rotations), the divergences above have a zero measure and do not cause divergences in the number of bosons $\langle 0|a_i^\dagger a_i|0\rangle$,[14] which measures the strength of the deviations from the classical state. In such a case, the Néel ordered state is stable with respect to quantum fluctuations, at least for large enough S.[15] On the other hand, if the number of such modes grows like N, the average number of bosons diverges and the spin-wave expansion breaks down (the initial assumption that $\langle 0|a_i^\dagger a_i|0\rangle$ is finite and small compared with S cannot be satisfied).

At this point, a route to obtaining a QSL appears to be to look for a lattice where the classical model has a sufficient number of "soft" modes, so that the zero-point motion of the spins restores the rotation invariance and destroys the long-range spin–spin correlations. This condition is realized on the kagome lattice, where, indeed, all numerical studies have concluded that there is no Néel order in this system (at least for $S = \frac{1}{2}$). However, the semiclassical spin wave theory described here breaks down. As discussed in the next sections, QSL states in Mott insulators possess some internal topological properties which are missed by the simple picture of a "disordered" state which would just be the quantum analogue of a high-temperature phase.

15.4 Lieb–Schultz–Mattis theorem, and Hastings's extension to $D > 1$: Ground state degeneracy in gapped spin liquids

The Lieb–Schultz–Mattis theorem [1] was originally derived for spin chains and spin ladders [18, 19] and was recently extended to higher dimensions in an important publication by Hastings [2] (see also [20] for an intuitive topological argument valid in any dimension, and [21] for a mathematically rigorous proof). It applies to spin Hamiltonians which are translation invariant in one direction (say x), and have a conserved magnetization $S^z_{\text{tot}} = \sum_i S^z_i$ and short-range interactions. In addition, the model must have periodic boundary conditions in the

[13] The general theory of the possible Jordan forms of $\mathcal{C}$ (dealing with the size and nature of the irreducible blocks) is in fact a result of *classical mechanics*, found by Williamson and outlined in [17].

[14] $\langle 0|a_i^\dagger a_i|0\rangle = \frac{1}{2}\langle 0|p_i^2 + q_i^2 - 1|0\rangle$ can be computed by expressing q_i and p_i in terms of $b_\alpha^\dagger$ and b_α, or in terms of the new positions and momenta $Q_\alpha = (1/\sqrt{2})(b_\alpha + b_\alpha^\dagger)$ and $P_\alpha = (1/\sqrt{2}i)(b_\alpha - b_\alpha^\dagger)$ Concentrating on the term $\langle 0|q_i^2|0\rangle$, q_i is a linear combination of the type $q_i = \sum_{\alpha=1}^N u_\alpha^i Q_\alpha + \sum_{\beta=1}^N v_\beta^i P_\beta$ (u and v are related to the eigenvectors of $\mathcal{C}$). From the fact that $|0\rangle$ is the vacuum of the b_α bosons, we have $\langle 0|P_i P_j|0\rangle = \langle 0|Q_i Q_j|0\rangle$ if $i \neq j$, and $\langle 0|P_i Q_j + Q_j P_i|0\rangle = 0\, \forall i, j$. Then the square of the spin deviation at site i (here the $\vec{x}_i$ component) is a linear combination of the zero-point fluctuations of the normal harmonic oscillators $\langle 0|q_i^2|0\rangle = \sum_\alpha (u_\alpha^i)^2 \langle 0|Q_\alpha^2|0\rangle + \sum_\alpha (v_\alpha^i)^2 \langle 0|P_\alpha^2|0\rangle$. Assuming regular behavior of the coefficients $(u_\alpha^i)^2$ and $(v_\alpha^i)^2$, $\langle 0|q_i^2|0\rangle$ is typically a sum of terms proportional to $\sim 1/\omega_\alpha$ when the mode frequency ω_α is small.

[15] This does not imply that the order should persist down to $S = \frac{1}{2}$.

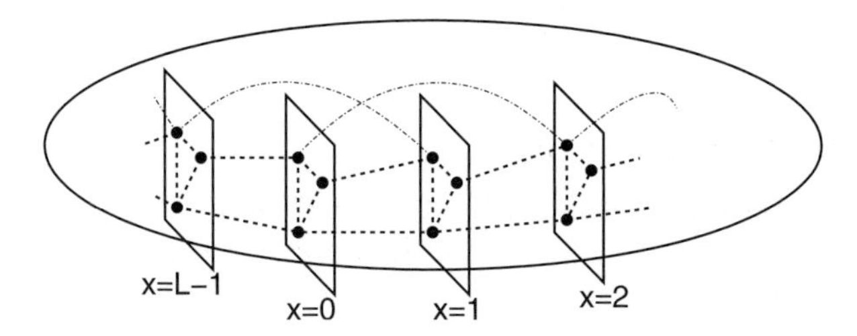

Fig. 15.3 A lattice model which is translation invariant and periodic in the x-direction can be viewed as a ring. The interactions J_{ij}, indicated by dashed lines, are invariant in the x-direction but otherwise arbitrary. In this example, each cross section has $C = 3$ sites.

x-direction. Although more general interactions can easily be considered,[16] we concentrate for simplicity on spin-S Heisenberg models, written as in eqn (15.1) (with $J_{i,j} = J_{i+x,j+x}$ to respect the translation invariance).

Following [20], we define a *cross section* as all of the sites sitting at a given value of x. By translation invariance, all cross sections are equivalent and contain C sites (Fig. 15.3). In a spin chain, each cross section contains a single site. In an n-leg spin ladder, $C = n$ sites. In a square lattice, $C = L_y$. On a D-dimensional lattice with n sites per unit cell, $C = nL^{D-1}$, etc. We denote the length of the system in the x-direction by L_x, and therefore CL_x is the total number of sites. Finally, we define $m^z = (1/CL_x)\langle 0|S^z_{\text{tot}}|0\rangle$ as the ground state magnetization per site.

The theorem says that if $C(S+m^z)$ is *not an integer*, either the ground state is degenerate or the spectrum has gapless excitations in the thermodynamic limit. In other words, if $C(S+m^z) \notin \mathbb{Z}$ the system *cannot have a unique ground state and a finite gap to excited states* in the thermodynamic limit. Although the proof in 1D [1] and Oshikawa's topological argument [20] (Section 15.4.1) are relatively simple, the proof appears quite involved for $D > 1$, and will not be discussed here.

What is the relation between the Lieb–Schultz–Mattis–Hastings (LSMH) theorem and QSLs? In most AF Heisenberg models on a finite-size lattice, $|0\rangle$ is a singlet and $m^z = 0$. If we focus on the case $S = \frac{1}{2}$, the theorem forbids a single ground state and a gap when C is *odd*. In particular, if the lattice is two-dimensional and describes a Mott insulator, the unit cell has an odd number n of sites, and any odd L_y can be chosen to get an odd $C = nL_y$ (note that the total number of sites is still even if L_x is even). If we assume that a *gapped* QSL is realized (for an example which fits the LSMH conditions, see for instance [22]), its ground state must be *degenerate* (with periodic boundary conditions). Usually, ground state degeneracies are the signature of some spontaneous symmetry breaking. However, by definition, a QSL respects all lattice symmetries. The degeneracy imposed by the LSMH theorem cannot be understood from this conventional point of view and is a hint that (gapped) QSL wave functions possess some interesting topological properties, which correspond to the notion of "topological order" introduced by Wen [23, 24] for spin systems and Wen and Niu [25] in the context of the fractional quantum Hall effect. As we will briefly discuss at the end of this chapter, this

[16] In particular, the interaction can be anisotropic, $S_i^z S_j^z + \Delta(S_i^x S_j^x + S_i^y S_j^y)$, and an external magnetic field parallel to the z-direction can be present.

topological degeneracy is deeply related to the exotic nature of the elementary excitations in a QSL.[17]

15.4.1 Oshikawa's topological argument

Oshikawa's argument is somewhat related to Laughlin's argument [26] for the quantization of the transverse conductivity in the quantum Hall effect. First, a "twisted" version of the Hamiltonian is introduced:

$$H_\theta = \frac{1}{2}\sum_{ij} J_{ij}\left[S_i^z S_j^z + \frac{1}{2}\left(e^{i\theta(x_i - x_j)/L_x} S_i^+ S_j^- + \text{H.c.}\right)\right], \tag{15.15}$$

where $0 \le x_i < L_x$ is the x-coordinate of site i (Fig. 15.4). It is simple to show that the spectra of H_0 and $H_{2\pi}$ are the same, since the unitary operator

$$U = \prod_i \exp\left(2i\pi \frac{x_i}{L_x} S_i^z\right) \tag{15.16}$$

maps H_0 onto $H_{2\pi}$:

$$U H_0 U^{-1} = H_{2\pi} \tag{15.17}$$

(the calculation simply uses $e^{i\theta S_i^z} S_i^+ e^{-i\theta S_i^z} = S_i^+ e^{i\theta}$).

Starting with a spectrum of H_0 which is gapped, we assume further that *the gap of H_θ remains finite when θ goes from 0 to 2π.*[18] On can follow the ground state of H_θ, which does not cross any other energy level as θ is varied. Assuming that the ground state $|0\rangle$ of H_0 is unique, and using the finite-gap hypothesis, it must evolve to the ground state of $H_{2\pi}$, denoted by $|2\pi\rangle$. Through eqn (15.17), both states are related: $|2\pi\rangle = U^{-1}|0\rangle$. However, the operator U does not always commute with the translation operator T and may change the momentum. The precise relation is

$$TU = UT \exp\left(2i\pi \frac{S^z_{\text{tot}}}{L_x}\right) \exp\left(2i\pi CS\right). \tag{15.18}$$

The first phase factor, equal to $2\pi C m^z$, comes from the shift by $2\pi/L_x$ of the local rotation angles after a translation. The second phase factor corrects the 2π jump of the rotation angle when passing from $x = L_x - 1$ to $x = 0$. This relation implies that the momentum k_0 of $|0\rangle$ (defined by $T|0\rangle = e^{ik_0}|0\rangle$) and the momentum $k_{2\pi}$ of $|2\pi\rangle = U^{-1}|0\rangle$ are related by

$$k_0 = k_{2\pi} + 2\pi C(S + m^z). \tag{15.19}$$

But H_θ is translation invariant (it commutes with T), and the momentum of each state (quantized for finite L_x) cannot change with θ. So $|0\rangle$ and $|2\pi\rangle$ have the same momentum, and $k_0 = k_{2\pi}[2\pi]$. From eqn (15.19), we find that $C(S + m^z)$ must be a integer.

[17] QSLs have "spinon" excitations which carry a spin $\frac{1}{2}$ (like an electron) but no electric charge.

[18] Hastings's argument does not directly use H_θ for a finite θ and does not rely on this assumption. This assumption is, however, reasonable because of the fact that, under an appropriate choice of gauge (frame), $H = H_0$ and H_θ differ only in the terms connecting the cross section at $x = L_x - 1$ to the cross section at $x = 0$ (boundary terms), and are identical in the bulk.

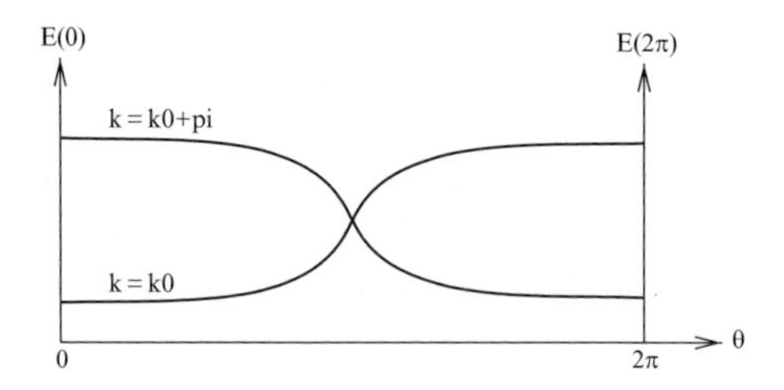

Fig. 15.4 Schematic spectrum of the twisted Hamiltonian (eqn (15.15)) as a function of the angle θ, in the case where $C(S+m^z)$ is a half-integer.

15.5 Anderson's short-range resonating-valence-bond picture

In the $1/S$ expansion, it is assumed that the spins experience small fluctuations about a well-defined direction and that the spin–spin correlations are long-ranged. This is, of course, incompatible with having a rotationally invariant QSL state. To gain some intuition about what a QSL wave function might look like, it is instructive to start from a completely opposite limit: a spin singlet state with extremely short-range correlations. A *short-range valence bond* (VB) state is such a wave function; it is the direct product of $S = 0$ states $|[ij]\rangle = (1/\sqrt{2})\,(|\uparrow_i\downarrow_j\rangle - |\downarrow_i\uparrow_j\rangle)$ on pairs of sites:

$$|\mathrm{VB}\rangle = |[i_0 i_1]\rangle \otimes |[i_2 i_3]\rangle \otimes |[i_4 i_5]\rangle \otimes \cdots |[i_{N-1} i_N]\rangle\,, \tag{15.20}$$

where each site of the lattice appears exactly once (Fig. 15.5). Such a VB state is said to be short-range if all pairs of sites coupled in a singlet are at a distance $|\mathbf{r}_{i_p} - \mathbf{r}_{i_{p+1}}|$ that is less than or equal to some fixed length r_{max} (much smaller than the lattice size). The simplest case is $r_{\mathrm{max}} = 1$, where each spin forms a singlet with one of its nearest neighbors.

In a VB state, the spin–spin correlations are short-ranged: $\langle \mathrm{VB}|\vec{S}_i \cdot \vec{S}_j|\mathrm{VB}\rangle = 0$ if $|\mathbf{r}_{i_p} - \mathbf{r}_{i_{p+1}}| > r_{\mathrm{max}}$. For a nearest-neighbor Heisenberg model on a bipartite lattice, one can compare the (expectation value of the) energy of a nearest-neighbor VB state with that of the simple two-sublattice Néel state $|\uparrow\downarrow\uparrow\downarrow\cdots\rangle$. The VB energy is $e_{\mathrm{VB}} = -J3/8$ per site and the Néel energy is $e_{\mathrm{N}} = -Jz/8$, where z is the number of nearest neighbors. If the lattice is not bipartite but admits a three-sublattice classical ground state (with spins pointing at 120 degrees from each other), the energy of a classical Néel state is $e_{\mathrm{N}} = -Jz/16$. From this, we observe, for instance, that the VB energy is *lower* than e_{N} on the kagome lattice. More

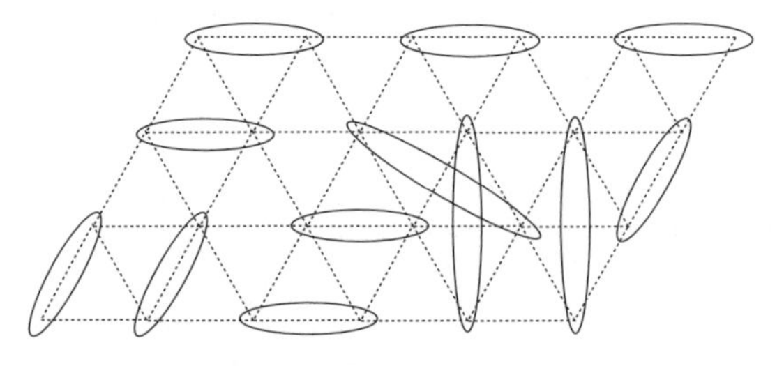

Fig. 15.5 A short-range valence bond state on the triangular lattice. The singlet pairs are marked with ellipses.

generally, this simple variational comparison shows that a low coordination z and frustrated interactions (which increase the number of sublattices in the classical ground state) tend to favor VB states, and thus possible QSL states.

In fact, there are many (frustrated and Heisenberg-like) toy models where some/the nearest-neighbor VB states are *exact* ground states. The best-known example is the Majumdar–Gosh model [27]. Consider a spin-$\frac{1}{2}$ Heisenberg chain with first- and second-neighbor couplings (J_1 and J_2, respectively). At $J_1 = 2J_2 > 0$, we have

$$H_{\mathrm{MG}} = 2\sum_i \vec{S}_i \cdot \vec{S}_{i+1} + \sum_i \vec{S}_i \cdot \vec{S}_{i+2} \tag{15.21}$$

and the (twofold degenerate) ground states are given exactly:

$$|a\rangle = \cdots \otimes |[01]\rangle \otimes |[23]\rangle \otimes |[46]\rangle \otimes \cdots , \tag{15.22}$$

$$|b\rangle = \cdots \otimes |[12]\rangle \otimes |[34]\rangle \otimes |[56]\rangle \otimes \cdots . \tag{15.23}$$

The proof can be done in three steps. First, the Heisenberg Hamiltonian on three sites $H_{ijk} = \vec{S}_i \cdot \vec{S}_j + \vec{S}_j \cdot \vec{S}_k + \vec{S}_k \cdot \vec{S}_i$ is written as $H_{ijk} = \frac{1}{2}(\vec{S}_i + \vec{S}_j + \vec{S}_k)^2 - \frac{9}{8}$. In this form, proportional to the square of the total spin, it is clear that the eigenvalues of H_{ijk} are $\frac{1}{2}S(S+1) - \frac{9}{8}$ with $S = \frac{1}{2}$ or $S = \frac{3}{2}$ (the only possible values of S for three spins $\frac{1}{2}$). So, if the sites ijk are in an $S = \frac{1}{2}$ state, they minimize H_{ijk} exactly. Second, the Majumdar–Gosh Hamiltonian is expressed as

$$H_{\mathrm{MG}} = \sum_i H_{i-1,i,i+1} \, . \tag{15.24}$$

Finally, we notice that the dimerized states $|a\rangle$ and $|b\rangle$ always have one singlet among the sites $i-1, i, i+1$, which are therefore in an $S = 1/2$ state. We conclude that $|a\rangle$ and $|b\rangle$ minimize all the terms in eqn (15.24) and are thus ground states of H_{MG}.

The Majumdar–Gosh model is the simplest model of a family of spin models where exact VB ground states can be found.[19] For instance, the Husimi cactus [30] is a lattice constructed as a tree (no loops) of corner-sharing triangles. Its geometry is locally similar to the kagome lattice but it has no closed loops (except, of course, for the triangles themselves). The argument above (writing the Hamiltonian as a sum of H_{ijk}) can be directly generalized to this case and shows that any nearest-neighbor VB state is a ground state. One can also mention the 2D Shastry–Sutherland Heisenberg model [31], where a particular nearest-neighbor VB state is the unique ground state, and which has an experimental realization in $SrCu_2(BO_3)_2$ [5].

So far, we do not yet have any gapped *liquid* state.[20] To obtain a qualitative idea of how VB states can be the building blocks of a gapped QSL, we will briefly explain the short-range resonating-valence-bond (RVB) picture proposed by Anderson [32]. If we exclude the toy

[19] There exists a general method for constructing an $\mathrm{SU}(2)$ symmetric spin model with short-range interactions such that all nearest-neighbor VB states are ground states [28]. Building on this idea, it was possible to construct $\mathrm{SU}(2)$ symmetric spin-$\frac{1}{2}$ models (with short-ranged interactions) with a gapped QSL ground state [29]. Although complicated, these models are among the very few examples where the ground state is well established to be a short-ranged resonating VB liquid.

[20] The ground states of H_{MG} spontaneously break the translation symmetry. On the Husimi cactus, the ground state is highly degenerate. The Shastry–Sutherland ground state does not break any symmetry (the ground state is unique), but the lattice has an *even* number of spins per unit cell and should be considered as a band insulator in our classification.

models discussed above, a VB state is generally not an eigenstate of the Heisenberg model. Starting from a nearest-neighbor VB state, the Heisenberg Hamiltonian will induce some dynamics among the VB states. If we take the kagome example, a nearest-neighbor VB state inevitably contains some "defect" triangles *without* any singlet.[21] While the term H_{ijk} leaves the VB state unchanged if the corresponding triangle has a singlet bond, the three VBs touching i, j, and k will be moved by H_{ijk} if (ijk) is a defect triangle. The ground state can be viewed as a linear combination of (many) VB configurations (not necessarily nearest-neighbor). Anderson suggested that, with appropriate interactions and lattice geometry, the ground state wave function could be "delocalized" over a large part of the subspace spanned by short-range VB states. By forming a linear superposition of a large number of very different VB states, the system may restore all the lattice symmetries (which are broken by an individual VB state) and form a QSL.

A more formal approach to this idea will be discussed in Section 15.6, but this picture can already be used to anticipate the nature of the magnetic excitations in such a short-range RVB liquid. To this end, we first consider a 2D model where one ground state is equal to (or dominated by) *one* particular VB state. Contrary to Anderson's RVB liquid, the wave function is localized in the vicinity of one particular VB state. This can be thought of as a 2D analogue of the Majumdar–Gosh chain, where the ground state is a spatially regular arrangement of singlet bonds. Many 2D models are known to realize such VB *crystals* (VBCs) [33]; we refer to [34] for a recent example where the exact ground states are known. In a VBC, a finite-energy excitation can be created by replacing a singlet bond by a triplet ($S = 1$), with an energy cost proportional to J. But is it possible to construct two *separated* spin-$\frac{1}{2}$ excitations in such a system? As a trial state, one can place two remote spins "up" (two spinon excitations) at sites 0 and i. Then, to minimize the energy, the regular VB structure of the ground state must be reconstructed as much as possible. However, owing to the spinons, the regular pattern cannot be fully reconstructed between 0 and i, and a "string" of misaligned VBs is unavoidable. The unpaired spins behave as a topological defect in the crystalline order. So, two remote spinons perturb the ordered VB background, not only in their vicinity but also all the way between them. They lead to an energy cost which is proportional to their separation.[22] So, isolated spinons are not finite-energy excitations in a VBC. The ordered VB background is a medium which confines the spinons in pairs. Since an RVB state should instead be viewed as a *liquid* (no broken symmetry, no long-range order), it is reasonable to expect the spinons to be able to propagate as independent particles. As we will see in the next section, the proper way to address this question of confinement and deconfinement of spinons is to understand the emergence of gauge degrees of freedom in these systems.

15.6 Schwinger bosons, large-$\mathcal{N}$ limit, and $\mathbb{Z}_2$ topological phase

15.6.1 Schwinger boson representation

The spin wave approach is a large-S approach and is unable to capture highly quantum states which are rotationally symmetric, such as RVB wave functions. From the discussion in Section 15.5, it is natural to look for a description in terms of *singlet fields* living on bonds, which will

[21] Whatever the nearest-neighbor VB state, exactly 1/4 of the triangles have no singlet bonds.

[22] The situation is very different in 1D. In the Majumdar–Gosh model, one can get a *finite-energy* state with two remote spinons by introducing a domain wall in the dimerization pattern between 0 and i.

be able to describe the presence or absence of a singlet between two sites. Such variables appear naturally when the Schwinger boson representation of the spin operators [35, 36] is used.

At each site, two types of bosons, carrying spins "up" and "down," are introduced, $a^\dagger_{i\uparrow}$ and $a^\dagger_{i\downarrow}$, and the spin operators are represented as bilinears in the boson creation and annihilation operators:

$$S^z_i = \frac{1}{2}\left(a^\dagger_{i\uparrow}a_{i\uparrow} - a^\dagger_{i\downarrow}a_{i\downarrow}\right), \quad S^+_i = a^\dagger_{i\uparrow}a_{i\downarrow}, \quad S^-_i = a^\dagger_{i\downarrow}a_{i\uparrow}. \tag{15.25}$$

With these relations, the commutation relations $[S^\alpha_i, S^\beta_i] = i\epsilon^{\alpha\beta\delta}S^\delta_i$ are automatically satisfied. The total spin reads $\vec{S}^2_i = (n_i/2)(n_i/2+1)$, where $n_i = a^\dagger_{i\uparrow}a_{i\uparrow} + a^\dagger_{i\downarrow}a_{i\downarrow}$ is the total number of bosons at site i. To fix the length of the spins, the following constraint must therefore be imposed on physical states:

$$a^\dagger_{i\uparrow}a_{i\uparrow} + a^\dagger_{i\downarrow}a_{i\downarrow} = 2S. \tag{15.26}$$

With this representation,[23] the Heisenberg interaction is of degree four in the boson operators and can be written

$$\vec{S}_i \cdot \vec{S}_j = S^2 - \frac{1}{2}(A_{ij})^\dagger A_{ij} \tag{15.27}$$

$$\text{with} \quad A_{ij} = a_{i\uparrow}a_{j\downarrow} - a_{i\downarrow}a_{j\uparrow}. \tag{15.28}$$

The bond operators $A^\dagger_{ij}$ behave as singlet creation operators: $A^\dagger_{ij}$, when applied to the boson vacuum, creates a spin singlet $|\uparrow_i\downarrow_j\rangle - |\downarrow_i\uparrow_j\rangle$ and, from eqn (15.27), $A^\dagger_{ij}A_{ij}$ is proportional to the number (0 or 1) of singlets between sites i and j. In addition, A_{ij} is invariant under rotations: redefining the bosons by an SU(2) matrix P, so that $\begin{bmatrix} a_\uparrow \\ a_\downarrow \end{bmatrix} \to P \begin{bmatrix} a_\uparrow \\ a_\downarrow \end{bmatrix}$, leaves A_{ij} unchanged.[24]

15.6.2 Mean-field approximation

Arovas and Auerbach [35] suggested an approximation in which the interaction is decoupled using mean-field expectation values,

$$A^\dagger_{ij}A_{ij} \longrightarrow A^\dagger_{ij}\langle A_{ij}\rangle + \langle A^\dagger_{ij}\rangle A_{ij} - |\langle A^\dagger_{ij}\rangle|^2, \tag{15.29}$$

and the constraint in eqn (15.26) is replaced by a condition on the *average* number of bosons per site,

$$\langle a^\dagger_{i\uparrow}a_{i\uparrow} + a^\dagger_{i\downarrow}a_{i\downarrow}\rangle = 2S. \tag{15.30}$$

By this replacement, the Hamiltonian becomes quadratic in the boson operator:

[23] Fermions can also be used, leading to other very interesting theories for (gapped or gapless) QSLs [12, 37, 38].

[24] A_{ij} can be written using the 2×2 antisymmetric tensor $\epsilon = \begin{bmatrix} 0 & -1 \\ 1 & 0 \end{bmatrix}$: $A_{ij} = \sum_{\sigma,\sigma'=\uparrow,\downarrow} \epsilon_{\sigma\sigma'}a_{i\sigma}a_{j\sigma'}$. The rotation invariance of A_{ij} follows from the fact that any $P \in \mathrm{SU}(2)$ satisfies $P^t\epsilon P = \epsilon$.

$$H \longrightarrow H_{\rm MF}[Q^0_{ij}, \lambda^0_j] = -\frac{1}{2} \sum_{ij} \left(A^\dagger_{ij} Q^0_{ij} + \bar{Q}^0_{ij} A_{ij} \right)$$
$$- \sum_i \lambda^0_i \left(a^\dagger_{i\uparrow} a_{i\uparrow} + a^\dagger_{i\downarrow} a_{i\downarrow} - 2S \right) + \text{cst}. \quad (15.31)$$

A chemical potential λ^0_i has been introduced at each site to tune the boson densities so that they satisfy eqn (15.30). The mean-field Hamiltonian $H_{\rm MF}$ (and thus its ground state $|0\rangle$) depends on the complex parameters Q^0_{ij} (one for each pair of sites ij, where $J_{ij} \neq 0$). These parameters have to be adjusted to satisfy the self-consistency conditions on each bond

$$Q^0_{ij} = \frac{1}{2} J_{ij} \langle 0 | a_{i\uparrow} a_{j\downarrow} - a_{i\downarrow} a_{j\uparrow} | 0 \rangle \,. \quad (15.32)$$

As in the spin wave approach, the Heisenberg model has been reduced to a quadratic boson model (here with some self-consistency conditions). However, the crucial difference is that the present formalism does not impose any preferred spin direction: giving a finite expectation value $A^0_{ij} \neq 0$ to the operator A_{ij} does not break the $\mathrm{SU}(2)$ symmetry, and this is a necessary condition to describe a QSL.

Generally speaking, two families of solutions can be found at this mean-field level. In the first class, favored when S is large, the Schwinger bosons *Bose-condense* into some particular mode. Because they carry a spin index, such a condensate state (spontaneously) breaks the $\mathrm{SU}(2)$ symmetry. These solutions describe Néel states with long-range spin–spin correlations. In such cases, the Schwinger boson mean-field theory is essentially equivalent to the spin wave approach (Section 15.3).

The second class corresponds to (mean-field) QSL states. There, the ground state is rotationally invariant, and the Bogoliubov quasiparticles obtained by diagonalizing $H_{\rm MF}$ are gapped. Since the corresponding creation operators, $b_{\uparrow,\alpha}$ and $b_{\downarrow,\alpha}$, are linear combinations of the original bosons, these excitations also carry a spin $\frac{1}{2}$. The most important question is whether the existence of these deconfined (free in the mean-field approximation) spinons is an artifact of the mean-field approximation, or whether they could survive in some Heisenberg spin model. In the first case, the inclusion of the fluctuations that were neglected would confine the spinons and would deeply change the nature of the ground state. The mean-field picture of a fully symmetric state with noninteracting spinon excitations is then qualitatively incorrect. Another possibility is that the spinons remains deconfined, even in the presence of fluctuations. In that case, the mean-field approximation is a very useful starting point. We will discuss in Section 15.6.5 a scenario where this is the case. But before we do that, we need to introduce the basic formalism that is needed to describe the fluctuations about the mean-field solution, and the emergence of gauge degrees of freedom in the system. The central question concerning the long-distance and low-energy properties of the system will be whether or not these gauge degrees of freedom confine the spinons.

15.6.3 Large $\mathcal{N}$, saddle point

To discuss the role of the fluctuations neglected in eqn (15.29), it is necessary to formulate the mean-field approximation as a saddle point approximation in the path integral formulation of the model. It will then be possible to identify the structure of the most important fluctuations about the saddle point. To do so, we duplicate the two species of bosons ($\uparrow$ and $\downarrow$) $\mathcal{N}$ times.

In addition to the site and up/down indices σ, the boson operators now carry an additional "flavor" index $m = 1, \cdots, \mathcal{N}$. The Hamiltonian and the constraint are then generalized to

$$H = -\frac{1}{2\mathcal{N}} \sum_{ij} J_{ij} A_{ij}^\dagger A_{ij} , \tag{15.33}$$

$$A_{ij} = \sum_{m=1}^{\mathcal{N}} a_{im\uparrow} a_{jm\downarrow} - a_{im\downarrow} a_{jm\uparrow} , \tag{15.34}$$

and

$$\sum_{m=1}^{\mathcal{N}} a_{im\uparrow}^\dagger a_{im\uparrow} + a_{im\downarrow}^\dagger a_{im\downarrow} = 2\mathcal{N}S . \tag{15.35}$$

For $\mathcal{N} = 1$, this model is the Heisenberg model with SU(2) symmetry. For $\mathcal{N} > 1$, this model has an enlarged symmetry given by the group Sp($\mathcal{N}$).[25] S is a parameter of the model, and is no longer related to a representation of SU(2) if $\mathcal{N} > 1$. The bond operator A_{ij} is a sum over all the flavors. For this reason, in the limit where $\mathcal{N}$ is very large, the fluctuations of A_{ij} become negligible compared with its expectation value, and the approximation made in eqn (15.29) becomes exact.

A formal way to establish this result is to adopt a formulation of the model where the partition function $\mathcal{Z} = \mathrm{Tr}\left[e^{-\beta H}\right]$ at temperature $T = \beta^{-1}$ is expressed as a coherent-state path integral over complex variables $z_{im\sigma}(\tau)$ (in correspondence with the boson operators $a_{im\sigma}$) which are periodic functions of an imaginary time $\tau \in [0, \beta[$. In this formalism, the partition function reads[26]

$$\mathcal{Z} = \int \mathcal{D}[z_{im\sigma}(\tau), \lambda_i(\tau)] \exp\left(- \int_0^\beta L_0 \, d\tau \right) , \tag{15.36}$$

[25] The symplectic group of $2\mathcal{N} \times 2\mathcal{N}$ matrices Sp($\mathcal{N}$) is the set of matrices P which satisfies $P^t \mathcal{J} P = \mathcal{J}$, where $\mathcal{J} = \begin{bmatrix} 0 & 1 & & & \\ -1 & 0 & & & \\ & & \ddots & & \\ & & & 0 & 1 \\ & & & -1 & 0 \end{bmatrix}$ generalizes the antisymmetric ϵ tensor.

[26] For an introduction to the path integral formalism in this context of quantum magnetism, see for instance [36]. We can sketch the main steps of the derivation in the case of a single bosonic mode $[a, a^\dagger] = 1$ as follows. For any complex number z, a coherent state $|z\rangle = e^{z a^\dagger}|0\rangle$ is defined. These states satisfy $a|z\rangle = z|z\rangle$, $\langle z|z'\rangle = e^{\bar{z}z'}$, and the resolution of the identity $(1/\pi)\int d^2z \, |z\rangle\langle z| e^{-|z|^2} = \mathbf{1}$. We write the partition function as a product over N_τ imaginary-time steps $\mathcal{Z} = \mathrm{Tr}\left[e^{-d\tau H} e^{-d\tau H} \cdots\right] = \lim_{N_\tau \to \infty} \mathrm{Tr}\left[(1 - d\tau H)(1 - d\tau H)\cdots\right]$ with $d\tau = \beta / N_\tau$. Then, the identity is inserted at each step: $\mathcal{Z} = \lim_{N_\tau \to \infty} \int \left(\prod_{\tau=1}^{N_\tau} d^2 z_\tau\right) e^{-|z_1|^2} \langle z_1 | 1 - d\tau H | z_{N_\tau}\rangle e^{-|z_{N_\tau}|^2} \langle z_{N_\tau} | 1 - d\tau H | z_{N_\tau - 1}\rangle \cdots e^{-|z_2|^2} \langle z_2 | 1 - d\tau H | z_1\rangle$. Next, we write $e^{-|z_i|^2}\langle z_i | 1 - d\tau H | z_{i-1}\rangle \simeq \exp\left[-\bar{z}_i(z_i - z_{i-1}) - d\tau H(\bar{z}_i, z_{i-1})\right]$, where the complex number $H(\bar{z}, z') = \langle z'|H|z\rangle$ is obtained by writing the Hamiltonian in a normal-ordered form and replacing $a^\dagger$ by $\bar{z}$ and a by z'. Taking the continuous-time limit $d\tau \to 0$ is formally written as $z_i - z_{i-1} \to \partial_\tau z(\tau)\, d\tau$ and leads finally to $\mathcal{Z} = \int \mathcal{D}[z] \exp(-\int_0^\beta L \, d\tau)$ with the Lagrangian $L = \bar{z}(\tau)\partial_\tau z(\tau) + H(\bar{z}(\tau), z(\tau))$.

$$L_0 = \sum_{i\,m\,\sigma} \bar{z}_{im\sigma} \partial_\tau z_{im\sigma} - \frac{1}{2\mathcal{N}} \sum_{ij} J_{ij} A_{ij}^\dagger A_{ij}$$
$$+ i \sum_{i\,m} \lambda_i \left(\bar{z}_{im\uparrow} z_{im\uparrow} + \bar{z}_{im\downarrow} z_{im\downarrow} - 2S \right), \tag{15.37}$$

$$A_{ij} = \sum_{m=1}^{\mathcal{N}} \left(z_{im\uparrow} z_{jm\downarrow} - z_{im\downarrow} z_{jm\uparrow} \right), \tag{15.38}$$

where a Lagrange multiplier λ has been introduced at each lattice site and each time step to enforce the constraint of eqn (15.35) exactly (to simplify the notation, the τ dependence of all fields is implicit).

Now, a Hubbard–Stratonovich transformation is performed:

$$\mathcal{Z} = \int \mathcal{D}[z_{im\sigma}(\tau), \lambda_i(\tau), Q_{ij}(\tau)] \exp\left(- \int_0^\beta L_1 \, d\tau \right), \tag{15.39}$$

$$L_1 = \sum_{i\,m\,\sigma} \bar{z}_{im\sigma} \partial_\tau z_{im\sigma} + \sum_{ij} \left(\frac{2\mathcal{N}}{J_{ij}} |Q_{ij}|^2 - \bar{Q}_{ij} A_{ij} - Q_{ij} \bar{A}_{ij} \right)$$
$$+ i \sum_{i\,m} \lambda_i \left(\bar{z}_{im\uparrow} z_{im\uparrow} + \bar{z}_{im\downarrow} z_{im\downarrow} - 2S \right). \tag{15.40}$$

This new formulation involves an additional complex field Q_{ij} on each bond. The equivalence of L_1 with the initial Lagrangian L_0 can be checked in a simple way by performing Gaussian integrations over $Q_{ij}(\tau)$ for each bond and each time step: $\int \mathcal{D}[Q_{ij}(\tau)] \exp\left(- \int_0^\beta L_1 \, d\tau \right) = \exp\left(- \int_0^\beta L_0 \, d\tau \right)$ (up to a multiplicative constant). At this point, the $\mathcal{N}$ flavors of particles are no longer coupled to each other, but are coupled to a common bond field Q_{ij}. So, for a fixed space–time configuration of Q, we have $\mathcal{N}$ independent copies of the same boson system. In addition, the Lagrangian L_1 is now quadratic in the variable z. We denote by $G_{Q,\lambda}^{-1}$ the corresponding quadratic form, a big matrix which has indices for space (i), time (τ), spin (σ), and complex conjugation (z versus $\bar{z}$) (but no flavor index), and depends on the auxiliary field Q and λ. L_1 is then given by

$$L_1 = \sum_{ij} \frac{2\mathcal{N}}{J_{ij}} |Q_{ij}|^2 - 2i\mathcal{N}S \sum_i \lambda_i$$
$$+ \sum_m \left[\bar{z}_{i\sigma}(\tau); z_{i\sigma}(\tau) \right] G_{Q,\lambda}^{-1} \begin{bmatrix} z_{j\sigma'}(\tau') \\ \bar{z}_{j\sigma'}(\tau') \end{bmatrix}. \tag{15.41}$$

Performing the Gaussian integral over the z fields is now simple, as it gives $(\det[G])^{\mathcal{N}}$, also equivalent to $e^{\mathcal{N}\mathrm{Tr}[\log(G)]}$. The partition function is now expressed as a path integral with the fields Q and λ only, but with a complicated non-Gaussian weight:

$$\mathcal{Z} = \int \mathcal{D}[z_{i\sigma}(\tau), \lambda_i(\tau), Q_{ij}(\tau)] \exp\left(-\mathcal{N} \int_0^\beta L_2 \, d\tau \right), \tag{15.42}$$

$$L_2 = + \sum_{ij} \frac{2}{J_{ij}} |Q_{ij}|^2 - 2iS \sum_i \lambda_i + \mathrm{Tr}[\log(G_{Q,\lambda})]. \tag{15.43}$$

Here, the flavor indices m have disappeared and $\mathcal{N}$ appears only as a global multiplicative factor in the action. With this formulation of the $\mathrm{Sp}(\mathcal{N})$ "spin" model, it is clear that, in the limit $\mathcal{N} \to \infty$, the partition function will be dominated by the configurations (Q^0, λ^0) which are *saddle points* of the action $\mathcal{S}[Q, \lambda] = \int_0^\beta L_2 \, d\tau$. In other words, *the fluctuations of Q_{ij} and λ_i are frozen when $\mathcal{N} \to \infty$*. Such saddle points are obtained by requiring

$$\left.\frac{\partial \mathcal{S}}{\partial \lambda_i(\tau)}\right|_{Q^0, \lambda^0} = 0 \,, \tag{15.44}$$

$$\left.\frac{\partial \mathcal{S}}{\partial Q_{ij}(\tau)}\right|_{Q^0, \lambda^0} = 0 \,, \tag{15.45}$$

and in most cases they are found to be time-independent, i.e. $Q^0_{ij}(\tau), \lambda^0_i(\tau) \to Q^0_{ij}, \lambda^0_i$. The equations above can then be shown to be equivalent to the self-consistency conditions of eqns (15.30) and (15.32), with $Q^0_{ij} = (J_{ij}/2\mathcal{N}) \sum_m \langle 0 | a_{im\uparrow} a_{jm\downarrow} - a_{im\downarrow} a_{jm\uparrow} | 0 \rangle$.

15.6.4 Fluctuations about a saddle point and gauge invariance

We are now ready to discuss the fluctuations that are present when $\mathcal{N}$ is finite, where the field $Q_{ij}(\tau)$ is able to fluctuate around its mean-field value Q^0_{ij}. Treating all of the possible fluctuations is certainly very difficult, as it would amount to solving the original spin problem. A possible approach is to compute perturbatively the first $1/\mathcal{N}$ corrections to the mean-field results [35]. However, this can miss some important effects (instabilities) which are not perturbative in $1/\mathcal{N}$, and will generally not shed light on the issue of spinon confinement that we are interested in. Instead, as in [39, 40], we will examine the qualitative structure of the fluctuation modes which are important for the long-distance properties of the system. In particular, we would like to know if some fluctuations could confine the spinons (in which case the mean-field picture is incorrect), or if the QSL state is stable at finite $\mathcal{N}$. As we will see, there are some fluctuation modes which are described by a gauge field [39, 40] and mediate some (possibly long-ranged) interaction between the spinons. The dynamics of this gauge field is therefore crucial to the physics of the spin system. In some cases this gauge field will be in a confining phase, and the $\mathcal{N} = \infty$ limit (where the fluctuations are frozen out) does not represent the physics of the finite-$\mathcal{N}$ models [39]. In some other situations, the gauge field has a deconfined phase and a QSL state with elementary spinon excitation is possible [40].

First, it should be noticed that the description of the spin operators using Schwinger bosons is redundant in the sense that an arbitrary local change of phase in the boson operators does not change the physical spin operators. In the path integral formulation, this becomes a full space–time gauge invariance. The Lagrangian L_1 (eqn (15.40)) is invariant under

$$z_{im\sigma}(\tau) \longrightarrow e^{i\Lambda_i(\tau)} z_{im\sigma}(\tau) \,, \tag{15.46}$$

$$Q_{ij}(\tau) \longrightarrow e^{i(\Lambda_i(\tau)+\Lambda_j(\tau))} Q_{ij}(\tau) \,, \tag{15.47}$$

$$\lambda_i(\tau) \longrightarrow \lambda_i(\tau) - \partial_\tau \Lambda_i(\tau) \,, \tag{15.48}$$

where $\Lambda_i(\tau)$ is some arbitrary angle at each site and time step.

However, this local $\mathrm{U}(1)$ gauge invariance is broken to a smaller invariance group in the vicinity of a saddle point (Q^0, λ^0). This can be illustrated in the simpler context of a classical ferromagnetic Heisenberg model. A ground state is magnetized in one particular direction and thus breaks the $O(3)$ symmetry of the Hamiltonian. The theory for the (transverse) spin *deviations* around this ferromagnetic state has an $O(2)$ symmetry, not $O(3)$. The situation is similar for the fluctuations of the bond field Q_{ij}. Although the model has a local $\mathrm{U}(1)$ gauge invariance, *the action describing the fluctuations around Q^0_{ij} has a lower invariance group*. In the ferromagnetic example, we look at the rotations under which the ground state is unchanged. Similarly, we look for the gauge transformations which leave Q^0_{ij} unchanged. These transformations form the *invariant gauge group* (IGG) of the saddle point, a concept introduced by X.-G. Wen [41]. A gauge transformation $i \mapsto \Lambda_i$ belongs to the IGG of Q^0_{ij} if it is static and satisfies

$$Q^0_{ij} = Q^0_{ij} e^{i(\Lambda_i+\Lambda_j)} . \tag{15.49}$$

If the lattice made by the bonds where Q^0_{ij} is nonzero is *bipartite*, it is easy to show that $\Lambda_i = \theta$ on sublattice A and $\Lambda_i = -\theta$ on sublattice B satisfy eqn (15.49) for any (global) angle θ. In such a case, the IGG is isomorphic to $\mathrm{U}(1)$. On the other hand, if the lattice of the bonds where $Q^0_{ij} \neq 0$ is *not* bipartite, the IGG is isomorphic to $\mathbb{Z}_2$, since $\Lambda_i = \pi$ and $\Lambda_i = 0$ are the only two solutions to eqn (15.49) when $Q^0_{ij} \neq 0$.

The general result [41] is that, among the fluctuations around the saddle point Q^0, some modes are described by a *gauge field*. with a gauge group given by the IGG. We will illustrate this result for the simple case IGG $= \mathbb{Z}_2$.[27]

15.6.5 $\mathbb{Z}_2$ gauge field

If the IGG is $\mathbb{Z}_2$, the important fluctuations turn out to be fluctuations of the *sign* of Q_{ij}. We therefore parameterize these fluctuations in the following way:

$$Q_{ij}(\tau) = Q^0_{ij}\, e^{i\mathcal{A}_{ij}(\tau)}, \quad \mathcal{A}_{ij}(\tau) \in \{0, \pi\}, \tag{15.50}$$

where the field $\mathcal{A}_{ij}$ will play the role of a "discrete" ($\mathbb{Z}_2$) vector potential living on the links of the lattice (pairs of sites where $Q^0_{ij} \neq 0$).

Doing the integration over all the other fluctuation modes (amplitude fluctuations of the bond field Q_{ij}, fluctuations of λ_i, etc.) in order to obtain an effective action for $\mathcal{A}_{ij}$ and the bosons $z_{i\sigma}$ only[28] is formally possible, but it is of course a very difficult task in practice. One can instead determine the symmetry constraints, and, in a Landau–Ginzburg type of approach, construct the simplest action compatible with these symmetries.

For this, we consider the (static) local gauge transformation $i \mapsto \Lambda_i$ with the restriction $\Lambda_i \in \{0, \pi\}$. Because $\mathcal{A}_{ij}$ is defined modulo 2π, $-\Lambda_j$ is equivalent to $+\Lambda_j$ and the transformation rules take the usual form (except for the discrete nature of $\mathcal{A}_{ij}$),

$$z_{i\sigma} \longrightarrow e^{i\Lambda_i} z_{i\sigma}, \tag{15.51}$$

$$\mathcal{A}_{ij} \longrightarrow \mathcal{A}_{ij} + \Lambda_i - \Lambda_j . \tag{15.52}$$

These local transformations form a very large symmetry group (2 to the power of the number of lattice sites) and severely constrain the effective Hamiltonian for these degrees of freedom. Because of this invariance, a term such as $\mathcal{A}_{ij}$, $\mathcal{A}^2_{ij}$, or even $\cos(\mathcal{A}_{ij})$ cannot appear as an

[27] The cases where IGG $= \mathrm{U}(1)$ are generically unstable saddle points: the gauge fluctuations lead to spinon confinement, and lattice symmetry breaking (VBC) when $S = \frac{1}{2}$ [39]. This will not be discussed here.

[28] From now on, we go back to $\mathcal{N} = 1$ and drop the flavor index m for simplicity.

energy term.[29] Instead, only the products of $e^{i\mathcal{A}_{ij}}$ on closed loops are gauge invariant. As a circulation of a vector potential, these loop terms are the analogue of the magnetic flux in electromagnetism. Such products can thus appear in an effective description of the fluctuations about the mean-field solution. Terms such as $\mathcal{E}_{ij} = \partial_\tau \mathcal{A}_{ij} + \lambda_i - \lambda_j$, which are equivalent to the electric field, are also gauge invariant. As for the couplings to the bosons, the couplings to $\mathcal{A}$ allowed by the gauge invariance (and spin rotations) are of the type $\bar{z}_{i\sigma} e^{i\mathcal{A}_{ij}} z_{j\sigma}$.

15.6.6 A simple effective model

We can combine the gauge-invariant terms above into a simple Hamiltonian which can phenomenologically, when IGG $= \mathbb{Z}_2$, describe the gauge fluctuations about a saddle point and their effect on the spinons:

$$\begin{aligned} H = & -K \sum_{\square} \sigma^z_{ij}\sigma^z_{jk}\sigma^z_{kl}\sigma^z_{li} - \Gamma \sum_{\langle ij \rangle} \sigma^x_{ij} \\ & - t \sum_{\langle ij \rangle, \sigma=\uparrow,\downarrow} \left(b^\dagger_{i\sigma}\, \sigma^z_{ij}\, b_{j\sigma} + \text{H.c.} \right) + \Delta \sum_{i\sigma} b^\dagger_{i\sigma} b_{i\sigma} \\ & + V \sum_i \left[\left(b^\dagger_{i\uparrow} b_{i\uparrow} + b^\dagger_{i\downarrow} b_{i\downarrow} - \frac{1}{2} \right)^2 - \frac{1}{4} \right]. \end{aligned} \tag{15.53}$$

The operator σ^z_{ij} has eigenvalues ± 1, like a pseudo-spin $\frac{1}{2}$, and corresponds to $e^{i\mathcal{A}_{ij}}$ in the path integral formulation (eqn (15.50)). σ^x_{ij} corresponds to the electric-field operator. In the path integral, $\mathcal{A}_{ij}$ and $\mathcal{E}_{ij}$ are conjugated. So σ^x_{ij} and σ^z_{ij} should not commute on the same bond. The natural choice in our discrete case is $\sigma^x_{ij}\sigma^z_{ij} = -\sigma^z_{ij}\sigma^x_{ij}$. So, σ^x_{ij} and σ^z_{ij} are the x- and z-components of the pseudo-spin $\frac{1}{2}$. The bosons represent the Bogoliubov quasiparticles (spinons) of the mean-field Hamiltonian. The first term (K) is a sum over all the elementary plaquettes (square here for simplicity) and corresponds to the magnetic energy of the gauge field. The second term (Γ) is the electric energy, which generates fluctuations in the magnetic flux. The third term (t) describes the hopping of spinons and their interaction with the gauge field. The last terms represent the energy cost $\Delta > 0$ to create a spinon (related to the spin gap of the spin model) and some (large) penalty V when more than one spinon is on the same site.

This model is, of course, not directly related to the original spin model but contains the same two important ingredients that have been identified in the large-$\mathcal{N}$ limit (spinons coupled to $\mathbb{Z}_2$ gauge field fluctuations) and can serve as a simplified, phenomenological description of a gapped QSL.

Because of the gauge symmetry, the physical Hilbert space of the model must be constrained to avoid spurious degrees of freedom: two states which differ by a gauge transformation correspond to a single physical state and should not appear twice in the spectrum. In the Hamiltonian formulation of gauge theories, the solution is to construct the operators U_{i_0} which generate the local gauge transformations, and impose the condition that all the physical states should be invariant under these transformations: $U_{i_0}|\text{phys.}\rangle = |\text{phys.}\rangle \ \forall i_0$. In the present

[29] In the same way, a mass term such as the square of the vector potential $A^2_{\mu\nu}$ is forbidden by gauge invariance in conventional electromagnetism.

case, an elementary gauge transformation at site i_0 changes the value $\sigma^z_{i_0 j}$ for all neighbors j of i_0 (denoted by $j \in +$). In addition, it changes the sign of the boson operators at i_0. This transformation is implemented by the following unitary operator:

$$U_{i_0} = \exp\left[i\pi(b^\dagger_{i_0\uparrow}b_{i_0\uparrow} + b^\dagger_{i_0\downarrow}b_{i_0\downarrow})\right]\prod_{j\in+}\sigma^x_{i_0 j}\,. \tag{15.54}$$

The constraint $U_{i_0} = 1$ is the lattice version of Gauss's law $\mathrm{div}\vec{E} = \rho$ in electromagnetism, and the spinons appear to play the role of "electric" charges.

Readers familiar with lattice gauge theories will have recognized the Hamiltonian formulation of a $\mathbb{Z}_2$ gauge theory [42]. However, to show that the ground state of this model realizes a topological phase (when Γ is small enough), we will show that it is very close to the toric-code model introduced by Kitaev [3].

15.6.7 Toric-code limit

One goal of these notes was to show that (gapped) QSLs in Mott insulators are topologically ordered states with emergent gauge degrees of freedom. To conclude, we will now take advantage of Kitaev's lectures on topological states of matter (Chapter 4 in this book), and show the close connection between the large-$\mathcal{N}$ description of a gapped QSL and Kitaev's toric code [3].

We consider the limit of eqn (15.53) when $t = 0$, $\Gamma = 0$, and $V = \infty$. In this limit, the bosons cannot hop any more, and there can only be zero or one per site: $n_i = b^\dagger_{i_0\uparrow}b_{i_0\uparrow} + b^\dagger_{i_0\downarrow}b_{i_0\downarrow} \in \{0,1\}$. Using $U_i = 1$ (eqn (15.54)), we find $e^{i\pi n_i} = \prod_{j\in+}\sigma^x_{ij}$, so that the boson occupation numbers can be expressed in terms of the (lattice divergence of the) electric-field operators: $2n_i = 1 - \prod_{j\in+}\sigma^x_{ij}$. The Hamiltonian can then be written as

$$H = -K\sum_{\square}\sigma^z_{ij}\sigma^z_{jk}\sigma^z_{kl}\sigma^z_{li} - \frac{1}{2}\Delta\sum_i\prod_{j\in+}\sigma^x_{ij}\,, \tag{15.55}$$

which is *exactly* the (solvable) toric-code Hamiltonian [3].

We can now import some results from the toric-code analysis. Although simple to derive in the framework of eqn (15.55), they are highly nontrivial from the point of view of the original spin model. First, the ground state breaks no symmetry, and the spinons (here at the sites i with $\prod_{j\in+}\sigma^x_{ij} = -1$) are free particles; they are not confined by the gauge field fluctuations. Secondly, the ground state is degenerate on a cylinder and on a torus (periodic boundary conditions), as required by the LSMH theorem. The ground states are topologically ordered in the sense that no local observable can distinguish the different ground states. Beyond the spinons, the model also has $\mathbb{Z}_2$-vortex excitations, which correspond to plaquettes with $\sigma^z_{ij}\sigma^z_{jk}\sigma^z_{kl}\sigma^z_{li} = -1$. These gapped excitations are singlet states in the original spin model since the bond field Q_{ij} and its sign fluctuations σ^z_{ij} are rotationally invariant.[30] These excitation have nontrivial mutual statistics with respect to the spinons, and a bound state of a spinon

[30] Reference [22] provides an example of a spin-$\frac{1}{2}$ model (with $\mathrm{U}(1)$ symmetry) where such a $\mathbb{Z}_2$ QSL is realized and where these vortex excitations, dubbed *visons*, have been studied. Visons have also been studied in the context of quantum dimer models, which are simplified models for the short-range VB dynamics in frustrated quantum antiferromagnets (see [33] for a brief introduction).

and a vison behaves as a fermion. Finally, the topological properties of the model (fractional excitations and topological degeneracy) are robust to perturbations, and should persist in the presence of a small Γ and small t (eqn (15.53)).

References

[1] E. H. Lieb, T. D. Schultz, and D. C. Mattis, *Ann. Phys. (N.Y.)* **16**, 407 (1961).
[2] M. Hastings, *Phys. Rev. B* **69**, 104431 (2004).
[3] A. Kitaev, *Ann. Phys.* **303**, 2 (2003).
[4] S. Taniguchi *et al.*, *J. Phys. Soc. Jpn.* **64**, 2758 (1995).
[5] H. Kageyama *et al.*, *Phys. Rev. Lett.* **82**, 3168 (1999).
[6] K. Takatsu, W. Shiramura, and H. Tanaka, *J. Phys. Soc. Jpn.* **66**, 1611 (1997); W. Shiramura *et al.*, *J. Phys. Soc. Jpn.* **66**, 1900 (1997).
[7] P. P. Shores *et al.*, *J. Am. Chem. Soc.* **127**, 13462 (2005); J. S. Helton *et al.*, *Phys. Rev. Lett.* **98**, 107204 (2007); P. Mendels *et al.*, *Phys. Rev. Lett.* **98**, 077204 (2007); A. Olariu *et al.*, *Phys. Rev. Lett.* **100**, 087202 (2008); T. Imai *et al.*, *Phys. Rev. Lett.* **100**, 077203 (2008).
[8] Z. Hiroi *et al.*, *J. Phys. Soc. Jpn.* **70**, 3377 (2001); F. Bert *et al.*, *Phys. Rev. Lett.* **95**, 087203 (2005).
[9] Y. Shimizu *et al.*, *Phys. Rev. Lett.* **91**, 107001 (2003).
[10] T. Itou *et al.*, *Phys. Rev. B* **77**, 104413 (2008).
[11] M. Ryuichi, K. Yoshitomo, and I. Hidehiko, *Phys. Rev. Lett.* **92**, 025301 (2004).
[12] W. Rantner and X.-G. Wen, *Phys. Rev. Lett.* **86**, 3871 (2001); M. Hermele *et al.*, *Phys. Rev. B* **70**, 214437 (2004); Y. Ran *et al.*, *Phys. Rev. Lett.* **98**, 117205 (2007); J. Alicea *et al.*, *Phys. Rev. Lett.* **95**, 241203 (2005).
[13] P. W. Anderson, *Phys. Rev.* **86**, 694 (1952).
[14] T. Holstein and H. Primakoff, *Phys. Rev. B* **58**, 1098 (1940).
[15] J. T. Chalker, P. C. Holdsworth, and E. F. Shender, *Phys. Rev. Lett.* **68**, 855 (1992).
[16] P. Chandra and B. Douçot, *Phys. Rev. B* **38**, 9335 (1988).
[17] V. I. Arnold, *Mathematical Methods of Classical Mechanics*, Graduate Texts in Mathematics, Springer-Verlag, 1989.
[18] I. Affleck, *Phys. Rev. B* **37**, 5186 (1988).
[19] M. Oshikawa, M. Yamanaka, and I. Affleck, *Phys. Rev. Lett.* **78**, 1984 (1997).
[20] M. Oshikawa, *Phys. Rev. Lett.* **84**, 1535 (2000).
[21] B. Nachtergaele and R. Sims, *Commun. Math. Phys.* **276**, 437 (2007).
[22] L. Balents, M. P. A. Fisher, and S. M. Girvin, *Phys. Rev. B* **65**, 224412 (2002); D. N. Sheng and L. Balents, *Phys. Rev. Lett.* **94**, 146805 (2005).
[23] X.-G. Wen, *Phys. Rev. B* **40**, 7387 (1989).
[24] X.-G. Wen, *Phys. Rev. B* **44**, 2664 (1991).
[25] X.-G. Wen and Q. Niu, *Phys. Rev. B* **41**, 9377 (1990).
[26] R. B. Laughlin, *Phys. Rev. B* **23**, 5632 (1981).
[27] C. K. Majumdar and D. K. Ghosh, *J. Math. Phys.* **10**, 1399 (1969).
[28] D. J. Klein, *J. Phys. A: Math. Gen.* **15**, 661 (1982).
[29] K. S. Raman, R. Moessner, and S. L. Sondhi, *Phys. Rev. B* **74**, 064413 (2005).
[30] P. Chandra and B. Douçot, *J. Phys. A: Math. Gen.* **27**, 1541 (1994).
[31] B. S. Shastry and B. Sutherland, *Physica (Amsterdam)* **108B**, 1069 (1981).

[32] P. W. Anderson, *Mater. Res. Bull.* **8**, 153 (1973).
[33] G. Misguich and C. Lhuillier, in *Frustrated Spin Systems*, ed. H. T. Diep, p. 229, World Scientific, Singapore, 2005 [also available at arXiv:cond-mat/0310405].
[34] A. Gellé *et al.*, *Phys. Rev. B* **77**, 014419 (2008).
[35] D. Arovas and A. Auerbach, *Phys. Rev. B* **38**, 316 (1988).
[36] A. Auerbach, *Interacting Electrons and Quantum Magnetism*, Springer-Verlag, 1994.
[37] I. Affleck, *Phys. Rev. Lett.* **54**, 966 (1985).
[38] I. Affleck and J. B. Marston, *Phys. Rev. B* **37**, 3774 (1988); J. B. Marston and I. Affleck, *Phys. Rev. B* **39**, 11538 (1989).
[39] N. Read and S. Sachdev, *Phys. Rev. Lett.* **62**, 1694 (1989).
[40] N. Read and S. Sachdev, *Phys. Rev. Lett.* **66**, 1773 (1991).
[41] X.-G. Wen, *Phys. Rev. B* **65**, 165113 (2002).
[42] J. B. Kogut, *Rev. Mod. Phys.* **51**, 659 (1979).

16

Superspin chains and supersigma models: A short introduction

H. SALEUR

Hubert Saleur,
IPhT, CEA Saclay,
91191 Gif-sur-Yvette, France,
and Physics Dept., USC,
Los Angeles, CA 90089-0484, USA.

16.1 Introduction

These lectures offer a short introduction to the field of superspin chains and supersigma models. By this I mean, more precisely, models whose target (be it the value of a lattice spin or a field) admits the action of a superalgebra or a supergroup, under which the physics exhibits some invariance. No confusion should arise with space–time supersymmetry of particle physics: the "super" is only in the target, and does not usually affect space–time.

In condensed matter physics, such target spaces arise in the study of noninteracting disordered systems [1]. For instance, the solution to the transition between plateaux in the integer quantum Hall effect boils down to identifying the strong-coupling limit of the $\mathrm{U}(1,1|2)/\mathrm{U}(1|1) \times \mathrm{U}(1|1)$ sigma model at $\theta = \pi$ (for a review and many useful remarks, see [2]).

Supergroups and supercoset targets appear in the description of strings in anti-de Sitter space, which play a major role on the string theory side of the AdS/CFT duality [3, 4]. For instance, strings on $AdS_3 \times S^3$ are related to the $PSL(2|2)$ sigma model.

Supergroups appeared in the context of geometrical problems in the pioneering paper of Parisi and Sourlas as early as 1980 [5]. Supersymmetry/disorder/random-geometry problems have a long, tangled history.

Supergroups are simple examples of noncommutative manifolds, and are natural objects to consider from a mathematical point of view. In fact, some of the early work on the subject occurred in the context of knot theory.

Ordinarily, "particle physics supersymmetry" usually makes things simpler: bosonic and fermionic diagrams come with opposite signs, so divergences cancel out in superstring theory, 2D $N = 2$ theories benefit from powerful nonrenormalization theorems so the exponents can be calculated readily from the Landau–Ginzburg action, etc. In our case, however, "supergroup supersymmetry" renders things very complicated instead. There are several reasons for this; one is that we have supersymmetry only in the target space. Another is that, for various reasons—which will be especially clear in statistical-mechanics applications—we do not impose unitarity requirements. As a result, we have to deal with several striking features:

- The theories are violently nonunitary: path integrals are naively divergent, and S-matrices cannot be unitary ($\sum p_i = 1$ but some $p_i > 1$!). This makes Bethe ansatz approaches (in particular, those to tackle nonperturbative RG flows) difficult to implement.
- The underlying representation theory involves disgusting problems of indecomposability and wilderness which mathematicians are willing to consider only at the most abstract level. In the field theories, this means that one has to deal with logarithmic conformal field theories (CFTs).
- The space of CFTs itself is hard to map out. In the case of ordinary compact groups, conformal invariance and global group symmetry imply Kac–Moody symmetry. WZW models are easily manageable. In the case of supergroups, WZW models form only a small class of possible CFTs. Other classes include:
    1. PCM on OSp($2n + 2|2n$) and PSL($n|n$), which are conformal with or without a WZW term and provide lines of CFTs.
    2. Supersphere sigma models OSp($2n + 2|2n$)/OSp($2n + 1|2n$), which also provide lines of CFTs.
    3. Strong-coupling limits of superprojective sigma models $\mathrm{U}(n+m|n)/\mathrm{U}(1) \times \mathrm{U}(n+m-1|n)$ at $\theta = \pi$.

4. Weak-coupling limits of supersphere sigma models in (Goldstone) spontaneously-broken-symmetry phases.
5. And God knows what else.

Progress has been slow but steady. Even if no pattern has truly emerged yet, at least some partial answers are now known. The purpose of these lectures is not to give a review of these answers or of the remaining problems, but simply to introduce the topic through a few very simple examples. The same philosophy applies to references: only very few are given, and their choice is at least partly a matter of taste.

16.2 Some mathematical aspects: The $\mathrm{gl}(1|1)$ case

16.2.1 The Lie superalgebra $\mathrm{gl}(1|1)$ and its representations

Defining relations. The Lie superalgebra $\mathfrak{g} = \mathrm{gl}(1|1)$ is generated by two bosonic elements E, N and two fermionic generators $\Psi^\pm$ such that E is central and the other generators obey

$$[N, \Psi^\pm] = \pm\Psi^\pm \quad \text{and} \quad \{\Psi^-, \Psi^+\} = E\,.$$

The even subalgebra is thus given by $\mathfrak{h}^{(0)} = \mathfrak{gl}(1) \oplus \mathfrak{gl}(1)$. Let us also fix the following Casimir element C for $\mathrm{gl}(1|1)$:

$$C = (2N-1)E + 2\Psi^-\Psi^+\,.$$

The choice of C is not unique, since we could add any function of the central element E. This has interesting consequences in field theory.

Finally, we recall the definition of the supertrace $\mathrm{Str}(.) = \mathrm{Tr}((-1)^F.)$. The superdimension is the supertrace of the identity, i.e. the number of bosons minus the number of fermions. The superdimension of $\mathrm{gl}(1|1)$ is zero.

Some remarks: while one goes from u(2) to sl(2) by factoring out a trivial u(1), the identity E occurs on the right-hand side of the commutation relations for $\mathrm{gl}(1|1)$. Since N, meanwhile, is a fermion-number-counting operator, it is really quite natural to consider $\mathrm{gl}(1|1)$ for field theory applications (or $\mathrm{psl}(1|1)$; see below), in contrast to the sl(2) case. Also, the notation $\Psi^\pm$ does not imply any kind of Hermitian conjugation (see below).

Irreducible representations. To begin with, we list the irreducible representations, which fall into two different series. There is one series of two-dimensional representations $\langle e, n\rangle$ which is labeled by pairs e, n with $e \neq 0$ and $n \in \mathbb{R}$. In these representations, the generators take the form $E = e\mathbf{1}_2$ and

$$N = \begin{pmatrix} n-1 & 0 \\ 0 & n \end{pmatrix}, \quad \Psi^+ = \begin{pmatrix} 0 & 0 \\ e & 0 \end{pmatrix}, \quad \Psi^- = \begin{pmatrix} 0 & 1 \\ 0 & 0 \end{pmatrix}.$$

These representations are the typical representations (long multiplets) of $\mathfrak{g}$ =gl(1|1). In addition, there is one series of atypical representations $\langle n\rangle$ (short multiplets). These are one-dimensional and are parameterized by the value $n \in \mathbb{R}$ of N. All other generators vanish.

In the representations $\langle e, n\rangle$, the fermionic generators appear on a somewhat different footing since Ψ^+ depends on the parameter e while Ψ^- does not. There exists another family of two-dimensional representations $\widetilde{\langle e,n \rangle}$, however, in which the roles of Ψ^- and Ψ^+ are interchanged:

$$N = \begin{pmatrix} n & 0 \\ 0 & n-1 \end{pmatrix}, \quad \Psi^+ = \begin{pmatrix} 0 & 1 \\ 0 & 0 \end{pmatrix}, \quad \Psi^- = \begin{pmatrix} 0 & 0 \\ e & 0 \end{pmatrix}.$$

As long as $e \neq 0$, the representations $\langle e, n \rangle$ and $\widetilde{\langle e,n \rangle}$ are equivalent. In fact, the isomorphism between the two representations may be implemented by conjugation with the matrices $W_e = e\sigma^+ + \sigma^-$, where $\sigma^\pm$ are the usual Pauli matrices.

Indecomposability—a first look. For the typical representations, we assume that the parameter e does not vanish. But it is still interesting to explore what happens when we set $e = 0$. The above matrices certainly continue to provide a representation of gl(1|1); the only difference is that this is no longer irreducible. In fact, we observe that the basis vector $|0\rangle = (1,0)^T$ generates a one-dimensional invariant subspace of the corresponding two-dimensional representation space. But one should not conclude that there exists an invariant complement. In fact, it is impossible to decouple the vector $|1\rangle = (0,1)^T$ from the representation, since $\Psi^-|1\rangle = |0\rangle$, independently of the choice of the parameter e. The representation $\langle 0, n \rangle$ is therefore indecomposable but it is not irreducible. We can think of $\langle 0, n \rangle$ as being built up from two atypical constituents, namely the representations $\langle n \rangle$ and $\langle n-1 \rangle$. To visualize the internal structure of $\langle 0, n \rangle$, we may employ the following diagram,

$$\langle 0, n \rangle : \quad \langle n-1 \rangle \longleftarrow \langle n \rangle .$$

Later we shall see much more complicated composites of atypical representations. It is therefore useful to become familiar with diagrammatic representations of indecomposables.

The isomorphism between $\langle e, n \rangle$ and $\widetilde{\langle e,n \rangle}$ does not survive in the limit $e \to 0$: the representations $\langle 0, n \rangle$ and $\widetilde{\langle 0,n \rangle}$ are inequivalent. $\widetilde{\langle 0,n \rangle}$ is also an indecomposable representation that is built up from the same atypical constituents as $\langle 0, n \rangle$, but this time the nonvanishing generator Ψ^+ maps us from $\langle n \rangle$ to $\langle n-1 \rangle$, i.e.

$$\widetilde{\langle 0,n \rangle} : \quad \langle n-1 \rangle \longrightarrow \langle n \rangle .$$

In many cases it is possible and convenient to consider the representations $\langle 0, n \rangle$ and $\widetilde{\langle 0,n \rangle}$ as limits of typical representations.

Projective covers. Having seen all of the irreducible representations $\langle e, n \rangle$ and $\langle n \rangle$ of gl(1|1) along with their limits as e goes to zero, our next task is to compute tensor products of typical representations $\langle e_1, n_2 \rangle$ and $\langle e_2, n_2 \rangle$. Here, we emphasize that we are dealing with graded tensor products; that is, when we pass a fermionic operator through a fermionic state, we generate an additional minus sign. We will follow the convention that $|0\rangle$ is bosonic and $|1\rangle$ is fermionic for the time being. It is of course possible to switch the Z_2 grading and decide that $0\rangle$ is fermionic, etc. As long as $e_1 + e_2 \neq 0$, the tensor product is easily seen to decompose into a sum of two typical representations,

$$\langle e_1, n_2 \rangle \otimes \langle e_2, n_2 \rangle = \langle e_1 + e_2, n_1 + n_2 - 1 \rangle \oplus \langle e_1 + e_2, n_1 + n_2 \rangle .$$

But when $e_1 + e_2 = 0$, we obtain a four-dimensional representation that cannot be decomposed into a direct sum of smaller subrepresentations! The representation matrices of these four-dimensional indecomposables $\mathcal{P}_n$ read as follows ($n \equiv n_1 + n_2 - 1$, $e \equiv 1$):

$$N = \begin{pmatrix} n-1 & 0 & 0 & 0 \\ 0 & n & 0 & 0 \\ 0 & 0 & n & 0 \\ 0 & 0 & 0 & n+1 \end{pmatrix}, \quad \Psi^{+} = \begin{pmatrix} 0 & 0 & 0 & 0 \\ -1 & 0 & 0 & 0 \\ 1 & 0 & 0 & 0 \\ 0 & 1 & 1 & 0 \end{pmatrix}, \quad \Psi^{-} = \begin{pmatrix} 0 & 1 & 1 & 0 \\ 0 & 0 & 0 & 1 \\ 0 & 0 & 0 & -1 \\ 0 & 0 & 0 & 0 \end{pmatrix}.$$

As we have seen before, it is useful to picture the structure of indecomposables. The form of N tells us that $\mathcal{P}_n$ is composed of the atypical irreducibles $\langle n-1\rangle, 2\langle n\rangle, \langle n+1\rangle$. The action of $\Psi^{\pm}$ relates these four representations as follows:

$$\mathcal{P}_n: \qquad \langle n\rangle \longrightarrow \langle n+1\rangle \oplus \langle n-1\rangle \longrightarrow \langle n\rangle\,, \tag{16.1}$$

or

$\mathcal{P}_n$: (16.2)

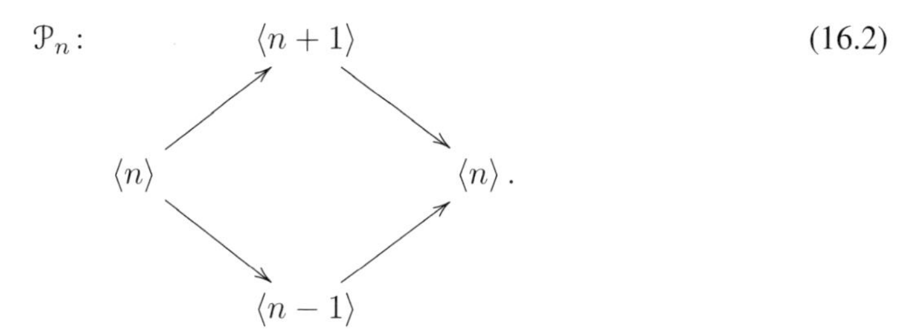

There are a few remarks we would like to make at this point. The first one concerns the form of the Casimir element C in the representations $\mathcal{P}_n$. It is straightforward to see that C maps the subspace $\langle n\rangle$ on the left onto the $\langle n\rangle$ on the right in the above diagram and that it is zero otherwise. This means that C cannot be diagonalized in $\mathcal{P}_n$. We shall return to this observation later on.

It is also obvious from the diagrammatic representation that $\mathcal{P}_n$ contains the indecomposables $\langle 0, n\rangle$ and $\langle \widetilde{0, n+1}\rangle$ as subrepresentations. In this sense, the latter are extendable into a larger indecomposable. For the representation $\mathcal{P}_n$, the situation is quite different: it may be shown (and is intuitively clear) that $\mathcal{P}_n$ is maximal in the sense that it can never appear as a subrepresentation of a larger indecomposable. In the mathematics literature, such representations are known as projective. Since the projective representation $\mathcal{P}_n$ contains the irreducible $\langle n\rangle$ as a true subrepresentation, $\mathcal{P}_n$ is also called the projective cover of $\langle n\rangle$.

Note that the term "projective" is used here with a very different meaning from what physicists call projective representations—that is, representations involving spurious phases. Such representations will not occur in the present context.

The typical representations $\langle e, n\rangle, e \neq 0$, along with the indecomposables $\mathcal{P}_n$, exhaust the set of finite-dimensional projectives of gl(1|1). An important property is that projective representations close under tensor products. In particular, tensor products of the representations $\mathcal{P}_n$ do not generate any new types of representations. That is not to say that there are not any others.

In fact, there is a large family of indecomposables ("zigzag modules") with arbitrarily large dimension, as described below. The following two subsections are borrowed from [6].

Zigzag modules. There exist two other different families of indecomposable representations $\mathcal{Z}^d(n)$ and $\widetilde{\mathcal{Z}}^d(n)$, which we shall call zigzag and antizigzag representations. They are parameterized by the eigenvalue $n \in \mathbb{C}$ of N with the largest real part and by the number $d =$

$1, 2, 3, \ldots$ of their atypical constituents. On a basis of eigenstates $|m\rangle, m = n, \ldots, n-d+1$, for the element N, the generators of the zigzag representations $\mathcal{Z}^d(n)$ read

$$N|m\rangle = m|m\rangle \quad , \quad \psi^{\pm}|m\rangle = \frac{1}{2}(1 + (-1)^{n-m})|m \pm 1\rangle \, , \tag{16.3}$$

and E vanishes identically. Here we adopt the convention that $|m\rangle = 0$ when m is outside the allowed range. Similarly, we can introduce antizigzag representations $\widetilde{\mathcal{Z}}^d(n)$ through

$$N|m\rangle = m|m\rangle \quad , \quad \psi^{\pm}|m\rangle = \frac{1}{2}(1 - (-1)^{n-m})|m \pm 1\rangle \, . \tag{16.4}$$

The only difference between the formulas (16.3) and (16.4) is in the sign between the two terms for the action of fermionic elements. Note that we have $\langle n\rangle \cong \mathcal{Z}^1(n) \cong \widetilde{\mathcal{Z}}^1(n)$.

Once more, we can display the structure of the zigzag and antizigzag modules through their associated diagram. In doing so, we shall separate two cases depending on the parity of d. When $d = 2p$ is even, we find

$$\mathcal{Z}^{2p}(n): \qquad \langle n-2p+1\rangle \longleftarrow \langle n-2p+2\rangle \longrightarrow \cdots \longleftarrow \langle n-2\rangle \longrightarrow \langle n-1\rangle \longleftarrow \langle n\rangle \, ,$$

$$\widetilde{\mathcal{Z}}^{2p}(n): \qquad \langle n-2p+1\rangle \longrightarrow \langle n-2p+2\rangle \longleftarrow \cdots \longrightarrow \langle n-2\rangle \longleftarrow \langle n-1\rangle \longrightarrow \langle n\rangle \, .$$

Observe that the leftmost atypical constituent is invariant for even-dimensional zigzag modules, a property that is not shared by the even-dimensional antizigzag representations, which, by construction, always possess an invariant constituent in their rightmost position. When $d = 2p+1$ is odd, on the other hand, the corresponding diagrams read

$$\mathcal{Z}^{2p+1}(n): \qquad \langle n-2p\rangle \longrightarrow \langle n-2p+1\rangle \longleftarrow \cdots \longleftarrow \langle n-2\rangle \longrightarrow \langle n-1\rangle \longleftarrow \langle n\rangle \, ,$$

$$\widetilde{\mathcal{Z}}^{2p+1}(n): \qquad \langle n-2p\rangle \longleftarrow \langle n-2p+1\rangle \longrightarrow \cdots \longrightarrow \langle n-2\rangle \longleftarrow \langle n-1\rangle \longrightarrow \langle n\rangle \, .$$

In this case, both ends of the antizigzag modules correspond to invariant subspaces. In analogy with the graphical presentation of the projective representations $\mathcal{P}_{\mathfrak{h}}(n)$, we can modify the diagrams for $\mathcal{Z}$ and $\widetilde{\mathcal{Z}}$ a little by moving the sources up such that all arrows run at a 45 degree angle. The resulting pictures explain the name "zigzag module."

Tensor products. We now discuss some tensor products of finite-dimensional representations of $\mathfrak{gl}(1|1)$. Obviously, there are quite a few cases to consider. For the tensor product of two typical representations, we find

$$\langle e_1, n_1\rangle \otimes \langle e_2, n_2\rangle = \begin{cases} \mathcal{P}_{n_1+n_2-1} & \text{for } e_1 + e_2 = 0 \, , \\ \bigoplus_{p=0}^{1} \langle e_1 + e_2, n_1 + n_2 - p\rangle & \text{for } e_1 + e_2 \neq 0 \, . \end{cases} \tag{16.5}$$

This formula should only be used when $e_1, e_2 \neq 0$.

Next we would like to consider the tensor products involving projective covers $\mathcal{P}_n$ in addition to typical representations. These are given by

$$\begin{aligned} \langle e, n\rangle \otimes \mathcal{P}_m &= \langle e, n+m+1\rangle \oplus 2 \cdot \langle e, n+m\rangle \oplus \langle e, n+m-1\rangle \, , \\ \mathcal{P}_n \otimes \mathcal{P}_m &= \mathcal{P}_{n+m+1} \oplus 2 \cdot \mathcal{P}_{n+m} \oplus \mathcal{P}_{n+m-1} \, , \end{aligned} \tag{16.6}$$

where we assume once more that $e \neq 0$ in the first line. We observe that typical representations and projective covers close under tensor products, in perfect agreement with the general behavior of projective representations.

Some remarks: the classification of indecomposable representations for super-Lie algebra is a difficult problem, usually believed to be "wild" (in a technical sense). Except for the simplest cases, one can glue diagrams together with arrows in various ways (as in the case of the zigzag representations) to build out of well-defined blocks, shapes that can be almost arbitrary. The few models well understood up to now, however, suggest that physics is simpler: only projective representations seem to play a role. These are much easier to characterize and classify. While there is plethora of general references on superalgebras, there are not so many cases worked out explicitly. My favorite pedestrian papers on the subject are [6–8], and the beautiful thesis [9]. In addition, super-Lie algebras admit quantum group deformations, and are related to R-matrix solutions of the Bethe ansatz. The gl(1|1) case is related in this way to the Alexander–Conway polynomial.

16.2.2 The supergroup U(1|1)

We now move on to supergroups. Since we now have fermionic generators, we need Grassmann numbers to multiply them. I will give a short summary of what's necessary here. Details (together with much information about particle physics SUSY) can be found in [10].

Usually, one starts with a finite number of symbols α_i, $i = 1, \ldots \mathcal{N}$, and the calculational rule that all of them anticommute pairwise (and thus square to zero). The Grassmann algebra $G_\mathcal{N}$ is the set of "numbers" which are polynomials in the real numbers and the α_i's with real coefficients. For instance, for $\mathcal{N} = 1$, it is the set of "supernumbers" $a = a_0 + a_1\alpha_1$, for $\mathcal{N} = 2$, the set of supernumbers $a = a_0 + a_1\alpha_1 + a_2\alpha_2 + a_{12}\alpha_1\alpha_2$, etc. In general, $G_\mathcal{N}$ is of dimension $2^\mathcal{N}$. The bosonic (commuting) component of these supernumbers is called the "body" (e.g. $a_0 + a_{12}\alpha_1\alpha_2$ for $\mathcal{N} = 2$), and the fermionic (anticommuting) component is called the "soul." It is convenient to consider that $\mathcal{N} = \infty$, and to forget the "super" and just talk about numbers.

One can of course take the coefficients a_i to be complex. As for the basis α_i, it is often assumed that it is "real" so that $\alpha_i^* = \alpha_i$. Complex conjugation is then defined to obey $(a + b)^* = a^* + b^*$, and $(ab)^* = b^* a^*$. Thus, for instance, $(\alpha_1\alpha_2)^* = \alpha_2\alpha_1 = -\alpha_1\alpha_2$ and hence $\alpha_1\alpha_2$ is "purely imaginary." It is also possible to decide that the basis is not real. In this case, one needs to make room for the conjugates α_i^*. In most aspects, this is really the same as doubling the number of species $\mathcal{N}$: complex conjugation is not really very useful for fermions, and some authors recommend forgetting it altogether.

We thus define U(1|1) as the group of transformations that leave the quadratic form $z^* z + \eta^* \eta$ invariant, where z is a complex bosonic number and η is fermionic. It is easy to see that these transformations can always be written in the form

$$g = \begin{pmatrix} e^{iu} & 0 \\ 0 & e^{iv} \end{pmatrix} \begin{pmatrix} 1 - \alpha\alpha^*/2 & i\alpha \\ i\alpha^* & 1 - \alpha^*\alpha/2 \end{pmatrix}. \tag{16.7}$$

Using, for a general supermatrix $G = \begin{pmatrix} A & B \\ C & D \end{pmatrix}$, the definition of the adjoint

$$G^t = \begin{pmatrix} A^t & C^t \\ B^t & D^t \end{pmatrix}, \tag{16.8}$$

we have

$$g^\dagger = \begin{pmatrix} 1-\alpha\alpha^*/2 & -i\alpha \\ -i\alpha^* & 1-\alpha^*\alpha/2 \end{pmatrix} \begin{pmatrix} e^{-iu} & 0 \\ 0 & e^{-iv} \end{pmatrix} \tag{16.9}$$

such that, of course, $g^\dagger g = g^\dagger g = 1$. This clearly shows the role of the two U(1) bosonic subgroups, and the fermionic part.

Recalling similarly the expression for the superdeterminant (or Berezinian) of a general supermatrix sdet $g = \det A/\det(D - CA^{-1}B)$, we find here

$$\text{sdet}\, g = e^{i(u-v)}\,. \tag{16.10}$$

The group can be considered as the (purely imaginary) exponential of the Lie algebra

$$g = e^{i(xE+yN)} e^{i(\eta_+\Psi^+ + \eta_-\Psi^-)}\,, \tag{16.11}$$

where

$$\begin{aligned} x &= \frac{nu-(n-1)v}{e}\,, \\ y &= v-u\,, \\ \eta_- &= \alpha\,, \\ \eta_+ &= \frac{\alpha^*}{e}\,. \end{aligned} \tag{16.12}$$

Other, more convenient forms are used in the literature, obtained by moving the exponentials around. In [11], for instance, we set $g = e^{i(\eta_+\Psi^+} e^{i(xE+yN)} e^{i\eta_-\Psi^-)}$ instead.

I need also to provide a little reminder about integrals over Grassmann variables. For some such number θ, any real function f expands as $f(\theta) = a + b\theta$. We set

$$\int d\theta = 0, \quad \int d\theta\, \theta = 1\,, \tag{16.13}$$

so $\int f(\theta)\, d\theta = b$: integration is the same as differentiation, etc. A well-known formula for integration over fermionic variables is

$$\int d\theta_1\, d\theta_1^* \ldots d\theta_n\, d\theta_n^* \exp\left(\sum_{i,j} \theta_i^* A_{ij} \theta_j\right) = \text{Det}\, A\,. \tag{16.14}$$

Since Grassmann integration is in fact differentiation, note that we do not have to worry about the sign of the eigenvalues A in this integral: it always converges. Also, if we think about it, nothing in this formula uses the fact that θ^* is or is not the conjugate of θ. We can treat these two variables as independent. We can equivalently set $\theta = \theta_1 + i\theta_2$ assuming $\theta_{1,2}$ to be real, make a change of variables, etc., and get the same result. A related remark is that $\theta\theta^*$ is not "positive" or "negative," although it is real (and it can be read as $2i\theta_2\theta_1$ as well).

The invariant metric for U(1|1) (such that $\int f(g)\, d\mu(g) = \int f(g_0 g)\, d\mu(g)$) is

$$d\mu = dx\, dy\, d\eta_-\, d\eta_+\,. \tag{16.15}$$

In principle, one has to be careful about the fact that bosonic supernumbers are not necessarily "without a soul," that is, they are not necessarily ordinary real or complex numbers. When one is integrating over them, however, it is usually the case that one can forget this potential problem—in a rather precise sense, integrals over bosonic supernumbers are better seen as contour integrals, which can be deformed to lie entirely within the ordinary numbers. As a result, the volume of the supergroup U(1|1) vanishes identically, since the fermion contribution is identically zero. The question of the volume of GL(1|1) is more delicate, as the integrals over the real coordinates diverge.

Let us now try to move in the direction of defining sigma models. Still using eqn (16.7), we can find the action of the principal chiral model (PCM):

$$\mathrm{Str}\left(\partial_\mu g \partial_\mu g^{-1}\right) = \left(\partial_\mu u\right)^2 - \left(\partial_\mu v\right)^2 + 2\partial_\mu \alpha \partial_\mu \alpha^* + i\left(\partial_\mu u - \partial_\mu v\right)\left(\alpha \partial_\mu \alpha^* - \partial_\mu \alpha \alpha^*\right). \tag{16.16}$$

Despite the presence of the factor i, the last term is real. Its presence shows that the model is not free.

Meanwhile, since we have opposite signs for the u, v terms in the real part, a path integral built using (16.16) will never be well defined. Note that the problem would be the same for GL(1|1)—obtained, for instance, by taking real exponentials of the Lie algebra—since making u, v purely imaginary will switch the two signs simultaneously.

This problem can been tackled in two possible ways. One is to try to modify the target into something that is not a group any longer—roughly, by making purely imaginary only those coordinates which come with the wrong sign in the action. The other is to proceed rather formally with supercurrent algebras, minisuperspace analysis, etc., and later on, to perform analytic continuations to get physically meaningful results. In any case, a crucial aspect is that it is not possible to deal with a compact target: even though the parameterization of U(1|1) allows for compactification of u, v coordinates, only the theory on the universal cover seems meaningful. There is thus not much difference, in the end, between GL(1|1) and U(1|1).

From the discussion of the superdeterminant, we see that SU(1|1) will be obtained simply by setting $u = v$. The resulting group has a trivial remaining U(1) in the factor, which is usually discarded to obtain what is usually called PSU(1|1): the latter is the group obtained by identifying elements of SU(1|1) that differ by a multiple of the identity. Note that there is no 2×2 supermatrix representation of this group: indeed, by taking the product of two matrices with $u = v = 0$ and fermionic parameters β, γ, one obtains a matrix with a fermionic parameter $\beta + \gamma$ and $u = v = (i/2)(\beta\gamma^* + \beta^*\gamma)$ (it is real, since the fermions are anticommuting).

The sigma model on PSU(1|1) is conveniently defined by restricting the supertrace to the leftover two fermionic generators,

$$\mathrm{Str}\left(g^{-1}\partial_\mu g \Psi^-\right) \mathrm{Str}\left(g^{-1}\partial_\mu g \Psi^+\right) = e \partial_\mu \alpha \partial_\mu \alpha^* . \tag{16.17}$$

Note that $\mathrm{Str}\left(g^{-1}\partial_\mu g E\right) \propto \mathrm{Str}\left(g^{-1}\partial_\mu g\right) = 0$, so the U(1) part of SL(1|1) does not contribute to the action at all, and represents a sort of gauge invariance: one can in fact think of the PSU(1|1) sigma model as the sigma model on the complex superprojective space $\mathrm{CP}^{0|1}$.

Some remarks: the U(1|1) WZW model has given rise to rather a lot of work, and is by now essentially understood [11]. Work is in progress on the perturbed WZW model and on the PCM model. The difficulties in defining the functional integrals for supersigma models have been discussed in [12]. Aspects of the analytic continuation necessary in the "naive" approach have been discussed in [13]. For relations to knot theory, etc., see [14].

16.3 The two simplest sigma models

16.3.1 psl(1|1) and symplectic fermions

We consider now a theory involving two fermionic fields $\psi^{1,2}$ with action [15]

$$S = \frac{1}{4\pi}\int d^2x\,\partial_\mu\psi^1\partial_\mu\psi^2 = \frac{1}{8\pi}\int d^2x\left(\partial_\mu\psi^1\partial_\mu\psi^2 - \partial_\mu\psi^2\partial_\mu\psi^1\right). \tag{16.18}$$

It is possible to think of the fermions as conjugate and use the notation $\psi, \psi^\dagger$ instead. This will introduce some factors of i into the action but, as commented on earlier, will not change any of the physical properties.

The main feature of this theory is that its partition function vanishes identically,

$$\int [d\psi^1 d\psi^2]\exp(-S) = 0\,, \tag{16.19}$$

because of the zero mode in the Laplacian. There are several ways to interpret this result; one of them is that this theory is a psl(1|1) current algebra (see below). The point for now is that it makes defining correlations in the theory a little tricky. The best way to proceed is to decide to define averages as functional integrals with no partition function normalization (as in Liouville theory):

$$\langle O\rangle \propto \int [d\psi^1 d\psi^2] O\exp(-S)\,. \tag{16.20}$$

It follows that, with proper normalization (the notation and formulas are borrowed from [15])

$$\begin{aligned}
\langle 1\rangle &= 0\,,\\
\langle \psi^\alpha(z_1,\bar z_1)\psi^\beta(z_2,\bar z_2)\rangle &= d^{\alpha\beta}\,,\\
\langle \psi^\alpha(1)\psi^\beta(2)\psi^\gamma(3)\psi^\delta(4)\rangle &\\
= d^{\alpha\beta}d^{\gamma\delta}(\Delta_{12}+\Delta_{34}) &- d^{\alpha\gamma}d^{\beta\delta}(\Delta_{13}+\Delta_{24}) + d^{\alpha\delta}d^{\beta\gamma}(\Delta_{14}+\Delta_{23})\,,\\
\ldots\,,
\end{aligned} \tag{16.21}$$

where $d^{12} = -d^{21} = 1$. The normalization is chosen such that the two-point function—which is a constant independent of $z_{1,2}$—is unity. The quantity Δ is the Gaussian propagator,

$$\Delta = \mathcal{Z} - \ln|z|^2\,, \tag{16.22}$$

where $\mathcal{Z}$ is a constant. Note that no coupling constant appears in these expressions. This is expected; fermions do not really have a "scale," and a coupling g_σ in front of the integral could readily be absorbed into a change of variable. To proceed, we introduce another bosonic ground state by sending the two fermionic fields to the same point, and subtracting (logarithmic) divergences in the correlators. This leads to a field $\omega \equiv :\psi^1\psi^2:$ with the additional propagators

$$\begin{aligned}
\langle\omega\rangle &= 1\,,\\
\langle\omega(1)\omega(2)\rangle &= -2\Delta(z_{12})\,,\\
\langle\psi^\alpha(1)\psi^\beta(2)\omega(3)\rangle &= -d^{\alpha\beta}\left(\Delta_{13}+\Delta_{23}-\Delta_{12}\right),\\
&\ldots\,.
\end{aligned} \tag{16.23}$$

Since the identity has a vanishing one-point function, it is often convenient to give it another name, and call it Ω. The operator product expansions (OPEs) follow, and the most important are

$$\begin{aligned}\psi^\alpha(z,\bar{z})\psi^\beta &= d^{\alpha\beta}\left(\omega + \Delta(z)\Omega\right),\\ \psi^\alpha(z,\bar{z})\omega &= -\Delta(z)\psi^\alpha,\\ \omega(z,\bar{z})\omega &= -\Delta(z)\left(2\omega + \Delta(z)\Omega\right),\end{aligned} \tag{16.24}$$

where fields with no argument are taken at the origin. The action is invariant under transformations

$$\begin{pmatrix}\psi^1\\ \psi^2\end{pmatrix} \to \begin{pmatrix} a & b\\ c & d\end{pmatrix}\begin{pmatrix}\psi^1\\ \psi^2\end{pmatrix}, \tag{16.25}$$

provided $ad - bc = 1$, i.e. there is an SP(2) symmetry. Introducing the Noether currents

$$\begin{aligned}J^0 &= \frac{1}{4}\left(\psi^1\partial\psi^1 + \psi^2\partial\psi^2\right),\\ J^1 &= \frac{1}{4}\left(-\psi^1\partial\psi^1 + \psi^2\partial\psi^2\right),\\ J^2 &= \frac{1}{4}\left(\psi^1\partial\psi^2 + \psi^2\partial\psi^1\right),\end{aligned} \tag{16.26}$$

and their $\bar{J}$ partners, we have $\bar{\partial}J + \partial\bar{J} = 0$, and the integrated charges

$$Q^a = \frac{1}{2i\pi}\int J^a\, dz - \bar{J}^a\, d\bar{z} \tag{16.27}$$

satisfy

$$\left[Q^a, Q^b\right] = f^{ab}_c Q^c \tag{16.28}$$

with $f_2^{01} = -1, f_1^{02} = 1, f_0^{12} = 1$. Nevertheless the currents do not satisfy Kac–Moody OPEs. For instance, by forming the combinations $J^\pm = J^0 \pm J^1$, we find

$$J^+(z,\bar{z})J^-(0) = \frac{\omega + \Omega}{4z^2} + \frac{\Omega\Delta(z)}{4z^2} + \ldots, \tag{16.29}$$

so that the $1/z^2$ term has unpleasant logarithmic terms (similar behavior appears in the $1/z$ term). Also, the left–right OPEs are nontrivial; for instance,

$$J^+(z,\bar{z})\bar{J}^-(0) = \frac{\Omega}{4z\bar{z}} + \ldots. \tag{16.30}$$

Nevertheless, when we compute correlators, the logarithmic terms disappear because $\langle\Omega\rangle = 0$, and one finds, for instance,

$$\begin{aligned}\langle J^+(z,\bar{z})J^-(0)\rangle &= \frac{1}{4z^2},\\ \langle J^+(z,\bar{z})\bar{J}^-(0)\rangle &= 0,\end{aligned} \tag{16.31}$$

which corresponds formally to a level $k = 1/2$ sl(2) current algebra. In the ordinary case, knowledge of these two-point functions would allow us to conclude that the OPEs are of

Kac–Moody type: continuous symmetry plus conformal invariance would imply Kac–Moody symmetry. This is not true here, because we have no unitarity: there are fields whose one- or two-point functions are zero, and yet these fields themselves do not vanish. The question of classifying "logarithmic deformations" or current algebras remains, sadly, unsolved.

The model also contains a sub-psl(1|1) current algebra generated by the two fermionic currents $\partial\psi^\alpha$. In fact, this is the most obvious symmetry of the problem when one considers the full field content of the theory. For instance, this model has four fields with conformal weight $h = 0$, namely $\Omega, \omega, \psi^1, \psi^2$, and under the action of the zero modes of the fermionic currents, they can be arranged into the ubiquitous four-dimensional projective representation of psl(1|1) (which is also a representation of gl(1|1)),

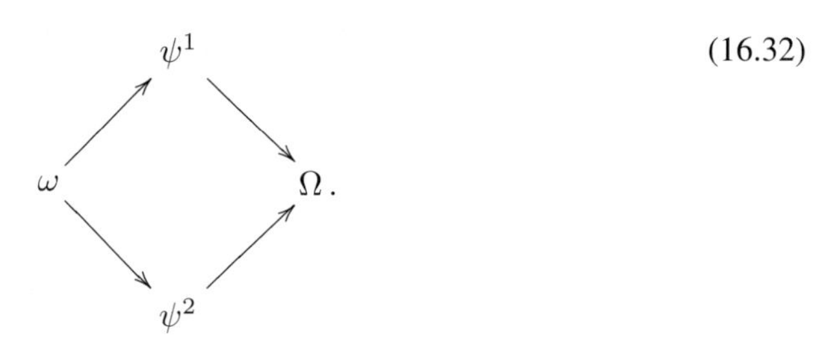

(16.32)

Meanwhile, the stress energy tensor of the theory $T \propto \partial\psi^1\partial\psi^2$ has a nontrivial OPE with the field ω and L_0 maps ω onto Ω, just like the Casimir element in the super-Lie algebra. This is but one characteristic feature of a logarithmic CFT.

Some remarks: symplectic fermions are related to the so-called $\eta\xi$ system. There have been some ambiguities in deciding how to handle the fact that the partition function vanishes—it is tempting, for instance, to decide to suppress the zero mode so that $\langle\psi^1\psi^2\rangle$ becomes nontrivial. This, however, does not really seem consistent. Two very useful papers on this and related topics are [16, 17].

16.3.2 The supersphere sigma model

In fact, by considering the antisymmetric combination of the two terms in J^2 as well as terms $\psi^1\psi^2\partial\psi^\alpha$, it is possible overall to collect eight fields with conformal weight $(1, 0)$, indicating a more complicated symmetry picture for the theory. A good way to think of this is to observe that symplectic fermions are the weak-coupling limit of a model with spontaneously broken symmetry, the supersphere sigma model.

One of the origins of supergroups in condensed matter physics was the introduction of a "supersymmetric" extension of the usual O(m) sigma model (in order to give a well-defined meaning to the $m \to 0$ limit). In general, one can define a nonlinear sigma model with a target space that is the supersphere $S^{2n+m-1,2n} = \mathrm{OSP}(2n+m|2n)/\mathrm{OSP}(2n+m-1|2n)$ as follows. We use as coordinates a real scalar field

$$\vec{\phi} \equiv (\phi^1, \ldots, \phi^{2n+m}, \psi^1, \ldots, \psi^{2n}) \tag{16.33}$$

and the bilinear form

$$\vec{\phi} \cdot \vec{\phi'} = \sum \phi^a (\phi')^a + \sum J_{\alpha\beta}\psi^\alpha(\psi')^\beta , \tag{16.34}$$

where $J_{\alpha\beta}$ is a symplectic form, which we take to consist of diagonal blocks $\begin{pmatrix} 0 & 1 \\ -1 & 0 \end{pmatrix}$. The $\cdot$ product is invariant under a supergroup of transformations called the orthosymplectic supergroup OSP($2n+m|2n$), which obviously contains the bosonic components $\mathrm{O}(2n+m) \times \mathrm{SP}(2n)$.

The unit supersphere is then defined by the constraint

$$\vec{\phi}.\vec{\phi} = 1 \,. \tag{16.35}$$

The action of the sigma model (the convention is that the Boltzmann weight is e^{-S}) reads

$$S = \frac{1}{2g_\sigma} \int d^2x \, \partial_\mu \vec{\phi}.\partial_\mu \vec{\phi} \,. \tag{16.36}$$

The perturbative β function depends only on m, to all orders:

$$\beta(g_\sigma) = (m-2)g_\sigma^2 + O(g_\sigma^3) \,. \tag{16.37}$$

We see that the model for g_σ positive flows to strong coupling for $m > 2$. This is the case of ordinary sigma models: the symmetry is restored at large length scales, and the field theory is massive. For $m < 2$, however, the model flows to weak coupling, and the symmetry is spontaneously broken. One expects this scenario to work for g_σ small enough, and the Goldstone phase to be separated from a nonperturbative strong-coupling phase by a critical point.

The possibility of spontaneous symmetry breaking in 2D goes against the Coleman–Mermin–Wagner theorem, but one should not forget that the latter is valid for ordinary, compact groups only. This can be seen more clearly by arguments other than by looking at the beta function. The point is that, in the usual O(m) models, fluctuations of the transverse degrees of freedom diminish the value of the two-point function of the order parameter in the searched-for broken-symmetry phase by logarithmic terms that diverge at large distances. To lowest order in g_σ,

$$\langle \vec{\phi}(z,\bar{z})\vec{\phi}(0)\rangle = \left[1 - \mathrm{cst}\, g_\sigma (m-2) \ln \frac{|z|}{a}\right]^{(m-1)/(m-2)} , \tag{16.38}$$

where we have not made explicit the cst term, which is normalization dependent (but positive and independent of m). For $m > 2$, we see indeed that fluctuations diminish the expectation value. But if $m < 2$, we see that they in fact increase this value. In the Goldstone phase, the supersphere sigma models turn out in fact to have correlations that diverge (logarithmically) at large distances.

The simplest case occurs when $m = -1, n = 1$, where the supersphere condition reads $\phi^2 + 2\psi^1\psi^2 = 1$. This can be solved for ϕ (up to a sign, but we will not worry about this here) via $\phi = 1 - \psi^1\psi^2$. The sigma model action is

$$S = \frac{1}{2g_\sigma} \int d^2x \left[(\partial_\mu \phi)^2 + 2\partial_\mu \psi^1 \partial_\mu \psi^2\right] . \tag{16.39}$$

We can check its invariance explicitly by looking at an infinitesimal OSP($1|2$) transformation

$$\begin{aligned}\delta\phi &= -\delta\xi^1\,\psi^1 + \delta\xi^2\,\psi^2\,,\\ \delta\psi^1 &= -\delta\xi^2\,\phi + \delta a\,\psi^1 + \delta c\,\psi^2\,,\\ \delta\psi^2 &= -\delta\xi^1\,\phi + \delta b\,\psi^1 - \delta a\,\psi^2\,,\end{aligned} \tag{16.40}$$

where $\delta\xi^1$, $\delta\xi^2$ are "small" fermionic deformation parameters and $\delta a, \delta c$ are small bosonic parameters. This change leaves $\phi^2 + 2\psi^1\psi^2$ invariant. The action can then be re-expressed in terms of the fermion variables only,

$$S = \frac{1}{g_\sigma}\int d^2x\,\left[\partial_\mu\psi^1\partial_\mu\psi^2 - \psi^1\psi^2\partial_\mu\psi^1\partial^\mu\psi_2\right], \tag{16.41}$$

where now the infinitesimal transformations are implemented through

$$\begin{aligned}\delta\psi^1 &= -\delta\xi^2(1-\psi^1\psi^2) + \delta a\,\psi^1 + \delta c\,\psi^2\,,\\ \delta\psi^2 &= -\delta\xi^1(1-\psi^1\psi^2) + \delta a\,\psi^1 - \delta a\,\psi^2\,.\end{aligned} \tag{16.42}$$

Notice that the relative normalization of the two terms can be changed at will by changing the normalization of the fermions. The relative sign can also be changed by switching the fermion labels $1 \to 2$. However, the sign of the four-fermion term cannot be changed, and determines whether the model is massive or massless in the IR. For g_σ positive, the model flows (perturbatively) to weak coupling in the IR. In that limit, indeed, we obtain symplectic fermions!

16.3.3 Application to a random geometrical problem: Trees and fermions

The discrete Laplacian is related to spanning trees in a beautiful way. Consider a loopless undirected graph with V vertices. We define the discrete Laplacian of G as a $V \times V$ matrix $M = \{m_{ij}\}$, with matrix elements, for $i \neq j$,

$$m_{ij} = -\text{sum over edges connecting } i \text{ and } j$$

and, for $i = j$,

$$m_{ii} = \text{number of edges incident on } i\,.$$

Whereas $\det M = 0$, the determinant of the matrix $M(i)$ obtained by removing the i-th row and the i-th column is independent of i, and

$$\det M(i) = \text{number of spanning trees of } G\,. \tag{16.43}$$

Recall that a tree is a connected graph having no cycles.

If we introduce the shorthand notation $[d\psi^1 d\psi^2] = \prod_i d\psi^1(i)d\psi^2(i)$, this allows us to reformulate Kirchhoff's theorem as

$$\int [d\psi^1 d\psi^2]\mathrm{e}^{\psi^1 M\psi^2} = \det M = 0\,, \tag{16.44}$$

$$\int [d\psi^1 d\psi^2]\psi^1(i)\psi^2(i)\mathrm{e}^{\psi^1 M\psi^2} = \text{number of spanning trees of } G\,. \tag{16.45}$$

We then introduce the two objects (n is the number of vertices)

$$Q_1 = \sum_{i=1}^{V} \psi^1(i)\psi^2(i)\,,$$

$$Q_2 = \frac{1}{2}\sum_{i,j=1}^{V} (-m_{ij})\psi^1(i)\psi^2(i)\psi^1(j)\psi^2(j)\,. \tag{16.46}$$

It is a beautiful recent result that

$$\int [d\psi^1 d\psi^2] \mathrm{e}^{a(Q_1-Q_2)} \mathrm{e}^{\psi^1 M \psi^2} = \sum_{p=1}^{\infty} a^p N(F_p) \equiv Z\,, \tag{16.47}$$

where $N(F_p)$ is the number of p-forests on G. Hence actions with four-fermion terms also have a geometrical interpretation (Fig. 16.1).

Moreover, one can also show, after a few manipulations, that this is the same as the partition function of the lattice OSP(1|2) sphere sigma model

$$\frac{a^V Z}{2^V} = \int_+ [d\phi\, d\psi^1 d\psi^2\, \delta(\phi^2 + 2\psi^1\psi^2 - 1)] \exp\left[-\frac{1}{a}\sum_{<ij>} (\vec{\phi}(i)\vec{\phi}(j) - 1)\right], \tag{16.48}$$

where we have used the standard result $\vec{\phi}(i)\vec{\phi}(j) = \phi(i)\phi(j) + \psi^1(i)\psi^2(j) + \psi^1(j)\psi^2(i)$. Some remarks: sadly I do not know much literature about superspheres; some aspects are discussed in [12]. The physics of sigma models is discussed in [18, 19]. The interpretation in terms of trees and forests appeared in [20].

16.4 From gl(N—N) spin chains to sigma models

16.4.1 The simplest gl(1|1) spin chain

We now consider an antiferromagnetic spin chain with gl(1|1) symmetry. To build it, we consider a space made of $2L$ alternating fundamental representations $\langle 1, 1\rangle \equiv \square$ and its conjugate $\langle -1, 0\rangle \equiv \bar{\square}$. The fermion operators f_i (we use this notation instead of Ψ^-) and $f_i^\dagger$, $i = 0, 1,$

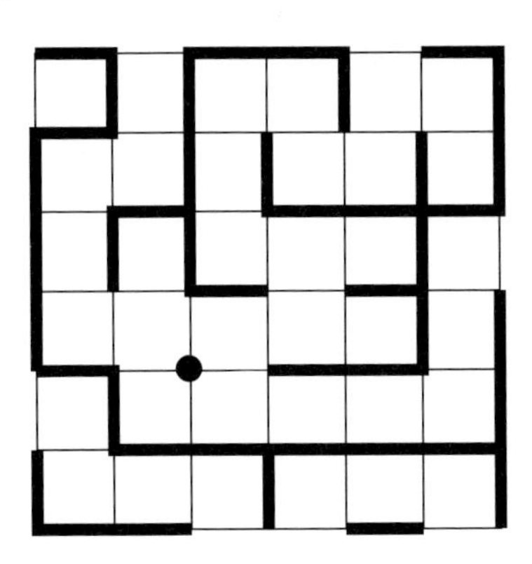

Fig. 16.1 Geometrical interpretation of actions with four-fermion terms.

..., $2L-1$, obey $\{f_i, f_{i'}\} = 0$, $\{f_i, f_{i'}^\dagger\} = (-1)^i \delta_{ii'}$. We define the fermion number at site i as $F_i = (-1)^i f_i^\dagger f_i$. In the natural basis of states $|0\rangle, |1\rangle$ indexed by fermion number, we then have

$$f_0 = \begin{pmatrix} 0 & 1 \\ 0 & 0 \end{pmatrix}, \quad f_0^\dagger = \begin{pmatrix} 0 & 0 \\ 1 & 0 \end{pmatrix},$$
$$f_1 = \begin{pmatrix} 0 & 1 \\ 0 & 0 \end{pmatrix}, \quad f_1^\dagger = \begin{pmatrix} 0 & 0 \\ -1 & 0 \end{pmatrix}. \tag{16.49}$$

We choose first as the coupling the nearest-neighbor interaction

$$e_i = (f_i^\dagger + f_{i+1}^\dagger)(f_i + f_{i+1}) = -(f_i + f_{i+1})f_i^\dagger + f_{i+1}^\dagger), \tag{16.50}$$

for $i = 0, \ldots, 2L-2$. With a basis ordered $|0,0\rangle, |0,1\rangle, |1,0\rangle, |1,1\rangle$, the interaction between the $i=0$ and $i=1$ sites, for instance, reads

$$e_0 = \begin{pmatrix} 0 & 0 & 0 & 0 \\ 0 & -1 & -1 & 0 \\ 0 & 1 & 1 & 0 \\ 0 & 0 & 0 & 0 \end{pmatrix}, \tag{16.51}$$

while e_1 in the corresponding basis reads as $-e_0$. Note that one might wish to perform a particle–hole transformation on the $\bar{\square}$ representations to have e look like an object contracting identical colors, etc. This will not be necessary, however.

The interaction is chosen to commute with gl(1|1): indeed, since the Casimir element takes the form $C = (2N-1)E + 2\Psi^-\Psi^+$, in the product of $\square$ and $\bar{\square}$, $E = 1 - 1 = 0$ and thus $C = -2e$.

We have so far refrained from using the notation Ψ and $\Psi^\dagger$ for the generators. It is certainly perfectly possible to define a scalar product invariant under gl(1|1) such that Ψ^+ is the adjoint of Ψ^-. Note, however, that while it is possible to have this scalar product positive definite in $\square$, in the case of $\bar{\square}$, the minus sign in the anticommutator of the fermions means that the two states in the representation necessarily have opposite norm squared. This means that in the tensor product, some states have zero norm squared. Let us start by deciding that $\langle 0,0|0,0\rangle = 1$. Then we have $\langle 0,1|0,1\rangle = \langle 0,0|f_1 f_1^\dagger|0,0\rangle = -1$, $\langle 1,0|1,0\rangle = 1$ and

$$\begin{aligned} \langle 1,1|1,1\rangle &= \left(f_1^\dagger f_0^\dagger|0,0\rangle, f_1^\dagger f_0^\dagger|0,0\rangle\right) \\ &= \langle 0,0|f_0 f_1 f_1^\dagger f_0^\dagger|0,0\rangle = \langle 0,0|f_0(-1 - f_1^\dagger f_1)\Psi_0^\dagger|0,0\rangle = -1\,. \end{aligned} \tag{16.52}$$

Thus the two states $|0,1\rangle$ and $\pm|1,0\rangle$ both have zero norm squared! Note that

$$\begin{aligned} e_0\left(|0,1\rangle - |1,0\rangle\right) &= 0\,, \\ e_0\left(|0,1\rangle + |1,0\rangle\right) &= 2\left(|0,1\rangle - |1,0\rangle\right), \end{aligned} \tag{16.53}$$

which shows that e_0 is not diagonalizable in the tensor product.

The elementary Hamiltonians e_i satisfy the Temperley–Lieb (TL) algebra relations

$$\begin{aligned} e_i^2 &= m e_i\,, \\ e_i\, e_{i\pm 1}\, e_i &= e_i\,, \\ e_i\, e_j &= e_j\, e_i \qquad (j \neq i,\ i \pm 1)\,, \end{aligned} \tag{16.54}$$

with, moreover, the parameter $m = 0$. They have a standard graphical interpretation, which is more naturally understood when one turns to a 2D point of view.

Note that the fact that we take an alternating fundamental and its conjugate can be reinterpreted by supposing that the edges carry a fixed orientation (this is reminiscent of the Chalker Coddington model): see Fig. 16.2. The graphical representation of the TL algebra then corresponds to having lines propagating on the edges, with two and only two possible splittings at every vertex of the lattice. This draws dense polymers on the square lattice (Fig. 16.3). The simplest Hamiltonian is $H = -\sum e_i$ (note that the sign does not matter here, as the algebra is invariant under $e_i \to -e_i$). It is in the same universality class as the 2D model where both splittings occur with equal weights.

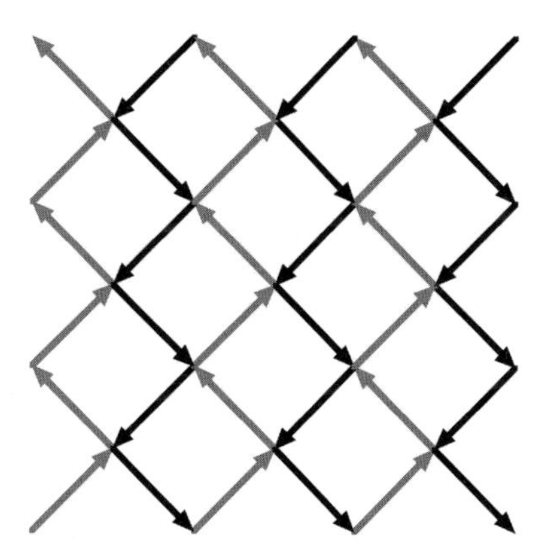

Fig. 16.2 Reinterpretation of alternating fundamental and conjugate.

Fig. 16.3 Graphical representation of TL algebra.

These e_i turn out to commute with more than gl(1|1): there is an enlarged symmetry that we shall call $\mathcal{A}_{1|1}$, generated by the following operators (this is discussed in great detail in [21]):

$$\Psi^- = \sum_i f_i \,, \tag{16.55}$$

$$\Psi^+ = \sum_i f_i^\dagger \,, \tag{16.56}$$

$$F_{(2)} = \sum_{i<i'} f_i f_{i'} \,, \tag{16.57}$$

$$F_{(2)}^\dagger = \sum_{i<i'} f_{i'}^\dagger f_i^\dagger \,, \tag{16.58}$$

$$N = \sum_i (-1)^i f_i^\dagger f_i - L \,. \tag{16.59}$$

The graded commutators of these five close on themselves, so they form a Lie superalgebra. Ψ^+, Ψ^- generate a Lie sub-superalgebra isomorphic to psl(1|1). $F_{(2)}$, $F_{(2)}^\dagger$, and N generate an sl_2 Lie subalgebra, where N is $2S_z$, and F, $F^\dagger$ transform as a doublet under this sl_2, and so form a Lie ideal.

The system can also be defined on a circle, in which case we need to add a coupling between the $i = 0$ and the $i = 2L - 1$ sites. We will restrict ourselves for the moment to periodic boundary conditions for the fermions. Note that in this case the naive extension of the $F_{(2)}$ generators vanishes; the issue of symmetries is then slightly more complicated.

It is convenient, in order to match with ordinary fermion chains, to make a transformation $f_{2i-1}^\dagger \to -f_{2i-1}^\dagger$, so the f's now satisfy $\{f_i, f_j^\dagger\} = \delta_{ij}$ independently of the parity of i, j. Then the Hamiltonian reads

$$H = -\sum_{i=0}^{2L-1} e_i = \sum_{j=0}^{L} 2\left(f_{2j}f_{2j}^\dagger - f_{2j-1}f_{2j-1}^\dagger\right) - f_{2j}f_{2j-1}^\dagger - f_{2j}f_{2j+1}^\dagger + f_{2j-1}f_{2j}^\dagger + f_{2j+1}f_{2j}^\dagger \tag{16.60}$$

and commutes with the total number of fermions

$$N = \sum f_i f_i^\dagger \,. \tag{16.61}$$

We diagonalize the Hamiltonian:

$$H = \sum_{p=1}^{p=L-1} \sin k \left(\chi_k^\dagger \chi_k - \eta_k^\dagger \eta_k\right) + 2 f_0 f_\pi^\dagger, \;\; k = \frac{p\pi}{L} \,, \tag{16.62}$$

where the χ, η are normalized fermionic modes, i.e. $\{\chi_k^\dagger, \chi_k'\} = \delta_{kk'}$ etc. Introducing the Fourier transform

$$f_k = \frac{1}{\sqrt{2L}} \sum_{j=0}^{2L-1} e^{ikj} f_j, \;\; \left\{f_k, f_{k'}^\dagger\right\} = \delta_{kk'} \,, \tag{16.63}$$

the proper modes follow from

$$
\begin{aligned}
f_k &= \frac{1}{\sqrt{2\tan(k/2)}}(\chi_k + \eta_k)\,, \\
f_k^\dagger &= \frac{\sqrt{\tan(k/2)}}{\sqrt{2}}(\chi_k^\dagger + \eta_k^\dagger)\,, \\
f_{k+\pi} &= \frac{\sqrt{\tan(k/2)}}{\sqrt{2}}(\chi_k - \eta_k)\,, \\
f_{k+\pi}^\dagger &= \frac{1}{\sqrt{2\tan(k/2)}}(\chi_k^\dagger - \eta_k^\dagger)\,.
\end{aligned} \tag{16.64}
$$

We see in particular that the superscript $\dagger$ for the χ, η modes is misleading, as, for example, $\chi, \chi^\dagger$ are not related by complex conjugation.

Note that we are summing momenta only in the interval $[0, \pi]$; momenta in $[\pi, 2\pi]$ are included in the fermion eigenmodes. We have commutation with the two gl(1|1) generators $f_0 \propto \sum f_i$ and $f_\pi^\dagger \propto \sum (-1)^i f_i^\dagger$, resulting in two zero modes. The term $f_0 f_\pi^\dagger$ is not diagonalizable. For $L = 2$, for instance, we have two modes at vanishing energy, one mode at energy ϵ ($= 1$ in our normalizations) and one at energy $-\epsilon$. There are thus three eigenvalues of H: ϵ and $-\epsilon$, degenerate four times, and 0, degenerate eight times. The ground state for arbitrary L will be degenerate four times. The appearance of powers of 2 is not an accident, and reveals a hidden affine sl(2) symmetry.

More detailed study shows that the continuum limit is described by a symplectic fermion theory.

16.4.2 Remark: The Yang–Baxter gl(1|1) spin chain

It is important to note that the model is not the one that would be obtained via the general quantum inverse scattering construction. In the latter, the same representation has to be carried by a horizontal or a vertical line, so the geometry is more like that in Fig. 16.4.

An integrable Hamiltonian can, however, be extracted which acts in the same Hilbert space of alternating $\square, \bar{\square}$. It reads in this case

$$
H_{\text{int}} \propto \sum e_i e_{i+1} + e_{i+1} e_i \tag{16.65}
$$

and is not in the same universality class. Its continuum limit is a bit mysterious, but can be tackled using the Bethe ansatz: it still does not depend on the sign of the Hamiltonian, although it is more difficult to understand why. The equations read symbolically

$$
\begin{aligned}
\left(\frac{\alpha - i}{\alpha + i}\right)^L &= \prod \frac{\alpha - \beta - 2i}{\alpha - \beta + 2i}\,, \\
\left(\frac{\beta - i}{\beta + i}\right)^L &= \prod \frac{\beta - \alpha - 2i}{\beta - \alpha + 2i}\,,
\end{aligned}
$$

with the energy

$$
E \propto \sum \frac{1}{1 + \alpha^2} + \frac{1}{1 + \beta^2}\,.
$$

It is found that the central charge is $c = -1$, and that the spectrum corresponds to the same symplectic fermion theory to which a noncompact free boson has been added. Recall that

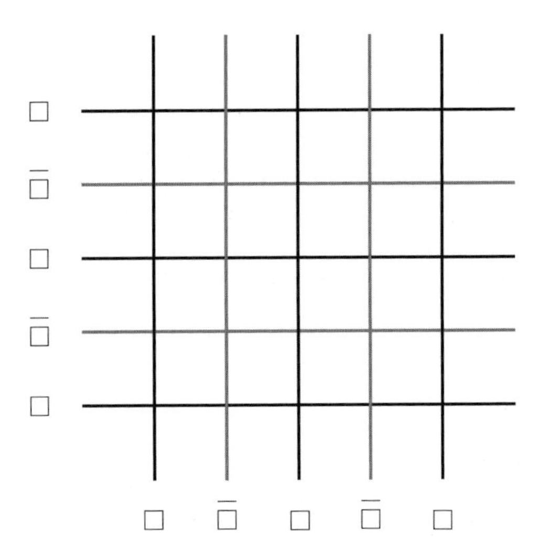

Fig. 16.4 Geometry obtained via general quantum inverse scattering construction.

in general in statistical mechanics, one considers bosons compactified on a circle. When the action is normalized as $S = (1/8\pi)\int(\partial_\mu\phi)^2$ and the field is compactified as $\phi \equiv \phi + 2\pi R$, the spectrum of dimensions is

$$x = \frac{e^2}{R^2} + \frac{m^2 R^2}{4}.$$

In the limit $R \to \infty$, the dimensions become degenerate and the spectrum can be considered as a continuum starting above $\Delta + \bar{\Delta} = 0$. To emphasize the latter point, consider the generating function of levels coming from the boson

$$\begin{aligned} Z_\phi = \mathrm{Tr}\, q^{L_0 - 1/24}\bar{q}^{\bar{L}_0 - 1/24} &= \frac{1}{\eta\bar{\eta}}\sum_{e,m} q^{(e/R+mR/2)^2/2}\bar{q}^{(e/R-mR/2)^2/2} \\ &= \frac{R}{\sqrt{2}}\frac{1}{\sqrt{\mathrm{Im}\,\tau}\eta\bar{\eta}}\sum_{m,m'}\exp\left(-\frac{\pi R^2|m\tau - m'|^2}{2\,\mathrm{Im}\,\tau}\right) \\ &\underset{R\to\infty}{\approx} \frac{R}{\sqrt{2}}\frac{1}{\sqrt{\mathrm{Im}\,\tau}\eta\bar{\eta}}, \end{aligned} \tag{16.66}$$

where $\eta(q) = q^{1/24}\prod_{n=1}^{\infty}(1-q^n) = q^{1/24}P(q)$. If we observe now that we can write

$$\frac{1}{\sqrt{\mathrm{Im}\,\tau}\eta\bar{\eta}} = 4\int_0^\infty ds\, \frac{q^{s^2}\bar{q}^{s^2}}{\eta\bar{\eta}},$$

this can be interpreted as an integral over a continuum of critical exponents $\Delta = \bar{\Delta} = s^2$.

In this limit, the radius R plays the role of the density of levels, and is proportional to the (diverging) size of the target space.

On the lattice, this translates into very large approximate degeneracies as the size of the system gets larger, with, for instance, more and more levels seemingly converging to the ground state in that limit. More detailed analytic studies of the equations confirms this scenario. It is tempting to think that the continuum limit of the spin chain should be an SL(1|1) WZW model, the "E" degree of freedom having become dynamical on that special line.

For a discussion of quantum inverse scattering in the context of supergroups and sigma models, see [22]. It is a little intriguing that the Bethe equations in this section do not obey the general pattern of correspondence with the Cartan matrix of the underlying Lie superalgebra. For a discussion of the continuous spectrum in this model, see [23].

16.4.3 The simplest $\mathrm{gl}(n|n)$ spin chain

The following construction is borrowed from [21].

We can generalize all this by taking the fundamental of the Lie superalgebra gl$(n|n)$ for i even, and its dual for i odd. Technically, the chain may be constructed using boson operators $b_{ia}^{\dagger}, \bar{b}_{ia}^{\dagger}$ for $a = 1, \ldots, n$, and fermion operators $f_{ia}^{\dagger}, \bar{f}_{i}^{a\dagger}$ for $a = n + 1, \ldots, 2n$.

For i even, we have $[b_i^a, b_{jb}^{\dagger}] = \delta_{ij}\delta_b^a$ $(a, b = 1, \ldots, n)$, and fermion operators $f_i^{\alpha}, f_{i\alpha}^{\dagger}$, $\{f_i^{\alpha}, f_{j\beta}^{\dagger}\} = \delta_{ij}\delta_{\beta}^{\alpha}$ $(\alpha, \beta = 1, \ldots, n)$; here, labels such as α on the fermion operators stand for $\alpha = a - n$ for a corresponding index a. For i odd, we have similarly boson operators $\bar{b}_{ia}, \bar{b}_i^{a\dagger}$, $[\bar{b}_{ia}, \bar{b}_j^{b\dagger}] = \delta_{ij}\delta_a^b$ $(a, b = 1, \ldots, n)$, and fermion operators $\bar{f}_{i\alpha}, \bar{f}_i^{\alpha\dagger}$, $\{\bar{f}_{i\alpha}, \bar{f}_j^{\beta\dagger}\} = -\delta_{ij}\delta_{\alpha}^{\beta}$ $(\alpha, \beta = 1, \ldots, n)$. Notice again the minus sign in the last anticommutator; since our convention is that the $\dagger$ stands for the adjoint, this minus sign implies that the norms squared of any two states that are mapped onto each other by the action of a single $\bar{f}_{i\alpha}$ or $\bar{f}_i^{\alpha\dagger}$ have opposite signs.

The space V is now defined as the subspace of states that obey the constraints

$$\sum_a b_{ia}^{\dagger} b_i^a + \sum_{\alpha} f_{i\alpha}^{\dagger} f_i^{\alpha} = 1 \quad (i \text{ even}), \tag{16.67}$$

$$\sum_a \bar{b}_i^{a\dagger} \bar{b}_{ia} - \sum_{\alpha} \bar{f}_i^{\alpha\dagger} \bar{f}_{i\alpha} = 1 \quad (i \text{ odd}). \tag{16.68}$$

The generators of the Lie superalgebra gl$(n|n)$ acting on each site of the chain are the bilinear forms $J_{ia}^b = b_{ia}^{\dagger} b_i^b$, $f_{i\alpha}^{\dagger} f_i^{\beta}$, $b_{ia}^{\dagger} f_i^{\beta}$, $f_{i\alpha}^{\dagger} b_i^b$ for i even, and similarly for i odd.

The tensor product of the fundamental and its conjugate in gl$(n|n)$ gives rise, as in the case of $n = 1$, to a big indecomposable block. The TL generators are constructed as follows. First, we note that for any two sites i (even), j (odd), the combinations

$$\sum_a \bar{b}_{ja} b_i^a + \sum_{\alpha} \bar{f}_{j\alpha} f_i^{\alpha}, \quad \sum_a b_{ia}^{\dagger} \bar{b}_j^{a\dagger} + \sum_{\alpha} f_{i\alpha}^{\dagger} \bar{f}_j^{\alpha\dagger}$$

are invariant under gl$(n|n)$. We then introduce, for each pair of neighbors i, $i+1$,

$$
\begin{aligned}
d_{V_{i+1}} &= \sum_a \bar{b}_{i+1,a} b_i^a + \sum_\alpha \bar{f}_{i+1,\alpha} f_i^\alpha \quad (i \text{ even}), \\
d_{V_{i+1}} &= \sum_a \bar{b}_{ia} b_{i+1}^a + \sum_\alpha \bar{f}_{i\alpha} f_{i+1}^\alpha \quad (i \text{ odd}), \\
b_{V_i} &= \sum_a b_{ia}^\dagger \bar{b}_{i+1}^{a\dagger} + \sum_\alpha f_{i\alpha}^\dagger \bar{f}_{i+1}^{\alpha\dagger} \quad (i \text{ even}), \\
b_{V_i} &= \sum_a b_{i+1,a}^\dagger \bar{b}_i^{a\dagger} + \sum_\alpha f_{i+1,\alpha}^\dagger \bar{f}_i^{\alpha\dagger} \quad (i \text{ odd}).
\end{aligned}
$$

Then the TL generators can be written as

$$
e_i = b_{V_i} d_{V_{i+1}}
$$

for all i.

Algebraically, we have gained very little by going to $n > 1$, at least at first sight. We can, in particular, decide to study the same Hamiltonian $H = -\sum e_i$ as before. The difference from the gl(1|1) case will then be that, because the group is larger, more invariants exist, and thus more correlation functions can be considered. In general, one expects the spectrum of one of these models with gl$(n|n)$ symmetry to be a subset of the spectrum of models with gl$(n'|n')$ symmetry when $n' > n$.

16.5 A conformal sigma model at $c = -2$

16.5.1 Adding couplings in the spin chain

We can also decide to study a more general situation, where other interactions preserving the gl$(n|n)$ symmetry are added. For instance, we can consider adding the term $P_{i,i+2}$, the graded permutation operator between representations on next-to-nearest neighbors. It is an easy exercise to check that in the case $n = 1$, this does not change the universality class, and we remain within symplectic fermions. But things are different for $n > 1$. To see what might happen, it is useful to get some inspiration from the usual mapping between spin chains and sigma models (the following discussion is borrowed from [21], in which more useful references can be found).

Let us recall what happens in the usual sl(2) case. The general strategy is to use geometrical or coherent-state quantization to describe individual spins via a path integral. Taking (antiferromagnetic) interactions into account, in the large-spin limit we get the $\mathrm{O}(3) = \mathrm{SU}(2)/\mathrm{U}(1)$ sigma model at $\theta = 2\pi s$. We also find that the bare coupling constant is $g_\sigma^2 \propto 1/s$ and that there is a flow to large coupling, so the physics at $s = \frac{1}{2}$ is described by the XXX antiferromagnetic spin chain, and thus the SU(2) WZW model at level one.

The extension of this to sl(m) and alternating representations (so as to have antiferromagnetic physics) can be done using coherent-state representation of the states on every site of the spin chain. This leads to matrices living in the group modulo the isotropy group of the highest-weight state, $Q = g|hw\rangle\langle hw|g^{-1}$. For representations $\square\square\square\ldots$ and their conjugates, this gives $\mathrm{SU}(m)/\mathrm{SU}(m-1) \times \mathrm{U}(1)$. One can show similarly that in the case at hand, the large-distance physics for our antiferromagnetic Hamiltonians should be described by the $\mathrm{U}(n+m|n)/\mathrm{U}(1) \times \mathrm{U}(n+m-1|n)$ or $\mathrm{CP}^{n+m-1|n}$ model at $\theta = \pi$, and that $m = 0$.

The fields can be represented by complex components z^a ($a = 1, \ldots, n+m$), η^α ($\alpha = 1, \ldots, n$), where z^a is commuting and η^α is anticommuting. In these coordinates, at each point in space–time, the solutions to the constraint $z_a^\dagger z^a + \eta_\alpha^\dagger \eta^\alpha = 1$ (conjugation $\dagger$ obeys $(\eta\xi)^\dagger = \xi^\dagger \eta^\dagger$ for any η, ξ), modulo U(1) phase transformations $z^a \mapsto e^{iB} z^a$, $\eta^\alpha \mapsto e^{iB} \eta^\alpha$, parameterize $\mathrm{CP}^{n+m-1|n}$. The Lagrangian density in two-dimensional Euclidean space–time is

$$\mathfrak{L} = \frac{1}{2g_\sigma}\left[(\partial_\mu - ia_\mu) z_a^\dagger (\partial_\mu + ia_\mu) z^a + (\partial_\mu - ia_\mu)\eta_\alpha^\dagger(\partial_\mu + ia_\mu)\eta^\alpha\right] + \frac{i\theta}{2\pi}(\partial_\mu a_\nu - \partial_\nu a_\mu)\,,$$

where a_μ ($\mu = 1, 2$) stands for

$$a_\mu = \frac{i}{2}[z_a^\dagger \partial_\mu z^a + \eta_\alpha^\dagger \partial_\mu \eta^\alpha - (\partial z_a^\dagger) z^a - (\partial \eta_\alpha^\dagger)\eta^\alpha]\,.$$

Fields are subject to the above constraint, and under the U(1) gauge invariance, a_μ transforms as a gauge potential; a gauge must be fixed in any calculation. This setup is similar to the nonsupersymmetric CP^{m-1} model

The coupling constants are g_σ (there is only one such coupling, because the target supermanifold is a supersymmetric space) and θ, the coefficient of the topological term, so θ is defined modulo 2π.

Now we finally get to the important points. First, the β function of the model obeys

$$\frac{dg_\sigma}{dl} = \beta(g_\sigma) = mg_\sigma^2 + O(g_\sigma^3)\,.$$

For $m = 0$, it vanishes to leading order. In fact, the β function is independent of n. Now, for $n = 1$, we recover our theory of symplectic fermions

$$\mathfrak{L} = \frac{1}{2g_\sigma}\partial_\mu \eta^\dagger \partial_\mu \eta\,,$$

for which we know that g_σ is redundant. Thus $\beta = 0$ to all orders. We have a conformal sigma model (I will not discuss the very interesting role of the topological angle), which should contain as a subset a theory of symplectic fermions, and exhibit other fields with g_σ-dependent features as well.

Getting a handle on such massless sigma models is one of the challenges of the field. We can make a little progress here by thinking of spin chains. So let us consider the Hamiltonian

$$H = -\sum e_i + w\sum P_{i,i+2}\,. \tag{16.69}$$

It is natural to expect that the continuum limit of this more general $H(w)$ corresponds to the sigma model, with g_σ a function of w (one can argue that $g_\sigma \to 0$ as $w \to \infty$). While the Hamiltonian does not seem solvable explicitly, one can certainly study it numerically. For this, it is very convenient to use a graphical representation of the operators: while the e_i describe contractions, the P describe the six leg crossings. These can be implemented elegantly in a model of a dense polymer on a triangular lattice.

The simplest invariant tensors correspond to observables called L-leg polymer operators (do not confuse this L with the size of the spin chains in the foregoing sections). Their conformal dimension is obtained by forcing L lines to propagate through the system without ever being contracted. The very small set of representations available guarantees that only the case $L = 2$ can occur in gl(1|1). Thus we expect that for $L = 0, 2$, the properties are independent of g_σ, but that for $L \geq 4$, they do depend on it.

16.5.2 Minisuperspace and weak-coupling limit

In the weak-coupling limit where g_σ is small, meanwhile, we can solve the sigma model using the minisuperspace (particle limit) approach.

Consider a sigma model on a cylinder of circumference r or, equivalently, at temperature $T = 1/r$. Doing a Wick rotation transforms the space into a basic circle $x \equiv x + r$, while the imaginary time runs to infinity along the axis of the cylinder. At small r, i.e. large temperature, it is reasonable to neglect the fluctuations of the fields in the transverse direction, and replace the fields $\phi^i(x, y)$ by $\phi^i(y)$.

To be more precise, let us describe the problem in a Hamiltonian formalism. In general, one has to deal with wave functions Ψ which are functions of the field configuration at a given time (or imaginary time), $\Psi[\phi^i(x)]$. In the minisuperspace limit, these become functions of the x independent approximations of the fields, i.e. functions on the target space itself. If the sigma model of interest is a model on a (super)group, the wave functions become functions on that (super)group. The Hamiltonian becomes a differential operator on these functions.

To see how this works and to fix the notation, consider briefly the O(2) action

$$A = \frac{1}{2g_\sigma} \int d^2x \left[(\partial_\mu \phi^1)^2 + (\partial_\mu \phi^2)^2 \right], \tag{16.70}$$

where φ is the angle of the vector $\vec{n}$, quantized on a circle of circumference 2π. The minisuperspace approximation should be valid in the limit of g large. This corresponds to small temperatures in the XY model, i.e. the limit where the free-floating vortex operators are strongly irrelevant.

So, in the minisuperspace approximation, the action is

$$A = \frac{1}{2g_\sigma} \int d\tau \, \dot{\varphi}^2 \tag{16.71}$$

and yields the quantized Hamiltonian

$$H = \frac{g_\sigma T}{2} \Pi^2 = -\frac{g_\sigma T}{2} \hat{\Delta}_1 \, . \tag{16.72}$$

Here Π is the canonical momentum associated with φ, the equal-time commutator is $[\Pi, \varphi] = 1/i$, and $\hat{\Delta}_1 = d^2/d\varphi^2$ is the Laplacian on the circle. The Hamiltonian H has eigenfunctions $\Psi_n(\phi) = e^{ni\phi}$ with eigenenergies $E_n = g_\sigma T n^2/2$. Again, this approximation should become good when g_σ is small, so these dimensions are small and accumulate near the ground state.

On the other hand, H reads, in the Virasoro formalism,

$$H = 2\pi T \left(L_0 + \bar{L}_0 - \frac{c}{12} \right).$$

The spectrum is thus, from the exact solution,

$$E = 2\pi T \left(\frac{e^2 g_\sigma}{4\pi} + \frac{\pi}{g_\sigma} m^2 - \frac{1}{12} \right) \tag{16.73}$$

and indeed coincides in the limit of small g_σ with the one obtained in the minisuperspace limit. Note that the central charge, being a contribution of order $O(1)$ to the spectrum, should not be visible in the minisuperspace approximation.

So, we can try to obtain estimates of the exponents of our massless sigma model by using the minisuperspace approach. The spectrum of the Laplacian on the ordinary projective space $\mathrm{CP}^m = \mathrm{U}(m+1)/[\mathrm{U}(m) \times \mathrm{U}(1)]$ is known to be of the form

$$E_l = 4l(l+m), \quad l \text{ integer}. \tag{16.74}$$

We set $m = -1$ in this equation to obtain what presumably is the spectrum for our superprojective space (for harmonic analysis on U(1|1), see [11]),

$$E_l = (2l-1)^2 - 1\,, \tag{16.75}$$

leading to a prediction for the exponents (where $2l \equiv L$ now)

$$h_L = \frac{g_\sigma}{2}\left[(L-1)^2 - 1\right], \tag{16.76}$$

and so $h_2 = 0, h_4 = 8g_\sigma^2, h_6 = 24g_\sigma^2, h_8 = 48g_\sigma^2, \ldots$. We now turn specifically to the case of free boundary conditions. In this case, it is known that when $w = 0$, the exponents read exactly

$$h_L = \frac{L(L-1)}{2}, \tag{16.77}$$

formally fitting the minisuperspace formula for $g_\sigma = 1$. It is thus tempting to speculate that this formula might be exact for all values of $w > 0$. Numerical results illustrating this speculation are presented in Fig. 16.5 (for technical reasons, the best results occur with an odd number of legs, for which one expects the general formula to apply as well, with L odd, though the sigma model description is slightly more involved), where the coupling g_σ has been extracted from the exponents for various values of L and w using eqn (16.76).

Many questions remain to be studied here, including the case of periodic boundary conditions, and the role of θ terms.

Some remarks: not all of the results in this section have yet been published, and they are part of joint work with C. Candu, J. Jacobsen, and N. Read. Conformal sigma models are

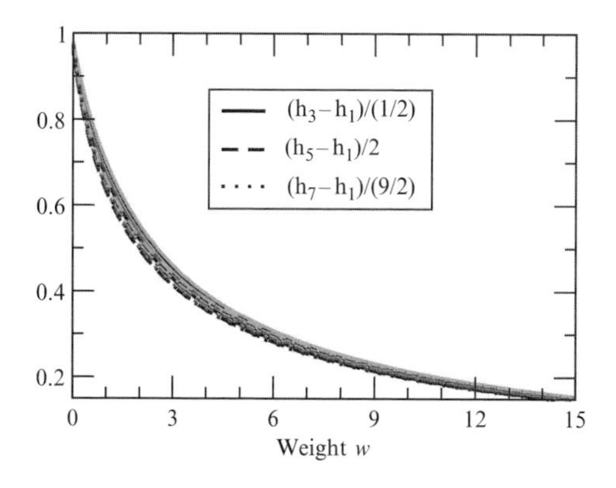

Fig. 16.5 Numerical results illustrating speculation about eqn (16.77) (see text).

certainly some of the most intriguing objects in this field: their action is known, they have a continuous symmetry, they are conformally invariant, they are most likely integrable—and, yet, there is no known way to solve them. For two inspiring references on the PSL(2|2) PCM, see [24, 25].

16.6 Conclusions

Of course, much remains to be said. As these lectures indicate, some progress has occurred in the understanding of CFTs and of the simplest massive sigma models. Of all the things that remain to be done, I feel that it would be most useful to solve—most likely using S-matrices and the Bethe ansatz—a nontrivial example of RG flow, such as the flow from the O(3) sigma model at $\theta = \pi$ to the $SU(2)_1$ WZW model in the nonsuper case.

Acknowledgments

I thank all the friends and collaborators who have shared my interest in supertarget physics over the years, particularly C. Candu, Y. Ikhlef, J. L. Jacobsen, L. Rozansky, B. Wehefritz-Kaufmann, and, especially, N. Read and V. Schomerus, from whom I learned a great deal of what is in these notes.

References

[1] K. B. Efetov, Supersymmetry and theory of disordered metals, *Adv. Phys.* **32**, 53 (1983).
[2] M. Zirnbauer, Conformal field theory of the integer quantum Hall plateau transition, arXiv:hep-th/9905054.
[3] R. R. Metsaev and A. A. Tseytlin, Type IIB superstring action in $\mathrm{AdS}(5) \times \mathrm{S}(5)$ background, *Nucl. Phys. B* **533**, 109 (1998) [arXiv:hep-th/9805028].
[4] N. Berkovits, C. Vafa, and E. Witten, Conformal field theory of AdS background with Ramond–Ramond flux, *JHEP* **9903**, 018 (1999) [arXiv:hep-th/9902098].
[5] G. Parisi and N. Sourlas, Self avoiding walk and supersymmetry, *J. Phys. Lett. (Paris)* **41**, L403 (1980).
[6] G. Gotz, T. Quella, and V. Schomerus, Representation theory of sl(2|1), arXiv:hep-th/0504234.
[7] M. Scheunert, W. Nahm, and V. Rittenberg, *J. Math. Phys.* **18**, 146 (1977).
[8] M. Marcu, *J. Math. Phys.* **21**, 1277 (1980); *J. Math. Phys.* **21**, 1284 (1980).
[9] J. Germoni, *Représentations indécomposables des superalgèbres de Lie spéciales linéaires*, Publications de l'Institut de Recherche Mathématique Avancée, Université de Strasbourg.
[10] P. Freund, *Introduction to Supersymmetry*, Cambridge University Press.
[11] V. Schomerus and H. Saleur, The GL(1|1)WZW model: From supergeometry to logarithmic CFT, *Nucl. Phys. B* **734**, 221–245 (2006) [arXiv:hep-th/0510032].
[12] M. Zirnbauer, Riemannian symmetric superspaces and their origin in random matrix theory, *J. Math. Phys.* **37**, 4986 (1996) [arXiv:math-ph/9808012].
[13] H. Saleur and V. Schomerus, On the SU(2|1) WZW model and its statistical mechanics applications, *Nucl. Phys. B* **775**, 312–340 (2007) [arXiv:hep-th/0611147].
[14] L. Rozansky and H. Saleur, Quantum field theory for the multivariable Alexander–Conway polynomial, *Nucl. Phys. B* **376**, 461 (1992).

[15] H. G. Kausch, Symplectic fermions, *Nucl. Phys. B* **583**, 513 (2000) [arXiv:hep-th/0003029].

[16] M. Flohr, Bits and pieces in logarithmic conformal field theory, *Int. J. Mod. Phys. A* **18**, 4497 (2003) [arXiv:hep-th/0111228].

[17] M. R. Gaberdiel, An algebraic approach to logarithmic conformal field theory, *Int. J. Mod. Phys. A* **18**, 4593 (2003) [arXiv:hep-th/0111260].

[18] H. Saleur and B. Wehefritz-Kaufmann, Integrable quantum field theories with OSP(m/2n) symmetries, *Nucl. Phys. B* **628**, 407 (2002) [arXiv:hep-th/0112095].

[19] J. L. Jacobsen and H. Saleur, The arboreal gas and the supersphere sigma model, *Nucl. Phys. B* **716**, 439–461 (2005) [arXiv: cond-mat/0502052].

[20] S. Caracciolo, J. L. Jacobsen, H. Saleur, A. D. Sokal, and A. Sportiello, Fermionic field theory for trees and forests, *Phys. Rev. Lett.* **93**, 080601 (2004) [arXiv:cond-mat/0403271].

[21] N. Read and H. Saleur, *Nucl. Phys. B* **613**, 409 (2001); *Nucl. Phys. B* **777**, 263 (2007).

[22] F. H. L. Essler, H. Frahm, and H. Saleur, Continuum limit of the integrable $sl(2|1)$ $3 - \bar{3}$ superspin chain, *Nucl. Phys. B* **712**, 513 (2005) [arXiv:cond-mat/0501197].

[23] Y. Ikhlef, J. Jacobsen, and H. Saleur, A staggered 6 vertex model with a non compact continuum limit, *Nucl. Phys. B* **789**, 483–524 (2007) [arXiv:cond-mat/0612037].

[24] M. Bershadsky, S. Zhukov, and A. Vaintrob, PSL(n|n) sigma model as a conformal field theory, *Nucl. Phys. B* **559**, 205 (1999) [arXiv:hep-th/9902180].

[25] G. Gotz, T. Quella, and V. Schomerus, The WZNW model on $PSU(1, 1|2)$, *JHEP* **0703**, 003 (2007) [arXiv:hep-th/0610070].

17
Integrability and combinatorics: Selected topics

P. ZINN-JUSTIN

Paul Zinn-Justin,
LPTMS (CNRS, UMR 8626), Université Paris-Sud,
91405 Orsay Cedex, France,
and LPTHE (CNRS, UMR 7589),
Université Pierre et Marie Curie, Paris 6,
75252 Paris Cedex, France.

Introduction

The purpose of these lectures is twofold. On the one hand, they will try to show how methods coming from modern physics and, more specifically, from quantum integrable models can be used to solve difficult problems of enumerative combinatorics. As the field of combinatorics is expanding rapidly, problems of enumeration are becoming more and more difficult, and direct combinatorial proofs often become extremely complicated and tedious. This is where physical ideas can come in to provide conceptually simple proofs. On the other hand, these lectures will try to define and explain some basic tools of combinatorics which may be of use to a physicist, especially one working in two-dimensional (or one+one-dimensional) systems.

The lectures will cover three subjects. Though they are in principle different, there will be some definite connections and the last part will provide ideas on how to reunite them. The first subject is free-fermionic methods. Though free fermions in two dimensions may seem excessively simple to the physicist, they already provide a wealth of combinatorial formulas. In fact, they have become extremely popular in the recent mathematical literature (see e.g. [33, 48]). This discussion of free fermions will also allow us to introduce Schur functions, which are omnipresent in any combinatorics computation. The second subject is the six-vertex model, and, in particular, the six-vertex model with domain wall boundary conditions. This is an example of a nontrivial model of two-dimensional statistical mechanics which is exactly solvable (i.e. quantum integrable in the correspondence between 2D statistical mechanics and 1D quantum systems). We shall apply this model to the enumeration of alternating-sign matrices. Finally, the last part discusses yet another model, this time made of loops. We shall discuss its combinatorial properties, and introduce a unifying equation, the quantum Knizhnik–Zamolodchikov equation, which will allow us to reconnect to the other two subjects.

17.1 Free-fermionic methods

17.1.1 Definitions

Operators and Fock space. Consider a fermionic operator $\psi(z)$,

$$\psi(z) = \sum_{k\in\mathbb{Z}+1/2} \psi_{-k} z^{k-1/2}\,, \qquad \psi^{\text{star}}(z) = \sum_{k\in\mathbb{Z}+1/2} \psi_k^{\text{star}} z^{k-1/2}\,, \tag{17.1}$$

with anticommutation relations

$$[\psi_r^{\text{star}}, \psi_s]_+ = \delta_{rs}\,, \qquad [\psi_r, \psi_s]_+ = [\psi_r^{\text{star}}, \psi_s^{\text{star}}]_+ = 0\,; \tag{17.2}$$

$\psi(z)$ and $\psi^{\text{star}}(z)$ should be thought of as generating series for ψ_k and ψ_k^{star}, so that z is just a formal variable (see also Section 17.1.2). What we have here is a complex (charged) fermion, with particles, and antiparticles which can be identified with holes in the Dirac sea. These fermions are one-dimensional, in the sense that their states are indexed by (half-odd) integers; ψ_k^{star} creates a particle (or destroys a hole) at location k, whereas ψ_k destroys a particle (or creates a hole) at location k.

We shall explicitly build the Fock space $\mathcal{F}$ and the representation of the fermionic operators now. We start from a vacuum $|0\rangle$ which satisfies

$$\psi_k\,|0\rangle = 0\,, \quad k>0\,, \qquad \psi_k^{\text{star}}\,|0\rangle = 0\,, \quad k<0\,; \tag{17.3}$$

that is, it is a Dirac sea filled up to location 0:

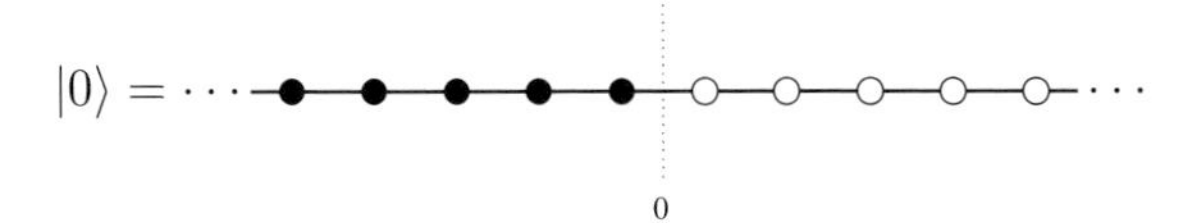

Then any state can be built by the action of ψ_k and ψ_k^{star} from $|0\rangle$. In particular, one can define more general vacua at level $\ell \in \mathbb{Z}$:

$$|\ell\rangle = \begin{cases} \psi^{\mathrm{star}}_{\ell-1/2}\psi^{\mathrm{star}}_{\ell-3/2}\cdots\psi^{\mathrm{star}}_{1/2}\,|0\rangle\,, & \ell > 0 \\ \psi_{\ell+1/2}\psi_{\ell+3/2}\cdots\psi_{-1/2}\,|0\rangle\,, & \ell < 0 \end{cases}$$

$$= \cdots \bullet\!\!-\!\!\bullet\!\!-\!\!\bullet\!\!-\!\!\bullet\!\!-\!\!\bullet\!\!\underset{\ell}{|}\!\!-\!\!\circ\!\!-\!\!\circ\!\!-\!\!\circ\!\!-\!\!\circ\!\!-\!\!\circ \cdots\,, \tag{17.4}$$

which will be useful in what follows. They satisfy

$$\psi_k\,|\ell\rangle = 0\,, \quad k > \ell\,, \qquad \psi_k^{\mathrm{star}}\,|\ell\rangle = 0\,, \quad k < \ell\,. \tag{17.5}$$

More generally, we define a *partition* to be a weakly decreasing finite sequence of non-negative integers $\lambda_1 \geq \lambda_2 \geq \cdots \geq \lambda_n \geq 0$. We usually represent partitions as *Young diagrams*: for example, $\lambda = (5, 2, 1, 1)$ is depicted as

$\lambda =$

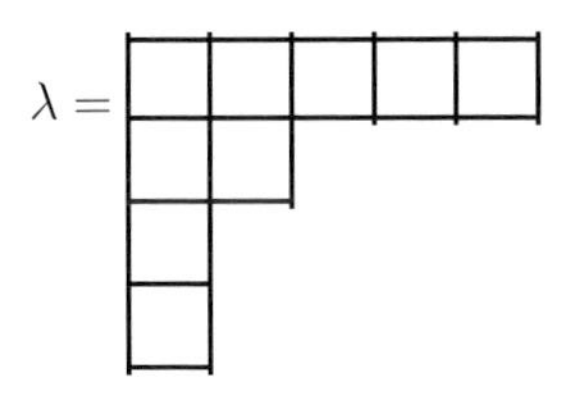

With each partition $\lambda = (\lambda_1, \ldots, \lambda_n)$ we associate the following state in $\mathcal{F}_\ell$:

$$|\lambda;\ell\rangle = \psi^{\mathrm{star}}_{\ell+\lambda_1-1/2}\psi^{\mathrm{star}}_{\ell+\lambda_2-3/2}\cdots\psi^{\mathrm{star}}_{\ell+\lambda_n-n+1/2}\,|\ell-n\rangle\,. \tag{17.6}$$

Note the important property that if one "pads" a partition with extra zeros, then the corresponding state remains unchanged. In particular, for the empty diagram $\varnothing$, $|\varnothing;\ell\rangle = |\ell\rangle$. For $\ell = 0$, we just write $|\lambda;0\rangle = |\lambda\rangle$.

This definition has the following nice graphical interpretation: the state $|\lambda;\ell\rangle$ can be described by numbering the edges of the boundary of the Young diagram, in such a way that the main diagonal passes between $\ell - \frac{1}{2}$ and $\ell + \frac{1}{2}$; then the occupied and empty sites correspond to vertical and horizontal edges, respectively. With the example above and $\ell = 0$, we find the following (only the occupied sites are numbered for clarity):

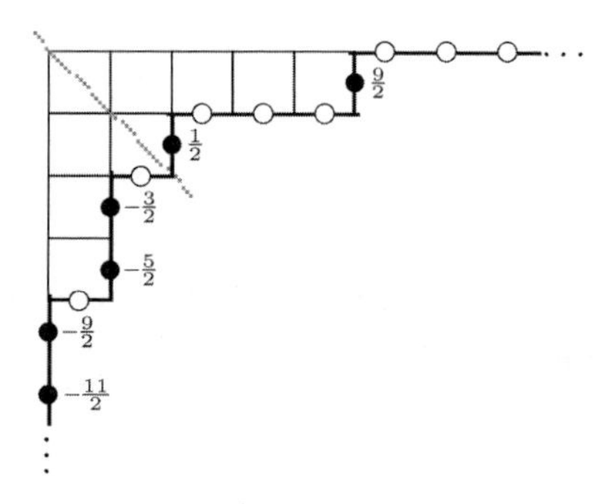

The $|\lambda;\ell\rangle$, where λ runs over all possible partitions (two partitions being identified if they can be obtained from each other by adding or removing zero parts), form an orthonormal basis of a subspace of $\mathcal{F}$, which we denote by $\mathcal{F}_\ell$. ψ_k and $\psi_k^{\rm star}$ are Hermitian conjugates of each other.

Note that eqn (17.6) fixes our sign convention for the states. In particular, this implies that when one acts with ψ_k or $\psi_k^{\rm star}$ on a state $|\lambda\rangle$ with a particle or a hole at k, one produces a new state $|\lambda'\rangle$ with a particle removed or added, respectively, at k times -1 to the power of the number of particles to the right of k.

The states λ can also be produced from the vacuum by acting with ψ to create holes; paying attention to the sign issue, we find

$$|\lambda;\ell\rangle = (-1)^{|\lambda|}\psi_{\ell-\lambda'_1+1/2}\cdots\psi_{\ell-\lambda'_m+m-1/2}\,|\ell+m\rangle\,, \tag{17.7}$$

where the λ'_i are the lengths of the columns of λ, $|\lambda|$ is the number of boxes in λ, and $m=\lambda_1$. This formula is formally identical to eqn (17.6) if we renumber the states from right to left, exchange ψ and $\psi^{\rm star}$, and replace λ with its *transpose* diagram λ' (this property is graphically clear). So the particle–hole duality translates into transposition of Young diagrams.

Finally, we introduce the normal ordering with respect to the vacuum $|0\rangle$,

$$:\psi_j^{\rm star}\psi_k: \;=\; -:\psi_k\psi_j^{\rm star}: \;=\; \begin{cases}\psi_j^{\rm star}\psi_k\,, & j>0\,,\\ -\psi_k\psi_j^{\rm star}\,, & j<0\,,\end{cases} \tag{17.8}$$

which allows us to get rid of trivial infinite quantities.

The relativistic chiral fermion. One possible interpretation of these operators is that they correspond to chiral massless relativistic charged free fermions, with eqn (17.1) solving implicitly the equations of motion. Indeed, if z is a complex variable which represents space–time, then ψ and $\psi^{\rm star}$ satisfy $\bar\partial\psi(z)=\bar\partial\psi^{\rm star}(z)=0$, which is the solution of the equations of motion for a free fermion with action $S=\int d^2z\,\psi^{\rm star}(z)\bar\partial\psi(z)$, or with Hamiltonian $H=\sum_k k:\psi_k^{\rm star}\psi_k:$.

This is not the point of view we shall adopt in the rest of these lectures, since the dynamics of our system will be given by a different type of Hamiltonian (though it will be quadratic, resulting in a free theory as well). In fact, it is more natural to think of k as being a position, and therefore of z as a momentum (with $|z|=1$), though this is to some extent a matter of taste.

Note, for example, that one can derive the following formula using solely the construction of the previous section (the proof is elementary and left to the reader—careful with the signs!):

$$\begin{aligned}
&\langle \ell |\, \psi(w_r) \cdots \psi(w_1) \psi^{\text{star}}(z_1) \cdots \psi^{\text{star}}(z_s)\, | m \rangle \\
&\quad = \delta_{\ell+r,m+s} \prod_{i=1}^{r} w_i^{-m} \prod_{j=1}^{s} z_i^{m} \, \frac{\prod_{1 \le i < j \le r}(w_j - w_i) \prod_{1 \le i < j \le s}(z_i - z_j)}{\prod_{1 \le i \le r, 1 \le j \le s}(w_i - z_j)} \,. \qquad (17.9)
\end{aligned}$$

But this is in fact the Wick theorem for free fermions with propagator $\langle \psi(w) \psi^{\text{star}}(z) \rangle = 1/(w - z)$.

$\mathfrak{gl}(\infty)$ *and* $\hat{\mathfrak{u}}(1)$ *action.* The bilinears $\psi^{\text{star}}(z)\psi(w)$ give rise to the Schwinger representation of $\mathfrak{gl}(\infty)$ on $\mathcal{F}$, whose usual basis is $:\psi_r^{\text{star}} \psi_s:$, $r, s \in \mathbb{Z} + \frac{1}{2}$, and the identity. In the first-quantized picture, this representation is simply the natural action of $\mathfrak{gl}(\infty)$ on the one-particle Hilbert space $\mathbb{C}^{\mathbb{Z}+1/2}$ and exterior products thereof. The electric charge $J_0 = \sum_r :\psi_r^{\text{star}} \psi_r:$ is a conserved number and classifies the irreducible representations of $\mathfrak{gl}(\infty)$ inside $\mathcal{F}$, which are all isomorphic. The highest-weight vectors are precisely our vacua $|\ell\rangle$, $\ell \in \mathbb{Z}$, so that $\mathcal{F} = \oplus_{\ell \in \mathbb{Z}} \mathcal{F}_\ell$, with $\mathcal{F}_\ell$ being the subspace in which $J_0 = \ell$.

The $\mathfrak{u}(1)$ current

$$j(z) = \,:\psi^{\text{star}}(z)\psi(z):\, = \sum_{n \in \mathbb{Z}} J_n z^{-n-1} \,, \qquad (17.10)$$

with $J_n = \sum_r :\psi_{r-n}^{\text{star}} \psi_r:$, forms a $\hat{\mathfrak{u}}(1)$ (Heisenberg) subalgebra of $\mathfrak{gl}(\infty)$:

$$[J_m, J_n] = m \delta_{m,-n} \,. \qquad (17.11)$$

Note, in particular, that positive modes commute among themselves. This allows us to define the general "Hamiltonian"

$$H[t] = \sum_{q=1}^{\infty} t_q J_q \,, \qquad (17.12)$$

where $t = (t_1, \ldots, t_q, \ldots)$ is a set of parameters ("times").

The J_q, $q > 0$, displace one of the fermions q steps to the left. This is expressed by the following formulas describing the time evolution of the fermionic fields:

$$\begin{aligned}
\epsilon H[t] \psi(z) \epsilon - H[t] &= \epsilon - \sum_{q=1}^{\infty} t_q z^q \psi(z) \,, \\
\epsilon H[t] \psi^{\text{star}}(z) \epsilon - H[t] &= \epsilon + \sum_{q=1}^{\infty} t_q z^q \psi^{\text{star}}(z)
\end{aligned} \qquad (17.13)$$

(proof: compute $[J_q, \psi^{[\text{star}]}(z)] = \pm z^q \psi^{[\text{star}]}(z)$ and exponentiate). Of course, similarly, J_{-q}, $q > 0$, moves one fermion q steps to the right.

17.1.2 Schur functions

Schur functions are undoubtedly the most important basis in the theory of symmetric functions (symmetric polynomials of an arbitrarily large number of variables). As we shall see, they are also closely connected to free fermions.

Definition. There are several ways to define Schur functions. We use the following. Let λ be a partition. Then the associated Schur function s_λ is

$$s_\lambda(x_1, \ldots, x_n) = \frac{\det_{1\le i,j\le n}(x_i^{\lambda_j+n-j})}{\prod_{i<j}(x_i - x_j)} . \tag{17.14}$$

This is sometimes called the Weyl formula. Here it is assumed that λ has been padded to the number of variables. If there are fewer variables than the number of nonzero parts of λ, then $s_\lambda = 0$. Note that the denominator is nothing but the numerator for $\lambda_j = 0$ (up to a sign, it is the Vandermonde determinant), so that $s_\varnothing = 1$. One can think of the numerator/denominator as the (Slater) wave function of n fermions in a first-quantized picture, the numerator being an excited state and the denominator being the ground state. The connection between Schur functions and free fermions that we shall establish later is essentially a refined, second-quantized version of this idea.

$s_\lambda(x_1, \ldots, x_n)$ is explicitly symmetric by permutation of its arguments. It is also not too hard to check that $s_\lambda(x_1, \ldots, x_n, 0, \ldots, 0) = s_\lambda(x_1, \ldots, x_n)$; that is, s_λ satisfies a stability property with respect to the number of variables.

Since the numerator vanishes when any two x_i coincide, the denominator divides the numerator, so that s_λ is a polynomial in the x_i. In fact, one can easily see that it is a homogeneous polynomial of degree the number of boxes of λ.

Remark: defined in terms of a fixed number n of variables, as in eqn (17.14), $s_\lambda(x_1, \ldots, x_n)$ has the following group-theoretic interpretation. The polynomial irreducible representations of $GL(n)$ are known to be indexed by partitions. More precisely, the representation on tensors of order k decomposes as a direct sum of isotypic representations corresponding to every partition λ such that $|\lambda| = k$. Then $s_\lambda(x_1, \ldots, x_n)$ is the character of the representation λ evaluated for the diagonal matrix $\mathrm{diag}(x_1, \ldots, x_n)$. Hence, the dimension of λ as a $GL(n)$ representation is given by $s_\lambda(\underbrace{1, \ldots, 1}_{n}) = \prod_{1\le i<j\le n}(\lambda_i - i - \lambda_j + j)/(j - i)$ (proof: use $x_i = q^i$, compute, and send q to 1).

Examples: $s_{\square} = \sum_i x_i$, $s_{\substack{\square\\\square}} = \sum_{i<j} x_i x_j$, $s_{\square\square} = \sum_{i\le j} x_i x_j$.

Power sums. Any symmetric polynomial in some variables x_i is in fact a polynomial in its power sums, i.e.

$$t_q = \frac{1}{q}\sum_i x_i^q , \qquad q \ge 1 .$$

If there are n variables, only n of the t_q are independent; however, for symmetric functions (i.e. for an arbitrarily large number of variables), they should be considered as independent.

We shall often use this parameterization for Schur functions; we denote it by $s_\lambda[t_1, \ldots, t_q, \ldots]$ or simply $s_\lambda[t]$.

Note that transposition of a diagram corresponds to changing the sign of even-power sums:

$$s_{\lambda'}[t_1, t_2, \ldots, t_q, \ldots] = s_\lambda[t_1, -t_2, \ldots, (-1)^{q-1}t_q, \ldots] .$$

Examples: $s_{\square} = t_1$, $s_{\substack{\square\\\square}} = \frac{1}{2}t_1^2 - t_2$, $s_{\square\square} = \frac{1}{2}t_1^2 + t_2$, $s_{\substack{\square\square\\\square}} = \frac{1}{3}t_1^3 - t_3$.

Remark: according to the Schur–Weyl duality, the coefficients of the expansion of s_λ in products of t's are known (the Frobenius formula): they are essentially characters of the symmetric group. As a corollary, the dimension of λ as a representation of the symmetric group is given by $|\lambda|!\, s_\lambda[1,0,\ldots,0,\ldots]$.

17.1.3 From free fermions to Schur functions

Basic relation. We now prove the following important identity:

$$\langle \ell |\, \epsilon H[t]\, |\lambda;\ell\rangle = s_\lambda[t]\,, \tag{17.15}$$

which shows that the map $|\Phi\rangle \mapsto \langle\ell|\,\epsilon H[t]\,|\Phi\rangle$ is an isomorphism from $\mathcal{F}_\ell$ to the space of symmetric functions.

Proof Owing to the obvious translational invariance of all the operators involved, we may as well set $\ell = n$. We use the definition (17.6) of $|\lambda\rangle$ and the commutation relations (17.13) to rewrite the left-hand side as

$$\langle n|\,\epsilon H[t]\,|\lambda;n\rangle$$
$$= \epsilon \sum_{q\geq 1} t_q \sum_{i=1}^{n} z_i^q \,\langle n|\,\psi^{\mathrm{star}}(z_1)\psi^{\mathrm{star}}(z_2)\cdots\psi^{\mathrm{star}}(z_n)\,|0\rangle\,\Big|_{z_1^{n+\lambda_1-1} z_2^{n+\lambda_2-2}\cdots z_n^{\lambda_n}}\,,$$

where $\big|_{\ldots}$ means picking one term in a generating series.

As a special case of eqn (17.9), we can evaluate the remaining bra–ket to be (we now use the $\ell = 0$ notation for the left-hand side)

$$\langle 0|\,\epsilon H[t]\,|\lambda\rangle = \epsilon \sum_{q\geq 1} t_q \sum_{i=1}^{n} z_i^q \prod_{1\leq i<j\leq n} (z_i - z_j)\Big|_{z_1^{n+\lambda_1-1} z_2^{n+\lambda_2-2}\cdots z_n^{\lambda_n}}\,.$$

We now write $t_q = (1/q)\sum_{j=1}^{n} x_j^q$ and note that $\epsilon\sum_{q\geq 1} t_q \sum_{i=1}^{n} z_i^q = \prod_{i,j=1}^{n}(1 - z_i x_j)^{-1}$. We recognize (part of) the Cauchy determinant:

$$\langle 0|\,\epsilon H[t]\,|\lambda\rangle = \frac{\det_{1\leq i,j\leq n}(1 - x_i z_j)^{-1}}{\prod_{i<j}(x_i - x_j)}\Big|_{z_1^{n+\lambda_1-1} z_2^{n+\lambda_2-2}\cdots z_n^{\lambda_n}}\,.$$

At this stage, we can just expand separately each column of the matrix $(1 - x_i z_j)^{-1}$ to pick the right power of z_j; we find

$$\langle 0|\,\epsilon H[t]\,|\lambda\rangle = \frac{\det_{1\leq i,j\leq n}(x_i^{\lambda_j+n-j})}{\prod_{i<j}(x_i - x_j)}\,,$$

which is our definition (17.14) of a Schur function. □

We shall now compute $\langle 0|\,\epsilon H[t]\,|\lambda\rangle$ in various ways. In fact, many of the methods used are equally applicable to the following more general quantity:

$$s_{\lambda/\mu}[t] = \langle \mu|\,\epsilon H[t]\,|\lambda\rangle\,, \tag{17.16}$$

where λ and μ are two partitions. It is easy to see that in order for $s_{\lambda/\mu}[t]$ to be nonzero, we must have $\mu \subset \lambda$, when λ and μ are considered as Young diagrams; in this case $s_{\lambda/\mu}$ is known as the *skew Schur function* associated with the skew Young diagram λ/μ. The latter is depicted as the complement of μ inside λ. This is appropriate because skew Schur functions factorize in terms of the connected components of the skew Young diagram λ/μ.

Examples: $s_{(2,1)/(1)} = s_{\square}^2 = t_1^2$, $s_{(3,2)/(1)} = \frac{5}{24}t_1^4 + \frac{1}{2}t_1^2 t_2 + \frac{1}{2}t_2^2 - t_1 t_3 - t_4$.

Wick theorem and Jacobi–Trudi identity. First, we apply the Wick theorem. We consider the following as the definition of the time evolution of fermionic fields:

$$\begin{aligned} \psi_k[t] &= \epsilon H[t] \psi_k \epsilon - H[t] \,, \\ \psi_k^{\text{star}}[t] &= \epsilon H[t] \psi_k^{\text{star}} \epsilon - H[t] \,. \end{aligned} \tag{17.17}$$

In fact, eqn (17.13) gives us the "solution" of the equations of motion in terms of the generating series $\psi(z)$ and $\psi^{\text{star}}(z)$.

Noting that the Hamiltonian is quadratic in the fields, we now state the Wick theorem:

$$\langle \ell | \, \psi_{i_1}[0] \cdots \psi_{i_n}[0] \psi_{j_1}^{\text{star}}[t] \ldots \psi_{j_n}^{\text{star}}[t] \, | \ell \rangle = \det_{1 \le p,q \le n} \langle \ell | \, \psi_{i_p}[0] \psi_{j_q}^{\text{star}}[t] \, | \ell \rangle \,. \tag{17.18}$$

Next, we start from the expression (17.16) for $s_{\lambda/\mu}[t]$: padding λ or μ with zeros so that they have the same number of parts n, we can write

$$s_{\lambda/\mu}[t] = \langle -n | \, \psi_{\mu_n - n + 1/2} \cdots \psi_{\mu_1 - 1/2} \epsilon H[t] \psi_{\lambda_1 - 1/2}^{\text{star}} \cdots \psi_{\lambda_n - n + 1/2}^{\text{star}} \, | -n \rangle$$

and apply the Wick theorem to find

$$s_{\lambda/\mu}[t] = \det_{1 \le p,q \le n} \langle -n | \, \psi_{\mu_p - p + 1/2} \epsilon H[t] \psi_{\lambda_q - q + 1/2}^{\text{star}} \, | -n \rangle \,.$$

It is easy to see that $\langle -n | \, \psi_i \epsilon H[t] \psi_j^{\text{star}} \, | -n \rangle$ does not depend on n and thus depends only on $j - i$. Let us denote it by

$$h_k[t] = \langle 1 | \, \epsilon H[t] \psi_{k+1/2}^{\text{star}} \, | 0 \rangle \,, \qquad \sum_{k \ge 0} h_k[t] z^k = \langle 1 | \, \epsilon H[t] \psi^{\text{star}}(z) \, | 0 \rangle = \epsilon \sum_{q \ge 1} t_q z^q \tag{17.19}$$

($k = j - i$; note that $h_k[t] = 0$ for $k < 0$).

The final formula that we obtain is

$$s_{\lambda/\mu}[t] = \det_{1 \le p,q \le n} \left(h_{\lambda_q - \mu_p - q + p}[t] \right) \tag{17.20}$$

or, for regular Schur functions,

$$s_\lambda[t] = \det_{1 \le p,q \le n} \left(h_{\lambda_q - q + p}[t] \right) . \tag{17.21}$$

This is known as the Jacobi–Trudi identity.

By using "particle–hole duality," we can find a dual form of this identity. We describe our states in terms of hole positions, parameterized by the lengths of the columns λ'_p and μ'_q, according to eqn (17.7):

$$s_{\lambda/\mu}[t] = (-1)^{|\lambda|+|\mu|} \langle m| \psi^{\mathrm{star}}_{-\mu'_m+m-1/2} \cdots \psi^{\mathrm{star}}_{-\mu'_1+1/2} \epsilon H[t] \psi_{-\lambda'_1+1/2} \cdots \psi_{-\lambda'_m+m-1/2} |m\rangle .$$

Again the Wick theorem applies, and expresses $s_{\lambda/\mu}$ in terms of the two point-function $\langle m| \psi^{\mathrm{star}}_i \epsilon H[t] \psi_j |m\rangle$, which depends only on $i - j = k$ and is given by

$$\begin{aligned} e_k[t] &= (-1)^k \langle -1| \epsilon H[t] \psi_{-k+1/2} |0\rangle , \\ \sum_{k \geq 0} e_k[t] z^k &= \langle -1| \epsilon H[t] \psi(-z) |0\rangle = \epsilon \sum_{q \geq 1} (-1)^{q-1} t_q z^q . \end{aligned} \tag{17.22}$$

The final formula takes the form

$$s_{\lambda/\mu}[t] = \det_{1 \leq p,q \leq n} \left(e_{\lambda'_q - \mu'_p - q + p}[t] \right) \tag{17.23}$$

or, for regular Schur functions,

$$s_\lambda[t] = \det_{1 \leq p,q \leq n} \left(e_{\lambda'_q - q + p}[t] \right) . \tag{17.24}$$

This is the dual Jacobi–Trudi identity, also known as the Von Nägelsbach–Kostka identity.

Schur functions and lattice fermions. In the remainder of this subsection (Section 17.1.3), we shall assume a fixed, finite number of variables n, and set as before $t_q = (1/q) \sum_{i=1}^n x_i^q$. In this case we can write

$$\epsilon H[t] = \prod_{i=1}^n \epsilon \phi_+(x_i) , \qquad \phi_+(x) = \sum_{q \geq 1} \frac{x^q}{q} J_q .$$

So we can think of the "time evolution" as a series of discrete steps represented by commuting operators $\exp \phi_+(x_i)$. In the language of statistical mechanics, these are transfer matrices (and the one-parameter family of transfer matrices $\exp \phi_+(x)$ is of course related to the integrability of the model). We now show that they have a very simple meaning in terms of lattice fermions.

Consider a two-dimensional square lattice, one direction being our space $\mathbb{Z} + \frac{1}{2}$ and one direction being time. In what follows, we shall reverse the arrow of time (that is, we shall consider that time flows upwards in the pictures), which makes the discussion slightly easier, since products of operators are read from left to right. The rule for going from one step to the next according to the evolution operator $\exp \phi_+(x)$ can be formulated either in terms of particles or in terms of holes:

- Each particle can either go straight or hop to the right as long as it does not reach the (original) location of the next particle. Each step to the right is given a weight of x.
- Each hole can either go straight or one step to the left as long as it does not bump into its neighbor. Each step to the left is given a weight of x.

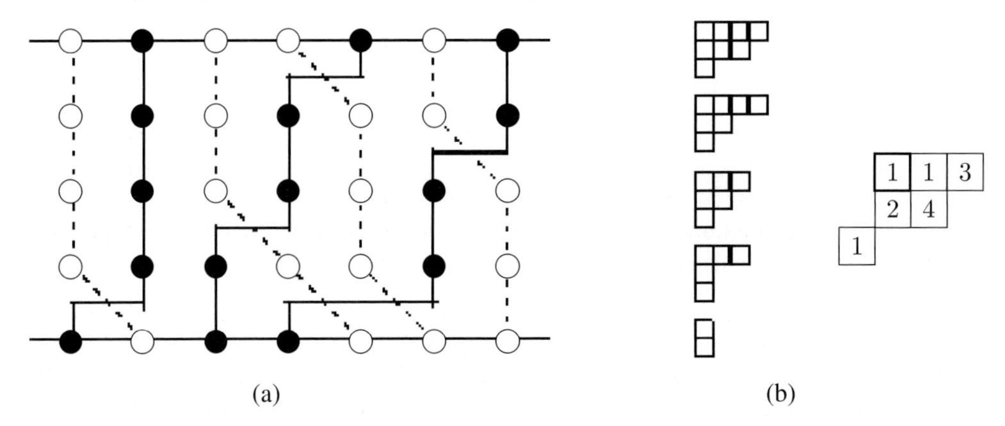

Fig. 17.1 A lattice fermion configuration and the corresponding (skew) SSYT.

Obviously, the second description is simpler. An example of a possible evolution of the system with given initial and final states is shown in Fig. 17.1(a).

The proof of these rules consists in computing $\langle\mu|\,\epsilon\phi_+(x)\,|\lambda\rangle$ explicitly by applying, say, eqn (17.23) for $t_q = (1/q)x^q$, and noting that in this case, according to eqn (17.22), $e_n[t] = 0$ for $n > 1$. This strongly constrains the possible transitions and produces the description above.

Relation to semistandard Young tableaux. A semistandard Young tableau (SSYT) of shape λ is a filling of the Young diagram of λ with elements of some ordered alphabet, in such a way that the rows are weakly increasing and the columns are strictly increasing.

We shall use here the alphabet $\{1, 2, \ldots, n\}$. For example, with $\lambda = (5, 2, 1, 1)$, one possible SSYT with $n \geq 5$ is

1	2	4	5	5
3	3			
4				
5				

It is useful to think of Young tableaux as time-dependent Young diagrams where the number indicates the step at which a given box was created. Thus, with the same example, we get

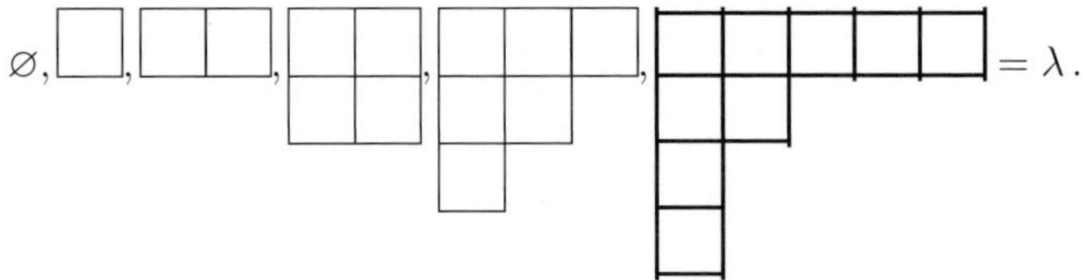

In other words, a Young tableau is nothing but a statistical configuration of our lattice fermions, where the initial state is the vacuum. Similarly, a skew SSYT is a filling of a skew Young diagram with the same rules; it corresponds to a statistical configuration of lattice fermions with arbitrary initial and final states. The correspondence is exemplified in Fig. 17.1(b).

Each extra box corresponds to a step to the right for particles or to the left for holes. The initial and final states are $\varnothing$ and λ, which is the case for Schur functions (see eqn (17.15)). We conclude that the following formula holds:

$$s_\lambda(x_1, \dots, x_n) = \sum_{T \in \mathrm{SSYT}(\lambda, n)} \prod_{b \text{ box of } T} x_{T_b} . \tag{17.25}$$

This is often taken as a definition of Schur functions. It is explicitly stable with respect to n in the sense that $s_\lambda(x_1, \dots, x_n, 0, \dots, 0) = s_\lambda(x_1, \dots, x_n)$. It is, however, not obvious from this definition that s_λ is symmetric by permutation of its variables. This fact is a manifestation of the underlying free-fermionic ("integrable") behavior. Of course, an identical formula holds for the more general case of skew-symmetric Schur functions.

Nonintersecting lattice paths and Lindström–Gessel–Viennot formula. The rules for evolution given earlier in this section strongly suggest the following explicit description of lattice fermion configurations. Consider the directed graphs in Fig. 17.2 (the graphs are in principle infinite to the left and right, but any given bra–ket evaluation involves only a finite number of particles and holes and therefore the graphs can be truncated to a finite part). Consider *nonintersecting lattice paths* (NILPs) on these graphs: these are paths with given starting points (at the bottom) and given ending points (at the top), which follow the edges of the graph, respecting the orientation of the arrows, and which are not allowed to touch at any vertices. One can check that the trajectories of holes and particles following the rules described earlier are exactly the most general NILPs on these graphs.

In this context, the Wick theorem (17.21) (i.e. the Jacobi–Trudi identity) is most naturally proved in the "functional integral" formalism. We recall it here.

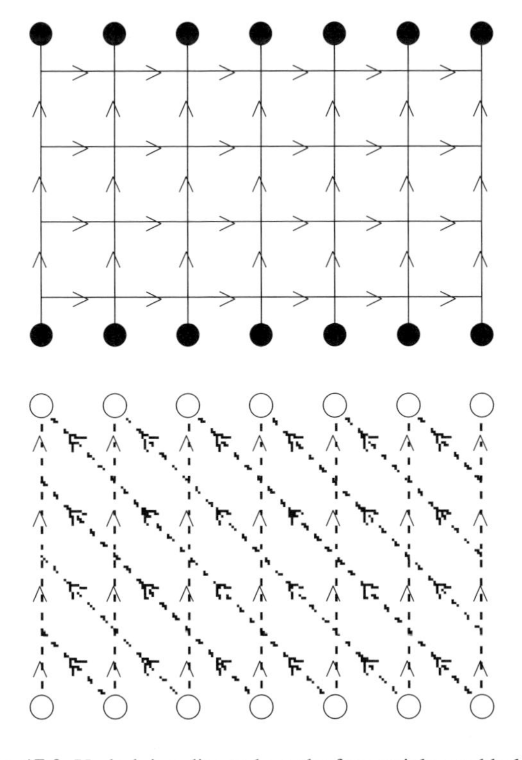

Fig. 17.2 Underlying directed graphs for particles and holes.

We assign Grassmann variables χ_i and χ_i^{star} to each vertex of the graph and consider the action

$$-S = \sum_i \chi_i \chi_i^{\text{star}} + \sum_{i \to j} w_{ij} \chi_i^{\text{star}} \chi_j ,$$

where the w_{ij} are arbitrary weights on the directed edges $i \to j$. Then we have the following identity: the weighted sum of paths from starting locations $i_1, \ldots, i_n$ to ending locations $j_1, \ldots, j_n$, where the weight of a path is the products of the weights of the edges, is given by

$$N(i_1, \ldots, i_n; j_1, \ldots, j_n) = \left\langle \chi_{i_1}^{\text{star}} \cdots \chi_{i_n}^{\text{star}} \chi_{j_n} \cdots \chi_{j_1} \right\rangle .$$

In this formula, the averaging is with respect to the measure $\exp(-S)$. The Wick theorem for Gaussian integrals then asserts that $\langle \chi_{i_1}^{\text{star}} \cdots \chi_{i_n}^{\text{star}} \chi_{j_n} \cdots \chi_{j_1} \rangle = \det_{p,q} \langle \chi_{i_p}^{\text{star}} \chi_{j_q} \rangle$, or

$$N(i_1, \ldots, i_n; j_1, \ldots, j_n) = \det_{p,q} N(i_p; j_q) . \tag{17.26}$$

This result is clearly valid for any planar directed acyclic graph (and with appropriate starting points and endpoints: no paths between starting points or between endpoints should be possible). In combinatorics, the formula (17.26) is known as the Lindström–Gessel–Viennot (LGV) formula [26, 39].

So, we need only to compute $N(i; j)$, the weighted number of paths from i to j. Let us do so in our problem.

In the case of particles (top part of Fig. 17.2), numbering the initial and final points from left to right, we find that the weighted sum of paths from i to j, where a weight x_i is given to each rightward move at time step i, depends only on $j - i$; if we denote this sum by $h_{j-i}(x_1, \ldots, x_n)$, we have the obvious generating-series formula

$$\sum_{k \geq 0} h_k(x_1, \ldots, x_n) z^k = \prod_{i=1}^{n} \frac{1}{1 - z x_i} .$$

Note that this formula coincides with the alternate definition (17.19) of $h_k[t]$ if we set as usual $t_q = (1/q) \sum_{i=1}^{n} x_i^q$. Thus, applying the LGV formula (17.26) and choosing the correct initial and final points for Schur functions or skew Schur functions, we recover eqns (17.20) and (17.21) immediately.

In the case of holes (bottom part of Fig. 17.2), numbering the initial and final points from right to left, we find once again that the weighted sum of paths from i to j, where a weight x_i is given to each leftward move at time step i, depends only on $j - i$; if we denote this sum by $e_{j-i}(x_1, \ldots, x_n)$, we have the equally obvious generating-series formula

$$\sum_{k \geq 0} e_k(x_1, \ldots, x_n) z^k = \prod_{i=1}^{n} (1 + z x_i) ,$$

which coincides with eqn (17.22), thus allowing us to recover eqns (17.23) and (17.24).

Relation to standard Young tableaux. A standard Young tableau (SYT) of shape λ is a filling of the Young diagram of λ with elements of some ordered alphabet, in such a way that both the rows and the columns are strictly increasing. There is no loss of generality in assuming that the alphabet is $\{1, \ldots, n\}$, where $n = |\lambda|$ is the number of boxes in λ. For example,

1	2	6	8	9
3	4			
5				
7				

is an SYT of shape $(5, 2, 1, 1)$.

Standard Young tableaux are connected to the representation theory of the symmetric group; in particular, the number of such tableaux with a given shape λ is the dimension of λ as an irreducible representation of the symmetric group, which is, up to a trivial factor, the evaluation of the Schur function s_λ at $t_q = \delta_{1q}$. In this case we have $H[t] = J_1$, and there is only one term contributing to the bra–ket $\langle\lambda|\, \epsilon H[t]\, |0\rangle$ in the expansion of the exponential:

$$s_\lambda[\delta_{1\cdot}] = \frac{1}{n!} \langle\lambda|\, J_1^n\, |0\rangle\,.$$

In terms of lattice fermions, J_1 has a direct interpretation as the transfer matrix for one particle hopping one step to the left. As the notion of the SYT is invariant by transposition, particles and holes play a symmetric role so that the evolution can be summarized by either of the following two rules:

- Exactly one particle moves one step to the right in such a way that it does not bump into its neighbor; all the other particles go straight.
- Exactly one hole moves one step to the left in such a way that it does not bump into its neighbor; all the other holes go straight.

An example of such a configuration is given in Fig. 17.3.

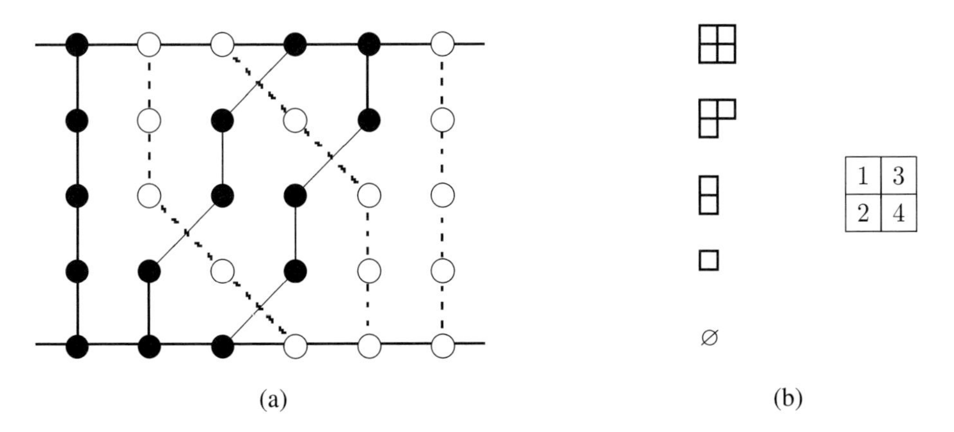

Fig. 17.3 (a) A lattice fermion configuration and (b) the corresponding SYT.

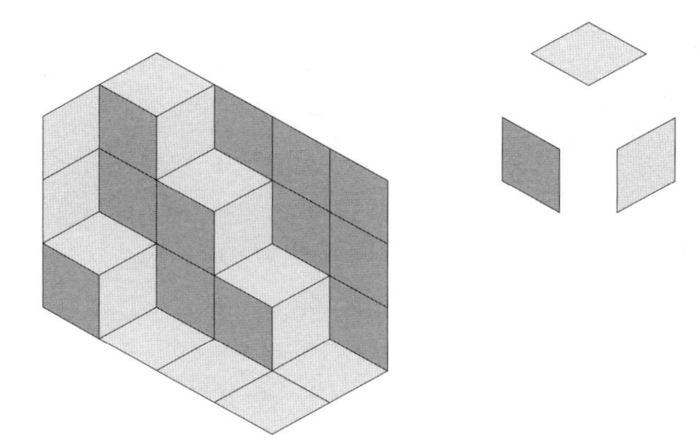

Fig. 17.4 A plane partition of size $2 \times 3 \times 4$.

17.1.4 Application: Plane partition enumeration

Plane partitions are an important class of objects in combinatorics. The name originates from the way they were first introduced [40] as two-dimensional generalizations of partitions. Here we shall directly define them graphically.

Definition. Intuitively, plane partitions are piles of boxes (cubes) in the corner of a room, subject to the constraints of gravity. An example is given in Fig. 17.4. Typically, we ask for the cubes to be contained inside a bigger box (parallelepiped) of given size.

Alternatively, one can project the picture onto a two-dimensional plane (which is inevitably what we do when we draw the picture on paper), and the result is a tiling of a region of the plane by lozenges (or rhombi) of three possible orientations, as shown on the right of the figure. If the cubes are inside a parallelepiped of size $a \times b \times c$, then, possibly drawing the walls of the room as extra tiles, we obtain a lozenge tiling of a hexagon with sides a, b, c, which is the situation we consider now.

MacMahon formula. In order to display the free-fermionic nature of plane partitions, we shall consider the following operation. In the 3D view, consider slices of the pile of boxes formed by planes that are parallel to two of the three axes and are located halfway between successive rows of cubes. In the 2D view, this corresponds to selecting two of the three orientations of the lozenges and building paths out of these. Figure 17.5 shows on the left the result of such an operation: a set of lines going from one side to the opposite side of a hexagon. The lines are, by definition, nonintersecting and can only move in two directions. Inversely, any set of such NILPs produces a plane partition.

At this stage, one could apply the LGV formula. But there is no need, since this is actually the case already considered in Section 17.1.3. Compare Figs. 17.5 and 17.1: the trajectories of holes are exactly our paths (it is left to the reader as an exercise to check that the trajectories of particles form another set of NILPs corresponding to another choice of two orientations of lozenges; what about the third choice?). If we attach a weight x_i to each dark gray lozenge at step i, we find that the weighted enumeration of plane partitions in an $a \times b \times c$ box is given by

$$N_{a,b,c}(x_1, \dots, x_{a+b}) = \langle 0 | \, \epsilon H[t] \, | b \times c \rangle = s_{b \times c}(x_1, \dots, x_{a+b}) \,,$$

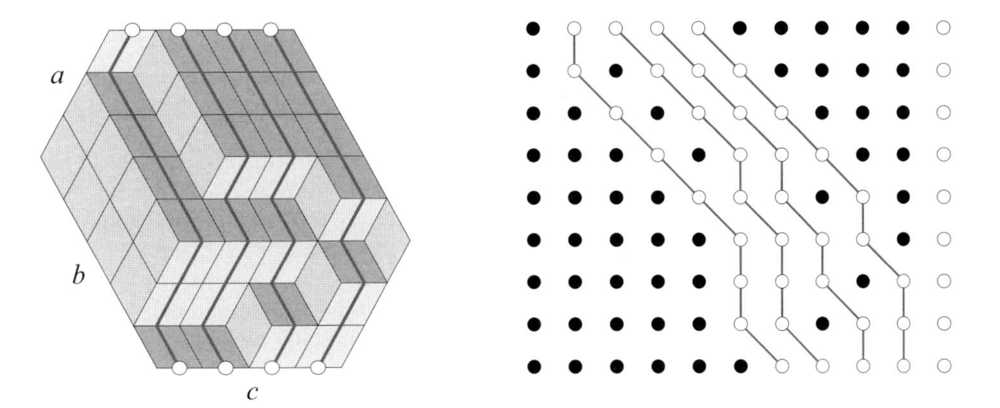

Fig. 17.5 NILPs corresponding to a plane partition.

where $b \times c$ is the rectangular Young diagram with height b and width c. In particular, the unweighted enumeration is the dimension of the Young diagram $b \times c$ as a $GL(a+b)$ representation:

$$N_{a,b,c} = \prod_{i=1}^{a} \prod_{j=1}^{b} \prod_{k=1}^{c} \frac{i+j+k-1}{i+j+k-2},$$

which is the celebrated MacMahon formula. But the more general formula provides various refinements. For example, one can assign a weight q to each cube in the 3D picture. It is left as an exercise to show that this is achieved by setting $x_i = q^{-i}$ (up to a global power of q). This way, we find the q-deformed formula

$$N_{a,b,c}(q) = \prod_{i=1}^{a} \prod_{j=1}^{b} \prod_{k=1}^{c} \frac{1-q^{i+j+k-1}}{1-q^{i+j+k-2}}.$$

Many more formulas can be obtained in this formalism. The reader may, for example, prove that

$$N_{a,b,c} = \sum_{\lambda:\lambda_1 \le c} s_\lambda(\underbrace{1,\ldots,1}_{a}) s_\lambda(\underbrace{1,\ldots,1}_{b})$$

or that

$$N_{a,b,c} = \det(1 + T_{c\times b} T_{b\times a} T_{a\times c})$$

(where $T_{y\times x}$ is a matrix with y rows and x columns and entries $\binom{i}{j}$, $i = 0,\ldots,y-1$, $j = 0,\ldots,x-1$), and also investigate their possible refinements. (For more formulas similar to the last one, see [19].)

Note that our description in terms of paths clearly breaks the threefold symmetry of the original hexagon. It strongly suggests that one should be able to introduce *three* series of parameters $x_1,\ldots,x_{a+b}$, $y_1,\ldots,y_{b+c}$, $z_1,\ldots,z_{a+b}$ to provide an even more refined counting of plane partitions. With two sets of parameters, this is in fact known in the combinatorial literature and is related to so-called double Schur functions (see for example [41] and references therein), which are essentially a form of supersymmetric Schur functions. The full three-parameter generalization is less well-known and appears in [59], making use of triple Schur functions (see [38] for a definition of multiple Schur functions).

Plane partitions exhibit interesting phenomena in the continuum limit, including spatial phase separation [8]; these are discussed in Kenyon's lectures.

Remark: as the name suggests, plane partitions are higher-dimensional versions of partitions, that is, of Young diagrams. After all, each slice that we have used to define our NILPs is also a Young diagram itself. However, these Young diagrams should not be confused with the ones obtained from the NILPs by the correspondence of Section 17.1.3.

Totally symmetric self-complementary plane partitions. Many more complicated enumeration problems have been addressed in the mathematical literature (see [53]). In particular, consider lozenge tilings of a hexagon of shape $2a \times 2a \times 2a$. We note that there is a group of transformations acting naturally on the set of configurations. We consider here the dihedral group of order 12, which consists of rotations by $\pi/3$ and reflections with respect to axes passing through opposite corners of the hexagon or through the centers of opposite edges. With each of its subgroups one can associate an enumeration problem.

Here we discuss only the case of maximal symmetry, i.e. the enumeration of plane partitions with dihedral symmetry. These are called in this case totally symmetric self-complementary plane partitions (TSSCPPs). The fundamental domain is a twelfth of the hexagon (see Fig. 17.6). Inside this fundamental domain, one can use the equivalence to NILPs by considering light gray and dark gray lozenges. However, it is clear that the resulting NILPs are not of the same type as those considered before for general plane partitions, for two reasons: (a) the starting and ending points are not on parallel lines, and (b) the endpoints are in fact free to lie anywhere on a vertical line. However, the LGV formula still holds. For future purposes, we provide below an integral formula for the counting of TSSCPPs [18], where a weight τ is attached to every dark gray lozenge in the fundamental domain.

Let us call the location of the endpoint of the j-th path r_j, numbered from top to bottom

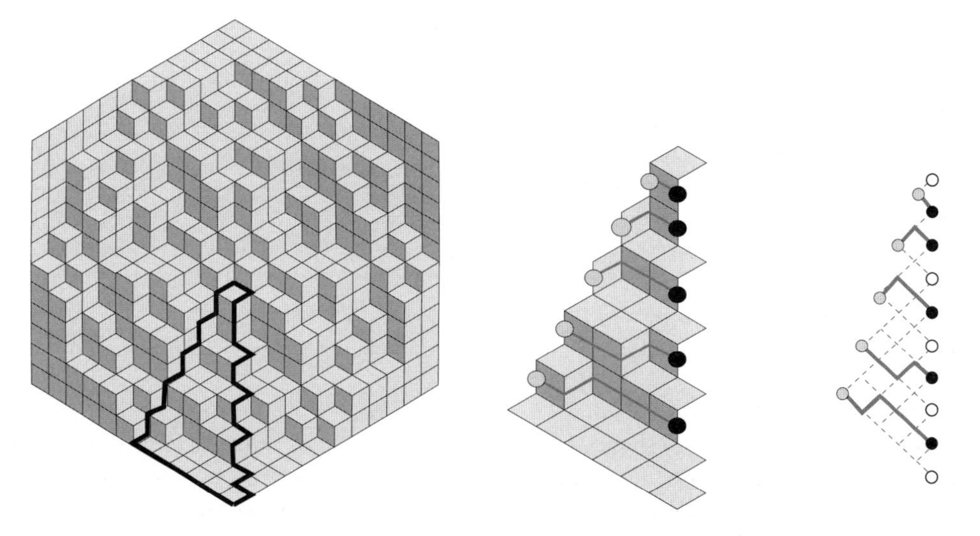

Fig. 17.6 A TSSCPP and the associated NILP.

starting at zero. We first apply the LGV formula to write the number of NILPs with given endpoints as $\det(N_{i,r_j})_{1\le i,j\le n-1}$, where $N_{i,r} = \tau^{2i-r-1}\binom{i}{2i-r-1} = (1+\tau u)^i|_{u^{2i-r-1}}$. Next we sum over them and obtain

$$\begin{aligned} N_n(\tau) &= \sum_{0\le r_1<r_2<\cdots<r_{n-1}} \det[(1+\tau u_i)^i u_i^{r_j}]\big|_{\prod_{i=1}^{n-1} u_i^{2i-1}} \\ &= \prod_{i=1}^{n-1}(1+\tau u_i)^i \sum_{0\le r_1<r_2<\cdots<r_{n-1}} \det(u_i^{r_j})\big|_{\prod_{i=1}^{n-1} u_i^{2i-1}}\,. \end{aligned} \tag{17.27}$$

We recognize the numerator of a Schur function; the summation is simply over all Young diagrams with n parts. At this stage, we use a classical summation formula to conclude that

$$N_n(\tau) = \prod_{1\le i<j\le n-1} \frac{u_j - u_i}{1-u_i u_j} \prod_{i=1}^{n-1} \frac{(1+\tau u_i)^i}{1-u_i}\Big|_{\prod_{i=1}^{n-1} u_i^{2i-1}}\,. \tag{17.28}$$

This formula can be used to generate these numbers efficiently by computer; we find

$$N_n(1) = 1, 2, 7, 42, 429\ldots,$$

which have only small prime factors; this allows us to conjecture a simple product form,

$$N_n(1) = \prod_{i=0}^{n-1} \frac{(3i+1)!}{(n+i)!} = \frac{1!4!\ldots(3n-2)!}{n!(n+1)!\ldots(2n-1)!}\,,$$

which was in fact proven in [2]. We shall later provide an actual derivation of this evaluation. See Fig. 17.7 for the first few TSSCPPs.

More general dynamics. The dynamics that we have used here are rather trivial (the simplest rules for hopping, and translational invariance). Many 1D discrete models turn out to be equivalent to free fermions, with possibly more general Hamiltonians. An example that has become popular is the totally asymmetric exclusion process (TASEP) (see Majumdar's lectures). Another possible variation is to consider particles on a half-line, which corresponds to a neutral fermion; see, for example, [27, 55].

However, the most interesting models usually involve interactions between fermions, and the methods presented here no longer apply. These models, however, often remain quantum integrable—typically, large classes of fermionic models with four-fermion interactions are integrable. The six-vertex model described in the next section can in fact be formulated in such a way. The general partial asymmetric exclusion process (PASEP) [14, 21, 51] is also of this type.

17.1.5 Classical integrability

The free-fermionic Fock space is also important for the construction of solutions of *classically integrable* hierarchies. We refer to [30] and references therein for details. Here we shall only mention in passing the basic idea, since lack of time prevented this from being discussed during the lectures. Recall the isomorphism $\Phi \mapsto \langle \ell | \, e H[t] \, |\Phi\rangle$ from $\mathcal{F}_\ell$ to the space of polynomials in the variables t_q (or, equivalently, to the space of symmetric functions if the t_q

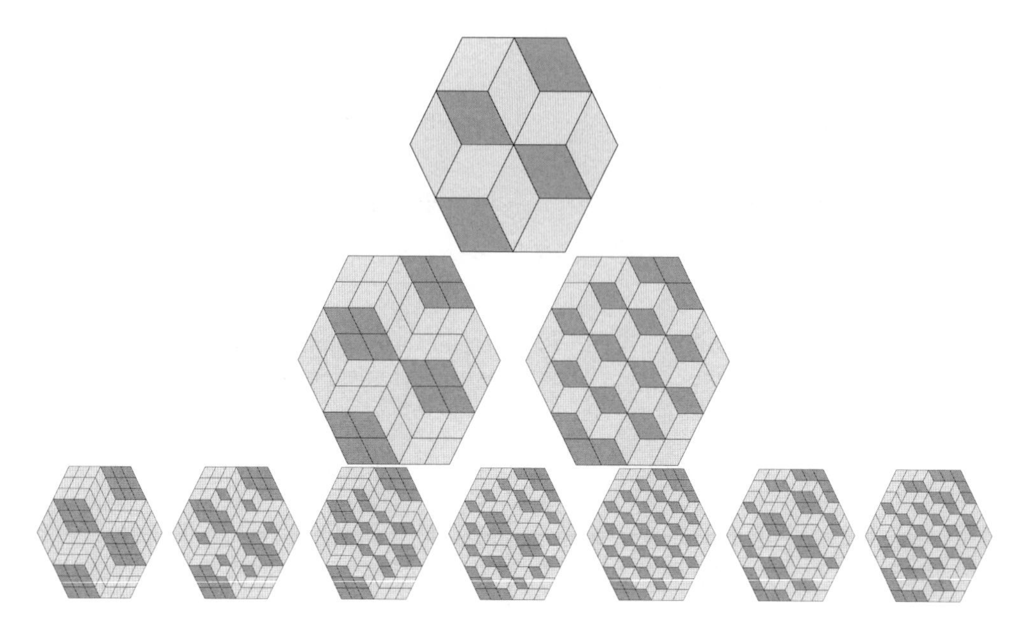

Fig. 17.7 All TSSCPPs of size 1, 2, 3.

are interpreted as power sums). The resulting symmetric function will be a τ-*function* of the Kadomtsev–Petiashvili (KP) hierarchy (as a function of the t_q) for appropriately chosen $|\Phi\rangle$. By "appropriately chosen" we mean the following.

In the first-quantized picture, the essential property of free fermions is the possibility to write their wave function as a Slater determinant; this amounts to considering states which are exterior products of one-particle states. Geometrically, this is interpreted as saying that the state (defined up to multiplication by a scalar) really lives in a subspace of the full Hilbert space called a Grassmannian. The equations defining this space (Plücker relations) are quadratic; these equations are differential equations satisfied by $\langle \ell | \, \epsilon H[t] \, |\Phi\rangle$. They are Hirota's form of the equations defining the KP hierarchy.

17.2 The six-vertex model

The six-vertex model was presented in great detail in Reshetikhin's lectures. Here we provide only a minimum information on this model.

17.2.1 Definition

Configurations. The six-vertex model is defined on a (subset of the) square lattice by putting arrows (with two possible directions) on each edge of the lattice, with the additional rule that at each vertex, there are as many incoming arrows as outgoing ones. See Fig. 17.8 for an example, and for an alternative description ("square ice"). Around a given vertex, there are only six configurations of edges which respect the arrow conservation rule (see Fig. 17.9), hence the name of the model.

Weights. Let us consider Boltzmann weights that are invariant under reversal of every arrow. The weights are assigned to the six vertices and are traditionally called a, b, c (see Fig. 17.9).

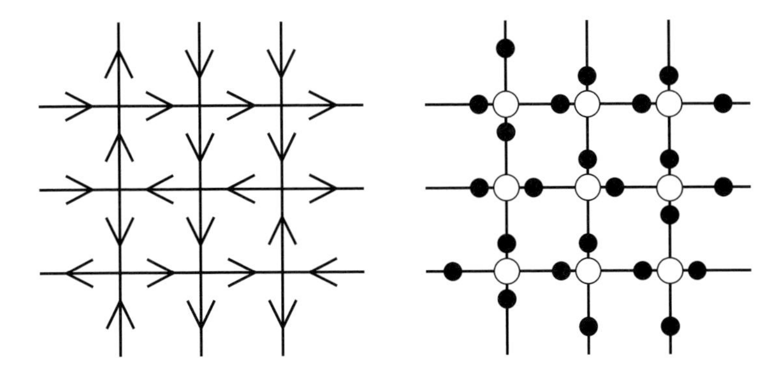

Fig. 17.8 A configuration of the six-vertex model.

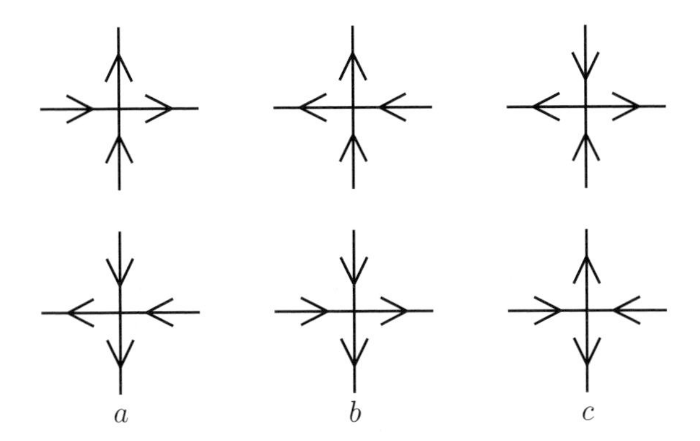

Fig. 17.9 Weights of the six-vertex model.

Thus the partition function is given by

$$Z = \sum_{\text{configurations}} \prod_{\text{vertex}} (\text{weight of the vertex})\,.$$

An additional remark is useful. With any fixed boundary conditions, one can show that the difference between the numbers of vertices of the two types c is constant (independent of the configuration). This means that one can actually give different weights c_1 and c_2 to them: only the product $c^2 = c_1 c_2$ will matter. This will be used in what follows.

There is another way to formulate the partition function, using a transfer matrix. In order to set up a transfer matrix formalism, we first need to specify the boundary conditions. Let us consider doubly periodic boundary conditions in the two directions of the lattice, so that the model is defined on a lattice of size $M \times L$ with the topology of a torus. Then one can write

$$Z = \operatorname{tr} T_L^M\,,$$

where T_L is the $2^L \times 2^L$ transfer matrix which corresponds to a periodic strip of size L. In other words, the indices of the matrix T_L are sequences of L up/down arrows. T_L can itself be expressed as a product of matrices which encode the vertex weights; in the case of integrable models, we usually denote this matrix by the letter R:

$$R = \begin{array}{c} \\ \rightarrow\uparrow \\ \rightarrow\downarrow \\ \leftarrow\uparrow \\ \leftarrow\downarrow \end{array} \begin{array}{c} \begin{array}{cccc} \rightarrow\uparrow & \rightarrow\downarrow & \leftarrow\uparrow & \leftarrow\downarrow \end{array} \\ \begin{pmatrix} a & 0 & 0 & 0 \\ 0 & b & c & 0 \\ 0 & c & b & 0 \\ 0 & 0 & 0 & a \end{pmatrix} \end{array} . \tag{17.29}$$

Then we have

$$T_L = \mathrm{tr}_0\, R_{0L} \cdots R_{02} R_{01} = \cdots \quad \cdots , \tag{17.30}$$

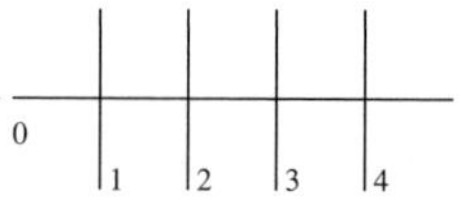

where R_{ij} means the matrix R acting on the tensor product of the i-th and j-th spaces, and 0 is an additional auxiliary space encoding the horizontal edges, as in the picture (note that the trace is over the auxiliary space and means graphically that the horizontal line reconnects with itself). In the picture, "time" flows upwards and to the right.

17.2.2 Integrability

Properties of the R-matrix. Let us now introduce the following parameterization of the weights:

$$\begin{aligned} a &= qx - q^{-1}x^{-1} , \\ b &= x - x^{-1} , \\ c &= q - q^{-1} ; \end{aligned} \tag{17.31}$$

x, q are enough to parameterize them up to a global scaling. Instead of q, one often uses

$$\Delta = \frac{a^2 + b^2 - c^2}{2ab} = \frac{q + q^{-1}}{2} .$$

In general, q or Δ is fixed whereas x is a variable parameter, called the spectral parameter. It can itself be thought of as a ratio of two spectral parameters attached to the lines crossing at a vertex.

The matrix $R(x)$ then satisfies the following remarkable identity (the Yang–Baxter equation):

$$R_{12}(x_2/x_1) R_{13}(x_3/x_1) R_{23}(x_3/x_2) = R_{23}(x_3/x_2) R_{13}(x_3/x_1) R_{12}(x_2/x_1) ,$$

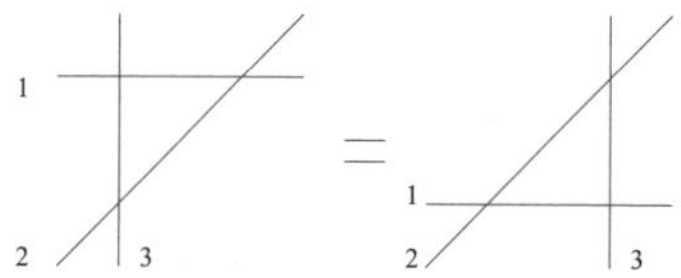

This is formally the same equation that is satisfied by S-matrices in an integrable field theory (field theory with factorized scattering, i.e. such that every S-matrix is a product of two-body S matrices).

The R-matrix also satisfies the unitarity equation

$$R_{12}(x)R_{21}(x^{-1}) = (qx - q^{-1}x^{-1})(qx^{-1} - q^{-1}x)I \,,$$

with $x = x_2/x_1$. The scalar function could, of course, be absorbed by appropriate normalization of R.

Commuting transfer matrices. Now consider the transfer matrix as a function of the spectral parameter x:

$$T_L(x) = \mathrm{tr}_0 \, R_{0L}(x) \cdots R_{02}(x) R_{01}(x) \,.$$

Then, using the Yang–Baxter equation repeatedly, we obtain the relation

$$[T_L(x), T_L(x')] = 0 \,.$$

We thus have an infinite family of commuting operators. In practice, for a finite chain, $T_L(x)$ is a Laurent polynomial in x, so there is a finite number of independent operators.

Note that we could have used the more general *inhomogeneous* transfer matrix

$$T_L(x_0; x_1, \ldots, x_L) = \mathrm{tr}_0 \, R_{0L}(y_L/x_0) \cdots R_{02}(y_2/x_0) R_{01}(y_1/x_0) \,,$$

where now we have spectral parameters y_i attached to each vertical line i and one more parameter x_0 attached to an auxiliary line. The same commutation relations then hold for fixed y_i and variable x_0.

Bethe ansatz. We shall not discuss the Bethe ansatz here, since it will be developed in other lectures. Roughly, the (algebraic) Bethe ansatz [22, 23] consists in considering states Ψ of the form shown in Fig. 17.10, and choosing appropriately the parameters $x_1, \ldots, x_k$ in such a way that Ψ is an eigenvector of the transfer matrix. This turns out to be equivalent to imposing some algebraic equations (Bethe equations) on the x_i.

In the case of the largest eigenvalue of the transfer matrix, one can solve the Bethe equations exactly in the limit where the size of the system goes to infinity. This gives access to the bulk free energy, which allows one to describe the phase diagram.

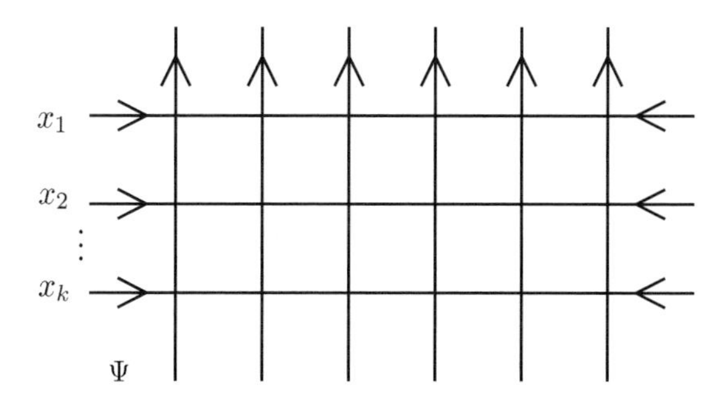

Fig. 17.10 A Bethe state.

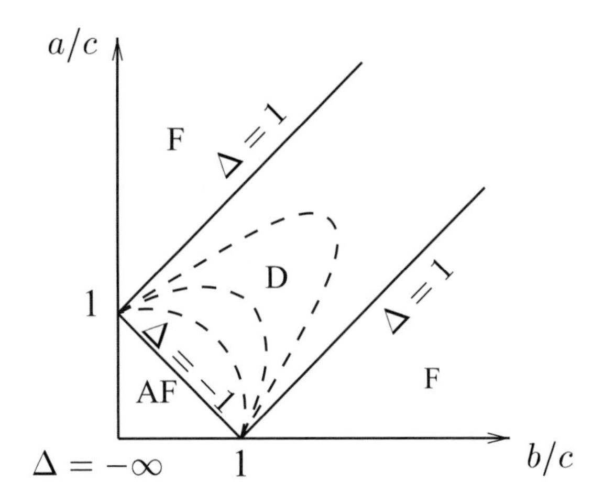

Fig. 17.11 Phase diagram of the six-vertex model.

17.2.3 Phase diagram

We find the following phase diagram, based on the exact solution of the model. The physical properties of the system depend only on $\Delta = (q + q^{-1})/2$, x playing the role of a lattice anisotropy parameter. We distinguish three phases (see Fig. 17.11):

- $\Delta \geq 1$: *ferroelectric phase.* This phase is noncritical. Furthermore, there are no local degrees of freedom: the system is frozen into regions filled with one of the vertices of type a or b (i.e. all arrows are aligned), and no local changes (that respect arrow conservation) are possible.
- $\Delta < -1$: *antiferroelectric phase.* This phase is also noncritical. This time, there is a finite correlation length. The ground state of the transfer matrix corresponds to a state with zero polarization (in the limit $\Delta \to -\infty$, it is simply an alternation of up and down arrows).
- $-1 \leq \Delta < 1$: *disordered phase.* This phase is critical. It possesses a continuum limit with conformal symmetry, and the limiting infrared conformal field theory is well known: it is simply the $c = 1$ theory of a free boson on a circle of radius R given by

$$R^2 = \frac{1}{2(1-\gamma/\pi)}, \qquad \Delta = -\cos\gamma\,.$$

  In other words, it is a bosonic field $\varphi(z, \bar{z})$ with $\varphi \equiv \varphi + 2\pi R$ and action $(1/2\pi) \int d^2z\, \partial\varphi\bar{\partial}\varphi$. The primary operators for the underlying chiral algebras are the electromagnetic vertex operators $\epsilon i((n/R)\varphi + mR\tilde{\varphi})$, with conformal weight $\Delta = (1/8)m^2R^2 + (1/2)(n^2/R^2)$. These are discussed in detail in Nienhuis's lectures, and we shall not go further in this direction.

Free-fermion point. Inside the disordered phase, there is a special point $\Delta = 0$. By combining the equivalences given in [32, 58], one can provide a NILP representation of the six-vertex model at $\Delta = 0$, thus showing it is a system of free fermions. This is described on Fig. 17.12. Note that the correspondence is not one-to-one: one of the c vertices corresponds to two possible local paths.

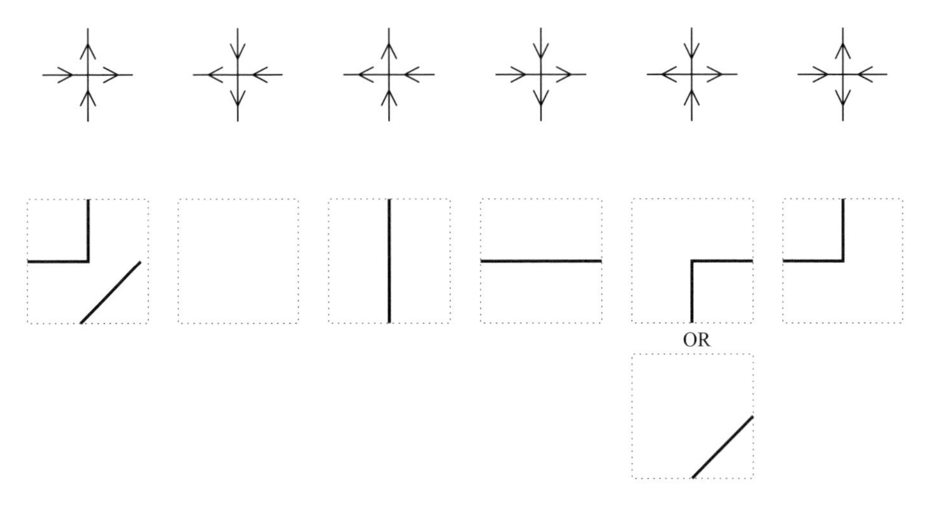

Fig. 17.12 From the $\Delta = 0$ six-vertex model to NILPs.

The directed graph of the NILPs is the basic pattern (with edge weights α, β, γ, ϵ, δ) repeated, with paths moving upwards and to the right, and with the weights indicated on the edges. Comparing the weights, we get the relations

$$\begin{aligned} a &= \alpha\beta\delta = 1\,, \\ b &= \alpha\gamma = \beta\epsilon\,, \\ c_1 &= \epsilon\gamma + \delta\,, \\ c_2 &= \alpha\beta\,. \end{aligned}$$

By combining these, we find that $a^2 + b^2 - c_1c_2 = 0$, so the correspondence only makes sense at $\Delta = 0$.

17.2.4 Equivalence to loop models

Here we follow the terminology of Nienhuis's lectures concerning loop models.

Completely packed loops. In order to go from the six-vertex model to completely packed loops (CPLs), one can first transform the six-vertex model into a height model and then apply the general spin-model–loop-model correspondence à la Pasquier [44]. Since this approach was emphasized by Nienhuis, we shall use here another (strictly equivalent) route. An example is shown in Fig. 17.13.

We start from a CPL configuration. We can introduce a local weight u for one of the two types of CPL vertices, say NE/SW loops. Furthermore, the (unoriented) loops carry a weight n. A convenient way to make the latter weight also local is to turn unoriented loops into *oriented loops*: each configuration is now expanded into $2^{\#\,\text{loops}}$ configurations, with every

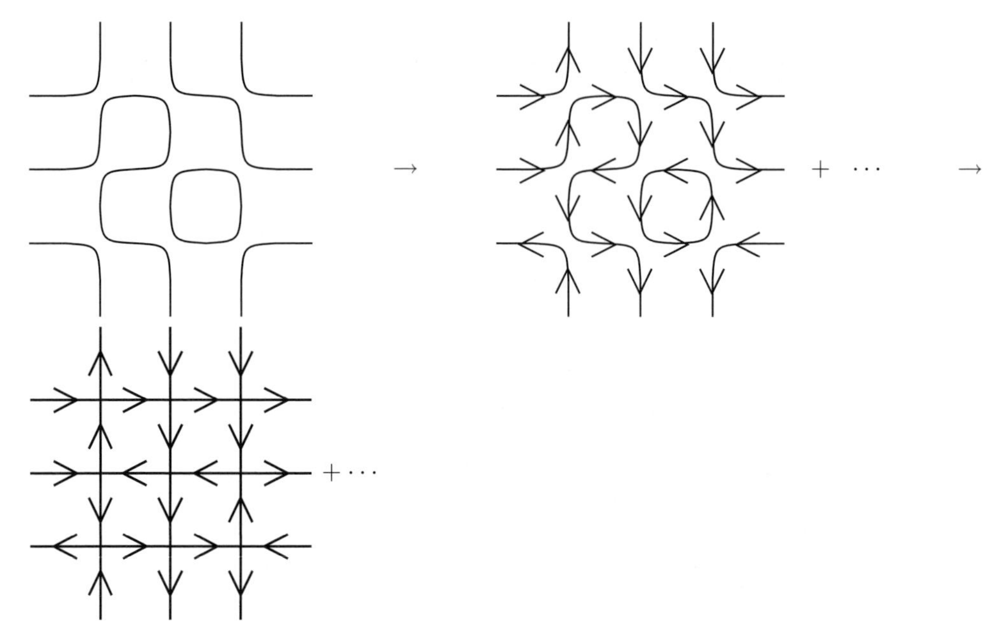

Fig. 17.13 From CPLs to the six-vertex model.

possible orientation of the loops. The weight of a 90 degree turn is chosen to be $\omega^{\pm 1/4}$, where $n = \omega + \omega^{-1}$.

Finally, we forget about the original loops, retaining only the arrows. We note that arrow conservation is automatically satisfied around each vertex: we thus obtain one of the six vertex configurations:

$a =$ … $=$ … $= u\,,$ $\qquad b =$ … $=$ … $= 1\,,$

$c_1 =$ … $+$ … $= u\,\omega^{1/2} + \omega^{-1/2}\,,$ $\qquad c_2 =$ … $+$ … $= \omega^{1/2} + u\,\omega^{-1/2}\,.$

Note that if $u = 1$, all weights become rotationally invariant and $a = b$, $c_1 = c_2$.

We can then check that the formula $\Delta = -n/2$ holds (or, equivalently, $q = -\omega$), where u plays the role of a spectral parameter. In particular, the critical phase $|\Delta| < 1$ corresponds to $|n| < 2$.

Remark: this construction works only in the plane. On a cylinder or on a torus, we have a problem: there are noncontractible loops which, according to the prescription above, get a weight of 2. We shall not discuss here how to correct this (see Section 17.3.1); we simply note that this explains the discrepancy in the central charges between six-vertex model ($c = 1$) and the CPL model ($c < 1$).

Fully packed loops. There is a more limited relation to the model of fully packed loops (FPLs). The limitation comes from the fact that one cannot assign an actual weight to the

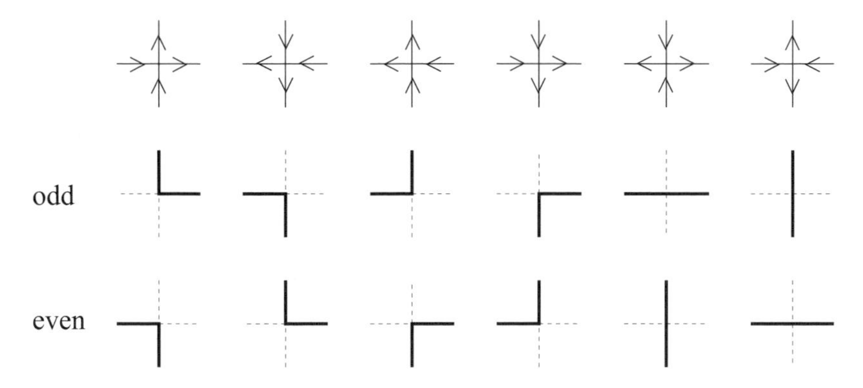

Fig. 17.14 From the six-vertex model to FPLs.

loops, so that we only obtain an $n = 1$ model. This time, the correspondence between configurations is one-to-one: starting from the six-vertex model side, we impose the condition that at every vertex, arrows pointing in the same direction should be in the same state (occupied or empty) on the FPL side. This forces us to distinguish between odd and even sublattices, and leads to the rules of Fig. 17.14.

For rotational invariance of the FPL weights, we must have $a = b$. c/a then has the meaning of a "rigidity" parameter of the loops.

17.2.5 Domain wall boundary conditions

Domain wall boundary conditions (DWBCs) are special boundary conditions which were originally introduced in order to study correlation functions of the six-vertex model [35]. However, they are also interesting in their own right.

Definition. DWBCs are defined on an $n \times n$ square grid: all external edges of the grid are fixed according to the rule that vertical edges are outgoing and horizontal edges are incoming. An example is given in Fig. 17.15.

With each horizontal and vertical line we associate a spectral parameter x_i and y_j, respectively. The partition function is thus

$$Z_n(x_1, \ldots, x_n; y_1, \ldots, y_n) = \sum_{\text{configurations}} \prod_{i,j=1}^{n} w(y_j/x_i)\,,$$

where $w = a, b, c$ depending on the type of vertex.

Korepin's recurrence relations. In [35], a way to compute Z_n inductively was proposed. This way is based on the following properties:

- $Z_1 = q - q^{-1}$.
- $Z_n(x_1, \ldots, x_n; y_1, \ldots .y_n)$ is a symmetric function of the $\{x_i\}$ and of the $\{y_i\}$ (separately). This is a consequence of the Yang–Baxter equation:

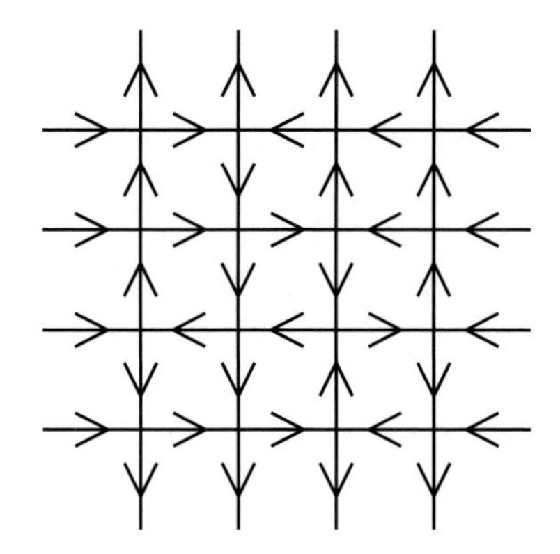

Fig. 17.15 An example of a configuration with domain wall boundary conditions.

$$(qy_{i+1}/y_i - q^{-1}y_i/y_{i+1})Z_n(\ldots, y_i, y_{i+1}, \ldots)$$

$$= (qy_{i+1}/y_i - q^{-1}y_i/y_{i+1})$$

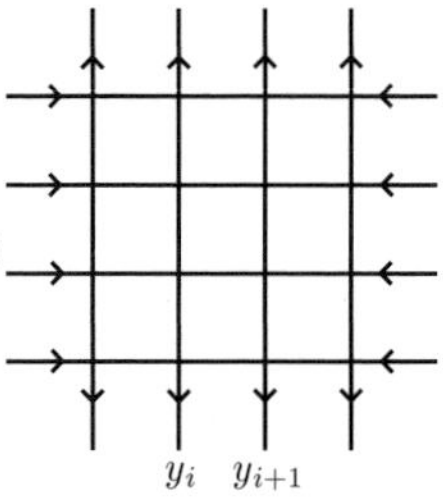

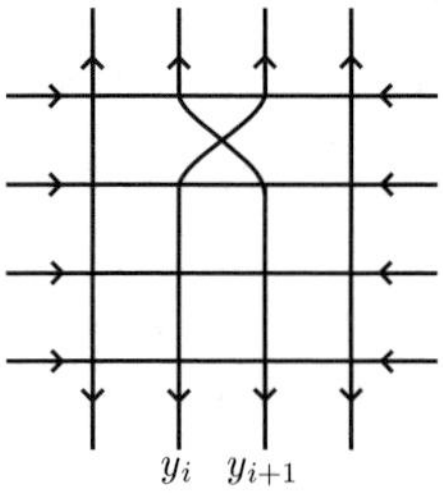

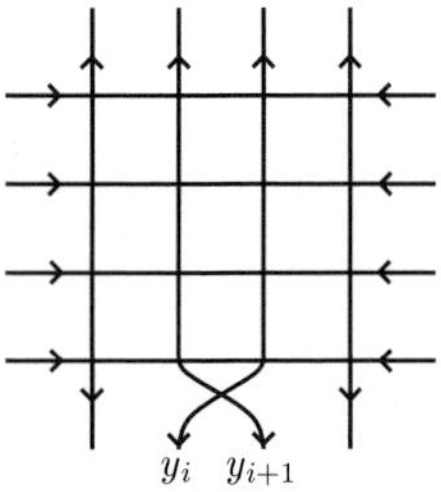

$y_i \quad y_{i+1}$

$$= \qquad = \qquad = \cdots =$$

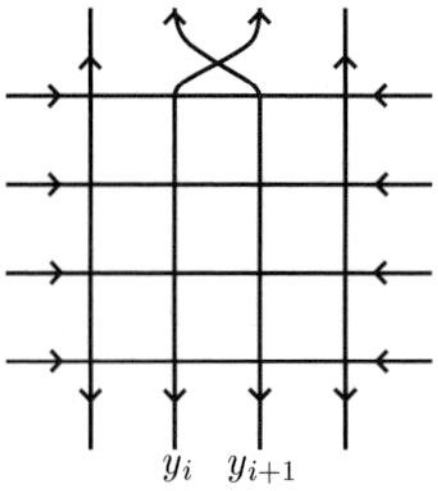

$y_i \quad y_{i+1}$ $\qquad$ $y_i \quad y_{i+1}$ $\qquad$ $y_i \quad y_{i+1}$

$$= (qy_{i+1}/y_i - q^{-1}y_i/y_{i+1})$$

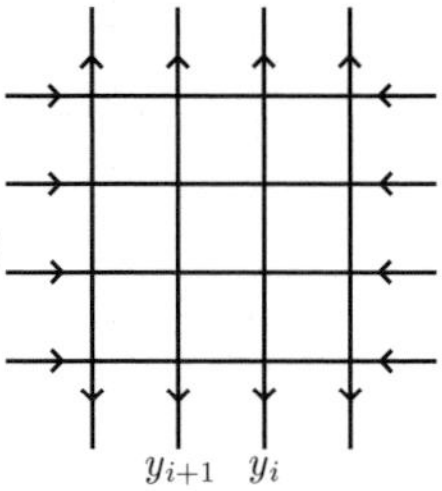

$y_{i+1} \quad y_i$

$$= (qy_{i+1}/y_i - q^{-1}y_i/y_{i+1})Z_n(\ldots, y_{i+1}, y_i, \ldots)$$

and similarly for the x_i.

- Z_n multiplied by x_i^{n-1} or y_i^{n-1} is a polynomial of degree at most $n-1$ in each variable x_i^2 or y_i^2, respectively. This is because (i) each variable, say x_i, appears only in row i; (ii)

a and b are linear combinations of x_i^{-1} and x_i, and c is a constant; and (iii) there is at least one vertex of type c on each row/column.

- The Z_n obey the following recursion relation:

$$
\begin{aligned}
Z_n(x_1, \ldots, x_n; y_1 = x_1, \ldots, y_n) \\
&= (q - q^{-1}) \prod_{i=2}^{n} \left(\frac{q x_1}{x_i} - \frac{q^{-1} x_i}{x_1} \right) \\
&\times \prod_{j=2}^{n} \left(\frac{q y_j}{x_1} - \frac{q^{-1} x_1}{y_j} \right) Z_{n-1}(x_2, \ldots, x_n; y_2, \ldots, y_n) \,. \qquad (17.32)
\end{aligned}
$$

Since $y_1 = x_1$ implies $b(y_1/x_1) = 0$, by inspection all configurations with nonzero weights are of the form shown in Fig. 17.16. This results in the identity.

Note that by the symmetry property, eqn (17.32) fixes Z_n at n distinct values of y_1: x_i, $i = 1, \ldots, n$. Since Z_n is of degree $n - 1$ in y_1^2, it is entirely determined by this equation.

Izergin's formula. Remarkably, there is a closed expression for Z_n due to Izergin [28, 29]. It is a determinant formula,

$$
\begin{aligned}
Z_n &= \frac{\prod_{i,j=1}^{n} (x_j/y_i - y_i/x_j)(q x_j/y_i - q^{-1} y_i/x_j)}{\prod_{1 \le i < j \le n} (x_i/x_j - x_j/x_i)(y_i/y_j - y_j/y_i)} \\
&\times \det \left(\frac{q - q^{-1}}{(x_j/y_i - y_i/x_j)(q x_j/y_i - q^{-1} y_i/x_j)} \right)_{i,j=1,\ldots,n} \,. \qquad (17.33)
\end{aligned}
$$

The hard part is finding the formula, but once it has been found, it is a simple check to prove that it satisfies all the properties of the previous section and, in particular, the recurrence relations (left as an exercise).

Relation to classical integrability and random matrices. The Izergin determinant formula is curious because it involves a simple determinant, which reminds us of free-fermionic models.

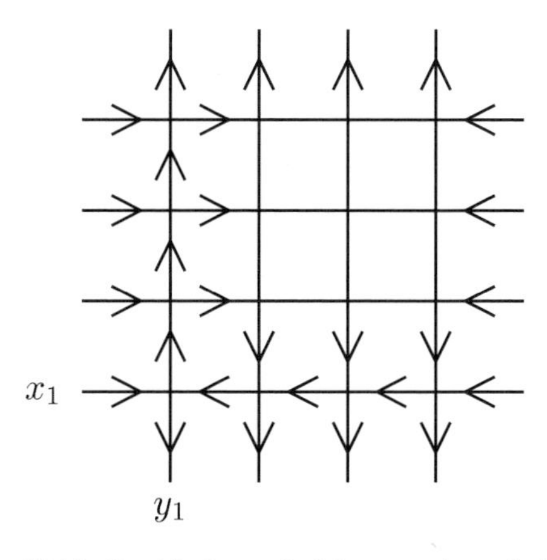

Fig. 17.16 Graphical proof of the recursion relation.

And, indeed, it turns out that it can be written in terms of free fermions or, equivalently, that it provides a solution to a hierarchy of classically integrable partial differential equations, in the present case the two-dimensional Toda lattice hierarchy. Owing to lack of time, this was not discussed during the lectures, and we refer to [58].

Thermodynamic limit. The six-vertex model suffers from a strong dependence on boundary conditions owing to the constraints imposed by arrow conservation. In particular, there is no thermodynamic limit in the usual sense (i.e. independently of boundary conditions). This was observed in [36, 58], where the bulk free energy with DWBCs was computed and found to be different from that with periodic boundary conditions. In [60], it was suggested that the six-vertex model undergoes spatial phase separation, similarly to the dimer models discussed in Kenyon's lectures. This was motivated by some numerical evidence, as well as by the exact result at the free-fermion point $\Delta = 0$, at which the arctic circle theorem [31] applies: the boundary between the phases is known exactly to be an ellipse (a circle for $a = b$) tangent to the four sides of the square.

Since then, there has been a considerable amount of work in this area. There has been more numerical work [1]. The results in [36] have been proven rigorously and extended using sophisticated machinery in the series of papers [4–6] by Bleher and others. Finally, the curve separating the phases has been studied in a series of papers by Colomo and Pronko [9, 10], and recently these authors proposed equations for this curve in the cases $a = b$, $\Delta = \pm 1/2$ [11].

Application: Alternating-sign matrices. Alternating-sign matrices are another important object in enumerative combinatorics. They are defined as follows. An alternating-sign matrix (ASM) is a square matrix made of 0s, 1s and -1s such that if 0s are ignored, 1s and -1s alternate in each row and column, starting and ending with 1s. For example,

$$\begin{matrix} 0 & 1 & 0 & 0 \\ 0 & 0 & 1 & 0 \\ 1 & 0 & -1 & 1 \\ 0 & 0 & 1 & 0 \end{matrix}$$

is an ASM of size 4. The enumeration of ASMs is a famous problem with a long history (see [7]). Here we simply note that ASMs are in fact in bijection with six-vertex model configurations with DWBCs [37]. The correspondence is quite simple and is summarized in Fig. 17.17. For example, Fig. 17.15 becomes the 4×4 ASM above.

We can therefore reinterpret the partition function of the six-vertex model with DWBCs as a weighted enumeration of ASMs. It it natural to set the weights of all zeros to be equal ($a = b$), which leaves us with only one parameter c/a, the weight of ± 1. In fact, here we shall

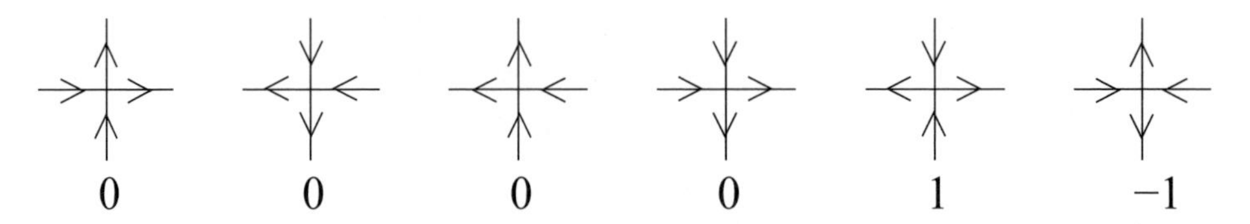

Fig. 17.17 From the six-vertex model to ASMs.

consider only the pure enumeration problem, that is, the problem with all weights equal. We thus compute $\Delta = 1/2$ and $q = \epsilon i\pi/3$, and then $x_i = q$, $y_j = 1$ so that the three weights are $w(x_i/y_j) = q - q^{-1}$.

At this stage, there are several options. We might try to evaluate directly the formula (17.33); since the determinant vanishes in the homogeneous limit where all the x_i or y_j coincide, this is a somewhat involved computation and is the content of Kuperberg's paper [37].

There is, however, a much easier way, discovered independently by Stroganov [54] and Okada [42]. It consists in identifying Z_n *at* $q = \epsilon i\pi/3$ with a Schur function. Consider the partition $\lambda^{(n)} = (n-1, n-1, n-2, n-2, \ldots, 1, 1)$, that is, the Young diagram

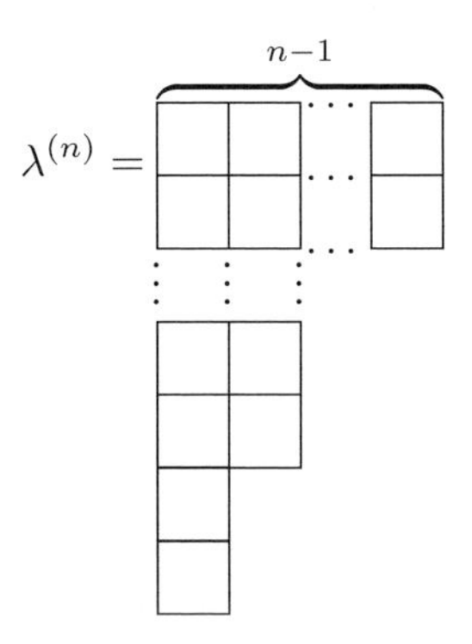

In this case, $s_{\lambda^{(n)}}(z_1, \ldots, z_{2n})$ is a polynomial of degree at most $n-1$ in each z_i (use eqn (17.25)) and satisfies the following:

$$s_{\lambda^{(n)}}(z_1, \ldots, z_j = q^{-2} z_i, \ldots, z_n) = \prod_{\substack{k=1 \\ k \neq i,j}}^{2n} (z_k - q^2 z_i) s_{\lambda^{(n-1)}}(z_1, \ldots, \hat{z}_i, \ldots, \hat{z}_j, \ldots, z_{2n}),$$

where the hats mean that these variables are skipped (start from eqn (17.14), find all the zeros as $z_j = q^2 z_i$, and then set $z_i = z_j = 0$ to find what is left).

This looks similar to the recursion relation (17.32). After appropriate identification, we find

$$\begin{aligned} Z_n(x_1, \ldots, x_n; y_1, \ldots, y_n)\big|_{q=\epsilon i\pi/3} &= (-1)^{n(n-1)/2}(q - q^{-1})^n \prod_{i=1}^{n} (q x_i y_i)^{-(n-1)} \\ &\quad \times s_{\lambda^{(n)}}(q^2 x_1^2, \ldots, q^2 x_n^2, y_1^2, \ldots, y_n^2). \end{aligned}$$

Note that Z_n possesses, at the point $q = \epsilon i\pi/3$, an enhanced symmetry in the whole set of variables $\{q x_1, \ldots, q x_n, y_1, \ldots, y_n\}$. Finally, setting $x_i = q^{-1}$ and $y_j = 1$ and remembering that this will give a weight of $(q - q^{-1})^{n^2}$ to each ASM, we conclude that the number of ASMs is given by

$$A_n = 3^{-n(n-1)/2} s_{\lambda^{(n)}}(\underbrace{1, \ldots, 1}_{2n}) = 3^{-n(n-1)/2} \prod_{1 \le i < j \le 2n} \frac{\lambda_i^{(n)} - i - \lambda_j^{(n)} + j}{j - i}.$$

Simplifying the product results in

$$A_n = \prod_{i=0}^{n-1} \frac{(3i+1)!}{(n+i)!} = 1, 2, 7, 42, 429 \ldots, \tag{17.34}$$

which is a sequence of numbers that we have encountered before! In fact, the first proof of the formula (17.34), due to Zeilberger [56], amounts to showing (nonbijectively) that the number of ASMs is the same as the number of TSSCPPs.

The following are the ASMs of size 1, 2, 3:

$$1$$

$$\begin{matrix}1&0\\0&1\end{matrix} \qquad \begin{matrix}0&1\\1&0\end{matrix}$$

$$\begin{matrix}1&0&0\\0&1&0\\0&0&1\end{matrix} \qquad \begin{matrix}1&0&0\\0&0&1\\0&1&0\end{matrix} \qquad \begin{matrix}0&1&0\\1&0&0\\0&0&1\end{matrix} \qquad \begin{matrix}0&1&0\\0&0&1\\1&0&0\end{matrix} \qquad \begin{matrix}0&1&0\\1&-1&1\\0&1&0\end{matrix} \qquad \begin{matrix}0&0&1\\1&0&0\\0&1&0\end{matrix} \qquad \begin{matrix}0&0&1\\0&1&0\\1&0&0\end{matrix}$$

17.3 Razumov–Stroganov conjecture

17.3.1 Some boundary observables for loop models

Here we return to a model which has already been mentioned (CPLs), but with some specific boundary conditions which will play an important role, since the observables that we shall compute live at the boundary. Several geometries are possible and lead to interesting combinatorial results [12, 20], but here we consider only the case of a cylinder.

Loop model on a cylinder. We consider the model of completely packed loops on a semi-infinite cylinder with a finite, even number of sites $L = 2n$ around the cylinder (see Fig. 17.18). It it convenient to draw the dual square lattice of that of the vertices, so that the cylinder is divided into *plaquettes*. Each plaquette can contain one of the two drawings  and 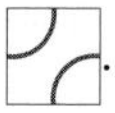 .

We now set $n = 1$; that is, we do not put any weights on the loops. There are no more nonlocal weights, and in fact each plaquette is independent of the other plaquettes. So we can reformulate this model as a purely probabilistic model, in which each plaquette is drawn independently at random, with, say, probability p for  and $1 - p$ for 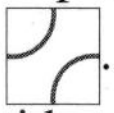 .

Finally, we define the observables that we are interested in. We consider the *connectivity* of the boundary points, i.e. the endpoints of loops (which in this case are not loops but paths) lying on the bottom circle. We encode them into connectivity patterns that are called *link patterns* in the literature. In the present context, they can be visualized as follows. We project the cylinder onto a disk in such a way that the boundaries coincide and the infinity is somewhere inside the disk. We remove all loops except the boundary paths. Up to deformation of these resulting paths, what we obtain is a noncrossing pairing of $2n$ points on the boundary of the disk. This is what we call a link pattern of size $2n$; we denote their set by P_n. (Exercise: show that the number of such link patterns is $c_n = (2n)!(n!(n+1)!)$, called the Catalan number.) So what we are computing in this model is simply the probability of occurrence of each link pattern. These probabilities can be encoded as one vector with c_n entries

$$|\Psi_L\rangle = \sum_{\pi \in P_n} \Psi_\pi \, |\pi\rangle \,,$$

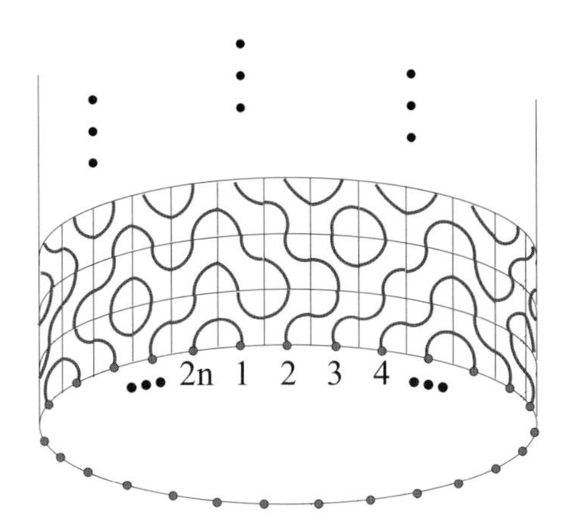

Fig. 17.18 The CPL model on a cylinder.

where P_n is the set of link patterns of size $2n$ and Ψ_π is the probability of link pattern π.

Markov process on link patterns. We now show that $|\Psi_L\rangle$ can be reinterpreted as the steady state of a Markov process on link patterns. This is easily understood by considering a transfer matrix formulation of the model. As before, we introduce the transfer matrix that creates one extra row on the semi-infinite cylinder, the important point being that the transfer matrix encodes not the actual plaquettes but the effect of the new plaquettes on the connectivity of the endpoints: that is, we define $T_{\pi,\pi'}(p)$ to be the probability that, starting from a configuration of the cylinder whose endpoints are connected via the link pattern π' and then adding a row of plaquettes, one obtains a new configuration whose endpoints are connected via the link pattern π. This form a $c_n \times c_n$ matrix $T_L(p)$.

This transfer matrix is actually stochastic in the sense that

$$\sum_{\pi \in P_n} T_{\pi,\pi'}(p) = 1 \qquad \forall \pi' , \tag{17.35}$$

which expresses the conservation of probability. This is of course a special feature of the transfer matrix at $n = 1$. Note that eqn (17.35) says that $T_L(p)^T$ has an eigenvector $(1, \dots, 1)$ with eigenvalue 1.

The matrix $T_L(p)$ has non-negative entries; it is easy to show that it is primitive (the entries of $T_L(p)^n$ are positive). These are the hypotheses of the Perron–Frobenius theorem. Therefore, $T_L(p)$ possesses a unique eigenvector $|\Psi_L\rangle$ with positive entries; the corresponding eigenvalue is positive and is larger in modulus than all other eigenvalues. Now the theorem also applies to $T_L(p)^T$ and, by uniqueness, we conclude that the largest eigenvalue of $T_L(p)$ and of $T_L(p)^T$ is 1. In conclusion, we find that the eigenvector with positive entries of $T_L(p)$, which with a bit of foresight we call $|\Psi_L\rangle$ again, satisfies

$$T_L(p) \, |\Psi_L\rangle = |\Psi_L\rangle \, . \tag{17.36}$$

(In fact, the whole reasoning in this paragraph is completely general and applies to any Markov process, $|\Psi_L\rangle$ being, up to normalization, the steady state of the Markov process defined by $T_L(p)$.)

Two more observations are needed. Firstly, eqn (17.36) is clearly satisfied by the vector of probabilities that we defined earlier (the semi-infinite cylinder being invariant under addition of one extra row); it is in fact defined uniquely up to normalization by eqn (17.36). This explains why we have used the same notation.

Secondly, $|\Psi_L\rangle$ is in fact independent of p. In the next section, we obtain this result by going back to the six-vertex model and showing that p plays the role of a spectral parameter, so that $[T_L(p), T_L(p')] = 0$. Another proof will be presented when we discuss the quantum Knizhnik–Zamolodchikov equation.

Equivalence to the six-vertex model revisited: The space of states. We now show that the transfer matrix that we have just defined is essentially the same as the one in Section 17.2.2 for the six-vertex model, up to a change of basis and issues of boundary conditions.

We start from the equivalence described in Section 17.2.4. The basic idea is to orient the loops. So we start from a link pattern and add arrows to each "loop" (pairing of points). Forgetting about the original link pattern, we obtain a collection of $2n$ up or down arrows, which form a state of the six-vertex model in the transfer matrix formalism. To assign weights, it is convenient to think of the points as being on a straight line with the loops emerging perpendicularly: this way, each loop can only acquire a weight of $\omega^{\pm 1/2}$, depending on whether it is moving to the right or to the left. For example, in the case of a size $L = 2n = 4$,

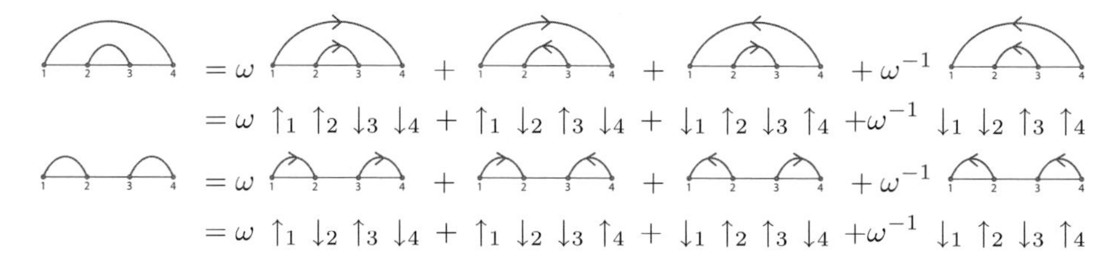

There is only one problem with this correspondence: it is not obviously compatible with periodic boundary conditions. We would like to identify a loop from i to j, $i < j$, and a loop from j to $i + L$, $j < i + L$. This is only possible if we impose *twisted boundary conditions* on the six-vertex model, i.e. we assume that $\uparrow_{i+L} = \omega \uparrow_i$, $\downarrow_{i+L} = \omega^{-1} \downarrow_i$. As was explained in Reshetikhin's lectures, this preserves the integrability of the model.

This mapping from the space of link patterns (of dimension c_n) to that of sequences of arrows (of dimension 2^{2n}) is injective, so the space of link patterns is isomorphic to a certain subspace $\mathbb{C}^{2^{2n}}$. Our claim, which we shall not prove in detail here but which is a natural consequence of the general formalism, is that the transfer matrix of the six-vertex model (defined in Section 17.2.2) with the additional twist above leaves this subspace invariant and is identical to the transfer matrix T_L of our loop model up to this isomorphism, the correspondence of weights being the same as in Section 17.2.4 (in particular, $\Delta = -n/2$).

All that we have said so far in this section in relation to the equivalence to the six-vertex model is valid for any values of the parameters. However, in the end we are only interested in the situation $n = 1$, which is the same as $\Delta = -1/2$ or $q = \epsilon 2\pi i/3$. Note that this is *not* the same value of Δ $(+1/2)$ that was discussed in Section 17.2.5 in relation to ASMs. p now plays the role of the spectral parameter (explicitly, $p = (qx - q^{-1}x^{-1})/(qx^{-1} - q^{-1}x)$). In particular, we conclude that $[T_L(p), T_L(p')] = 0$, as announced.

17.3.2 Properties of the steady state

Some empirical observations. We begin with an example of size $L = 2n = 8$. By brute-force diagonalization of the stochastic matrix T_L, we can obtain the vector $|\Psi_L\rangle$ of probabilities,

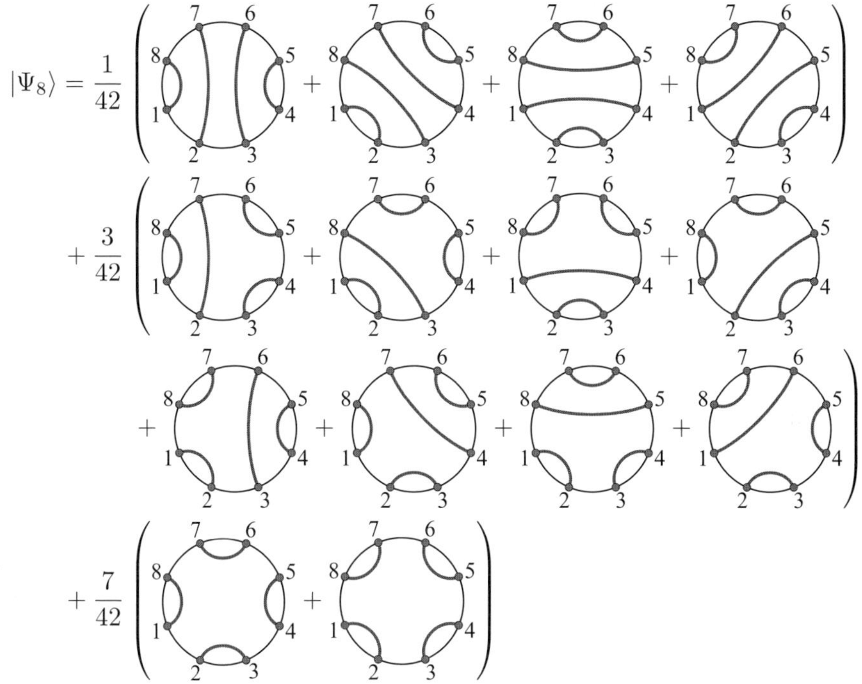

$$|\Psi_8\rangle = \frac{1}{42}\left(\cdots + \cdots + \cdots + \cdots \right) + \frac{3}{42}\left(\cdots + \cdots + \cdots + \cdots + \cdots + \cdots + \cdots + \cdots \right) + \frac{7}{42}\left(\cdots + \cdots \right)$$

We recognize some of our favorite numbers A_n, namely 7 and 42.

In fact, Batchelor, de Gier and Nienhuis [3] conjectured the following properties for all system sizes $L = 2n$:

(1) The smallest probabilities correspond to all pairings being parallel, and are equal to $1/A_n$.

(2) All probabilities are integer multiples of the smallest probability.

(3) The largest probabilities correspond to nearest neighbors being paired, and are equal to A_{n-1}/A_n.

All of these properties have now been proven [17, 18], as will be discussed in Section 17.3.3.

The general conjecture. A question, however, remains: according to property 2 above, if we multiply the probabilities by A_n, we obtain a collection of integers. The smallest is 1 and the largest is A_{n-1}, but what can we say about the others?

Recall that A_n also counts the number of six-vertex model configurations with DWBCs. Furthermore, we showed that there is a one-to-one correspondence between six-vertex model configurations and FPL configurations (see Section 17.2.4). The 42 FPLs of size 4×4 are drawn explicitly in Fig. 17.19.

Note that DWBCs translate into the fact that every other external edge is occupied. Interestingly, we find that the reformulation in terms of FPLs allows us to introduce once again a notion of connectivity. Indeed, there are $2n$ occupied edges on the exterior square and they are

Fig. 17.19 The 42 FPLs of size 4×4.

paired by the FPL model. We can therefore count separately FPLs with a given link pattern π; let us denote the resulting number by A_π. The Razumov–Stroganov conjecture [49] then states that

$$\Psi_\pi = \frac{A_\pi}{A_n},$$

thus relating two different models of loops (CPL and FPL) with completely different boundary conditions. And, even though both models are equivalent to the six-vertex model, the values of Δ are also different (they differ by a sign).

The Razumov–Stroganov (RS) conjecture remains open, although some special cases have been proved (e.g. in [59]).

The relation to the conjectured properties listed previously is as follows. It is easy to show that if π is a link pattern with all pairings parallel, then there exists a unique FPL configuration with connectivity π. Thus the RS conjecture implies property (1). Furthermore, since all A_π are integers, it obviously implies property (2). Property (3), however, remains nontrivial, since even assuming the RS conjecture it amounts to saying that $A_\pi = A_{n-1}$ in the case of the two link patterns π that pair nearest neighbors, which has not been proven.

17.3.3 The quantum Knizhnik–Zamolodchikov equation

We now introduce a new equation whose solution will correspond roughly to a double generalization of the ground state eigenvector $|\Psi_L\rangle$ of the loop model introduced above: (i) it contains inhomogeneities, and (ii) it is a continuation of the original eigenvector to an arbitrary value of q, the original value being $q = \epsilon 2\pi i/3$.

Temperley–Lieb algebra. First we need to define the Temperley–Lieb algebra and its action on the space of link patterns (a vector space with a canonical basis equal to the $|\pi\rangle$ indexed by link patterns).

The Temperley–Lieb algebra of size L with parameter τ is given by generators e_i, $i = 1, \dots, L-1$, and the relations

$$e_i^2 = \tau e_i\,, \qquad e_i e_{i\pm 1} e_i = e_i\,, \qquad e_i e_j = e_j e_i\,, \qquad |i-j| > 1\,.$$

In order to define the action of the Temperley–Lieb generators e_i on link patterns, it is simpler to view them graphically as $e_i =$; then the relations of the Temperley–Lieb algebra and the representation in the space of link patterns become natural graphically. For example, we find

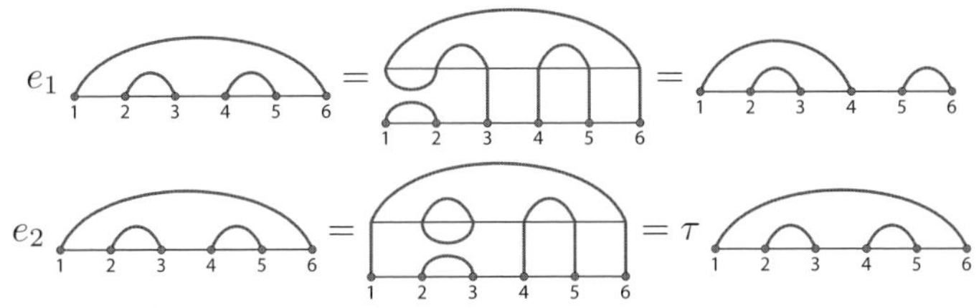

where, for convenience, we have flattened link patterns to pairings inside the upper half-plane of points on a line. The role of the parameter τ is that each time a closed loop is formed, it can be erased at the cost of a multiplication by τ.

In what follows, we set $\tau = -(q + q^{-1})$; q is thus a free parameter.

Definition. We introduce once again the R-matrix, but this time it is rotated by 45 degrees and acts a little differently than before. Namely, it acts on the vector space spanned by link patterns, in the following way:

$$\check{R}_i(z) = \frac{(q^{-1} - qz)I + (1-z)e_i}{q^{-1}z - q}\,.$$

Redrawing e_i slightly as $e_i =$, and similarly $I =$, we recognize the two (rotated) CPL plaquettes. In this section, it is convenient to use spectral parameters z that are the *squares* of our old spectral parameters x. Indeed, using the equivalence to the six-vertex model described in Section 17.3.1, which amounts to the representation

$$e_i = \begin{pmatrix} 0 & 0 & 0 & 0 \\ 0 & -q & 1 & 0 \\ 0 & 1 & -q^{-1} & 0 \\ 0 & 0 & 0 & 0 \end{pmatrix}$$

for the Temperley–Lieb generators (acting on the i-th and $(i+1)$-th spaces), we essentially recover the R-matrix of the six-vertex model after the change of variables $z = x^2$:

$$\check{R}(z) = \frac{1}{qx^{-1} - q^{-1}x} \begin{pmatrix} qx - q^{-1}x^{-1} & 0 & 0 & 0 \\ 0 & (q-q^{-1})x^{-1} & x - x^{-1} & 0 \\ 0 & x - x^{-1} & (q-q^{-1})x & 0 \\ 0 & 0 & 0 & qx - q^{-1}x^{-1} \end{pmatrix}$$

on condition that we perform the following transformations: $\check{R}(z) \propto \mathcal{P}x^{\kappa/2}R(x)x^{-\kappa/2}$, where $\mathcal{P}$ permutes the factors of the tensor product, and $\kappa = \mathrm{diag}(0, 1, -1, 0)$.

Consider now the following system of equations for $|\Psi_L\rangle$, a vector-valued function of the $z_1, \ldots, z_L, q, q^{-1}$ $(i = 1, \ldots, L-1)$:

$$\check{R}_i(z_{i+1}/z_i)\,|\Psi_L(z_1, \ldots, z_L)\rangle = |\Psi_L(z_1, \ldots, z_{i+1}, z_i, \ldots, z_L)\rangle\,, \tag{17.37}$$

$$\rho\,|\Psi_L(z_1, \ldots, z_L)\rangle = c\,|\Psi_L(z_2, \ldots, z_L, sz_1)\rangle\,, \tag{17.38}$$

where ρ rotates link patterns in the following way,

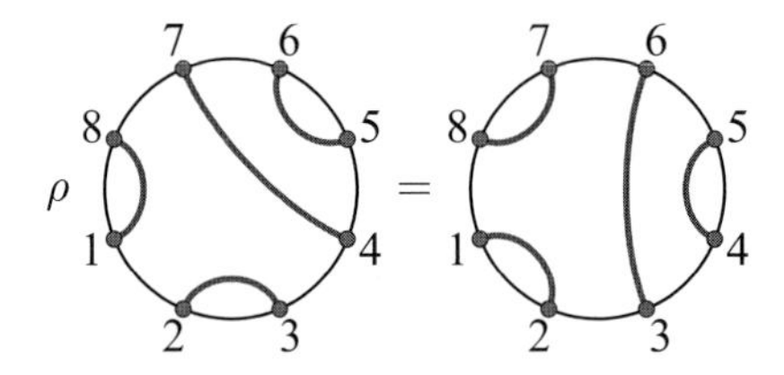

and c is a constant that is needed for homogeneity. s is a parameter of the equation: if we set $s = q^{2(k+\ell)}$ with $k = 2$ (technically, this is the dual Coxeter number of the underlying quantum group), then ℓ is called the *level* of the quantum Knizhnik–Zamolodchikov (qKZ) equation.

This equation first appeared in [52] in a study of form factors in integrable models. It is not what is usually called the quantum Knizhnik–Zamolodchikov equation; the latter was introduced in [25] as a q-deformation of the Knizhnik–Zamolodchikov (KZ) equation (the qKZ equation is to quantum affine algebras what the KZ equation is to affine algebras). The qKZ equation is of the form

$$S_i(z_1, \ldots, z_L)\,|\Psi_L\rangle\,(z_1, \ldots, z_i, \ldots, z_L) = |\Psi_L\rangle\,(z_1, \ldots, sz_i, \ldots, z_L)\,, \tag{17.39}$$

where S_i can be defined pictorially as

$$S_i = \cdots \qquad \cdots\,,$$

i

and where an empty box is just the "face" graphical representation of the R-matrix (dual to the "vertex" representation that we used for the six-vertex model):

$$= \frac{q^{-1} - qz}{q^{-1}z - q} \qquad + \frac{1 - z}{q^{-1}z - q} \qquad .$$

The spectral parameters z to be used in S_i are as follows: for the box numbered j, z_j/z_i if $j > i$ or $z_j/(sz_i)$ if $j < i$. Loosely, S_i is the "scattering matrix for the i-th particle."

Alternatively, S_i can be expressed as a product of $\check{R}_i$ and ρ:

$$S_i = \check{R}_{i-1}(z_{i-1}/(sz_i)) \cdots \check{R}_2(z_2/(sz_i))\check{R}_1(z_1/(sz_i))\rho\check{R}_{L-1}(z_L/z_i)\cdots \\ \times \check{R}_{i+1}(z_{i+2}/z_i)\check{R}_i(z_{i+1}/z_i)\,.$$

It is a simple exercise to check, using this expression, that the system (17.37) and (17.38) implies the qKZ equation (17.39). Naively, the converse is untrue. However, one can show that, up to some transformations, a solution of eqn (17.39) can always be reduced to a solution of eqns (17.37) and (17.38) (the details are beyond the scope of these lectures).

In what follows, we shall only be interested in *polynomial* solutions of the system (17.37) and (17.38).

Relation to the affine Hecke algebra. There is an equivalent point of view, which was advocated by Pasquier [45] in the context of the Razumov–Stroganov conjecture.

We start from the qKZ system (eqns (17.37) and (17.38)) and rewrite it in such a way that the action on the finite-dimensional part (on the space of link patterns) is separated from the action on the variables (on the space of polynomials of L variables). Equation (17.38) is actually already of this form; eqn (17.37) needs to be rewritten slightly:

$$(qz_i - q^{-1}z_{i+1})\partial_i\,|\Psi_L\rangle = (e_i + q + q^{-1})\,|\Psi_L\rangle\,, \tag{17.40}$$

where $\partial_i \equiv (1/(z_{i+1}-z_i))(\tau_i - 1)$, and τ_i is the operator that switches variables z_i and z_{i+1}, so that the left-hand side acts only on the polynomial part of $|\Psi_L\rangle$, whereas the right-hand side acts only on link patterns.

The operators $t_i = (q^{-1}z_{i+1} - qz_i)\partial_i$ acting on polynomials (we must check that t_i acting on a polynomial produces a polynomial) form a representation of the *Hecke algebra* (with parameter $-\tau$); in other words, they satisfy the relations

$$t_i^2 = -\tau t_i\,, \qquad t_i t_{i+1} t_i - t_i = t_{i+1} t_i t_{i+1} - t_{i+1}\,, \qquad t_i t_j = t_j t_i\,, \quad |i-j|>1\,.$$

Equivalently, $t_i + \tau$ satisfies the same relations, with $-\tau$ replaced with τ. Note the important fact that the Temperley–Lieb algebra is a quotient of the Hecke algebra (it is easy to check that the e_i satisfy all the relations of the Hecke algebra).

We can add an extra operator on the space of polynomials—the one that appears in the right-hand side of eqn (17.38), namely the cyclic shift r of spectral parameters $z_1 \mapsto z_2 \mapsto \cdots \mapsto z_L \mapsto sz_1$. The t_i together with r generate a representation of the *affine Hecke algebra*.

We can now interpret eqns (17.38)–(17.40) as follows: we have on the one hand a representation of the affine Hecke algebra on the space of link patterns (with generators e_i and ρ), and on the other hand a representation of the same algebra on polynomials of L variables (the $t_i + \tau$ and r). $|\Psi_L\rangle$ provides a bridge between these two representations; it is essentially an invariant object in the tensor product of the two; that is, it provides a subrepresentation of the space of polynomials (explicitly, the span of the Ψ_π) which is isomorphic to the dual of the (irreducible) representation on the link patterns.

So, the search for polynomial solutions of eqns (17.37) and (17.38) is equivalent to finding irreducible subrepresentations of the action of the affine Hecke algebra on the space of polynomials.

Remark: the direct relation between the qKZ system and representations of an appropriate affine algebra works only for the A_n series of algebras, i.e. the affine Hecke algebra. For more complicated situations such as the BWM algebra, it fails because one cannot separate the two different actions [46].

Polynomial solution. On general grounds, we expect only polynomial solutions for integer values of the level. Here we shall need only a solution at level 1, that is $s = q^6$. Note that the KZ equation at level 1 is essentially connected to free fermions (coming in two species), so that what we shall produce is essentially a q-deformed version of free fermions (the other difference being that we use the basis of link patterns and not the usual basis of the six-vertex model). By comparison with free-fermionic formulas, we expect this solution to be of degree $n(n-1)$.

We shall build this solution in several steps. First, we use a "nice" property of our basis of link patterns, that is, the fact that eqn (17.40) can be written as a *triangular* linear system in the components of $|\Psi_L\rangle$. This requires us to define an order on link patterns, which is most conveniently described as follows. We draw link patterns once again as pairings of points on a line and consider the operation described in Fig. 17.20. This gives a bijection between link patterns of size $2n$ and Young diagrams inside the staircase diagram $(n-1, n-2, \ldots, 1)$. Then the order is that of inclusion of Young diagrams. The smallest element, corresponding to the empty Young diagram, is denoted by π_0; it connects i and $L+1-i$ (note that it is one of the link patterns with all pairings parallel, which correspond to the smallest probability, $1/A_n$, in the loop model). We consider now the exchange equation (17.40) and write it in components; we find two possibilities:

- i and $i+1$ are not paired. Then we find that eqn (17.40) involves only Ψ_π, and implies that $qz_i - q^{-1}z_{i+1}$ divides Ψ_π and, furthermore, $\Psi_\pi/(qz_i - q^{-1}z_{i+1})$ is symmetric under exchange of z_i and z_{i+1}.
- i and $i+1$ are paired. Then

$$(qz_i - q^{-1}z_{i+1})\partial_i\Psi_\pi = \sum_{\pi' \neq \pi, e_i \cdot \pi' = \pi} \Psi_{\pi'};$$

  that is, eqn (17.40) involves a sum over preimages of π by e_i viewed as acting on the set of link patterns. It turns out there are two types of preimages of a given π: in terms of Young diagrams, there is the Young diagram obtained from π by adding one box at $i, i+1$ (which is always possible unless π is the largest element), and there are other Young diagrams that are included in π. So we can write the equation

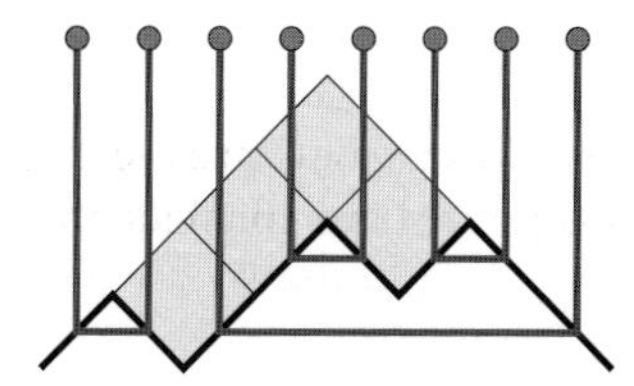

Fig. 17.20 From a Young diagram to a link pattern: in this example, from the partition $(2,1,1)$ to the pairings $(1,2)$, $(3,8)$, $(4,5)$, $(6,7)$.

$$\Psi_{\pi+\text{one box at }(i,i+1)} = (qz_i - q^{-1}z_{i+1})\partial_i \Psi_\pi - \sum_{\substack{\pi' \subset \pi \\ e_i \cdot \pi' = \pi}} \Psi_{\pi'} ,$$

which has the desired triangular structure and allows us to build the Ψ_π one by one by adding boxes to the corresponding Young diagram. However, there is no equation for Ψ_{π_0}. In fact, this triangular system can be explicitly solved [34] (see also [13]) in the sense that every Ψ_π can be written as a series of operators acting on Ψ_{π_0}. We shall not need this here.

From the discussion above, we find that all we need is to fix Ψ_{π_0}. We use the following simple observation, which generalizes the first case in the dichotomy above: *if there are no pairings between points $i, i+1, \dots, j$ in π, then $\prod_{i\le p<q\le j}(qz_p - q^{-1}z_q)$ divides Ψ_π.* (One can prove this by induction on $j - i$.)

In the case of π_0, we find $n(n-1)$ factors, which exhausts the expected degree. We therefore make the minimality assumption that Ψ_{π_0} is just

$$\Psi_{\pi_0} = \prod_{1\le i<j\le n} (qz_i - q^{-1}z_j) \prod_{n+1\le i<j\le 2n} (q^{-1}z_j - qz_i) , \tag{17.41}$$

where we recall that the system size is $L = 2n$.

It remains a nontrivial fact that with such a choice of Ψ_{π_0}, eqn (17.38) is satisfied, with $s = q^6$. We refer to [16, 45] for details.

Connection to the loop model. In general, the two problems of diagonalizing the transfer matrix and finding solutions of the qKZ equation are unrelated. However, there is exactly one value of q where a solution of the qKZ equation does in fact provide an eigenvector of the transfer matrix. This is when the parameter s is equal to 1, which occurs here when $q = \epsilon 2\pi i/3$ (other sixth roots of unity are possible but they either are trivial or give the same result as the one that we have picked). Note that in this case eqn (17.38) becomes a simple rotational-invariance condition. Furthermore, the real qKZ equation (17.39) becomes an eigenvector equation for the scattering matrices:

$$S_i(z_1,\dots,z_L)\,|\Psi_L(z_1,\dots,z_L)\rangle = |\Psi_L(z_1,\dots,z_L)\rangle .$$

These scattering matrices do not involve any extra shifts of the spectral parameters, and, as is well known in the case of Bethe ansatz, they are just specializations of the inhomogeneous transfer matrix. Indeed, if we define $T_L(z; z_1,\dots,z_L)$ to be simply

$$T_L = z \; \boxed{}_{z_1}\boxed{}_{z_2} \cdots \boxed{}_{z_L}$$

(with periodic boundary conditions), we can then observe that $S_i(z_1,\dots,z_L) = T(z_i; z_1,\dots, z_L)$. By a Lagrange interpolation argument, we conclude that

$$T_L(z; z_1,\dots,z_L)\,|\Psi_L(z_1,\dots,z_L)\rangle = |\Psi_L(z_1,\dots,z_L)\rangle ,$$

i.e. $|\Psi_L\rangle$ is, up to normalization, the steady state of the inhomogeneous Markov process defined by $T_L(z; z_1,\dots,z_L)$. In order to recover the original homogeneous Markov process, we simply set all $z_i = 1$.

Note that the normalization is fixed by specifying the value of Ψ_{π_0}; in particular, if all $z_i = 1$, using eqn (17.41) we find that $\Psi_{\pi_0} = 3^{n(n-1)/2}$, to be compared with the (conjectured) probability $1/A_n$ associated with π_0. In other words, with this normalization, we should have $\sum_{\pi \in P_n} \Psi_\pi = 3^{n(n-1)/2} A_n$.

Integral formulas. Using the formalism of the qKZ equation allows us to prove the properties discussed in Section 17.3.2, as well as to reconnect the three models that we have found in which the same numbers A_n appear. Recently, a particularly useful tool to exploit these solutions of the qKZ equation has been introduced [18, 24, 50]: it consists in writing integral formulas for them.

Here we shall give only the main points, and let those who are interested read the relevant papers. We concentrate on one single quantity: the sum of all components of Ψ_π. We claim that the following formula holds:

$$\sum_{\pi \in P_n} \Psi_\pi = (-1)^{n(n-1)/2} \prod_{1 \le i < j \le 2n} (q z_i - q^{-1} z_j) \oint \cdots \oint \prod_{\ell=0}^{n-1} \frac{dw_\ell (q w_\ell - z_{2\ell+1})}{2\pi i}$$
$$\times \frac{\prod_{0 \le \ell < m \le n-1} (w_m - w_\ell)(q w_\ell - q^{-1} w_m)}{\prod_{\ell=0}^{n-1} \prod_{1 \le i \le 2\ell+1} (w_\ell - z_i) \prod_{2\ell+1 \le i \le 2n} (q w_\ell - q^{-1} z_i)},$$

where the contours surround the z_i counterclockwise, but not the $q^{-2} z_i$.

At this stage, one can do two things. On the one hand, one can check directly [24] that $\sum_{\pi \in P_n} \Psi_\pi$, at $q = \epsilon 2\pi i/3$, equals the partition function, up to prefactors, of the six-vertex model with DWBCs at $q = \epsilon i\pi/3$ or, equivalently, the Schur function $s_{\lambda^{(n)}}$:

$$\sum_{\pi \in P_n} \Psi_\pi(z_1, \ldots, z_{2n})\big|_{q=\epsilon 2\pi i/3} = s_{\lambda^{(n)}}(z_1, \ldots, z_{2n}).$$

We conclude that the steady-state probability of the link pattern π_0 is, in the homogeneous case $z_i = 1$, $\Psi_{\pi_0}/(\sum_{\pi \in P_n} \Psi_\pi) = 1/A_n$. More generally, we can find the probability of link pattern π_0 for arbitrary inhomogeneities.

On the other hand, one can try to set $z_i = 1$ directly in the integral formula; it is convenient to send the two poles (1 and q^{-2}) to zero and infinity, respectively, by the homographic transformation $u_\ell = (w_\ell - 1)/(q w_\ell - q^{-1})$. We find

$$\frac{1}{\Psi_{\pi_0}} \sum_{\pi \in P_n} \Psi_\pi(1, \ldots, 1) = \oint \cdots \oint \prod_{\ell=0}^{n-1} \frac{(1 + u_\ell) du_\ell}{2\pi i \, u_\ell^{2\ell+1}} \prod_{0 \le \ell < m \le n-1} (u_m - u_\ell)(1 + \tau u_m + u_\ell u_m),$$

where $\tau = -q - q^{-1}$, and the contours surround zero. This can be rewritten

$$\frac{1}{\Psi_{\pi_0}} \sum_{\pi \in P_n} \Psi_\pi(1, \ldots, 1) = \prod_{\ell=0}^{n-1} (1 + u_\ell) \prod_{0 \le \ell < m \le n-1} (u_m - u_\ell)(1 + \tau u_m + u_\ell u_m)\Big|_{\prod_{i=1}^{n-1} u_i^{2i}}.$$

In fact, we can simply set $u_0 = 0$ to get

$$\frac{1}{\Psi_{\pi_0}} \sum_{\pi \in P_n} \Psi_\pi(1, \ldots, 1) = \prod_{\ell=1}^{n-1} (1 + u_\ell) \prod_{1 \le \ell < m \le n-1} (u_m - u_\ell)(1 + \tau u_m + u_\ell u_m)\Big|_{\prod_{i=1}^{n-1} u_i^{2i-1}}, \tag{17.42}$$

which looks strikingly similar to the formula for the weighted enumeration of TSSCPPs (17.28).

There is, however, an important difference. Whereas eqn (17.28) just contains a product of functions of one variable times an antisymmetric function of the u_i (which, ultimately, comes from the free-fermionic nature of the model), eqn (17.42) does not possess any particular symmetry with respect to exchange of its variables, owing to the factors $1 + \tau u_m + u_\ell u_m$ (which come from the q-deformed Vandermonde product in the original integral formula). One can, however, antisymmetrize the formula (17.42) [24, 57], and one then recovers eqn (17.28). The result is a very nontrivial connection which goes beyond the value $q = \epsilon 2\pi i/3$: for generic q, the sum of the components of the solution of the qKZ equation reproduces the weighted enumeration of TSSCPPs with weight $\tau = -q - q^{-1}$, proving a conjecture formulated in [15].

References

[1] D. Allison and N. Reshetikhin, Numerical study of the 6-vertex model with domain wall boundary conditions, arXiv:cond-mat/0502314.

[2] G. E. Andrews, Plane partitions V: The TSSCPP conjecture, *J. Combin. Theory A* **66** (1994), 28–39.

[3] M. T. Batchelor, J. de Gier, and B. Nienhuis, The quantum symmetric XXZ chain at $\Delta = -1/2$, alternating sign matrices and plane partitions, *J. Phys. A* **34** (2001), L265–L270; arXiv:cond-mat/0101385.

[4] P. Bleher and V. Fokin, Exact solution of the six-vertex model with domain wall boundary conditions. Disordered phase, *Commun. Math. Phys.* **268** (2006), 223–284; arXiv: math-ph/0510033.

[5] P. Bleher and K. Liechty, Exact solution of the six-vertex model with domain wall boundary conditions. Ferroelectric phase, arXiv:0712.4091.

[6] P. Bleher and K. Liechty, Exact solution of the six-vertex model with domain wall boundary conditions. Critical line between ferroelectric and disordered phases, arXiv: 0802.0690.

[7] D. Bressoud, *Proofs and Confirmations: The Story of the Alternating Sign Matrix Conjecture*, Mathematical Association of America (1999).

[8] H. Cohn, M. Larsen, and J. Propp, The shape of a typical boxed plane partition, *N. Y. J. Math.* **4** (1998), 137–165; arXiv:math/9801059.

[9] F. Colomo and A. G. Pronko, The arctic circle revisited, arXiv:0704.0362.

[10] F. Colomo and A. G. Pronko, Emptiness formation probability in the domain-wall six-vertex model, arXiv:0712.1524.

[11] F. Colomo and A. G. Pronko, The limit shape of large alternating sign matrices, arXiv: 0803.2697.

[12] J. de Gier, Loops, matchings and alternating-sign matrices, *Discr. Math.* **298** (2005), 365–388; arXiv:math.CO/0211285.

[13] J. de Gier and P. Pyatov, Factorised solutions of Temperley–Lieb qKZ equations on a segment, arXiv:0710.5362.

[14] B. Derrida, M. Evans, V. Hakim, and V. Pasquier, Exact solution of a 1D asymmetric exclusion model using a matrix formulation, *J. Phys. A* **26** (1993), 1493–1517.

[15] P. Di Francesco, Totally symmetric self-complementary plane partitions and quantum Knizhnik–Zamolodchikov equation: A conjecture, *J. Stat. Mech.* (2006) P09008; arXiv: cond-mat/0607499.

[16] P. Di Francesco and P. Zinn-Justin, Quantum Knizhnik–Zamolodchikov equation, generalized Razumov–Stroganov sum rules and extended Joseph polynomials, *J. Phys. A* **38** (2005), L815–L822; arXiv:math-ph/0508059.

[17] P. Di Francesco and P. Zinn-Justin, Around the Razumov–Stroganov conjecture: Proof of a multi-parameter sum rule, *Electron. J. Combin.* **12** (2005) R6; arXiv:math-ph/0410061.

[18] P. Di Francesco and P. Zinn-Justin, Quantum Knizhnik–Zamolodchikov equation, totally symmetric self-complementary plane partitions and alternating sign matrices, *Theor. Math. Phys.* **154**(3) (2008), 331–348; arXiv:math-ph/0703015.

[19] P. Di Francesco, P. Zinn-Justin, and J.-B. Zuber, Determinant formulae for some tiling problems and application to fully packed loops, *Ann. Inst. Fourier (Grenoble)* **55**(6) (2005), 2025–2050; arXiv:math-ph/0410002.

[20] P. Di Francesco, P. Zinn-Justin, and J.-B. Zuber, Sum rules for the ground states of the $O(1)$ loop model on a cylinder and the XXZ spin chain, *J. Stat. Mech.* (2006) P08011; arXiv:math-ph/0603009.

[21] F. Essler and V. Rittenberg, Representations of the quadratic algebra and partially asymmetric diffusion with open boundaries, *J. Phys. A* **29** (1996), 3375–3407; arXiv:cond-mat/9506131.

[22] L. D. Faddeev, How algebraic Bethe ansatz works for integrable model, Les Houches lecture notes (1995), arXiv:hep-th/9605187.

[23] L. D. Faddeev, E. K. Sklyanin, and L. A. Takhtajan, *Theor. Math. Phys.* **40** (1979), 194.

[24] T. Fonseca and P. Zinn-Justin, On the doubly refined enumeration of alternating sign matrices and totally symmetric self-complementary plane partitions, *Electron. J. Combin.*, in press; arXiv:0803.1595.

[25] I. B. Frenkel and Yu. B. Reshetikhin, Quantum affine algebras and holonomic difference equations, *Commun. Math. Phys.* **146** (1992), 1–60; http://projecteuclid.org/euclid.cmp/1104249974.

[26] I. M. Gessel and X. Viennot, Binomial determinants, paths and hook formulae, *Adv. Math.* **58** (1985), 300–321.

[27] J. Harnad and A. Yu. Orlov, Fermionic construction of tau functions and random processes, arXiv:0801.0066.

[28] A. G. Izergin, Partition function of a six-vertex model in a finite volume, *Sov. Phys. Dokl.* **32** (1987), 878–879.

[29] A. G. Izergin, D. A. Coker, and V. E. Korepin, Determinant formula for the six-vertex model, *J. Phys. A* **25** (1992), 4315.

[30] M. Jimbo and T. Miwa, Solitons and infinite dimensional Lie algebras, *Publ. RIMS* **19** (1983), 943–1001.

[31] W. Jockush, J. Propp, and P. Shor, Random domino tilings and the arctic circle theorem, arXiv:math.CO/9801068.

[32] K. Johansson, The arctic circle boundary and the Airy process, *Ann. Probab.* **33** (2005), 1–30.

[33] K. Johansson, Random matrices and determinantal processes, Les Houches lectures, arXiv:math-ph/0510038.

[34] A. Kirillov Jr. and A. Lascoux, Factorization of Kazhdan–Lusztig elements for Grassmanians, *Adv. Stud.* **28** (2000), 143–143; arXiv:math.CO/9902072.

[35] V. E. Korepin, Calculation of norms of Bethe wave functions, *Commun. Math. Phys.* **86** (1982), 391.

[36] V. Korepin and P. Zinn-Justin, Thermodynamic limit of the six-vertex model with domain wall boundary conditions, *J. Phys. A* 33 (2000), 7053–7066; arXiv:cond-mat/0004250.

[37] G. Kuperberg, Another proof of the alternating sign matrix conjecture, *Int. Math. Res. Not.* (1996) 139–150; arXiv:math/9712207.

[38] A. Lascoux, Symmetric functions and combinatorial operators on polynomials, *CBMS 99*, American Mathematical Society (2001).

[39] B. Lindström, On the vector representations of induced matroids, *Bull. London Math. Soc.* **5** (1973), 85–90.

[40] P. MacMahon, *Combinatory Analysis*, Cambridge University Press (1915).

[41] A. I. Molev and B. E. Sagan, A Littlewood–Richardson rule for factorial Schur functions, *Trans. Am. Math. Soc.* **351**(11) (1999), 4429–4443; arXiv:q-alg/9707028.

[42] S. Okada, Enumeration of symmetry classes of alternating sign matrices and characters of classical groups, *J. Algebraic Combin.* **23** (2006), 43–69; arXiv:math/0408234.

[43] A. Okounkov and N. Reshetikhin, Correlation function of Schur process with application to local geometry of a random 3-dimensional Young diagram, *J. Am. Math. Soc.* **16** (2003), 581–603; arXiv:math.CO/0107056.

[44] V. Pasquier, Two-dimensional critical systems labelled by Dynkin diagrams, *Nucl. Phys. B* **285** (1987), 162.

[45] V. Pasquier, Quantum incompressibility and Razumov Stroganov type conjectures, *Ann. Henri Poincaré* **7** (2006), 397–421; arXiv:cond-mat/0506075.

[46] V. Pasquier, Incompressible representations of the Birman–Wenzl–Murakami algebra, arXiv:math/0507364.

[47] A. V. Razumov and Yu. G. Stroganov, Spin chain and combinatorics, *J. Phys. A* **34** (2001), 3185–3190; arXiv:cond-mat/0012141.

[48] A. V. Razumov and Yu. G. Stroganov, Spin chains and combinatorics: Twisted boundary conditions, *J. Phys. A* **34** (2001), 5335–5340; arXiv:cond-mat/0102247.

[49] A. V. Razumov and Yu. G. Stroganov, Combinatorial nature of ground state vector of $O(1)$ loop model, *Theor. Math. Phys.* **138** (2004), 333–337; arXiv:math.CO/0104216.

[50] A. V. Razumov, Yu. G. Stroganov, and P. Zinn-Justin, Polynomial solutions of qKZ equation and ground state of XXZ spin chain at $\Delta = -1/2$, *J. Phys. A* **40**(39) (2007), 11827–11847; arXiv:0704.3542.

[51] G. Schutz and E. Domany, Phase transitions in an exactly soluble one-dimensional exclusion process, *J. Stat. Phys.* **72** (1993), 277; arXiv:cond-mat/9303038.

[52] F. A. Smirnov, A general formula for soliton form factors in the quantum sine–Gordon model, *J. Phys. A* **19** (1986), L575–L578.

[53] R. Stanley, A baker's dozen of conjectures concerning plane partitions, in *Combinatoire Enumérative*, eds. G. Labelle and P. Leroux, Lecture Notes in Mathematics, Vol. 1234, Springer-Verlag (1986), pp. 285–293.

[54] Yu. Stroganov, A new way to deal with Izergin–Korepin determinant at a third root of unity, *Teor. Mat. Fiz.* **1** (2006), 65–76; arXiv:math-ph/0204042.

[55] J. W. van de Leur and A. Yu. Orlov, Random turn walk on a half line with creation of particles at the origin, arXiv:0704.1157.

[56] D. Zeilberger, Proof of the alternating sign matrix conjecture, *Electron. J. Combin.* **3**(2) (1996), R13; arXiv:math/9407211.

[57] D. Zeilberger, Proof of a conjecture of Philippe Di Francesco and Paul Zinn-Justin related to the qKZ equations and to Dave Robbins' two favorite combinatorial objects, http://www.math.rutgers.edu/ zeilberg/mamarim/mamarimhtml/diFrancesco.html, 2007.

[58] P. Zinn-Justin, Six-vertex model with domain wall boundary conditions and one-matrix model, *Phys. Rev. E* (3) (2000), 3411–3418; arXiv:math-ph/0005008.

[59] P. Zinn-Justin, Proof of the Razumov–Stroganov conjecture for some infinite families of link patterns, *Electron. J. Combin.* **13**(1) (2006), R110; arXiv:math/0607183.

[60] P. Zinn-Justin, The influence of boundary conditions in the six-vertex model, arXiv:cond-mat/0205192.

Part III

Seminars

18

A rigorous perspective on Liouville quantum gravity and the KPZ relation

B. Duplantier

Bertrand Duplantier,
Institut de Physique Théorique, CNRS/URA 2306,
Bât. 774, Orme des Merisiers,
CEA/Saclay,
F-91191 Gif-sur-Yvette Cedex,
France.

Dedication

This lecture is dedicated to the memory of ODED SCHRAMM, who died in a tragic hiking accident on September 1, 2008. It describes joint work with SCOTT SHEFFIELD, from the Department of Mathematics at the Massachusetts Institute of Technology, formerly at the Courant Institute for Mathematical Sciences at NYU.

Preface

We present here a (mathematically rigorous) probabilistic and geometrical proof of the KPZ relation between scaling exponents in a Euclidean planar domain D and in Liouville quantum gravity. It uses a properly regularized quantum area measure $d\mu_{\gamma,\varepsilon} = \varepsilon^{\gamma^2/2} e^{\gamma h_\varepsilon(z)} dz$, where dz is the standard 2D Euclidean (i.e. Lebesgue) measure on D, γ is a real parameter, $0 \leq \gamma < 2$, and $h_\varepsilon(z)$ denotes the mean value on a circle of radius ε centered at z of an instance h of the Gaussian free field on D. The proof extends to the boundary geometry. The singular case $\gamma > 2$ is shown to be related to the quantum measure $d\mu_{\gamma'}$, $\gamma' < 2$, by the fundamental duality $\gamma\gamma' = 4$.

This lecture is essentially an expanded version of a recently published Letter [38], cast in a language accessible to theoretical physicists. A more complete and mathematical version can be found in [37].

18.1 Introduction

18.1.1 Historical perspective

One of the major theoretical advances in physics over the past thirty years has been the realization in gauge theory and string theory that transition amplitudes require to be summed over random surfaces, which replaces the traditional sums over random paths, i.e. the celebrated Feynman path integrals of quantum mechanics and quantum field theory. Polyakov [70] first understood that the summation over random Riemannian metrics involved could be represented mathematically by the now celebrated *Liouville theory of quantum gravity*.

The latter can be simply described as follows. Consider a bounded planar domain $D \subset \mathbb{C}$ as the parameter domain of a random Riemannian surface, and consider an instance h of the Gaussian free field (GFF) on D, with Dirichlet energy

$$(h, h)_\nabla := \frac{1}{2\pi} \int_D \nabla h(z) \cdot \nabla h(z)\, dz\,. \tag{18.1}$$

The quantum area is then (formally) defined by $\mathcal{A} = \int_D e^{\gamma h(z)} dz$, where dz is the standard 2D Euclidean (i.e. Lebesgue) measure and $e^{\gamma h(z)}$ is the *random* conformal factor of the Riemannian metric, with a constant $0 \leq \gamma < 2$. The quantum Liouville action is then

$$S(h) = \frac{1}{2}(h, h)_\nabla + \lambda \mathcal{A}\,, \tag{18.2}$$

where $\lambda \geq 0$ is the so-called "cosmological constant." The corresponding Boltzmann–Gibbs weight is thus

$$\exp\left[-S(h)\right] = \exp\left[-\frac{1}{2}(h, h)_\nabla\right] \exp(-\lambda \mathcal{A})\,, \tag{18.3}$$

to be integrated over a "flat" functional measure $\mathcal{D}h$ on h, defined heuristically as a "uniform measure on the space of all functions." (Of course, the latter makes perfect sense if one considers only a finite-dimensional vector space of functions, such as real-valued functions defined on the vertices of a lattice, or functions whose Fourier coefficients beyond a certain frequency threshold are identically zero—in this case $\mathcal{D}h$ would be the Lebesgue measure on the vector space.)

Kazakov introduced the key idea of placing (critical) statistical models on random planar lattices when he solved the Ising model exactly [52]. This anticipated the breakthrough by Knizhnik, Polyakov, and Zamolodchikov in [57], who predicted that corresponding critical exponents (i.e. conformal weights x) of any critical statistical model in the Euclidean plane and in quantum gravity (Δ) would obey the KPZ relation [57]

$$x = \frac{\gamma^2}{4}\Delta^2 + \left(1 - \frac{\gamma^2}{4}\right)\Delta\,. \tag{18.4}$$

The positive inverse of the relation (18.4) is

$$\Delta_\gamma := \frac{1}{\gamma}\left(\sqrt{4x + \left(\frac{2}{\gamma} - \frac{\gamma}{2}\right)^2} - \left(\frac{2}{\gamma} - \frac{\gamma}{2}\right)\right). \tag{18.5}$$

In the critical continuum limit, the statistical system borne by the random lattice is described by a conformal field theory (CFT) with a central charge $c \leq 1$, which *fixes* the value of γ as a function of c [57]:

$$\gamma = \frac{1}{\sqrt{6}}\left(\sqrt{25 - c} - \sqrt{1 - c}\right) \leq 2, \quad c \leq 1\,. \tag{18.6}$$

With this identification, we find the standard form of the KPZ relation,

$$\Delta = \frac{\sqrt{24x + 1 - c} - \sqrt{1 - c}}{\sqrt{25 - c} - \sqrt{1 - c}}\,. \tag{18.7}$$

The first check of the KPZ relation was provided by the independent, earlier result by Kazakov on the Ising model on a random lattice. This fact then strongly suggested that discrete random lattices could provide a regularization of continuum Liouville quantum gravity. (See the recent historical note [73].) Another confirmation soon followed with the resolution of the complete spectrum of a polymer on a randomly triangulated lattice [35]. The original derivation by KPZ used the so-called light-cone gauge. It was followed by a (still formal) derivation in the conformal gauge of Liouville quantum gravity [20, 25].

This provides the core continuous model of "2D quantum gravity," whose deep and manifold connections to string theory, conformal field theory, random planar lattice models, random matrix theory, and stochastic Loewner evolution (SLE) are often still conjectural from a mathematical perspective (see [5, 21, 23, 32, 33, 45, 69, 71, 78, 83] and references therein).

Elaborate techniques have been developed for calculating correlation functions in Liouville field theory [26, 44, 46, 48, 62, 75, 81, 82, 84, 85]. A recent combinatorial approach studies random planar maps via bijections to sets of labeled trees [11, 12, 14, 24, 77]. This approach is well suited to the study of *geodesics* [13, 15–17, 66] and allows a detailed mathematical construction of continuum scaling limits [65, 68] and of higher-genus maps [67].

In the latter case, the associated topological expansion has recently seen significant progress [40, 42, 43].

The KPZ relation has been checked in multiple ways using explicit calculations in geometrical models on random planar lattices [3, 19, 27, 35, 36, 41, 52, 53, 60, 61, 63, 64]. (See also Kostov's chapter in this volume.) 2D quantum gravity and the KPZ relation have also been used to predict the harmonic measure and rotation multifractal spectra for any conformally invariant curve in the plane [28–32, 34].

Despite its great importance for conformal field theory, the KPZ relation (18.4) has never been proven rigorously, nor has its range of validity been properly defined, and not even its geometrical meaning has been fully understood. The first proof appeared in [37]. The aim of this lecture is to present such a proof in a minimal, yet rigorous way, which closely follows [38].

In this geometrical and probabilistic approach, we start from the *critical* Liouville gravity, with the action S (18.2) taken at $\lambda = 0$, i.e. a *free-field* action. The role of the cosmological constant λ in the weight (18.3) is to control the expectation of the whole quantum area; it thus acts as an "infrared" regulator, while the KPZ relation is a local scaling property of the quantum measure, hence an "ultraviolet" phenomenon, and therefore independent of the value of λ.

We define a properly regularized quantum area measure, which allows a transparent probabilistic understanding of the KPZ relation (18.4) for any scaling fractal set in D, as a direct consequence of the underlying *Brownian stochastic properties* of the two-dimensional GFF. We also prove the boundary analogue of the KPZ relation for fractal subsets of the boundary ∂D.

One striking and important consequence of our perspective is that the KPZ relation appears to hold in a much broader context than the original CFT realm, which relates γ to c, i.e. it appears to hold for *any fractal structure as measured with the quantum random measure* $e^{\gamma h(z)}dz$, *and for any* $0 \leq \gamma < 2$. For instance, it predicts that the set of Euclidean exponents x of a random or a self-avoiding walk (a CFT with $c = 0$) obeys eqn (18.4) with $\gamma = \sqrt{8/3}$ in pure gravity ($c = 0$), but also with $\gamma = \sqrt{3}$ on a random lattice equilibrated with Ising spins ($c = 1/2$). This *central charge mixing* yields new KPZ exponents Δ_γ in eqn (18.5) or (18.7), settling theoretically an issue raised earlier but inconclusively in numerical simulations [6, 51].

Note that eqn (18.6) gives only values of the parameter γ in the range $\gamma \leq 2$. Our probabilistic approach also allows us to explain the *duality* property of Liouville quantum gravity: for $\gamma > 2$, the *singular* quantum measure can be properly defined in terms of the *regular* γ' quantum measure, for the dual value $\gamma' = 4/\gamma < 2$, establishing the existence of the so-called "other branch" of the γ-KPZ relation and its correspondence to the standard γ'-KPZ relation for $\gamma' < 2$, as argued long ago in [54–56].

The following approach is essentially based on [38]. An extended mathematical version of that work can be found in [37], along with outstanding open problems relating discrete models and SLE to Liouville quantum gravity, which we hope will be solved by the methods introduced here. Several follow-up works exist, both at the rigorous level [9, 76] and the heuristic one [22].

18.1.2 Quantum measure

For concreteness, let h be an instance of a centered GFF on a bounded simply connected domain D with zero boundary conditions. This means that $h = \sum_n \alpha_n f_n$, where the α_n are i.i.d. zero-mean, unit-variance, normal random variables, and the f_n are an orthonormal basis, with respect to the Dirichlet inner product

$$(f_1, f_2)_\nabla := \frac{1}{2\pi} \int_D \nabla f_1(z) \cdot \nabla f_2(z)\, dz\,,$$

of the Hilbert space closure $H(D)$ of the space $H_s(D)$ of C^∞ real-valued functions compactly supported on D. Although this sum diverges pointwise almost surely, it does converge almost surely in the space of distributions on D, and one can also make sense of the mean value of h on various sets. (See [79] for a detailed account of this construction of the GFF.)

Given an instance h of the Gaussian free field on D, let $h_\varepsilon(z)$ now denote the mean value of h on a circle of radius ε centered at z (where $h(z)$ is defined to be zero for $z \in \mathbb{C} \setminus D$). As we shall see in Section 18.3, this provides a convenient regularization of the Liouville quantum measure, defined as

$$d\mu_{\gamma,\varepsilon} := \varepsilon^{\gamma^2/2} e^{\gamma h_\varepsilon(z)} dz\,, \tag{18.8}$$

where the extra power of ε is required for the limit measure to exist when $\varepsilon \to 0$. We then have the following proposition.

Proposition 18.1 *Fix $\gamma \in [0, 2)$ and define h and D as above. Then it is (almost surely) the case that as $\varepsilon \to 0$ (along powers of two), the measures $d\mu_{\gamma,\varepsilon} = \varepsilon^{\gamma^2/2} e^{\gamma h_\varepsilon(z)} dz$ converge weakly to a limiting measure, which we denote by $d\mu_\gamma := e^{\gamma h(z)} dz$.*

This proposition, rigorously proven in [37], mathematically defines Liouville quantum gravity. This construction will be explained in the next sections.

For each $z \in D$, we now denote by $C(z; D)$ the *conformal radius* of D viewed from z. That is, $C(z; D) = |\phi'(z)|^{-1}$, where $\phi : D \to \mathbb{D}$ is a conformal map to the unit disk with $\phi(z) = 0$. The following gives an equivalent definition of the (regularized) quantum measure $d\mu_{\gamma,\varepsilon}$ (18.8):

$$d\mu_{\gamma,\varepsilon} = \exp\left(\gamma h_\varepsilon(z) - \frac{\gamma^2}{2} \operatorname{Var} h_\varepsilon(z) + \frac{\gamma^2}{2} \log C(z; D)\right) dz\,, \tag{18.9}$$

and the continuum quantum measure $d\mu_\gamma$ is the (weak) limit for $\varepsilon \to 0$ of these measures. This identity will be obtained in Section 18.3 from a simple geometrical analysis of the properties of the GFF, which shows in particular that the variance of the circular average $h_\varepsilon(z)$ is, explicitly,

$$\operatorname{Var} h_\varepsilon(z) = -\log \varepsilon + \log C(z; D)\,.$$

A standard property of the expectation of the exponential of a Gaussian random variable Y is

$$\mathbb{E} \exp Y = \exp\left(\mathbb{E} Y + \frac{1}{2} \operatorname{Var} Y\right),$$

where here $Y := h_\varepsilon(z)$ and $\mathbb{E}Y = \mathbb{E}h_\varepsilon(z) = 0$ (see Section 18.3). The *expectation* of the quantum measure (18.9) therefore reads

$$\mathbb{E}\,\mu_{\gamma,\varepsilon}(A) = \int_A C(z;D)^{\gamma^2/2}dz =: \mathbb{E}\,\mu_\gamma(A)\,,$$

independently of ε, and for each measurable subset $A \subset D$.

Intuitively, we interpret the pair (D, μ_γ) as describing a "random surface" $\mathcal{M}$ parameterized conformally by D, with an area measure given by μ_γ. The term "random metric" is often used as well; however, we stress that, since we have not endowed D with a two-point distance function, "random metric" in the Liouville quantum gravity context does not mean "random metric space," but "random measure." The *fluctuations* of this random measure are at the heart of the KPZ relation, to which we turn now.

18.1.3 Euclidean and quantum scaling exponents

Definition 18.2 *For any fixed measure μ_γ on D (which we call the "quantum" measure), we let $B^\delta(z)$ be a Euclidean ball centered at z whose radius is chosen so that $\mu_\gamma(B^\delta(z)) = \delta$. (If there does not exist a unique δ with this property, we take the radius to be* $\sup\{\varepsilon : \mu_\gamma(B_\varepsilon(z)) \leq \delta\}$.*) We refer to $B^\delta(z)$ as the quantum ball of area δ centered at z. In particular, if $\gamma = 0$ then μ_0 is the Lebesgue measure and $B^\delta(z)$ is $B_\varepsilon(z)$, where $\delta = \pi\varepsilon^2$.*

Given a subset $X \subset D$, we denote the ε neighborhood of X by

$$B_\varepsilon(X) = \{z : B_\varepsilon(z) \cap X \neq \emptyset\}\,.$$

We also define the *quantum δ neighborhood* of X by

$$B^\delta(X) = \{z : B^\delta(z) \cap X \neq \emptyset\}\,.$$

Translated into probability language, the *KPZ formula* is a quadratic relationship between the expectation fractal dimension of a random subset of D defined in terms of the Euclidean measure (which is the Liouville gravity measure with $\gamma = 0$) and the corresponding expectation fractal dimension of X defined in terms of Liouville gravity with $\gamma \neq 0$.

We say that a (deterministic or random) fractal subset X of D has a *Euclidean expectation dimension* $2-2x$ and *Euclidean scaling exponent* x if the expected area of $B_\varepsilon(X)$ decays like $\varepsilon^{2x} = (\varepsilon^2)^x$, i.e.

$$\lim_{\varepsilon\to 0} \frac{\log \mathbb{E}\mu_0(B_\varepsilon(X))}{\log \varepsilon^2} = x\,.$$

In other words, if z is chosen uniformly in D, and independently of X, then the probability of intersection of the Euclidean ball $B_\varepsilon(z)$ and X (as illustrated in Fig. 18.1) scales as

$$\mathbb{P}\{B_\varepsilon(z) \cap X \neq \emptyset\} \asymp \varepsilon^{2x}\,,$$

in the sense that

$$\lim_{\varepsilon\to 0} \frac{\log \mathbb{P}\{B_\varepsilon(z) \cap X \neq \emptyset\}}{\log \varepsilon} = 2x\,.$$

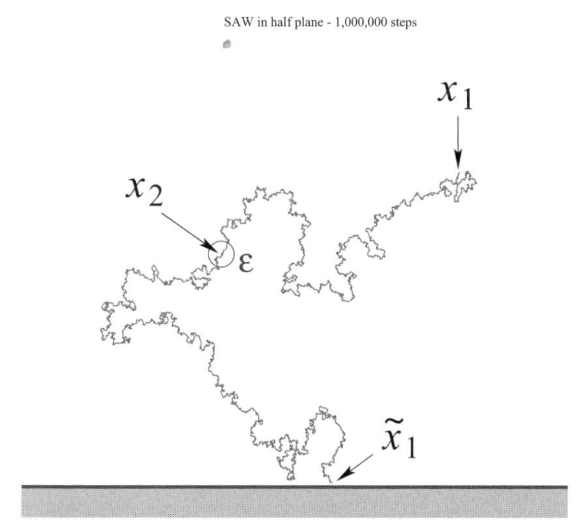

Fig. 18.1 A fractal path X (here a self-avoiding walk) intersecting a Euclidean ball B_ε of radius ε. The intersection probability scales as $\mathbb{P}\{B_\varepsilon(z) \cap X \neq \emptyset\} \asymp \varepsilon^{2x}$, with a fractal exponent $x = x_2$ for typical intersections in the bulk of path X, and $x = x_1$ at the tip of X. The boundary exponent is $\tilde{x}_1$ for the path extremity hitting the (Dirichlet) boundary. *[Courtesy of T. Kennedy (University of Arizona).]*

We fix $\gamma \in [0, 2)$. We say that X has a *quantum scaling exponent* Δ if, when X and μ_γ (as defined above) are chosen independently, we have

$$\lim_{\delta \to 0} \frac{\log \mathbb{E}\mu_\gamma(B^\delta(X))}{\log \delta} = \Delta .$$

For z chosen uniformly in D, and independently of X and μ_γ, the joint probability of intersection of the quantum ball $B^\delta(z)$ and X then scales as

$$\mathbb{P}\{B^\delta(z) \cap X \neq \emptyset\} \asymp \delta^\Delta ,$$

in the sense that

$$\lim_{\delta \to 0} \frac{\log \mathbb{P}\{B^\delta(z) \cap X \neq \emptyset\}}{\log \delta} = \Delta .$$

18.1.4 Box formulation of Liouville quantum gravity

One can alternatively define quantum scaling dimensions using boxes instead of balls. We define a *dyadic square* to be a closed square (including its interior) of one of the grids $2^{-k}\mathbb{Z}^2$ for some integer k. Let μ be any measure on $\mathbb{C}$. For $\delta > 0$, we define a (μ, δ) *box* S to be a dyadic square S with $\mu(S) < \delta$ and $\mu(S') \geq \delta$, where S' is the dyadic parent of S. Clearly, if a point $z \in \mathbb{C}$ does not lie on a boundary of a dyadic square—and it satisfies $\mu(\{z\}) < \delta < \mu(\mathbb{C})$—then there is a unique (μ, δ) box containing z, which we denote by $S^\delta(z)$. Let $\mathcal{S}_\mu^\delta$ be the set of all (μ, δ) boxes. The boxes in $\mathcal{S}_\mu^\delta$ do not overlap one another except at their boundaries. Thus, they form a tiling of $\mathbb{R}^2$ (see Figs. 18.2 and 18.3 for an illustration of this construction for the Liouville quantum gravity measure on a discrete torus).

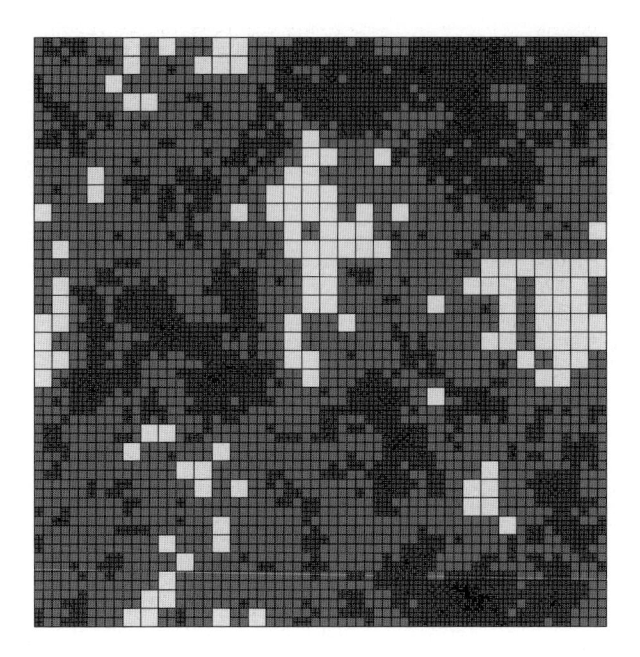

Fig. 18.2 (μ_γ, δ) boxes of the random measure $\mu_\gamma = e^{\gamma h} dz$, where $\gamma = 1$ and h is a (discrete) Gaussian free field on a very fine (1024×1024) grid on the torus, dz is the counting measure on the vertices of that grid, and δ is 2^{-12} times the total mass of μ_γ. (We view μ_γ as a discrete approximation of the continuum Liouville quantum gravity measure.) One way to construct this figure is to view the entire torus as a square, then subdivide each square whose μ_γ measure is at least δ into four smaller squares, and repeat until all squares have a μ_γ measure less than δ. The squares shown have roughly the same μ_γ size, in the sense that each square has a μ_γ measure less than δ but each square's dyadic parent has a μ_γ measure greater than δ. (From [37].)

When ε is a power of 2, we define $S_\varepsilon(z)$ analogously to be the dyadic square containing z with edge length ε, i.e. Euclidean area ε^2. Likewise, we define

$$S_\varepsilon(X) = \{z : S_\varepsilon(z) \cap X \neq \emptyset\},$$

$$S^\delta(X) = \{z : S^\delta(z) \cap X \neq \emptyset\},$$

i.e. we define them as the unions of (μ_0, ε^2) *Euclidean* boxes or (μ_γ, δ) *quantum* boxes, respectively, intersected by X (Fig. 18.4). The following gives the equivalence of the definition of the scaling dimension when boxes are used instead of balls.

Proposition 18.3 *Fix $\gamma \in [0, 2)$ and let X be a random subset of a deterministic compact subset of D. Let $N_\gamma(\delta, X)$ be the number of (μ_γ, δ) boxes intersected by X, and let $N_0(\varepsilon^2, X)$ be the number of dyadic squares intersecting X that have edge length ε (a power of 2). Then X has a Euclidean scaling exponent $x \geq 0$ if and only if*

$$\lim_{\varepsilon \to 0} \frac{\log \mathbb{E}[\mu_0(S_\varepsilon(X))]}{\log \varepsilon^2} = \lim_{\varepsilon \to 0} \frac{\log \mathbb{E}[\varepsilon^2 N_0(\varepsilon^2, X)]}{\log \varepsilon^2} = x$$

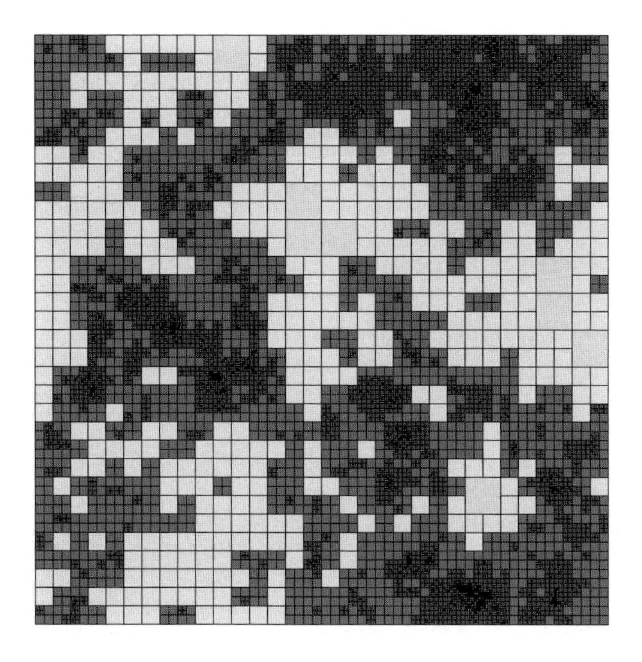

Fig. 18.3 Analogue of Fig. 18.2 with $\gamma = 3/2$, using the same instance h of the GFF. (From [37].)

or, equivalently,

$$\lim_{\varepsilon \to 0} \frac{\log \mathbb{E}[N_0(\varepsilon^2, X)]}{\log \varepsilon^2} = x - 1 .$$

Similarly, X has a quantum scaling exponent Δ if and only if, when X and μ_γ (as defined above) are chosen independently, we have

$$\lim_{\delta \to 0} \frac{\log \mathbb{E}[\mu_\gamma(S^\delta(X))]}{\log \delta} = \lim_{\delta \to 0} \frac{\log \mathbb{E}[\delta\, N_\gamma(\delta, X)]}{\log \delta} = \Delta \tag{18.10}$$

or, equivalently,

$$\lim_{\delta \to 0} \frac{\log \mathbb{E}[N_\gamma(\delta, X)]}{\log \delta} = \Delta - 1 . \tag{18.11}$$

Equivalently, for z chosen uniformly in D, and independently of X and μ_γ, the joint probability of intersection of the quantum square $S^\delta(z)$ and X scales as

$$\mathbb{P}\{S^\delta(z) \cap X \neq \emptyset\} \asymp \delta^\Delta ,$$

in the sense that

$$\lim_{\delta \to 0} \frac{\log \mathbb{P}\{S^\delta(z) \cap X \neq \emptyset\}}{\log \delta} = \Delta .$$

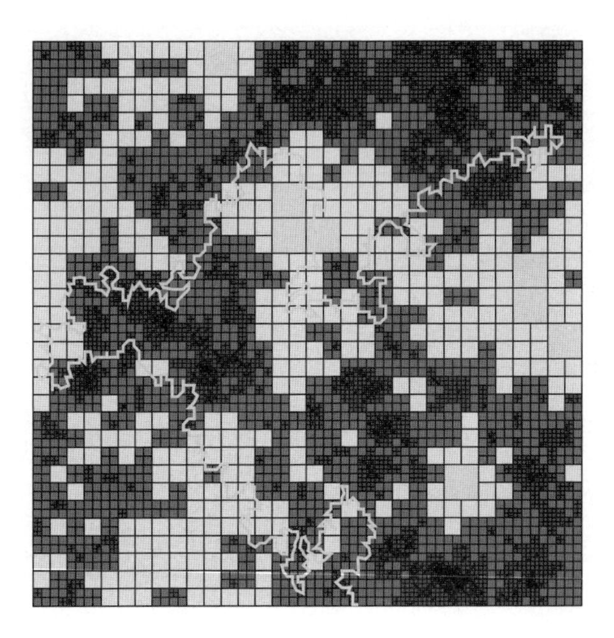

Fig. 18.4 A fractal set X (here a self-avoiding walk) intersecting the quantum grid $S^{\delta}_{\mu_\gamma}$ made of (μ_γ, δ) boxes for $\gamma = 3/2$ (Fig. 18.3). The number $N_\gamma(\delta, X)$ of (μ_γ, δ) boxes intersected by X scales as $N_\gamma(\delta, X) \asymp \delta^{\Delta-1}$, where Δ is the quantum scaling exponent of X. Equivalently, the joint probability of intersection of a dyadic quantum box $S^{\delta}(z)$ and X scales as $\mathbb{P}\{S^{\delta}(z) \cap X \neq \emptyset\} \asymp \delta^{\Delta}$.

18.1.5 Statement of the KPZ relation

The following is the KPZ scaling-exponent relation. To avoid boundary technicalities, we restrict our attention here to a compact subset of D. The case of boundary exponents will be dealt with in Section 18.5.

Theorem 18.4 *Fix $\gamma \in [0, 2)$ and a compact subset $\tilde{D}$ of D. If $X \cap \tilde{D}$ has a Euclidean scaling exponent $x \geq 0$ then it has a quantum scaling exponent $\Delta = \Delta_\gamma$, where Δ_γ is the non-negative solution (18.5) to*

$$x = \frac{\gamma^2}{4}\Delta_\gamma^2 + \left(1 - \frac{\gamma^2}{4}\right)\Delta_\gamma\,. \tag{18.12}$$

Note that the expectation in the above is with respect to both random variables, X and μ_γ.

18.2 GFF regularization

In this section, we present a mathematically convenient regularization of the Gaussian free field in the continuum [37].

18.2.1 GFF circular average

Let h be a centered Gaussian free field on a bounded simply connected domain D with Dirichlet zero boundary conditions. As already remarked in [72], special care is required to make

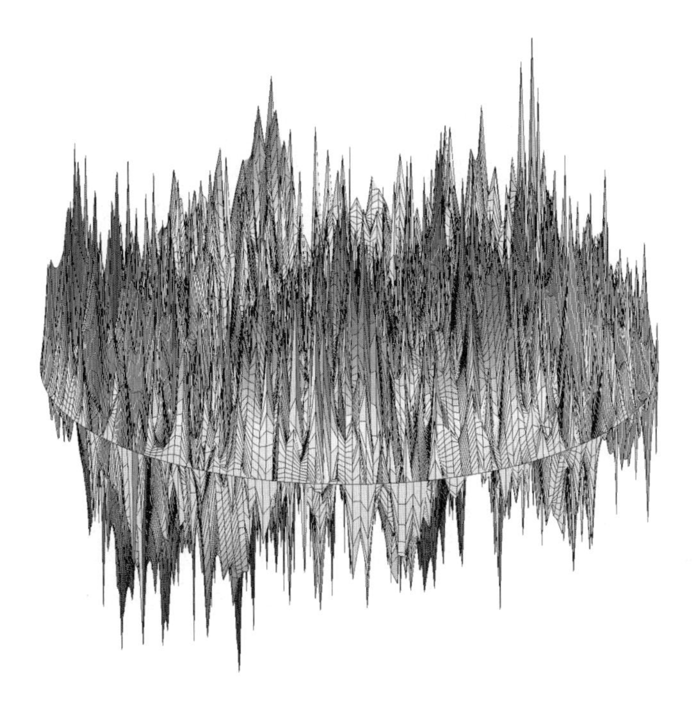

Fig. 18.5 GFF with Dirichlet boundary conditions on a disk. A very fine regularization leads to a proliferation of spikes. *[Courtesy of N.-G. Kang (Caltech).]*

sense of the quantum gravity measure, since the GFF is a distribution and not a function (it typically oscillates between $\pm\infty$) (see e.g. [79] and Fig. 18.5).

For each $z \in D$, we define the ball $B_\varepsilon(z) := \{w : |w - z| < \varepsilon\}$, and let ε_0^z be the largest ε such that $B_\varepsilon(z) \subset D$ (Fig. 18.6). When $\varepsilon \leq \varepsilon_0^z$, we write $h_\varepsilon(z)$ for the average value of h on the circle $\partial B_\varepsilon(z) := \{w : |w - z| = \varepsilon\}$. (Similar averages were considered in [8].) We denote by $\rho_\varepsilon^z(y)$ the uniform (Dirac-like) density (of total mass one) localized on the circle $\partial B_\varepsilon(z)$,

$$\rho_\varepsilon^z(y) := \frac{1}{2\pi} \int_0^{2\pi} d\theta \, \delta(y - z - \varepsilon e^{i\theta}) \,, \tag{18.13}$$

such that one can write $h_\varepsilon(z)$ as the scalar product on D

$$h_\varepsilon(z) = (h, \rho_\varepsilon^z) := \int_D h(y) \rho_\varepsilon^z(y) \, dy \,. \tag{18.14}$$

The density ρ_ε^z is naturally associated with a Newtonian potential f_ε^z. We define a function $f_\varepsilon^z(y)$, for $y \in D$, as

$$f_\varepsilon^z(y) := -\log \max(\varepsilon, |z - y|) - \tilde{G}_z(y) \,, \tag{18.15}$$

where $\tilde{G}_z(y)$ is the *harmonic extension* of $-\log|z - y|$ to D, i.e. the harmonic function of $y \in D$ with a boundary value equal to the restriction of $-\log|z - y|$ to $y \in \partial D$. By construction, this $f_\varepsilon^z(y)$ satisfies Dirichlet boundary conditions and the Poisson equation

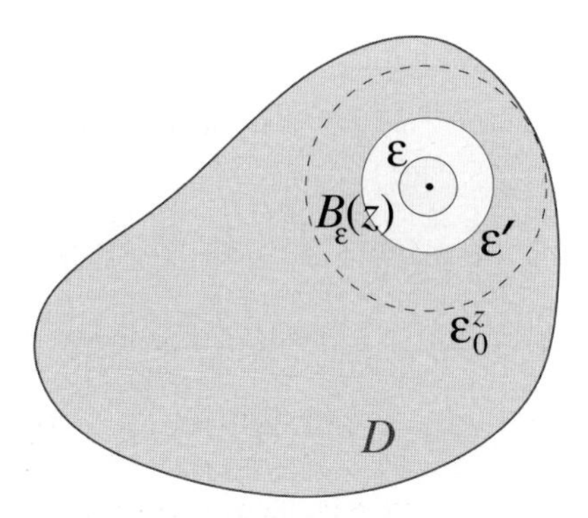

Fig. 18.6 Nested balls $B_\varepsilon(z)$ centered at z. The value ε_0^z is the largest radius ε such that $B_\varepsilon(z) \subset D$. A uniform (Dirac-like) density (of total mass one) is localized on the circle $\partial B_\varepsilon(z)$.

$$-\Delta f_\varepsilon^z = 2\pi \rho_\varepsilon^z \,. \tag{18.16}$$

This (regular) potential function is represented in Fig. 18.7.

Substituting eqn (18.16) into eqn (18.14), integrating by parts, and using Dirichlet boundary conditions for h, we immediately have the following:

$$\begin{aligned} h_\varepsilon(z) = (h, \rho_\varepsilon^z) &= \frac{1}{2\pi}(h, -\Delta f_\varepsilon^z) \\ &= \frac{1}{2\pi}\int_D \nabla h(y) \cdot \nabla f_\varepsilon^z(y)\, dy \,, \end{aligned}$$

which we write as

$$h_\varepsilon(z) = (h, f_\varepsilon^z)_\nabla \,, \tag{18.17}$$

in terms of the Dirichlet inner product

$$(f_1, f_2)_\nabla := \frac{1}{2\pi}\int_D \nabla f_1(y) \cdot \nabla f_2(y)\, dy \,, \tag{18.18}$$

i.e. the *interaction energy of the fields associated with potentials* $f_{i=1,2}$.

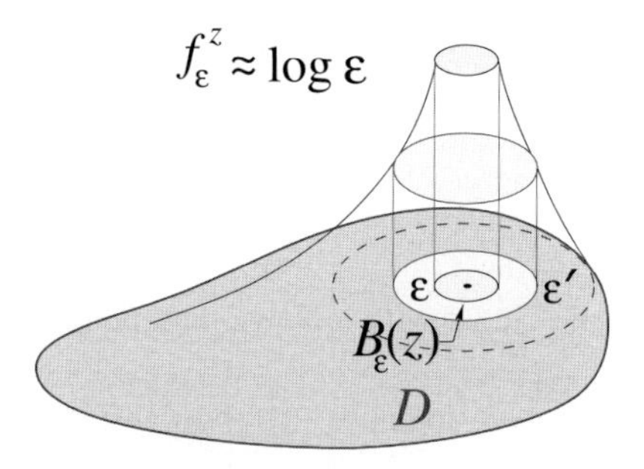

Fig. 18.7 The potential f_ε^z (18.15) created by a uniform mass distribution localized on the circle $\partial B_\varepsilon(z)$; its leading constant value inside the disk $B_\varepsilon(z)$ is $-\log \varepsilon$.

18.2.2 GFF covariance

In fact, the random variables $(h, f)_\nabla$ are zero-mean Gaussian random variables for each f, with the fundamental covariance property

$$\mathrm{Cov}\big((h, f_1)_\nabla, (h, f_2)_\nabla\big) = (f_1, f_2)_\nabla\,, \tag{18.19}$$

where the covariance of two random variables A, B is defined as usual as

$$\mathrm{Cov}(A, B) := \mathbb{E}(A\,B) - \mathbb{E}(A)\,\mathbb{E}(B)\,.$$

From this and eqn (18.17), we immediately deduce the covariance of the averaged fields $h_\varepsilon(z)$ on two arbitrary circles (Fig. 18.8):

$$\mathrm{Cov}\big(h_{\varepsilon_1}(z), h_{\varepsilon_2}(y)\big) = (f^z_{\varepsilon_1}, f^y_{\varepsilon_2})_\nabla\,,$$

which is precisely the *Newtonian interaction energy* of the unit mass distributions localized on the two circles $\partial B_{\varepsilon_1}(z)$ and $\partial B_{\varepsilon_2}(y)$.

18.2.3 GFF circular average and Brownian motion

Consider *nested circles* centered at z (Fig. 18.6). We obtain the covariance of the averaged $h_\varepsilon(z)$ fields (18.17) on those circles,

$$\mathrm{Cov}\big(h_\varepsilon(z), h_{\varepsilon'}(z)\big) = (f^z_\varepsilon, f^z_{\varepsilon'})_\nabla\,,$$

an interaction energy which is easily evaluated by using Gauss's theorem. By integrating by parts, and using eqn (18.16), the definition (18.13), and the explicit potential (18.15) or Gauss's theorem, the Newtonian interaction energy of two nested circles can be written as

$$\begin{aligned}(f^z_\varepsilon, f^z_{\varepsilon'})_\nabla &= (2\pi)^{-1}(f^z_\varepsilon, -\Delta f^z_{\varepsilon'}) = (f^z_\varepsilon, \rho^z_{\varepsilon'})\\ &= f^z_{\max(\varepsilon,\varepsilon')}(z) = -\log\max(\varepsilon, \varepsilon') - \tilde{G}_z(z)\,,\end{aligned} \tag{18.20}$$

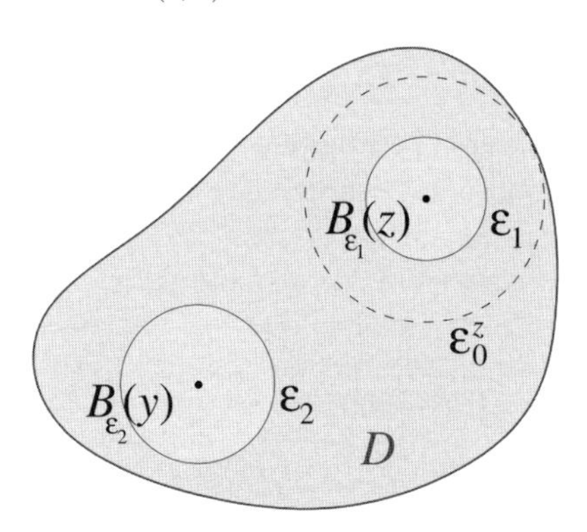

Fig. 18.8 Balls $B_{\varepsilon_1}(z)$ and $B_{\varepsilon_2}(y)$ with unit masses localized on their respective boundaries $\partial B_{\varepsilon_1}(z)$ and $\partial B_{\varepsilon_2}(y)$.

where we have used the fact that the function $\tilde{G}_z$ is harmonic, and hence has all averages over concentric circles equal to its value at their common center z: $(\tilde{G}_z, \rho^z_{\varepsilon'}) = \tilde{G}_z(z)$. This gives the explicit form of the covariance

$$\mathrm{Cov}\big(h_\varepsilon(z), h_{\varepsilon'}(z)\big) = \mathbb{E}\big(h_\varepsilon(z) h_{\varepsilon'}(z)\big) = -\log \max(\varepsilon, \varepsilon') - \tilde{G}_z(z)\,, \tag{18.21}$$

with $\mathbb{E}\, h_\varepsilon(z) = 0$ here for Dirichlet boundary conditions.

Specializing the above results to $\varepsilon' = \varepsilon$, we obtain in particular the *variance* of the field $h_\varepsilon(z)$,

$$\mathrm{Var}\, h_\varepsilon(z) = (f^z_\varepsilon, f^z_\varepsilon)_\nabla = f^z_\varepsilon(z) = -\log \varepsilon - \tilde{G}_z(z)\,. \tag{18.22}$$

The harmonic extension $\tilde{G}_z(y)$, taken at the point $y = z$, satisfies the identity $\tilde{G}_z(z) = -\log C(z; D)$, in terms of the conformal radius C of D viewed from z, a smooth function of z.

Brownian motion. From eqns (18.21) and (18.22), and the identity

$$\mathrm{Var}\big[h_\varepsilon(z) - h_{\varepsilon'}(z)\big] = \mathrm{Var}\, h_\varepsilon(z) + \mathrm{Var}\, h_{\varepsilon'}(z) - 2\, \mathrm{Cov}\big(h_\varepsilon(z), h_{\varepsilon'}(z)\big)\,,$$

we thus obtain the two important variances

$$\mathrm{Var}\, h_\varepsilon(z) = -\log \varepsilon + \log C(z; D)\,, \tag{18.23}$$

$$\mathrm{Var}\big[h_\varepsilon(z) - h_{\varepsilon'}(z)\big] = |\log \varepsilon - \log \varepsilon'|\,. \tag{18.24}$$

The interpretation of eqns (18.23) and (18.24) is immediate: for fixed z, the Gaussian random variable $h_\varepsilon(z)$ is a *one-dimensional standard Brownian motion* $\mathcal{B}_t$ *when parameterized by a time* $t := -\log[\varepsilon/C(z; D)]$ [37].

Note that eqn (18.23) mathematically quantifies the unboundedness of the GFF fluctuations in the continuum as $\varepsilon \to 0$, as already illustrated in Fig. 18.5. Moreover, for fixed z, the identity in law

$$h_\varepsilon(z) \overset{\text{(in law)}}{=} \mathcal{B}_t\,, \tag{18.25}$$

$$-\log[\varepsilon/C(z; D)] = t\,, \tag{18.26}$$

$$\mathbb{E}\, h_\varepsilon(z) = \mathbb{E}\, \mathcal{B}_t = 0 \tag{18.27}$$

shows that $h_\varepsilon(z)$ does not converge to a limit as $\varepsilon \to 0$ (which would be $h(z)$ if h were a proper continuous function, and not a distribution), but instead it wanders indefinitely like Brownian motion in the time parameter as $t \to \infty$.

Note that in the parameterization (18.26) we have $t \in [t^z_0, \infty)$, where

$$t^z_0 := -\log\left[\frac{\varepsilon^z_0}{C(z; D)}\right], \quad \varepsilon^z_0 = \sup\{\varepsilon : B_\varepsilon(z) \subset D\}\,.$$

Thanks to the *Koebe theorem* [see e.g. [74]], we have

$$\frac{1}{4} C(z; D) \leq \varepsilon^z_0 \leq C(z; D)\,,$$

and hence $0 \leq t^z_0 \leq 2 \log 2$.

A perhaps more convenient variant of the above parameterization then consists in taking the radius $\varepsilon_0 := \varepsilon_0^z$ as a reference point, and simply writing [37]

$$h_\varepsilon(z) - h_{\varepsilon_0}(z) \overset{\text{(in law)}}{=} \mathcal{B}_t\,, \tag{18.28}$$

$$-\log(\varepsilon/\varepsilon_0) = t \in [0,\infty),\ \varepsilon \leq \varepsilon_0\,, \tag{18.29}$$

$$\mathcal{B}_0 = 0\,. \tag{18.30}$$

This Brownian property of the circular average $h_\varepsilon(z)$ will play an essential role in the sequel, in which we prove the validity of the KPZ relation.

18.3 Random measure and Liouville quantum gravity

18.3.1 Gaussian exponential averages

Recall, first, that if Y is a Gaussian random variable with mean $\mathbb{E}Y = a$ and variance $\mathrm{Var}Y = b$, then $\mathbb{E}\,e^Y = e^{a+b/2}$. Applying this to the Gaussian variable $h_\varepsilon(z)$, we have the following exponential expectation from eqn (18.23), since $\mathbb{E}\,h_\varepsilon(z) = 0$:

$$\mathbb{E}\,e^{\gamma h_\varepsilon(z)} = e^{\mathrm{Var}[\gamma h_\varepsilon(z)]/2} = \left[\frac{C(z;D)}{\varepsilon}\right]^{\gamma^2/2}. \tag{18.31}$$

Since eqn (18.31) ultimately *diverges* for $\varepsilon \to 0$, we are led to *regularize* Liouville quantum gravity by defining the random measure (18.8)

$$d\mu_{\gamma,\varepsilon} = M_{\gamma,\varepsilon}(z)\,dz\,, \quad M_{\gamma,\varepsilon}(z) := \varepsilon^{\gamma^2/2} e^{\gamma h_\varepsilon(z)}\,, \tag{18.32}$$

in a way similar to the so-called Wick normal ordering (see e.g. [80]). Owing to eqn (18.23), this definition coincides with the one (18.9) given in the introduction. In [37], it is shown that for $\gamma \in [0,2)$, the limit of this regularized measure exists as $\varepsilon \to 0$, which mathematically defines Liouville quantum gravity (see also [1, 47] for earlier work).

18.3.2 GFF sampling and random measure

We now consider a measure on pairs (z,h), where h is a Gaussian free field and, given h, the point z is chosen from the regularized quantum area measure $e^{\gamma h_\varepsilon(z)}\,dz$. Such a measure has the form

$$e^{\gamma h_\varepsilon(z)} dh\,dz = \exp\left[-\frac{1}{2}(h,h)_\nabla + \gamma h_\varepsilon(z)\right]\mathcal{D}h\,dz\,, \tag{18.33}$$

where dh represents the whole GFF measure and $\mathcal{D}h$ the flat functional measure. The total action of the GFF *Liouville weighted measure* is thus the quadratic combination

$$S_\gamma(h) := \frac{1}{2}(h,h)_\nabla - \gamma h_\varepsilon(z)\,. \tag{18.34}$$

Owing to eqn (18.17), the latter can be rewritten as

$$\begin{aligned} S_\gamma(h) &= \frac{1}{2}(h,h)_\nabla - \gamma(h, f_\varepsilon^z)_\nabla = \frac{1}{2}(h - \gamma f_\varepsilon^z, h - \gamma f_\varepsilon^z)_\nabla - \frac{\gamma^2}{2}(f_\varepsilon^z, f_\varepsilon^z)_\nabla \\ &= \frac{1}{2}(h', h')_\nabla - \frac{\gamma^2}{2}\operatorname{Var} h_\varepsilon(z), \qquad h' := h - \gamma f_\varepsilon^z\,, \end{aligned} \tag{18.35}$$

where we have used $\operatorname{Var} h_\varepsilon(z) = (f_\varepsilon^z, f_\varepsilon^z)_\nabla$ from eqn (18.22). In the substitution $h' := h - \gamma f_\varepsilon^z$, h' is a *standard* GFF. The probability weight involved in our random measure defined by eqns (18.33)–(18.35) can finally be written, thanks to eqn (18.31), as

$$\begin{aligned} \exp\left[-S_\gamma(h)\right] &= \exp\left[-\frac{1}{2}(h,h)_\nabla + \gamma h_\varepsilon(z)\right] \\ &= \exp\left[-\frac{1}{2}(h', h')_\nabla\right] \mathbb{E}\, e^{\gamma h_\varepsilon(z)}\,, \end{aligned} \tag{18.36}$$

where the second factor on the right-hand side is the *marginal distribution density* of z. The meaning of eqn (18.36) is that, after z has been sampled from its marginal distribution, the law of h *weighted* by $e^{\gamma h_\varepsilon(z)}$ is identical to that of the original GFF h' *plus* the deterministic function γf_ε^z (18.15):

$$h \overset{\text{(in law)}}{=} h' + \gamma f_\varepsilon^z\,. \tag{18.37}$$

18.3.3 Random measure and Brownian motion with drift

For fixed z, the identity in the law (18.37) can be averaged on the circle $\partial B_\varepsilon(z)$ with the density ρ_ε^z (18.13) to yield

$$\begin{aligned} h_\varepsilon(z) &\overset{\text{(in law)}}{=} h'_\varepsilon(z) + \gamma(f_\varepsilon^z, \rho_\varepsilon^z) \\ &= h'_\varepsilon(z) + \gamma f_\varepsilon^z(z) \\ &= h'_\varepsilon(z) - \gamma \log\left[\frac{\varepsilon}{C(z;D)}\right], \end{aligned} \tag{18.38}$$

where use has been made of the result (18.20). Since h' is a standard Gaussian free field, one can apply to its circular average $h'_\varepsilon(z)$ the identity in law given by eqns (18.25) and (18.26) to standard Brownian motion $\mathcal{B}_t$, $t = -\log[\varepsilon/C(z;D)]$, and rewrite eqn (18.38) as

$$h_\varepsilon(z) \overset{\text{(in law)}}{=} \mathcal{B}_t + \gamma t\,. \tag{18.39}$$

The statement of eqn (18.39) is the following variant of a statement in [37].

Proposition 18.5 *For fixed z, when the law of h is weighted by the Liouville conformal factor $e^{\gamma h_\varepsilon(z)}$ as in eqn (18.33) or (18.36), the law of the Gaussian random variable $h_\varepsilon(z)$ is identical to that of one-dimensional standard Brownian motion $\mathcal{B}_t$ with drift γt, when parameterized by a time $t = -\log[\varepsilon/C(z;D)]$.*

Alternatively, we can use the second parameterization (18.29) for $h'_\varepsilon(z)$ and the identity in law (18.28) to standard Brownian motion $\mathcal{B}_t$, $t = -\log(\varepsilon/\varepsilon_0)$, and rewrite eqn (18.38) as in [37]:

$$h_\varepsilon(z) - h_{\varepsilon_0}(z) \overset{\text{(in law)}}{=} h'_\varepsilon(z) - h'_{\varepsilon_0}(z) - \gamma \log\left(\frac{\varepsilon}{\varepsilon_0}\right)$$
$$\overset{\text{(in law)}}{=} \mathcal{B}_t + \gamma t, \ \mathcal{B}_0 = 0 \,. \tag{18.40}$$

The identities in law (18.39) and (18.40) above differ simply by the choice of time origin.

18.4 Proof of the KPZ relation

18.4.1 Quantum balls and scaling

It can be shown [37] that when ε is small, the regularized stochastic quantum measure $\mu_{\gamma,\varepsilon}(B_\varepsilon(z))$ (18.8) or (18.32) of the Euclidean ball $B_\varepsilon(z)$ and the limit quantum measure $\mu_\gamma(B_\varepsilon(z))$ of Proposition 18.1 are very well approximated by the simple form $\mu_\gamma(B_\varepsilon(z)) \simeq \pi\varepsilon^2\varepsilon^{\gamma^2/2}e^{\gamma h_\varepsilon(z)}$, i.e. the Lebesgue measure of the ball times the regularized conformal factor at the ball's center z. Let us therefore write

$$\mu_\gamma(B_\varepsilon(z)) = \pi \exp\left[\gamma h_\varepsilon(z) + \gamma Q_\gamma \log \varepsilon\right], \tag{18.41}$$
$$Q_\gamma := \frac{2}{\gamma} + \frac{\gamma}{2}, \tag{18.42}$$

and, in the simplified perspective of this lecture, take eqn (18.41) to be the *definition* of $\mu_\gamma(B_\varepsilon(z))$. That is, we view μ_γ as a function on balls of the form $B_\varepsilon(z)$, defined by eqn (18.41), rather than a fully defined measure on D. In accordance with Definition 18.2 in Section 18.1.3, we then define the *quantum ball* $\tilde{B}^\delta(z)$ of area δ centered at z as the (largest) Euclidean ball $B_\varepsilon(z)$ whose radius ε is chosen so that

$$\mu_\gamma(\tilde{B}^\delta(z)) = \delta \,. \tag{18.43}$$

According to the definitions given in Section 18.1.3, we say that a (deterministic or random) fractal subset X of D has a *Euclidean scaling exponent* x (and Euclidean dimension $2-2x$) if, for z chosen uniformly in D and independently of X, the probability

$$\mathbb{P}\{B_\varepsilon(z) \cap X \neq \emptyset\} \asymp \varepsilon^{2x}\,, \tag{18.44}$$

in the sense that $\lim_{\varepsilon\to 0} \log \mathbb{P}/\log\varepsilon = 2x$.

Similarly, we say that X has a *quantum scaling exponent* Δ if, when X and (z,h), sampled with weights given by eqns (18.33) and (18.36), are chosen independently, we have

$$\mathbb{P}\{\tilde{B}^\delta(z) \cap X \neq \emptyset\} \asymp \delta^\Delta\,. \tag{18.45}$$

18.4.2 Stopping time

According to Proposition 18.5, $h_\varepsilon(z)$ in eqn (18.41) sampled with weights given by eqns (18.33) and (18.36) has the same law (18.39) as Brownian motion with drift, or, equivalently,

$h_\varepsilon(z) - h_{\varepsilon_0}(z)$ has the law (18.40). Substituting $t = -\log(\varepsilon/\varepsilon_0)$, we can rewrite eqn (18.41) in the normalized form

$$\frac{\mu_\gamma(B_\varepsilon(z))}{\mu_\gamma(B_{\varepsilon_0}(z))} \overset{\text{(in law)}}{=} \exp\left[\gamma \mathcal{B}_t + \gamma(\gamma - Q_\gamma)t\right]. \tag{18.46}$$

To simplify the notation, from now on we shall retain the normalization radius ε_0 in the expression for t, and the normalization quantum area $\mu_\gamma(B_{\varepsilon_0}(z))$, as in eqn (18.46). This amounts to a redefinition of δ in eqn (18.43) by a (random) change of units of the Euclidean length and quantum area.

The equality of renormalized eqn (18.43) to eqn (18.46) then relates the Euclidean radius ε *stochastically* to the given quantum area δ, and, thanks to eqn (18.46), characterizes the (smallest) time $t = -\log \varepsilon/\varepsilon_0$ such that

$$\exp\left[\gamma(\mathcal{B}_t - a_\gamma t)\right] = \delta\,, \tag{18.47}$$

where we have introduced the parameter

$$a_\gamma := Q_\gamma - \gamma = \frac{2}{\gamma} - \frac{\gamma}{2} > 0\,,\ \gamma \in (0,2)\,. \tag{18.48}$$

The Euclidean radius is thus given in terms of the *stopping time* (Fig. 18.9)

$$-\log \varepsilon_A := T_A := \inf\{t : -\mathcal{B}_t + a_\gamma t = A\}\,, \tag{18.49}$$

$$A := -(\log \delta)/\gamma > 0\,. \tag{18.50}$$

At time $t = 0$, the Brownian motion $\mathcal{B}_t$ is started at $\mathcal{B}_0 = 0$, in accordance with eqn (18.40).

The probability (18.44) that the Euclidean ball $B_{\varepsilon_A}(z)$ intersects X scales as $\varepsilon_A^{2x} = e^{-2xT_A}$. Computing its expectation

$$\mathbb{E}\,\varepsilon_A^{2x} = \mathbb{E}\left[\exp\left(-2xT_A\right)\right] \tag{18.51}$$

with respect to the random stopping time T_A will give the quantum probability (18.45)

$$\mathbb{P}\{\tilde{B}^\delta(z) \cap X \neq \emptyset\} \asymp \mathbb{E}\left[\exp\left(-2xT_A\right)\right], \tag{18.52}$$

in the sense that the ratio of the logarithms of these quantities tends to 1 as $\delta \to 0$, $A \to \infty$.

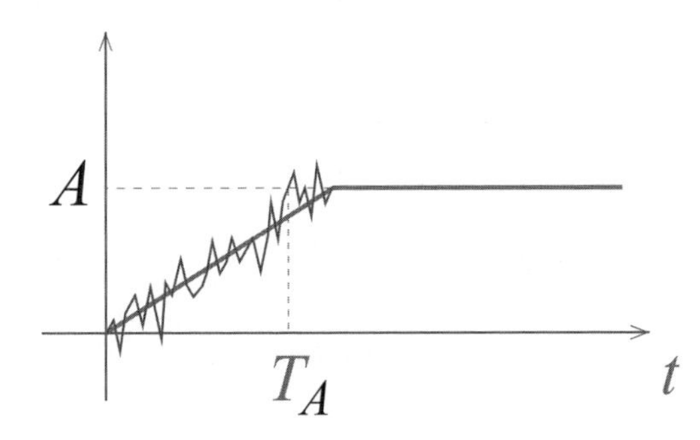

Fig. 18.9 Illustration of the definition (18.49) of the stopping time T_A. The drift term $a_\gamma t$ gives a constant positive slope, on which the Brownian motion $\mathcal{B}_t$ is superimposed.

18.4.3 Brownian martingale

Now consider, for any β, the standard *Brownian exponential martingale*

$$\mathbb{E}\left[\exp\left(-\beta\mathcal{B}_t - \frac{\beta^2 t}{2}\right)\right] = \mathbb{E}\left[\exp(-\beta\mathcal{B}_0\right] = 1\,, \tag{18.53}$$

valid for $0 \le t < \infty$ and $\mathcal{B}_0 = 0$.

When $0 \le t \le T_A$, we have $-\mathcal{B}_t + a_\gamma t - A \le 0$ by the definition of T_A, so the argument of the martingale (18.53) remains bounded for $\beta > 0$, because $\exp(\beta A - (\beta a_\gamma + \beta^2/2)t) \le \exp(\beta A)$, and hence it is bounded by a fixed constant. We can then apply the martingale (18.53) at the stopping time T_A, where $T_A < \infty$ and $\mathcal{B}_{T_A} = a_\gamma T_A - A$. We thus get

$$\mathbb{E}\left[\exp\left[\beta A - \left(\beta a_\gamma + \frac{\beta^2}{2}\right) T_A\right] 1_{T_A<\infty}\right] = 1\,,$$

which can be written as

$$\mathbb{E}[\exp(-2xT_A)1_{T_A<\infty}] = \exp(-\beta A)\,, \tag{18.54}$$

$$2x := \beta a_\gamma + \frac{\beta^2}{2}\,. \tag{18.55}$$

This is valid for $\beta \ge 0$, which is thus equal to the positive root of eqn (18.55),

$$\beta = \beta_\gamma(x) := (a_\gamma^2 + 4x)^{1/2} - a_\gamma \tag{18.56}$$

(see e.g. [10]). Thanks to the definition (18.50) of A, we thus obtain the scaling behavior

$$\mathbb{E}[\exp(-2xT_A)1_{T_A<\infty}] = \exp(-\beta_\gamma(x)A) = \delta^{\Delta_\gamma}\,, \tag{18.57}$$

$$\Delta_\gamma := \frac{\beta_\gamma(x)}{\gamma} = \frac{(a_\gamma^2 + 4x)^{1/2} - a_\gamma}{\gamma}\,. \tag{18.58}$$

For $x = 0$, we find in particular

$$\mathbb{P}(T_A < \infty) = \mathbb{E}[1_{T_A<\infty}] = 1\,, \tag{18.59}$$

since $\beta_\gamma(0) = (|a_\gamma| - a_\gamma)/\gamma = 0$ for $a_\gamma > 0, \gamma \in (0,2)$, so that the conditioning on $T_A < \infty$ can actually be *omitted* for $\gamma \in (0,2)$.

We thus obtain, with eqns (18.52) and (18.57), the expected quantum scaling behavior of eqn (18.45), together with $\Delta = \Delta_\gamma$ (eqn (18.58)), which is, thanks to eqn (18.48),

$$\Delta_\gamma = \gamma^{-1}\left(\sqrt{4x + \left(\frac{2}{\gamma} - \frac{\gamma}{2}\right)^2} - \left(\frac{2}{\gamma} - \frac{\gamma}{2}\right)\right), \tag{18.60}$$

i.e., the expected *inverse KPZ* function (18.5). Note also that substituting $\beta_\gamma = \gamma\Delta_\gamma$ into eqn (18.55) gives

$$2x = \left(2 - \frac{\gamma^2}{2}\right)\Delta_\gamma + \frac{\gamma^2}{2}\Delta_\gamma^2\,, \tag{18.61}$$

i.e. the KPZ relation (18.4), QED.

It is interesting to note that in eqn (18.55), which is in essence the KPZ relation, the term linear in β corresponds in the Brownian martingale (18.54) to the intersection time $T'_A := A/a_\gamma$ of the linear drift term $a_\gamma t$ with the level A in Fig. 18.9. This explains the origin of the term linear in Δ_γ in eqn (18.61). The second term, $\beta^2/2$, in eqn (18.55) corresponds to pure Brownian fluctuations in the martingale (18.54) and in Fig. 18.9. These fluctuations thus generate the distinctive quadratic term in the KPZ formula (18.61).

18.4.4 Probability distribution and GFF points

The inverse Laplace transform $P_A(t)$ of eqn (18.57), with respect to $2x$, is the probability density of $T_A = -\log \varepsilon_A$ such that $P_A(t)\, dt := \mathbb{P}\,(T_A \in [t, t+dt])$. Its explicit form is [37]

$$P_A(t) = \frac{A}{\sqrt{2\pi t^3}} \exp\left[-\frac{1}{2t}(A - a_\gamma t)^2\right]. \tag{18.62}$$

This probability density is associated with the variations of the Euclidean area $\pi\varepsilon^2 := \pi e^{-2t}$ for a given quantum area $\delta := e^{-\gamma A}$. Its integral equals

$$\int_0^\infty dt P_A(t) = \exp\left[-A(|a_\gamma| - a_\gamma)\right] \tag{18.63}$$

$$= 1, \qquad a_\gamma \geq 0\,, \tag{18.64}$$

$$= \exp\left(2Aa_\gamma\right) < 1, \quad a_\gamma < 0\,. \tag{18.65}$$

The fist case, $a_\gamma > 0$, corresponds to $\gamma \in [0, 2)$, so that P_A is a normalized probability density, in accordance with eqn (18.59). The case $a_\gamma < 0$ for $\gamma > 2$ will be interpreted in Section 18.6.

From eqn (18.62), we deduce that for A large (i.e. δ and ε small), the variable $A/T_A = (\log \delta)/(\gamma \log \varepsilon_A)$ is concentrated in eqn (18.57),

$$\mathbb{E}[\exp(-2xT_A)1_{T_A<\infty}] = \int_0^\infty dt\, P_A(t) \exp(-2xt)\,, \tag{18.66}$$

near the saddle point value $A/t := a_\gamma + \gamma\Delta_\gamma$. Reverse engineering to the Brownian motion $\mathcal{B}_t$ via eqn (18.49), we find that $-\mathcal{B}_t/t = A/t - a_\gamma$ is concentrated near $\gamma\Delta_\gamma$, and, finally, using the identity in law (18.39), that the GFF circle average $h_\varepsilon/\log \varepsilon^{-1} \overset{\text{(in law)}}{=} \mathcal{B}_t/t + \gamma$ has a typical value $\gamma - \gamma\Delta_\gamma$. We thus conclude that a point z that is typical with respect to the quantum measure is an *α-thick point* of h [49]:

$$\alpha := \lim_{\varepsilon \to 0} \frac{\log h_\varepsilon(z)}{\log \varepsilon^{-1}}\,,$$

with a typical value

$$\alpha = \gamma - \gamma\Delta_\gamma \tag{18.67}$$

for a fractal of quantum scaling dimension Δ_γ [37, 38].

18.5 Boundary KPZ relation

18.5.1 Boundary quantum measure

Most of the results about random measures on D have straightforward analogues about random measures on the boundary ∂D. The proofs are essentially identical, but we will sketch the differences in the arguments here.

Suppose that D is a domain with a piecewise linear boundary ∂D and that h is an instance of the GFF on D, now with *free* boundary conditions, and normalized to have mean zero. This means that $h = \sum_n \alpha_n f_n$, where the α_n are i.i.d. zero-mean, unit-variance, normal random variables and the f_n are an orthonormal basis, with respect to the inner product

$$(f_1, f_2)_\nabla = (2\pi)^{-1} \int_D \nabla f_1(z) \cdot \nabla f_2(z)\, dz\,,$$

of the Hilbert space closure $H(D)$ of the space of C^∞ of bounded real-valued (but *not* necessarily compactly supported) functions on D with mean zero.

Suppose that f is a smooth function that is *not* compactly supported on D but has a gradient that vanishes on ∂D in the normal direction to ∂D, and we write $-\Delta f = 2\pi\rho$. Then integration by parts implies that $(h, \rho) = (h, f)_\nabla$ with variance $\mathrm{Var}(h, \rho) = \mathrm{Var}(h, f)_\nabla = (f, f)_\nabla$, as in the Dirichlet boundary case.

We can also make sense of the statement that $h_\varepsilon(z)$, for a point $z \in \partial D$ on the boundary of D, is the mean value of $h(z)$ on a *semicircle* $\partial B_\varepsilon(z) \cap D$ of radius ε centered at z and contained in the domain D. In this case, if $\varepsilon \leq \varepsilon_0$, where ε_0 is small enough so that $B_{\varepsilon_0}(z)$ contains no corners of D and exactly one semidisk of $B_{\varepsilon_0}(z)$ lies in D, we can consider the geometry of D to be locally equivalent to that of the upper half-plane $\mathbb{H}$.

By integration by parts, we can then write

$$h_\varepsilon(z) - h_{\varepsilon_0}(z) = (h, \hat{f}^z_\varepsilon)_\nabla\,, \tag{18.68}$$

where $\hat{f}^z_\varepsilon$ is a continuous function which is equal, on the half-annulus $\mathbb{H} \cap [B_{\varepsilon_0}(z) \setminus B_\varepsilon(z)]$, to

$$\hat{f}^z_\varepsilon(y) = 2\log|y - z|, \;\; y \in \mathbb{H} \cap \{y : \varepsilon \leq |y - z| < \varepsilon_0\}\,, \tag{18.69}$$

and is constant outside of the half-annulus: $\hat{f}^z_\varepsilon(y) = 2\log\varepsilon$ for $|y - z| \leq \varepsilon$ and $\hat{f}^z_\varepsilon(y) = 2\log\varepsilon_0$ for $|y - z| \geq \varepsilon_0$. (The $2\log|\cdot - z|$ in place of $\log|\cdot - z|$ comes from the fact that we are taking an average over half the circle $\partial B_\varepsilon(z)$.)

The variance of $h_\varepsilon(z) - h_{\varepsilon_0}(z)$ is given by the Dirichlet energy

$$\mathrm{Var}[h_\varepsilon(z) - h_{\varepsilon_0}(z)] = (\hat{f}^z_\varepsilon, \hat{f}^z_\varepsilon)_\nabla = -2\log\left(\frac{\varepsilon}{\varepsilon_0}\right), \tag{18.70}$$

which is twice as big as before (given the factor of 2 in the definition of $\hat{f}^z_\varepsilon$ and the fact that the integral is only over half as much area).

From eqn (18.70) we then have, for fixed $z \in \partial D$, the result that the Gaussian variable $h_\varepsilon(z) - h_{\varepsilon_0}(z)$ *is a standard Brownian motion* $\mathcal{B}_{2t}$ *in a time* $2t := -2\log(\varepsilon/\varepsilon_0)$ *for* $\varepsilon \leq \varepsilon_0$, i.e. the identity in law

$$z \in \partial D,\;\; h_\varepsilon(z) - h_{\varepsilon_0}(z) \overset{\text{(in law)}}{=} \mathcal{B}_{2t}\,, \tag{18.71}$$

$$-\log(\varepsilon/\varepsilon_0) = t\,, \tag{18.72}$$

$$0 = \mathcal{B}_0\,. \tag{18.73}$$

Thus, at a point on the interior of one of the boundary lines of D, the variance of $h_\varepsilon(z)$ scales like $-2\log\varepsilon$ instead of $-\log\varepsilon$. We then define the *regularized boundary quantum measure*

$$d\mu_{\gamma,\varepsilon}^{B} := \varepsilon^{\gamma^2/4} e^{\gamma h_\varepsilon(z)/2}\, dz\,, \tag{18.74}$$

where in this case dz is the Lebesgue measure on the boundary ∂D of D, with the conformal factor $e^{\gamma h_\varepsilon(z)/2}$ needed for integrating a quantum length instead of an area; as before, the power of ε chosen makes the expectation $\varepsilon^{\gamma^2/4}\mathbb{E}_h e^{\gamma h_\varepsilon(z)/2}$ regular in ε (compare with eqn (18.32)). In other words, the factor preceding dz in the definition (18.74) is an exponential martingale in time $t = -\log \varepsilon$.

We define the boundary quantum measure $d\mu_\gamma^B$ to be the (weak) limit as $\varepsilon \to 0$ of the measures $d\mu_{\gamma,\varepsilon}^B$. The existence of this limit when $0 \leq \gamma < 2$ is ensured by a complete analogue of Proposition 18.1 for the boundary case [37].

18.5.2 Boundary scaling and KPZ relation

For $z \in \partial D$, we write $\hat{B}_\varepsilon(z) := B_\varepsilon(z) \cap \partial D$ and we define $\hat{B}^\delta(z)$ to be the (largest) set $\hat{B}_\varepsilon(z)$ whose μ_γ^B measure is δ.

Likewise, we define

$$\hat{B}_\varepsilon(X) = \{z \in \partial D : \hat{B}_\varepsilon(z) \cap X \neq \emptyset\}$$

and

$$\hat{B}^\delta(X) = \{z \in \partial D : \hat{B}^\delta(z) \cap X \neq \emptyset\}\,.$$

We say that a (deterministic or random) fractal subset X of the boundary of D has a *Euclidean expectation dimension* $1 - \tilde{x}$ and a *Euclidean scaling exponent* $\tilde{x}$ in the boundary sense if the expected measure of $\hat{B}_\varepsilon(X)$ decays like $\varepsilon^{\tilde{x}}$, i.e.

$$\lim_{\varepsilon \to 0} \frac{\log \mathbb{E}\, \mu_0(\hat{B}_\varepsilon(X))}{\log \varepsilon} = \tilde{x}\,.$$

In other words, if z is chosen uniformly in ∂D, and independently of X, then the probability of boundary intersection of the Euclidean segment $\hat{B}_\varepsilon(z)$ and X scales as

$$\mathbb{P}\{\hat{B}_\varepsilon(z) \cap X \neq \emptyset\} \asymp \varepsilon^{\tilde{x}}\,, \tag{18.75}$$

in the sense that

$$\lim_{\varepsilon \to 0} \frac{\log \mathbb{P}\{\hat{B}_\varepsilon(z) \cap X \neq \emptyset\}}{\log \varepsilon} = \tilde{x}\,.$$

We say that X has a *boundary quantum scaling exponent* $\tilde{\Delta}$ if, when X and μ_γ^B (as defined above) are chosen independently, we have

$$\lim_{\delta \to 0} \frac{\log \mathbb{E}\, \mu_\gamma^B(\hat{B}^\delta(X))}{\log \delta} = \tilde{\Delta}\,.$$

For z chosen uniformly in ∂D, and independently of X and μ_γ^B, the joint probability of intersection of the boundary quantum interval $\hat{B}^\delta(z)$ and X then scales as

$$\mathbb{P}\{\hat{B}^\delta(z) \cap X \neq \emptyset\} \asymp \delta^{\tilde{\Delta}}\,, \tag{18.76}$$

in the sense that

$$\lim_{\delta \to 0} \frac{\log \mathbb{P}\{\hat{B}^{\delta}(z) \cap X \neq \emptyset\}}{\log \delta} = \tilde{\Delta}.$$

The validity of the KPZ relation (18.4) for the pair $(\tilde{x}, \tilde{\Delta})$, as anticipated in [32], is ensured by the following theorem [37].

Theorem 18.6 *Fix $\gamma \in [0, 2)$ and let $\hat{\partial D}$ be a closed subinterval of one of the boundary line segments of ∂D. If $X \cap \hat{\partial D}$ has a Euclidean scaling exponent $\tilde{x} \geq 0$ then it has a quantum scaling exponent $\tilde{\Delta} = \tilde{\Delta}_\gamma$, where $\tilde{\Delta}_\gamma$ is the non-negative solution to*

$$\tilde{x} = \frac{\gamma^2}{4} \tilde{\Delta}_\gamma^2 + \left(1 - \frac{\gamma^2}{4}\right) \tilde{\Delta}_\gamma. \tag{18.77}$$

The proofs in the boundary case proceed in exactly the same way as in the interior point case, up to factors of 2 in various places. We sketch the proof in order to indicate where those factors of 2 appear.

18.5.3 Boundary quantum ball

As in Section 18.4.1, when ε is small, the regularized stochastic quantum measure $\mu^B_{\gamma,\varepsilon}(\hat{B}_\varepsilon(z))$ (18.74) of the interval $\hat{B}_\varepsilon(z)$, the intersection of the Euclidean ball $B_\varepsilon(z)$ with ∂D, and the limit quantum measure $\mu^B_\gamma(\hat{B}_\varepsilon(z))$ are very well approximated by the simple form $\mu^B_\gamma(\hat{B}_\varepsilon(z)) \simeq 2\varepsilon \times \varepsilon^{\gamma^2/4} e^{\gamma h_\varepsilon(z)/2}$, i.e. the Lebesgue measure of the interval times the regularized conformal factor at the ball's center z. Let us then write

$$\mu^B_\gamma(\hat{B}_\varepsilon(z)) = 2 \exp\left[\frac{\gamma}{2} h_\varepsilon(z) + \frac{\gamma}{2} Q_\gamma \log \varepsilon\right], \tag{18.78}$$

$$Q_\gamma = \frac{2}{\gamma} + \frac{\gamma}{2},$$

and, in the simplified perspective of this lecture, take eqn (18.78) to be the *definition* of the quantum boundary measure $\mu^B_\gamma(\hat{B}_\varepsilon(z))$.

Let $\hat{B}^\delta(z)$ now be the boundary intersection of the largest Euclidean ball $B_\varepsilon(z)$ in D centered at $z \in \partial D$ for which eqn (18.78) is equal to the quantum length δ:

$$\mu^B_\gamma(\hat{B}^\delta(z)) = \delta. \tag{18.79}$$

18.5.4 GFF sampling and boundary quantum measure

We can repeat the whole analysis of Sections 18.3.2 and 18.3.3 in the case of the boundary weight $e^{\gamma h_\varepsilon(z)/2}$. Because of eqn (18.68), the deterministic potential f^z_ε (18.15) is now replaced by $\hat{f}^z_\varepsilon$ (18.69). In analogy to the result (18.37), we have that, after sampling $z \in \partial D$ from its marginal distribution, the law of h *weighted* by $e^{\gamma h_\varepsilon(z)/2}$ is identical to that of the original GFF h' *plus* the deterministic function $(\gamma/2)\hat{f}^z_\varepsilon$ (18.69) (up to the ε_0-dependent constant term):

$$h \overset{\text{(in law)}}{=} h' + \frac{\gamma}{2} \hat{f}^z_\varepsilon. \tag{18.80}$$

For fixed $z \in \partial D$, the identity in law (18.80) can be averaged over the *half-circle* $\partial B_\varepsilon(z) \cap \mathbb{H}$, as in eqn (18.68), to yield

$$\begin{aligned} h_\varepsilon(z) - h_{\varepsilon_0}(z) &\overset{\text{(in law)}}{=} h'_\varepsilon(z) - h'_{\varepsilon_0}(z) + \frac{\gamma}{2}(\hat{f}^z_\varepsilon, \hat{f}^z_\varepsilon)_\nabla, \quad z \in \partial D\,, \\ &= h'_\varepsilon(z) - h'_{\varepsilon_0}(z) - \gamma \log(\varepsilon/\varepsilon_0)\,. \end{aligned} \tag{18.81}$$

18.5.5 Boundary Brownian martingale

Since h' is a standard Gaussian free field, we can apply to its half-circular average $h'_\varepsilon(z)$ the identity in law given by eqns (18.71) and (18.72) to the standard Brownian motion $\mathcal{B}_{2t}$, $t = -\log(\varepsilon/\varepsilon_0)$, and rewrite eqn (18.81) as

$$z \in \partial D\colon \; h_\varepsilon(z) - h_{\varepsilon_0}(z) \overset{\text{(in law)}}{=} \mathcal{B}_{2t} + \gamma t \;\; (\mathcal{B}_0 = 0)\,, \tag{18.82}$$

where $\mathcal{B}_{2t}$ evolves as a Brownian motion with twice the variance of standard Brownian motion, because of the free boundary conditions on ∂D.

The (relative) quantum boundary measure (18.78) can thus be written as

$$\begin{aligned} \frac{\mu^B_\gamma(\hat{B}_\varepsilon(z))}{\mu^B_\gamma(\hat{B}_{\varepsilon_0}(z))} &= \exp\left[\frac{\gamma}{2}\mathcal{B}_{2t} + \frac{1}{2}\gamma^2 t - \frac{\gamma}{2}Q_\gamma t\right], \\ Q_\gamma &= \frac{2}{\gamma} + \frac{\gamma}{2}\,. \end{aligned} \tag{18.83}$$

Hereafter, we shall insert in eqn (18.79) the reference quantum measure of the fixed interval $\hat{B}_{\varepsilon_0}(z)$ as in eqn (18.83), which amounts to a redefinition of the quantum length δ.

Thus, equating eqn (18.83) to the *quantum boundary length* δ, as in the definition (18.79) of the quantum interval $\hat{B}^\delta$, corresponds in eqn (18.83) to the smallest t for which $\gamma\mathcal{B}_{2t} + \gamma^2 t - \gamma Q_\gamma t = 2\log\delta$. That is, $-\mathcal{B}_{2t} + (Q_\gamma - \gamma)t = -2(\log\delta)/\gamma$. If we set $A := -(\log\delta)/\gamma$, such that $A > 0$, this smallest time is a stopping time T_A such that

$$T_A = \inf\{t : -\mathcal{B}_{2t} + a_\gamma t = 2A\}, \; a_\gamma = Q_\gamma - \gamma = \frac{2}{\gamma} - \frac{\gamma}{2} > 0\,. \tag{18.84}$$

As above, we consider a two-part experiment in which we first sample T_A and then sample z and check to see whether a ball of radius $\varepsilon_A = e^{-T_A}$ intersects X on the boundary. The ratio of the logarithms of this probability $\mathbb{P}(\hat{B}_{\varepsilon_A}(z) \cap X \neq \emptyset)$ and of $\mathbb{E}\left[(\varepsilon_A)^{\tilde{x}}\right] = \mathbb{E}\left[\exp(-\tilde{x}T_A)\right]$ tends to 1 as $\delta \to 0$, $A \to \infty$, which we write

$$\mathbb{P}\{\hat{B}^\delta(z) \cap X \neq \emptyset\} \asymp \mathbb{E}\left[\exp(-\tilde{x}T_A)\right]. \tag{18.85}$$

We next consider, for any β, the exponential martingale $\exp(-(\beta/2)\mathcal{B}_{2t} - (\beta^2/4)t)$, such that

$$\mathbb{E}\left[\exp\left(-\frac{\beta}{2}\mathcal{B}_{2t} - \frac{\beta^2}{4}t\right)\right] = \mathbb{E}\left[\exp\left(-\frac{\beta}{2}\mathcal{B}_0\right)\right] = 1\,.$$

For $\beta A \geq 0$, and hence $\beta \geq 0$, we can apply this at the stopping time T_A, where $T_A < \infty$ and $\mathcal{B}_{2T_A} = a_\gamma T_A - 2A$. We thus get

$$\mathbb{E}\left[\exp\left[\beta A - \left(\frac{\beta}{2}a_\gamma + \frac{\beta^2}{4}\right)T_A\right]1_{T_A<\infty}\right] = 1\,,$$

which can be written as

$$\mathbb{E}[\exp(-\tilde{x}T_A)1_{T_A<\infty}] = \exp(-\beta A)\,, \tag{18.86}$$

$$\tilde{x} := \frac{\beta}{2}a_\gamma + \frac{\beta^2}{4}\,. \tag{18.87}$$

This is valid for $\beta \geq 0$, which is thus equal to the positive root of eqn (18.87),

$$\beta_\gamma(\tilde{x}) = (a_\gamma^2 + 4\tilde{x})^{1/2} - a_\gamma\,, \tag{18.88}$$

which is the same function as in eqn (18.56) in the bulk case. Thanks to the definition of A, we thus obtain

$$\mathbb{E}[\exp(-\tilde{x}T_A)1_{T_A<\infty}] = \exp(-\beta_\gamma(\tilde{x})A) = \delta^{\tilde{\Delta}_\gamma}\,, \tag{18.89}$$

$$\tilde{\Delta}_\gamma := \frac{\beta_\gamma(\tilde{x})}{\gamma} = \frac{(a_\gamma^2 + 4\tilde{x})^{1/2} - a_\gamma}{\gamma}\,. \tag{18.90}$$

As in eqn (18.59), the conditioning on $T_A < \infty$ can actually be omitted, since $a_\gamma > 0$ for $\gamma \in [0, 2)$.

Equations (18.85), (18.89), and (18.90) establish the validity of the quantum boundary scaling (18.76), together with the *boundary KPZ relation* (18.77). The latter and eqn (18.90) are the same as Theorem 18.4 and eqn (18.58) in the bulk case, QED.

We remark, finally, that a procedure similar to that above allows us to make sense of measures restricted to lines in the interior of the domain.

18.6 Liouville quantum duality

18.6.1 $\gamma' = 4/\gamma$ Liouville duality

For $\gamma > 2$, the Liouville measure (18.32) corresponds to the so-called "other" gravitational dressing of the Liouville potential [7, 54–56]. The corresponding random surface is meant to be the scaling limit of random simply connected surfaces with large amounts of area cut off by small bottlenecks [2, 4, 18, 39, 58, 59] (see also [50]). This surface turns out to be a tree-like foam of Liouville quantum bubbles of *dual* parameter

$$\gamma' := \frac{4}{\gamma},\ \gamma' < 2 < \gamma\,, \tag{18.91}$$

("baby universes"), connected to each other at "pinch points" and rooted at a "principal bubble" parameterized by D. A precise mathematical description requires additional machinery and will appear elsewhere. Here we relate γ to γ' formally.

The definition of quantum balls in eqns (18.41) and (18.43) makes sense when $\gamma > 2$. Noting that $Q_{\gamma'} = Q_\gamma =: Q$, we have

$$\mu_{\gamma'}(B_\varepsilon(z)) = \pi\varepsilon^{\gamma' Q}e^{\gamma' h_\varepsilon(z)} = \mu_\gamma(B_\varepsilon(z))^{\gamma'/\gamma} = \mu_\gamma^{4/\gamma^2} \tag{18.92}$$

(up to an irrelevant power of π)—i.e. a γ-quantum ball of size δ has a γ'-quantum size

$$\delta' := \delta^{4/\gamma^2}\,. \tag{18.93}$$

(Intuitively, the ball contains about a fraction δ' of the total γ'-quantum area but only a fraction $\delta < \delta'$ of the γ-quantum area because the latter also includes points on nonprincipal bubbles.) The number of γ'-quantum-size-δ' balls (i.e. γ-quantum-size-δ balls) needed to cover the principal bubble D thus scales as

$$N_{\gamma'}(\delta') \asymp \delta'^{-1} = \delta^{-4/\gamma^2}\,. \tag{18.94}$$

If the fractal random subset $X \subset D$ has a scaling exponent x, then eqn (18.45), established for $\gamma' < 2$, essentially says (see eqns (18.10) and (18.11)) that the expected number $N_{\gamma'}(\delta', X)$ of γ'-quantum-size-δ' balls (i.e. the number $N_\gamma(\delta, X)$ of γ-quantum-size-δ balls) required to cover X, when normalized by $N_{\gamma'}(\delta')$, scales, as in eqn (18.10), as

$$\frac{N_{\gamma'}(\delta', X)}{N_{\gamma'}(\delta')} \asymp \delta'^{\Delta_{\gamma'}}\,. \tag{18.95}$$

From eqns (18.93) and (18.94) there follows the scaling identity

$$N_{\gamma'}(\delta', X) \asymp \delta'^{\Delta_{\gamma'}-1} = \delta^{\Delta_\gamma - 1} \asymp N_\gamma(\delta, X)\,, \tag{18.96}$$

with the scaling exponent Δ_γ defined by the *duality relation*

$$\Delta_\gamma - 1 := \frac{4}{\gamma^2}(\Delta_{\gamma'} - 1),\ \ \gamma' = \frac{4}{\gamma}\,. \tag{18.97}$$

Equation (18.96) is thus the generalization of the quantum scaling (18.11) to the case $\gamma > 2$.

Rewriting the KPZ relation (18.4) as

$$x - 1 = \frac{\gamma^2}{4}(\Delta - 1)^2 + \left(1 + \frac{\gamma^2}{4}\right)(\Delta - 1)\,,$$

we note that the triple $(x, \gamma, \Delta_\gamma)$, thanks to eqn (18.97), satisfies eqn (18.4) since $(x, \gamma', \Delta'_\gamma)$ does. The quantum scaling exponent Δ_γ, as defined by the duality relation (18.97), then *coincides* with the quantum scaling exponent in eqns (18.58) and (18.60), *generalized to* $\gamma > 2$:

$$\begin{aligned} \Delta_\gamma &= \gamma^{-1}[(a_\gamma^2 + 4x)^{1/2} - a_\gamma]\,, \\ \gamma > 2,\ a_\gamma &= \frac{2}{\gamma} - \frac{\gamma}{2} < 0\,. \end{aligned} \tag{18.98}$$

The duality relation (18.97) can indeed be written as

$$\gamma(\Delta_\gamma - 1) = \gamma'(\Delta_{\gamma'} - 1)\,, \tag{18.99}$$

which is easily checked using eqn (18.98). Dual quantum exponents appear in [7, 32, 55, 56]. In [32], the duality relation (18.97) was introduced in the natural context of boundary exponents associated with the *dense phase* of the $O(N)$ model or with the stochastic Loewner evolution SLE_κ for $\kappa \geq 4$. (See also Kostov's chapter in this book.)

18.6.2 Distribution of pinch points

A rigorous construction of the singular quantum measures $\mu_{\gamma>2}$ will be presented elsewhere. We shall simply mention here that it uses the dual measure $\mu_{\gamma'<2}$ and the additional randomness of sets of point masses $\{z_k\}$, where *finite* amounts $\mu_\gamma(z_k)$ of quantum area are *localized* and which are such that $\{z_k, \mu_\gamma(z_k)\}$ is distributed as a Poissonian point process of intensity $\mu_{\gamma'} \times \Lambda^\nu$, where $\Lambda^\nu(d\eta) := \eta^{-\nu-1} d\eta$, with parameter $\nu := 4/\gamma^2 \in (0, 1)$. The characteristic Laplace transform of the γ-quantum area $\mu_\gamma(\mathcal{D})$ of any subdomain $\mathcal{D} \subset D$ then obeys the *duality identity*

$$\mathbb{E} \exp[-\lambda \mu_\gamma(\mathcal{D})] = \exp[\Gamma(-\nu)\lambda^\nu \mu_{\gamma'}(\mathcal{D})]$$

for any $\lambda > 0$, where $\Gamma(-\nu) = -\Gamma(1-\nu)/\nu < 0$ is the usual Euler Γ-function. This duality identity implies that the dual cosmological constant associated with $\mu_{\gamma'}$ is exactly

$$\lambda' := -\Gamma(-\nu)\lambda^\nu, \ \nu = \frac{4}{\gamma^2}.$$

In other words, given $\mu_{\gamma'}$, the typical size of $\mu_\gamma(\mathcal{D})$ is $\mu_{\gamma'}(\mathcal{D})^{\gamma^2/4}$, in agreement with eqn (18.92) and with the scaling expected from the physics literature. The relations above are the probabilistic formulation in Liouville quantum gravity of the *Legendre transform* relating the free energies of dual Liouville theories, as discovered in [55]. (See also Kostov's chapter in this book.)

18.6.3 Brownian approach to duality

When $\gamma > 2$, the ε-*regularized* measures $M_\varepsilon(z)\, dz$ (18.32) converge to zero. If we choose the pair (z, h) from the weighted measure $M_\varepsilon(z)\, dh\, dz$ as in eqn (18.36) and consider the Brownian description (18.49), we find that $a_\gamma < 0$ for $\gamma > 2$, i.e. the drift term runs in a direction opposite to $A > 0$, so that $T_A = \infty$ for large A. The weighted measure is thus *singular*; i.e. there is a quantum area of at least δ *localized* at z for small enough δ.

The Brownian-martingale result (18.57) still holds for $\gamma > 2$:

$$\mathbb{E}[\exp(-2xT_A)1_{T_A<\infty}] = \delta^{\Delta_\gamma}, \qquad (18.100)$$
$$\frac{(a_\gamma^2 + 4x)^{1/2} - a_\gamma}{\gamma} = \Delta_\gamma .$$

For $x = 0$, it gives the probability, at a given z, for T_A to be finite:

$$\mathbb{P}(T_A < \infty) = \mathbb{E}[1_{T_A<\infty}] = \delta^{\Delta_\gamma(0)} = \delta^{1-4/\gamma^2} = \frac{\delta}{\delta'} < 1, \qquad (18.101)$$

where $\Delta_\gamma(0) = (|a_\gamma| - a_\gamma)/\gamma = 1 - 4/\gamma^2$; this is in agreement with the probability integral (18.65).

We may define a δ-*regularized* measure $M^\delta(z)\, dh\, dz$ as $M_{\varepsilon_A}(z)\, dh\, dz$ restricted to the event $T_A < \infty$. This means that there is *strictly less* than the quantum area δ localized at the point z. Replacing γ with $\gamma' = 4/\gamma$ and δ with δ' has the same effect as multiplying $M^\delta(z)$ by $\delta'/\delta = \delta^{4/\gamma^2-1}$, so

$$\delta^{4/\gamma^2-1} M^\delta(z)\, dz = \frac{\delta'}{\delta} M^\delta(z)\, dz =: d\mu_{\gamma',\varepsilon_A} \qquad (18.102)$$

does converge to $d\mu_{\gamma'}$ with $\gamma' < 2$.

From eqns (18.100) and (18.101), we deduce the *conditional* expectation

$$\frac{\mathbb{E}[\exp(-2xT_A)1_{T_A<\infty}]}{\mathbb{E}[1_{T_A<\infty}]} = \delta^{\Delta_\gamma} \times \frac{\delta'}{\delta} = \delta'^{\,\Delta_{\gamma'}}\,, \tag{18.103}$$

which scales now in terms of the γ'-quantum area δ' with the *dual* quantum scaling exponent $\Delta_{\gamma'}$, $\gamma' < 2$, in complete agreement with the definition of the measure (18.102) above.

18.6.4 Properties of Liouville duality

Using eqn (18.98) and $a_{\gamma'} = -a_\gamma$, we obtain the duality identity

$$x = \Delta_\gamma \Delta_{\gamma'}, \;\; \gamma\gamma' = 4\,,$$

as anticipated in [32]. Thanks to eqn (18.99), the typical GFF thickness (18.67)

$$\alpha = \gamma(1 - \Delta_\gamma) = \gamma'(1 - \Delta_{\gamma'})$$

is *invariant* under duality and obeys the Seiberg bound [78]

$$\alpha \le Q_\gamma\,.$$

The *string susceptibility exponent*

$$\gamma_{\text{str}} = 2 - \frac{2}{\gamma} Q_\gamma = 1 - \frac{4}{\gamma^2} \tag{18.104}$$

obeys the expected duality relation

$$(1 - \gamma_{\text{str}})(1 - \gamma'_{\text{str}}) = 1\,,$$

with

$$\gamma_{\text{str}} > 0 > \gamma'_{\text{str}}, \;\; \gamma > 2 > \gamma' = \frac{4}{\gamma}\,.$$

Liouville quantum gravity is thus expected to describe, for $\gamma > 2$, the continuum limit of critical discrete matrix models where a nonstandard string susceptibility exponent $\gamma_{\text{str}} > 0$ appears; this happens precisely at special values of the couplings in those models, where a fine tuning of the fugacity controlling the occurrence of bottlenecks leads to a critical proliferation of "baby universes". (See [2, 4, 7, 18, 32, 39, 54–56, 58, 59].)

18.6.5 The other branch of the KPZ relation

When $\gamma \ge 2$, the positive root Δ_γ (see eqns (18.5) and (18.98)) of the relation (18.4) is

$$\Delta_\gamma = \frac{1}{\gamma}\left(\sqrt{4x + \left(\frac{2}{\gamma} - \frac{\gamma}{2}\right)^2} + \left|\frac{2}{\gamma} - \frac{\gamma}{2}\right|\right), \;\; \gamma \ge 2\,, \tag{18.105}$$

which can no longer be written in the standard KPZ form given in eqns (18.6) and (18.7).

If we use the standard conformal-field-theory parameterization (18.6) with a central charge $c \leq 1$ for the dual value $\gamma' = 4/\gamma \leq 2$,

$$\gamma' = \frac{1}{\sqrt{6}}\left(\sqrt{25-c} - \sqrt{1-c}\right) \leq 2, \ \ c \leq 1\,, \tag{18.106}$$

we get for $\gamma \geq 2$ the other parameterization

$$\gamma = \frac{4}{\gamma'} = \frac{1}{\sqrt{6}}\left(\sqrt{25-c} + \sqrt{1-c}\right) \geq 2, \ \ c \leq 1\,. \tag{18.107}$$

With this identification, we find for eqn (18.105) the nonstandard form, i.e. the "other branch" of the KPZ relation

$$\Delta_{\gamma \geq 2} = \frac{\sqrt{24x+1-c} + \sqrt{1-c}}{\sqrt{25-c} + \sqrt{1-c}}, \ \ c \leq 1\,, \tag{18.108}$$

in agreement with [54–56].

The string susceptibility exponent (18.104) for $\gamma' \leq 2$, when parameterized as in eqn (18.106), has the usual form

$$\gamma'_{\mathrm{str}} = 1 - \frac{4}{\gamma'^2} = \frac{1}{12}\left(c - 1 - \sqrt{(25-c)(1-c)}\right) \leq 0, \ \ c \leq 1\,, \tag{18.109}$$

while in the $\gamma \geq 2$ domain, the other parameterization (18.107) gives its dual value

$$\gamma_{\mathrm{str}} = 1 - \frac{4}{\gamma^2} = \frac{1}{12}\left(c - 1 + \sqrt{(25-c)(1-c)}\right) \geq 0, \ \ c \leq 1\,. \tag{18.110}$$

Finally, for the SLE_κ process, $\gamma = \sqrt{\kappa}$ [37], so that the present Liouville duality $\gamma\gamma' = 4$ and the SLE duality $\kappa\kappa' = 16$ [30, 31] coincide.

These lecture notes have explained the KPZ relation for continuum Liouville quantum gravity. Some outstanding open problems relate discrete models and SLE to Liouville quantum gravity, as described in [37]. We hope they will be solved by the methods described here.

Acknowledgments

It is a pleasure to thank Michael Aizenman, Omer Angel, Jacques Franchi, Peter Jones, Igor Klebanov, Ivan Kostov, Greg Lawler, Jean-François Le Gall, Andrei Okounkov, Tom Spencer, Paul Wiegmann, Michel Zinsmeister, and the late Oded Schramm for useful discussions. We also thank Emmanuel Guitter for his help with the figures. Support from the ANR project "GranMa" (ANR-08-BLAN-0311-01) is gratefully acknowledged. This research has made use of NASA's Astrophysics Data System.

References

[1] S. Albeverio, G. Gallavotti, and R. Hoegh-Krohn. Some results for the exponential interaction in two or more dimensions. *Commun. Math. Phys.*, 70:187–192, 1979.

[2] L. Alvarez-Gaumé, J. L. F. Barbón, and Č. Crnković. A proposal for strings at $D > 1$. *Nucl. Phys. B*, 394:383–422, 1993.

[3] J. Ambjorn, K. N. Anagnostopoulos, U. Magnea, and G. Thorleifsson. Geometrical interpretation of the Knizhnik–Polyakov–Zamolodchikov exponents. *Phys. Lett. B*, 388:713–719, 1996.

[4] J. Ambjorn, B. Durhuus, and T. Jonsson. A Solvable 2d Gravity Model with $\gamma > 0$. *Mod. Phys. Lett. A*, 9:1221–1228, 1994.

[5] J. Ambjørn, B. Durhuus, and T. Jonsson. *Quantum Geometry: A Statistical Field Theory Approach*, Cambridge Monographs on Mathematical Physics. Cambridge University Press, Cambridge, 1997.

[6] K. Anagnostopoulos, P. Bialas, and G. Thorleifsson. The Ising model on a quenched ensemble of $c = -5$ gravity graphs. *J. Stat. Phys.*, 94:321–345, 1999.

[7] J. L. F. Barbón, K. Demeterfi, I. R. Klebanov, and C. Schmidhuber. Correlation functions in matrix models modified by wormhole terms. *Nucl. Phys. B*, 440:189–214, 1995.

[8] M. Bauer. *Aspects de l'invariance conforme*. Ph.D. thesis, Université Paris 7, 1990.

[9] I. Benjamini and O. Schramm. KPZ in one dimensional random geometry of multiplicative cascades. *Commun. Math. Phys.*, 46–56, 2009.

[10] A. N. Borodin and P. Salminen. *Handbook of Brownian Motion*, 2nd edn. Birkhäuser Verlag, Basel, 2000, p. 295, paragraphs 2.0.1 and 2.0.2.

[11] M. Bousquet-Mélou and G. Schaeffer. The degree distribution in bipartite planar maps: Applications to the Ising model. In *Proceedings of FPSAC 03 (Formal Power Series and Algebraic Combinatorics)*, Vadstena, Sweden, June 2003, eds. K. Eriksson and S. Linusson, pp. 312–323; arXiv:math/0211070, 2003.

[12] J. Bouttier, P. Di Francesco, and E. Guitter. Census of planar maps: From the one-matrix model solution to a combinatorial proof. *Nucl. Phys. B*, 645:477–499, 2002.

[13] J. Bouttier, P. Di Francesco, and E. Guitter. Geodesic distance in planar graphs. *Nucl. Phys. B*, 663:535–567, 2003.

[14] J. Bouttier, P. Di Francesco, and E. Guitter. Blocked edges on Eulerian maps and mobiles: Application to spanning trees, hard particles and the Ising model. *J. Phys. A*, 40:7411–7440, 2007.

[15] J. Bouttier and E. Guitter. Statistics of geodesics in large quadrangulations. *J. Phys. A*, 41:145001, 2008.

[16] J. Bouttier and E. Guitter. The three-point function of planar quadrangulations. *J. Stat. Mech.*, 7:P07020, 2008.

[17] J. Bouttier and E. Guitter. Confluence of geodesic paths and separating loops in large planar quadrangulations. *J. Stat. Mech.*, P03001, 2009.

[18] S. R. Das, A. Dhar, A. M. Sengupta, and S. R. Wadia. New critical behavior in $d = 0$ large-N matrix models. *Mod. Phys. Lett. A*, 5:1041–1056, 1990.

[19] J.-M. Daul. Q-states Potts model on a random planar lattice. arXiv:hep-th/9502014, 1995.

[20] F. David. Conformal field theories coupled to 2-D gravity in the conformal gauge. *Mod. Phys. Lett. A*, 3:1651–1656, 1988.

[21] F. David. Simplicial quantum gravity and random lattices. In *Gravitation et quantifications (Les Houches, Session LVII, 1992)*, eds. B. Julia and J. Zinn-Justin, pp. 679–749. Elsevier, Amsterdam, 1995.

[22] F. David and M. Bauer. Another derivation of the geometrical KPZ relations. *J. Stat. Mech.*, 3:P03004, 2009.

[23] P. Di Francesco, P. Ginsparg, and J. Zinn-Justin. 2D gravity and random matrices. *Phys. Rep.*, 254:1–133, 1995.

[24] P. Di Francesco and E. Guitter. Geometrically constrained statistical systems on regular and random lattices: From folding to meanders. *Phys. Rep.*, 415(1):1–88, 2005.

[25] J. Distler and H. Kawai. Conformal field theory and 2D quantum gravity. *Nucl. Phys. B*, 321:509–527, 1989.

[26] H. Dorn and H.-J. Otto. Two- and three-point functions in Liouville theory. *Nucl. Phys. B*, 429:375–388, 1994.

[27] B. Duplantier. Random walks and quantum gravity in two dimensions. *Phys. Rev. Lett.*, 81:5489–5492, 1998.

[28] B. Duplantier. Harmonic measure exponents for two-dimensional percolation. *Phys. Rev. Lett.*, 82:3940–3943, 1999.

[29] B. Duplantier. Two-dimensional copolymers and exact conformal multifractality. *Phys. Rev. Lett.*, 82:880–883, 1999.

[30] B. Duplantier. Conformally invariant fractals and potential theory. *Phys. Rev. Lett.*, 84:1363–1367, 2000.

[31] B. Duplantier. Higher conformal multifractality. *J. Stat. Phys.*, 110:691–738, 2003. Special issue in honor of Michael E. Fisher's 70th birthday.

[32] B. Duplantier. Conformal fractal geometry & boundary quantum gravity. In *Fractal Geometry and Applications: A Jubilee of Benoît Mandelbrot*, Part 2, Proceedings of Symposia on Pure Mathematics, Vol. 72, pp. 365–482. American Mathematical Society, Providence, RI, 2004.

[33] B. Duplantier. Conformal random geometry. In *Mathematical Statistical Physics (Les Houches Summer School, Session LXXXIII, 2005)*, eds. A. Bovier, F. Dunlop, F. den Hollander, A. van Enter, and J. Dalibard, pp. 101–217. Elsevier, Amsterdam, 2006.

[34] B. Duplantier and I. A. Binder. Harmonic measure and winding of conformally invariant curves. *Phys. Rev. Lett.*, 89:264101, 2002.

[35] B. Duplantier and I. Kostov. Conformal spectra of polymers on a random surface. *Phys. Rev. Lett.*, 61:1433–1437, 1988.

[36] B. Duplantier and I. K. Kostov. Geometrical critical phenomena on a random surface of arbitrary genus. *Nucl. Phys. B*, 340:491–541, 1990.

[37] B. Duplantier and S. Sheffield. Liouville quantum gravity and KPZ. arXiv:0808.1560, 2008.

[38] B. Duplantier and S. Sheffield. Duality and KPZ in Liouville quantum gravity. *Phys. Rev. Lett.*, 102:150603, 2009.

[39] B. Durhuus. Multi-spin systems on a randomly triangulated surface. *Nucl. Phys. B*, 426:203–222, 1994.

[40] B. Eynard. Large N expansion of convergent matrix integrals, holomorphic anomalies, and background independence. *J. High Energy Phys.*, 3:3, 2009.

[41] B. Eynard and C. Kristjansen. Exact solution of the $O(n)$ model on a random lattice. *Nucl. Phys. B*, 455:577–618, 1995.

[42] B. Eynard and N. Orantin. Invariants of algebraic curves and topological expansion. arXiv:0702.045, *Commun. Number Theor. Phys.*, 1:347–452, 2007.

[43] B. Eynard and N. Orantin. Topological expansion and boundary conditions. *J. High Energy Phys.*, 6:37, 2008.

[44] V. Fateev, A. B. Zamolodchikov, and A. B. Zamolodchikov. Boundary Liouville field theory I. Boundary state and boundary two-point function. arXiv:hep-th/0001012, 2000.
[45] P. Ginsparg and G. Moore. Lectures on 2d gravity and 2d string theory (TASI 1992). In *Recent Directions in Particle Theory, Proceedings of the 1992 TASI*, eds J. Harvey and J. Polchinski. World Scientific, Singapore, 1993.
[46] M. Goulian and M. Li. Correlation functions in Liouville theory. *Phys. Rev. Lett.*, 66:2051–2055, 1991.
[47] R. Hoegh-Krohn. A general class of quantum fields without cut-offs in two space–time dimensions. *Commun. Math. Phys.*, 21:244–255, 1971.
[48] K. Hosomichi. Bulk-boundary propagator in Liouville theory on a disc. *J. High Energy Phys.*, 11:44, 2001.
[49] X. Hu, J. Miller, and Y. Peres. Thick points of the Gaussian free field. arXiv:0902.3842, 2009.
[50] S. Jain and S. D. Mathur. World-sheet geometry and baby universes in 2D quantum gravity. *Phys. Lett. B*, 286:239–246, 1992.
[51] W. Janke and D. A. Johnston. The wrong kind of gravity. *Phys. Lett. B*, 460:271–275, 1999.
[52] V. A. Kazakov. Ising model on a dynamical planar random lattice: Exact solution. *Phys. Lett. A*, 119:140–144, 1986.
[53] V. A. Kazakov and I. K. Kostov. Loop gas model for open strings. *Nucl. Phys. B*, 386:520–557, 1992.
[54] I. R. Klebanov. Touching random surfaces and Liouville gravity. *Phys. Rev. D*, 51:1836–1841, 1995.
[55] I. R. Klebanov and A. Hashimoto. Non-perturbative solution of matrix models modified by trace-squared terms. *Nucl. Phys. B*, 434:264–282, 1995.
[56] I. R. Klebanov and A. Hashimoto. Wormholes, matrix models, and Liouville gravity. *Nucl. Phys. B Proc. Suppl.*, 45:135–148, 1996.
[57] V. G. Knizhnik, A. M. Polyakov, and A. B. Zamolodchikov. Fractal structure of 2D-quantum gravity. *Mod. Phys. Lett. A*, 3:819–826, 1988.
[58] G. P. Korchemsky. Loops in the curvature matrix model. *Phys. Lett. B*, 296:323–334, 1992.
[59] G. P. Korchemsky. Matrix model perturbed by higher order curvature terms. *Mod. Phys. Lett. A*, 7:3081–3100, 1992.
[60] I. K. Kostov. $O(n)$ vector model on a planar random lattice: Spectrum of anomalous dimensions. *Mod. Phys. Lett. A*, 4:217–226, 1989.
[61] I. K. Kostov. Boundary correlators in 2D quantum gravity: Liouville versus discrete approach. *Nucl. Phys. B*, 658:397–416, 2003.
[62] I. K. Kostov. Boundary ground ring in 2D string theory. *Nucl. Phys. B*, 689:3–36, 2004.
[63] I. K. Kostov. Boundary loop models and 2D quantum gravity. *J. Stat. Mech.*, 08:P08023, 2007.
[64] I. K. Kostov, B. Ponsot, and D. Serban. Boundary Liouville theory and 2D quantum gravity. *Nucl. Phys. B*, 683:309–362, 2004.
[65] J.-F. Le Gall. The topological structure of scaling limits of large planar maps. *Invent. Math.*, 169:621–670, 2007.
[66] J.-F. Le Gall. Geodesics in large planar maps and the Brownian map. *Acta. Math.* (to

appear) arXiv:0804.3012, 2008.

[67] G. M. Miermont. Tessellations of random maps of arbitrary genus. *Ann. Éc. Norm. Supér.*, 42(5):725–781, 2009. arXiv:0712.3688, 2007.

[68] G. M. Miermont. On the sphericity of scaling limits of random planar quadrangulations. *Electron. Commun. Probab.*, 13:248–257, 2008.

[69] Y. Nakayama. Liouville field theory. *Int. J. Mod. Phys. A*, 19:2771–2930, 2004.

[70] A. M. Polyakov. Quantum geometry of bosonic strings. *Phys. Lett. B*, 103:207–210, 1981.

[71] A. M. Polyakov. *Gauge Fields and Strings*. Harwood Academic, Chur, 1987.

[72] A. M. Polyakov. Quantum gravity in two-dimensions. *Mod. Phys. Lett. A*, 2:893, 1987.

[73] A. M. Polyakov. From quarks to strings. arXiv:0812.0183, 2008.

[74] C. Pommerenke. *Boundary Behaviour of Conformal Maps*, Grundlehren Math. Wiss., Vol. 299. Springer-Verlag, Berlin, 1992.

[75] B. Ponsot and J. Teschner. Boundary Liouville field theory: Boundary three-point function. *Nucl. Phys. B*, 622:309–327, 2002.

[76] R. Rhodes and V. Vargas. KPZ formula for log-infinitely divisible multifractal random measures. arXiv:0807.1036, 2008.

[77] G. Schaeffer. *Conjugaison d'arbres et cartes combinatoires aléatoires*. Ph.D. thesis, University of Bordeaux I, Talence, 1998.

[78] N. Seiberg. Notes on quantum Liouville theory and quantum gravity. *Prog. Theor. Phys. Suppl.*, 102:319–349, 1990.

[79] S. Sheffield. Gaussian free fields for mathematicians. *Probab. Theor. Rel. Fields*, 139:521–541, 2007.

[80] B. Simon. *The* $P(\Phi)_2$ *Euclidean (Quantum) Field Theory*. Princeton University Press, Princeton, NJ, 1974.

[81] J. Teschner. On the Liouville three-point function. *Phys. Lett. B*, 363:65–70, 1995.

[82] J. Teschner. Liouville theory revisited. *Class. Quantum Grav.*, 18:R153–R222, 2001.

[83] J. Teschner. From Liouville theory to the quantum geometry of Riemann surfaces. In *Prospects in Mathematical Physics*, Contemporary Mathematics, Vol. 437, pp. 231–246. American Mathematical Society, Providence, RI, 2007.

[84] A. B. Zamolodchikov and A. B. Zamolodchikov. Structure constants and conformal bootstrap in Liouville field theory. *Nucl. Phys. B*, 477:577–605, 1996.

[85] A. B. Zamolodchikov. Higher equations of motion in Liouville field theory. *Int. J. Mod. Phys. A*, 19:510–523, 2004.

19
Topologically protected qubits based on Josephson junction arrays

M. V. FEIGEL'MAN

Mikhail V. Feigel'man,
L. D. Landau Institute for Theoretical Physics,
Kosygina 2, Moscow, 119334,
Russia.

Preface

In this lecture, I provide a rather detailed review of an approach to the physical implementation of topologically protected qubits in Josephson junction arrays developed in several papers (Ioffe and Feigel'man 2002, Douçot *et al.* 2003, Douçot *et al.* 2005, Protopopov and Feigel'man 2004). At the end of these notes I also provide a brief account of the most recent achievements in this direction (Douçot and Ioffe 2009, Gladchenko *et al.* 2009)

19.1 Introduction

Quantum computing (Ekert and Jozsa 1996, Steane 1998) is in principle a very powerful technique for solving classic "hard" problems such as factorizing large numbers (Shor 1994) and sorting large lists (Grover 1996). The remarkable discovery of quantum error correction algorithms (Shor 1995) shows that there is no problem of principle involved in building a functioning quantum computer. However, implementation still seems dauntingly difficult: the essential ingredient of a quantum computer is a quantum system with 2^K (with $K \gg 100$) quantum states which are degenerate (or nearly so) in the absence of external perturbations and are insensitive to the "random" fluctuations which exist in every real system, but which may be manipulated by controlled external fields with errors less than 10^{-6}. Moreover, the standard schemes for error correction (assuming an error rate of order 10^{-6}) require very big system sizes, $K \sim 10^4$–10^6, to correct the errors (i.e. the total number of all qubits exceeds by a factor of 100–1000 the number of qubits needed to perform a computational algorithm in the "ideal" condition of no errors). If frequency of errors could be reduced by orders of magnitude, the conditions for residual error corrections would become much less stringent, and the total size of the system, K, would be much smaller (Preskill 1998).

Insensitivity to random fluctuations means that any coupling to the external environment neither induces transitions among these 2^K states nor changes the phase of one state with respect to another. Mathematically, this means that one requires a system whose Hilbert space contains a 2^K-dimensional subspace, called the "protected subspace" (Wen and Niu 1990, Wen 1991, Kitaev 2003), within which any local operator $\hat{O}$ has (to a high accuracy) only state-independent diagonal matrix elements: $\langle n|\hat{O}|m\rangle = O_0\delta_{mn} + o(\exp(-L))$, where L is a parameter such as the system size that can be made as large as desired. It has been very difficult to design a system which meets these criteria. Many physical systems (for example, spin glasses (Mezard *et al.* 1997)) exhibit exponentially many distinct states, so that the off-diagonal matrix elements of all physical operators between these states are exponentially small. In such systems, the longitudinal relaxation of a superposition of these states is very slow. The absence of a transverse relaxation, which is due to different diagonal matrix elements O_{mm} and O_{nn}, is a different matter: it is a highly nontrivial requirement that is not satisfied by the usual physical systems (such as spin glasses) and which puts systems that satisfy it in a completely new class.

One very attractive possibility, proposed in a seminal paper by Kitaev (2003) and developed further by Bravyi and Kitaev (2000) and Freedman *et al.* (2000), involves a protected subspace created by a topological degeneracy of the ground state. Typically, such a degeneracy happens if the system has a conservation law such as the conservation of the *parity* of the number of "particles" along some long contour, and the absence of any local order parameter. Physically, it is clear that two states that differ only by the parity of some big number that cannot be obtained from any local measurement are very similar to each other. A cartoon example of this idea can be presented as follows. Consider two locally flat surfaces, one with the topology of a simple cylinder, and another with the topology of a Möbius strip, and imagine an observer moving on one of these surfaces. Clearly, the only way to decide which surface the observer is located on is to walk a whole loop around the strip, and then the observer will be either at the same point (if the surface is a cylinder) or on the other side of the surface (if it is a Möbius strip).

The model (called the "toric code") proposed by Kitaev (2003) has been shown to exhibit many properties of an ideal quantum computer; here we review several attempts to propose physical realizations of this type of mathematical model, which seem to be possible if one deals with arrays of submicron Josephson junctions. The first such an attempt was made by Ioffe *et al.* (2002). Further developments of these kind of ideas were proposed by Ioffe and Feigel'man (2002) and Douçot *et al.* (2003). The relevant degrees of freedom of this new array are described by a model very much analogous to the one proposed by Kitaev (2003).

In physical terms, the array proposed in these papers may exist in two different phases: (i) a topological superconductor; that is, a superfluid of $4e$ composite objects; and (ii) a topological insulator; that is, a superfluid of superconductive vortices with a flux quantum $\Phi_0 = hc/2e$. The topological degeneracy of the ground state (i) arises because $2e$ excitations have a gap. In such a system with the geometry of an annulus, one extra Cooper pair injected at the inner boundary can never escape it; on the other hand, it is clear that two states differing by the parity of the number of Cooper pairs at the boundary are practically indistinguishable by a local measurement. In the ground state (ii), the lowest excitation is a "half-vortex" (i.e. a vortex with a flux $\Phi_0/2$), and a topological double degeneracy appears owing to the possibility of putting a half-vortex inside the opening, without paying any energy.

Below, we first describe the physical array, in the "topological superconductor" state, and identify its relevant low-energy degrees of freedom and the mathematical model which describes their dynamics. We then show how protected states appear in this model, derive the parameters of the model, and identify various corrections appearing in a real physical system and their effects. Then we discuss a generalization of this array that is needed to obtain in a controllable way the second phase of the "topological insulator." Finally, we discuss how one can manipulate quantum states in a putative quantum computer based upon those arrays, and the physical properties expected in small arrays of this type.

We remark that the properties of the excitations and topological order parameter exhibited by the system we propose here are in many respects similar to the properties of the ring exchange and frustrated-magnets models discussed, for example, by Wen and Niu (1990), Wen (1991), Read and Chakraborty (1989), Read and Sachdev (1991), Kivelson (1989), Senthil and Fisher (2001*a*,*b*), Balents *et al.* (2002), Misguich *et al.* (2002), Motrunich and Senthil (2002), Fendley *et al.* (2002), and Ioselevich *et al.* (2002).

We also present a discussion of a simpler (in terms of practical implementation) example of a Josephson array, which does not seem to belong to the family of Kitaev toric-code models, but has been theoretically shown (Douçot *et al.* 2005) to possess a high degree of robustness with respect to noise and disorder.

19.2 Topological superconductor

The basic building block of the array is a rhombus made of four Josephson junctions with each side of the rhombus containing one Josephson contact; these rhombi form a hexagonal lattice, as shown in Fig. 19.1. We denote the centers of the hexagons by letters $a, b, \ldots$ and the individual rhombi by $(ab), (cd) \ldots$, because each rhombus is in one-to-one correspondence with the link (ab) between the sites of the triangular lattice dual to the hexagonal lattice. The lattice is placed in a uniform magnetic field so that the flux through each rhombus is $\Phi_0/2$. The geometry is chosen in such a way that the flux Φ_s through each star of David is a half-

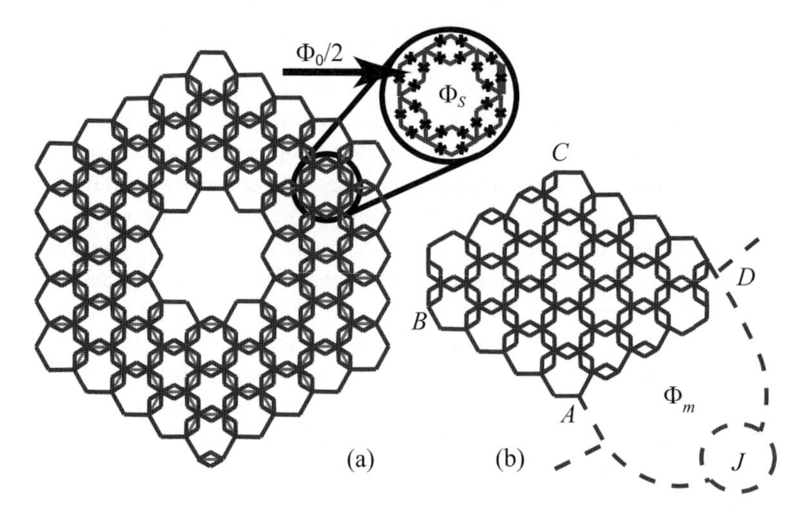

Fig. 19.1 Examples of the proposed Josephson junction array. The thick lines show superconductive wires; each wire contains one Josephson junction, as shown in the detailed view of one hexagon. The width of each rhombus is such that the ratio of the area of the star of David to the area of one rhombus is an odd integer. The array is put in a magnetic field such that the flux through each elementary rhombus and through each star of David (inscribed in each hexagon) is a half-integer. The thin lines show the effective bonds formed by the elementary rhombi. The Josephson coupling provided by these bonds is π-periodic. (a) Array with one opening; generally, the effective number of qubits K is equal to the number of openings. The choice of boundary condition shown here makes the superconducting phase unique along the entire length of the outer or inner boundary; the state of the entire boundary is described by a single degree of freedom. The topological order parameter controls the phase difference between the inner and outer boundaries. Each boundary includes one rhombus to allow experiments with flux penetration; the magnetic flux through the opening is assumed to be $(\Phi_0/2)(1/2 + m)$ for any integer m. (b) With this choice of boundary circuits, the phase is unique only inside the sectors AB and CD of the boundary; the topological degree of freedom controls the difference between the phases of these boundaries. This allows a simpler setup of the experimental test for the signatures of the ground state described in the text, for example by a SQUID interference experiment, sketched here, that involves a measuring loop with a flux Φ_m and a very weak junction J balancing the array.

integer multiple of Φ_0: $\Phi = (n_s + \frac{1}{2})\Phi_0$.[1] Finally, the lattice contains globally a number K of large openings (the size of the opening is much larger than the lattice constant; a lattice with $K = 1$ is shown in Fig. 19.1(a)). The dimension of the protected space will be shown to be equal to 2^K. The system is characterized by the Josephson energy, $E_J = (\hbar/2e)I_c$, of each contact and by the capacitance matrix of the islands (the vertices of the lattice). We shall assume (as is usually the case) that the capacitance matrix is dominated by the capacitances of individual junctions; we write the charging energy as $E_C = e^2/2C$. The "phase regime" of this network implies that $E_J > E_C$. The whole system is described by the Lagrangian

$$\mathcal{L} = \sum_{(ij)} \frac{1}{16E_C}(\dot{\phi}_i - \dot{\phi}_j)^2 + E_J \cos(\phi_i - \phi_j - a_{ij}), \tag{19.1}$$

where the ϕ_i are the phases of the individual islands and the a_{ij} are chosen to produce the correct magnetic fluxes. The Lagrangian (19.1) contains only gauge-invariant phase differences $\phi_{ij} = \phi_i - \phi_j - a_{ij}$, so it will be convenient sometimes to treat them as independent variables satisfying the constraint $\sum_\Gamma \phi_{ij} = 2\pi\Phi_\Gamma/\Phi_0 + 2\pi n$, where the sum is taken over a closed loop Γ and n is an arbitrary integer.

As will become clear below, it is crucial that the degrees of freedom at the boundary have dynamics identical to the dynamics of those in the bulk. To ensure this, one needs to add additional superconducting wires and Josephson junctions at the boundary. There are a few ways to do this, and two examples are shown in Fig. 19.1: a type I boundary (the entire length of the boundary in Fig. 19.1(a) and parts AB and CD in Fig. 19.1(b)), and a type II boundary (parts BC and AD in Fig. 19.1(b)). For both types of boundaries, one needs to include in each boundary loop a flux equal to $Z_b * \pi/2$, where Z_b is the coordination number of the dual triangular-lattice site. For instance, for four-coordinated boundary sites one needs to enclose an integer flux in these contours. In a type I boundary, the entire boundary corresponds to one degree of freedom (the phase at some point), while a type II boundary includes many rhombi, so it contains many degrees of freedom.

Note that each (inner and outer) boundary shown in Fig. 19.1(a) contains one rhombus; we have included it to allow flux to enter and exit through the boundary when it is energetically favorable.

19.3 Ground state, excitations, and topological order

In order to identify the relevant degrees of freedom in this highly frustrated system, we consider first an individual rhombus. As a function of the gauge-invariant phase difference between the far ends of the rhombus, the potential energy is

$$U(\phi_{ij}) = -2E_J(|\cos(\phi_{ij}/2)| + |\sin(\phi_{ij}/2)|). \tag{19.2}$$

This energy has two equivalent minima, at $\phi_{ij} = \pm\pi/2$, which can be used to construct an elementary unprotected qubit (Blatter *et al.* 2001). In each of these states, the phase changes

[1] The flux Φ_s can be also chosen so that it is an integer multiple of Φ_0: this would not change the final results significantly but would change the intermediate arguments and make them longer, so for clarity we discuss in detail only the half-integer case here. Note, however, that the main quantitative effect of this alternative choice of the flux is beneficial: it pushes up the phase transition line separating the topological and superconducting phases shown in Fig. 19.3 for the half-integer case.

by $\pm\pi/4$ in each junction, going clockwise around the rhombus. We denote these states as $|\uparrow\rangle$ and $|\downarrow\rangle$, respectively. In the limit of large Josephson energy, the space of low-energy states of the full lattice is described by these binary degrees of freedom; the set of operators acting on these states is given by the Pauli matrices $\sigma_{ab}^{x,y,z}$. We now combine these rhombi into hexagons, forming the lattice shown in Fig. 19.1. This gives another condition: the sum of phase differences around a hexagon should be equal to the flux, Φ_s through each star of David inscribed in this hexagon. The choice $\phi_{ij} = \pi/2$ is consistent with a flux Φ_s that is equal to a half-integer number of flux quanta. This state minimizes the potential energy (19.2) of the system. This is not, however, the only choice. Although flipping the phase of one dimer changes the phase flux around the star by π and thus is prohibited, flipping two, four, and six rhombi is allowed; generally, the low-energy configurations of $U(\phi)$ satisfy the constraint

$$\hat{P}_a = \prod_b \sigma_{ab}^z = 1\,, \tag{19.3}$$

where the product runs over all neighbors, b, of site a. The number of (classical) states satisfying the constraint (19.3) is still huge: the corresponding configurational entropy is extensive (proportional to the number of sites). We now consider the charging energy of the contacts, which results in the quantum dynamics of the system. We show that it reduces this degeneracy to a much smaller number, 2^K. The dynamics of an individual rhombus is described by a simple Hamiltonian $H = \tilde{t}\sigma_x$, but the dynamics of a rhombus embedded in the array is different because individual flips are not compatible with the constraint (19.3). The simplest dynamics compatible with eqn (19.3) contains flips of three rhombi belonging to the elementary triangle (a,b,c), $\hat{Q}_{(abc)} = \sigma_{ab}^x\sigma_{bc}^x\sigma_{ca}^x$ and therefore the simplest quantum Hamiltonian operating on the subspace defined by eqn (19.3) is

$$H = -r\sum_{(abc)} Q_{(abc)}\,. \tag{19.4}$$

We discuss the derivation of the coefficient r in this Hamiltonian and the correction terms and their effects below, but first we solve the simplified model given by eqns (19.3) and (19.4) and show that its ground state is "protected" in the sense described above and that excitations are separated by a gap.[2]

Clearly, it is very important that the constraint is imposed on all sites, including boundaries. Evidently, some boundary hexagons are only partially complete, but the constraint should be still imposed on the corresponding sites of the corresponding triangular lattice. This is ensured by additional superconducting wires that close the boundary hexagons in Fig. 19.1.

We note that the constraint operators commute not only with the full Hamiltonian but also with individual operators $\hat{Q}_{(abc)}$: $[\hat{P}_a, \hat{Q}_{(abc)}] = 0$. The Hamiltonian (19.4) without any constraint has an obvious ground state, $|0\rangle$, in which $\sigma_{ab}^x = 1$ for all rhombi. This ground state, however, violates the constraint. This can be fixed by noting that since the operators $\hat{P}_a$ commute with the Hamiltonian, any state obtained from $|0\rangle$ by acting on it by $\hat{P}_a$ is also a ground state. We can now construct a true ground state satisfying the constraint by taking

[2]In a rotated basis $\sigma^x \to \sigma^z$, $\sigma^z \to \sigma^x$, this model is reduced to a special case of the Z_2 lattice gauge theory (Wegner 1971, Balian *et al.* 1975) which contains only a magnetic term in the Hamiltonian, with the constraint (19.3) playing the role of a gauge invariance condition.

$$|G\rangle = \prod_a \frac{1 + \hat{P}_a}{\sqrt{2}} |0\rangle \,. \tag{19.5}$$

Here $(1 + \hat{P}_a)/\sqrt{2}$ is a projector onto the subspace that satisfies the constraint at site a and preserves the normalization.

Obviously, the Hamiltonian (19.4) commutes with any product of $\hat{P}_a$ which is equal to a product of operators σ^z_{ab} around a set of closed loops. These integrals of motion are fixed by the constraint. However, for a topologically nontrivial system, there appear a number of other integrals of motion. For a system with K openings, a product of operators σ^z_{ab} along a contour γ that begins at one opening and ends at another (or at the outer boundary; see Fig. 19.2),

$$\hat{T}_q = \prod_{(\gamma_q)} \sigma^z_{ab} \,, \tag{19.6}$$

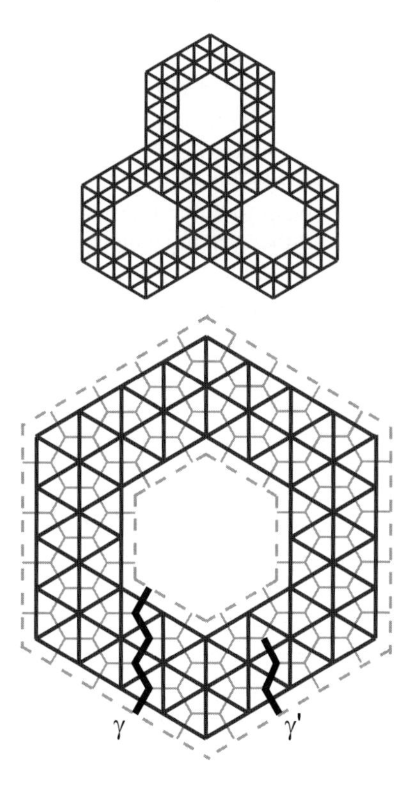

Fig. 19.2 Top: a lattice with $K = 3$ openings; the ground state of the Josephson junction array on this lattice is $2^K = 8$-fold degenerate. Bottom: location of the discrete degrees of freedom responsible for the dynamics of the Josephson junction array shown in Fig. 19.1. The spin degrees of freedom describing the state of the elementary rhombi are located on the bonds of the triangular lattice (shown by thick lines), while the constraints are defined on the sites of this lattice. The dashed line indicates the boundary condition imposed by the physical circuitry shown in Fig. 19.1(a). The contours γ and γ' are used in the construction of the topological order parameter and excitations.

commutes with the Hamiltonian and is not fixed by the constraint. Physically, these operators count the parity of "up" rhombi along such a contour. The presence of these operators results in the degeneracy of the ground state. Note that multiplying such an operator by an appropriate $\hat{P}_a$ gives a similar operator defined on a shifted contour, so all topologically equivalent contours give one new integral of motion. Further, multiplying two operators defined along contours beginning at the same (e.g. outer) boundary and ending in different openings A and B is equivalent to an operator defined on a contour leading from A to B, so independent operators can be defined by, for example, a set of contours that begin at one opening and end at the outer boundary. The state $|G\rangle$ constructed above is not an eigenstate of these operators, but this can be fixed by defining

$$|G_f\rangle = \prod_q \frac{1 + c_q \hat{T}_q}{\sqrt{2}} |G\rangle \,, \tag{19.7}$$

where $c_q = \pm 1$ is the eigenvalue of the operator $\hat{T}_q$ defined on the contour γ_q. Equation (19.7) is the final expression for the ground state eigenfunctions.

The construction of the excitations is similar to the construction of the ground state. First, we notice that since all operators $\hat{Q}_{abc}$ commute with each other and with the constraints, any state of the system can be characterized by the eigenvalues ($Q_{abc} = \pm 1$) of $\hat{Q}_{abc}$. The lowest excited state corresponds to only one Q_{abc} being -1. Notice that a simple flip of one rhombus (by an operator $\sigma^z_{(ab)}$) somewhere in the system changes the sign of *two* eigenvalues Q_{abc}, corresponding to the two triangles to which it belongs. To change only one Q_{abc}, we need to consider a continuous string of these flip operators starting from the boundary: $|(abc)\rangle = v_{(abc)}|0\rangle$ with $v_{(abc)} = \prod_{\gamma'} \sigma^z_{(cd)}$, where the product is over all rhombi (cd) that belong to a path γ' that begins at the boundary and ends at (abc) (see Fig. 19.2, which shows one such path). This operator changes the sign of only one Q_{abc}, the one that corresponds to the "last" triangle. This construction does not satisfy the constraint, so we have to apply the same "fix" as for the ground state construction above,

$$|v_{(abc)}\rangle = \prod_q \frac{1 + c_q \hat{T}_q}{\sqrt{2}} \prod_a \frac{1 + \hat{P}_a}{\sqrt{2}} v_{(abc)}|0\rangle \,, \tag{19.8}$$

to get the final expression for the lowest-energy excitations. The energy of each excitation is $2r$. Note that a single-flip excitation at a rhombus (ab) can be viewed as a combination of two elementary excitations located at the centers of the triangles to which the rhombus (ab) belongs, and has twice their energy. Generally, all excited states of the model (19.4) can be characterized as a number of elementary excitations (19.8), so they give an exact quasiparticle basis. Note that creation of a quasiparticle at one boundary and moving it to another is equivalent to the operator $\hat{T}_q$, so this process acts as τ^z_q in the space of the 2^K degenerate ground states. As will be shown below, in the physical system of Josephson junctions, these excitations carry charge $2e$ so that the process τ^z_q is equivalent to a transfer of charge $2e$ from one boundary to another.

Consider now the matrix elements $O_{\alpha\beta} = \langle G_\alpha|\hat{O}|G_\beta\rangle$ of a local operator $\hat{O}$ between two ground states, for example of an operator that is composed of a small number of σ_{ab}. To evaluate this matrix element, we first project a general operator onto the space that satisfies the constraint: $\hat{O} \to \mathcal{P}\hat{O}\mathcal{P}$, where $\mathcal{P} = \prod_a (1 + \hat{P}_a)/2$. The new (projected) operator is also local; it has the same matrix elements between ground states but it commutes with all $\hat{P}_a$. Since it is local, it can be represented as a product of operators σ_z and $\hat{Q}$, which implies that it also commutes with all $\hat{T}_q$. Thus, its matrix elements between different states are exactly zero. Further, using the fact that it commutes with $\hat{P}_a$ and $\hat{T}_q$, we can write the difference between its diagonal elements evaluated between states that differ by parity over the contour q as

$$O_+ - O_- = \langle 0| \prod_i \frac{1 + \hat{P}_i}{\sqrt{2}} \hat{T}_q \hat{O} |0\rangle \,. \tag{19.9}$$

This equation can be viewed as a sum of products of σ_z operators. Clearly, to get a nonzero contribution, each σ^z should enter an even number of times. Each $\hat{P}$ contains a closed loop of six σ^z operators, so any product of these terms is also a collection of a closed loops of σ^z. In contrast, the operator $\hat{T}_q$ contains a product of σ^z operators along the loop γ, so the product of them contains a string of σ_z operators along the contour that is topologically equivalent to γ. Thus, we get a nonzero $O_+ - O_-$ only for operators $\hat{O}$ which contain a string of σ^z operators along a loop that is topologically equivalent to γ, which is impossible for a local operator. Thus, we conclude that for this model, all nondiagonal matrix elements of a local operator are *exactly* zero, while all diagonal elements are *exactly* equal.

19.4 Effect of physical perturbations

We now come back to the original physical system described by the Lagrangian (19.1), and derive the parameters of the model (19.4) and discuss the most important corrections to it and their effect. We begin with the derivation. In the limit of small charging energy, a flip of three rhombi occurs by a virtual process in which the phase ϕ_i at one (6-coordinated) island i changes by π. In the quasi-classical limit, the phase differences at the individual junctions are $\phi_{\text{ind}} = \pm\pi/4$; the leading quantum process changes the phase at one junction by $3\pi/2$ and that of others by $-\pi/2$, changing the phase across the rhombus by $\phi \to \phi + \pi$. The phase differences ϕ satisfy the constraint that the sum of them over closed loops remains $2\pi(n + \Phi_s/\Phi_0)$. The simplest such process preserves the symmetry of the lattice, and changes simultaneously the phase differences in the three rhombi containing island i, keeping all other phases constant. The action for such a process is three times the action of the elementary transitions in the individual rhombi, S_0:

$$r \approx E_J^{3/4} E_C^{1/4} \exp(-3S_0), \quad S_0 = 1.61\sqrt{\frac{E_J}{E_C}} \,. \tag{19.10}$$

In an alternative process, the phase differences between i and other islands change in turn, via a high-energy intermediate state in which one phase difference has changed while the others remain close to their original values. An estimate of the action for this process shows that it is larger than $3S_0$, so eqn (19.10) gives the dominating contribution. There are in fact many processes that contribute to this transition: the phase of island i can change by $\pm\pi$ and, in addition, in each rhombus one can choose arbitrarily the junction in which the phase changes by $\pm 3\pi/2$; the amplitudes of all these processes should be added. This does not change the result qualitatively unless these amplitudes exactly cancel each other, which happens only if the charge of the island is exactly a half-integer (because phase and charge are conjugate, the amplitude difference of processes that are different by 2π is $\exp(2\pi i q)$). We assume that in a generic case, this cancelation does not occur. External electrical fields (created by, for example, stray charges) might induce noninteger charges on each island, which would lead to a randomness in the phase and amplitude of r. The phase of r can be eliminated by a proper gauge transformation $|\uparrow\rangle_{ab} \to e^{i\alpha_{ab}}|\uparrow\rangle_{ab}$ and has no effect at all. The amplitude variations result in a position-dependent quasiparticle energy.

We now consider the corrections to the model (19.4). One important source of corrections is the difference of the actual magnetic flux through each rhombus from the ideal value $\Phi_0/2$. If this difference is small, it leads to a bias of "up" versus "down" states; their energy difference becomes

$$2\epsilon = 2\pi\sqrt{2}\frac{\delta\Phi_d}{\Phi_0}E_J\,. \tag{19.11}$$

Similarly, differences in the actual flux through the stars of David and differences in the Josephson energies of individual contacts lead to the following interaction between "up" states:

$$\delta H_1 = \sum_{(ab)} V_{ab}\sigma^z_{ab} + \sum_{(ab),(cd)} V_{(ab),(cd)}\sigma^z_{ab}\sigma^z_{cd}\,, \tag{19.12}$$

where $V_{ab} = \epsilon$ for a uniform field that deviates slightly from the ideal value, and $V_{(ab),(cd)} \neq 0$ for rhombi belonging to the same hexagon. Consider now the effect of perturbations described by δH_1 given by eqn (19.12). These terms commute with the constraint but do not commute with the main term H, so the ground state is no longer $|G_\pm\rangle$. In other words, these terms create excitations (19.8) and give them kinetic energy. In the leading order of the perturbation theory, the ground state becomes $|G_\pm\rangle + (\epsilon/4r)\sum_{(ab)}\sigma^z_{(ab)}|G_{i\pm}\rangle$ Qualitatively, this corresponds to the appearance of virtual pairs of quasiparticles in the ground state. The density of these quasiparticles is ϵ/r. As long as these quasiparticles do not form a topologically nontrivial string, all previous conclusions remain valid. However, there is a nonzero amplitude to form such a string—it is now exponential in the system size. With exponential accuracy, this amplitude is $(\epsilon/2r)^L$, which leads to an energy splitting of the two ground state levels and of the matrix elements of typical local operators of the same order $E_+ - E_- \sim O_+ - O_- \sim (\epsilon/2r)^L$.

The physical meaning of the $v_{(abc)}$ excitations become clearer if we consider the effect of the addition of one σ^z operator to the end of the string defining the quasiparticle: this results in a charge transfer of $2e$ across the last rhombus. To prove this, note that the wave function of a superconductor corresponding to a state which is a symmetric combination of $|\uparrow\rangle$ and $|\downarrow\rangle$ is periodic with period π and thus corresponds to a charge which is a multiple of $4e$, while an antisymmetric combination corresponds to a charge $(2n+1)2e$. The action of one σ_z induces a transition between these states and thus transfers a charge $2e$. Thus, these excitations carry charge $2e$. Note that continuous degrees of freedom are characterized by long-range order in $\cos(2\phi)$ and thus correspond to the condensation of pairs of Cooper pairs. In other words, this system superconducts with an elementary charge $4e$ and has a gap, $2r$, relative to the excitations carrying charge $2e$. A similar pairing of Cooper pairs was shown to occur in a chain of rhombi in a recent paper (Douçot and Vidal 2002); the formation of a classical superconductive state with an effective charge $6e$ in a frustrated kagome wire network was predicted by Park and Huse (2001).

The model (19.4) completely ignores processes that violate the constraint at each hexagon. Such processes might violate the conservation of the topological invariants $\hat{T}_q$ and thus are important for the long-time dynamics of the ground state manifold. In order to consider these processes, we need to go back to a full description involving the continuous superconducting phases ϕ_i. Since the potential energy (19.2) is periodic with a period π, it is convenient to separate the degrees of freedom into continuous parts (defined modulo π) and discrete parts. The continuous parts have long-range order: $\langle \cos(2\phi_0 - 2\phi_r) \rangle \sim 1$. The elementary excitations of the continuous degrees of freedom are harmonic oscillations and vortices. The harmonic oscillations interact with the discrete degrees of freedom only through the local currents that they generate; furthermore, the potential (19.2) is very close to quadratic, so we conclude that they are practically decoupled from the rest of the system. In contrast to this, vortices have an important effect. By construction, an elementary vortex carries a flux π in this problem. Consider the structure of these vortices in greater detail. The superconducting phase should change by 0 or 2π when one moves round a closed loop. In the case of a half-vortex, this is achieved if a gradual change of π is compensated (or augmented) by a discrete change of π on a string of rhombi which costs no energy. Thus, from the viewpoint of discrete degrees of freedom, the position of a vortex is a hexagon where the constraint (19.3) is violated. The energy of the vortex is found from the usual arguments:

$$E_v(R) = \frac{\pi E_J}{4\sqrt{6}} \left(\ln(R) + c\right), \quad c \approx 1.2; \tag{19.13}$$

it is logarithmic in the vortex size R. The process that changes the topological invariant $\hat{T}_q$ is one in which a half-vortex completes a circle around an opening. The amplitude of such a process is exponentially small: $(\tilde{t}/E_v(D))^\Lambda$, where $\tilde{t}$ is the amplitude to flip one rhombus and Λ is the length of the shortest path around the opening. In the quasi-classical limit, the amplitude $\tilde{t}$ can be estimated analogously to eqn (19.10): $\tilde{t} \sim \sqrt{E_J E_Q} \exp(-S_0)$. Half-vortices would appear in a realistic system if the flux through each hexagon were systematically different from the ideal half-integer value. The presence of free vortices destroys topological invariants, so either a realistic system should be not too large (so that deviations of the total flux do not induce free vortices) or these vortices should be localized in prepared traps (e.g. stars of David with fluxes slightly larger or smaller than Φ_s). In the absence of half-vortices, the model is equivalent to the Kitaev model (Kitaev 2003) placed on a triangular lattice in the limit of infinite energy of an excitation that violates the constraints.

Quantitatively, the expression for the parameters of the model (19.4) become exact only if $E_J \gg E_C$. One expects, however, that the qualitative conclusions will remain the same and that the formulas derived above will provide good estimates of the scales even for $E_J \sim E_C$, provided that the charging energy is not so large as to result in a phase transition to a different phase. One expects this transition to occur at $E_c^* = \eta E_J$ with $\eta \sim 1$, the exact value of which can be reliably determined only from numerical simulations. Practically, since the perturbations induced by flux deviations from Φ_0 are proportional to $(\delta\Phi/\Phi_0)(E_J/r)$ and r becomes exponentially small at small E_C, the optimal choice of the parameters for a physical system is $E_C \approx E_c^*$. We show a schematic illustration of the phase diagram in Fig. 19.3. We assume here that the transition to the insulating phase is direct; another alternative is an intermediate phase in which the energy of a vortex becomes finite instead of being logarithmic. If this phase indeed exists, it is likely to have properties more similar to the one discussed in

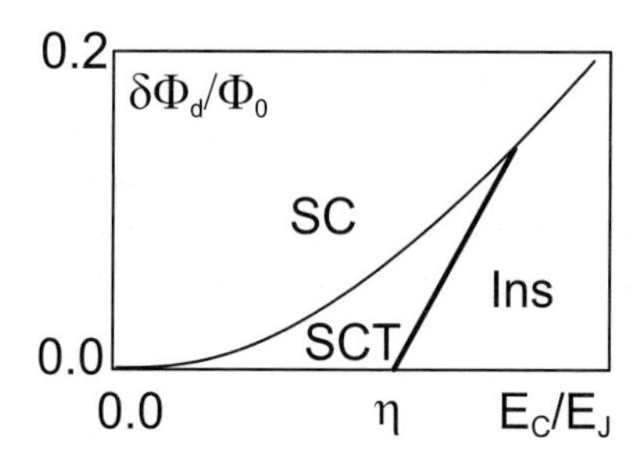

Fig. 19.3 Schematic illustration of the phase diagram for half-integer Φ_s at low temperatures: $\delta\Phi_d$ is the deviation of the magnetic flux through each rhombus from its ideal value. "SC" stands for the usual superconducting phase, and "SCT" for the phase with the $\cos(2\phi)$ long-range order of the continuous degrees of freedom and the discrete topological order parameter discussed extensively in the text of this chapter. The SCT and SC phases are separated by a 2D quantum Ising phase transition.

Kitaev (2003) (in the next section we consider a generalized Josephson junction array, where such an intermediate phase does exist). The "topological" phase is stable in a significant part of the phase diagram. The phase transition between the "topological superconductor" and the usual type of superconductor belongs to the class of quantum spin-$\frac{1}{2}$ 2D Ising models on a hexagonal lattice, placed in a transverse field:

$$H_{\text{Ising}} = -\epsilon \sum_{(ij)} s_j^x s_i^x - r \sum_i s_i^z \,. \tag{19.14}$$

Here i, j denote sites of a hexagonal lattice; the eigenvalue of the operator s_i^z measures the parity of the number of Cooper pairs n_i on the i-th island, where $n_i (\text{mod } 2) = \frac{1}{2}(1 - s_i^z)$; the parameters r and ϵ are defined in eqns (19.10) and (19.11). As long as the ratio $\lambda = r/\epsilon$ is larger than some critical value λ_c, the ground state has an even number of Cooper pairs on each island, which corresponds to our "topological superconductor" phase. The values of λ_c for square, triangular, and cubic lattices have been found via quantum Monte Carlo simulations (Bishop *et al.* 2000); in particular, $\lambda_c^{\text{triang}} \approx 4.6 \pm 0.3$, and $\lambda_c^{\text{square}} \approx 2.7 \pm 0.3$. There is no available data for a hexagonal lattice; based upon the above results, one could estimate $\lambda_c^{\text{hex}} \approx 2 \pm 0.5$.

Furthermore, since the vortex excitations have logarithmic energy, we expect that this phase will survive at finite temperatures as well. In the thermodynamic limit, at $T \neq 0$ one gets a finite density of $2e$-carrying excitations ($n_v \sim \exp(-2r/T)$), but vortices remain absent as long as the temperature is below a BKT-like depairing transition for half-vortices.

19.5 Topological insulator

Generally, increasing the charging energy in a Josephson junction array makes it an insulator. This transition is due to an increase of the phase fluctuations in the original array and the resulting appearance of free vortices, which form a superfluid of their own. A new situation arises in a topological superconductor because it allows half-vortices. Two scenarios are now possible. The "conventional" scenario would involve condensation of half-vortices, since they are conjugate to charges $4e$. In this case we get an insulator with elementary excitations carrying a charge $4e$. An alternative is condensation of full vortices (pairs of half-vortices), with a finite gap to half-vortices. In this case the elementary excitations are charge $2e$ objects. A similar fractionalization was discussed in the context of high-T_c superconductors by Senthil and Fisher (2000) and Senthil and Fisher (2001b), and in the context of spin or quantum dimer systems by Moessner *et al.* (2002), Balents *et al.* (2002), Senthil and Motrunich (2002), and Misguich *et al.* (2002). Such an insulator acquires interesting topological properties on a lattice with holes because each hole leads to a new binary degree of freedom which describes the presence or absence of a half-vortex. The energies of these states are equal up to corrections which vanish exponentially with the size of the holes. These states cannot be distinguished by local measurements and have all of the properties expected for a topological insulator. They can be measured, however, if the system is adiabatically brought into the superconductive state by changing some controlling parameter. Here we discuss a modification of the "topologically superconductive" array, proposed by Douçot *et al.* (2003), that provides such a control parameter and, at the same time, allows us to solve the model and compute the properties of the topological insulator. The key idea of this modification is to allow full vortices (of flux Φ_0) to move with a large amplitude between plaquettes of the hexagonal lattice, so that they lower their energy owing to delocalization, and eventually Bose-condense, while half-vortices are still kept (almost) localized.

Consider the array shown in Fig. 19.4, which contains rhombi with junctions characterized by Josephson and charging energies $E_J > E_C$ and weak junctions with $\epsilon_J \ll \epsilon_C \ll E_C$. Each rhombus encloses half of a flux quantum, leading to an exact degeneracy between the two states of opposite chirality of the circulating current. This degeneracy is a consequence of the symmetry operation which combines a reflection about the long diagonal of the rhombus and a gauge transformation needed to compensate the change of the flux $\Phi_0/2 \to -\Phi_0/2$. This gauge transformation changes the phase difference along the diagonal by π. This Z_2 symmetry implies the conservation of the parity of the number of pairs at each site of the hexagonal lattice and is the origin of the Cooper pair binding. We assume that each elementary hexagon contains exactly k such junctions: if each link contains one weak junction, $k = 6$, but generally it can take any value $k \geq 1$. As will be shown below, the important condition is the number of weak junctions that one needs to cross in an elementary loop. Qualitatively, a value $k \geq 1$ ensures that it costs a little to put a vortex in any hexagon.

For the general arguments that follow below, the actual construction of the weak links is not important; however, it is difficult practically to vary the ratio of the capacitance to the Josephson energy, so a weaker Josephson contact usually implies a larger Coulomb energy. This can be avoided if the weak contact is made from a Josephson junction loop frustrated by a magnetic field. The charging energy of this system is half the charging energy of the individual junction, while the effective Josephson junction strength is $\epsilon_J = 2\pi(\delta\Phi/\Phi_0)E_0$, where E_0 is the Josephson energy of each contact and $\delta\Phi = \Phi - \Phi_0/2$ is the difference of the flux from half a flux quantum. This construction also allows one to control the system by varying the magnetic field.

Under these conditions, the whole array is insulating. Assume that ϵ_J sets the lowest energy scale in this problem (the exact condition will be discussed below). The state of the array is controlled by discrete variables $u_{ij} = 0, 1$ that describe the chiral state of each rhombus and by continuous phases ϕ_{ij} that specify the state of each weak link (here and below, i, j

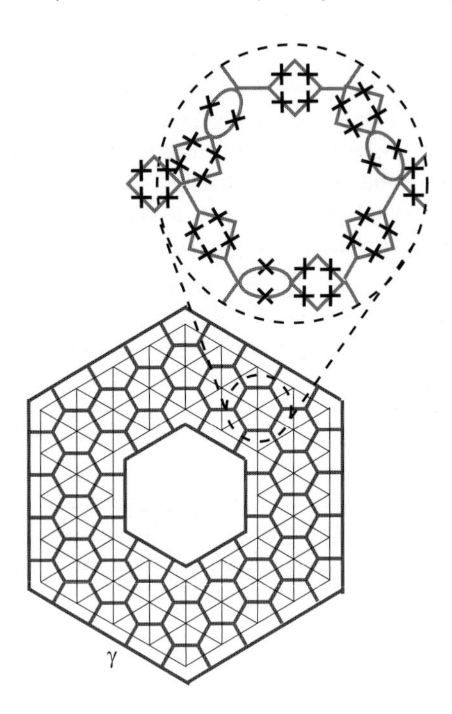

Fig. 19.4 Schematic illustrations of an array. Main figure: global structure of the array. Discrete variables controlling the low-energy properties are defined on the links of the hexagonal lattice. Generally, the lattice might have K large holes; here we show an example with $K = 1$. Zoom in: each inner bond of the lattice contains a rhombus made of four Josephson junctions; some bonds also contain an effective weaker link made of two Josephson junctions so that each hexagon of the lattice contains $k = 3$ such links. The flux through each rhombus is half a flux quantum, $\Phi_0/2$, and the flux through a loop constituting a weak link is close to half a flux quantum, $\Phi = \Phi_0/2 + \delta\Phi$. The boundary of the lattice contains rhombi and weak links so that each boundary plaquette has the same number, k, of weak links as the bulk hexagon.

denote the sites of the hexagonal lattice). If the Josephson coupling ϵ_J is identically equal to 0, different islands are completely decoupled and the potential energy does not depend on the discrete variables u_{ij}. For small ϵ_J, we can evaluate its effect using perturbation theory:

$$V(u) = -W \cos\left(\pi \sum_{\text{hex}} u_{ij}\right), \qquad W = \frac{k^k}{k!}\epsilon_J \left(\frac{\epsilon_J}{8\epsilon_C}\right)^{k-1}. \tag{19.15}$$

This potential energy lowers the relative energy of classical configurations of u_{ij} that satisfy the constraint $\sum_{\text{hex}} u \equiv 0 \bmod 2$ but it does not prohibit configurations with $\sum_{\text{hex}} u \equiv 1 \bmod 2$.

Consider now the dynamics of the discrete variables. Generally, two types of tunneling processes are possible. In the first type, the phase changes by π across each of the three rhombi that have a common site. This is the same process as that which gives the leading contribution to the dynamics of the superconducting array studied by Ioffe and Feigel'man

(2002); its amplitude is given (in the quasi-classical approximation) by the formula (19.10) above. In the second type of process, the phase changes across one rhombus and across one weak junction. Because the potential energy of the weak junction is assumed to be very small, the main effect of the weak junction is to change the kinetic energy. The total kinetic energy for this process is the sum of the terms due to the phase across the rhombi and across the weak link. Assuming that these phase variations are equal and opposite in sign, the former is about $E_C^{-1}\dot{\phi}^2$ while the latter is about $\epsilon_C^{-1}\dot{\phi}^2$, so the effective charging energy of this process is $\widetilde{E_C} = \left(E_C^{-1} + \epsilon_C^{-1}\right)^{-1}$. For $\epsilon_C \ll E_C$, this charging energy is small and such processes are suppressed. Thus, under these conditions the dominating process is the simultaneous flip of three rhombi as in the superconducting case. In the following, we restrict ourselves to this case. Further, we shall assume that $r \gg W$ so that, in the leading order, one can neglect the potential energy compared with the kinetic energy corresponding to the flip of three rhombi. As W is increased by turning on ϵ_J, the continuous phase ϕ_{ij} orders and the transition into the superconducting state happens at $\epsilon_J \sim \epsilon_C$. At larger ϵ_J, W becomes ϵ_J and, with a further increase of ϵ_J, for $\epsilon_J \gg r$ vortices completely disappear from the low-energy spectrum and the array becomes equivalent to the one studied by Ioffe and Feigel'man (2002).

The low-energy states are the ones that minimize the kinetic energy corresponding to simultaneous flip processes,

$$H_T = -r \sum_i \prod_{j(i)} \tau_{ij}^x \,. \tag{19.16}$$

Here $j(i)$ denotes the nearest neighbors of site i, τ_{ij}^x is the operator that flips the discrete variables u_{ij}, and r is given by eqn (19.10). The states minimizing this energy satisfy the gauge invariance condition

$$\prod_{j(i)} \tau_{ij}^x |\Psi\rangle = |\Psi\rangle \,. \tag{19.17}$$

The Hilbert space of states that satisfy the condition (19.17) is still huge. If all weaker terms in the Hamiltonian are neglected, all states that satisfy eqn (19.17) are degenerate. These states can be visualized in terms of half-vortices positioned on the sites of the dual lattice a, b. Indeed, a convenient way to describe the different states that satisfy eqn (19.17) is to note that the operator $\prod_{j(i)} \tau_{ij}^x$ does not change the value of $\sum_{\text{hex}} u_{ij}$ [mod(2)]. Thus, one can fix the values of $\sum_{\text{hex}} u_{ij} = v_a$ on all hexagonal plaquettes a and impose the constraint (19.17). In physical terms, the binary values $v_i = 0, 1$ describe the positions of half-vortices on the dual (triangular) lattice. This degeneracy between different states is lifted when the subdominant terms are taken into account. The main contribution to the potential energy of these half-vortices comes from eqn (19.15); it is simply proportional to their number. The dynamics of these vortices is due to processes in which only one rhombus changes its state, and a corresponding flip of the phase across the weak junction. The amplitude of this process is

$$\widetilde{r} \approx E_J^{3/4} E_C^{1/4} \exp(-\widetilde{S_0}), \quad \widetilde{S_0} = 1.61 \sqrt{\frac{E_J}{\widetilde{E_C}}} \,.$$

In this approximation, the effective Hamiltonian controlling these vortices becomes (cf. eqn (19.14))

$$H_v = -\widetilde{r} \sum_{(ab)} \sigma_a^x \sigma_b^x - W \sum_a \sigma_a^z \,, \tag{19.18}$$

where the operators σ_a act in the usual way on the states with and without half-vortices at plaquette a, and the first sum runs over adjacent plaquettes (ab). This Hamiltonian describes an Ising model in a transverse field. For small $W/\widetilde{r} < \lambda_c \sim 1$, its ground state is "disordered", i.e. $\langle \sigma^z \rangle = 0$ but $\langle \sigma^x \rangle \neq 0$, while for $W/\widetilde{r} > \lambda_c$ it is "ordered", i.e. $\langle \sigma^z \rangle \neq 0$, $\langle \sigma^x \rangle = 0$. The critical value of the transverse field is known from extensive numerical simulations (Bishop *et al.* 2000): $\lambda_c \approx 4.6 \pm 0.3$ for a triangular lattice. The "disordered" state corresponds to a liquid of half-vortices, while in the "ordered" state the density of free half-vortices vanishes, i.e. the ground state contains an even number of half-vortices so the total vorticity of the system is zero. To prove this, we start from the state $|\uparrow\rangle$, which is the ground state at $\widetilde{r}/W = 0$, and consider the effect of $\widetilde{r}\sum_{(ab)} \sigma_b^x \sigma_a^x$ in perturbation theory. Higher-energy states are separated from the ground state by the gap W, so each order is finite. Furthermore, in each order the operator $\sigma_a^x \sigma_b^x$ creates two more half-vortices, proving that the total number of half-vortices remains even in each order. The states with an odd number of half-vortices have a gap $\Delta(\widetilde{r}/W)$, which remains nonzero for $W/\widetilde{r} > \lambda_c$.

In terms of the original discrete variables defined on the rhombi, the Hamiltonian (19.18) becomes

$$H_u = -\widetilde{r} \sum_{(ij)} \tau_{ij}^x - W \sum_i \prod_{j(i)} \tau_{ij}^z \,, \tag{19.19}$$

where the τ operators act on the state of each rhombus. This Hamiltonian commutes with the constraint (19.17) and is in fact the simplest Hamiltonian of the lattice Z_2 gauge theory. The disordered regime corresponds to a confined phase of this Z_2 gauge theory, leading to elementary charge $4e$ excitations, and the ordered regime corresponds to the deconfined phase.

Consider now a system with a nontrivial topology, for example with a hole. In this case the set of variables v_a is not sufficient to determine uniquely the state of the system; one has to supplement it by the variable $v_0 = \sum_L u_{ab}$, where the sum is taken over a closed contour L that goes around the hole. Physically, this variable describes the presence/absence of a half-vortex in the hole. The effective Hamiltonian of this additional variable has only a kinetic part because the presence or absence of a half-vortex in a hole which has l weak links in its perimeter gives a potential energy $W_0 = c\epsilon_J (\epsilon_J/\epsilon_C)^l$, which is exponentially small for $l \gg 1$. The kinetic part is similar to the other variables: $H_0 = -\widetilde{r}\sum_{a \in I} \sigma_a^x \sigma_0^x$, which describes a process in which a half-vortex jumps from the hole to the inner boundary, I, of the system. In a state with $\langle \sigma^z \rangle \neq 0$, this process increases the energy of the system by $\tilde{W}(\widetilde{r}/W)$ ($\tilde{W}(0) = W$ and $\tilde{W}(\lambda_c) = 0$). In a state with $\langle \sigma^x \rangle \neq 0$, it costs nothing. Thus, a process in which a half-vortex jumps from the hole into the system and another half-vortex exits into the outside region appears in the second order of the perturbation theory; the amplitude of this process is $t_v = \widetilde{r}^2 \sum_{i \in I, j \in O} g_{ab}$, where the sum is performed over all sites of the inner (I) and outer (O) boundaries and $g_{ab} = \langle \sigma_a^x (1/(H - E_0) \sigma_b^x \rangle$ has the physical meaning of the amplitude for tunneling of a half-vortex from the inner to the outer boundary. At small $\widetilde{r}/W$, we can estimate g_{ab} using a perturbation expansion in $\widetilde{r}/W$: the leading contribution appears in the $|a-b|$-th order of the perturbation theory, and thus $g_{ab} \propto (\widetilde{r}/W)^{|a-b|}$. Thus for small $\widetilde{r}/W$, the tunneling amplitude of the half-vortex is exponentially small in the distance L from the outer to the inner boundary; we expect that it will remain exponentially small for all $\widetilde{r}/W < \lambda_c$. For $\widetilde{r}/W > \lambda_c$, this amplitude is of the order of $\widetilde{r}^2/W$ and therefore is significant.

In a different language, in a system with a hole we can construct a topological invariant $\mathcal{P} = \prod_\gamma \tau_{ij}^x$ (the contour γ is shown in Fig. 19.4), which can take values ± 1 . Note that

Table 19.1 Duality between topological superconductor and topological insulator

	Topological superconductor	Topological insulator
Ground state	Condensate of charges $4e$	Condensate of 2π phase vortices
Fluxons	Gapful, charge $2e$	Gapful, π phase vortices
Pseudocharges	Half fluxes with energy $\epsilon \sim E_J \log(L)$	Charge $2e$ with $\epsilon = 2r + \beta\epsilon_C Lg$
Ground state degeneracy	Charge on the inner boundary mod $4e$	Number (mod 2) of π vortices in the hole
Ground state splitting	$\left(\frac{\delta\Phi}{\Phi_0}\frac{E_J}{r}\right)^L$	$(\widetilde{r}/W)^L$

the contour now goes via triangular lattice sites (where vortices are defined), whereas in the first (superconductive array) version the corresponding path was drawn via sites of the basic hexagonal lattice. The same arguments as used for the superconducting array show that any dynamics consistent with constraint (19.17) preserves $\mathcal{P}$. Thus, formally, the properties of the topological insulator are very similar to the properties of the topological superconductor discussed by Ioffe and Feigel'man (2002) if one replaces the words "Cooper pair" by "half-vortex" and vice versa. We summarize this duality in Table 19.1, where $\beta \sim 1$, $Lg = \min\left[\log L, \log(c/c_0)\right]$, c is the capacitance of a weak link, and c_0 is the self-capacitance of an island.

Note that at small $\widetilde{r}/W \to 0$, the ground state of the Hamiltonian (19.19) satisfies the condition (19.17) and minimizes the second term in eqn (19.19), i.e. it satisfies the condition $\prod_{j(i)} \tau^z_{ij}|\Psi\rangle = |\Psi\rangle$; it can be written explicitly as

$$|0\rangle_{\text{ins}} = \prod_i \frac{1}{2}\left(1 + \prod_{j(i)} \tau^z_{ij}\right) \prod_{kl} | \rightarrow\rangle_{kl} \,. \tag{19.20}$$

This state is a linear superposition of the degenerate states with $\mathcal{P} = 1$ and $\mathcal{P} = -1$:

$$|0\rangle_{\text{ins}} = \frac{1}{\sqrt{2}}\left(|+\rangle_{\text{ins}} + |-\rangle_{\text{ins}}\right). \tag{19.21}$$

It coincides with the ground state $|G\rangle$ of the discrete variables in the superconducting array (cf. eqn (19.5)). The orthogonal superposition of $\mathcal{P} = \pm 1$ states,

$$|1\rangle_{\text{ins}} = \frac{1}{\sqrt{2}}\left(|+\rangle_{\text{ins}} - |-\rangle_{\text{ins}}\right), \tag{19.22}$$

corresponds to a half-vortex inside the hole. The energy difference between the above two states $E_1 - E_0$ is exponentially small in the insulating state of the array, whereas it is large in the superconductive state.

19.6 Quantum manipulations

We now discuss the manipulation of the protected states formed in this system. We start from the *superconductive* version of the array.

First, we note that here the topological invariant $\hat{T}_q$ has a simple physical meaning—it measures the total phase difference (modulo 2π) between the inner and outer boundaries. In an array with an even number of rhombi between the internal and outer boundaries, a state with eigenvalue $c_q = 1$ has a phase difference 0, whereas a state with eigenvalue $c_q = -1$ has a phase difference π. This means that measuring this phase difference measures the state of a qubit in the same basis as that in which $\hat{T}_q$ is diagonal. For the following discussion, we define a set of Pauli matrices $\Sigma_q^{x,y,z}$ acting in the 2×2 qubit space, such that $\Sigma_q^z \equiv T_q$.

The introduction of a weak coupling between these boundaries by a very weak Josephson circuit (characterized by a small energy e_J) would change the phase of these states in a controllable manner, for example by a unitary transformation

$$U^z = \exp(ie_J t\Sigma_q^z)\,. \tag{19.23}$$

A transformation coupling two qubits can be obtained if one introduces a weak Josephson circuit that connects two different inner boundaries (corresponding to different qubits). Namely, it will produce an operation

$$U_{p,q}^z = \exp(ie_J t\Sigma_q^z\Sigma_p^z)\,. \tag{19.24}$$

Analogously, the virtual process involving half-vortex motion around the opening gives the tunneling amplitude ϵ_t between topological sectors, for example for the unitary transformation $U^x = \exp(it\epsilon_t\Sigma_q^x)$. This tunneling can be controlled by a magnetic field if the system is prepared with some number of vortices that are pinned in the idle state in a special plaquette where the flux is an integer. A slow (adiabatic) change of this flux to a normal (half-integer) value would release the vortex and result in the transition between topological sectors with $\epsilon_t \sim \tilde{t}/D^2$.

These operations are analogous to the usual operations on a qubit and are prone to the usual sources of errors. This system, however, allows another type of operation that is naturally discrete. As we have shown above, the transmission of an elementary quasiparticle across the system changes its state by Σ_q^z. This implies that the discrete process of the transfer of one pair across the system is equivalent to a Σ_q^z transformation. Similarly, a controlled process in which a vortex is moved around a hole results in a discrete Σ_q^x transformation. Moreover, this system allow one to make discrete transformations such as $\sqrt{\Sigma^{x,z}}$. Consider, for instance, a process in which, by changing the total magnetic flux through the system, one half-vortex is placed in the center of the system shown in Fig. 19.1(b) and then released. It can escape through the left or through the right boundary; in one case the state does not change, but in the other it changes by Σ_x. The amplitudes add, resulting in the operation $(1 + i\Sigma^x)/\sqrt{2}$. Analogously, using electrostatic gate(s) to pump one charge $2e$ from one boundary to the island in the center of the system and then releasing it results in a $(1 + i\Sigma^z)/\sqrt{2}$ transformation. This type of process allows a straightforward generalization to an array with many holes: in that case, an extra half-vortex or charge should be placed at equal distances from the inner and outer boundaries.

The degenerate ground states in the *insulating* array can be manipulated in the same way as in the superconductor, up to duality (half-vortex $\rightarrow$ Cooper pair and vice versa). As mentioned above, these states $|0\rangle_{\text{ins}}$ and $|1\rangle_{\text{ins}}$ correspond to the absence or presence of a half-vortex inside the hole. We denote by $\tilde{\Sigma}^{x,y,z}$ the Pauli operators acting in the space spanned by $|0\rangle_{\text{ins}}$ and $|1\rangle_{\text{ins}}$. An adiabatic change of the local magnetic field that drags one half-vortex across the system flips the state of the system, providing us with an implementation of the operator $\tilde{\Sigma}^x$ acting on the state of the qubit. Analogously, the motion of an elementary charge $2e$ around the hole changes the relative phase of the states $|0\rangle_{\rm ins}$ and $|1\rangle_{\rm ins}$ by π, providing us with the operator $\tilde{\Sigma}^z$. The operators $\sqrt{\tilde{\Sigma}^{x,z}}$ can be realized in a way similar to that described for the superconductive array. Rotation by an arbitrary angle $U^x = \exp(i\alpha\tilde{\Sigma}^x)$, which is an analogue of the operator (19.23), can achieved by modifying (during a time t) the parameter $\tilde{r}$ in such a way as to produce a non-negligible amplitude A for tunneling of a half-vortex across the system: $\alpha = At$. In the same way, a two-qubit entanglement operation can be realized, which is an analogue of the operator (19.24)):

$$\tilde{U}^x_{p,q} = \exp(itA_{pq}\tilde{\Sigma}^x_p\tilde{\Sigma}^x_q)\,, \tag{19.25}$$

where in this case A_{pq} is the amplitude for tunneling of a half-vortex between holes p and q.

19.7 Physical properties of small arrays

19.7.1 Superconductive array

Even without these applications for quantum computation, the physical properties of this array are remarkable: it exhibits long-range order in the square of the usual superconducting order parameter, $\langle\cos(2(\phi_0-\phi_r))\rangle \sim 1$, without the usual order $\langle\cos(\phi_0-\phi_r)\rangle = 0$; and the charge transferred through the system is quantized in units of $4e$. This can be tested in an interference experiment, sketched in Fig. 19.1(b); as a function of the external flux Φ_m, the supercurrent through the loop should be periodic with half the usual period. This simpler array can also be used for a kind of "spin echo" experiment: applying two consecutive operations $(1+i\Sigma_x)/\sqrt{2}$ of the kind described above should give a unique classical state again, whereas applying only one of them should result in a quantum superposition of two states with equal weight.

This echo experiment can be used to measure the decoherence time in this system. Generally, one distinguishes between processes that flip the classical states and processes that change the relative phases of the states. In the NMR literature, the former are referred to as transverse relaxation, and the latter as longitudinal relaxation. Transverse relaxation occurs when a vortex is created and then moved around the opening by external noise. Assuming thermal noise, we estimate the rate of this process $\tau_\perp^{-1}$ to be of the order of $\tilde{t}\exp(-E_V(L)/T)$. Similarly, the transfer of a quasiparticle from the outer to the inner boundary changes the relative phase of the two states, leading to longitudinal relaxation. The rate for this is proportional to the density of quasiparticles: $\tau_\parallel^{-1} = R\exp(-2r/T)$. The coefficient R depends on the details of the physical system. In an ideal system with some nonzero uniform value of ϵ (see eqn (19.12)), quasiparticles are delocalized and $R \sim \epsilon/L^2$. Random deviations of fluxes Φ_r from half-integer values produce randomness in ϵ, in which case one expects Anderson localization of quasiparticles owing to *off-diagonal* disorder, with a localization length of the order of the lattice spacing, and thus $R \sim \bar{\epsilon}\exp(-cL)$ with $c \sim 1$, and where $\bar{\epsilon}$ is a typical value of ϵ. Stray charges induce randomness in the values of r, i.e. add some *diagonal* disorder. When the random part of r, δr, becomes larger than $\bar{\epsilon}$, the localization becomes stronger: $R \sim \bar{\epsilon}(\bar{\epsilon}/\bar{\delta r})^L$, where $\bar{\delta r}$ is a typical value of δr. Upon further increase of the stray charge field, there appear rare sites where r_i is much smaller than the average value. Such sites act as additional openings in the system. If the density of these sites is significant, the effective length that controls the decoherence becomes the distance between these sites. For a typical $E_V(L) \approx E_J \approx 2K$, the transverse relaxation time reaches seconds for $T \sim 0.1\,\mathrm{K}$, whereas a realistic value of $\epsilon/r \sim 0.1$ implies that, owing to quasiparticle localization in the random case the longitudinal relaxation reaches the same scale for systems of size $L \sim 10$; note that the temperature T has only to be somewhat lower than the excitation gap $2r$ in order to make the longitudinal rate low.

Most properties of the array are only weakly sensitive to the effect of stray charges: as discussed above, they result in a position-dependent quasiparticle potential energy, which has very little effect because these quasiparticles have no kinetic energy and are localized anyway. The direct effect of stray charges on the topologically protected subspace can also be physically described as the effect of the electrostatic potential on states with even and odd charges at the inner boundary; since the absolute value of the charge fluctuates strongly, this effect is exponentially weak.

19.7.2 Insulating array

The signature of the topological insulator is the persistence of a trapped half flux quantum inside the central hole (see Fig. 19.4), which can be observed by cycling the magnetic field so as to drive the system back and forth between the insulating and superconducting states. This trapping is especially striking in the insulator. Experimentally, this can be revealed by driving the array slowly into the superconducting state and then measuring the phase difference between opposite points such as those in the outer ring in Fig. 19.4. In the state with a half-vortex, the phase difference is $\pi/2 + \pi n$, whereas it is πn in the other state. The contribution πn is due to vortices of the usual kind, which get trapped in the large hole. This slow transformation can be achieved by changing the strength of the weak links using the external magnetic field as a control parameter. The precise nature of the superconductive state is not essential, because a phase difference π between points can be interpreted as being due to a full vortex trapped in a hole in a conventional superconductor or due to a π periodicity in a topological superconductor, which makes no essential difference. These flux-trapping experiments are similar to the ones proposed for high-T_c cuprates (Senthil and Fisher 2001*a*,*b*), but with a number of important differences: the trapped flux is half of Φ_0, the cycling does not involve the temperature (avoiding problems with excitations), and the final state can be either a conventional or a topological superconductor.

19.8 XZ array

It is well known that stable degeneracy of quantum levels is almost always due to a high degree of symmetry in the system. Examples are numerous: time inversion invariance ensures the degeneracy of states with half-integer spin, rotational symmetry results in the degeneracy of states with nonzero momentum, etc. In order for the degeneracy to be stable with respect to local noise, one needs to ensure that sufficient symmetry remains even if part of the system is excluded. The simplest example is provided by six Josephson junctions connecting four superconducting islands, so that each island is connected to every other (Feigel'man *et al.* 2004*a*,*b*). In this miniarray, all islands are equivalent, so it is symmetric under all transformations of the permutation group S_4. This group has a two-dimensional representation, and thus pairs of exactly degenerate states. With an appropriate choice of parameters, one can make these doublets the ground state of the system. Noise acting on one superconducting island reduces the symmetry to the permutation group of three elements, which still has two-dimensional representations. So, this system is protected from noise to first order ($n = 2$). The goal of this lecture is to discuss designs giving systems that are protected from noise to higher orders. Note that systems with higher symmetry groups, such as five junctions connected by ten junctions (group S_5), typically do not have two-dimensional representations, so in these systems one can typically obtain much higher degeneracy but not higher protection. Here we review the results of Douçot *et al.* (2005), where a somewhat different approach to the construction of highly protected states was proposed.

Generally, one gets degenerate states if there are two symmetry operations described by operators P and Q that commute with the Hamiltonian but do not commute with each other. If $[P, Q]|\Psi\rangle \neq 0$ for any $|\Psi\rangle$, all states are at least doubly degenerate. A local noise term is equivalent to adding other terms in the Hamiltonian which might not commute with these operators, thereby lifting the degeneracy. Clearly, in order to preserve the degeneracy, one needs to have two (*sets* of n elements each), of noncommuting operators $\{P_i\}$ and $\{Q_i\}$, so that any given local noise field does not affect some of them; furthermore, preferably, any given local noise should affect at most *one* P_i and Q_i. In this case, the effect of the noise appears when n noise fields act *simultaneously,* i.e. in the n-th order in the noise strength. Another important restriction comes from the condition that these symmetry operators should not result in a higher degeneracy of the ideal system. For two operators P and Q, this implies that $[P^2, Q] = 0$ and $[P, Q^2] = 0$. One can construct degenerate eigenstates of the Hamiltonian by starting with the eigenstate $|0\rangle$ of H and Q and acting on this state with P. The resulting state, $|1\rangle$, should be different from the original state because P and Q do not commute: $[P, Q]\Psi \neq 0$ for any Ψ. In a doubly degenerate system, by acting again on this state with the operator P, one should get back the state $|1\rangle$, so $[P^2, Q] = 0$. For a set of operators, the same argument implies that in order to get a double degeneracy (and not more), one requires that $[P_i P_j, Q] = 0$ and $[P, Q_i Q_j] = 0$ for any i, j. In this case, one can diagonalize simultaneously the sets of operators $\{Q_i\}$, $\{Q_i Q_j\}$, and $\{P_i P_j\}$. Consider a ground state $|0\rangle$ of the Hamiltonian which is also an eigenstate of all these operators. By acting on it with, say, P_1 we get a new state, $|1\rangle$, but since $|1\rangle \propto (P_i P_1) P_1 |0\rangle = P_i (P_1 P_1)|0\rangle \propto P_i|0\rangle$, none of the other operators of the same set would produce a new state.

19.8.1 "Quantum compass" model

The conditions discussed above are fully satisfied by a spin $S = 1/2$ model on a square $n \times n$ array described by the Hamiltonian

$$H = -J_x \sum_{i,j} \sigma^x_{i,j} \sigma^x_{i,j+1} - J_z \sum_{i,j} \sigma^z_{i,j} \sigma^z_{i+1,j} \,. \tag{19.26}$$

Here the σ are Pauli matrices; note that the first term couples spins in same row of the array, while the second couples them along the columns. It is not important for the following discussion whether the boundary conditions are periodic or free, but since the latter are much easier to implement in hardware, we shall assume them in the following. Furthermore, the signs of the couplings are irrelevant because for a square lattice, one can always change them by choosing a different spin basis on one sublattice. For the sake of argument, we have assumed that the signs of the couplings are ferromagnetic; this is also a natural sign for the Josephson junction implementations described in Section 19.3. The Hamiltonian (19.26) was first introduced by Kugel and Khomskii (1982) as a model for the anisotropic exchange interaction in transition metal compounds, but its properties remain largely unclear.

The Hamiltonian (19.26) has two sets of integrals of motion, $\{P_i\}$ and $\{Q_i\}$, with n operators each:

$$P_i = \prod_j \sigma^z_{i,j}\,,$$
$$Q_j = \prod_i \sigma^x_{i,j}\,,$$

i.e. each P_i is the row product of $\sigma^z_{i,j}$ and Q_j is the column product of $\sigma^x_{i,j}$. Consider the operator P_i first. It obviously commutes with the second term in the Hamiltonian and, because the first term contains two $\sigma^z_{i,j}$ operators in the same row, P_i contains either none of them or both of them and since different Pauli matrices anticommute, P_i commutes with each term in the Hamiltonian (19.26). Similarly, $[Q_i, H] = 0$. Clearly, different P_i commute with themselves, i.e. $P_i^2 = 1$, and similarly $[Q_i, Q_j] = 0$ and $Q_i^2 = 1$, but they do not commute with each other:

$$\begin{aligned} \{P_i, Q_j\} &= 0\,, \\ [P_i, Q_j]^2 &= 4\,, \end{aligned} \tag{19.27}$$

so $[P_i, Q_j]|\Psi\rangle \neq 0$ for any $|\Psi\rangle$, and thus, in this model, all states are at least doubly degenerate. Further, because P_iP_j contains two $\sigma^z_{i,j}$ in any column, this product commutes with all operators Q_k, and similarly $[Q_kQ_l, P_i] = 0$. Thus, we conclude that in this model all states are doubly degenerate, there is no symmetry reason for a larger degeneracy, and that this degeneracy should be affected by noise only in the n-th order of perturbation theory.

To estimate the effect of noise (which appears in this high order), one needs to know the energy spectrum of the model and what its low-energy states are. All states of the system can be characterized by the set $\{\lambda_i = \pm 1\}$ of the eigenvalues of the operators P_i (or alternatively by the eigenvalues of the operators Q_j). The degenerate pairs of states are formed by the two sets $\{\lambda_i\}$ and $\{-\lambda_i\}$, and each operator Q_j interchanges these pairs: $Q_i\{\lambda_i\} = \{-\lambda_i\}$. We believe that different choices of $\{\lambda_i = \pm 1\}$ exhaust *all* low-energy states in this model, i.e. that there are exactly 2^n low-energy states. Note that this is a somewhat unusual situation; normally one expects n^2 modes in a 2D system and thus 2^{n^2} low-energy states. The number 2^n of low-energy states is natural for a one-dimensional system and would also appear natural in two-dimensional systems if these states were associated with the edge. Here, however, we cannot associate them with edge states, because they do not disappear for periodic boundary conditions. We cannot prove our conjecture in the general case, but we can see that it is true when one coupling is much larger than the other, and we have verified it numerically when the couplings are of the same order of magnitude. We start with an analytic treatment of the $J_z \gg J_x$ case.

When one coupling is much larger than others, it is convenient to start with a system where these others are absent and then treat them as small perturbations; in the limit $J_x = 0$, all columns are independent and the ground state of each column is an Ising ferromagnet. The ground state of each column is doubly degenerate: $|1\rangle_j = \prod_i |\uparrow\rangle_{ij}$ and $|2\rangle_j = \prod_i |\downarrow\rangle_{ij}$, giving us 2^n degenerate states in this limit. The excitations in each column are static kinks against the background of these states, and each kink has an energy $2J_z$. If we now include the J_x coupling, we see that it creates two kinks in each of the neighboring columns, thereby increasing the energy of the system by $8J_z$, so the lowest order of the perturbation theory is small in $J_x/8J_z$. A splitting between the 2^n states occurs owing to high-order processes which flip all spins in two columns. In the leading approximation, one can calculate the amplitude of this process ignoring other columns. Thus, for this calculation we can consider a model with only two columns that can be mapped onto a single Ising chain in a transverse field in the following manner. The ground state of two independent columns belongs to the sector of the Hilbert space characterized by all $P_i P_j = 1$; it is separated from the rest of the spectrum by a gap of the order of $2J_z$. Furthermore, the Hamiltonian does not mix these sectors with different P_i, so in order to find the low-energy states, it is sufficient to diagonalize the problem in the sector $P_i = 1$. In this sector, only two states are allowed in each row: $|\uparrow\uparrow\rangle$ and $|\downarrow\downarrow\rangle$; in the basis of these states, the Hamiltonian is reduced to

$$H_{\text{col}} = -2J_z \sum_i \tau_i^z \tau_{i+1}^z - J_x \sum_i \tau_i^x \, , \tag{19.28}$$

where the τ are Pauli matrices acting in the space of the $|\uparrow\uparrow\rangle$ and $|\downarrow\downarrow\rangle$ states. One can show that this leads to a splitting $2\Delta \approx (J_x/2J_z)^n(2J_z)$ between the symmetric and antisymmetric combinations of the two ferromagnetic chains in the problem. Thus, we conclude that the effective Hamiltonian of the low-energy states in the full system is

$$H = \Delta \sum_j \widehat{\tau}_j^x \widehat{\tau}_{j+1}^x \, ,$$

where the $\widehat{\tau}$ are Pauli matrices acting in the space of the $|1\rangle$ and $|2\rangle$ states describing the global state of the whole column. This effective low-energy model also describes a ferromagnetic chain in which the excitations (static kinks) are separated from the degenerate ground state by a gap 2Δ. In the basis of these 2^n low-energy states, the operators $Q_j = \widehat{\tau}_j^x$.

We conclude that in the limit $J_z \gg J_x$, 2^n low-energy states form a narrow (of the order of Δ) band inside a much larger gap, J_z, characterized by different eigenvalues of the operators Q_j and by one value $P_i P_j = 1$. In the opposite limit $J_x \gg J_z$, the low-energy states form a narrow band inside a gap of size J_x characterized by different eigenvalues of operators P_j and by the same value $Q_i Q_j = 1$. Consider now the effect of weak noise in the former limit. To be more specific, we consider the effect of additional single-site fields

$$H_n = \sum_{i,j} h_{i,j}^z \sigma_{i,j}^z + h_{i,j}^x \sigma_{i,j}^x \, .$$

The first term shifts (up or down) the energies of each ferromagnetic column by $H_i^z = \sum_i h_{i,j}^z$, while the second term gives transitions between up and down states in each column. These transitions appear only in the order n of the perturbation theory, so their amplitude is exponentially small: $H_j^x = (\prod_i h_{ij}^x/J_z)J_z$. Thus, when projected onto the low-energy subspace, this noise part becomes

$$H_n = \sum_j H_j^z \widehat{\tau}_j^z + H_j^x \widehat{\tau}_j^x \, .$$

The effect of the first term on the ground state degeneracy appears in the n-th order of the perturbation theory in H_j^z/Δ and so it is much bigger than that of the second term, because Δ becomes exponentially small as $n \to \infty$ for $J_x \ll J_z$. Note that although the effect of the $h_{i,j}^z \sigma_{i,j}^z$ noise appears only in a large order of the perturbation theory, it is not small, because of the small energy denominator in this parameter range. Similarly, we expect that in the opposite limit, $J_z \ll J_x$, the low-lying states will be characterized by the set of eigenvalues of the operators P_i; the effect of the $h_{i,j}^x \sigma_{i,j}^x$ noise grows rapidly while the effect of the $h_{i,j}^z \sigma_{i,j}^z$ noise decreases with increasing J_x. We conclude that in the limits when one coupling is much larger than the other ($J_z \gg J_x$ or $J_z \ll J_x$), the gap closes very quickly (exponentially) and the nonlinear effect of the appropriate noise grows rapidly with the system size. These qualitative conclusions should remain valid for all couplings except at a special isotropic point ($J_x = J_y$) unless the system undergoes a phase transition near this point (at some $J_x/J_y = j_c \sim 1$).

In order to check these conclusions, we have numerically diagonalized small spin systems containing up to 5×5 spins subjected to a small random field $h_{i,j}^z$ flatly distributed in the interval $(-\delta/2, \delta/2)$. We see that indeed the gap closes rather fast away from the special point $J_x = J_y$ (Fig. 19.5) but remains significant near $J_z = J_x$, where it clearly has a much weaker size dependence. Interestingly, the gap between the lowest 2^n states and the rest of the spectrum expected in the limits $J_z \gg J_x$ and $J_z \ll J_x$ appears only at $J_x/J_z > j_c$, with a practically size-independent $j_c \approx 1.2$. We also see that the condition $P_i = 1$ eliminates all low-lying states in the limit $J_z \ll J_x$, where the lowest excited state in the $P_i = 1$ sector is separated from the ground state by a large gap and, in fact, provides a lower bound for all high-energy states. The special nature of this state appears only at $J_x/J_z > j_c$. Clearly, the system behaves quite differently near the isotropic point and away from it, but the size limitations do not allow us to conclude whether these different regimes correspond to two different phases (with the "isotropic" phase restricted to a small range of parameters $j_c^{-1} < J_x/J_z < j_c$) or whether this is a signature of a critical region which becomes narrower as the size increases. Although we do not see any appreciable change in j_c with the system size, our numerical data do not allow us to exclude the possibility that j_c tends to unity in the thermodynamic limit. We conclude that numerical data *favor* the intermediate-phase scenario. In contrast to this, the analytical results for two and three chains indicate that a transition occurs only at the point $J_z = J_x$. Namely, both the two- and the three-chain models with periodic boundary conditions in the transverse direction can be mapped onto solvable models with a transition at $J_z = J_x$: in the case of two chains, the problem is mapped onto the exactly solvable Ising model in a transverse field as described above, while the three-chain model is mapped onto the four-state Potts model in a similar way. The latter is not exactly solvable in the whole parameter range but it obeys an exact duality that allows one to determine its critical point (Baxter 1982, Wu 1982); furthermore, its exponents can be determined from conformal field theory (Dotsenko and Fateev 1984). The mapping of the three-chain problem onto the Potts model is possible because the number of states of a three-spin rung for a given value of the conserved operator P is 4, while the number of different terms in the Hamiltonian that couples the adjacent rungs is 3. For a larger number of chains, the number of states in each rung grows exponentially, while the number of terms in the Hamiltonian grows only linearly, making such mappings impossible. In this sense, the two- and three-chain models are exceptional and it is fairly possible that an intermediate phase appears in models with a larger number of chains.

Finally, we have checked the effect of $h_{i,j}^z \sigma_{i,j}^z$ on the ground state degeneracy splitting;

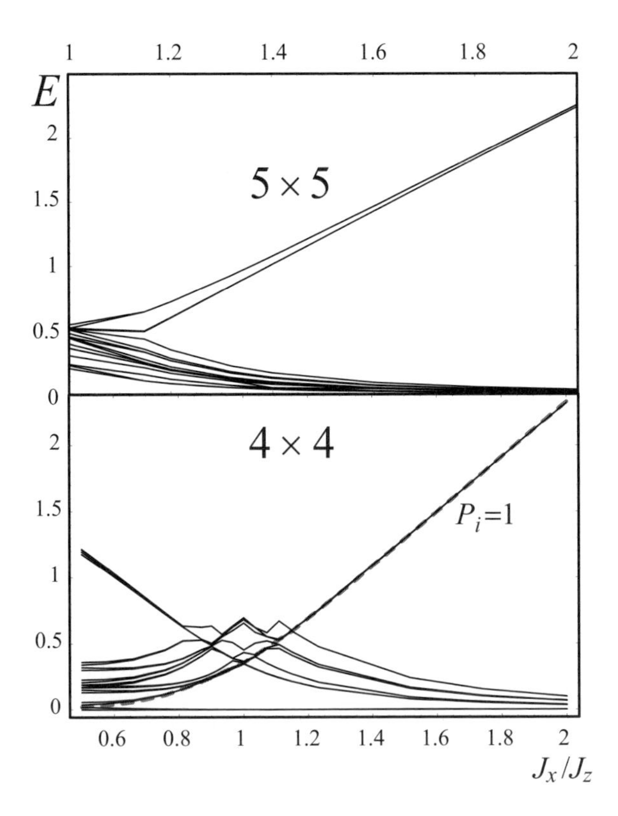

Fig. 19.5 Energy spectrum of 5×5 and 4×4 systems in units of the coupling J_z as a function of J_x/J_z. We show the energies of the lowest 40 states for the 5×5 system (upper panel) and the lowest 20 states for the 4×4 system (lower panel). One can clearly see that at large anisotropy, a well-defined low-energy band is formed, which contains 2^5 states for the 5×5 system and 2^4 states for the 4×4 one. In order to verify that the low-energy states are in one-to-one correspondence with the eigenvalues of P_i for large J_x/J_z, we have calculated the second lowest eigenstate in the $P_i = 1$ sector (the first one is the ground state). As shown in the lower panel, this state indeed has a large gap for $J_x/J_z \gtrsim 1.2$.

our results are shown in Fig. 19.6. We see that, as expected, this disorder becomes relevant for $J_x \ll J_z$, while in the opposite limit its effect quickly becomes unobservable. We conclude that at (and perhaps near) the isotropic point, the gap closes slowly with the system size and the effect of even significant disorder ($\delta = 0.1$) becomes extremely small for medium-sized systems.

Although it is not clear how fast the gap closes in the thermodynamic limit (if it closes exponentially fast, the system never becomes truly protected from noise, because the effect of the high-order terms might become very large), our numerical results clearly indicate that medium-size (4×4 or 5×5) systems provide extremely good protection from noise, suppressing its effect by many orders of magnitude. This should be enough for all practical purposes. Furthermore, if it is possible to construct systems where $P_i P_j = 1$ (in other words, with an additional term in the Hamiltonian $H_P = -\Delta \sum_{ij} P_i P_j$ with a significant Δ), this would eliminate the dangerous low-energy states, leading to good protection for all coupling

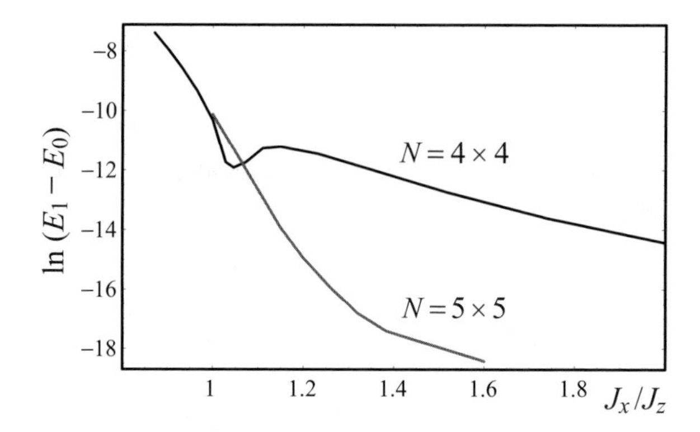

Fig. 19.6 Ground state splitting by a random field in the z-direction for 5×5 and 4×4 systems. The random field acted on each spin and was randomly distributed in the interval $(-0.05, 0.05)$. Note that the effect of the random field in the z-direction becomes larger for $J_x \ll J_z$, as expected (see text). Because near the $J_x = J_z$ isotropic point the gap for the 5×5 system is significantly smaller than the gap for the 4×4 system, this relatively large disorder has almost the same effect on these systems at $J_x \sim J_z$. We have verified numerically that a decrease of the disorder by a factor of two leads to dramatically smaller effects for the 5×5 system, confirming the scaling $E_1 - E_0 \propto \delta^n$ discussed in the text; for $J_x/J_z > 1.2$, the difference $E_1 - E_0$ is difficult to resolve numerically.

strengths $J_x \geq J_z$. Indeed, in this case, we can repeat the previous analysis and conclude that the effects of noise appear only in the n-th order and that now the perturbation theory in the "dangerous" noise, H_j^z, implies an expansion in H_j^z/Δ where Δ is no longer exponentially small but is the coefficient in the Hamiltonian H_P. Thus, in this case these higher-order terms become small.

Recently, a number of studies (Dorier *et al.* 2005, Chen *et al.* 2007, Orus *et al.* 2008) have been performed to elucidate the nature of the ground state and low-lying states, and of the quantum phase transition(s) in the quantum compass model as a function of the ratio J_x/J_z. In particular, Chen *et al.* and Orus *et al.* concluded the existence of a single first-order transition at $J_x = J_z$.

19.8.2 Josephson junction implementations

The basic ingredient of any spin-$1/2$ implementation in a Josephson junction array is an elementary block that has two (nearly degenerate) states. One of the simplest implementations is provided by a four-Josephson-junction loop (shown as a rhombus in Fig. 19.7(a)) penetrated by a magnetic flux $\Phi_0/2$. Classically, this loop is frustrated and its ground state is degenerate: it corresponds to phase differences $\pm\pi/4$ across each junction constituting the loop. Two states (spin "up" and "down") then correspond to states with a phase difference $\pm\pi/2$ across the rhombus. For an isolated rhombus, a nonzero (but small) charging energy $E_C = e^2/2C$ would result in transitions between these two states with an amplitude

$$r \approx E_J^{3/4} E_C^{1/4} e^{-s\sqrt{E_J/E_C}}, \tag{19.29}$$

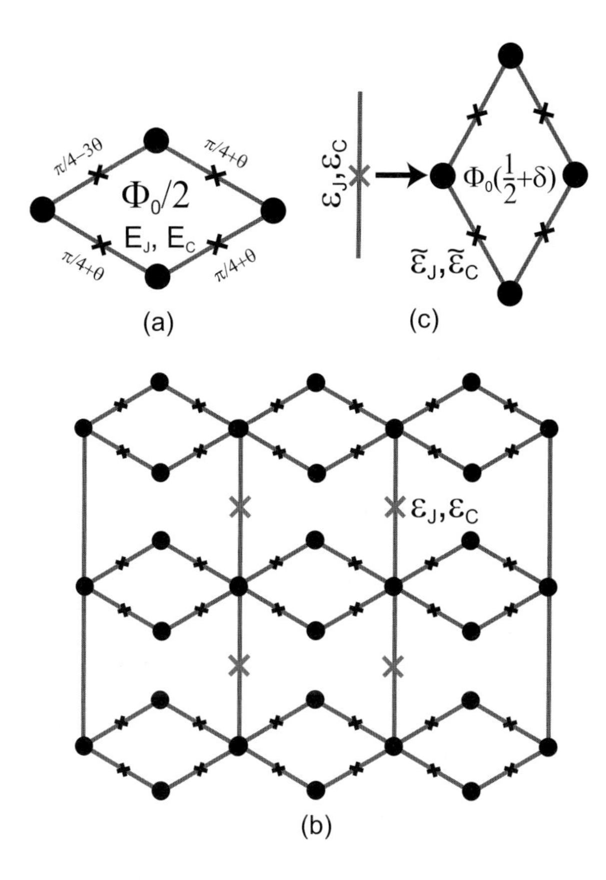

Fig. 19.7 Schematic illustrations of an array equivalent to the spin model with the interaction (19.30) in the vertical direction. (a) The main element of the array, a superconducting rhombus frustrated by a magnetic flux $\Phi_0/2$. The Josephson energy of each rhombus is minimal for $\theta = 0$ and $\theta = -\pi/2$. A significant charging energy induces transitions $\theta = 0 \leftrightarrow \theta = -\pi/2$ between these energy minima. (b) The array geometry. The superconducting boundary conditions chosen here ensure that $P_i P_j = 1$, thereby eliminating all low-lying states in the appropriate regime (see text). (c) The requirement that the continuous phase does not fluctuate much while the discrete variables have large fluctuations is easier to satisfy in a very big array ($L > 20$) if one replaces the vertical links by rhombi with junctions with $\tilde{\epsilon_J}, \tilde{\epsilon_C}$ frustrated by a flux $\Phi_0(\delta + 1/2)$.

thereby lifting this degeneracy. Here s is a numerical coefficient of the order of unity, which was found by Douçot *et al.* (2003) to be approximately equal to 1.61, and E_J is the Josephson energy of each junction.

Simplest Josephson junction array. We begin with a Josephson junction array that has two sets of the integrals of motion $\{P_i\}$ and $\{Q_i\}$ discussed above, which is shown in Fig. 19.7. This array contains rhombi with junctions characterized by Josephson and charging energies $E_J \gtrsim E_C$, and weaker vertical junctions characterized by energies ϵ_J, ϵ_C. As we explain below, although this array preserves the integrals of motion $\{P_i\}$ and $\{Q_i\}$, it maps onto a spin model that differs from (19.26); we consider a more complicated array that is completely equivalent to spin model (19.26) later. The state of the system is fully characterized by the state of each rhombus (described by an effective spin $1/2$) and by the small deviations of the continuous superconducting phase across each junction from its equilibrium (classical) value. Ignoring the continuous phase for the moment, we see that the potential energy of the array shown in Fig. 19.7 is given by

$$H_z = -\epsilon_J \sum_{i,j} \tau^z_{i,j} \tau^z_{i+1,j}, \qquad \tau^z_{i,j} \equiv \prod_{k<j} \sigma^z_{i,k}\,. \tag{19.30}$$

Physically, the variable $\tau^z_{i,j}$ describes the phase of the rightmost corner of each rhombus with respect to the left (grounded) superconducting wire modulo π. The right superconducting wire (which connects the rightmost corners of the rhombi in the last column) ensures that the phase differences along all rows are equal. In the limit of a large phase stiffness, this implies that the numbers of rhombi with a phase difference $\pi/2$ should be equal for all rows modulo 2. This constraint does not allow an individual rhombus flip; instead, a flip of one rhombus must always be followed by a flip of another in the same row. If, however, the phase stiffness is low, a flip of one rhombus can also be compensated by continuous phase deformations in the other rhombi constituting this row; we derive the conditions under which we can exclude these processes below. The simplest allowed process is a simultaneous flip of two rhombi in one row,

$$H_x = -\sum_{i,j,k} \widetilde{J_x}(j-k)\sigma^x_{i,j}\sigma^x_{i,k}\,, \tag{19.31}$$

where $\widetilde{J_x}(k)$ is the amplitude to flip two rhombi a distance k apart. Both the potential energy (19.30) and the kinetic energy (19.31) commute with the integrals of motion $\{P_i\}$ and $\{Q_j\}$, so that we expect that the main feature of this model, namely the existence of protected doublets, will be preserved by this array.

As explained in the previous section, in order to achieve really good protection one needs to eliminate all low-energy states (except for the degenerate ground state) characterized by different values of the operators $\{P_i\}$ and $\{Q_j\}$. The array shown in Fig. 19.7 has a boundary conditions implying $P_i P_j = 1$ for any i, j because, in this array, the sums along each row of the phases across individual rhombi should be equal for all rows. Thus, for a sufficiently large tunneling amplitude $\widetilde{J_x}(k)$, this array should have two degenerate ground states separated from the rest of the spectrum by a large gap. Physically, these two states correspond to two different values of the phase difference along each row. The quantitative condition that ensures that the tunneling amplitude $\widetilde{J_x}(k)$ is large enough depends on the range of $\widetilde{J_x}(k)$. The simplest situation is realized if only nearest-neighbor rhombi flip with a significant amplitude J_x. Because a flip of two nearest-neighbor rhombi is equivalent to a flip of the phase on the island between them, in this case the spin model given in eqns (19.30) and (19.31) is equivalent to a collection of independent vertical Ising chains with Hamiltonian

$$H = -\sum_{i,j} \epsilon_J \tau^z_{i,j}\tau^z_{i+1,j} - J_x \tau^x_{i,j}\,.$$

For $J_x \gg 2\epsilon_J$, each chain described by this Hamiltonian has a unique ground state separated by $\Delta = 2J_x$ from the rest of the spectrum. As the ratio ϵ_J/J_x grows, the gap decreases.

In the opposite limiting case of a very long range, $\widetilde{J_x}(k) = J_x$, one can treat the interaction (19.31) in the mean-field approximation

$$H_x = -J_x L_x \langle \sigma^x_{i,k} \rangle \sum_{i,j} \sigma^x_{i,j} \,. \tag{19.32}$$

At large J_x, the ground state of this system is also a doublet (characterized by $\langle \sigma^x_{i,k} \rangle = \pm 1$), with all other excitations separated by a gap $\Delta = 2L_x J_x$ from the rest of the spectrum. As we increase the vertical coupling ϵ_J, the gap to the excitations gets smaller. At very large ϵ_J, the Hamiltonian is dominated by the ferromagnetic coupling in the vertical direction, so in this regime there are many low-energy states corresponding to two possible magnetizations of each column. The magnitude of ϵ_J for which the gap decreases significantly can be estimated from the first-order correction in ϵ_J. The dominant contribution comes from transitions involving rhombi in the outermost rows. They occur with amplitude ϵ_J and lead to states with energy Δ, so we expect that as long as $\epsilon_J \lesssim \Delta$, the system has a doubly degenerate ground state separated from the other states by gap of the order of Δ.

The amplitude $\widetilde{J_x}(k)$ for the simultaneous flip of two rhombi can be found from the same calculation that was used to calculate a single-rhombus flip and a simultaneous flip of three rhombi. If $\epsilon_C \gg E_C$, the contribution of the vertical links to the total kinetic energy of the superconducting phase is small and can be treated as a small perturbation; in this case

$$\widetilde{J_x}(k) \approx E_J^{3/4} E_C^{1/4} e^{-2s\sqrt{E_J/E_C}(1+ckE_C/\epsilon_C)} \,, \tag{19.33}$$

where $c \sim 1$. Here, a factor 2 appears in the exponential because, in this process, the phases across two neighboring rhombi change simultaneously. Note that although the relative change in the action due to vertical links is always small, their contribution might suppress the flips of all rhombi except nearest neighbors if $E_J E_C/\epsilon_C^2 \gg 1$. However, even a relatively large ϵ_C (so that $E_C/\epsilon_C \ll 1$) can be sufficient to suppress the processes involving non-nearest neighbors. We conclude that the low-energy states become absent as long as

$$\epsilon_J < L_{\text{eff}} J_x \,, \tag{19.34}$$

where $L_{\text{eff}} = 1/2$ if $E_J E_C/\epsilon_C^2 \gg 1$ and $L_{\text{eff}} \approx \min(\epsilon_C/\sqrt{E_J E_C}, L)$ if $E_J E_C/\epsilon_C^2 \ll 1$. These estimates assume that the main contribution to the capacitance comes from the junctions and ignores the contribution from the self-capacitance. If the self-capacitance is significant, processes involving more than one island become quickly suppressed.

We now consider the effect of the continuous fluctuations of the superconducting phase. Generally, a finite phase rigidity allows a single-rhombus flip, described by the $\sum_{ij} \tilde{t}\sigma^x_{ij}$ term in the effective spin Hamiltonian. This term does not commute with the integrals of motion P_i, and thereby destroys the protected doublets. However, for a significant phase rigidity, the energy of a state formed by a single rhombus flip, U_{sf}, is large. If, further, the amplitude $\tilde{t}$ of these processes is small, i.e. $\tilde{t} \ll U_{\text{sf}}$, the states corresponding to single flips can be eliminated from the effective low-energy theory and the protection is restored. If $\tilde{t} > U_{\text{sf}}$, the protection is lost.

We thus begin our analysis of the effects of a finite phase rigidity with a consideration of the dangerous single-rhombus flips. Generally, the continuous phase can be represented as a sum of two parts: one that it is due to vortices, and a spin-wave part which does not change the winding numbers of the phase. As is usual in XY systems, it is the vortex part that is the most relevant for the physical properties. In particular, in these arrays it is the vortex part that controls the dynamics of the discrete subsystem. Note that, unlike conventional arrays, arrays containing rhombi allow two types of vortices: half-vortices and full vortices, because of the double periodicity of each rhombus. A flip of an individual rhombus is equivalent to the creation of a pair of half-vortices. If the ground state of the system contains a liquid of half-vortices, these processes become real and the main feature of the Hamiltonian, namely the existence of two sets of anticommuting variables, is lost. We now estimate the potential energy of a half-vortex and of a pair associated with a single flip, U_{sf}, and the amplitude to create such a pair, $\widetilde{t}$. We begin with the potential energy, which is different in different limits. Let us consider the simpler limiting case where rhombus flips do not affect the rigidity in the vertical direction; it remains ϵ_J. Further, we have to distinguish between the case of a very large size in the horizontal direction and a moderate size because the contribution from the individual chains can be dominant in a moderate system if $E_J \gg \epsilon_J$. In a very large system of linear size L with rigidity ϵ_J in the vertical direction, the potential energy of one vortex is

$$E_v = \pi\sqrt{E_J\epsilon_J}\ln(L)\,, \tag{19.35}$$

while the energy of a vortex–antivortex pair at a large distance R from each other is

$$U_v(R) = \pi\sqrt{E_J\epsilon_J}\ln(R)\,. \tag{19.36}$$

These formulas can be derived by noting that at large scales the superconducting phase changes slowly, which allows one to use a continuous approximation for the energy density, $E = \frac{1}{2}E_J(\partial_x\phi)^2 + \frac{1}{2}\epsilon_J(\partial_z\phi)^2$. If we then rescale the x-coordinate by $x \to \widetilde{x}\sqrt{E_J/\epsilon_J}$, we get an isotropic energy density $E = \frac{1}{2}\sqrt{E_J\epsilon_J}(\nabla\phi)^2$. The continuous approximation is valid if both rescaled coordinates $\widetilde{x}, z \gtrsim 1$. Thus, in a system with $E_J \sim \epsilon_J$, the formulas (19.35) and (19.36) remain approximately correct even at small distances $R \sim 1$, so a flip of a single rhombus creates a half-vortex–antihalf-vortex pair with energy $E_p \approx E_J$, but the formulas become parametrically different in a strongly anisotropic system. Consider first the limit $\epsilon_J = 0$. Here the chains of rhombi are completely decoupled and the energy of two half-vortices separated by one rhombus in the vertical direction (the configuration created by a single flip) is due to the phase gradients in only one chain, $U_{\text{sf}}^{(0)} = \pi^2 E_J/(2L)$, which appear because the ends of the chain have a phase fixed by the boundary superconductor. A very small coupling between the chains adds $U_{\text{sf}}^{(1)} = (2/\pi)\epsilon_J L$ to this energy, so the total potential energy of a single flip inside the array is

$$U_{\text{sf}} = \frac{\pi^2 E_J}{2L} + \left(\frac{4}{\pi}\right)\epsilon_J L, \qquad L \ll \sqrt{\frac{E_J}{\epsilon_J}}\,. \tag{19.37}$$

This formula is correct as long as the second term is much smaller than the first one; these terms become comparable at $L = \sqrt{E_J/\epsilon_J}$, and at larger L the potential energy associated with the single flip saturates at

$$U_{\text{sf}} = \gamma\sqrt{E_J\epsilon_J}, \qquad L \gg \sqrt{\frac{E_J}{\epsilon_J}}\,, \tag{19.38}$$

where $\gamma \approx 3.3$. Qualitatively, a single flip leads to a continuous phase configuration where the phase gradients are significant in a narrow strip in the x-direction of length $\sqrt{E_J/\epsilon_J}$ and width ~ 1. The phase configuration resulting from such a process is shown in Fig. 19.8.

These formulas assume that the rigidity of the superconducting phase remains ϵ_J, which is, strictly speaking, only true if the discrete variables are perfectly ordered in the vertical direction. Indeed, the coupling in the vertical direction contains $\epsilon_J \cos(\phi)\tau^z_{i,j}\tau^z_{i+1,j}$, which renormalizes to $\epsilon_J \cos(\phi)\langle\tau^z_{i,j}\tau^z_{i+1,j}\rangle$ in a fluctuating system. In the opposite limit of strongly fluctuating rhombi, the average value of $\langle\tau^z_{i,j}\tau^z_{i+1,j}\rangle$ becomes small, and we can estimate it from the perturbation theory expansion in ϵ_J, which sets the lowest energy scale of the problem: $\langle\tau^z_{i,j}\tau^z_{i+1,j}\rangle \approx \epsilon_J/(L_{\text{eff}}J_x)$, which renormalizes the value of ϵ_J:

$$\epsilon_J \rightarrow \widetilde{\epsilon_J} = \frac{\epsilon_J^2}{L_{\text{eff}}J_x} .$$

This renormalized value of ϵ_J should be used in the estimates of the vortex energy in eqns (19.37) and (19.38). This does not affect the estimates much unless the system is deep in the fluctuating regime. Unlike the potential energy, the single-flip processes occur with an amplitude

$$\widetilde{t} = E_J^{3/4} E_C^{1/4} e^{-s\sqrt{E_J/E_C}} \tag{19.39}$$

in all regimes. This formula neglects the contribution of the continuous phase to the action of the tunneling process. The reason is that both the potential energy (19.38) of the half-vortex and the kinetic energy required to change the continuous phase are much smaller than the corresponding energies for an individual rhombus, E_J, E_C. In order to estimate the kinetic energy, consider the contribution of the vertical links (the horizontal links give an equal contribution). There are roughly $\sqrt{E_J/\epsilon_J}$ such links, so their effective charging energy is about $e_C\sqrt{\epsilon_J/E_J}$. If all junctions in this array are made with the same technology, their Josephson energies and capacitances are proportional to their areas, so $\epsilon_J/E_J = E_C/\epsilon_C \equiv \eta$; in the

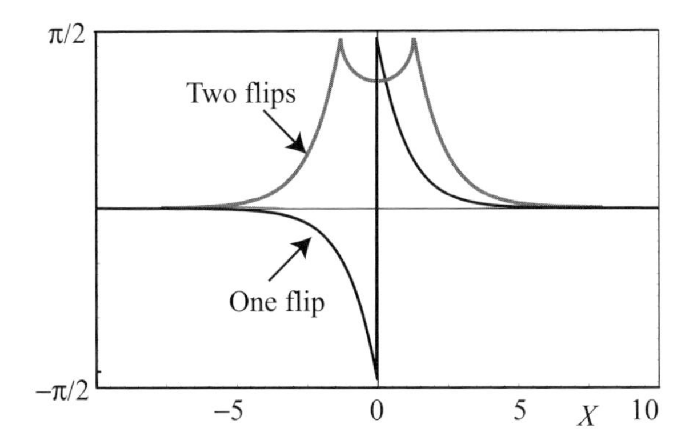

Fig. 19.8 Phase variation along the horizontal axis after a flip of a single rhombus and after consecutive flips of two rhombi located at a distance from each other of twice the core size of each rhombus, $\sqrt{E_J/\epsilon_J}$. The horizontal axis shows the distance $X = \sqrt{\epsilon_J/E_J}x$ measured in units of the vortex core size.

following we shall refer to such junctions as *similar*. In this case the array is characterized by two dimensionless parameters, $\eta \ll 1$ and $E_J/E_C \gg 1$, and the additional contribution to the charging energy, $\eta^{1/2}E_C^{-1}$, coming from the vertical links is smaller than that of the individual rhombi, E_C^{-1}, and thus does not change the dynamics.

We conclude that the dangerous real single-flip processes become forbidden if $\widetilde{t} \ll U_{\rm sf}$, where $\widetilde{t}$ is given by eqn (19.39) and $U_{\rm sf}$ by eqn (19.37) or (19.38). This condition is not difficult to satisfy in a real array, because the amplitude $\widetilde{t}$ is typically much smaller than E_J. Further, for moderately sized arrays (with $L = 5$–10, which already provides very good protection), the energy of a single-rhombus flip is only slightly smaller than E_J, so the condition $\widetilde{t} \ll U_{\rm sf}$ is not really restrictive. Note, however, that in order to eliminate low-energy states of the discrete subsystem, we also need to satisfy the condition (19.34), which implies that the tunneling processes should occur with a significant amplitude. While this might be difficult in an infinite array made from similar junctions (with the same product of charging and Josephson energies), this is not really a restrictive condition for moderately sized arrays. One could choose, for instance, a system of $L \times L$ rhombi with $L = 5$–10 Josephson contacts, with $E_J = 10E_C$. This choice would give $\widetilde{t} \approx 0.35E_C$ and $J_x \approx 0.2E_C$ for a system of disconnected horizontal chains. The condition $\widetilde{t} \ll U_{\rm sf}$ is well satisfied. If we now choose vertical links with $\epsilon_J = 0.5E_C$ and, correspondingly, $\epsilon_C = 20E_C$, we get $L_{\rm eff} \approx 5$, so that the condition (19.34) is satisfied as well and there are no low-energy states. It is more difficult, however, to eliminate the low-energy states in the infinite array of coupled chains shown in Fig. 19.7 and to satisfy the condition $\widetilde{t} \ll U_{\rm sf}$ at the same time, especially if all junctions are to be "similar" in the sense defined above. This can be achieved, however, by replacing the vertical links by rhombi frustrated by a flux $\Phi = (1/2+\delta)\Phi_0$ with $\delta \ll 1$, with each junction characterized by $\widetilde{\epsilon_J} \lesssim E_J$ and $\epsilon_C \gtrsim E_C$. These rhombi would provide a significant rigidity to the continuous phase fluctuations (with effective rigidity $\widetilde{\epsilon_J}$) but only weak coupling ($\epsilon_J = \delta\widetilde{\epsilon_J}$) between the discrete degrees of freedom.

Finally, we discuss the effect of the finite phase rigidity on the amplitude of the two-rhombus processes, $J_x(k)$. The condition that real single-flip process do not occur does not exclude virtual processes that flip two rhombi consecutively in the same chain. This would lead to an additional contribution to $\widetilde{J_x}(k)$ (eqn (19.33)). To estimate this contribution, we note that immediately after two flips, the continuous phase has the configuration shown in Fig. 19.8, which is associated with an energy $\sim U_{\rm sf}$. The amplitude for two such consecutive flips is $\widetilde{t}^2/U_{\rm sf}$; it can become of the order of $\widetilde{J_x}(k)$ in a large system (where $U_{\rm sf}$ is small). However, the amplitude of the full process involves an additional action, which further suppresses this amplitude. This happens because the two consecutive flips lead to the high-energy virtual state sketched in Fig. 19.8, and in order to get back to the low-energy state the resulting continuous phase has to evolve dynamically. To estimate the action corresponding to this evolution, we note that its dynamics is controlled by $\sqrt{E_J/\epsilon_J}$ junctions with charging energy ϵ_C. For the purpose of this estimate, we can replace these junctions by a single junction with capacitive energy $\epsilon_C\sqrt{\epsilon_J/E_J}$. Thus, the final stage of this process leads to an additional term in the action $\delta S \sim E_J/\epsilon_C = \eta(E_J/E_C)$. Depending on the parameter $\sqrt{E_J E_C}/\epsilon_C = \eta\sqrt{E_J/E_C}$ this additional contribution to the action is smaller or larger than the total action, but even if it is smaller, it is still large compared with unity if $E_J \gg \epsilon_C$. In this case, the processes that do not change the continuous phase dominate. We emphasize again that in any case, transitions involving two flips in the same row commute with both integrals of motion P, Q and thus do not affect the qualitative conclusions of the previous section. For practically important similar junctions, this means the following. If $\eta \gg \sqrt{E_C/E_J}$, only nearest-neighbor rhombi flip with the amplitude J_x given by eqn (19.33). If $\sqrt{E_C/E_J} \gg \eta \gg E_C/E_J$, flips occur for rhombi in the same row if they are closer than $L_{\text{eff}} < \eta^{-1}\sqrt{E_C/E_J}$. Finally, for $\eta \ll E_C/E_J$, the distance between flipped rhombi exceeds the size of a half-vortex, and two-rhombus flips in a large ($L \gg \sqrt{E_J/\epsilon_J}$) array happen via virtual half-vortices in the continuous phase.

In the discussion above, we have implicitly assumed a superconducting boundary condition such as that shown in Fig. 19.7. Such boundary conditions imply that in the absence of significant continuous phase fluctuations, $P_i P_j = 1$. Physically, this means that if the array as a whole is a superconductor, it still has two states characterized by a phase difference $\Delta\phi = 0$ or π between the left and right boundaries even in the regime where the individual phases in the middle fluctuate strongly between the values 0 and π. In this regime of strong discrete phase fluctuations, the external fields are not coupled to the global degree of freedom $\Delta\phi$ describing the array as a whole. In principle, it is also possible to have a similar array with open boundary conditions, but in this case it is more difficult to eliminate low-lying states because there is no reason for the constraint $P_i P_j = 1$ in this case.

Array equivalent to a spin model with local interactions. In order to construct an array equivalent to the spin model (19.26), we need to couple the rhombi in such a way that transitions involving only one rhombus in a row are not allowed but the superconducting phase varies significantly between one rhombus and the next in the row. This is achieved if the rhombi are connected in a chain by a weak Josephson link, characterized by a Josephson energy e_J and Coulomb energy e_C, so that $E_J \gtrsim e_J$ and $e_C \ll E_C$, as shown in Fig. 19.9(b). In this case, a simultaneous tunneling of two rhombi which does not change the phase difference across the weak junction is not affected; its amplitude J_x is still given by eqn (19.33).

In this array, the flip of a single rhombus may be due to two alternative mechanisms. The first one involves the creation of a half-vortex (as discussed earlier) and does not involve a change in the phase across a weak junction. Alternatively, the flip can be due to a phase flip by π across a weak junction. Because $e_C \ll E_C$, this process is slow and its amplitude is low:

$$t = E_J^{3/4} e_C^{1/4} e^{-s'\sqrt{E_J/e_C}},$$

where $s' \approx 1$. Furthermore, this process increases the energy of the system by e_J, so if its amplitude is small, so that $t \ll e_J$, it can be completely neglected. Normally, junctions with a smaller Josephson energy, $e_J \lesssim E_J$, have a bigger charging energy, not a smaller one as required here. To avoid this problem, we note that these weak junctions can be implemented in practice as a two-junction SQUID loop containing a flux $\Phi = (1/2 + \delta)\Phi_0$ as shown in Fig. 19.9(c). The effective Josephson coupling provided by such a loop is $e_J = \sin(2\pi\delta)E_J^{(0)}$ (where $E_J^{(0)}$ is the energy of the individual junction), while its effective charging energy is $e_C = E_J^{(0)}/2$. This allows one to use bigger junctions for these weak junctions, and another advantage is that it provides an additional controlling parameter in the system. This construction seems somewhat similar to the partially frustrated rhombi that one needs to introduce as vertical links in the large arrays discussed previously (Fig. 19.7c), but it serves a completely opposite purpose: it increases the density of the full vortices while keeping the discrete variables coupled. The partially frustrated rhombi, on the other hand, suppress the fluctuations of the continuous phase while allowing independent fluctuations of the discrete variables in

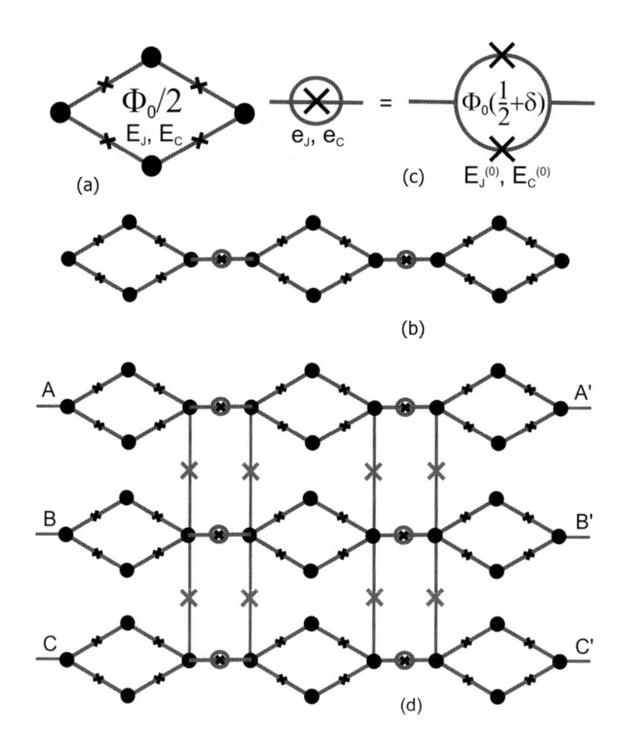

Fig. 19.9 Schematic illustrations of an array. (a) The elementary Josephson circuit, emulating a spin $1/2$, consists of a four-junction loop penetrated by a magnetic flux $\Phi_0/2$. (b) Implementation of a spin chain by Josephson junction loops. Here, elementary rhombi are connected by weak links; the appropriate parameters for these links can be obtained if each link in fact consists of two elementary Josephson junctions as shown in (c). (d) Full array implementing the spin model (19.26). The natural boundary conditions for this array are periodic, i.e. the point A should be connected to A', B to B', etc.

different rows. Such a dramatic difference is made possible by a combination of two reasons. First, the frustration has a different effect on these elements: in the case of the SQUID loops, a half flux eliminates the Josephson coupling completely, while in the case of the rhombi, it leads to an exact degeneracy between two discrete states and to a rigidity of the continuous phase. Second, the values of the charging energies are rather different: in the case of a SQUID loop, the charging energy of its junctions completely dominates all processes in which the phase changes across this loop, thereby prohibiting single-rhombus flips. In the case of partially frustrated rhombi, their charging energy is of the order of the charging energy of the contacts in the horizontal rhombi and thus it suppresses double flips of distant rhombi in one chain but has a relatively minor effect on the nearest rhombi.

The J_z interaction between the spins is provided by pairs of weak Josephson contacts as shown in Fig. 19.9(d), with a Josephson energy $\epsilon_J \lesssim E_J$ and a charging energy $\epsilon_C \gtrsim E_C$. These junctions do not affect the tunneling process of each rhombus but provide a weak interaction between them of the form (19.26) with a strength

$$J_z = \epsilon_J \, . \tag{19.40}$$

Similarly to the array discussed in the previous subsection, we have to choose the parameters so that the energy of a half-vortex and the states resulting from a single flip is sufficiently high compared with the amplitude of single flips. The previous discussion carries over to this array. The only change is that the energy of a vortex in an infinite system contains the weakest link in the horizontal direction, i.e. e_J instead of E_J:

$$\widetilde{t} \ll U_{\rm sf}\,, \tag{19.41}$$

$$\begin{aligned} U_{\rm sf} &= \pi^2 e_J/(2L) + (4/\pi)\epsilon_J L, \qquad & L &\ll \sqrt{e_J/\epsilon_J}\,, \\ U_{\rm sf} &= \gamma\sqrt{e_J\epsilon_J}, & L &\gg \sqrt{e_J/\epsilon_J}\,. \end{aligned} \tag{19.42}$$

Finally, we have to ensure that the phases of consecutive rhombi are decoupled and the interaction between discrete degrees of freedom is purely local. This is satisfied if the continuous phase across weak junctions fluctuates strongly, i.e. the energy of the usual type of vortex (as opposed to a half-vortex) is smaller than the kinetic energy: $\sqrt{e_J\epsilon_J} \ll e_C$. Physically, this means that the array as a whole is an insulator, similarly to the topological insulator considered by Douçot *et al.* (2003), owing to the full vortices that move in a vertical direction, thereby decoupling different columns of rhombi. This condition does not contradict the condition $\widetilde{t} \ll U_{\rm sf}$, because the latter involves the exponentially small amplitude for flipping a single rhombus. If both conditions are satisfied, the absolute value of the phase on each island constituting a rhombus fluctuates but the difference across the rhombus remains a slow variable taking two discrete values, and it flips only simultaneously with another one in the same row. Note that the interaction between discrete variables belonging to one column is due to a loop formed by these rhombi and two vertical junctions; it is therefore always local by construction, and its value is given by eqn (19.40). Repeating the previous arguments, we see that in order to suppress the simultaneous flips of distant rhombi in the same row, one needs also to satisfy the condition $E_J E_C/\epsilon_C^2 \gg 1$, but in contrast to the case of the regime discussed there, here the conditions on the vertical junctions are not difficult to satisfy, because one does not need to keep the long-range order in the continuous phase. Under these conditions, the dynamics of the array is described by the Hamiltonian (19.26).

Although in this regime the system as a whole is an insulator, it does not allow a half-vortex to move across it. So, physically, the two states of the global system can be observed in the array with periodic boundary conditions shown in Fig. 19.9(d): here, the two different states correspond to a half-vortex being trapped or not trapped inside the big loop formed by the array as a whole owing to the periodic boundary conditions.

19.9 Rhombus chain: Quantitative analysis

In this section we present a brief review of recent results (Protopopov and Feigel'man 2004, 2006) from detailed theoretical studies of the one-dimensional chains made out of frustrated "rhombi" used as a major building block in the previous discussion. Originally, the problem of such a chain was formulated by Douçot and Vidal (2002) in the approximation of a local Coulomb interaction (i.e. assuming that the major capacitances in the problem are those which correspond to individual islands). In fact, experimentally the capacitances C of the interisland junctions are typically much larger (by a factor of the order of 100) than the self-capacitances of the islands C_0. Therefore, in order to account for the specific features of experimentally realizable situations, a detailed analysis of the superconducting properties of rhombus chains was conducted in the limit $C_0/C \to 0$.

This case is also simpler for theoretical treatment, since the Lagrangian of the system becomes a sum of terms, such that each of them belongs to an individual rhombus only. The only source of coupling between different rhombi is the periodic boundary condition along the chain. A method to treat a similar problem was developed by Matveev *et al.* (2002). These authors considered a simple chain of N Josephson junctions in a closed-ring geometry, and reduced the calculation of the supercurrent in the large-N limit to the solution of a Schrödinger equation for a particle moving in a periodic potential $\sim \cos x$, with appropriate boundary condition. Matveev *et al.* assumed (we will do the same) that the Josephson energy E_J of a junction was large compared with its charging energy $E_C = e^2/2C$. Protopopov and Feigel'man (2004) generalized the method of Matveev *et al.* in order to use it for the case of a ring of rhombi. In this case, the fictitious particle of the theory is still moving in a cos-like potential, but it now acquires a large *spin* $S = \frac{1}{2}N$, where N is the number of rhombi in the ring. In the maximally frustrated case $|\Phi_r - \Phi_0/2| \equiv \delta\Phi = 0$, the x-projection of the spin is an integral of motion, which must be chosen to minimize the total energy. As a result, $S_x = \pm\frac{1}{2}N$ and the whole problem reduces to the one studied by Matveev *et al.* up to a trivial redefinition of the parameters. In this situation, the ground state energy and the supercurrent (which is proportional to the derivative of the ground state energy with respect to the total flux Φ_c) are periodic functions of Φ_c with period $\Phi_0/2$, i.e. $4e$ transport takes place. A nonzero flux deviation $\delta\Phi$ produces a longitudinal field h_z coupled to the z-component of the spin of the fictitious particle, which now acquires nontrivial dynamics. It was shown that in the limit of a sufficiently long rhombus chain, the whole problem can be analyzed in terms of the semiclassical dynamics of a particle with a large spin and a spin-dependent potential barrier. In general, there are two tunneling trajectories; one of them corresponds to the usual $2e$ transport, whereas the other corresponds to $4e$ transport. By comparing the actions of these trajectories for different $\delta\Phi$, the critical flux deflection $\delta\Phi_c$ was found as a function of the ratio $E_J/E_C \gg 1$.

The resulting amplitudes of the supercurrent components are determined primarily by the classical actions on corresponding trajectories:

$$I(\gamma) = I_{2e}\sin\gamma + I_{4e}\sin(2\gamma)\,, \tag{19.43}$$

where

$$I_{4e} = A_{4e}\exp(-S_E^{4e}), \quad I_{2e} = A_{2e}\exp(-S_E^{2e})\,, \tag{19.44}$$

γ is the phase difference between the endpoints of the chain, and S_E^{4e} and S_E^{2e} are the values of the tunneling actions on trajectories of the first and second type, respectively. Both S_E^{4e} and S_E^{2e} are large in the region of strong fluctuations $Nw \gg 1$, and thus the total supercurrent will in general be dominated by least-action processes.

In Protopopov and Feigel'man (2004), the actions S_E^{4e} and S_E^{2e} were calculated in order to find the dominant charge transport mechanism in different regions of the parameters. At $d \equiv \sqrt{2}\pi\upsilon/(E_J\delta^{3/2}) \ll 1$, i.e. at sufficiently large flux deflections δ, the dominant process was found to be the usual $2e$ transfer. The supercurrent amplitude is

$$I_{2e} \approx 32\cdot 2^{1/4} I_c^0\sqrt{N}\left(\frac{\upsilon}{E_J\sqrt{\delta}}\right)^{3/2}\exp\left\{-\frac{8\sqrt{2}}{\pi}\frac{N\upsilon}{E_J\sqrt{\delta}}\right\}. \tag{19.45}$$

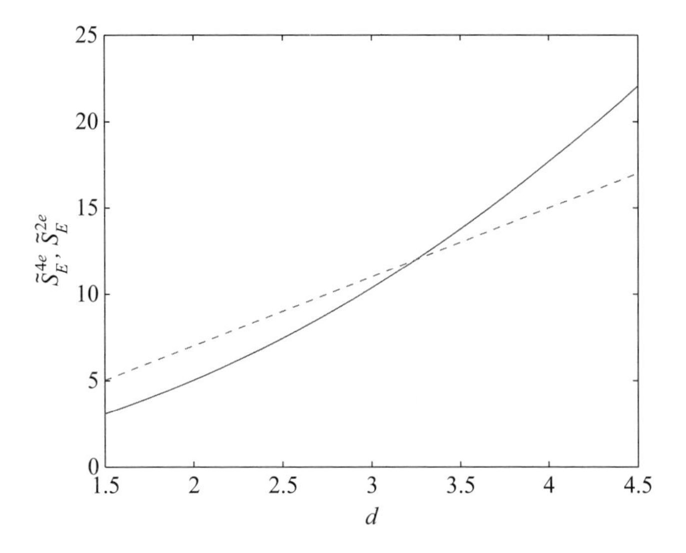

Fig. 19.10 Results of numerical evaluation of $\widetilde{S}_E^{2e}(d)$ (solid line) and $\widetilde{S}_E^{4e}(d)$ (dashed line).

At small flux deflections δ, the parameter $d \gg 1$ and the dominant trajectory is then the $4e$ one. The current is then obtained in the form

$$I_{4e} \approx 128 \cdot 2^{1/4} I_c^0 \sqrt{N} \left(\frac{v}{E_J \sqrt{\delta}} \right)^{3/2} \exp \left\{ -\frac{32\sqrt{2}}{\pi} \frac{Nv}{E_J \sqrt{\delta}} + \frac{8N\delta}{\pi^2} \right\} \tag{19.46}$$

at $(1/\delta^{3/2})(v/E_J) \gg 1$.

In the intermediate region of $d \sim 1$, we analyzed the saddle-point equations for the action numerically. First, we wrote S_E^{4e} and S_E^{2e} in the form $S_E^{4e} = h\widetilde{S}_E^{4e}(d)$ and $S_E^{2e} = h\widetilde{S}_E^{2e}(d)$. Then the functions $\widetilde{S}_E^{4e}(d)$ and $\widetilde{S}_E^{2e}(d)$, depending on a single parameter d, were evaluated numerically. The result is presented in Fig. 19.10. The actions for the two types of trajectories are equal at $d = d_0 \approx 3.2$, where we have $\widetilde{S}_E^{4e}(d_0) = \widetilde{S}_E^{2e}(d_0) \approx 11.9$. Thus the crossover between the $4e$ regime and $2e$ regime takes place at

$$\delta\Phi = \delta\Phi^c = \left(\frac{v^2}{4\pi d_0^2 E_J^2} \right)^{1/3} \Phi_0 \approx 0.2 \left(\frac{v}{E_J} \right)^{2/3} \Phi_0 \,. \tag{19.47}$$

By varying the flux Φ_r in some vicinity of the crossover point (19.47), one can find both $2e$ and $4e$ components of the supercurrent, but their relative weight is expected to vary strongly with $\Phi_r - \Phi_r^c$, in analogy with phase coexistence near a first-order phase transition.

The above results were obtained for a regular rhombus chain, neglecting all kinds of disorder (random deviations of the areas of different rhombi and disparity of the Josephson critical currents, as well as random offset charges). In Protopopov and Feigel'man (2006), all these effects were treated quantitatively and it was shown that the major results presented above survive in a realistic situation of reasonable weak disorder.

19.10 Recent developments

The practical realization of the fully topologically protected arrays discussed in Sections 19.3–19.8 above is complicated owing to the very large number of Josephson junctions involved. On the other hand, simplified solutions such as rhombi chains (for an account of the first experiments with such chains, see Pop *et al.* (2008) are not sufficient, owing to the exponential decrease of the effective stiffness with the length of the chain, which introduces excessively strong phase fluctuations between the endpoints of the chain. The first attempt at a compromise between these two limits was discussed in the Section 19.9 above.

Recently, another approach was developed by Douçot and Ioffe (2009). These authors proposed to construct a moderate-size Josephson junction array made out of frustrated rhombi, in such a way that the current–phase relation of the array as a whole would be mostly π-periodic, with the disparity between 0 and π states being suppressed with respect to the case of single rhombi. Such a construction can then be repeated in a hierarchical way. A simple example of one step of such a transformation is shown in Fig. 19.11 (copied from Douçot and Ioffe 2009). The idea is to connect several (say, K) rhombus chains (each made out of N rhombi) in parallel. The relatively large N provides a sufficient amount of quantum phase slip (switching of current directions in different rhombi) to provide "screening" of local disorder, whereas the presence of $K > 1$ chains in parallel suppresses the tunneling of half-vortices across the whole array. The general idea for the optimization of such an array is to maximize simultaneously two basic energy scales in the problem: the effective Josephson energy E_2 of the whole array, and the energy gap Δ_0 with respect to excitations beyond the protected subspace. These two energies evolve in opposite directions: E_2 decreases whereas Δ_0 grows with increase of the ratio E_C/E_J. Therefore, the optimal parameters fall into the range of $E_C/E_J \sim 1$. Detailed analytical and numerical studies, reported by Douçot and Ioffe (2009), lead to the following choice of optimal parameters: $K = 2$, $N = 3$, and $E_C/E_J \approx 0.3$.

The first successful approach to increasing the robustness with respect to noise and disorder using "quantum protection" was reported by Gladchenko *et al.* (2009). An array containing 12 rhombi, composed of two combined units, each of the type shown in Fig. 19.11, was fabricated and measured. It was shown that the range of tolerated deviations $\delta\Phi$ of the magnetic flux through each rhombus (i.e. the domain of $\delta\Phi$ where the effective current–phase relation $I(\phi)$ is mostly π-periodic) is much wider for such an array (protected up to the fourth order against local errors) than for a simple two-rhombus chain. Another important result obtained by Gladchenko *et al.* (2009) was a demonstration of the sensitivity of the effective $I(\phi)$

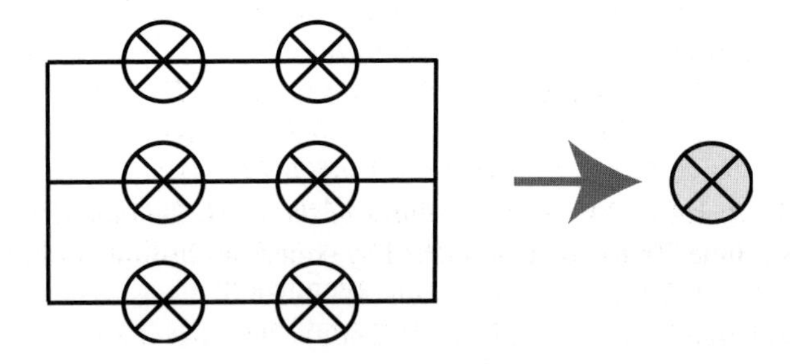

Fig. 19.11 Hierarchical construction with $K = 3$ and $N = 2$.

with respect to a nearby electric gate, which is a signature of the phase-coherent nature of the quantum phase slips in the array.

19.11 Conclusion

We have shown that Josephson junction arrays of two special types (shown in Figs. 19.1 and 19.4) have degenerate ground states described by a topological order parameter. The manifold of these states is protected in the sense that local perturbations have an exponentially weak effect on the relative phases and transition amplitudes of the states. The simpler array in Fig. 19.1 possesses a superconductive state with topological order. The minor modification shown in Fig. 19.4 may be brought into both superconductive and insulating "topological" phases in a controllable manner. Both versions of the array are expected to demonstrate very long coherence times in "spin-echo"-type experiments and to be promising basic elements for scalable quantum computers.

It is important to note that the topologically degenerate ground state is realized in large arrays over a finite range of magnetic flux (through the rhombi) near full frustration; see the phase diagram in Fig. 19.3. Therefore it is possible to study experimentally the range of stability of such phases via the global current–phase relation, similarly to the first experiment reported by Gladchenko *et al.* (2009) with simplified small-scale arrays made out of a few parallel rhombi chains.

Acknowledgments

I am grateful to my co-authors Benoit Douçot, Lev Ioffe, Alexei Ioselevich, and Ivan Protopopov, without whom the research reviewed here would not have been done. I am also glad to acknowledge useful discussions with Gianni Blatter, Vadim Geshkenbein, Dmitry Ivanov, Alexei Kitaev, Sergei Korshunov, Anatoly Larkin, and Bernard Pannetier.

References

L. Balents, S. M. Girvin, and M. P. A. Fisher (2002). *Phys. Rev. B* **65**, 224412.
R. Balian, J. M. Droufle, and C. Itzykson (1975). *Phys. Rev. D* **11**, 2098.
R. J. Baxter (1982). *Exactly Solvable Models in Statistical Mechanics*, Chapter 12. Academic Press, London.
R. F. Bishop, D. J. J. Farrel, and M. L. Ristig (2000). *Int. J. Mod. Phys. B* **14**, 1517.
G. Blatter, V. B. Geshkenbein, and L. B. Ioffe (2001). *Phys. Rev. B* **63**, 174511.
S. B. Bravyi and A. Kitaev (2000). Preprint http://xxx.lanl.gov/abs/quant-ph/0003137.
H.-D. Chen, C. Fang, J. Hu, and H. Yao (2007). *Phys. Rev. B* **75**, 144401.
J. Dorier, F. Becca, and F. Mila (2005). *Phys. Rev. B* **72**, 024448.
V. S. Dotsenko and V. A. Fateev (1984). *Nucl. Phys. B* **240**, 312.
B. Douçot and L. B. Ioffe (2009). *Proceedings of the L. D. Landau Memorial Conference* (Chernogolovka, June 2008), to be published by American Institute of Physics.
B. Douçot and J. Vidal (2002). *Phys. Rev. Lett.* **88**, 227005.
B. Douçot, M. V. Feigel'man, and L. B. Ioffe (2003). *Phys. Rev. Lett.* **90**, 107003.
B. Douçot, M. V. Feigel'man, L. B. Ioffe, and A. S. Ioselevich (2005). *Phys. Rev. B* **71**, 024505.
A. Ekert and R. Jozsa (1996). *Rev. Mod. Phys.* **68**, 733.

M. V. Feigel'man, L. B. Ioffe, V. B. Geshkenbein, P. Dayal, and G. Blatter (2004*a*). *Phys. Rev. Lett.* **92**, 098301.
M. V. Feigel'man, L. B. Ioffe, V. B. Geshkenbein, P. Dayal, and G. Blatter (2004*b*). *Phys. Rev. B* **70**, 224524.
P. Fendley, R. Moessner, and S. L. Sondhi (2002). *Phys. Rev. B* **66**, 214513.
M. H. Freedman, A. Kitaev, and Z. Wang (2000). Preprint http://xxx.lanl.gov/abs/quant-ph/0001071.
S. Gladchenko, D. Olaya, E. Dupont-Ferrier, B. Douçot, L. B. Ioffe, and M. E. Gershenson (2009). *Nature Phys.* **5**, 48; arXiv:0802.2295.
L. Grover (1996). In *Proceedings of the Annual ACM Symposium on the Theory of Computing*. ACM Press, New York.
L. B. Ioffe and M. V. Feigel'man (2002). *Phys. Rev. B* **66**, 224503.
L. B. Ioffe, M. V. Feigel'man, A. Ioselevich, D. Ivanov, M. Troyer, and G. Blatter (2002). *Nature* **415**, 503.
A. S. Ioselevich, D. A. Ivanov, and M. V. Feigel'man (2002). *Phys. Rev. B* **66**, 174405.
A. Yu. Kitaev (2003). *Ann. Phys.* **303**, 2–30; arXiv:quant-ph/9707021.
S. Kivelson (1989). *Phys. Rev. B* **39**, 259.
K. I. Kugel and D. I. Khomskii (1982). *Sov. Phys. Uspekhi* **25**, 231.
K. A. Matveev, A. I. Larkin, and L. I. Glazman (2002). *Phys. Rev. Lett.* **89**, 096802.
M. Mezard, G. Parisi, and M. Virasoro (1997). *Spin Glass Theory and Beyond.* World Scientific, Singapore.
G. Misguich, D. Serban, and V. Pasquier (2002). *Phys. Rev. Lett.* **89**, 137202.
R. Moessner, S. L. Sondhi, and E. Fradkin (2002). *Phys. Rev. B* **65**, 24504.
O. Motrunich and T. Senthil (2002). *Phys. Rev. Lett.* **89**, 277004.
R. Orus, A. C. Doherty, and G. Vidal (2008). Preprint arXiv:0809.4068.
K. Park and D. A. Huse (2001). *Phys. Rev. B* **64**, 134522.
I. M. Pop, K. Hallesbach, O. Buisson, W. Guichard, B. Pannetier, and I. V. Protopopov (2008). *Phys. Rev. B* **78**, 104504.
J. Preskill (1998). *Proc. Roy. Soc. A* **454**, 385.
I. V. Protopopov and M. V. Feigel'man (2004). *Phys. Rev. B* **70**, 184519.
I. V. Protopopov and M. V. Feigel'man (2006). *Phys. Rev. B* **74**, 064516.
N. Read and B. Chakraborty (1989). *Phys. Rev. B* **40**, 7133.
N. Read and S. Sachdev (1991). *Phys. Rev. Lett.* **66**, 1773.
T. Senthil and M. P. A. Fisher (2000). *Phys. Rev. B* **62**, 7850.
T. Senthil and M. P. A. Fisher (2001*a*). *Phys. Rev. Lett.* **86**, 292.
T. Senthil and M. P. A. Fisher (2001*b*). *Phys. Rev. B* **63**, 134521.
T. Senthil and O. Motrunich (2002). *Phys. Rev. B* **66**, 205104.
P. W. Shor (1994). In *Proceedings of the 35th Symposium on the Foundations of Computer Science*. IEEE Press, Los Alamitos, CA.
P. W. Shor (1995). *Phys. Rev. A* **52**, 2493.
A. Steane (1998). *Rep. Prog. Phys.* **61**, 117.
F. Wegner (1971). *J. Math. Phys.* **12**, 2259.
X.-G. Wen (1991). *Phys. Rev. B* **44**, 2664.
X.-G. Wen and Q. Niu (1990). *Phys. Rev. B* **41**, 9377.
F. Y. Wu (1982). *Rev. Mod. Phys.* **54**, 235.

20

On some quantum Hall states with negative flux

T. JOLICOEUR

Thierry Jolicoeur,
LPTMS, Université Paris-Sud,
Building 100-A,
91405 Orsay, France.

20.1 Introduction

The fractional quantum Hall effect (FQHE) is a remarkable state of electronic matter that occurs in two dimensions under a strong magnetic field. From a practical point of view, one observes experimentally a wealth of new thermodynamic phases in this special regime. While there is a theory for the most prominent states (in a sense given below), recent experiments have given evidence for more complexity, for which there is as yet no simple and universal theory. This seminar gives an overview of the situation as of the end of 2008 and focuses on some recent proposals for wave functions describing these new quantum Hall states.

For the sake of completeness, let us first set the stage for the FQHE. It is known to occur for particles confined in two spatial dimensions and subjected to a perpendicular magnetic field. In this setup, the one-body spectrum is drastically affected by the magnetic field: the kinetic energy is frozen, and there is a set of exactly degenerate energy levels called Landau levels (LLs). These levels are separated by the cyclotron energy $\hbar\omega_c$, where $\omega_c = eB/m$. Their degeneracy is given by eB/h times the area A of the sample. Increasing the field B leads to more degenerate levels and also to a larger separation between Landau levels. Imagine now that we have some particles in this situation and we look for the ground state. If we have N_e electrons and we increase B, at some point there will be enough states in the lowest Landau level (LLL) to accommodate every electron. This will happen first when $N_e = eB_1/h \times A$ and hence for a density $n = N_e/A = eB_1/h$. Beyond that value of B, there will be more states available and the problem of putting the electrons in the one-body orbitals becomes exponentially degenerate since the number of configurations is given by a binomial coefficient. If the cyclotron gap between the levels is large enough, it is the mutual interactions between electrons that will determine the structure of the ground state. There is no longer any interplay between kinetic energy and potential energy. The FQHE is a "pure" interaction effect. Typical two-dimensional electron gases have densities of the order of 10^{11} cm^{-2}. The field required to put every electron in the LLL is thus $B_1 \approx 4 \times n(10^{11})$ tesla. The number of occupied orbitals in the LLL is called in what follows the filling factor ν, which is the ratio of the number of electrons to the number of one-body quantum states in the LLL; $\nu = nh/eB$.

Let us note what would be the reasoning of a condensed matter physicist before the 1980s if confronted with this situation. The exactly degenerate Landau levels are of course an idealized situation, and the disorder present in real-world samples broadens the levels, which retain nevertheless a high density of states. These broadened levels are partially filled for $B > B_1$ so, as a function of the strength of the interactions, there are two plausible guesses: the first is a Fermi liquid in a partially filled band. Since the band is narrow, one can also envision the relevance of a Mott insulating state, i.e. a crystalline state of electrons. This second possibility is in fact realized at very small filling factors, $\nu \lesssim 1/7$. This is the so-called Wigner crystal. However, the Fermi liquid phase does not happen, and is replaced by a new kind of liquid state supporting fractionalization of quantum numbers. In fact, there is a whole series of such liquids as a function of the filling factor ν. Historically, the first state for which a satisfactory theoretical description was given, by Laughlin [1], was $\nu = 1/3$. The "composite fermion" scheme developed by Jain [2] gave a description of many other FQHE states. We will briefly give an overview of these theoretical approaches in Section 20.2.

Finally, note that we will not discuss the transport properties of these FQHE states. It is certainly true that the study of an FQHE state starts with a measurement of resistivity. However, the description of transport is a subject in itself and has been described in several previous Les Houches lectures [3]. For our purposes, we just need to know that the FQHE states are liquids without any obvious local symmetry breaking, they possess a gapped excitation spectrum, and, amongst the excitations, there are unconventional quasiparticles with fractional charge and statistics. Experiments are very often limited to gap measurements. The "most prominent" states are the ones with the largest gaps.

20.2 Classical hierarchies

We now describe briefly the existing theory of the most prominent FQHE states [4–6]. If we use the symmetric gauge for the external applied magnetic field $\mathbf{A} = \frac{1}{2}\mathbf{B} \times \mathbf{r}$, then a basis of the one-body eigenstates for the LLL is given by

$$\phi_m(z) = \frac{1}{\sqrt{2\pi 2^m m!}} z^m \mathrm{e}^{-|z|^2/4\ell^2}\,, \tag{20.1}$$

where m is a positive integer, $\ell = \sqrt{\hbar/eB}$ is the magnetic length, and $z = x + iy$ is the complex coordinate in the plane. In the absence of interactions and an external potential, all these states have exactly the same eigenenergy $\frac{1}{2}\hbar\omega_c$. These states have a definite chirality, since m is positive. The density associated with such a state is nonzero in the neighborhood of a ring centered on the origin (owing to the gauge choice), of radius $\ell\sqrt{2m}$. We now consider the many-body problem with the electrons all residing in the LLL. The states that we describe are also spin polarized, i.e. the Zeeman energy is large enough for this to be the case. This is not true in general, of course, and there are many interesting FQHE states involving the spin degree of freedom, but the states at very high field presumably do not involve the spin. Each electron is described by a complex variable, and thus the many-body wave function is of the form

$$\Psi(z_1,\ldots,z_N) = f(z_1,\ldots,z_N)\,\mathrm{e}^{-\sum_i |z_i|^2/4\ell^2}\,. \tag{20.2}$$

In general, contemplation of the full many-body wave function is not a very useful way to understand a physical system, but in FQHE physics this has proven to be the best approach (so far). Let us first understand what happens when the LLL is full of electrons, i.e. when $\nu = 1$ or $B = B_1$ as defined in the introduction. We imagine a cylindrical, infinite wall of a given radius R. This has the effect of sending to infinity all states in eqn (20.1) with m larger than $R^2/2\ell^2$. Only the states close to the boundary will have wave functions different to those given by eqn (20.1). This is a negligible effect for large systems. The number of states in the LLL is thus finite and equal to $M = R^2/2\ell^2 = eB/h \times$ (area of cylinder). If we fill exactly all these one-body states $\propto z^0, z^1, \ldots, z^M$ with fermions, there is a unique possible state for $B = B_1$, which is simply the Slater determinant of all occupied orbitals:

$$\Psi_{\nu=1} = \mathrm{Det}\left[z_i^{j-1}\right]\,\mathrm{e}^{-\sum_i |z_i|^2/4\ell^2}\,, \tag{20.3}$$

where the indices i, j run from 1 to N. This determinant is called the Vandermonde determinant and can be computed in closed form: $\mathrm{Det}\left[z_i^{j-1}\right] = \prod_{i<j}(z_i - z_j)$. While this leads to a nice wave function, this is not a very fascinating one. Since all orbitals are occupied, this is the exact ground state for any interactions between electrons if we neglect the possibility of transitions towards higher-lying LLs (note that when there is spin in the game, things are now nontrivial—we will not consider this important point). The density distribution of $\psi_{\nu=1}$ can

be guessed easily: since all orbitals are occupied in a uniform manner and these orbitals fill concentric rings of increasing radius with the same peak density, we have uniform coverage of a circular region. The state looks like a flat pancake, which decays to zero on a length scale given by ℓ close to the boundary only. Encouraged by this understanding, let us try to guess a candidate ground state for a filling factor less than unity. This was the approach used by Laughlin. He proposed to take simply the cube of the Vandermonde determinant. Since all factors $(z_i - z_j)$ becomes $(z_i - z_j)^3$, this repels the electrons and flattens the pancake of the charge distribution. It is easy to guess that the new wave function is still pancake-like but now with a density which is $1/3$ of the previous case $\nu = 1$, and hence it is a state with $\nu = 1/3$:

$$\Psi_{\nu=1/3} = \prod_{i<j} (z_i - z_j)^3 \, \mathrm{e}^{-\sum_i |z_i|^2/4\ell^2} \,. \tag{20.4}$$

Why is this considered as being close to the truth for interacting electrons at $\nu = 1/3$? Contrary to the case of $\nu = 1$, this is *not* an exact eigenstate of the Coulomb problem. We first need to make a detour by looking at the two-body problem in the case of the LLL. If we consider the kinetic energies of two charged particles, then there is the following identity:

$$\frac{1}{2m}(\mathbf{p}_1 + e\mathbf{A}_1)^2 + \frac{1}{2m}(\mathbf{p}_2 + e\mathbf{A}_2)^2 = \frac{1}{2M}(\mathbf{P}_{cm} + 2e\mathbf{A}_{cm})^2 + \frac{1}{2\mu}(\mathbf{p}_r + \frac{e}{2}\mathbf{A}_r)^2 \,, \tag{20.5}$$

where we have a separation of the motion of the center of mass $M = 2m$ and the relative particle motions. The "relative particles" also live in Landau levels, so their kinetic energy is frozen. To find the eigenenergies of the two-body problem is now trivial: we just have to take expectation values of the interaction potential $V(\mathbf{r}_1 - \mathbf{r}_2)$ for the eigenstates of the relative particles given by eqn (20.1), with $8\ell^2$ in the exponential instead of $4\ell^2$ since $e/2$ appears in the relative kinetic energy. These eigenenergies are thus $V_m \equiv \langle \phi_m | V | \phi_m \rangle$ for any (rotationally invariant!) potential (forgetting the cyclotron energy, which is independent of m). The exponent m appearing in the relative-particle wave function is positive, i.e. the relative angular momentum is always positive. So, any two-body interaction in the LLL is parameterized fully by the coefficients V_m, called pseudopotentials after the work of Haldane [4]. This peculiarity of the two-body problem also means that one can write the interaction Hamiltonian in a very special way:

$$\mathcal{H} = \sum_{i<j} \sum_{m} V_m \mathcal{P}_{ij}^m \,, \tag{20.6}$$

where $\mathcal{P}_{ij}^m$ projects the pair of particles i, j onto the relative angular momentum m and the sum over m is restricted to positive odd integers for spinless fermions owing to Pauli statistics. For a repulsive Coulomb potential, these pseudopotentials decrease with increasing values of m. There is a very fundamental property of the Laughlin wave function: it is the unique smallest-degree homogeneous polynomial which is a zero-energy eigenvalue of the special model where only V_1 is nonzero. This allows one to understand why we believe that the Laughlin wave function captures the correct physics. When we vary the pseudopotentials between a hard-core model with only nonzero V_1 and the true Coulomb problem, there is clear numerical evidence that nothing dramatic happens, i.e. we do not cross any phase boundary. This has been shown by Haldane by exact diagonalization of small systems [4].

The Laughlin wave function can only describe liquids with filling factors $1/m$, m odd, whereas there are many more FQHE states in the real world. We start by trying to rewrite the Laughlin state as a determinant. We like determinants since Slater determinants are the simplest way to get an efficient description of atoms, molecules, nuclei, and solids. With Slater determinants, we can make particle–hole excitations and hence construct excited states on top of the ground state. This is a very desirable theoretical tool. We do the following manipulation:

$$\Psi_{\nu=1/3} = c = \prod_{i \neq j}(z_i - z_j) \times \prod_{i<j}(z_i - z_j)\,, \tag{20.7}$$

where the ubiquitous exponential factor is not written, for clarity. The last factor is the Vandermonde determinant. We note that

$$\prod_{i \neq j}(z_i - z_j) = \prod_{j \neq 1}(z_1 - z_j) \cdots \prod_{j \neq N}(z_N - z_j) \equiv J_1 \ldots J_N\,, \tag{20.8}$$

where we have defined the so-called Jastrow factors J_i. These factors can be distributed along the columns of the Vandermonde determinant:

$$\Psi_{\nu=1/3} = J_1 \ldots J_N \times \prod_{i<j}(z_i - z_j) = \begin{vmatrix} J_1 & \ldots & J_N \\ z_1 J_1 & \ldots & z_N J_N \\ \vdots & \vdots & \vdots \\ \vdots & \vdots & \vdots \\ z_1^{N-1} J_1 & \ldots & z_N^{N-1} J_N \end{vmatrix}\,. \tag{20.9}$$

So this *is* a Slater determinant provided we change the rules in the following way: instead of using orbitals z^m which are bona fide one-body orbitals, we now use pseudo-orbitals $z^m J$ where J effectively repels all the other particles. This is not really a one-body object, but within this construct we can perform the usual Slater-like construction of excited states and so on. The first appearance of the Vandermonde determinant in eqn (20.7) is suggestive of a flux reduction effect arising from correlations. The Vandermonde determinant is the ground state at $\nu = 1$. It is as if the correlation factors $J_1 \ldots J_N$ reduce the magnetic field from B to $B_{\text{eff}} = B - 2nh/e$ so that $\nu = 1/3$ becomes $\nu = 1$. If we have

$$\Phi_\nu = \prod_{i<j}(z_i - z_j)^2 \Phi_{\nu^*}\,, \tag{20.10}$$

then the two filling factors are related by $1/\nu = 2 + 1/\nu^*$. This is intuitively reasonable: the Jastrow factor repels the particles and flattens the pancake, i.e. the charge distribution of the wave function. Let us now reason in terms of the effective filling factor ν^*. If ν is greater than $1/3$, this implies that the effective ν^* is now larger than one. We thus have to occupy Landau levels higher than the LLL in the Slater determinant. There is nothing wrong with that, provided we project back to the LLL. There will be a special filling when there is filling of an integer number p of LLs. It is natural, but not immediately obvious, to expect that such wave functions will have to do with incompressible FQHE states. The candidate "composite fermion" (CF) states [2, 6] are thus

$$\Phi_{\nu=p/(2p+1)} = \mathcal{P}_{\text{LLL}} \prod_{i<j}(z_i - z_j)^2 \Phi_{\nu^*=p}\,. \tag{20.11}$$

For example, the case $p = 2$ involves the second LL, which is spanned by the one-body orbitals $z^* z^m$ with $m \geq 0$. We can make a Slater determinant by putting half of the electrons in the pseudo-LLL and the other half in the second pseudo-LL:

$$\Phi_{\nu^*=2} = \begin{vmatrix} 1 & \dots & 1 \\ z_1 & \dots & z_N \\ \vdots & \vdots & \vdots \\ \vdots & \vdots & \vdots \\ z_1^{N/2-1} & \dots & z_N^{N/2-1} \\ z_1^* & \dots & z_N^* \\ z_1^* z_1 & \dots & z_N^* z_N \\ \vdots & \vdots & \vdots \\ \vdots & \vdots & \vdots \\ z_1^* z_1^{N/2-1} & \dots & z_N^* z_N^{N/2-1} \end{vmatrix}. \tag{20.12}$$

This is essentially the original Jain proposal, which has proven extremely successful. However, manipulation of this wave function is inconvenient in practice owing to the projection that one has to perform after multiplication by the Jastrow factor. Jain and Kamilla have shown that this scheme may be slightly altered to make it much more tractable while retaining all its good quantitative properties. The idea is again to distribute all factors J in the determinant as before, but now we project *before* computing the determinant. This means that we just have to project factors such as $z^* z^m J$ onto the LLL. This is done by replacing each z^* by $\partial/\partial z$. Doing the derivations is now trivial, and after some elementary manipulations we have

$$\Phi_{\nu=2/5} = \prod_{i<j}(z_i - z_j)^2 \begin{vmatrix} 1 & \dots & 1 \\ z_1 & \dots & z_N \\ \vdots & \vdots & \vdots \\ \vdots & \vdots & \vdots \\ z_1^{N/2-1} & \dots & z_N^{N/2-1} \\ \Sigma_1 & \dots & \Sigma_N \\ z_1 \Sigma_1 & \dots & z_N \Sigma_N \\ \vdots & \vdots & \vdots \\ \vdots & \vdots & \vdots \\ z_1^{N/2-1}\Sigma_1 & \dots & z_N^{N/2-1}\Sigma_N \end{vmatrix}, \tag{20.13}$$

where $\Sigma_i = \sum_{j \neq i} 1/(z_i - z_j)$. This is a typical example of the CF scheme. Note that one can change the partitioning between the two LLs involved by putting, say, N_1 electrons in the LLL and N_2 in the second LL, with $N_1 + N_2 = N$. All these states are observed in exact diagonalizations in an unbounded disk geometry. The reason why we believe in the CF state is slightly different in the case of the Laughlin state. Here, there is no simple Hamiltonian for which such states are unique exact ground states. But the spectroscopy of low-lying levels is correctly reproduced: we find good quantum numbers and a good ordering of levels in the CF scheme when compared with exact-diagonalization results. This CF scheme gives a reasonable description of FQHE states at $\nu = p/(2p + 1)$. By multiplying by extra powers of the Vandermonde determinant, it is easy to generate candidates for $p/(4p+1), p/(6p+1), \dots$ without new ideas. In the CF folklore, we say that CFs are electrons dressed by two flux tubes, which means that there is a Jastrow factor squared that appears in the trial wave function.

Note now that for B less than $2nh/e$, the effective field is *negative*. This happens for fractions between $\nu = 1$ and the limiting case $\nu = 1/2$ (which is a compressible state). This is easily included in the CF scheme, since changing the sign of B amounts to complex conjugation. We just have to use p filled CF levels with negative fluxes in eqn (20.11). Although there are more derivative operators, this also leads to satisfactory wave functions describing fractions, now at $p/(2p-1)$, $p/(4p-1)$, and so on. This CF construction also explains naturally the occurrence of a gapless compressible state at $\nu = 1/2$: this is the value for which the net effective flux is zero, suggestive of freely moving CF particles. For many years it has been known that one fraction, with $\nu = 5/2$, does not fit into the CF scheme. This FQHE state is in the second LL and is in fact a state with a partial filling $1/2$ of the second LL. It is reasonable to expect that a filled LLL with two spin values, i.e. $\nu = 2$, plays the role of an inert dielectric medium renormalizing the interactions, and hence we should find a replica of the FQHE phenomenon in the second LL if the interactions are not dramatically altered. In fact, there is a big difference, which is the appearance of an FQHE at $\nu = 5/2$ with an *odd* denominator, forbidden by the CF/hierarchical constructions. The most successful candidate to describe this state is the Moore–Read Pfaffian state [7, 8], which is a state with pairing between the effective CF particles. This state has fractionally charged excitations with charge $e/4$, for which there is some experimental evidence. It is described below.

Recent experiments [9] have uncovered states displaying the FQHE at filling factors $\nu = 4/11, 5/13, 4/13, 6/17$, and $5/17$ that do not belong to the primary FQHE sequences. In addition, there is also evidence for two new *even*-denominator fractions $\nu = 3/10$ and $3/8$. This is very unusual, since the only previously known example of an even-denominator fraction was the elusive $\nu = 5/2$ state. The state $3/8$ has also been observed [10] in the $N = 1$ LL at a total filling factor $\nu = 2 + 3/8$. The new odd-denominator fractions can be explained by hierarchical reasoning in the spirit of the original Halperin–Haldane hierarchy. For example, at $\nu = 4/11$, the CFs have an effective filling factor $\nu_{\mathrm{CF}} = 1 + 1/3$. If the interactions between the CFs have a repulsive short-range core then it is plausible that they themselves will form a standard Laughlin liquid at filling factor $1/3$ within the second CF Landau level. It should be pointed out that this construction of a "second generation" of composite fermions is part of the standard lore of the hierarchical view of FQHE states, since the CF construction and the older Halperin–Haldane hierarchy can be related by a change of basis in the lattice of quantum numbers [11]. Since the even-denominator fractions require clustering, they do not fit naturally into this picture. There is no natural "Grand Unification" of all these new fractions in the hierarchical constructs.

It is possible [12] to make a construction based on the idea of composite bosons that now carry an *odd* number of flux quanta, i.e. a Jastrow factor to an odd power in the trial wave function. This has to be contrasted with the previous CF construction, where we used only even powers of the Jastrow factor. The fluxes may be positive or negative. One can then exploit the possibility of clustering of bosons in the LLL. Indeed, it has been suggested [13] that incompressible liquids of Bose particle may form at fillings $\nu = k/2$ with integer k. We will now write down spin-polarized FQHE wave functions in disk and spherical geometries. By construction, they reside entirely in the LLL and have a filling factor $\nu = k/(3k \pm 2)$. Whereas the positive-flux series has already appeared in the work of Read and Rezayi [14, 15], the negative-flux series is new. These series produce candidate wave functions for all the states observed by Pan *et al.* [9] beyond the main CF sequences, thus unifying even- and odd-denominator fractions. For the fraction $3/7$, the negative-flux candidate wave function has an excellent overlap with the Coulomb ground state obtained by exact diagonalization on a sphere for $N = 6$ electrons.

The first observation is that some of the new fractions described in [9] are of the form $p/(3p \pm 1)$. This would be natural for the FQHE of *bosons*, where one expects the formation of composite fermions with an *odd* number of flux tubes, i.e. ^{1}CFs and ^{3}CFs. The ^{1}CFs lead to a series of Bose fractions at $\nu = p/(p+1)$ which has nothing to do with the present problem. But if the ^{3}CFs fill an integer number of pseudo-Landau levels, then this leads to magic filling factors $p/(3p\pm 1)$. Indeed, there is evidence from theoretical studies of bosons in the LLL with dipolar interactions [16] that such ^{3}CFs do appear. This suggests that composite bosons may form in the case of an electronic system consisting of three flux tubes bound to one electron, ^{3}CBs; the attachment may occur with a statistical flux parallel to or against the applied magnetic field. If ν stands for the electron filling factor and ν^* for the ^{3}CB filling factor, they are related by $1/\nu = 3 + 1/\nu^*$. The relationship between the wanted electronic trial wave function and the CB wave function is

$$\Psi_\nu^{\text{Fermi}}(\{z_i\}) = \mathcal{P}_{\text{LLL}} \prod_{i<j} (z_i - z_j)^3 \; \Phi_{\nu^*}^{\text{Bose}}(\{z_i, z_i^*\}) \,. \tag{20.14}$$

The Laughlin–Jastrow factor $\prod_{i<j}(z_i - z_i)^3$ transforms bosons into fermions and adequately takes into account the Coulomb repulsion. The next step is to find candidates for the trial state $\Phi_{\nu^*}^{\text{Bose}}$. It has been suggested [13] that bosons in the LLL may form incompressible states for $\nu^* = k/2$. There is some evidence that they are described by Read–Rezayi parafermionic states [14, 15] with clustering of k particles,

$$\Phi_{\nu^*=k/2}^{\text{RR}} = \mathcal{S}\left[\prod_{i_1<j_1} (z_{i_1} - z_{j_1})^2 \cdots \prod_{i_k<j_k} (z_{i_k} - z_{j_k})^2\right]. \tag{20.15}$$

In this equation, the symbol $\mathcal{S}$ means symmetrization of the product of Laughlin–Jastrow factors over all partitions of N particles into subsets of N/k particles (N being divisible by k) (the ubiquitous exponential factor appearing in all LLL states has been omitted for clarity). While the relevance of such states to bosons with contact interactions is not clear, it has been shown that longer-range interactions such as dipolar interactions may help stabilize these states [16]. Since the CBs are composite objects, it is likely that their mutual interaction also has some long-range character. It is thus natural to try the ansatz $\Phi_{\nu^*}^{\text{Bose}} = \Phi_{\nu^*=k/2}^{\text{RR}}$ in eqn (20.14). This leads to a series of states with electron filling factor $\nu = k/(3k+2)$, which is in fact the $M = 3$ case of the generalized (k, M) states constructed by Read and Rezayi. In this construction, the flux attached to the boson is positive. It is also natural to construct wave functions with negative flux [17] attached to the CBs. Now the Bose function depends only upon the antiholomorphic coordinates:

$$\Phi_{\nu^*}^{\text{Bose}}(\{z_i^*\}) = (\Phi_{\nu^*=k/2}^{\text{RR}}(\{z_i\}))^* \,. \tag{20.16}$$

The projection onto the LLL in eqn (20.14) means that the electronic wave function can be written as

$$\Psi_\nu^{\text{Fermi}}(\{z_i\}) = \Phi_{\nu^*=k/2}^{\text{RR}}\left(\left\{\frac{\partial}{\partial z_i}\right\}\right) \prod_{i<j}(z_i - z_j)^3 \,. \tag{20.17}$$

The filling factor of this new series of states is now $\nu = k/(3k - 2)$. These states can be written in spherical geometry with the help of the spinor components $u_i = \cos(\theta_i/2)e^{i\varphi_i/2}$, $v_i = \sin(\theta_i/2)e^{-i\varphi_i/2}$ ($\{\theta_i, \phi_i\}$ being standard polar coordinates), by making the following substitutions:

$$z_i - z_j \to u_i v_j - u_j v_i\,, \quad \partial_{z_i} - \partial_{z_j} \to \partial_{u_i}\partial_{v_j} - \partial_{v_i}\partial_{u_j}\,. \tag{20.18}$$

This construction leads to wave functions that have zero total angular momentum $L = 0$, as expected for liquid states. On the sphere, these two series of states have a definite relation between the number of flux quanta through the surface and the number of electrons. The positive-flux series has $N_\phi = N/\nu - 5$, while the negative-flux series has $N_\phi = N/\nu - 1$. Even when these states have the same filling factor as the standard-hierarchy/composite-fermion states, the shift (the constant term in the N_ϕ–N relation) is in general different. The positive-flux series starts with the Laughlin state for $\nu = 1/5$ at $k = 1$, the $k = 2$ state is the known Pfaffian state [7, 8] at $\nu = 1/4$, at $k = 3$ there is a state with $\nu = 3/11$ which competes with

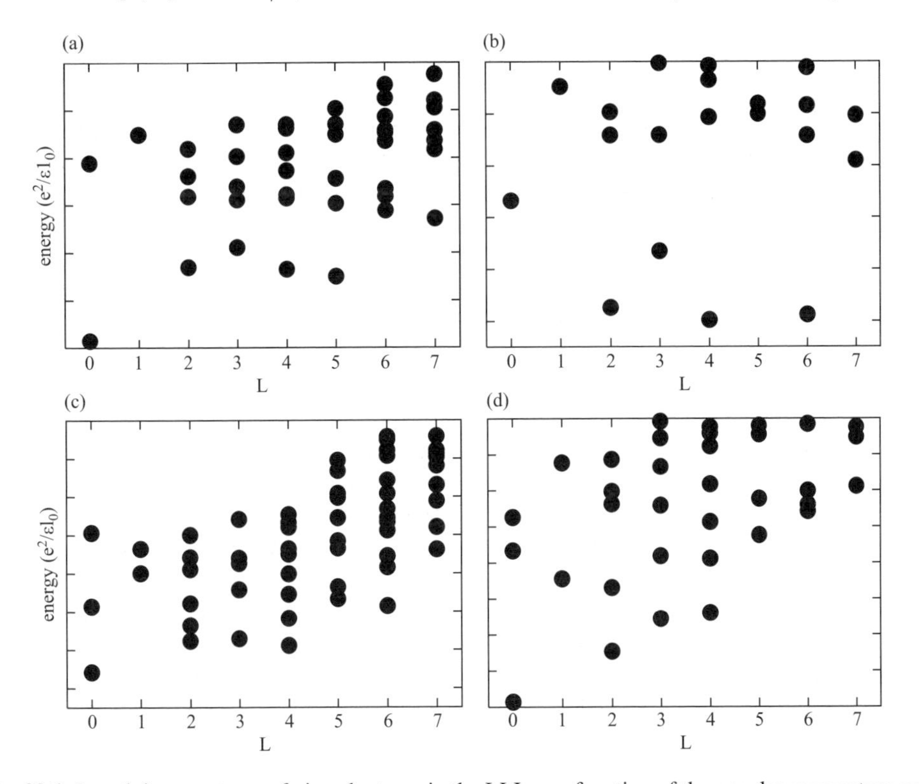

Fig. 20.1 Low-lying spectrum of nine electrons in the LLL as a function of the angular momentum on a sphere. Top panel: Coulomb interaction. (a) At $N_\phi = 16$, the flux needed for the $3/7$ CF state, there is a singlet ground state and a branch of collective excitations. (b) At the flux needed for the candidate state, there is no evidence of an FQHE state. Bottom panel: weakened potential with $V_1 = 0.7V_1^{\mathrm{Coulomb}}$. (c) The $3/7$ CF state is now compressible. (d) There is a possible new FQHE state with the shift required by eqn (20.17).

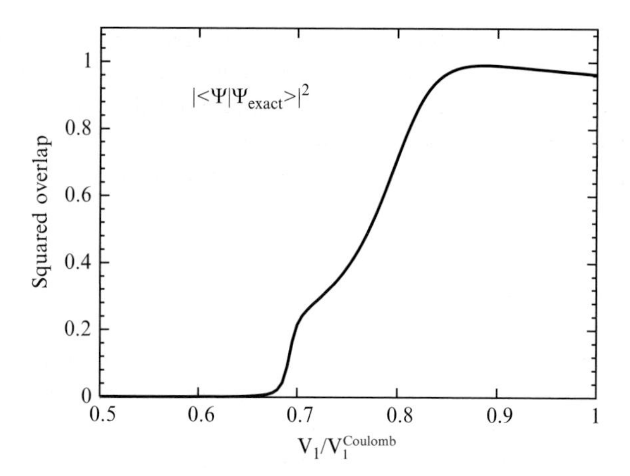

Fig. 20.2 The squared overlap for $N = 6$ electrons at $N_\phi = 13$ between the candidate wave function at $\nu = 3/7$ and the exact ground state computed by varying the pseudopotential V_1 with respect to the Coulomb value.

the ^{4}CF state with negative flux, and at $k = 4$ the competition is with the similar $\nu = 2/7$ ^{4}CF state. This series also contains $5/17$ at $k = 5$, $3/10$ at $k = 6$, and $4/13$ at $k = 8$. The negative-flux series starts with a filled Landau level at $k = 1$ and contains, notably, $5/13$ ($k = 5$), $3/8$ ($k = 6$), $4/11$ ($k = 8$), and $6/17$ ($k = 12$). It is not likely that these states will compete favorably with the main sequence CF states in view of the remarkable stability of the latter. However, the situation is open concerning the exotic even-denominator and the unconventional odd-denominator states. Also, the CF states may be destabilized by slightly tuning the interaction potential. It is known, for example, that there is a window of stability for a non-Abelian $\nu = 2/5$ state in the $N = 1$ LL [18], which is obtained by slightly decreasing the pseudopotential component V_1 with respect to its Coulomb value.

A similar phenomenon seems to happens at $\nu = 3/7$ in the LLL. The conventional CF state at this filling factor is a member of the principal sequence of states. It is realized for $N = 9$ electrons at $N_\phi = 16$ in the spherical geometry. There is a singlet ground state and a well-defined branch of neutral excitations for $L = 2, 3, 4, 5$ (see Fig. 20.1(a)). The negative-flux state in eqn (20.17) requires $N_\phi = 20$ for the same number of particles. At this flux, for a pure Coulomb interaction, there is simply a set of nearly degenerate states with no evidence for an incompressible state (see Fig. 20.1(b)). If the pseudopotential V_1 is decreased from its Coulomb LLL value, the CF state is quickly destroyed (Fig. 20.1(c)), but there is the appearance of a possibly incompressible state precisely at the special shift predicted above (Fig. 20.1(d)). There is an $L = 0$ ground state and a branch of excited states for $L = 2, 3, 4$. To check if this state has anything to do with the new negative-flux state proposed above, the overlap between the candidate wave function for $k = 3$ in eqn (20.17) and the numerically obtained ground state is displayed in Fig. 20.2 for $N = 6$ electrons at $N_\phi = 13$. Even for the pure Coulomb interaction, the squared overlap is 0.9641, and it rises to 0.99054 for $V_1 = 0.885 V_1^{\mathrm{Coulomb}}$.

More numerical evidence may be found in [12] concerning some of the other fractions in these series such as $3/8$ and $3/11$. It is difficult to study states with high-order k-clustering, since they require at least $2k$ particles. It should be noted that, so far, these new series of states have not been derived from correlators of a conformal field theory (CFT). It is known that the Read–Rezayi states can all be derived from expectation values of fields of parafermionic CFTs. A proof that the new states that we described above can be derived from a unitary CFT would be an indication that they describe incompressible candidate FQHE states [19].

References

[1] R. B. Laughlin, *Phys. Rev. Lett.* **50**, 1395 (1983).

[2] J. K. Jain, *Phys. Rev. Lett.* **63**, 199 (1989).

[3] A. H. MacDonald, in *Macroscopic Quantum Physics, Les Houches 1994*, E. Akkermans, G. Montambaux, J.-L. Pichard, and J. Zinn-Justin (eds.); S. M. Girvin and M. Shayegan, in *Topological Aspects of Low-Dimensional Systems, Les Houches 1998*, A. Comtet, T. Jolicoeur, S. Ouvry, and F. David (eds.), EDP Sciences/Springer-Verlag, Les Ulis/Berlin, 1999.

[4] R. E. Prange and S. M. Girvin (eds.), *The Quantum Hall Effect*, Springer-Verlag, 1990.

[5] S. Das Sarma and A. Pinczuk (eds.), *Perspectives in Quantum Hall Effect*, Wiley, New York, 1996.

[6] O. Heinonen (ed.), *Composite Fermions: A Unified View of the Quantum Hall Regime*, World Scientific, Singapore, 1998.

[7] G. Moore and N. Read, *Nucl. Phys. B* **360**, 362 (1991).

[8] M. Greiter, X. G. Wen, and F. Wilczek, *Nucl. Phys. B* **374**, 567 (1992).

[9] W. Pan, H. L. Stormer, D. C. Tsui, L. N. Pfeiffer, K. W. Baldwin, and K. W. West, *Phys. Rev. Lett.* **90**, 016801 (2003).

[10] J. S. Xia, W. Pan, C. L. Vincente, E. D. Adams, N. S. Sullivan, H. L. Stormer, D. C. Tsui, L. N. Pfeiffer, K. W. Baldwin, and K. W. West, *Phys. Rev. Lett.* **93**, 176809 (2004).

[11] N. Read, *Phys. Rev. Lett.* **65**, 1502 (1990).

[12] T. Jolicoeur, *Phys. Rev. Lett.* **99**, 036805 (2007).

[13] N. R. Cooper, N. Wilkin, and M. Gunn, *Phys. Rev. Lett.* **87**, 120405 (2001).

[14] N. Read and E. H. Rezayi, *Phys. Rev. B* **54**, 16864 (1996).

[15] N. Read and E. H. Rezayi, *Phys. Rev. B* **59**, 8084 (1999).

[16] E. H. Rezayi, N. Read, and N. R. Cooper, *Phys. Rev. Lett.* **95**, 160404 (2005).

[17] G. Möller and S. H. Simon, *Phys. Rev. B* **72**, 045344 (2005).

[18] E. H. Rezayi and N. Read, Non-Abelian quantized Hall states of electrons at filling factors $12/5$ and $13/5$ in the first excited Landau level, arXiv:cond-mat/0608346.

[19] N. Read, *Phys. Rev. B* **79**, 045308 (2009).

21

Supersolidity and what soluble models can tell us about it

D. THOULESS

David Thouless,
University of Washington,
Box 351560, Seattle, WA 98195, U.S.A.

21.1 Introduction

Theoretical work in the late 1960s suggested that solid helium might have an equilibrium concentration of vacancies, and that those vacancies should Bose condense at sufficiently low temperatures and lead to superfluid flow of matter in the crystal. Fruitless experimental searches for such supersolid properties were apparently eventually rewarded by the observation of a reduced moment of inertia below $0.2\,\mathrm{K}$ by the Pennsylvania State University group of Moses Chan in 2004. The experimental results are more complicated than were originally expected by theorists, and have provided a rich field of theoretical speculation. Many of us believe that the apparent flow occurs in networks of lower-dimensional features such as grain boundaries or dislocations. I explore some of the information we have on such systems from exact solutions of simple models.

21.2 Some old theory

Nearly forty years ago, it was suggested independently by Andreev and Lifshitz (1969) and Chester (1970) that a bosonic quantum solid could act in some respects like a superfluid. More detailed discussion of this was given by Leggett (1970). I have some documentary evidence to support my memory that Geoffrey Chester told me about his ideas on this topic when I was visiting Cornell in late 1968, because, to my surprise, I found that I had discussed the concept of supersolidity at that time in an otherwise irrelevant paper (Thouless 1969). In a classical solid, the energy of a vacancy (relative to the chemical potential) is the energy cost of removing one atom from the interior and taking it to the surface of the solid. Similarly, the energy of an interstitial is the energy cost of taking an atom from the surface to an interstitial position in the interior. In the zero-temperature limit of a stable classical solid or fermionic solid, the energy to create a vacancy is positive. For a bosonic solid, a vacancy has an additional negative energy because it can move freely through the lattice by changing places with the neighboring atoms, and its lowest energy is obtained by taking the $\mathbf{k} = 0$ state of the vacancy, the zero-momentum state. If this negative delocalization energy has a large enough magnitude to cancel the positive classical vacancy energy, it is energetically favorable to form vacancies in the ground state. If the interaction energy between vacancies is positive (repulsive), only a finite concentration of vacancies will occur, and the ground state of such vacancies will be a Bose condensate. This ideal supersolid should have a mass density different from the value expected from the density one would deduce from the lattice spacing. The superfluid flow of vacancies (or interstitials) should give rise to an anomalous mass flow, free from dissipation, at low velocities. There should be a specific-heat signature of a normal-to-superfluid transition at a temperature proportional to the low-temperature density of vacancies. In a recent paper, however, Pollet *et al.* (2008) argue that the interaction between vacancies is attractive, which should lead to phase separation rather than to Bose condensation.

Various experimental searches for such phenomena were made over the next 35 years without convincing success. In addition, there seems to be strong evidence from the theoretical side that the ideal solid should not be a supersolid, because good experimental and theoretical estimates for the vacancy and interstitial energies have given values which were considerably larger than estimates of the delocalization energy. There are some contrary opinions; for example, Galli *et al.* (2005) have done a simulation that gives a very small amount of off-diagonal long-range order, but that could be due to boundary effects in their small samples.

21.3 Some recent experimental results

Recently Kim and Chan (2004*a*,*b*, 2005, 2006), at Pennsylvania State University, found evidence of anomalous mass flow in samples of solid helium, both in bulk samples and in samples in porous media. These results were found by using a torsional oscillator to measure the moment of inertia of helium confined in a cylindrical container; this is a much-used modification of a method first developed by Andronikashvili more than fifty years ago. A cylindrical container is filled with helium under pressure and cooled down to below its freezing temperature; its pressure must be above 25 bar in order for it to solidify. The experiments found that there is a significant drop in the moment of inertia of the helium when the temperature falls below about $0.2\,\mathrm{K}$. This is reminiscent of the behavior of liquid helium, for which the drop in the moment of inertia is proportional to the ratio of the superfluid density to the total density, since the superfluid component has irrotational flow, and does not follow the rotational motion of the container if the container has axial symmetry. Kim and Chan found drops in the moment of inertia as large as 2% in some cases. The effect persists over a wide range of pressures, from the melting pressure up to 140 bar. The critical speed at which the effect begins to disappear is of the order of six microns per second, but some anomalous inertia continues up to about $100\,\mu\mathrm{m/s}$. The temperature at which the effect appears remains around $0.2\,\mathrm{K}$ through most of the pressure range.

A particularly notable feature of these observations is that blocking the channel around the axis along one line removes almost all the observed reduction in moment of inertia. Again, this is what one would expect if the mechanism is quantized superfluid flow around the axis of the cylinder.

Independent work in other places has found similar effects. At Cornell, Rittner and Reppy (2006, 2007) made extensive studies, and found even larger changes in the moment of inertia. At Keio University, Kondo *et al.* (2007) found results similar to those of Kim and Chan. At Rutgers, Aoki *et al.* (2007) had their sample in a compound pendulum which could be vibrated at two frequencies differing by a factor of around 2, and they found that the measured drop in the apparent moment of inertia was the same in both cases, which seems to exclude explanations of the change in the resonant frequency of the oscillator in terms of a change in the rigidity of the rod rather than a change in the moment of inertia of the sample.

Particularly interesting observations of a different sort were made by Sasaki *et al.* (2006) at ENS, Paris. These authors placed an inverted beaker so as to span an interface between solid and liquid helium, adjusting the pressure so that there was an excess pressure inside the beaker and the solid–liquid interface was higher inside the beaker than outside it. The system was observed by optical microscopy. It was found that where there were visible grain boundaries in the interior of the beaker, the interface dropped steadily until it reached the same level inside as outside, falling at a rate of about $6\,\mu\mathrm{m/s}$. When no grain boundaries were visible, the pressure difference between the inside and outside was maintained for four hours. This observation for crystals of lower quality is very reminiscent of the flow of helium films and of helium in capillaries or cracks, where flow is maintained at a rate independent of the magnitude of the pressure head, as was observed by Kamerlingh Onnes, by Rollin, and by many others in the early days of work on liquid helium. There have been other, unsuccessful, attempts to measure critical flow in the solid, by Greywall (1977) and by Day *et al.* (2005).

Recently, Ray and Hallock (2008) have devised a method similar to that of Sasaki *et al.* (2006) which works at pressures significantly different from the melting pressure. By feeding liquid helium through a porous plug into two well-separated points in helium, they showed that an initial pressure difference between the two points fell linearly with time until the pressures were almost equal. This again suggests a critical current, independent of driving pressure, flowing through the solid to equalize the chemical potentials of the helium at the two points.

Almost all subsequent work has shown up a feature that was not apparent in the earliest work by Kim and Chan, which is that lowering the concentration of ^{3}He in the sample (the ^{3}He isotope is a natural contaminant of helium gas, which stays in solution at low temperatures) and annealing the sample both have the effect of lowering the reduction in moment of inertia. There is no evidence that I know of to suggest that there is an anomalous moment of inertia in a good crystal of ^{4}He. Attention has therefore shifted to explanations of such anomalous behavior in disordered crystals, not in perfect crystals.

There is another phenomenon that seems to be associated with the reduction in moment of inertia at low temperatures. Day and Beamish (2007), at the University of Alberta, have found that there is also an increase in the rigidity modulus at low temperatures. For one sample, there was an 8% increase in the shear modulus for strains of around 2×10^{-8} or less, with the shear modulus rising in the temperature range from 0.15 K to 0.05 K. This effect is also reduced by annealing of the sample or by reducing the ^{3}He concentration.

21.4 Classical and nonclassical inertia

For a solid described in terms of quantum mechanics, uniform motion of the solid with velocity $\mathbf{v}$ can be described in terms of the Galilean boosted wave function

$$\Psi_{\mathbf{v}}(\mathbf{r}_j) = \Psi_0(\mathbf{r}_j - \mathbf{v}t) \exp\left\{ im \left[\sum_{j=1}^{N} \mathbf{r}_j \cdot \mathbf{v} - \frac{Nv^2}{2} \right] \right\}, \tag{21.1}$$

where Ψ_0 is the ground state wave function. Up to a point, a similar construction can be used for a steady rigid-body rotation, with the phase factor $m\mathbf{r}_j \cdot \mathbf{v}$ replaced by $m \int d\mathbf{r}_j \cdot \mathbf{v}$. For a classical solid, this construction more or less works, but for a quantum liquid it does not work, because the phase defined in this way is not single-valued. Only if the flow pattern is irrotational, such that $\mathbf{v}$ is the gradient of a scalar function of position, is this construction unambiguous. This is the argument that led early workers on superfluidity to the idea of irrotational flow of a Bose fluid, and led Onsager (1949) to explain the observed rigid-body flow of bulk helium in terms of an array of quantized vortices. For such an irrotational quantum fluid, the kinetic energy $\int d^3r\, \rho v^2/2$, where ρ is the mass density, is much lower than it is for rigid-body rotation.

The reason that ordinary solids behave classically but Bose-condensed fluids have irrotational flow is that in the fluid, a single particle can go right round a loop without much disturbance of the other particles, so there is a one-particle phase, the phase of the condensate wave function, that must be single-valued modulo 2π, while for a regular solid all the particles move together, and the only obvious phase is associated with the N-particle system as a whole. More than forty years ago, I did a study of exchange in quantum solids inspired by Phil Anderson, mostly directed towards the fermion solid ^{3}He, which confirmed that exchange was a rather small perturbation in solid helium, whose magnitude was shown later by experiment to be of the order of 10^{-7} eV (Thouless 1965). I therefore argue that any nonclassical inertial effects found in the solid must be due to departures from the ideal solid, perhaps due to point defects, as Andreev and Lifshitz (1969) proposed, or to a quantum fluid associated with extended defects such as dislocations or grain boundaries. This is also the conclusion of Prokof'ev and Svistunov (2005), who give reasons that I find compelling.

Anderson (2008) argued that, contrary to these ideas, quantized vortices can exist in a quantum solid, just as they can exist in a cuprate superconductor above the transition temperature, and that a "vortex fluid" may exist in a quantum solid. Although I agree that vortices exist in superfluids above the transition temperature, and, as Anderson observes, this was implicit in the papers of Kosterlitz and Thouless (1972, 1973) on two-dimensional superfluidity, and I am happy with the idea that vortices exist above the critical temperature in cuprates, I do not think they can exist in solids except through association with free vacancies or interstitials. My main reason for this is that in a neutral system, we associate a phase gradient with a mass current. In a charged system, it can be associated alternatively with the electromagnetic vector potential. It is my view that, in our present state of understanding, there can be no vortices in the absence of a gauge field without real circulation of the matter current.

In the next section, I look for the information on such problems that soluble models can give us.

21.5 One-dimensional models

It was pointed out by Girardeau (1960) that there is a simple transformation between the eigenstates of identical noninteracting fermions and the eigenstates of a system of bosons whose only interaction is a hard-core repulsion that prevents them from overlapping or passing one another. If we consider a set of N identical particles, each of mass m, confined to a line of length L, with a zero-range hard-core interaction between them, it makes little difference what the statistics of the particles is. The wave function must vanish whenever the coordinates of two particles coincide, and this is satisfied by the wave function for noninteracting spinless fermions. The only difference that statistics makes is that for fermions, the derivative of the wave function is continuous in the neighborhood of $x_n = x_{n+1}$, whereas for bosons it has a discontinuous sign change. The wave function therefore has the form

$$\Psi \propto \det \psi_j(x_n) \;\text{ for }\; 0 < x_1 < \cdots < x_n < x_{n+1} < \cdots < x_N < L + x_1 \,, \tag{21.2}$$

where the $\psi_j(x)$ are a set of independent single-particle wave functions satisfying appropriate boundary conditions. The wave function for other orderings of the particles is determined by the appropriate representation of the permutation group.

When the particles are confined to a loop (periodic boundary conditions), rather than to a line, the situation is a little more complicated, since cyclic permutations such as

$$x_n \to x_{n+1} \;\text{ for }\; 0 < n < N \,, \;\; x_N \to x_1 + L \,, \tag{21.3}$$

are now dynamically possible. For fermions or for an odd number of bosons, this requires the usual periodic boundary conditions $\psi_j(x + L) = \psi_j(x)$, while for an even number of bosons it requires the antiperiodic conditions $\psi_j(x + L) = -\psi_j(x)$. If the potential energy is constant round the loop then we have $\psi_j(x) = \exp(ik_j x)$, where k_j is an integer multiple of $2\pi/L$ for fermions or an odd number of bosons, and is a half-integer multiple of $2\pi/L$ for an even number of bosons. In this situation the total momentum (regarded as an algebraic scalar along the local direction of the loop) is an integer multiple of Nh/L for bosons and for an odd number of fermions, so that the circulation is an integer multiple of h/m. For an even number of fermions, the center-of-mass momentum is a half-integer multiple of Nh/L, and the circulation is a half-integer multiple of h/m. For distinguishable but dynamically identical particles with a hard-core interaction, all that is required is that the wave function returns to its original value when *all* the particles are simultaneously translated all the way around the loop, so the total momentum can be any integer multiple of h/L. In this model, a system of identical particles cannot have a low translational velocity in response to a slow motion of the container, because the quantized circulation does not allow an average velocity between zero and h/mL.

The case of a zero-range hard-core potential is only one of a number of one-dimensional models for which exact solutions have been known for a long time. A hard core of finite range c can be solved by replacing x_n by $y_n = x_n - nc$, and the only effects of the excluded volume are to replace L by $L - Nc$ in the expressions for the quantized circulation and the energies of the ground state and the gap. The case of a finite-strength δ-function interaction between bosons is well known to be equivalent to the Luttinger model for identical fermions (Mattis and Lieb 1965), and so this maps onto the Luttinger liquid.

So far, this model bears no relation to the situation of a regular quantum solid constrained by the boundaries of a moving container or to the possible situation of frictionless flow in a fixed container. The lines of atoms in a regular solid are subject to a periodic potential due to their neighbors, and the irregularities of the inner surface of the container will superpose disorder on the potential that lines or loops of atoms are subject to. We therefore need, at least, to consider the effects of a periodic containing potential and of comoving disorder on the motion of the atoms relative to a moving boundary.

The periodic potential by itself, if it is commensurate with the spacing between the atoms, is enough to destroy the possibility of nonzero circulation relative to the container. In this case the single-particle wave functions ψ_j are Bloch waves completely filling a band, and then, if the periodic potential moves, the full band just follows the motion of the potential, as happens for a full electron band in a moving insulator. The Green's functions for such a system are localized, with a localization length that corresponds to the exponential decay length for the middle of the energy gap. An incommensurate periodic potential would give a partially filled band, and the holes or particles in this partially filled band could give a nonclassical contribution to the inertia, just as the carriers in a doped semiconductor serve as current carriers with an anomalous mass.

It is well known that for a one-dimensional system any disorder localizes all the energy levels of a system of noninteracting particles, whether fermions or bosons. For white noise disorder, the theory of this was worked out by Berezinskii (1973), and the localization length is related to what would be calculated as the classical mean free path of the particles (Thouless 1973). Again, just as for a commensurate periodic potential, this exponential localization ensures that the particles follow the motion of the disorder potential, just as it does for a classical solid held in a solid cryostat with irregular boundaries. For the combination of a commensurate periodic potential and weak disorder, one can argue that, provided the disorder is weak enough not to shift energy levels across the band gap, the energy levels remain exponentially localized, and will follow the movement of the container that provides the periodic potential and the disorder (Niu and Thouless 1984).

For this one-dimensional loop model, anomalous inertial responses are completely destroyed both by a commensurate periodic potential and by disorder, and there seems to be no vestige of quantized vorticity left.

In the absence of disorder, and with a periodic potential for which the number of periods is somewhat different from N, the bosonic wave functions correspond to fermionic wave functions with an incompletely filled lowest band or with a partially filled second band. In this case the filled energy band contributes nothing to the circulation, and only the holes in the lowest band or the extra particles in the upper band need be considered. The reduced number of these particles or holes can greatly reduce the anomalous momentum which they contribute if the periodic substrate is set in motion.

In order to make comparison with some of the realistic mechanisms which have been suggested for anomalous inertia in quantum solids, such as channeling along a network of intersecting dislocations, as suggested by Boninsegni *et al.* (2007), one needs to be interested in long channels with disordered periodicity, containing a dense collection of interacting atoms in the channel. A more detailed version of this theory has been published by Pollet *et al.* (2008).

We would like to know something about one-dimensional models which involve both a two-body interaction (a short-ranged repulsion is likely to be particularly important) and an external potential displaying the properties of both periodicity or near-periodicity, and disorder. I know of very few such models which are soluble, but there is at least one to hand. For spinless fermions we know that the wave function vanishes wherever two particles coincide, so a zero-range hard-core interaction does not change the eigenstates or energies whatever the one-particle substrate potential is. There are various soluble one-dimensional models, such as the tight-binding model, the Krönig–Penney model (one-dimensional δ functions), models with periodic square wells, and models with white noise disorder (Frisch and Lloyd 1960, Halperin 1965, Berezinskii 1973). Even without exact solutions, there are strong arguments, some rigorous (Goldsheid *et al.* 1977, for example), to support the idea that all eigenstates are exponentially localized for one-dimensional disordered or nearly periodic substrate potentials. We also know that for a noninteracting many-fermion system with filled bands, the many-body states are exponentially localized.

Using the boson–fermion transformation, we can immediately see that similar statements can be made about the many-boson system with a zero-range hard core. The effect of the external potential on its many-body wave functions and on its energy levels is much the same as for the fermion system. The energy levels in a moving frame of reference can be obtained by a Galilean boost from those in a stationary frame in the same way as in eqn (21.1). This gives

$$\Psi_v(x_j) = \Psi_0(x_j - vt) \exp\left\{ im \left[\sum_{j=1}^{N} x_j v - \frac{Nv^2}{2} \right] \right\}, \tag{21.4}$$

and, to preserve the symmetry under cyclic permutation of the particles along a loop of length L, we need v to be a multiple of h/mL, and the momentum to be a multiple of Nh/L. For a one-dimensional model like this, the existence of a quantized circulation of the center-of-mass motion does not lead to stable circulating currents, because the boson version of a transition from a left-going state to a right-going state can change the circulation by one quantum unit h/m.

Although the one-dimensional models studied by exact methods lead to the absence of any flow in the zero-temperature limit, there are arguments that show that an attractive interaction between fermions can lead to a metallic state for fermions in the presence of weak disorder (Giamarchi and Schulz 1988). The Giamarchi and Schultz paper argues that the attractive interaction promotes pairing, and so the Bose condensation of pairs can compete with the tendency towards localization. This is also supported by simulations of such systems (Schmitt-eckert *et al.* 1998).

21.6 Two-dimensional flow

Experimental evidence from a wide variety of sources indicates that solid helium, away from its single-crystal equilibrium state, and with ^{3}He impurity at a level of several parts per million or more, may display nonclassical behavior in a rotating container, and may also display nondissipative flow. A survey of our theoretical knowledge of one-dimensional fluids makes us suspect that separated one-dimensional channels are not likely to give us this sort of behavior. However, Kim and Chan (2005) found similar phenomena for solid helium trapped in porous gold, which can be regarded as an interconnected network of narrow channels; and superfluidity of the liquid in porous media has been studied for many years, and the general theory is well understood. It is also possible that superfluidity could occur on interconnected surfaces such as grain boundaries. In this section I discuss some of the possibilities, where connections with exact solutions are much harder to come by than in the case of one-dimensional systems.

There has been a recent article by Balibar and Caupin (2008) that surveys the experimental and theoretical situation. This article focuses on two important uncertainties in the experimental evidence for superfluid behavior in the solid. One of the crucial experiments that appears to demonstrate anomalous mass flow is the control experiment done by Kim and Chan (2004*b*), in which the oscillating chamber was blocked by a barrier, so that there was no longer a continuous loop surrounding the axis of the system. This appeared to destroy most of the anomalous moment of inertia, although not quite as much as one would expect. Balibar and Caupin argue that this needs to be done more precisely, and with blocked and unblocked chambers that are more directly comparable in their dimensions and in the densities of the samples. A second crucial experiment is the observation of what appears to be a critical mass flow under pressure in the experiment of Sasaki *et al.* (2006)—Balibar and Caupin were also authors of that paper. In this experiment, the pressure of the solid was necessarily very close to the melting pressure. Balibar and Caupin now think that the flow that was observed could have been due to ordinary superfluid liquid along a line at which a grain boundary met the wall of the container, and argue that this observation needs more careful exploration. The work of Ray and Hallock (2008) extends this a little above the melting pressure, but other experiments which have attempted to detect critical flow at pressures much higher than the melting pressure have not given positive results, although nonclassical inertia has been measured at much higher pressures.

Gaudio *et al.* (2008) have introduced a model of transport along grain boundaries, in which the corners where four edges meet are pinned by ^{3}He impurities. The point defects which are inevitable where incommensurately ordered crystals meet may provide possible carriers for two-dimensional superfluid flow along the grain boundaries. In this work, the rounding of the temperature dependence of the transition to the low-temperature phase and its dependence on the ^{3}He concentration were ascribed to the length-scale dependence of the superfluid transition in two dimensions, which was derived by Ambegaokar *et al.* (1980).

21.7 Conclusions

Experimental work done in recent years gives strong, but not compelling, evidence for fluid-like flow in solid ^{4}He under some conditions, but gives no support for the existence of supersolidity in a good single crystal. If such flow is real, it seems most likely to be confined to grain boundaries, grain edges, dislocations, or the interface between the solid and the container. Exact solutions of simple models give us some idea of what should not happen, but leave a number of possibilities open. We can hope that further experiments and more detailed theories will soon tell us what is happening.

Acknowledgments

I wish to thank numerous people who have guided and stimulated my thoughts on this topic, particularly Phil Anderson, Sebastien Balibar, Geoffrey Chester, Moses Chan, and John Reppy. I thank Wayne Saslow for pointing out some errors in my original text.

References

Ambegaokar, V., Halperin, B. I., Nelson, D. R., and Siggia, E. D. (1980). *Phys. Rev. B* **21**, 1806.
Anderson, P. W. (2008). *Phys. Rev. Lett.* **100**, 215301.
Andreev, A. F. and Lifshitz, I. M. (1969). *Zh. Eksp. Teor. Fiz.* **56**, 2057.
Aoki, Y., Graves, J. C., and Kojima, H. (2007). *Phys. Rev. Lett.* **99**, 015301.
Balibar, S. and Caupin, F. (2008). *J. Phys.: Condens. Matter* **20**, 173201.
Berezinskii, V. L. (1973). *Zh. Eksp. Teor. Fiz.* **65**, 1251.
Boninsegni, M., Kuklov, A. B., Pollet, L., Prokof'ev, N. V., Svistunov, B. V., and Troyer, M. (2007). *Phys. Rev. Lett.* **99**, 035301.
Chester, G. V. (1970). *Phys. Rev. A* **2**, 256.
Day, J. and Beamish, J. (2007). *Nature* **450**, 853.
Day, J., Herman, T., and Beamish, J. (2005). *Phys. Rev. Lett.* **95**, 035301.
Frisch, H. L. and Lloyd, S. P. (1960). *Phys. Rev.* **120**, 1175.
Galli, D., Rossi, M., and Reatto, L. (2005). *Phys. Rev. B* **71**, 140506.
Gaudio, S., Cappelluti, E., Rastelli, G., and Pietrenero, L. (2008). *Phys. Rev. Lett.* **101**, 075301.
Giamarchi, T. and Schulz, H. J. (1988). *Phys. Rev. B* **37**, 325.
Girardeau, M. (1960). *J. Math. Phys.* **1**, 516.
Goldsheid, I. Ya., Molchanov, S. A., and Pastur, L. A. (1977). *Func. Anal. Appl.* **11**, 1.
Greywall, D. S. (1977). *Phys. Rev. B* **16**, 1291.
Halperin, B. I. (1965). *Phys. Rev.* **139A**, 104.
Kim, E. and Chan, M. H. W. (2004*a*). *Nature* **427**, 225.
Kim, E. and Chan, M. H. W. (2004*b*). *Science* **305**, 1941.
Kim, E. and Chan, M. H. W. (2005). *J. Low Temp. Phys.* **138**, 859.
Kim, E. and Chan, M. H. W. (2006). *Phys. Rev. Lett.* **97**, 115302.
Kondo, M., Takada, S., Shibayama, Y., and Shirahama, K. (2007). *J. Low Temp. Phys.* **148**, 695.
Kosterlitz, J. M. and Thouless, D. J. (1972). *J. Phys. C* **5**, L124.
Kosterlitz, J. M. and Thouless, D. J. (1973). *J. Phys. C* **6**, 1181.

Leggett, A. J. (1970). *Phys. Rev. Lett.* **25**, 1543.
Mattis, D. C. and Lieb, E. H. (1965). *J. Math. Phys.* **6**, 304.
Niu, Q. and Thouless, D. J. (1984). *J. Phys. A* **17**, 2453.
Onsager, L. (1949). *Nuovo Cimento* **6**, Suppl. 2, 249.
Pollet, L., Boninsegni, M., Kuklov, A. B., Prokof'ev, N. V., Svistunov, B. V., and Troyer, M. (2008). *Phys. Rev. Lett.* **101**, 097202.
Prokof'ev, N. and Svistunov, B. (2005). *Phys. Rev. Lett.* **94**, 155302.
Ray, M. W. and Hallock, R. B. (2008). *Phys. Rev. Lett.* **100**, 235301.
Rittner, A. S. C. and Reppy, J. D. (2006). *Phys. Rev. Lett.* **97**, 165301.
Rittner, A. S. C. and Reppy, J. D. (2007). *Phys. Rev. Lett.* **98**, 175302.
Sasaki, S., Ishiguro, R., Caupin, F., Maris, H., and Balibar, S. (2006). *Science* **313**, 1098.
Schmitteckert, P., Schulze, T., Schuster, C., Schwab, P., and Eckern, U. (1998). *Phys. Rev. Lett.* **80**, 560.
Thouless, D. J. (1965). *Proc. Phys. Soc.* **86**, 893.
Thouless, D. J. (1969). *Ann. Phys.* **52**, 403.
Thouless, D. J. (1973). *J. Phys. C* **6**, L49.